Cell Biology

A LABORATORY HANDBOOK

Second Edition

Volume Two

Second Edition
Volume Two

Cell Biology

A LABORATORY HANDBOOK

Edited by

JULIO E. CELIS

Department of Medical Biochemistry
University of Aarhus, Denmark

ACADEMIC PRESS

San Diego   London   Boston   New York   Sydney   Tokyo   Toronto

Cover photograph for Volume 2: A GTPase-deficient mutant of Rab5, a regulator of endocytic transport, transiently expressed in BHK cells using the vaccinia T7 virus system (see chapter by Stenmark and Zerial) causes the formation of large vacuoles produced by an exaggerated homotypic fusion between early endosomes. Provided by Marino Zerial.

Academic Press
a division of Harcourt Brace & Company
525 B Street, Suite 1900, San Diego, California 92101-4495, USA
http://www.apnet.com

Academic Press Limited
24-28 Oval Road, London NW1 7DX, UK
http://www.hbuk.co.uk/ap/

Library of Congress Card Catalog Number: 97-80300

International Standard Book Number: 0-12-164727-7

PRINTED IN THE UNITED STATES OF AMERICA
97 98 99 00 01 02 DO 9 8 7 6 5 4 3 2 1

PART 3

ORGANELLES AND CELLULAR STRUCTURES

PART 5

ANTIBODIES

SECTION A—Production of Antibodies   379

A COMPLETE SUBJECT INDEX APPEARS IN VOLUME 4.

CONTENTS OF OTHER VOLUMES

VOLUME 1

VOLUME 3

PART 10
INTRACELLULAR MEASUREMENTS

PART 11
CYTOGENETICS AND *In Situ* HYBRIDIZATION

Numbers in parentheses indicate the volume (bold face) and page on which the authors' contribution begin.

Ruedi Aebersold (**4**, 514), Department of Molecular Biology, University of Washington, Seattle, Washington 98195.

Ueli Aebi (**3**, 277, 292), M.E. Müller Institute for Microscopy, Biozentrum, University of Basel, Basel CH-4056 Switzerland.

Gudrun Ahnert-Hilger (**4**, 103), Institut für Anatomie, Universitatsklinikum Charite Humboldt-Universitat zu Berlin, Berlin 10115 Germany.

David J. Allan (**1**, 327), School of Life Sciences, Queensland University of Technology, Brisbane 04000 Australia.

Robert C. Allen (**4**, 421), Laboratory of Biochemical Genetics, National Institute of Mental Health, Washington, DC 20032 and Department of Pathology, University of South Carolina, Charlston, South Carolina 29425.

Terence D. Allen (**3**, 346), CRC Department of Structural Cell Biology, University of Manchester, Paterson Institute for Cancer Research, Manchester M20 9BX United Kingdom.

Adam Amsterdam (**3**, 502), Department of Biology, Institute of Technology, Cambridge, Massachusetts 02139.

Søren S. L. Andersen (**2**, 205), Department of Cell Biology, EMBL, Heidelberg D-69117 Germany.

K. I. Anderson (**2**, 372), Institute of Molecular Biology, Austrian Academy of Sciences, Institute of Molecular Biology, Salzburg A-5020 Austria.

Michael J. Anderson (**3**, 405), Ludwig Institute for Cancer Research, La Jolla, California 92093.

Helena Andersson (**4**, 218), Bioscience at Novum, Karolinska Institutet, Huddinge S-141 57 Sweden.

James M. Angelastro (**1**, 244), Department of Pathology, Columbia University College of Physicians and Surgeons, New York, New York 10032.

Geert Angenon (**4**, 92), Department of Genetics, Flanders Interuniversity Institute for Biotechnology (VIB), Universität Gent, Gent B-9000 Belgium.

Wilhelm Ansorge (**4,** 23, 57), EMBL, Heidelberg D-69117 Germany.

Nobukazu Araki (**2,** 495), Department of Anatomy, Ehime University School of Medicine, Shigenobu, Ehime 791-02 Japan.

Anthony J. Ashford (**2,** 205), Department of Cell Biology, EMBL, Heidelberg D-69117 Germany.

Kathryn R. Ayscough (**2,** 477), Department of Biochemistry, Medical Sciences Institute, University of Dundee, DD1 4HN Scotland, United Kingdom.

Ivan C. Baines (**1,** 413), Laboratory of Cell Biology, National Heart, Lung, and Blood Institute, National Institutes of Health, Bethesda, Maryland 20892.

William E. Balch (**2,** 258), Department of Cell Biology, The Scripps Research Institute, La Jolla, California 92037.

Shoshana Bar-Nunn (**4,** 458), Department of Biochemistry, Tel-Aviv University, Ramat-Aviv, Tel-Aviv 69978 Isreal.

Udo Baron (**4,** 230), Zentrum für Molekulare Biologie, Universität Heidelberg, Heidelberg D-69120 Germany.

Jiri Bartek (**4,** 489), Division for Cancer Biology, Danish Cancer Society, Copenhagen DK-2100 Denmark.

Bodil Basse (**2,** 429; **4,** 375, 404), Department of Medical Biochemistry and Danish Center for Human Genome Research, Aarhus University, Aarhus C Denmark.

Philippe I. H. Bastiaenes (**3,** 136), Department of Molecular Biology, Max Planck Institute for Biophysical Chemistry, Göttingen D-37018 Germany.

Paul J. Battista (**1,** 137), Cell Culture Research and Development, Life Technologies, Incorporated, Grand Island, New York 14072.

Peter Beard (**1,** 493), Swiss Institute for Experimental Cancer Research, Epalinges 01066 Switzerland.

Mary C. Beckerle (**4,** 441), Department of Biology, University of Utah, Salt Lake City, Utah 84112.

Sven-Erik Behrens (**2,** 174), Justus-Liebig-Universität Giessen, Institute für Virologie (FB18), D-35392 Giessen Germany.

Jürgen Bereiter-Hahn (**3,** 54), Johann-Wolfgang-Gothe Universität, Biozentrum, Frankfurt am Main, D-60439 Germany.

Michael W. Berns (**2,** 193), Beckman Laser Institute and Medical Clinic, University of California, Irvine, Irvine, California 92612.

Andrew Bett (**1,** 500), Merck Human Genetics, Merck and Company, Incorporated, West Point, Pennsylvania 19486.

Hartmut Beug (**1,** 107), Research Institute of Molecular Pathology, Vienna A-1030 Austria.

Shyamala Bhaskaran (**1,** 472; **4,** 176), Department of Soil and Crop Science, Texas A & M University, College of Agriculture, College Station, Texas 77843-2474.

Roland Bilang (**4,** 171), Institute of Plant Sciences, Swiss Federal Institute of Technology, Zurich CH-8092 Switzerland.

R. Curtis Bird (**1,** 205, 209), Department of Pathobiology, Auburn University, Auburn, Alabama 36849.

Lucille Bitensky (**3,** 238), Department of Medicine, Charing Cross and Westminster Medical School, London W6 8RF England.

Guenther Boeck (**2**, 63), Institute for General and Experimental Pathology, University of Innsbruck Medical School, Innsbruck A-6020 Austria.

Bryan J. Bolton (**1**, 178), European Collection of Cellcultures, Wiltshire SP4 0JG United Kingdom.

Lars Bolund (**3**, 437), Institute for Human Genetics, University of Aarhus, Aarhus C DK-8000 Denmark.

Jesper Bonde (**1**, 164), Departments of Medicine, Hematology, Medical Microbiology, and Immunology, Aarhus University Hospital, Aarhus C DK-8000 Denmark.

Michel Bornens (**2**, 111), Institute Curie, Section Recherche CRNS UMR 144, Paris, Cedex 05 75248 France.

Alan Boyde (**1**, 142; **3**, 179), Department of Anatomy, University College, London WC1E 6BT United Kingdom.

Savile Bradbury (**3**, 34), Pembroke College, Oxford OX4 1PD United Kingdom.

Rodrigo Bravo (**3**, 460), Oncology, Exploratory, and Drug Discovery Research, Bristol-Myers Squibb, Princeton, New Jersey 08543.

Andreas Bremer (**3**, 277, 292), Bellevue Asset Management AG, CH-6301 Zug Switzerland.

John W. Breneman, III (**3**, 444), Life Science Group, BIO-RAD Laboratories, Hercules, California 94547.

Heather L. Brownell (**4**, 64, 75), Microbiology and Immunology and Pathology, Queen's University, Kingston, Ontario K7L 3N6 Canada.

Josef Brunner (**4**, 495), Laboratorium für Biochemie II, ETH-Zentrum, Zurich CH-8092 Switzerland.

Nils Brünner (**1**, 16, 291), The Finsen Laboratory, Rigshospitalet, Copenhagen Denmark.

Hermann Bujard (**4**, 230), Zentrum fur Molekulare Biologie, Universität Heidelberg, Heidelberg D-69120 Germany.

Debra Burdick (**2**, 239), Margaret M. Dyson Vision Research Institute, Department of Ophthalmology, Cornell University Medical College, New York, New York 10021.

Ian M. Caldicott (**1**, 398), Marana, Arizona 85653.

Sophie Calvet (**4**, 111), Centre National de la Recherche Scientifique, Ecole Normale Supérieure, Cedex 05 Paris 75230 France.

Keith H. S. Campbell (**3**, 487), Roslin Institute, Midlothian EH25 9PS United Kingdom.

Marie-France Carlier (**2**, 359), Laboratorie d'Enzymologie de CNRS, Gir-sur-Yvette, Cedex 91198 France.

Edwin Carmack (**4**, 539), Department of Molecular Biotechnology, University of Washington, Seattle, Washington 98195.

Daniel Carrasco (**3**, 460), Onocology, Exploratory, and Drug Discovery Research, Bristol-Myers Squibb, Princeton, New Jersey 08543.

Nigel P. Carter (**3**, 421), The Sanger Centre, Wellcome Trust, Genome Campus, Hinxton, Cambridge CB10 1SA United Kingdom.

Doris Cassio (**3**, 399), Universitie Paris Sud, Orsay, Cedex 91405 France.

Ariana Celis (**1**, 5; **2**, 392; **4**, 375), Department of Medical Biochemistry and Danish Centre for Human Genome Research, Aarhus University, Aarhus C DK-8000 Denmark.

Julio E. Celis (**1**, 5; **2**, 392, 429; **4**, 187, 225, 361, 375, 404, 409, 450, 454, 482), Department of Medical Biochemistry and Danish Centre for Human Genome Research, Aarhus University, Aarhus C DK-8000 Denmark.

Victoria E. Centonze (**3**, 149), University of Wisconsin, Madison, Wisconsin 53706.

Roberta Charpentier (**1**, 321), The Center for Gerontological Research, Allegheny University of the Health Sciences, Philadelphia, Pennsylvania 19129.

Joseph Chayen (**3**, 238), Department of Medicine, Charing Cross and Westminster Medical School, London W6 8RF England.

Jonathan D. Chesnut (**4**, 239), Invitrogen, San Diego, California 92008.

Piotr Chomczynski (**2**, 221), Molecular Research Center, Incorporated, Cincinnati, Ohio 45212.

Teresa Christianson (**4**, 474), COGNIS, Incorporated, Santa Rosa, California 95407.

Valentina Ciccarone (**4**, 145), Life Technologies, Rockville, Maryland 20849.

Mark S.F. Clarke (**4**, 49), Universities Space Research Association, Division of Space Life Sciences, Houston, Texas 77058.

Stephen Cohen (**3**, 510), EMBL, Heidelberg D-69117 Germany.

Frank R. Collart (**1**, 233), Centre for Mechanistic Biology and Biotechnology, Argonne National Laboratory, Argonne, Illinois 60493.

Carl W. Cotman (**1**, 65), Organized Research Unit in Brain Aging and Psychobiology and Neurobiology, University of California, Irvine, Irvine, California 92717.

Aaron W. Crawford (**4**, 441), Department of Biological Sciences, Union College, Schenectady, New York 12308.

G. Joseph Creed (**4**, 421), Laboratory of Biochemical Genetics, National Institutes of Health, Washington, DC 20032.

Vincent J. Cristofalo (**1**, 321), The Center for Gerontological Research, Allegheny University of the Health Sciences, Philadelphia, Pennsylvania 19129.

R. A. Cross (**2**, 317), Molecular Motors Group, Marie Curie Research Institute, Oxted, Surrey RH8 0TL United Kingdom.

Matthew E. Cunningham (**1**, 244), Department of Pathology, Columbia University College of Physicians and Surgeons, New York, New York 10032.

Rolf H. Dahl (**2**, 12), Cellular and Structural Biology, University of Colorado, Health Sciences Center, Denver, Colorado 80262.

Dorthe Danielsen (**3**, 518), Laboratory of Gene Expression and Department of Molecular Biology, University of Aarhus, Aarhus C DK-8000 Denmark.

Zbigniew Darzynkiewicz (**1**, 261, 341), The Cancer Research Institute, New York Medical College, Elmsford, New York 10523.

Richard W. Davies (**4**, 569), Amersham International PLC, Buckinghamshire HP7 9NA United Kingdom.

Massimo De Felici (**1**, 73), Dipartimento di Sanita Pubblica e Biologia Cellulare, Sezione di Istologia e Embriologia, Universita Di Roma "Tor Vergata", Rome 00173 Italy.

Meltsje J. de Hoop (**1**, 154), Cell Biology Program, EMBL, Heidelberg D-69012 Germany.

Kurt Dejgaard (**2**, 392; **4**, 482), Institute of Medical Biochemistry, Danish Centre for Human Genome Research, Aarhus C DK-8000 Denmark.

Robert T. Dell'Orco (**1**, 133), ProhibiTech, Edmond, Oklahoma 73034.

Nicolas Demaurex (**3**, 380), Department of Cell Biology, The Hospital for Sick Children, Toronto, Ontario M5G 1X8 Canada.

Daniele Derossi (**4**, 111), Centre National de la Recherche Scientifique, Ecole Normale Supérieure, 75230 Paris Cedex 05 France.

Michel Desjardins (**2**, 75), Departement d'Anatomie, Universite de Montreal, Montreal, Quebec H3C 1J4 Canada.

Peter N. Devreotes (**1**, 431), Department of Biological Chemistry, Johns Hopkins University School of Medicine, Baltimore, Maryland 21205.

T. Michael Dexter (**4**, 281), Paterson Institute for Cancer Research, Manchester M20 4BX United Kingdom.

Willy Dillen (**4**, 92), Department of Genetics, Flanders Interuniversity Institute for Biotechnology (VIB), Universität Gent, Gent B-9000 Belgium.

Roeland W. Dirks (**3**, 453), Department of Cytochemistry and Cytometry, Syluius Laboratories, Leichen University, Al Leiden 02333 The Netherlands.

Bernhard Dobberstein (**2**, 265; **4**, 495), Zentrum für Molekulare Biologie, Universität Heidelberg, Heidelberg D-69120 Germany.

Carlos G. Dotti (**1**, 154), Cell Biology Program, EMBL, Heidelberg D-69117 Germany.

Julian A.T. Dow (**2**, 518), Division of Molecular Genetics, Institute of Biomedical and Life Sciences, Glasgow G11 6NU Scotland, United Kingdom.

David G. Drubin (**2**, 477), Department of Molecular and Cellular Biology, University of California, Berkeley, Berkeley, California 94720.

Graham A. Dunn (**3**, 44), MRC Muscle and Cell Motility Unit, The Randall Institute, Kings College London, London WC2B 5RL United Kingdom.

Suzanne Eaton (**3**, 170), EMBL, Heidelberg D-69012 Germany.

Christoph Eckerskorn (**4**, 304), Protein Analysis, Max Planck Institute for Biochemistry, Martinsried D-2152 Germany.

Roy Edward (**1**, 197), DYNAL A.S. Skoyen, Oslo NO212 Norway.

Richard Egel (**1**, 421), Department of Genetics, Institute of Molecular Biology, University of Copenhagen, Copenhagen-K, DK-1353 Denmark.

Lars Ekblad (**2**, 5), Department of Biochemistry, Center for Chemistry and Chemical Engineering, Lund S-221 00 Sweden.

Hugh Y. Elder (**3**, 340), Institute of Biomedical and Life Sciences, University of Glasgow, Glasgow G12 8QQ Scotland.

Susannah Eliott (**1**, 431), Department of Biological Sciences, Macquarie University, Sidney NSW 2109 Australia.

Maria Ekström (**4**, 218), Bioscience at Novum, Karolinska Institutet, Huddinge S-141 57 Sweden.

Jimmy K. Eng (**4**, 539), Department of Molecular Biotechnology, University of Washington, Seattle, Washington 98195.

Anne-Marie Engel (**1**, 285), Bartholin Instituttet, Kobenhavns Kommunehospital, Denmark.

David Enshell (**2**, 446), Cell Research and Immunology, Tel-Aviv University, Ramat-Aviv, Tel-Aviv 69978 Isreal.

Patricia A. Estes (**1**, 521), Department of Pediatric Hematology/Oncology, University of Colorado School of Medicine, Denver, Colorado 80262.

Martin Evans (**1,** 86), Institute of Cancer and Developmental Biology, and Department of Genetics, University of Cambridge, Cambridge United Kingdom.

Dariush Fahimi (**2,** 87), Department of Anatomy and Cell Biology (II), University of Heidelberg, Heidelberg, D-69120 Germany.

Stephen E. Farinelli (**1,** 244), Department of Pathology, Columbia University College of Physicians and Surgeons, New York, New York 10032.

Marcus Fechheimer (**1,** 431), Department of Zoology, University of Georgia, Athens, Georgia 30602.

Patricia M. Ferrier (**3,** 487), Roslin Institute, Midlothian EH25 9PS United Kingdom.

Irene Fialka (**1,** 107), Research Institute of Molecular Pathology, Vienna A-1030 Austria.

Jörgen Finneman (**3,** 518), Laboratory of Gene Expression and Department of Molecular Biology, University of Aarhus, Aarhus C CK-8000 Denmark.

Richard A. Firtel (**1,** 431), Center for Molecular Genetics, University of California, San Diego, San Diego, California 92093-0634.

Kevin L. Firth (**4,** 75), Ask Science Products, Incorporated, Kingston, Ontario, K7L 3Z8 Canada.

Ronald Frank (**2,** 398), Cell Biology and Immunology, GBF-National Research Center for Biotechnology, Braunschweig D-38124 Germany.

Mark Frattini (**1,** 513), Pritzker School of Medicine, University of Chicago, Illinois 60637.

Sabine Freundlieb (**4,** 230), Zentrum für Molekulare Biologie, Universität Heidelberg, Heidelberg, D-69120 Germany.

Irving B. Fritz (**1,** 186), Molecular and Cellular Physiology, AFRC Institute of Animal Physiology and Genetics Research, Cambridge United Kingdom.

Yoshio Fukui (**4,** 31), Department of Cell and Molecular Biology, Northwestern University Medical School, Chicago Illinois 60611.

David C. Fung (**2,** 350), Department of Biochemistry and Beckman Center, Stanford University School of Medicine, Stanford, California 94305.

Ruth Furukawa (**1,** 431), Department of Zoology, University of Georgia, Athens, Georgia 30602.

Barbara Galazkiewicz (**4,** 398), Department of Molecular Biology, Austrian Academy of Sciences, Salzburg A-5020 Autria.

Robert L. Garcea (**1,** 521), Department of Pediatric Hematolgoy/Oncology, University of Colorado School of Medicine, Denver, Colorado 80262.

Henrik Garoff (**1,** 534; **4,** 218), Department of Bioscience at Novum, Karolinska Institutet, Huddinge S-141 57 Sweden.

Benjamin Geiger (**2,** 457), Chemical Immunology, The Weizmann Institute of Science, Rehovot 76100 Isreal.

Jonathan M. Gershoni (**2,** 446; **4,** 458), Cell Research and Immunology, Tel-Aviv University, Ramat-Aviv, Tel-Aviv 69978 Israel.

Alasdair J. Gibb (**1,** 359), Department of Pharmacology, University College London, London WC1E GBT England.

Beth L. Gillece-Castro (**4,** 547), Genentech Incorporated, San Francisco, California 94080.

Mario Gimona (**4,** 265, 398), Institute of Molecular Biology, Austrian Academy of Sciences, Salzburg A-5020 Austria.

Sophie Giorgetti-Peraldi (**4,** 75), Joslin Diabetes Center and Department of Medicine, Harvard Medical School, Boston, Massachusetts 02215.

Martin W. Goldberg (**3,** 346), CRC Department of Structural Cell Physiology, Paterson Institute for Cancer Research, Manchester M20 9BX United Kingdom.

Jon W. Gordon (**3,** 473), Geriatrics and Adult Development, Mt. Sinai School of Medicine, New York, New York 10029.

Angelika Görg (**4,** 386), Technical University of Munich, Freising-Weihenstephan D-85350 Germany.

Adolf Graessmann (**4,** 11), Institut für Molekularebiologie und Biochemie, Freie Universität Berlin, 14195 Berlin.

Monika Graessmann (**4,** 11), Institut fur Molecularebiologie und Biochemie, Freie Universität Berlin, D-14195 Berlin.

Roland Graf[1] (**4,** 495), Laboratorium für Biochemie II, ETH-Zentrum, Zurich CH-8092 Switzerland.

Frank L. Graham (**1,** 500), Departments of Biology and Pathology, McMaster University, Hamilton, Ontario L8S 4K1 Canada.

Paola Grandi (**2,** 159), Institut für Biochemie I, University of Heidelberg, Heidelberg, D-69120 Germany.

Colin Gray (**1,** 142; **3,** 179), Department of Anatomy, University College, London WC1E 6BT United Kingdom.

John C. Gray (**2,** 81, 286), Department of Plant Sciences, University of Cambridge, Cambridge CB2 3EA United Kingdom.

Anna Greco (**2,** 135), Faculte de Medecini Lyon-Laennec, Universite Claude Bernard Lyon I, Lyon, Cedex 08 69372 France.

Lloyd A. Greene (**1,** 244), Department of Pathology, Columbia University College of Physicians and Surgeons, New York, New York 10032.

Gregory Gregoriadis (**4,** 131), Centre for Drug Delivery Research, School of Pharmacy, London WC1N 1AX United Kingdom.

Gareth Griffiths (**3,** 332), European Molecular Biology Laboratory, Heidelberg D-69012 Germany.

Rudolf Grimm (**4,** 304), Hewlett Packard, Waldbronn D-76337 Germany.

Sergio Grinstein (**3,** 380), Department of Cell Biology, The Hospital for Sick Children, Toronto, Ontario M5G 1X8 Canada.

Pavel Gromov (**4,** 187, 225, 375, 409, 450, 454), Department of Medical Biochemistry and Danish Centre for Human Genome Research, Aarhus University, Aarhus C DK-8000 Denmark.

Dale F. Gruber (**1,** 19, 547), Cell Culture Research and Development, Life Technologies Incorporated, Grand Island, New York 14072.

Jean Gruenberg (**2,** 63, 248), Department of Biochemistry, Sciences II, University of Geneva, 1211-Geneva-4 Switzerland.

Martin Guttenberger (**4,** 295), Botanisdres Institut, Universität Tübingen, Tübingen D-72076 Germany.

---

[1] Deceased.

Michaela Haberfellner (**2**, 63), Forschunginstitut für Molekulare Pathologie Ges.m.b.h.H, Research Institute of Molecular Pathology, Wien, Vienna A-1030 Austria.

Einar Hallberg (**2**, 152), Department of Biochemistry, Stockholm University, Stockholm S-106-91 Sweden.

Fiona C. Halliday (**1**, 359), Department of Pharmacology, University College London, London WC1E GBT England.

Allister O. Hamilton (**1**, 27), Institute of Virology, Glasgow University, G11 5JR Scotland, United Kingdom.

Ian N. Hampson (**4**, 281), Department of Obstetrics and Gynecology, St. Mary's Hospital, Manchester M13 0JH United Kingdom.

Markus Häner (**3**, 277, 292), M.E. Müller Institute for Microscopy, Biozentrum, University of Basel, CH-4056 Switzerland.

Brian V. Harmon (**1**, 327), School of Life Sciences, Queensland University of Technology, Brisbane 04000 Australia.

Stefanie Hauser (**2**, 265), Zentrum für Molekulare Biologie, Universität Heidelberg, Heidelberg D-69120 Germany.

Pamela Hawley-Nelson (**4**, 145), Life Technologies, Rockville, Maryland 20849.

Robert J. Hay (**1**, 35, 43, 553), American Type Culture Collection, Rockville, Maryland 20852.

Izumi Hayashi (**1**, 393), Division of Neurosciences Beckman Research Institute of the City of Hope, Duarte, California 91010.

Rebecca Heald (**2**, 326), Department of Molecular and Cell Biology, University of California, Berkeley, Berkeley, California 94720.

Johannes W. Hell (**2**, 102), Department of Pharmacology, University of Washington School of Medicine, Seattle, Washington 98195.

Stefan Herr (**4**, 57), Biochemical Instrumentation Programme, EMBL, Heidelberg D-69117 Germany.

Theresa Higgins (**1**, 275), Imperial Cancer Research Fund, London WC2A 3PX United Kingdom.

Johnny Hindkjær (**3**, 437), Department of Gynecology, Aarhus University Hospital, Aarhus C DK-8000 Denmark.

Mary Hitt (**1**, 500), Departments of Biology and Pathology, McMaster University, Hamilton, Ontario L8S 4K1 Canada.

James P. Hoeffler (**4**, 239), Invitrogen, San Diego, California 92008.

Aldebaran M. Hofer (**3**, 363), Biomedical Sciences, Universita Degli Studi di Padova, Padova 35121 Italy.

Robert M. Hoffman (**1**, 377), AntiCancer, Incorporated, San Diego California 92111.

Hans Jürgen Hoffmann (**4**, 450), Medical Biochemistry and Danish Centre for Human Genome Research, Aarhus University, Aarhus C DK-8000 Denmark.

Marianne Hokland (**1**, 164), Department of Medical Microbiology and Immunology, Aarhus University Hospital, Aarhus C DK-8000 Denmark.

Peter Hokland (**1**, 164), Department of Medicine and Hematology, Aarhus University Hospital, Aarhus C DK-8000 Denmark.

Matthew Holley (**2**, 404), Department of Physiology, University of Bristol School of Medical Sciences, Bristol BS8 1TD United Kingdom.

Steen Holmberg (**1**, 421), Department of Genetics, Institute of Molecular Biology, University of Copenhagen, Copenhagen-K DK-1353 Denmark.

Claus Holst-Hansen (**1**, 16, 291), The Finsen Laboratory, Rigshospitalet, Copenhagen Denmark.

Nancy Hopkins (**3**, 502), Department of Biology, Massachusetts Institute of Technology, Cambridge, Massachusetts 02139.

David Hopwood (**3**, 221), Department of Molecular and Cellular Pathology, Ninewells Hospital, Dundee DD1 9SY Scotland.

Kathryn E. Howell (**2**, 12), Cellular and Structural Biology, University of Colorado, Health Sciences Center, Denver, Colorado 80262.

Chih-Lin Hsieh (**3**, 391), Norris Cancer Center, University of Southern California, Los Angeles, California 90033.

Lukas A. Huber (**1**, 107; **2**, 56, 63), Forschungsinstitut für Molekulare Pathologie Ges.m.b.H., Research Institute of Molecular Pathology, Wien, Vienna A-1030 Austria.

Eliezer Huberman (**1**, 233), Center for Mechanistic Biology and Biotechnology, Argonne National Laboratory, Argonne, Illinois 60493.

Christian Huet (**2**, 381, 421), C.N.R.S. Délégation Ile-de-France Sud, Bureau Formation, Gif-Sur-Yvette 91198 Cedex, France.

Norman Hui (**2**, 46), Cell Biology Laboratory, Imperial Cancer Research Fund, London WC2A 3PX United Kingdom.

Tony Hunter (**4**, 310), Microbiology and Virology Laboratory, Salk Institute, La Jolla, California 92186.

Eduard C. Hurt (**2**, 159), Institut für Biochemie I, University of Heidelberg, Heidelberg D-69120 Germany.

Wieland B. Huttner (**2**, 93), Institute for Neurobiology, University of Heidelberg, D-69120 Heidelberg Germany.

Anthony A. Hyman (**2**, 205, 326), Department of Cell Biology, EMBL, Heidelberg D-69117 Germany.

Marko Hyvönen (**4**, 255), EMBL, Heidelberg D-69012 Germany.

Kazuo Ikeda (**1**, 393), Division of Neurosciences, Beckman Research Institute of the City of Hope, Duarte, California 91010.

Elina Ikonen (**2**, 229), Cell Biology Program, EMBL, Heidelberg D-69012 Germany.

Susanne Isenberg (**2**, 486), Max Planck Institute for Biophysical Chemistry, Göttingen, D-37018 Germany.

Mitsuo Ishigami (**1**, 466), Department of Biology, Shiga University, Ohtsu 520 Japan.

Ryoki Ishikawa (**1**, 466), Department of Pharmacology, Gunma University School of Medicine, Maebashi, Gunma 371 Japan.

Sunita Iyengar (**2**, 165), Department of Molecular Biology, Vanderbuilt University, Nashville Tennessee 37235.

Marty R. Jacobson (**4**, 5), Cell Biology Group, Worcester Foundation for Biomedical Research, Shrewsbury, Massachusetts 01545.

Reinhard Jahn (**2**, 102), Department of Neurobiology, Max Planck Institute for Biophysical Chemistry, D-37077 Göttingen Germany.

David W. Jayme (**1**, 19, 27, 547), Cell Culture Research and Development, Life Technologies, Incorporated, Grand Island, New York 14072.

Niels A. Jensen (**4**, 375), Department of Medical Biochemistry and Danish Center for Human Genome Research, Aarhus University, Aarhus C Denmark.

Bengt Jergil (**2**, 5), Department of Biochemistry, Center for Chemistry and Chemical Engineering, Lund S-221 00 Sweden.

Keith Jermyn (**1**, 431), Clare Hall Laboratories, The Imperial Cancer Research Fund, Herts EN6 3LD United Kingdom.

He Jiang (**2**, 336), Laboratory of Molecular Cardiology, National Heart, Lung, and Blood Institute, National Institutes of Health, Bethesda, Maryland 20892-1762.

Andrew D. Johnson (**3**, 340), Institute of Biomedical and Life Sciences, University of Glasgow, G12 8QQ Scotland.

G. R. Johnson[2] (**1**, 172), Leukaemia Foundation of Queensland, Daikyo Research Unit, Queensland Institute of Medical Research, Brisbane, Queensland 04029 Australia.

Kirby L. Johnson (**3**, 444), Biology and Biotechnology Research Program, Lawrence Livermore National Laboratory, Livermore, California 94551.

Alain Joliot (**4**, 111), Centre National de la Recherche Scientifique, Ecole Normale Supérieure, Paris Cedex 05 France.

Sheila J. Jones (**1**, 142; **3**, 179), Department of Anatomy, University College London, London WC1E 6BT United Kingdom.

Steven M. Jones (**2**, 12), Department of Cellular and Structural Biology, University of Colorado, Health Sciences Center, Denver, Colorado 80262.

Thomas M. Jovin (**3**, 136, 355), Department of Molecular Biology, Max Planck Institute for Biophysical Chemistry, Göttingen D-37018 Germany.

Gloria Juan (**1**, 261, 341), The Cancer Research Institute, New York Medical College, Elmsford, New York 10523.

Yasufumi Kaneda (**4**, 123), Institute for Molecular and Cellular Biology, Osaka University, Osaka 565 Japan.

Sirpa Kärenlampi (**2**, 410), Department of Biochemistry and Biotechnology, University of Kuopio, Kuopio 70211 Finland.

Christina Karlsson (**4**, 246), Zoology Wellcome/CRC Institute, Cambridge CB2 1QR United Kingdom.

Eric Karsenti (**2**, 326), Department of Cell Biology, EMBL, Heidelberg D-69117 Germany.

Berthold Kastner (**2**, 174), Insituit für Molekularbiologie und Tumorfoschung, Marburg an der Lahn D-35037 Germany.

Mariko Katoh (**1**, 406), Institute of Biological Sciences, The University of Tsukuba, Tsukuba, Ibaraki 00305 Japan.

Madeleine Kihlmark (**2**, 152), Department of Biochemistry, Stockholm University, Stockholm, 106 91 Sweden.

Julie R. Kikkert (**4**, 157), Horticultural Sciences, Hedrick Hall, Cornell University, Geneva, New York 14456.

Lars Kilaas (**1**, 197), Industrial Chemistry, Norwegian University of Science and Technology, Trondheim N7034 Norway.

[2] Deceased.

John V. Kilmartin (**2,** 120), MRC, Laboratory of Molecular Biology, Cambridge CB2 2QH England.

Linda A. King (**4,** 204, 211), Insect Virus Research Group, School of Biological and Molecular Sciences, Oxford Brookes University, Oxford OX3 0BP England.

D. Kirubi (**4,** 176), Soil and Crop Science, Texas A & M University, College of Agriculture, College Station, Texas 77843-2474.

T. S. Ko (**4,** 176), Department of Soil and Crop Science, Texas A & M University, College of Agriculture, College Station, Texas 77843-2474.

Jørn Koch (**3,** 437), Department of Cytogenetics, Danish Cancer Society, Aarhus C DK-8000 Denmark.

Kazuhiro Kohama (**1,** 466), Department of Pharmacology, Gnuma University School of Medicine, Maebashi, Gunma 371 Japan.

Harri Kokko (**2,** 410), Department of Biochemistry and Biotechnology, University of Kuopio, Kuopio 70211 Finland.

Steen Kølvraa (**3,** 437), Institute for Human Genetics, University of Aarhus, Aarhus C DK-8000 Denmark.

Edward D. Korn (**1,** 413), Laboratory of Cell Biology, National Heart, Lung, and Blood Institute, National Institutes of Health, Bethesda, Maryland 20892.

Gabriela Krockmalnic (**2,** 184), Department of Biology, Massachusetts Institute of Technology, Cambridge Massachusetts 02139.

John Kunich (**3,** 444), Biology and Biotechnology Research Program, Lawrence Livermore National Laboratory, Livermore, California 94551.

Yasuhiro Kurasawa (**1,** 406), Institute of Biological Sciences, The University of Tsukuba, Tsukuba, Ibaraki 00305 Japan.

Ilpo Kuronen (**2,** 410), Department of Biochemistry and Biotechnology, University of Kuopio, Kuopio 70211 Finland.

Adam Kuspa (**1,** 431), Biology Center for Molecular Genetics, University of California San Diego, San Diego, California 92093.

Sergei A. Kuznetsov (**2,** 344), Fachbereich Bilogie, Institut für Zoologie, Lehrstuhl für Tierphysiologie, Universität Rostock, Rostock D-18055 Germany.

Lance A. Ladic (**3,** 189), Department of Physiology, University of British Columbia, Vancouver, British Columbia V6T 1Z3 Canada.

Frank Lafont (**2,** 229), Cell Biology Program, EMBL, Heidelberg D-69012 Germany.

Laimonis A. Laimins (**1,** 513), Department of Microbiology and Immunology, Penn State Medical School, Hershey, Pennsylvania 17033.

Thomas E. Lallier (**1,** 302), Louisiana State University Medical Center, School of Dentistry, Department of Anatomy, New Orleans, Louisiana 70119.

Amale Laouar (**1,** 233), Centre for Mechanistic Biology and Biotechnology, Argonne National Laboratory, Argonne, Illinois 60493.

Frank Larsen (**1,** 197), DYNAL A.S., Skoyen, Oslo NO212 Norway.

Martin R. Larsen (**4,** 556), Department of Molecular Biology, Odense University, Odense M DK-5230 Denmark.

Pamela L. Larsen (**1,** 398), Gerontology, University of Southern California, Los Angeles, California 90089-0191.

T. J. Last (**1,** 253), Department of Cell Biology, University of Massachusetts Medical Center, Worcester, Massachusetts 01655.

Valérie Laurent (**2,** 359), Laboratorie d'Enzymologie de CNRS, Gir-sur-Yvette, Cedex 91198 France.

Jette B. Lauridsen (**2,** 149, 429; **4,** 375), Department of Medical Biochemistry and Danish Center for Human Genome Research, Aarhus University, Aarhus C, Denmark.

André Le Bivic (**4,** 341), Marseille II, CNRS, Universite D'Aix, Marseille, Cedex 09 13288 France.

Wallace M. LeStourgeon (**2,** 165), Department of Molecular Biology, Vanderbilt University, Nashville, Tennessee 37235.

Margaret Leversha (**3,** 428), The Sanger Centre, Wellcome Trust Genome Campus, Hinxton, Cambridge CB10 1SA United Kingdom.

Chung L. Li (**1,** 172), Leukaemia Foundation of Queensland Daikyo Research Unit, Queensland Institute of Medical Research, Brisbane, Queensland 04029 Australia.

Jane B. Lian (**1,** 253), Department of Cell Biology, University of Massachusetts Medical Center, Worcester, Massachusetts 01655.

Hong Liang (**2,** 193), Beckman Laser Institute and Medical Center, University of California, Irvine, Irvine, California 92612.

Robert Lindner (**2,** 70), Department of Immunobiology, University of Bonn, Bonn 53117 Germany.

Andreas Lingnau (**2,** 398), Department of Molecular Microbiology, Washington University Medical School, St. Louis, Missouri 63110.

Maarten H. K. Linskens (**4,** 275), GBB, Cell Engineering Facility, University of Groningen, Groningen 9747 AG The Netherlands.

Yagang Liu (**2,** 193), Beckman Instruments, Incorporated, Brea, California 92622-8000.

Leslie M. Loew (**3,** 375), Department of Physiology, University of Connecticut Health Center, Farmington, Connecticut 06030.

Deryk T. Loo (**1,** 65), Pharmaceutical Research Institute, Bristol-Myers Squibb, Seattle, Washington 98121.

Friedrich Lottspeich (**4,** 304), Protein Analytics, Max-Planck Institute for Biochemistry, Martinsried D-82152 Germany.

Jette Lovmand (**1,** 528), Department of Microbiology and Immunology, University of Aarhus, Aarhus C Denmark.

Reinhard Lührmann (**2,** 174), Insituit für Molekularbiologie und Tumorforschung, Marburg an der Lahn D-35037 Germany.

Jiri Lukas (**4,** 489), Division of Cancer Biology, Danish Cancer Society, Copenhagen DK-2100 Denmark.

Anders H. Lund (**1,** 528), Department of Molecular and Structural Biology, University of Aarhus, Aarhus C, Denmark.

Kunxin Luo (**4,** 310), Berkeley National Laboratory, Berkeley, California 94720.

Hans Lyon (**3,** 232), Department of Pathology, University of Copenhagen, Hvidovre DK-2650 Denmark.

Karol Mackey (**2,** 221), Molecular Research Center, Incorporated, Cincinnati, Ohio 45212.

Jean-Jacques Madjar (**2,** 135), Faculte de Medecini Lyon-Laennec, Universite Claude Bernard Lyon I, Lyon Cedex 08 69372 France.

Peder Madsen (**4,** 187, 225), Department of Medical Biochemistry and Danish Center for Human Genome Research, Aarhus University, Aarhus C Denmark.

Sandra K. O. Mann (**1,** 431), Department of Biology, Center for Molecular Genetics, University of California San Diego, San Diego, California 92093.

Ahmed Mansouri (**3,** 478), Max-Planck Institute of Biophysical Chemistry, Göttingen, Am Fassberg 37077 Germany.

Paul A. Marks (**1,** 239), Cell Biology Program, Memorial Sloan Kettering Cancer Center and the Sloan Kettering Division of the Graduate School of Medicine, Cornell University, New York, New York 10021.

Bruno Martoglio (**2,** 265; **4,** 495), Zentrum für Molekulare Biologie, Universität Heidelberg, Heidelberg D-69120 Germany.

Susan A. Marlow (**4,** 204, 211), Biological and Molecular Sciences, Oxford Brookes University, Oxford OX3 0BP United Kingdom.

Alan D. Marmorstein (**4,** 341), Margaret M. Dyson Vision Research Institute, Department of Ophthalmology, Cornell University Medical Center, New York, New York 10021.

Ona C. Martin (**2,** 507), Carnegie Institution, Baltimore, Maryland 21210.

Bruno Martoglio (**2,** 265; **4,** 495), Zentrum für Molekulare Biologie, Universität Heidelberg, Heidelberg D-69120 Germany.

Kyomu Matsumoto (**3,** 444), Biology and Biotechnology Research Program, Lawrence Livermore National Laboratory, Livermore, California 94551.

Glenn Matthews (**4,** 37, 190), Department of Surgery, School of Medicine, University of Birmingham, Birmingham B15 2TH United Kingdom.

Arvid B. Maunsbach (**3,** 249, 260, 268), Department of Cell Biology, Institute of Anatomy, Aarhus University, Aarhus C DK-8000 Denmark.

James G. McAfee (**2,** 165), Department of Molecular Biology, Vanderbilt University, Nashville, Tennessee 37235.

Laura McCabe (**1,** 253), Department of Cell Biology, University of Massachusetts Medical Center, Worcester, Massachusetts 01655.

Brenda McCormack (**4,** 131), Centre for Drug Delivery Research, School of Pharmacy, London WC1N 1AX United Kingdom.

Ann McGinty (**4,** 351), Centre for Peptide and Protein Engineering, School of Biology and Biochemistry, The Queen's University of Belfast, Belfast BT9 7BL Northern Ireland.

Paul L. McNeil (**4,** 49), Cellular Biology and Anatomy, Medical College of Georgia, Augusta, Georgia 30912.

Carl R. Merril (**4,** 421), Laboratory of Biochemical Genetics, National Institutes of Mental Health, Washington, DC 20032.

Karin Meyer (**1,** 164), Department of Medicine and Hematology, Aarhus University Hospital, Aarhus C DK-8000 Denmark.

Craig Meyers (**1,** 513), Department of Microbiology and Immunology, Penn State Medical School, Hershey, Pennsylvania 17033.

Liane Meyn (**1,** 154), Cell Biology Program, EMBL, Heidelberg D-69117 Germany.

Laurée Salamin Michel (**3,** 323), Laboratorie Analyse Ultrastlect, Universite Lausanne, Dorigny 01015 Switzerland.

Brigitte Mies (**2**, 469), Institute of Molecular Biology, Department of Cell Biology, Salzburg A-5020 Austria.

Timothy J. Mitchison (**3**, 127), Department of Cellular and Molecular Pharmacology, University of California, San Francisco, San Francisco, California 94143.

Didier Montarras (**1**, 226), Laboratoire Développement Cellulaire, Département de Biologie Moléculaire, Paris 75015 France.

Yvonne Morrison (**4**, 131), Centre for Drug Delivery Research, School of Pharmacy, London WC1N 1AX United Kingdom.

Mohammed Moudjou (**2**, 111), Institut Curie, Section Recherche CRNS UMR 144, Cedex 05 Paris, 75428 France.

Ruth M. Mould (**2**, 81, 286), Department of Plant Sciences, University of Cambridge, Cambridge CB2 3EA United Kingdom.

Anne Muesch (**2**, 239), Margaret M. Dyson Vision Research Institute, Department of Ophthalmology, Cornell University Medical College, New York, New York 10021.

Andrew Murray (**2**, 326), Department of Pharmacology, University of California, San Francisco, San Francisco, California 94143.

Nobuhiro Nakamura (**2**, 46), Cell Biology Laboratory, Imperial Cancer Research Fund, London WC2A 3PX United Kingdom.

Axl Alois Neurauter (**1**, 197), DYNAL A.S. Skoyen Oslo NO212 Norway.

Jeffrey A. Nickerson (**2**, 184), Department of Cell Biology, University of Massachusetts Medical School, Worcester, Massachusetts 01655.

Garth L. Nicolson (**1**, 296), Department of Tumor Biology, The University of Texas M. D. Anderson Cancer Center, Houston, Texas 77030.

Kirsten Niebuhr (**2**, 398), Institut Pasteur, Unité de Pathogénie Microbienne Moléculaire, Paris 75724 Cédex 15 France.

Markus Niederreiter (**4**, 398), Department of Molecular Biology, Austrian Academy of Sciences, Salzburg A-5020 Austria.

Heinz Nika (**4**, 514), Hewlett Packard Company, Palo Alto, California.

Akinori Noma (**1**, 125), Department of Physiology, Kyoto University, Kyoto 606-01 Japan.

Eckhard Nordhoff (**4**, 556), Department of Molecular Biology, Odense University, Odense M DK-5230 Denmark.

Osamu Numata (**1**, 406), Institute of Biological Sciences, The University of Tsukuba, Tsukuba, Ibaraki 00305 Japan.

William M. Nuttley (**2**, 295), Department of Anatomy and Cell Biology, University of Toronto, Toronto, Ontario M5S 1A8 Canada.

Bronwyn A. O'Brien (**1**, 327), School of Life Sciences, Queensland University of Technology, Brisbane 04000 Australia.

Tracy J. O'Connor (**1**, 149), NeuroSpheres Limited, Calgary, Alberta T2N 4N1 Canada.

Louis A. Obosi (**4**, 211), Insect Virus Research Group, School of Biological and Molecular Sciences, Oxford Brookes University, Oxford OX3 0BP England.

Martin Oft (**1**, 107), Research Institute of Molecular Pathology, Vienna A-1030 Austria.

Phil Oh (**2**, 34), Department of Pathology, Harvard Medical School, Boston, Massachusetts 02215.

Ronald Jowett Oldfield (**3**, 5), School of Biological Sciences, Macquarie University, Sydney, NSW, 2109 Australia.

Ey∂finnur Olsen (**4**, 361), Institute of Medical Biochemistry, Danish Centre for Human Genome Research, Aarhus C. DK-8000 Denmark.

Michael G. Ormerod (**1**, 351; **4**, 88), Scientific Consultant, Reigate RH2 ODE England.

Mary Osborn (**2**, 462, 486), Max Planck Institute for Biophysical Chemistry, Göttingen D-37018 Germany.

Thomas A. Owen (**1**, 253), Pfizer Central Research, Centeral Research Division, Groton, Connecticut 06340.

Rosalind J. Packer (**1**, 178), European Collection of Cell Cultures, Center for Applied Microbiology and Research, Wiltshire SP4 0JG United Kingdom.

Christian Paech (**4**, 474), Genencor International, Palo Alto, California 94304.

Richard E. Pagano (**2**, 507), Mayo Clinic and Foundation, Department of Biochemistry and Molecular Biology, Rochester, Minnesota 55905.

Carole A. Parent (**1**, 431), Department of Biological Chemistry, Johns Hopkins University School of Medicine, Baltimore, Maryland 21205.

Sung Hun Park (**4**, 176), Soil and Crop Science, Texas A & M University, College of Agriculture, College Station, Texas 77843.

Bryce M. Paschal (**2**, 305), Center for Cell Signaling, University of Virginia, Charlottesville, Virginia 22908.

James B. Pawley (**3**, 149), University of Wisconsin, Madison, Wisconsin 53706.

Finn Skou Pedersen (**1**, 528), Department of Molecular and Structural Biology and Microbiology and Immunology, University of Aarhus, Aarhus C, Denmark.

Thoru Pederson (**4**, 5), Cell Biology Group, Worcester Foundation for Biomedical Research, Shrewsbury, Massachusetts 01545.

Sheldon Penman (**2**, 184), Department of Biology, Massachusetts Institute of Technology, Cambridge, Massachusetts 02139.

Rainer Pepperkok (**4**, 23, 57), Imperial Cancer Research Fund, WC2A 3PX London.

Hedvig Perlmann (**2**, 438), Department of Immunology, Stockholm University, Stockholm 106 91 Sweden.

Peter Perlmann (**2**, 438), Department of Immunology, Stockholm University, Stockholm 106 91 Sweden.

Mark Pershouse (**3**, 411), Department of Neuro-Oncology, The Brain Tumor Center, The University of Texas M.D. Anderson Cancer Center, Houston, Texas 77030.

Paul D. Phillips (**1**, 321), Department of Biology, West Chester University, West Chester, Pennsylvania 19383.

Lía I. Pietrasanta (**3**, 355), Instituto de Investigaciones Bioquimicas, INIBIBB, Bahia Blanca 08000 Argentina.

Jonathon Pines (**4**, 246), Department of Zoology, Wellcome/CRC Institute, Cambridge CB2 1QR United Kingdom.

Christian Pinset (**1**, 226), Department of Biologie Moleculaire, Laboratoire de Developpement Cellulaire, Paris 75015 France.

Martin Poot[3] (**2**, 513), Department of Biosciences, Molecular Probes, Incorporated, Eugene, Oregon 97402.

Robert D. Possee (**4**, 204, 211), NERC, Institute of Virology and Environmental Microbiology, Oxford OX1 3SR United Kingdom.

Ingo Potrykus (**1**, 478; **4**, 171), Institute of Plant Sciences, Swiss Federal Institute of Technology, Zurich CH-8092 Switzerland.

Trevor Powell (**1**, 125), University Laboratory of Physiology, Oxford OX1 3PT United Kingdom.

Tullio Pozzan (**3**, 363), Biomedical Sciences, Universita Degli Studi di Padova, Padova 35121 Italy.

Ludvik Prevec (**1**, 500), Departments of Biology and Pathology, McMaster University, Hamilton, Ontario L8S 4K1 Canada.

Alain Prochiantz (**4**, 111), Developpement Et Evolution Du Systeme Nerveux, Cedex 05, Paris, 75230 France.

Anton K. Raap (**3**, 453), Department of Cytochemistry and Cytometry, Syluius Laboratories, Leichen University, Al Leiden 02333 The Netherlands.

Henrik Rahbek-Nielsen (**4**, 556), Department of Molecular Biology, Odense University, Odense M DK-5230 Denmark.

Marilyn J. Ramsey (**3**, 444), Biology and Biotechnology Research Program, Lawrence Livermore National Laboratory, Livermore, California 94551.

Leda Raptis (**4**, 64, 75), Microbiology and Immunology and Pathology, Queen's University, Kingston, Ontario K7L 3N6 Canada.

Hanne H. Rasmussen (**4**, 505), Department of Medical Biochemistry, University of Aarhus, Aarhus C DK-8000 Denmark.

Gitte Ratz (**4**, 375, 404), Department of Medical Biochemistry and Danish Center for Human Genome Research, Aarhus University, Aarhus C Denmark.

Ernst Reichmann (**1**, 107), Institute Suissede Recherches Experimentales sur le Cancer, Epalinges/Lausanne CH-1066 Switzerland.

Sigrid Reinsch (**3**, 170), EMBL, Heidelberg 69012 Germany.

Siegfried Reipert (**3**, 346), Vienna Biocentre, Institute of Biochemistry and Moleculare Cell Biology, Vienna A-1030 Austria.

Brent A. Reynolds (**1**, 149), NeuroSpheres Limited, Calgary, Alberta T2N 4N1 Canada.

Victoria M. Richon (**1**, 239), Cell Biology Program, Memorial Sloan Kettering Cancer Center and the Sloan Kettering Division of the Graduate School of Medicine, New York, New York 10021.

Donald L. Riddle (**1**, 398), Division of Biological Sciences, University of Missouri, Columbia, Missouri 65211.

Richard A. Rifkind (**1**, 239), Cell Biology Program, Memorial Sloan Kettering Cancer Center and the Sloan Kettering Division of the Graduate School of Medicine, New York, New York 10021.

William A. Ritchie (**3**, 487), Roslin Institute, Roslin, Midlothian EH25 9PS United Kingdom.

Linda J. Robinson (**2**, 248), Department of Biochemistry, Sciences II, University of Geneva, 1211-Geneva-4 Switzerland.

---

[3] Current address: Department of Pathology, University of Washington, Seattle, Washington 98195.

Enrique Rodriguez-Boulan (**2,** 239; **4,** 341), Margaret M. Dyson Vision Research Institute, Department of Ophthalmology, Cornell University Medical College, New York, New York 10021.

Peter Roepstorff (**4,** 556), Department of Molecular Biology, Odense University, Odense M DK-5230 Denmark.

Robert Romanek (**3,** 380), Division of Cell Biology, Hospital for Sick Children, Toronto M5G 1X8 Canada.

Norbert Roos (**3,** 332), Electronmicroscopical Unit for Biological Sciences, University of Oslo, Blindern, N-0316 Oslo Norway.

S. Rose (**4,** 176), Department of Soil and Crop Science, Texas A & M University, College of Agriculture, College Station, Texas 77843-2474.

Sabine Rospert (**2,** 277), Bio Sentrum University of Basel, Basel CH-4056 Switzerland.

Ori D. Rotstein (**3,** 380), Department of Surgery, Toronto General Hospital, Toronto M5G 2C4 Canada.

Klemens Rottner (**2,** 469), Institute of Molecular Biology, Department of Cell Biology, A-5020 Salzburg, Austria.

Michael P. Rout (**2,** 120, 143), Rockefeller University, New York, New York 10021.

Tony Rowe (**2,** 258), Department of Cell Biology, The Scripps Research Institute, La Jolla, California 92037.

Enrique Rozengurt (**1,** 275), Imperial Cancer Research Fund, London WC2A 3PC United Kingdom.

Michael J. Rudick (**4,** 337), Department of Biology, Texas Woman's University, Denton, Texas 76204-3799.

Shahin M. Saati (**4,** 275), Geron Corporation, Menlo Park, California 94025.

Yoshinaga Saeki (**4,** 123), Institute for Molecular and Cellular Biology, Osaka University, Osaka 565 Japan.

Daniel Safer (**4,** 371), Department of Cell and Developmental Biology, University of Pennsylvania, Philadelphia, Pennsylvania 19104.

Roghieh Saffie (**4,** 131), Centre for Drug Delivery Research, School of Pharmacy, London WC1N 1AX United Kingdom.

Rainer Saffrich (**4,** 23, 57), EMBL, Heidelberg 69117 Germany.

Roland Sahli (**1,** 493), Institute of Microbiology, Centre Hospitalier Universitaire, Vaudois, Lausanne 01011 Switzerland.

Hans Peter Saluz (**4,** 429), Hans-Knoll-Institüt für Natursoff-Forschung, Jena D-07745 Germany.

Paul M. Salvaterra (**1,** 393), Division of Neurosciences, Beckman Research Institute of the City of Hope, Duarte California 91010.

Jeremy Sanderson (**3,** 15), Sir William Dunn School of Pathology, University of Oxford, Oxford OX1 3RE United Kingdom.

Matti Sarasti (**4,** 255), EMBL, Heidelberg D-69012 Germany.

Nathalie Sartori (**3,** 323), Laboratoire Analyse Ultrastlect, Universite Lausanne, Dorigny 01015 Switzerland.

Kenneth E. Sawin (**3,** 127), Cell Cycle Club, ICRF, Lincoln's Inn Fields, London WC2A 3PX United Kingdom.

Achim Schaper (**3,** 355), Department of Molecular Biology, Max Planck Institute for Biophysical Chemistry, Göttingen D-37018 Germany.

Gottfried Schatz (**2,** 277), Bio Sentrum, University of Basel, Basel CH-4056 Switzerland.

Leif Schauser (**3,** 518), Laboratory of Gene Expression and Department of Molecular Biology, University of Aarhus, Aarhus C DK-8000 Denmark.

Wim J. J. M. Scheenen (**3,** 363), Department of Cellular Animal Physiology, ED Nijmegen 6525, The Netherlands.

Karen L. Schmeichel (**4,** 441), Ernest Orlando Lawrence Berkeley National Laboratory, Berkeley, California 94720.

Jan E. Schnitzer (**2,** 34), Department of Pathology, Harvard Medical School, Boston, Massachusetts 02215.

Morten Schou (**1,** 285), Bartholin Instituttet, Kobenhavns Kommunehospital, Denmark.

Sandra Scianimanico (**2,** 75), Departement d'Anatomie, Universite de Montreal, Montreal, Quebec H3C 1J4 Canada.

Robert E. Scott (**1,** 221), Department of Pathology, University of Tennessee College of Medicine, Memphis, Tennessee 38163.

Antonio Sechi (**2,** 366; **3,** 285), GBF, Gesellschaft für Biotechnologische Forschung, Department of Cell Biology and Immunology, Braunschweig D-38124 Germany.

Bartholomew M. Sefton (**4,** 310), Molecular Biology and Virology Laboratory, Salk Institute, La Jolla, California 92186.

Jeffrey Segall (**1,** 431), Anatomy and Structural Biology, Albert Einstein College of Medicine, Bronx, New York 10461.

Per O. Seglen (**1,** 119), Department of Cell Biology, Institute for Cancer Research, Montelbello, Oslo 3 N-0310 Norway.

James R. Sellers (**2,** 336), Laboratory of Molecular Cardiology, National Heart, Lung, and Blood Institute, National Institutes of Health, Bethesda, Maryland 20892.

Nicholas J. Severs (**1,** 125; **3,** 299), Department of Cardiac Medicine, National Heart and Lung Institute, Imperial College of Science, Technology, and Medicine, London SW3 6NP United Kingdom.

Deborah A. Shackelford (**4,** 466), Department of Neurosciences, University of California, San Diego, San Diego, California 92093.

Gad Shaulsky (**1,** 431), Department of Biology, Center for Molecular Genetics, University of California, San Diego, San Diego, California 92093.

Peter J. Shaw (**3,** 206), John Innes Centre, Colney, Norwich NR4 7UH United Kingdom.

David M. Shotton (**2,** 26, 213; **3,** 73, 85, 299, 310), Cell Biology Research Group, Department of Zoology, University of Oxford, Oxford OX1 3PS United Kingdom.

Kai Simons (**1,** 101; **2,** 56, 229), Cell Biology Program, EMBL, Heidelberg D-69012 Germany.

Symeon Siniossoglou (**2,** 159), Institüt für Biochemie I, University of Heidelberg, Heidelberg D-69120 Germany.

Mathilda Sjöberg (**1,** 534), Department of Biosciences at Novum, Karolinska Institutet, Huddinge S-141-57 Sweden.

Philip Skehan (**1,** 313), Andes Pharmaceuticals, Incorporated, Los Altos, California 94023.

Paul Slusarewicz (**2,** 46), Biosciences Division, Unilever Research, Colworth House, Sharnbrook, Bedford MK44 1LQ United Kingdom.

J. Victor Small (**2**, 366, 372, 469; **3**, 285), Institute of Molecular Biology, Austrian Academy of Sciences, Salzburg A-5020 Austria.

Roberta H. Smith (**1**, 472, 485; **4**, 176), Department of Soil and Crop Science, Texas A & M University, College of Agriculture, College Station, Texas 77843.

Carl Soderland (**1**, 137), Cell Culture Research and Development, Life Technologies, Incorporated, Grand Island, New York 14072.

Gregory J. Sonek (**2**, 193), Beckman Laser Institute and Medical Clinic, University of California, Irvine, Irvine, California 92612.

Karen J. Sorensen (**3**, 444), Biology and Biotechnology Research Program, Lawrence Livermore National Laboratory, Livermore, California 94551.

German Spangenberg (**1**, 478; **4**, 162), Plant Sciences and Biotechnology, Agriculture Victoria, La Trobe University, Bundoora, Victoria 3083 Australia.

M. Srivatanakul (**4**, 176), Department of Soil and Crop Science, Texas A & M University, College of Agriculture, College Station, Texas 77843.

Eric J. Stanbridge (**3**, 405), Microbiology and Molecular Genetics, University of California, Irvine, Irvine, California 92717.

Peter Steck (**3**, 411), Department of Neuro-Oncology, The Brain Tumor Center, The University of Texas M.D. Anderson Cancer Center, Houston, Texas 77030.

Gary S. Stein (**1**, 253), Department of Cell Biology, University of Massachusetts Medical Center, Worcester, Massachusetts 01655.

Janet L. Stein (**1**, 253), Department of Cell Biology, University of Massachusetts Medical Center, Worcester, Massachusetts 01655.

Peter Steinlein (**2**, 63), Forschunginstitüt für Molekulare Pathologie Ges.m.b.h.H., Research Institute of Molecular Pathology, Wien, Vienna A-1030 Austria.

Ernst Stelzer (**3**, 170), EMBL, Heidelberg D-69012 Germany.

Harald Stenmark (**1**, 540; **4**, 141, 201), Department of Biochemistry, The Norwegian Radium Hospital, Montelbello, Oslo N-0310 Norway.

Baruch Stern (**2**, 446), Cell Research and Immunology, Tel-Aviv University, Ramat-Aviv, Tel-Aviv 69978 Isreal.

Jane C. Stinchcombe (**2**, 93), MRC Laboratory for Molecular Cell Biology, University College of London, London WC1E 6BT United Kingdom.

Jens Stougaard (**3**, 518), Laboratory of Gene Expression and Department of Molecular Biology, University of Aarhus, Aarhus C DK-8000 Denmark.

Caterina Strambio-de-Castillia (**2**, 143), Laboratory of Cell Biology, Rockefeller University, New York, New York 10021.

Elton Stubblefield (**3**, 411), Department of Neuro-Oncology, The Brain Tumor Center, The University of Texas M.D. Anderson Cancer Center, Houston, Texas 77030.

Damir Sudar (**3**, 109), Life Sciences Division, Lawrence Berkely National Laboratory, Berkeley, California 94720.

Joel A. Swanson (**2**, 495), Department of Cell Biology, University of Michigan Medical School, Ann Arbor, Michigan 48109.

Nobuyuki Tanahashi (**2**, 129), The Tokyo Metropolitan Institute of Medical Scences, Bunkyo-ku, Toyoko 113 Japan.

Keiji Tanaka (**2**, 129), The Tokyo Metropolitan Institute of Medical Scences, Bunkyo-ku, Toyoko 113 Japan.

Hans J. Tanke (**3,** 64), Department of Cytochemistry and Cytometry, Leiden University, Al Leiden 02333 The Netherlands.

Kenneth K. Teng (**1,** 244), Division of Hematology/Onocology, Department of Medicine, Cornell University Medical College, New York, New York 10021.

Mark Terasaki (**2,** 501), Department of Physiology, University of Connecticut Health Center, Farmington, Connecticut 06032.

Julie A. Theriot (**2,** 350; **3,** 127), Department of Biochemistry and Beckman Center, Stanford University School of Medicine, Stanford, California 94305.

Thomas Thykjær (**3,** 518), Laboratory of Gene Expression and Department of Molecular Biology, University of Aarhus, Aarhus C DK-8000 Denmark.

Mary Lynn Tilkins (**4,** 145), Life Technologies, Incorporated, Grand Island, New York 14072.

Leath A. Tonkin (**4,** 275), Geron Corporation, Menlo Park, California 94025.

Régis Tournebize (**2,** 326), Department of Cell Biology, EMBL, Heidelberg D-69117 Germany.

James D. Tucker (**3,** 444), Biology and Biotechnology Research Program, Lawrence Livermore National Laboratory, Livermore, California 94551.

Pierre S. Tung (**1,** 186), Scarboro, Ontario M1W 3W1 Canada.

John Ugelstad[4] (**1,** 197; **2,** 12), Department of Industrial Chemistry, Norwegian University of Science and Technology, Trondheim N7034 Norway and Cellular and Structural Biology, University of Colorado, Health Sciences Center, Denver, Colorado 80262.

Mariëtte P. C. van der Corput (**3,** 453), Department of Cytochemistry and Cytometry, Syluius Laboratories, Leichen University, Al Leiden 02333 The Netherlands.

Peter van der Geer (**4,** 310), Department of Chemistry and Biochemistry, University of California, San Diego, San Diego, California 92093.

Marc Van Montagu (**4,** 92), Department of Genetics, Flanders Interuniversity Institute for Biotechnology (VIB), Universiteit Gent, Gent B-9000 Belgium.

Lucas J. van Vliet (**3,** 109), Faculty of Applied Physics, Delft University of Technology, Delft CJ-2628 The Netherlands.

Joël Vandekerckhove (**4,** 505), Institute of Biotechnology and Department of Biochemistry, University of Gent, Gent B-9000 Belgium.

Philip H. Vardy (**1,** 431), Department of Biological Sciences, University of Western Sydney-Nepean, Westmead NSW 2145 Australia.

Isabelle Vernos (**2,** 326), Department of Cell Biology, EMBL, Heidelberg D-69117 Germany.

Angelo L. Vescovi (**1,** 149), NeuroSpheres Limited, Calgary, Alberta T2N 4N1 Canada.

Pavel Vesely (**3,** 54), Czechoslavak Academy of Sciences, Institute of Molecular Genetics, 16637 Praha 6, Chech Republic.

Hanne Vestergaard (**1,** 164), Department of Medical Microbiology and Immunology, University of Aarhus, Aarhus C 8000 Denmark.

Hilkka Virta (**1,** 101), Department of Cell Biology, EMBL, Heidelberg D-69012 Germany.

Ann-Mari Voie (**3,** 510), EMBL, Heidelberg D-69117 Germany.

Tova Volberg (**2,** 457), Chemical Immunology, The Weizmann Institute of Science, Rehovot 76100 Israel.

[4] Deceased.

Alfred Völkl (**2,** 87), Department of Anatomy and Cell Biology (II), University of Heidelberg, Heidelberg D-69120 Germany.

Hironao Wakabayashi (**1,** 296), Department of Tumor Biology, The University of Texas M. D. Anderson Cancer Center, Houston, Texas 77030.

Brian Walker (**4,** 351), Centre for Peptide and Protein Engineering, School of Biology and Biochemistry, The Queen's University of Belfast, Belfast BT9 7BL Northern Ireland.

Andrew Wallace (**4,** 429), Medical Biology Centre, School of Biology and Biochemistry, The Queen's University of Belfast, Belfast BT9 7BL Northern Ireland.

Hanlin Wang (**1,** 221), Department of Pathology, University of Tennessee College of Medicine, Memphis, Tennessee 38163.

Yu-Li Wang (**4,** 5), Cell Biology Group, Worcester Foundation for Biomedical Research, Shrewsbury, Massachusetts 01545.

Zeng-Yu Wang (**1,** 478; **4,** 162), Plant Sciences and Biotechnology, Agriculture Victoria, La Trobe University, Bundoora, Victoria 3083 Australia.

Graham Warren (**2,** 46), Cell Biology Laboratory, Imperial Cancer Research Fund, London WC2A 3PX United Kingdom.

Yoshio Watanabe (**1,** 406), Jobu University, Isesaki City, Gunmma 372, Japan.

Fiona M. Watt (**1,** 113), Keratinocyte Laboratory, Imperial Cancer Research Fund, London WC2A 3PX United Kingdom.

Jürgen Wehland (**2,** 398), Department of Cell Biology and Immunology, GBF-National Research Center for Biotechnology, Braunschweig D-38124 Germany.

Dieter G. Weiss (**2,** 344; **3,** 99), Fachbereich Bilogie, Insitiüt für Zoologie, Lehrstuhl fur Tierphysiologie, Universitat Rostock, Rostock D-10855 Germany.

Walter Weiss (**4,** 386), Technical University of Munich, Freising-Weihenstephan D-85350 Germany.

Ulrich Weller (**4,** 103), Institut für Medizinische Mikrobiologie und Hygiene, Johannes-Gutenberg University-Mainz, Mainz D-55101 Germany.

Martin Wendland (**2,** 295), Marl D-45770 Germany.

Michael Whitaker (**3,** 121), Department of Physiological Sciences, University of Newcastle Medical School, United Kingdom.

Stefan Wiemann (**4,** 57), Deutshces Krebsforschungszentrum, Heidelberg D-69120 Germany.

Jeffrey Williams (**1,** 431), Clare Hall Laboratories, The Imperial Cancer Research Fund, Herts EN6 3LD United Kingdom.

Keith L. Williams (**1,** 431), Department of Biological Sciences, Macquarie University, Sidney NSW 2109 Australia.

Louise E. Wilson (**4,** 204), Biological and Molecular Sciences, Oxford Brookes University, Oxford OX3 0BP United Kingdom.

Stuart M. Wilson (**3,** 340), Department of Child Health, Ninewells Hospital and Medical School, Dundee DD1 9S4 Scotland.

Clay M. Winterford (**1,** 327), Department of Pathology, University of Queensland Medical School, Brisbane 04006 Australia.

John R. Yates, III (**4,** 529, 539), Department of Molecular Biotechnology, University of Washington, Seattle, Washington 98195.

Charles Yeaman (**2**, 239), Margaret M. Dyson Vision Research Institute, Department of Ophthalmology, Cornell University Medical College, New York, New York 10021.

Ian T. Young (**3**, 109), Faculty of Applied Physics, Delft University of Technology, Delft CJ-2628 The Netherlands.

Brahim Zadi (**4**, 131), Centre for Drug Delivery Research, School of Pharmacy, London WC1N 1AX United Kingdom.

Cecilia Zapata (**1**, 485; **4**, 176), Department of Soil and Crop Sciences, Texas A & M University, College of Agriculture, College Station, Texas 77843.

Marino Zerial (**1**, 540; **4**, 141, 201), EMBL, Heidelberg D-69102 Germany.

Daniel Zicha (**3**, 44), MRC Muscle and Cell Motility Unit, The Randall Institute, Kings College London, London WC2B 5RL United Kingdom.

Chiara Zurzolo (**4**, 341), Biologia e Patologia Cellulare e Molecolare, Universita Federico II, Napoli 80131 Italy.

The second edition of *Cell Biology: A Laboratory Handbook* has been revised and expanded, taking into consideration the feedback from the scientific community. As a result, the *Handbook* has grown to four volumes and contains over 240 chapters flowing from cells to proteins. Volume 1 deals with tissue culture and associated techniques, discusses viruses, and includes appendices. Volume 2 covers organelles and cellular structures, assays, antibodies, immunocytochemistry, vital staining of cells, and Internet Resources. Volume 3 presents light microscopy and contrast generation, electron microscopy, intracellular measurements, cytogenetics and *in situ* hybridization, and transgenics and gene knockouts. Volume 4 completes the set with topics such as transfer of macromolecules and small molecules, expression systems, differential gene expression, and proteins and also includes appendices.

As in the previous edition, we have kept to an explicit and reader-friendly format in an effort to facilitate the execution of the protocols. As mentioned by Professor Thomas Pollard in his review of the first edition: "Short of having an expert at your side, or a collection of smaller books in tissue culture, microscopy, and routine biochemical methods, this is probably the best way to introduce new methods in a cell biology laboratory." Since new techniques are emerging rapidly, we very much welcome additional comments and suggestions for future editions.

In addition to the authors, who went out of their way to keep the pressing deadlines, there are a few other people who contributed to the making of the second edition of *Cell Biology: A Laboratory Handbook*. Among them I would like to thank my son Juan Pablo for preparing the artwork for a few chapters and my secretaries, Lene Svith and Inge Detlefsen, for giving me a helping hand at various stages of the project, in particular, with the mailing of the final version of the manuscripts to Academic Press. I also thank my family for providing me with much needed support during hectic times.

Finally, I extend my deepest appreciation to the Associate Editors, who actively participated in the planning of the book as well as in the cross-referencing of the various chapters. Also my deepest gratitude goes to Craig Panner of Academic Press for his enthusiastic and professional handling of the project. Many thanks also go to Suzanne Miller and all of the staff in the Editorial, Production, and Marketing departments at Academic Press for having produced these handsome volumes on schedule.

Julio E. Celis
Editor

PART 3

Organelles and **C**ellular **S**tructures

Plasma Membrane and Cytoplasmic Organelles

Purification of Rat Liver Plasma Membranes by Affinity Partitioning

Lars Ekblad and Bengt Jergil

I. INTRODUCTION

Affinity partitioning in aqueous polymer two-phase systems is an alternative technique for the fractionation of biological material. Plasma membranes, for instance, may be purified using this technique by applying wheat germ agglutinin (WGA) or other lectins as affinity ligands (Persson *et al.*, 1991; Persson and Jergil, 1992). Fractionation by affinity partitioning is advantageous by being highly selective and much more rapid than conventional membrane preparation protocols. In addition, the aqueous polymer environment is gentle to membrane structure and function.

Aqueous two-phase systems will form when solutions of two structurally different polymers are mixed at sufficiently high concentrations, and each phase will be enriched in one of the polymers (for a review see Albertsson, 1986). Polyethylene glycol 3350 and dextran T500 is a commonly used polymer pair, and these will form an aqueous two-phase system above 5.4% (w/w) of each polymer. Biological material, including membranes, can be fractionated in such conventional two-phase systems. The partitioning of material between the phases is strongly dependent on, for instance, polymer concentration and salt contents in the system (Gierow *et al.*, 1986). The distribution of the material can be manipulated by altering these factors, but the selectivity obtained is often insufficient for the facile separation of membranes.

More selective conditions may be obtained by introducing an affinity ligand into one of the phases. By conjugating a suitable ligand to one of the phase polymers, the resulting polymer–ligand adduct will partition in the corresponding polymer phase, selectively pulling the target membranes into this phase. Ideally, the experimental conditions are chosen so that all membranes present partition in one of the phases in the absence of affinity ligand. When introduced, the ligand conjugated to the second-phase polymer will pull the target membranes only from the first phase into this phase.

This article describes the purification of plasma membranes by affinity partitioning in a polyethylene glycol/dextran two-phase system using WGA conjugated to dextran as the affinity ligand. WGA binds *N*-acetylglucosamine and sialic acid residues exposed on the extracellular face of plasma membranes, thereby discriminating these membranes from other cellular membranes. Two protocols are presented. The first one deals with the purification of plasma membranes from crude rat liver membranes by affinity partitioning (Persson *et al.*, 1991); the second protocol describes the purification of plasma membranes from rat liver tissue by homogenization and partitioning in a conventional two-phase system followed by affinity partitioning (Persson and Jergil, 1992). Although these protocols have been developed for rat liver membranes, they should be easy to adapt to plasma membranes from other animal tissues as well.

II. MATERIALS AND INSTRUMENTATION

Dextran T-500 (Cat. No. 17-0320) is from Pharmacia Biotech and polyethylene glycol 3350 (PEG) is from Union Carbide. The dextran is dissolved in water and freeze-dried prior to use. Dimethyl sulfoxide (DMSO) (Cat. No. 802912), triethylamine (Cat. No. 808352), dichloromethane (Cat. No. 822271), sodium dihydrogen phosphate monohydrate (Cat. No. 6346), sodium chloride (Cat. No. 106404), boric acid (Cat. No. 100160), and sulfuric acid (Cat. No. 732) are from Merck. DMSO, triethylamine, and dichloromethane are dried with molecular sieves prior to use. Lectin WGA (Cat. No. 1359029) is from Boehringer-Mannheim and 2,2,2-trifluoroethanesulfonyl chloride (tresyl chloride) is from Synthelec AB. Tris (Cat. No. T-1378) and N-acetyl-D-glucosamine (Cat. No. A-8625) are from Sigma. Sucrose (Cat. No. 10274) is from BDH. Molecular sieves (Cat. No. 1-6264) are from KEBO Lab.

The Filtron Omegacell 150 is from Filtron Technology Corp. Two bench centrifuges are used: Jouan B 3-11 equipped with a swinging bucket rotor for volumes larger than 5 ml and Wifug DOCTOR for smaller volumes. The Dounce homogenizer is from Kontes Glass Co.

III. PROCEDURES

A. Preparation of WGA–Dextran

The affinity ligand WGA is coupled to dextran in a two-step process. First, dextran T500 is activated by treatment with tresyl chloride, and then WGA is conjugated to the activated dextran. These steps can be separated in time, as activated dextran is sufficiently stable to be stored.

Solutions

1. *Coupling buffer:* 0.1 M sodium phosphate, 0.5 M NaCl, pH 7.5. To make 200 ml, dissolve 2.76 g $NaH_2PO_4 \cdot H_2O$ and 5.84 g NaCl in 160 ml distilled water. Adjust pH to 7.5 with NaOH and bring to 200 ml.

2. *0.4 M Tris–HCl, pH 7.5:* To make 200 ml, dissolve 9.68 g Tris in 160 ml distilled water. Adjust pH to 7.5 with HCl and bring to 200 ml.

Steps

Activation of Dextran

1. Dissolve 5 g of freeze-dried dextran T500 in 25 ml DMSO at room temperature. Add dropwise 1 ml of triethylamine followed by 5 ml dichloromethane (should take ca. 10 min). Chill on ice. Add 0.35 g tresyl chloride dropwise while stirring vigorously with a magnetic stirrer. Stir gently for another 30 min on ice, for 60 min at 4°C, and overnight at 20°C.

2. Add 50 ml dichloromethane to terminate the reaction and to precipitate the dextran. Wash the precipitate with several 25-ml portions of dichloromethane until firm consistency, each time thoroughly kneading with a glass rod. Dissolve the washed tresyl dextran in 30 ml water and dialyze against 2 × 5 l water, at least for 4 hr each time or overnight. All this is done at room temperature. Freeze-dry and store dry at −20°C. Tresyl dextran is stable for several months under these conditions. The recovery of dextran is about 95%.

Conjugation of WGA

1. Dissolve 2 g of tresyl dextran in 10 ml of coupling buffer. Add dropwise under vigorous stirring 10 mg of WGA dissolved in 1 ml of the same buffer and incubate overnight at 4°C with gentle agitation. Terminate by adding 10 ml of 0.4 M Tris–HCl, pH 7.5, and incubate for at least 1 hr to inactivate unreacted tresyl groups.

2. Remove uncoupled ligand and salts by repeated ultrafiltration in a Filtron Omegacell (150 ml in size, filter cutoff 100 kDa). Add the terminated incubation mixture and water to

150 ml in the cell and ultrafilter to a volume of ca. 15 ml. Repeat the addition of water followed by ultrafiltration another five times. Freeze-dry and store dry at $-20°C$.

3. Determine the amount of ligand coupled by protein analysis (Bradford, 1976) using WGA as standard. Usually 4 mg of WGA is found per gram of freeze-dried product. The recovery of dextran exceeds 90%. WGA–dextran can be stored for at least 6 months at $-20°C$ without losing its separation properties.

B. Purification of Rat Liver Plasma Membranes from a Crude Membrane Preparation

Plasma membranes may be purified from any crude membrane preparation, e.g., a microsomal fraction, by two-phase affinity partitioning (Persson *et al.*, 1991). The two-phase system is prepared by weighing the required amounts of WGA–dextran, polyethylene glycol and dextran stock solutions, and buffer into a test tube suitable for low-speed centrifugation. After admixing the crude membrane suspension, phases are separated to obtain affinity-purified plasma membranes in the dextran-rich bottom phase, whereas other membranes remain in the PEG-rich top phase. The system described here has a total weight of 2.0 g, a final concentration of 6.0% (w/w) of each phase polymer, and is suitable for up to 2 mg of membrane protein. If necessary, the system may be scaled up or down, retaining its separation characteristics.

Solutions

1. *Stock solution of dextran at 20% (w/w):* Dissolve 20 g of freeze-dried dextran T500 in distilled water and make up to a total weight of 100 g. Store in suitable aliquots at $-20°C$.

2. *Stock solution of PEG 3350 at 40% (w/w):* Dissolve 40 g of PEG 3350 in distilled water and make up to a total weight of 100 g. Store in suitable aliquots at $-20°C$.

3. *0.2 M borate buffer:* To make 200 ml, dissolve 2.48 g boric acid in 160 ml distilled water and adjust to pH 7.8 with a concentrated solution of Tris in distilled water. Bring to 200 ml.

4. *Preequilibrated top phase:* Weigh 600 mg 20% dextran stock solution, 300 mg 40% PEG stock solution, 150 mg borate buffer, and 950 mg distilled water into a clear test tube. Mix and leave at 4°C for at least 8 hr. Siphon off the top phase prior to use.

5. *0.1 M N-acetylglucosamine solution:* To make 100 ml, dissolve 2.21 g N-acetylglucosamine, 8.56 g sucrose, and 60.6 mg Tris in 80 ml distilled water. Adjust pH to 8.0 with HCl and bring to 100 ml.

6. *5 mM Tris–HCl, pH 8.0:* To make 100 ml, dissolve 60.6 mg Tris in 80 ml distilled water. Adjust pH to 8.0 with HCl and bring to 100 ml.

Steps

1. Weigh an amount of WGA–dextran containing 100 μg WGA (-a mg WGA-dextran) into a clear 3-ml test tube. Add 300 mg PEG stock solution, 600-5a mg dextran stock solution, 150 μl of 0.2 M borate buffer, and distilled water to 1.80 g. Finally, add 0.20 g membrane suspension (containing up to 2 mg membrane protein) in 5 mM Tris–HCl buffer, pH 8.0.

2. Bring the two-phase system to 4°C, preferably in a well-tempered coldroom overnight. It is important to maintain this temperature during the entire two-phase procedure, as phase separation is critically dependent on temperature.

3. Mix the system thoroughly by inverting the tube 20 times, vortexing, and inverting the tube another 20 times.

4. Separate the phases by gentle (150 g) centrifugation (Wifug Doctor at 1500 rpm) for 5 min. Siphon off the PEG-rich top phase immediately, saving the interphase and the plasma membrane-containing bottom phase.

5. Reextract the bottom phase by adding the same amount of fresh preequilibrated top phase as the top phase removed in step 4. Mix thoroughly and separate the phases by repeating steps 3 and 4.

6. Dilute the resulting bottom phase 10-fold in 0.1 M N-acetylglucosamine. Mix and centrifuge for 60 min at 100,000 g to sediment membranes. Resuspend the membrane pellet to the desired concentration in 5 mM Tris–HCl, pH 8.0, or another suitable buffer. Test the purity of plasma membranes by marker enzyme analyses (see Section IV).

C. Rapid Purification of Rat Liver Plasma Membranes

This procedure combines homogenization and phase separation in a conventional two-phase system with affinity partitioning to rapidly produce highly purified liver plasma membranes in a high yield (Persson and Jergil, 1992). All steps are performed at 4°C.

Solutions

1. 40% PEG stock solution, 20% dextran stock solution, 0.2 M borate buffer, and 0.1 M N-acetylglucosamine solution: See Section B.

2. 5 mM Tris–HCl, 0.25 M sucrose: To make 200 ml, dissolve 121 mg Tris and 17.1 g sucrose in 160 ml distilled water. Adjust pH to 8.0 and bring to 200 ml.

3. 0.2 M Tris–H$_2$SO$_4$, pH 7.8: To make 200 ml, dissolve 4.84 g Tris in 160 ml distilled water and adjust pH to 7.8 with H$_2$SO$_4$. Add water to 200 ml.

4. Homogenization two-phase system: 5.7/5.7 % (w/w) dextran/PEG in 15 mM Tris–H$_2$SO$_4$, pH 7.8. Weigh 11.4 g 20% dextran stock solution, 5.70 g 40% PEG stock solution, 3.00 g 0.2 M Tris–H$_2$SO$_4$, pH 7.8, and 14.9 g distilled water directly into a 40-ml Dounce homogenizer. Mix and leave at 4°C overnight. This gives a weight of 35 g which adds up to 40 g with the addition of 5 g of tissue sample (see step 1).

5. Preequilibrated top phase I: Weigh 11.4 g 20% dextran stock solution, 5.70 g 40% PEG stock solution, 3.00 g 0.2 M Tris–H$_2$SO$_4$, pH 7.8, and 19.9 g distilled water into a 50-ml measuring cylinder. Mix and leave at 4°C overnight. Collect the top phase immediately before use.

6. Preequilibrated affinity bottom phase: 6.0/6.0 % (w/w) dextran/PEG in 15 mM borate buffer, pH 7.8, with 200 μg WGA as WGA–dextran. Weigh an amount of WGA–dextran containing 200 μg WGA (-a mg WGA-dextran), 3000-5a mg 20% dextran stock solution, 1.5 g 40% PEG stock solution, 750 mg 0.2 M borate buffer, and distilled water to 10 g into a 35-ml centrifuge tube. Mix thoroughly to dissolve the WGA–dextran and leave at 4°C overnight. Collect the bottom phase by siphoning off the top phase immediately prior to use.

7. Preequilibrated top phase II: Weigh 15.0 g 20% dextran stock solution, 7.50 g PEG stock solution, 3.75 g 0.2 M borate buffer, and 23.75 g distilled water into a 50-ml measuring cylinder. Mix and leave at 4°C overnight. Collect the top phase immediately before use.

Steps

1. Perfuse a rat liver with 5 mM Tris–HCl, 0.25 M sucrose to remove as much blood as possible. Squash the liver through a garlic press and transfer a maximum of 5 g to a homogenization two-phase system. If less than 5 g liver is used, add distilled water to a total system weight of 40 g.

2. Homogenize by 20 up-and-down strokes with a loose-fitting Dounce A pestle.

3. Transfer the homogenization system to a centrifuge tube and spin at 150 g for 5 min in a swing-out rotor to speed up phase separation. Collect the top phase and measure its volume.

4. Reextract the bottom phase by adding an equal amount of preequilibrated top phase I to the top phase removed in step 3. Mix thoroughly by 20 inversions of the tube, vortexing, and another 20 inversions.

5. Separate the phases by centrifugation as in step 3. Collect the top phase and combine with the first one.

6. Add the combined top phases to a preequilibrated affinity bottom phase. Mix thoroughly as in step 4 and separate the phases as in step 3. Siphon off the top phase and measure its volume.

7. Reextract the bottom phase by adding an equal amount of preequilibrated top phase II. Mix and separate the phases as just described. Siphon off the top phase.

8. Reextract the bottom phase with preequilibrated top phase II as in step 7.

9. Dilute the final bottom phase 10-fold with 0.1 *M* N-acetylglucosamine solution, or some other suitable buffer. Mix and pellet the membranes by centrifugation for 60 min at 100,000 *g*.

10. Suspend the resulting membrane pellet in 10 mM Tris-HCl buffer, pH 7.5, or another suitable buffer.

IV. COMMENTS

A. Preparation of WGA–Dextran

Dextran T500, as obtained from the manufacturer, dissolves very slowly, whereas freeze-dried dextran is readily soluble. It is therefore convenient to use dextran freeze-dried from a water solution. In the activation step, triethylamine and dichloromethane should be added slowly to avoid precipitation of dextran. After these additions, the mixture is rather viscous and it should be ascertained that the tresyl chloride is thoroughly admixed upon addition.

After conjugation of WGA and dextran the removal of salt by ultrafiltration is crucial. Salt remaining from this step will affect the partitioning of membranes during the affinity step and may cause unwanted membranes to distribute in the WGA–dextran-containing bottom phase. The amount of WGA coupled may vary and should be determined by protein analysis using WGA as standard.

The separation properties of each batch of WGA–dextran should be tested. This is done by varying the amount of WGA–dextran in a series of affinity partitionings as described in steps 2–4 in Section IIIB. A series containing 0–100 μg WGA in 2.0-g systems is suitable (cf. Fig.1), and crude membranes, e.g., microsomes, may be used as a membrane source. After phase separation the distribution in top and bottom phases of a plasma membrane marker enzyme, e.g., 5'-nucleotidase or alkaline phosphodiesterase I, is determined. The specificity of the affinity separation is ascertained by analyzing marker enzymes for other membrane compartments. Such markers should remain in the top phase at all WGA–dextran concentrations (Fig.1). Usually, 100 μg WGA as WGA–dextran will pull more than 90% of plasma membrane markers into the dextran-rich bottom phase in a 2-g system, leaving more than 90% of other membranes in the PEG-rich top phase.

B. Purification of Plasma Membranes from Crude Membranes

The salt conditions chosen for affinity partitioning are crucial for a successful outcome. Borate will cause membranes in general to partition in the PEG-rich top phase by distributing preferentially in the dextran-rich bottom phase, thereby repelling negatively charged membranes. This repulsion is overcome at a narrow borate concentration interval by the interaction of plasma membranes with WGA–dextran.

Additional salts should be avoided as much as possible as they might affect the partitioning properties of the system negatively, e.g., by causing membranes in general to distribute unselectively in the bottom phase or by keeping plasma membranes too firmly in the top phase, counteracting WGA-dependent redistribution. The extreme sensitivity of the partitioning process to salt (even in the millimolar range) should be kept in mind. Improper control of salt conditions is a common cause when two-phase separations do not work according to specification.

As the separation of membranes in one step is incomplete, the bottom phase should be

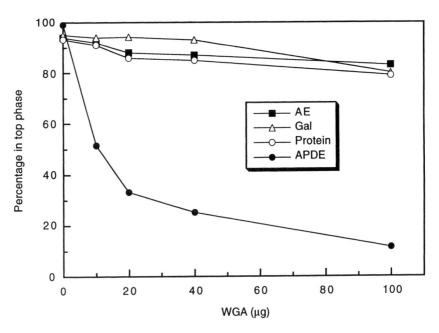

FIGURE 1 Affinity partitioning of rat liver microsomes using WGA–dextran. Rat liver microsomes were partitioned with increasing amounts of WGA–dextran in an affinity system containing 6.0% (w/w) each of PEG and dextran in 15 mM borate buffer, pH 7.8. Protein content was measured as well as markers for plasma membranes, alkaline phosphodiesterase (APDE); endoplasmic reticulum, arylesterase (AE); and Golgi membranes, galactosyltransferase (Gal). Reproduced with permission of The Biochemical Society and Portland Press.

reextracted with a fresh top phase to remove contaminants, thereby increasing the purity of the plasma membranes. A second reextraction is usually unnecessary, but may in some cases increase the purity further. In the final step, *N*-acetylglucosamine is added to facilitate the dissociation of WGA–dextran from membranes. This treatment may be omitted if the presence of WGA–dextran in the preparation does not disturb.

C. Rapid Purification of Plasma Membranes

In the protocol presented here the tissue is gently homogenized with a hand-driven Dounce homogenizer. The homogenization method is not critical, however, and can be modified according to particular experimental requirements.

The protocol includes a conventional two-phase separation followed by reextraction of the bottom phase with a fresh top phase. The reextraction will increase the plasma membrane yield by ca. 30%, but is not necessary to include when speed is important. More than 95% of mitochondrial and endoplasmic reticulum membranes are removed in the bottom phase. Plasma membranes are enriched in the combined top phases and are affinity extracted into a smaller volume of WGA-containing bottom phase. This latter phase is reextracted with a fresh top phase to increase plasma membrane purity; the second reextraction may be included to increase purity further, if deemed necessary. The affinity steps remove Golgi membranes, lysosomes, and remaining endoplasmic reticulum, whereas the final centrifugation removes soluble material still present after the extractions.

Preequilibrated phases should always be used for the affinity extraction and the reextractions. This is important as cross-extraction of the polymers occurs between the phases. Failure to use properly preequilibrated phases may therefore result in a one-phase system.

It is important to follow the distribution of wanted as well as unwanted membranes by marker enzyme analyses for each separation step, i.e., the first two-phase separation, the reextraction step, and after affinity partitioning, to ascertain that the various steps work properly. The final preparation should be enriched 30- to 40-fold over the homogenate (sum of combined top phases and bottom phase after the conventional two-phase steps) in markers for both apical and basolateral plasma membrane domains, with a recovery of 50–70% (Persson and Jergil, 1992). Only traces of other subcellular material should remain.

D. Membrane Marker Enzymes

The following marker enzymes are convenient in affinity partitioning procedures: Plasma membranes, apical domain–5′-nucleotidase (Avruch and Wallach, 1971) and alkaline phosphodiesterase I (Smith and Peters, 1980); plasma membranes, basolateral domain– asialoorosomucoid binding (Hubbard *et al.*, 1983; Gierow *et al.*, 1986); endoplasmic reticulum– arylesterase; Golgi membranes–*N*-acetylglucosamine galactosyltransferase (Fleischer and Smigel, 1978); lysosomes–*N*-acetyl-β-glucosaminidase (Touster *et al.*, 1970); mitochondria– succinate-cytochrome c reductase (Sottocasa *et al.*, 1967); and cytosol–lactate dehydrogenase (Bergmeyer and Bernt, 1970).

V. PITFALLS

1. Membranes distribute unselectively in the WGA-containing phase: This is usually caused by too much salt in the two-phase system, due either to incomplete desalting of WGA–dextran in the ultrafiltration step or to the incorrect addition of salt.

2. Membranes do not distribute as expected in conventional two-phase partitioning (steps 3 and 4 in Section IIIC): Check that salt conditions, polymer concentrations, and temperature are correct. Check the total recovery of enzyme activities. Pipetting errors due to the high viscosity of dextran-rich phases result in erroneously low activities.

3. Two-phase system does not form: Check that polymer concentrations are correct and that properly preequilibrated phases at the correct temperature are used.

Acknowledgment

This work was supported by the Swedish Natural Science Research Council.

References

Albertsson, P.-Å. (1986). "Partition of Cell Particles and Macromolecules." Wiley, New York.

Avruch, J., and Wallach, D. F. H. (1971). Preparation and properties of plasma membrane and endoplasmic reticulum fragments from isolated rat fat cells. *Biochim. Biophys. Acta* **233**, 334–347.

Bergmeyer, H. U., and Bernt, E. (1970). UV-test mit pyruvat und NADH. *In* "Methoden der Enzymatischen Analyse" (H. U. Bergmeyer, ed.), pp. 533–538. Verlag Chemie, Weinheim/Bergstrasse.

Bradford, M. M. (1976). A rapid and sensitive method for the quantitation of microgram quantities of protein utilizing the principle of protein dye binding. *Anal. Biochem.* **72**, 248–254.

Fleischer, B., and Smigel, M. (1978). Solubilization and properties of galctosyl transferase and sulfotransferase activities of Golgi membranes in Triton X-100. *J. Biol. Chem.* **253**, 1632–1638.

Gierow, P., Sommarin, M., Larsson, C., and Jergil, B. (1986). Fractionation of rat liver plasma membrane regions by two-phase partitioning. *Biochem. J.* **235**, 685–691.

Hubbard, A. L., Wall, D. A., and Ma, A. (1983). Isolation of rat hepatocyte plasma membranes I. Presence of the 3 major domains. *J. Cell Biol.* **96**, 217–229.

Persson, A., and Jergil, B. (1992). Purification of plasma membranes by aqueous two-phase affinity partitioning. *Anal. Biochem.* **204**, 131–136.

Persson, A., Johansson, B., Olsson, H., and Jergil, B. (1991). Purification of rat liver plasma membranes by wheat-germ-agglutinin affinity partitioning. *Biochem. J.* **273**, 173–177.

Smith, G. D., and Peters, T. J. (1980). Analytical subcellular fractionation of rat liver with special reference to the localization of putative plasma membrane marker enzymes. *Eur. J. Biochem.* **104**, 305–311.

Sottocasa, G. L., Kuylenstierna, B., Ernster, L., and Bergstrand, A. (1967). An electron-transport system associated with the outer membrane of liver mitochondria. *J. Cell Biol.* **32**, 415–438.

Touster, O., Aronson, N. N., Dulaney, J. T., and Hendrickson, H. (1970). Isolation of rat liver plasma membranes: Use of nucleotide pyrophosphatase and phosphodiesterase I as marker enzymes. *J. Cell Biol.* **47**, 604–618.

Immunoisolation of Organelles Using Magnetic Solid Supports

Steven M. Jones, Rolf H. Dahl, John Ugelstad[1], and Kathryn E. Howell

I. INTRODUCTION

Cell fractionation is one of the common tools used by cell and molecular biologists in determining the localization of specific molecules and the functional capacity of individual organelles. Classical fractionation methods rely on density and sedimentation velocity of the organelles or membranes derived from the organelles after homogenization of the cell (for review see Beaufay and Amar-Costesec, 1976). Immunoisolation utilizes the high specificity of antibodies to isolate the compartment(s) containing the cognate antigen. Many isolation protocols utilize a combination of the classical methods and immunoisolation. Reviews of applications of immunoisolation are described in Bailyes *et al.*, (1987) and in Howell *et al.* (1988, 1989).

This article illustrates the principles of immunoisolation using a system our group has employed extensively to study the molecules required for the budding of exocytic transport vesicles from the *trans*-Golgi network (TGN) of rat hepatocytes (Salamero *et al.*, 1990; Jones *et al.*, 1993). We begin by isolation of relatively intact, stacked Golgi fractions from rat liver by the procedure described by Leelavathi *et al.* (1970). Modifications of that procedure are described in the article by Norman Hui *et al.*, volume 2, "Purification of Rat Liver Golgi Stacks." The fraction is not completely homogenous, as it also contains individual cisternae, which have dissociated from the stack, and some vesicles, which were either associated with the Golgi in the cell or produced upon homogenization of the cell. This fraction is immobilized on magnetic beads by immunoisolation to provide relatively intact organelle with which to study the process of vesicle budding. In the immunoisolation step, components that have dissociated from the stacks during homogenization or the gradient fractionation will be isolated in proportion to the distribution of the antigen used for the isolation. The system is illustrated using two different antigens to isolate the stacked Golgi fraction: the polymeric IgA receptor (pIgA-R), a molecule en route to the plasma membrane that is relatively evenly distributed throughout the membranes of the entire Golgi complex, and TGN38, a transmembrane protein localized to the *trans*-cisternae and the TGN tubules (Sztul *et al.*, 1985; Luzio *et al.*, 1990).

In order to carry out immunoisolation, it is essential to identify an appropriate antigen and have a high-affinity antibody that recognizes the antigen in its native (undenatured) state. The following characteristics are essential for the antigen. (1) The distribution of the antigen should be limited to the compartment that you wish to isolate. In some cases the antigen may be more broadly distributed within the cell. This may not be a problem if a prefraction-

---

[1] Deceased.

ation step effectively removes the secondary compartment from the input fraction used for the immunoisolation. (2) The antigen should be of high enough abundance so that after homogenization of the cell, all vesicles derived from the compartment of interest contain the antigen. (3) The antigenic site is available for antibody binding, i.e., it is on the correct side of the membrane and not masked by other molecules. In general, the cytoplasmic domains of transmembrane proteins make the best antigens, but tightly associated peripheral membrane proteins also can be effective antigens for immunoisolation (e.g., antibodies against the heavy chain of clathrin efficiently isolate clathrin-coated vesicles).

Antibodies are often raised against denatured proteins and will not recognize the native protein and, therefore, will not function for immunoisolation. In fact, in our experience this has been the case more often than not. Monoclonal antibodies are preferred because they provide enough of the same antibody to complete the required series of experiments. We actually screen hybridomas for an appropriate clone by immunoisolation. Polyclonal antibodies usually need to be affinity purified to provide a high enough concentration of specific antibody on the bead surface for efficient isolation.

Many solid supports are available for immunoisolation and their properties are reviewed in Howell *et al.* (1988). We have focused on the use of magnetic beads because separation is based on a principle other than sedimentation and retrieval is rapid and efficient. Sedimentation and resuspension of a pellet of any fraction will result in the fragmentation of vesicles or organelles. The bead designed for immunoisolation has a thin shell of nonmagnetic material on the surface that minimizes interactions between beads and reduces nonspecific binding and fragmentation of bound organelles.

II. MATERIALS AND INSTRUMENTATION

Sodium tetraborate (Cat. No. B-9876), boric acid (Cat. No. B-0252), potassium phosphate, monobasic (Cat. No. P-0662), sodium phosphate, dibasic (Cat. No. S-0876), sodium chloride (Cat. No. S-9888), and Tween 20 (Cat. No. P-1379) are from Sigma. Tris base (Cat. No. 604 207) and bovine serum albumin (BSA, Cat. No.100 018) are from Boehringer-Mannheim. Dynabeads M-500, subcellular (Cat. No. 150 01), are from Dynal.

Magnets that fit various sizes and numbers of tubes and an adjustable, low rpm rotator are also available from Dynal. A low speed centrifuge is optional. Basic equipment and supplies for conventional cell fractionation and morphological and biochemical analysis of subcellular fractions will be required.

A specific type of monodisperse magnetic particles has been developed for the purpose of isolating subcellular compartments. The preparation of the monodisperse porous particles and subsequent introduction of magnetic iron is done according to the method described in this series (Larsen and Ugelstad). However, Dynabead M-450 or M-28 particles referred to there, which are successfully applied for selective isolation of cells, are not satisfactory for the present purpose. To obtain an effective and selective isolation of subcellular compartments, particles with an exceptionally smooth, hydrophilic surface are required to eliminate nonspecific attachment of cellular material. This surface also provides for sufficiently gentle handling of the fragile compartments so that they retain their *in vivo* functions after isolation. These special particles, Dynabead M-500, are prepared by treating the magnetized particles with monomers, in several steps. The monomers mutually react to build a cross-linked polymer that fills all pores within the particle, resulting in a layer of pure polymer at the surface of the bead. This procedure gives the required smooth and hydrophilic surface and ensures that no iron oxide is exposed at the surface. In the final step, a strongly hydrophilic hydroxy compound is covalently attached to the surface. By reaction of these hydroxyl groups with tosyl chloride, tosylated groups are formed that react with amino groups to covalently couple antibodies to the surface of the particles (see Howell *et al.*, 1989).

III. PROCEDURES

A. Preparation of Immunoadsorbant

For all the experiments described in this article, we used an immunoadsorbant consisting of a solid support (magnetic beads) with a linker antibody (an antibody against the Fc region of mouse IgG) covalently coupled to the surface of the bead. The isolation antibody, subsequently immunobound to the linker antibody, forms the completed immunoadsorbant (Fig. 1).

Solutions

1. *Borate buffer:* 0.1 M borate buffer, pH 9.5. Dissolve 19.07 g sodium tetraborate in 500 ml distilled H_2O. Adjust to pH 9.5 with 0.1 M boric acid and 1.2 g boric acid in 200 ml distilled H_2O.

2. *Bead wash:* 0.05 M Tris base, pH 7.8, 0.1 M NaCl, 0.01% BSA, and 0.1% Tween 20. Dissolve 3.03 g Tris base, 2.92 g NaCl in 450 ml distilled H_2O. Adjust to pH 7.8 with 1 N HCl. Add 0.5 ml Tween 20 and 50 mg BSA and stir until BSA is dissolved; adjust to a final volume of 500 ml.

3. *Isolation buffer:* phosphate-buffered saline and (PBS) 5 mg/ml BSA. Dissolve 4 g of NaCl, 0.1 g KCl, 0.72 g Na_2HPO_4, and 0.12 g of KH_2PO_4 in 450 ml distilled H_2O. Adjust the pH to 7.4 with HCl. Add 2.5 g BSA and stir until dissolved. Adjust to a final volume of 500 ml.

Steps

Covalent Coupling of Linker Antibody

1. Dynabeads M-500 are preactivated by the manufacturer and are stable for at least a year.

2. Pellet the beads using either a magnet or a centrifuge (1500 g for 3 min) to remove the storage buffer (typically 1 mM HCl).

3. Carry out covalent coupling of the linker antibody in borate buffer. Mix activated beads overnight at 20°C with 5–15 μg linker/2 mg beads/ml. Preferably the linker is an IgG fraction

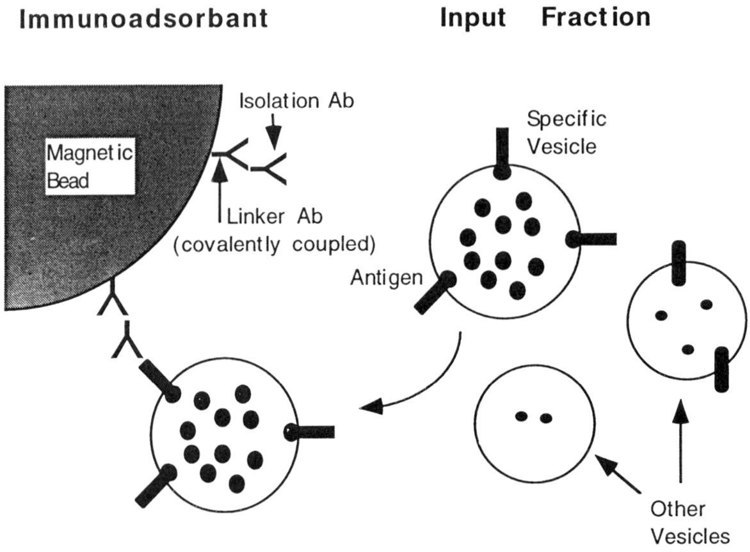

FIGURE 1 Immunoisolation using magnetic solid supports. Isolation antibodies, which recognize an epitope (antigen) on the cytoplasmic domain of a vesicle (or organelle), are bound to a magnetic bead via linker antibodies. The linkers are affinity-purified antibodies against the Fc fragment of mouse IgG and are covalently coupled to the surface of the bead. The complex (solid support, linker antibodies, and isolation antibodies) forms the immunoadsorbant which is mixed with the input fraction for the isolation. The isolation is carried out at 4°C in PBS and 5 mg/ml BSA, which is used to quench nonspecific binding to the immunoadsorbant. After an appropriate time (about 2 hr), the beads plus bound fraction (isolated fraction) are retrieved using a magnet and the nonisolated fraction is removed. The isolated fraction is washed, and the efficiency of the isolation is followed using morphological, immunological, biochemical, and functional assays. The specificity of the isolation is assessed two ways. The distribution of a "marker" in the input fraction that is not present in the compartment to be isolated is followed and the nonspecific association a "marker" of the compartment to be isolated on a control immunoadsorbant (containing an irrelevant antibody) is determined.

from an αFc antisera specific for the species of the isolation antibody, (i.e., α-mouse, α-rabbit, etc.). An αFc will orientate the isolation antibody and assure that its binding region is accessible. Other linkers can be used (e.g., streptavidin can be used to link biotinylated isolation antibodies).

4. The following day, pellet the beads to remove unbound linker antibody. Retain the unbound linker antibody and determine the amount either by OD_{280} or protein assay. The μg linker bound/mg beads is calculated, typically 3–5 μg bind/mg beads. Unbound linker can be reused.

5. Wash the beads three times, 10 min each, with bead wash, followed by two washes, 10 min each, with PBS/BSA. During the 10-min intervals, rotate the beads end over end at 2 rpm.

6. The bead with linker antibody can be stored in PBS/BSA at 4°C with 0.02% sodium azide for at least a month.

Alternative procedure: In some cases it is not necessary to use a linker antibody, and the isolation antibody can be covalently attached to the surface of the bead. In addition to the advantages already discussed, linker antibodies provide the following benefits: (1) the added length of the antibody complex gives additional flexibility to allow binding of the isolation antibody to its antigen; (2) the linker provides orientation of the isolation antibody, making it more efficient; (3) the specific binding sites are amplified; and (4) the linker provides versatility so that many different immunoadsorbants can be prepared by simply binding different isolation antibodies.

Binding of Isolation Antibody

Steps

1. Mix beads with covalently coupled linker antibodies with isolating antibodies at a final concentration of 10–20 μg /2 mg beads/ml in PBS/BSA at 4°C overnight with end-over-end rotation. We usually use total hybridoma supernatant, adjusted to the above concentrations, without purification of the IgG fraction. If polyclonal serum is used, affinity purification is usually required.

2. Wash the beads three times, 10 min each, with PBS/BSA. Because the binding is carried out in 5 mg/ml BSA, it is difficult to estimate how much specific antibody is bound. When setting up these assays, we have used radiolabeled antibodies for this determination. A microtiter capture assay is an alternative method that can be used.

3. The completed immunoadsorbant may be stored at 4°C in PBS/BSA with 0.02% sodium azide for at least a month, depending on the stability of the isolating antibodies.

4. A control immunoadsorbant should be prepared using an irrelevant antibody of the same Ig class as the isolation antibody.

Alternative procedure: For isolation, it is faster and more efficient to incubate the completed immunoadsorbant with the input fraction. However, this protocol requires that the isolation antibody immobilized on the bead recognizes its antigen with the same efficiency as it does in free solution. If this is not the case, the isolation antibody should be bound to the organelle or vesicles in suspension; subsequently, the complex is captured with the linker antibody/solid support. The disadvantage of the latter protocol is that an extra step is necessary to remove the unbound isolation antibody from the input fraction before capture with the linker antibody/solid support.

B. Preparation of Input Fraction

Solutions are not described here because these will vary considerably depending on the subcellular component to be isolated. Instead, guidelines for preparing an input fraction for immunoisolation are described.

The critical step to all fractionation procedures, including immmunoisolation, is homogenization of the cell or tissue. Decisions on how to homogenize the cell must be made based on the objectives of the experiment. For example, in classical fractionation, the cell is homogenized so that large, single-copy organelles like the endoplasmic reticulum and Golgi complex are homogenized into small, regular-sized vesicles, and the vesicles derived from these organelles are subsequently separated. This is ideal if the goal is to analyze the composition of the membrane or content of each compartment, but is less than ideal for cell-free assays used to study the function of the organelle. When the goal is to isolate the organelle relatively intact, as in our example in this article, the homogenization must be much more gentle and carefully controlled.

In both examples, the usual procedure is to first remove nuclei and cell debris from the homogenate and work only with a postnuclear supernatant (PNS). It is important to homogenize the cell so that nuclei remain intact and can be efficiently removed from the homogenate by a low-speed centrifugation step. If nuclei are broken and DNA is released into the homogenate, the entire homogenate tends to clump, resulting in a total inability to isolate any compartment in a specific manner. The damage to the homogenate resulting from broken nuclei and released DNA is complex and irreversible; DNase treatment will not result in the unclumping necessary to isolate components of the homogenate. In protocols where the homogenization step is gentle (i.e., for isolation of a stacked Golgi fraction), the loss of Golgi membranes to the nuclear pellet will be much greater (up to 50%) than with more vigorous homogenization protocols where the Golgi is totally vesicularized ($\sim$25%).

Before initiating immunoisolation experiments, the input fractions should be characterized both morphologically and biochemically. It is important to confirm that it contains a major amount of the component to be immunoisolated in a optimal condition. In many cases, isolation from the PNS is efficient and the nonspecific binding is minimal. There are two examples when it is advisable to start with a more enriched fraction. The first example is when the antigen used for the isolation is distributed in multiple subcellular compartments. This is the situation with the antigens used in the example we present. The pIgA-R is a plasma membrane receptor that trafficks through all compartments of the secretory and transcytotic pathways (Sztul *et al.*, 1985). It is present in the membranes of the biosynthetic pathway, especially the Golgi complex, in amounts higher than almost any other transmembrane protein en route to its final destination. The receptor is cleaved and secreted when it reaches the apical plasma membrane and is constantly replenished. The second example is when the nonspecific binding is very high; often an enrichment step using classical gradient fractionation is effective in reducing the background.

As already mentioned, we will illustrate the technique, using an entire organelle, as the example because each step requires more gentle handling than if a population of uniform sized vesicles were to be isolated. The stacked Golgi fraction is isolated from rat liver by the method of Leelavathi *et al.* (1970) and is refloated on a second gradient to deplete cytosol. This step is not necessary for immunoisolation but was essential for our studies to identify the components in cytosol necessary for the budding of pIgA-R containing exocytic vesicles from the TGN and provided a further enrichment of Golgi markers.

C. Characterization of the Input Fraction

1. Morphology

Standard EM procedures are used for morphological characterization of all fractions, including input, isolated, and nonisolated fractions. For the experiments presented here, fractions were fixed in suspension with 2% glutaraldehyde in 0.1 M cacodylate buffer, pH 7.35, and then pelleted (centrifuge for the input and nonbound fractions and magnet for the immunoisolated fraction). The pellets were washed (without resuspension) and postfixed with 1% OsO_4 in 0.8% potassium ferricyanide and 0.1 M cacodylate buffer, pH 7.35. After dehydration the samples were infiltrated and embedded in Spurr's resin.

An electron micrograph of a thin section of the stacked Golgi fraction used as the input fraction in the following experiments is shown in Fig. 2A. In an optimal section, many stacked Golgi cisternae are evident. The homogenization procedure results in an unequal fracture of

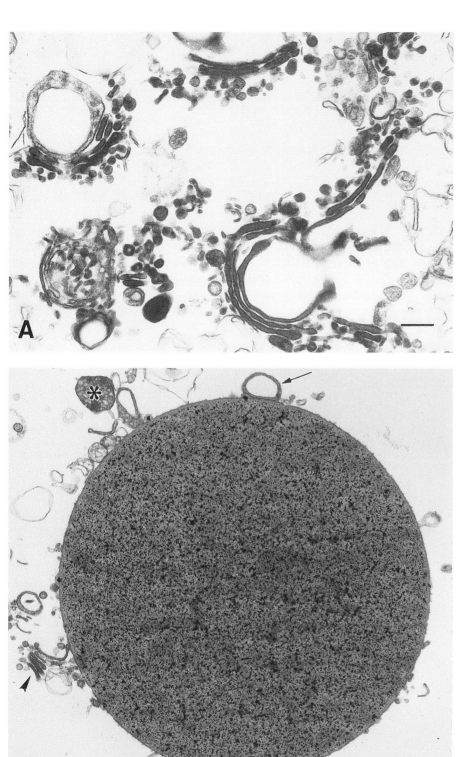

FIGURE 2 Morphological characterization of input and immunoisolated fractions. The fractions were fixed in suspension, pelleted, and prepared for EM using standard procedures. The thin sections provide overviews of the (A) input and (B) immunoisolated fractions. (A) The stacked Golgi fraction isolated from rat liver, the input fraction for immunoisolation, contains relatively intact stacks of Golgi cisternae. Cisternae and vesicles not apparently associated with a stack are also present. (B) The immunoisolated fraction bound to the immunoadsorbant shows the detail of both the bound fraction and the magnetic bead, solid support. Note that the surface of the bead is continuous and unbroken by crevices or pores. A stacked Golgi (arrowhead), single cisterna (arrow), and lipoprotein-filled cisternae (star) are all isolated on the same bead. Bars equal 200 nm.

the Golgi ribbon, resulting in both short and long segments. Often, within the isolated segments, Golgi cisternae appear to circularize. In addition to Golgi stacks the fraction contains dissociated cisternae and vesicles.

2. Biochemistry

The input fraction was characterized by determining the enrichment from the PNS of a number of antigens shown in Table I. The transmembrane proteins of the Golgi cisternae assayed included: p28, *cis* (Subramanian *et al.*, 1995); MG160, medial (Gonatas *et al.*, 1989); galactosyltransferase enzymatic activity (Bretz and S̆taubli, 1977); and TGN38, *trans*, and TGN (Luzio *et al*, 1990). All are 350- to 500-fold enriched. The liver plasma membrane protein, pp120/HA4 (Margolis *et al.*, 1988), and the rough and smooth endoplasmic reticulum marker, cytochrome P450 (Fleischer *et al.*, 1971), are de-enriched 0.13- and 0.12-fold, respectively. Molecules in transit through the Golgi complex, including pIgA-R, secretory proteins, transferrin, and apoE, are enriched 10- to 30-fold. These data, together with morphological data, indicate that the homogenization procedure resulted in breaking the Golgi ribbon into reasonably even segments that could be isolated intact. The lower enrichment of the *cis* and *trans* markers compared with the medial marker indicates that the ends of the Golgi stack are more labile to homogenization than the central cisternae. These data characterizing the input faction provide the baseline for the evaluation of the immunoisolation procedure.

D. Immunoisolation of Specific Fraction

Solutions

1. *PBS:* Dissolve 4 g of NaCl, 0.1 g KCl, 0.72g Na_2HPO_4, and 0.12g KH_2PO_4 in 450 ml distilled H_2O. Adjust the pH to 7.4 with HCl and to a final volume of 500 ml.

2. *PBS/BSA:* See Section IIIA.

Steps

1. Wash beads once with PBS/BSA. Resuspend beads at a concentration of 5 mg beads (completed immunoadsorbant)/ml and distribute 200 μl or 1 mg immunoadsorbant into

TABLE I Markers[a]

Marker	Compartment	Enrichment
TGN38	TGN[b]	325
Galactosyltransferase	*trans*	350
MG160	Medial Golgi[b]	500
p28	*cis*-Golgi[b]	350
bCOP	Golgi coat protein	15
p23	Endoplasmic reticulum[b]	ND[c]
Cytochrome P450 reductase	Endoplasmic reticulum[b]	0.04
HA4	Plasma membrane[b]	0.13
pIgA-R	Plasma membrane[b]	10
Transferrin	Secretory protein	11
ApoE	Secretory protein	31
VSV-G	Viral glycoprotein[b]	—

[a] Enrichment was determined by quantitative immunoblot, except for galactosyltransferase, which was measured enzymatically (Bretz and Stäubli, 1977).
[b] Transmembrane protein.
[c] Not determined.

each tube. Add the input fraction (10–250 μg protein) and adjust to a final volume of 1 ml with PBS/BSA.

2. Incubate with slow end-over-end rotation of the tubes for 1–2 hr at 4°C. **Do not vortex**, as this will sheer the bound fraction from the beads.

3. Recover the immunoadsorbant plus bound fraction using a magnet; pelleting in a microfuge often results in the fraction being sheared from the beads. The immunoadsorbant plus bound fraction may be washed in PBS/BSA, but each round of washing will result in some shearing of the specifically isolated fraction. Depending on the method used for the analysis of the isolated fraction, it is often advantageous to wash the isolated fraction with PBS minus BSA as the large amount of BSA in the fractions interferes with many subsequent assays.

4. The immunoadsorbant plus bound fraction can be resuspended in many solutions depending on the type of analysis to be carried out.

5. Parameters to vary to increase the efficiency of the isolation:

    a. Ratio of immunoadsorbant to input fraction in a constant volume: Optimal results are obtained when the minimal amount of immunoadsorbant is used to isolate the fraction of interest from the input fraction. Unoccupied bead surface has a greater chance of binding nonspecific components of the input fraction. Smaller volumes, increasing the concentration of the antibody/antigen, may decrease time required for binding, but also will increase nonspecific binding. Larger volumes, decreasing the concentration of antibody/antigen, have the opposite effect, but may also make analyzing the unbound fractions more difficult.

    b. Binding time: Because unoccupied bead surface has a greater chance of nonspecific binding, that chance becomes greater with increasing time of incubation. Therefore it is best to keep the incubation as short as possible. There is a more significant problem associated with increasing the time of incubation. Because the fractions to be isolated are small and fragile compared to the size of the magnetic bead, collisions of magnetic beads while mixing cause damage to the bound organelle or vesicle, resulting in the fragmentation of membranes (i.e., effectively producing microsomes) and the release of soluble luminal content. Note also that too vigorous a mixing of beads in washing will have the same consequences.

E. Typical Experiment: Isolation of Stacked Golgi Fraction

The optimal amount of input fraction necessary to isolate the stacked Golgi fraction with high efficiency and minimal nonspecific binding must be determined first. Initial experiments used a constant amount of immunoadsorbant (1 mg) and an increasing concentration of input fraction to assess the concentration where saturation occurs (Fig. 3). Both specific and control immunoadsorbants were used. In this experiment, the immunoadsorbants used to isolate the Golgi contained antibodies against the cytoplasmic domains of TGN38 and the pIgA-R. The control immunoadsorbant contained antibodies against the cytoplasmic domain of a viral protein, not present in the cells [the surface glycoprotein of vesicular stomatitis virus (VSV-G)]. Five concentrations of input fraction, from 0 to 200 μg protein, were incubated with 1 mg immunoadsorbant for 2 hr at 4°C.

The total amount of antigen isolated with each of the different immunoadsorbants was evaluated by quantitative immunoblot (Fig. 3B), and amounts (in PhosphorImager units) were plotted against the concentration of input fraction (Fig. 3A). The isolations are linear at low concentrations of input and saturate at the higher concentrations. The best efficiency is obtained just before saturation is reached (at 100 μg). The isolation efficiencies achieved with TGN38 and pIgA-R immunoadsorbants for the respective antigens were 95 and 80% (Fig. 4). The control immunoadsorbant (VSV-G) did not isolate compartments containing either TGN38 or pIgA-R, even at the highest concentration of input fraction.

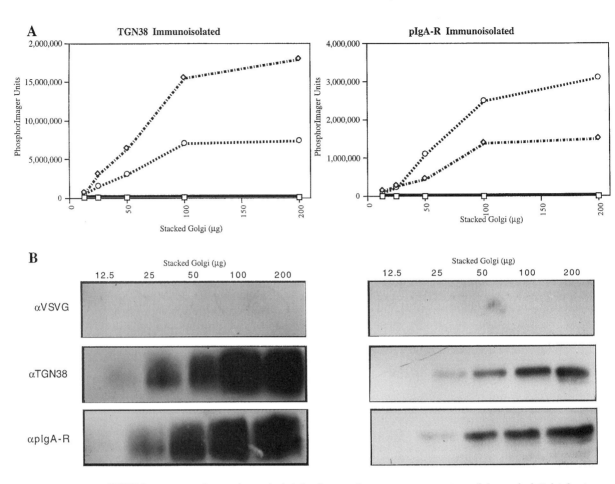

FIGURE 3 Immunoisolation of a stacked Golgi fraction. Increasing concentrations of the stacked Golgi fraction (25–200 μg protein; the input fraction) were mixed by rotation at 2 rpm with three different immunoabsorbants (1 mg/assay) for 2 hr at 4 °C. Immunoadsorbants used monoclonal antibodies against the cytoplasmic domains of (1) VSV-G, the control; (2) TGN38; and (3) the pIgA-R, both isolation antibodies (see Table I for details). After 2 hr the bound fractions were retrieved on a magnet and washed three times with PBS by gentle resuspension and retrieval with a magnet. The final fractions and aliquots of the input and nonisolated fractions were solubilized in SDS–PAGE sample buffer, resolved on a SDS gel, and transferred to a nitrocellulose filter. Filters were immunoblotted with antibodies against TGN38 and pIgA-R, detected with [^{125}I]protein A by autoradiography and quantitation by PhosphorImager analysis. Only data for the immunoisolation fractions are shown. Recoveries of each antigen from the input fraction in the immunoisolated and nonisolated fraction were between 90 and 95%. (A) The amount of antigen detected in the immunoisolated fractions, TGN38 (left) and pIgA-R (right), is plotted versus the amount of input fraction for each immunoabsorbant: αVSV-G ($\square$), αTGN38 ($\Diamond$) and αpIgA-R ($\bigcirc$). Quantitative data were obtained from the immunoblots in B. (B) Immunoblots of the total fractions isolated on different immunoadsorbants [αVSV-G , top lanes; αTGN38, middle lanes; and αpIgA-R, lower lanes] detected with antibodies against TGN38 (left) and pIgA-R (right). (Note that there is a slight misregistration of the bands in the TGN38 immunoisolated fractions.) There is no detectable isolation of either TGN38- or pIgA-R containing compartments on the control immunoadsorbant. The isolation of both TGN38- and pIgA-R-containing compartments is linear to approximately 100 μg protein, at which point the immunoadsorbant is saturated. (Note that the signal for TGN38 is much higher than that for pIgA-R, reflecting the difference in the amount of proteins in the input fraction.)

1. Morphological Analysis

A micrograph of an entire bead with the fraction isolated on the TGN38 immunoadsorbant is shown in Fig. 2B. Small stacks of cisternae and what appear as individual cisternae and vesicles are isolated (what appears to be a vesicle may well be a cisterna in cross section). Some of the vesicles are filled with lipoproteins, a major secretory product of liver. Even though, biochemically, the immunoadsorbant is saturated, morphologically, in a thin section, unoccupied regions are observed on the bead surface.

The thin section electron micrograph optimally shows the structure of the monodisperse magnetic beads. The lighter gray area is the original porous polymer; small black dots are

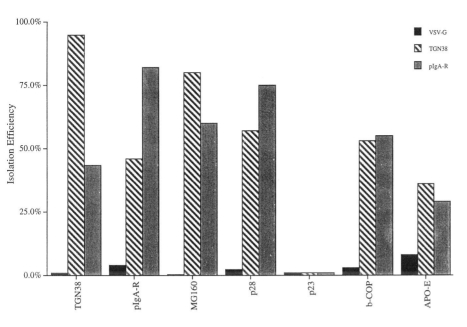

FIGURE 4 Efficiency of isolation of TGN38 and pIgA-R compared with numerous other 'markers' present in the input fraction. The immunoisolated fractions shown in Fig. 3 were further analyzed for the distribution of other markers associated with the Golgi complex. The same filters were stripped and reblotted with antibodies against TGN38, pIgA-R, MG160, p28, p23, β-COP, and apoE (see Table I for a description of the antigen). Isolation efficiency was calculated as the amount of antigen specifically isolated compared with the amount in the input fraction. The isolation efficiency of each antigen is shown for only the 100-μg (input) sample for each of the immunoadsorbants: (1) αVSV-G, the control, (black bars); (2) αTGN38 (striped bars); and (3) αpIgA-R (gray bars). All proteins predominately localized to the Golgi complex are present in the immunoisolated fraction, but their isolation efficiencies vary considerably. TGN38 and pIgA-R, the antigens used for the isolations, are isolated with 95 and 80% efficiency, respectively. p28 and MG160 *cis* and medial Golgi markers are recovered with 60 to 80% efficiency. β-COP, a cytosolic protein that associates with the Golgi membrane to become part of the COP I-coated vesicle, has a lower isolation efficiency (~55%), whereas apoE, an apo-lipoprotein of rat VLDL and HDL, is recovered at a much lower efficiency of ~30%. This could be the result of a wider distribution of apoE in all compartments of the secretory pathway and/or leakage of content lipoproteins during the isolation. The only marker not isolated to any significant extent is p23, which remained in the nonisolated fractions. The KDEL receptor (p23) is predominately localized in the ER but cycles into the *cis*-Golgi. In this fractionation the KDEL receptor was absent from the *cis*-cisternae isolated with either immunoadsorbant.

the maghemite that has been precipitated within the original polymer. Darker areas are the polymer that has been used to fill and seal the bead, providing a very thin layer at the surface (a nonmagnetic shell). These features are more clearly seen at higher magnifications in Figs. 4, 5, and 6.

Examples of fractions isolated with the two different isolation antibodies and viewed in different orientations plus the control immunoadsorbant are shown in Figs. 5 and 6. Total Golgi stacks are immunoisolated with antibodies against the cytoplasmic domains of TGN38 (Fig. 5), pIgA-R (Figs. 6A–C), and VSV-G (the control) (Fig. 6D). The isolation antibodies against TGN38 and the pIgA-R isolate morphologically similar stacks, cisternae, and vesicles whereas the control beads are empty. A stack can be bound to the beads with interactions resulting in many different orientations. Not all antigens of the fraction directly interact with the isolation antibodies bound to the bead surface. The Golgi stack show in Fig. 5A appears to interact with the bead surface only at the tip of a single cisterna, whereas others show more extensive interactions with single (Fig. 5C) or multiple cisternae (Fig. 6A). In some examples, a shorter segment of the ribbon appears to be isolated (Fig. 5A), whereas in others a more extensive region is isolated, comprising both compact and noncompact regions of the Golgi ribbon (Figs. 6B and 6C). In addition, the isolated organelle can appear to be quite distant from the surface of the bead, with the binding site in another plane of section. This would be the case in the example of Fig. 6B, if the section were taken in another plane.

One obtains a better understanding of the interactions that occur during the isolation from a morphological evaluation of the isolated fractions. First, sometimes the cisternae, by

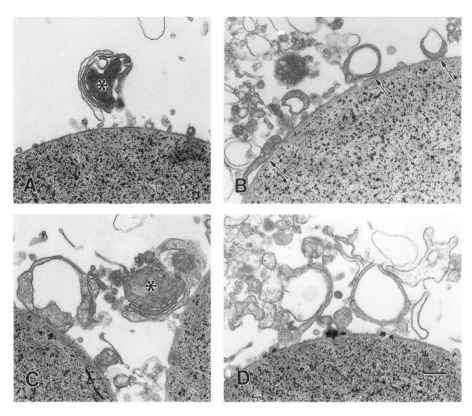

FIGURE 5 Morphological characterization of fractions immunoisolated with antibodies against the cytoplasmic domain of TGN38. (A and C) Examples of the different binding orientations and sections through relatively intact Golgi stacks. Some cisternae appear filled with a dense material, whereas others are packed with lipoprotein particles (asterisk), similar to the *in vivo* characteristics of rat liver Golgi. (B and D) The binding of individual cisternae filled with dense material to the surface of the bead (arrows). Bar equals 200 nm.

unknown forces, distend or balloon, as shown in Figs. 7A, 7C, and 7D. Other times the distended cisternae remain stacked (Figs. 7A and 7D), but others dissociate and account for some of the free cisternae in the fraction. More often, a single cisterna distends whereas others remain compact (Fig. 7C). A ballooned cisterna can fragment and release luminal content, as observed by lipoprotein particles spilling out of the damaged cisterna in Fig. 7D. Second, a single cisterna or stack may bind to more than one bead (Fig. 7B). In such cases the washing steps may sheer the stack or cisterna so that some remains with each bead, resulting in the release of luminal content. The cross-linking of an organelle to more than one bead can be minimized by adjusting the fraction to volume ratio. Occasionally the antigens within the bound region of the compartment will appear to cap and zipper to the surface of the bead (Fig. 7D).

2. Biochemical Analysis

Many assays can be used to analyze the results of immunoisolation experiments. The input, bound, and unbound fractions should be analyzed and a balance sheet produced to monitor activation/inactivation of enzymes or functions and loss of isolated fraction during the washing steps. The authors will illustrate with an evaluation of the immunoisolated Golgi fraction discussed earlier by following distribution of specific markers by quantitative immunoblotting. Additional assays may include other immunological assays, enzymatic activities, ligand blotting, distribution of labeled markers (i.e., internalization from the cell surface of radiolabeled LDL), and tests of specific functions.

An evaluation of other "markers" using quantitative immunoblotting is summarized in Fig. 4. The efficiency of the isolation (on the *x* axis) and antigens described in Table I (on the *y* axis) are reported in bar graphs. As with data for the saturation curve, no significant

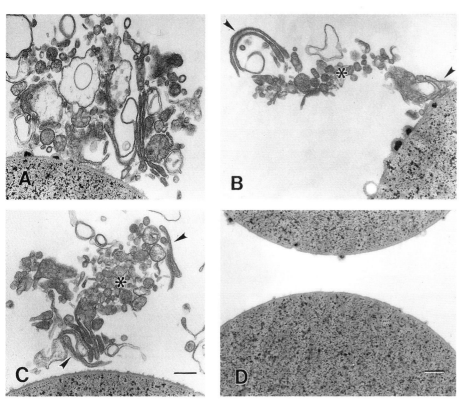

FIGURE 6 Morphological characterization of fractions immunoisolated with antibodies against the cytoplasmic domains of pIgA-R and VSV-G. (A–C) Examples of different binding orientations and sections through relatively intact Golgi stacks. (D) An example of a control immunoadsorbant demonstrating low nonspecific binding and the very regular, even surface of the magnetic beads. In both B and C, two stacked regions (arrowheads) appear to be held together by a less organized region (asterisk). This is similar to the compact and noncompact regions of the Golgi ribbon that have maintained their integrity throughout the isolation. Bars equal 200 nm.

amount of any antigen is isolated with the control (αVSV-G) immunoadsorbant. Each isolation immunoadsorbant (αTGN38 and αpIgA-R) was most efficient at the isolation of its specific antigen. pIgA-R is relatively uniformly distributed in all membranes of the Golgi stack whereas TGN38 is only localized to membranes of the *trans*-cisternae and some TGN tubules. Therefore, if all the stacks were retained together, both immunoadsorbants would isolate all antigens with the same efficiency. If the appropriate antigen is present, the dissociated stacks and vesicles will be isolated as well. Both biochemical data and morphological data indicate that the *cis* and TGN dissociate to some extent during the isolation of the stacked Golgi fraction and subsequent immunoisolation. For these reasons, variations in the isolation efficiencies of the two antigens are observed. We have not illustrated data on the unbound fraction, but recoveries from the input fraction are between 90 and 95%.

These data show that it is possible to obtain isolation efficiencies higher than with other available fractionation procedures (up to 95%) with minimal background. The two controls are both negative. Compartments not containing the antigen are not isolated with the isolation immunoadsorbants (e.g., p23), and the specific fraction is only isolated with the isolation antibodies and does not nonspecifically bind to an irrelevant immunoadsorbant.

IV. COMMENTS

We have presented only one application of immunoisolation and from the data a few important points are illustrated. Many different protocols for the isolation procedure and evaluation of the experiment are available. Cell sorting also can be carried out using the same basic technology (see Volume 1, Neurauter *et al.*, "Immunomagnetic Separation of Animal Cells"

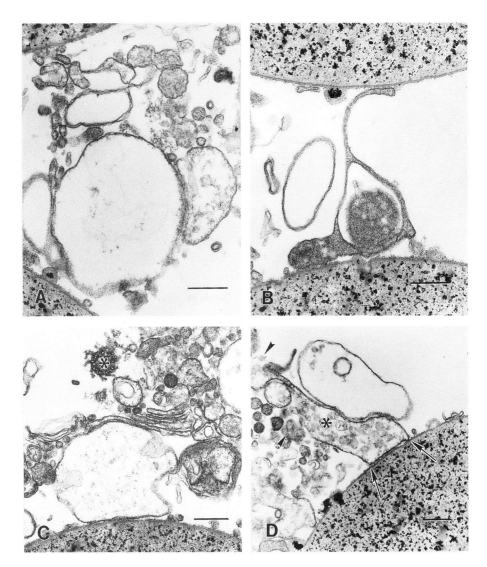

FIGURE 7 Examples of cisternal "ballooning," cross-linking, and capping in fractions immunoisolated using the TGN38 immunoadsorbant. (A, C, and D) Examples of cisternal "ballooning." In A and D, more than one cisterna has become distended with adjacent cisternae remaining attached. In B, only a single cisternae is distended; this cisternae is involved in the binding to the immunoadsorbant. D illustrates that when a ballooning cisternae (star) ruptures, the content lipoproteins (arrowheads) are often released. This phenomenon results in underrecovery of contents as compared with membrane. The ruptured cisternae appears bound to the bead by multiple antibody/antigen interactions, displaying the capping and zippering phenomenae (between arrows). (B) A single cisternae bound at its opposite ends to two different beads. It is likely that during additional washing steps this cisternae would rupture, resulting in the leakage of contents. *In vivo* the centrosome or microtubule organizing center is localized within the Golgi complex and is also observed within the isolated stacked Golgi fraction shown in C (asterisk). This is not an infrequent observation and confirms that this isolation procedure is gentle enough to isolate large organelles relatively intact. Bar equals 200 nm.

for additional information). The efficient isolation of the specific compartment and the lack of significant background are truly favorable properties of the M-500 magnetic particles. Other solid supports, such as porous beads or cellulose fibers, have problems with entrapment and fragmentation of bound fractions, which result in higher backgrounds and reduced efficiencies. Optimally employed, immunoisolation on M-500 beads avoids the harsher procedure of repeated pelleting and resuspension of organelles that are susceptible to further rupture and division into smaller vesicles. Finally, an organelle bound to a solid support offers a distinct advantage for cell-free assays in measuring the function of that compartment. The organelle bound to the beads can be introduced into and retrieved from successive reaction mixtures expeditiously and without significant damage to the isolated fraction.

V. PITFALLS

1. The success of the immunoisolation technique is dependent on the ability to produce a monodisperse input fractions from the tissue or cells of interest. Any amount of clumping of the organelles and vesicles in the cytosol will increase the level of nonspecific components isolated and washing will not be effective in reducing this.

2. Relatively large quantities of an antibody that will recognize its antigen within the membrane of the organelle to be isolated are required. Useful antibodies for immunoisolation may need to be specifically generated.

3. A large-scale preparation of immunoisolated fractions using the M-500 magnetic particles becomes expensive due to the cost of both the antibodies and the beads.

Acknowledgments

This work was supported by National Institutes of Health Grant GM 42629 to K. E. Howell. We acknowledge the use of Hepatobiliary Center, Cell Biology Core (NIH, P30 DK-34914), and the Cancer Center Monoclonal Antibody Facility (P30 CA-46934). SMJ thanks the American Liver Foundation for predoctoral fellowship support. We thank the following colleagues for providing antibodies for these studies: V. N. Subramaniam and W. Hong, αp28 and αp23; N. Gonatas, αMG160; G. Banting, αTGN38 luminal domain monoclonal; J.-P. Kraehenbuhl, αpIgA-R cytoplasmic domain monoclonal; and T. Kreis, αVSV-G cytoplasmic domain monoclonal and $\alpha\beta$-COP.

References

Bailyes, E. M., Richardson, P. J., and Luzio, J. P. (1987). Immunological methods applicable to membranes. *In* "Biological Membranes: A Practical Approach" (J. B. C. Findley and W. H. Evans, eds.), pp. 73–98. IRL Press, Oxford.

Beaufay, H., and Amar-Costesec, A. (1976). Cell fraction techniques. *Meth. in Membr. Biol.* **6**, 1–99.

Bretz, R., and Stäubli, W. (1977). Detergent influence on rat-liver galactosyltransferase activities toward different receptors. *Eur. J. Biochem.* **77**, 181–92.

Fleischer, S., Fleischer, B., Azzi, A., and Chance, B. (1971). Cytochrome b5 and P-450 in liver cell fractions. *Biochim. Biophys. Acta.* **225**, 194–200.

Gonatas, J. O., Mexitis, S. G. E., Stieber, A., Fleischer, B., and Gonatas, N. K. (1989). MG-160: A novel sialoglycoprotein of the medial cisternae of the Golgi apparatus. *J. Biol. Chem.* **264**, 646–653.

Howell, K. E., Gruenberg, J., Ito, A., and Palade, G. E. (1988). Immuno-isolation of subcellular components. *In* "Cell-Free Analysis of Membrane Traffic" (D. J. Morré, K. E. Howell, and G. M. W. Cook, eds.), pp. 77–90. A. R. Liss, New York.

Howell, K. E., Schmid, R., Ugelstad, J., and Gruenberg, J. (1989). Immuno-isolation using magnetic solid supports: Subcellular fractionation for cell free functional studies. *In* "Methods in Cell Biology" (A. M. Tartakoff, ed.), Vol. 31A, pp. 265–292. Academic Press, San Diego.

Jones, S. M., Crosby, J. R., Salamero, J., and Howell, K. E. (1993). A cytoplasmic complex of p62 and rab6 associates with TGN38/41 and is involved in budding of exocytic vesicles from the trans-Golgi network. *J. Cell Biol.* **122**, 775–788.

Leelavathi, D. E., Estes, L. W., Feingold, D. S., and Lombardi, B. (1970). Isolation of a Golgi-rich fraction from rat liver. *Biochem. Biophys. Acta* **211**, 124–138.

Luzio, P. J., Brake, B., Banting, G., Howell, K. E., Braghetta, P., and Stanley, K. K. (1990). Identification, sequencing and expression a TGN integral membrane protein (TGN38). *Biochem. J.* **270**, 97–102.

Margolis, R. N., Taylor, S. I., Seminar, D., and Hubbard, A. L. (1988). Identification of a pp120, an endogenous substrate for the hepatocyte insulin receptor tyrosine kinase, as an integral membrane glycoprotein of the bile canalicular domain. *Proc. Natl. Acad. Sci. USA* **85**, 7256–7259.

Salamero, J., Sztul, E. S., and Howell, K. E. (1990). Exocytic transport vesicles generated *in vitro* from the TGN carry secretory and plasma membrane proteins. *Proc. Natl. Acad. Sci. USA.* **87**, 7717–7721.

Subramaniam, V. N., Krijnse-Locker, J., Tang, B. L., Ericsson, M., Yusoff, A. R. B. M., Griffiths, G., and Hong, W. (1995). Monoclonal antibody HFD9 identifies a novel 28 kDa integral membrane protein of the cis-Golgi. *J. Cell Sci.* **108**, 2405–2414.

Sztul, E. S., Howell, K. E., and Palade, G. E. (1985). Biogenesis of the polymeric IgA receptor in rat hepatocytes. I. Kinetic studies of its intracellular forms. *J. Cell Biol.* **100**, 1248–1254.

Preparation of Human Erythrocyte Ghosts

David M. Shotton

I. INTRODUCTION

The human erythrocyte (red blood cell) is a remarkable cell in that it has no internal membranes. It is thus the ideal cell for many studies of plasma membrane structure and function, as erythrocytes are easy to obtain and one can prepare large quantities of pure erythrocyte plasma membrane from them by simple hypotonic osmotic lysis and washings to remove residual cytoplasm. It is for this reason primarily, and not because the erythrocyte membrane is in any way "typical," that it has been perhaps the most widely studied of all biological membranes, to the point where it is the only multifunctional plasma membrane whose functions and biophysical properties are reasonably well understood at the molecular level.

Under the influence of the hormone erythropoietin, erythrocytes develop from multipotent stem cells in the bone marrow through a series of well-defined stages. The penultimate stage is the reticulocyte, in which the internal membranous organelles are disassembled and the nucleus is physically expelled. Each of the erythrocytes thus formed is composed of a single external membrane, the plasma membrane, surrounding a cytoplasm made up of a concentrated solution of hemoglobin, some glycolytic enzymes, and minor amounts of other proteins and enzymes. The fully differentiated erythrocytes, now devoid of all biosynthetic and repair capabilities, are at this stage released into the circulation, where they survive for approximately 120 days before being desialated and removed from circulation.

Purified erythrocyte plasma membranes consist of approximately 52% protein, 40% lipids, and 8% carbohydrates, the latter being in the form of oligosaccharides distributed ~92% on glycosylated integral membrane proteins and ~8% on glycolipids. The lipids, mainly phospholipids, but including a high proportion of cholesterol, form a continuous closed noncompressible insulating bilayer, which separates the cytoplasm from the outside world, and through which the integral proteins, primarily the anion transporter and glycophorin A, facilitate the cell's interactions with that world. Evenly covering and tightly laminated to the cytoplasmic surface of this phospholipid bilayer is the membrane skeleton, an isotropic two-dimensional meshwork of peripheral membrane proteins, which is elastic and slightly contractile. It is this laminated complex of phospholipid bilayer and membrane skeleton that together constitutes the erythrocyte membrane.

The erythrocyte membrane is highly specialized in four particular aspects for its functional role of transporting oxygen and carbon dioxide between the lungs and the tissues. First, it imparts to the erythrocyte its typical flattened biconcave shape. Indeed, after hypotonic lysis, the residual empty plasma membrane retains this shape, and for this reason is referred to as an erythrocyte ghost. Although the cell has a diameter of about 8 μm, this flattened morphology results in a diffusion path distance from the cell surface to any point in the cytoplasm of only ~1 μm, greatly speeding the rate of diffusion and equilibration of oxygen and bicarbonate ions throughout the cell (Fig. 1a).

Second, the membrane is both very strong and easily deformable, permitting the cell to withstand without rupture the high sheer forces it experiences when being pumped through the heart and major arteries, and enabling it to distort into a cup shape and pass easily through small capillaries only ~3 μm in diameter, only to spring back to a biconcave disk once it enters more spacious surroundings. Both of these properties are primarily due to the membrane skeleton.

Third, the membrane is highly permeable to both oxygen and bicarbonate ions; the latter is formed when carbon dioxide dissolves in water. All membranes are freely permeable to uncharged oxygen molecules. However, the high permeability to bicarbonate ions is due to an extraordinarily high density in the membrane of the erythrocyte anion transporter, a large dimeric glycosylated globular integral protein, known for historical reasons as "Band 3," which exchanges bicarbonate and chloride ions across the membrane down their concentration gradients. There are about 600,000 Band 3 anion transporter dimers per cell, representing about 25% of the total membrane protein and occupying about 20% of the total plasma membrane area.

Finally, the erythrocyte has a very high negative surface charge density due to the presence of glycophorin A, the second major erythrocyte integral protein. The amino-terminal extracellular domain of this small protein molecule is very heavily glycosylated with oligosaccharides bearing negatively charged terminal sialic acid sugar residues. The consequence of the negative surface charge contributed by ~100,000 glycophorin A molecules per cell is a high degree of electrostatic repulsion between individual erythrocytes. Because about 40% of the total volume of blood is occupied by "solids" (the erythrocytes), it is thought that this charge repulsion serves to greatly reduce the blood viscosity, which otherwise would be too high to permit the rapid peripheral circulation typical of mammalian life. Rare individuals lacking glycophorin A compensate by hypersialation of the anion transporter.

The plasma membrane of the human erythrocyte is conveniently isolated by lysing erythrocytes in low salt buffers, such that the cell contents are lost, leaving a colorless empty

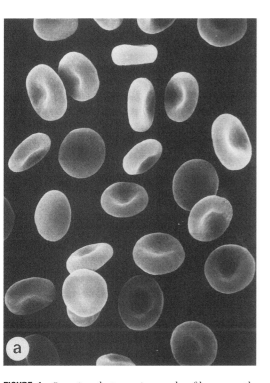 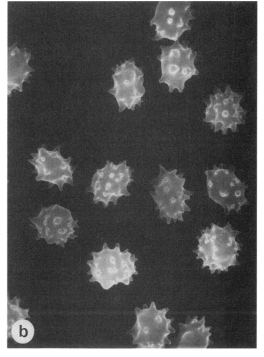

FIGURE 1 Scanning electron micrographs of human erythrocytes: (a) Normal biconcave disks, and (b) echinocytes. Reproduced from Lovrein and Anderson (1980) *J. Cell Biol.* **85**, 534–548, by copyright permission of The Rockefeller University Press. Cell diameter approximately 8 μm.

membrane sac, erythrocyte ghost, resembling in shape the cell from which it was derived. The method described in this article is based on that described by Dodge *et al.* (1963), and has been used routinely and successfully in the author's laboratory for many years, both for research purposes and for undergraduate practical classes.

II. MATERIALS AND INSTRUMENTATION

A. Buffers and Reagents

The volumes of buffers given are suitable for processing 20 ml of fresh blood, and should be scaled up for larger volumes.

1. EDTA anticoagulant: Mix 3.362 g disodium ethylenediaminetetraacetic acid (EDTA) anhydrous (MW 336.2 g) or 3.722 g dihydrate (MW 372.2 g) with 0.906 g NaCl (anhydrous) (MW 58.44 g). Add distilled water to give 100 ml. Use 1 ml of this solution of 100 mM EDTA and 155 mM NaCl per 9 ml of fresh blood.

2. Citrate anticoagulant: Mix 2.20 g trisodium citrate ($Na_3C_6H_5O_7 \cdot 2H_2O$, MW 294.10 g), 0.80 g citric acid ($C_6H_8O_7 \cdot H_2O$, MW 210.14 g) and D-glucose (2.45 g $C_6H_{12}O_6 \cdot H_2O$, (MW 198.16 g); **or** 2.20 g $C_6H_{12}O_6$ anhydrous, (MW 180.16 g)). Add distilled water to 100 ml (Source: "Handbook of Clinical Laboratory Data," 2nd Ed., p. 238, "Solution A"). Use 1 ml per 7 ml of fresh blood. This is better than EDTA anticoagulant if blood is to be stored before use. Blood may be stored in this for 3 weeks.

3. Isotonic 155 mM (310 mOsM) sodium chloride: 9.06 g/liter NaCl (anhydrous) (MW 58.44 g). Make 2 liters.

4. 310 mOsM sodium dihydrogen phosphate stock solution: 24.19 g/liter $NaH_2PO_4 \cdot 2H_2O$ (MW 156.08 g). Make 500 ml.

5. 20 mOsM sodium dihydrogen phosphate stock solution: Dilute 100 ml of the 310 mOsM sodium dihydrogen phosphate stock solution 1:15.5 with water to give 1,550 ml.

6. 310 mOsM disodium hydrogen phosphate stock solution: 14.67 g/liter Na_2HPO_4 (anhydrous) (MW 141.96) **or** 37.01 g/liter $Na_2HPO_4 \cdot 12H_2O$ (MW 358.14). Make 2 liters.

7. 20 mOsM disodium hydrogen phosphate stock solution: Dilute 200 ml of the 310 mOsM disodium hydrogen phosphate stock solution 1:15.5 with water to give 3,100 ml.

8. Isotonic 310 mOsM sodium phosphate buffer, pH 7.6 ("310 PB"): Titrate the remaining 1800 ml of the basic 310 mOsM disodium hydrogen phosphate solution with the acidic 310 mOsM monosodium dihydrogen phosphate solution (about 200 ml will need to be added) until the pH is 7.6.

9. 20 mOsM sodium phosphate buffer, pH 7.6 ("20 PB"): Titrate 1800 ml of the basic 20 mOsM disodium hydrogen phosphate solution with the acidic 20 mOsM monosodium dihydrogen phosphate solution (about 200 ml will need to be added) until the pH is 7.6.

B. Equipment

Magnetic stirrer, vortex mixer, and water pump aspirator with Büchner flask trap.

Refrigerated medium-speed centrifuge (up to 20,000 rpm) and suitable rotors and tubes. For example, using the Sorvall RC5B centrifuge, one requires: Sorvall SS34 rotor (8 × 50 ml), fixed angle 34° to vertical (3000 rpm = 1086 g_{max}, 16,000 rpm = 30,877 g_{max}), or equivalent rotor. Use with Sorvall Oak Ridge 50-ml nominal (35 ml actual) screw top tubes (polycarbonate 03934, polysulfone 03936, three-part sealing cap assembly 03277). Alternatively, flanged, straight-sided tubes (polycarbonate 50 ml 03146, polysulfone 03605) may be used if care is taken not to overfill. For large volumes of erythrocytes, use a Sorvall GS3 rotor (6 × 500 ml), fixed angle 20° to vertical (2500 rpm = 1,057 g_{max}). Use with Sorvall 500-ml nominal (350 ml actual) screw-top bottles (polycarbonate 03944, three-part sealing cap assembly 03280).

III. PROCEDURES

A. Notes on the Osmotic Lysis of Human Erythrocyte Membrane

Erythrocytes are said to lyse when their membranes break to release the cytoplasmic contents into solution (often termed hemolysis because it results in the release of hemoglobin). Lysis can be brought about either by osmotically induced swelling of the cell to its bursting point or by chemically dissolving away the semipermeable barrier created by the cell membrane's phospholipid bilayer.

1. Isotonic Conditions

In 155 mM NaCl, which is isotonic with human blood, erythrocytes are in osmotic equilibrium, with no internal turgor pressure (i.e., they are not inflated like a balloon). Cells isolated from fresh blood will show a normal biconcave disk shape, whereas cells prepared from blood bank blood, which has been stored for several days, may show shape changes, usually becoming round and crenated (i.e., having star-like projections), and are called echinocytes (Fig. 1). This is due to changes in the phosphorylation state of their membrane proteins and in the internal calcium concentration, consequent upon depletion of their internal ATP, and it can to some extent be reversed by incubation for several hours at 4°C in 10 mM glucose and 10 mM adenosine, which enables them to synthesize more ATP.

2. Hypertonic Solutions

Solutions of a higher ionic strength than that of blood plasma are referred to as hypertonic. When an erythrocyte is placed in a hypertonic solution, the lower external water potential causes water to leave the cell through the cell membrane. The cell shrinks and often adopts a characteristically crenated shape.

3. Hypotonic Lysis

Conversely, when an erythrocyte is placed in a hypotonic solution of lower ionic strength than blood, the higher external water potential causes water to enter the cell through its cell membrane. If the external solution is only slightly hypotonic, this process continues until the internal osmotic strength is lowered to equal that of the external solution, at which point net influx of water ceases. However, when an erythrocyte is placed in more dilute solutions, the membrane is ruptured by the increasing internal hydrostatic pressure before this equilibrium situation can be achieved. The cell then undergoes hypotonic lysis, losing its cytoplasmic contents to the solution.

4. Molarity and Osmolarity

Because the osmotic strength of a solution is determined by the total concentration of ions in that solution, it is often convenient, when dealing with osmotically sensitive systems, to calculate the strength of buffers in terms of their osmolarity rather than their molarity. For example, 0.1 M solutions of NaH_2PO_4 and Na_2HPO_4 will contain different numbers of ions and thus possess different osmotic strengths. The osmolarity of a solution is defined as the product of its molarity and the number of ions into which the substance dissociates. Thus 155 mM NaH_2PO_4 has an osmolarity of 310 mOsM, which is isotonic with human blood, as it dissociates into two ions. So does 103.3 mM Na_2HPO_4, as it dissociates into three ions. Although they differ in molarity, 310 mOsM solutions of these two phosphate salts can thus be mixed in any proportion to give a buffer of intermediate pH, without changing the osmotic strength of the buffer. This is the basis upon which the buffers used in the following procedures are prepared.

B. Obtaining Human Blood and Safety Considerations

There are two normal possibilities for obtaining human blood, either by taking fresh blood from a laboratory volunteer or by obtaining a bag of donor blood from a local hospital or regional blood bank. The choice between these two possibilities may be influenced by proximity to such a hospital or blood bank, the volume of blood required, the importance of having fresh blood rather than blood that has been stored for several days, the availability of staff qualified to take blood from a laboratory donor, and the restrictions of local and national safety regulations, which must be learned about and strictly adhered to.

The principal potential hazard from working with human blood is the risk of contracting a serious viral infection. For this reason, blood should be used only from donors who have been tested and shown to be free of HIV and hepatitis virus infections. As an additional precaution, laboratory coats and latex surgical gloves should be worn at all stages of the preparation of erythrocyte ghosts, and mouth pipetting must be scrupulously avoided. All procedures should be carried out on sealed plastic surfaced benches or on benches covered with "Benchcote" (Whatman Ltd.). All potentially contaminated solutions generated during the preparation, including supernatants from centrifugation washes and any excess blood, should be poured into a container previously filled half full with a 2× strength solution of an antiviral disinfectant, such as 2% Virkon (Antec International Ltd.), and left for 24 hr before disposal to drains. Contaminated glassware, centrifuge tubes, and apparatus should be rinsed thoroughly immediately after use in a 1× strength solution of the same disinfectant (e.g., 1% Vircon) and then left to soak in this for 24 hr before being washed normally. Gloves, tissues, disposable plasticware, and empty blood bags should be placed in the autoclave bag immediately after use and autoclaved as soon as is conveniently possible. At the end of a preparation yielding creamy-white washed erythrocyte ghosts, the chances of any viruses remaining in the ghosts are minimal.

A separate potential hazard of causing harm while extracting blood from a laboratory volunteer can arise from the use of a nonsterile hypodermic syringe or as a result of incorrect technique. For these reasons, blood should only be taken by properly trained qualified staff.

While sonicating erythrocyte ghosts to promote the subsequent release of spectrin from the membrane (See volume 2, Shotton, "Purification of Human Erythrocyte Spectrin and Actin" for additional information), ear protectors should be worn.

C. Method of Taking Fresh Venous Blood from a Volunteer Donor

Although large volumes of blood are most conveniently obtained from a blood bank in units of 500 ml, small volumes (5 to 50 ml) of fresh blood may (safety regulations permitting) be conveniently obtained from a laboratory volunteer, most easily from the vein that is situated just below the skin on the inner aspect of the elbow:

Steps

1. Seat the volunteer with the left arm extended, palm up, and supported beneath the elbow and the wrist.

2. Ask the volunteer to firmly grasp a cylindrical object approximately 25 mm in diameter with the left hand, and restrict the returning veinous blood supply from this arm using an inflatable or elastic pressure cuff secured around the upper left arm.

3. Sterilize the skin of the inner left elbow by swabbing with 95% ethanol.

4. While keeping the needle in its protective cover, remove the end cap from the cover of a sterile 20-gauge $1\frac{1}{2}$-in. (0.95 × 40 mm) hypodermic needle and attach the needle to a sterile syringe of the desired capacity. Free the initial adhesion of the plunger in the barrel of the syringe by moving it slightly in and out, and then expel all the air by pushing the plunger fully in.

5. While standing facing and on the left side of the donor, remove the protective cover from the needle and gently but firmly insert the needle, at an angle of about 30° to the surface of the skin and bevel tip down, through the skin and fully into the distended vein.

6. While holding the needle steady in the vein with one hand, slowly withdraw the plunger of the syringe with the other to permit the blood to flow into the syringe. Continue until the desired volume of blood has been extracted.

As an alternative to using a hypodermic needle attached to a syringe, one can use a needle attached to a special sterile-evacuated tube already containing anticoagulant ("Vacutainer," from Beckton-Dickinson Ltd.). In this case, the needle of the Vacutainer assembly is first inserted into the vein as described, and then the Vacutainer tube is pushed forward relative to the needle so that the sharp back end of the needle penetrates the rubber membrane sealing the evacuated tube, allowing the vacuum to suck blood into the tube. When flow has ceased, continue as described below.

7. Keeping the needle steadily in place, release the pressure cuff and cover the site of needle entry into the skin with a small sterile cotton swab. Having done this, remove the syringe needle from the arm, and request the donor to hold the swab firmly in place while raising the arm vertically above the shoulder.

8. While the donor is recovering in this position, remove the needle from the syringe or Vacutainer, disposing of it into a "sharps" container that will later be autoclaved or incinerated. If a syringe was used, express the blood from the syringe barrel into a tube or flask of suitable size, previously prepared and containing an appropriate volume of anticoagulant. Mix thoroughly, seal, and stand in ice in an ice bucket. If a Vacutainer was used, simply mix thoroughly and stand in ice.

9. Finally, cover the donor's needle wound with a sterile dressing or adhesive plaster, and request that (s)he maintain pressure on the wound and keep the arm elevated for a little longer in the unlikely event of any residual bleeding.

D. Preparation of Washed Erythocytes

All procedures should be conducted at 4°C or on ice.

Steps

1. Wash each volume of blood containing anticoagulant with 12 to 15 vol of isotonic 155 mM NaCl by mixing. For 20 ml of donor blood, which will yield about 8 ml of packed erythrocytes, this is most conveniently done by placing 2.5 ml of the blood in each of eight 35-ml Oak Ridge-style screw-top polycarbonate centrifuge tubes on ice and then squirting the cold isotonic NaCl from a wash bottle into each individual tube while holding it at an angle on the rotating head of a vortex mixer. When the tube contains about 20 ml the vortex will fail. At this point, place the tube in the ice bucket and make up the volume to approximately 35 ml with NaCl solution, balancing tubes in pairs by eye ready for subsequent centrifugation, and seal with screw caps. For larger volumes of blood, the initial washes may be conducted in 500-ml centrifuge bottles with screw tops, inverting the bottles several times to achieve thorough mixing of the contents.

2. Spin the tubes (or bottles) in a precooled angle rotor of a refrigerated medium-speed centrifuge at 4°C and 1000 g for 10 min to sediment the blood cells. When the rotor has come to a standstill, carefully remove each tube, slowly rotating it by 180° about its long axis in the process while maintaining the tube at an angle of ~45° to the horizontal , and place the tube on ice at this angle with the loose cell pellet undisturbed on the *lower inside surface* of the base of the tube.

3. Observe the cell pellet closely. Upon the surface of the red erythrocytes which form the bulk of the pellet, you will see a wispy whitish film of leukocytes (white blood cells) and platelets, the so-called "buffy coat." Because these cells are rich in proteolytic enzymes that might damage erythrocyte membrane proteins at later stages of the work, it is important to remove this buffy coat as thoroughly as possible, without losing too many of the underlying erythrocytes. This is done by carefully aspirating the buffy coat with moderate suction only, using a long glass Pasteur pipette connected via a Büchner flask containing 2% Vircon to a water suction pump before removing the supernatant. Attempt to remove most but not all of the buffy coat, realizing that there will be subsequent opportunities to remove any missed on the first attempt. Once the major part of the buffy coat has been removed, aspirate away the rest of the straw-colored plasma-rich supernatant using the same Pasteur pipette, taking care not to remove any erythrocytes.

4. Resuspend the erythrocytes in 12 to 15 vol of fresh cold isotonic saline, resediment the cells, and again aspirate the supernatant, as in steps 1 to 2 above. Repeat this washing procedure a third time, but this time resuspending the cells in isotonic 310 mOsM sodium phosphate buffer, pH 7.6 (recipe below). At each aspiration stage, try to remove more of the residual buffy coat without aspirating too many erythrocytes. You should aim to lose no more than 5% of the total volume of erythrocytes during all three aspirations of the buffy coat. Any residual leucocytes will be effectively removed after subsequent lysis of the washed erythrocytes. At the end of these three washes, you will be left with about 1 ml of deep red packed erythrocytes for every 2.5 ml of whole blood washed. Remove a small sample of the washed erythrocytes at this stage and store on ice for subsequent microscopy or gel analysis, if required.

E. Hypotonic Lysis of Erythrocytes and Purification of Erythrocyte Ghosts

Steps

1. After the third wash of the erythrocytes in isotonic sodium phosphate buffer, pH 7.6, lyse the erythrocytes by resuspending as before, but this time using 12 to 15 vol of 20 mOsM sodium phosphate buffer pH 7.6. Pool the resuspended and lysing erythrocyte preparations in a beaker, add a magnetic stir bar, cover, and stir gently for 20 min on ice using a magnetic stirrer.

The lysis of the erythrocytes at this stage will be apparent from the change of the suspension from opaque to transparent, due to the loss of light scattering as hemoglobin is lost from the cells. Before lysis, the erythrocytes are dense objects of high refractive index floating in a colorless liquid. As such, much of the light passing through the suspension is refracted and scattered, and even quite dilute suspensions of erythrocytes appear opaque. After lysis, one is left with colorless empty erythrocyte ghosts of low refractive index floating in a uniformly red solution. Much less light is scattered under these conditions, even though the total cell and hemoglobin concentrations are the same, and so the lysed suspension is now transparent.

2. Divide the suspension into 35-ml capacity screw-top centrifuges tubes, seal, and spin at 30,000 g for 10 min at 4°C. When the rotor has come to a standstill, remove the tubes in turn onto ice by rotating them by 180° about their long axis while maintaining their angle at ~45° to the horizontal, exactly as before, taking particular care not to disturb the tube contents. Each tube will now contain a deep red supernatant above a large pale red layer of erythrocyte ghosts, under which will be a small red and white pellet of lysed leukocytes, platelets and other debris.

3. This next step is the only one of the whole procedure that is technically tricky, where lack of care may result in loss of yield. With a long Pasteur pipette and gentle suction from the water suction pump, first carefully aspirate away the red hemoglobin-rich supernatant while maintaining the tube at an angle of ~45°, as before. The principal difference from performing this maneuver with intact erythrocytes is that the ghost pellet is much looser and more easily disturbed, whereas the supernatant is so deeply colored that it is difficult to see the boundary between supernatant and pellet. Overenthusiasm at this stage will result in loss of many of the ghosts, so undertake the aspiration slowly, carefully advancing the tip of the pipette down the side of the tube toward the pellet. Once about 70% of the supernatant has been removed in this way, do *not* advance the pipette tip any further. Rather, to remove the final 30% of the supernatant, tilt the tube slowly toward the horizontal so that the meniscus between the supernatant and the ghosts moves up the side of the tube toward the tip of the Pasteur pipette. This is best seen if the tube is initially wipe free of any external condensation or moisture, and is now viewed against bright room lights from below. This view can be achieved by carefully holding the tube above your head or, if the tubing from the water pump is of insufficient length (as is usually the case!), by crouching down and squinting upwards. Alternatively one may look down into a mirror placed flat on the workbench beneath the tube. With care, it is possible to remove all the supernatant without loss of any erythrocyte ghosts.

4. If the centrifuge tube is now rotated by 180° about its long axis, while holding it at an angle of ~45° to the horizontal, the large loose pellet of erythrocyte ghosts will flow away under gravity to the lower part of the tube, leaving the small red and white pellet of lysed leukocytes, platelets and debris sticking to the upper inner surface of the tube. After waiting 30 sec for the ghosts to flow away, increase the suction pressure from the water pump and carefully suck away this sticky pellet completely from the upper inner surface of the end of the tube, leaving only the erythrocyte ghosts in the bottom of the tube.

5. Using the procedures previously described, wash the erythrocyte ghosts twice more by vigorous vortex mixing with 12 to 15 vol of 20 mOsM phosphate buffer, centrifugation at 30,000 g, and aspiration at each stage, attempting to remove the progressively smaller residual sticky pellet beneath the ghosts. After the second wash, the ghosts should appear pale under a pale red supernatant and, after removal of the third clear supernatant, the tubes should finally contain about 1 ml of pale creamy white erythrocyte ghosts, without a trace of pink, for every 1 ml of washed erythrocytes lysed. Remove a small sample of the erythrocyte ghosts at this stage and store on ice for subsequent microscopy or gel analysis, if required. (See volume 2, Shotton, "Purification of Human Erythrocyte Spectrin and Actin" for details of polyacrylamide gel electrophoresis patterns.)

If fresh blood was used, both the intact erythrocytes and the ghosts should appear as typically flattened biconcave disks by phase-contrast light microscopy using a 40× dry objective. The erythrocytes should be highly refractile and orange in appearance, and the ghosts are seen as colorless gray outlines. If the blood has been stored for any length of time before use, many of the cells will have become crenated due to the depletion of internal ATP levels, being more rounded with undulating or spiky protrusions, and the ghosts made from them will retain these shapes (Fig. 1).

Reference

Dodge, J. T., Mitchell, C., and Hanahan, D. J. (1963). The preparation and chemical characterisation of haemoglobin-free ghosts of human erythrocytes. *Arch. Biochem. Biophys.* **100,** 119–130.

Isolation and Subfractionation of Plasma Membranes to Purify Caveolae Separately from Glycosyl-phosphatidylinositol-Anchored Protein Microdomains

Phil Oh and Jan E. Schnitzer

I. INTRODUCTION

It is becoming quite clear that the plasma membrane consists of distinct microdomains with their own unique molecular topographies. This article describes several protocols for first purifying plasma membranes from tissue and cultured cells and then subfractionating them to purify their caveolae separately from other plasmalemmal microdomains rich in glycosyl-phosphatidylinositol (GPI)-anchored proteins. Caveolae (also called noncoated plasmalemmal vesicles) are specialized flask-shaped invaginated microdomains located on the cell surface membrane of many types of cells. Little has been known about their function primarily because of a lack of means to study these structures. Past investigations were limited predominately to morphological studies looking at presumed vesicular trafficking and the occasional discovery of molecules localized to caveolae. Directed research, which is less dependent on happenstance, has evolved largely because of newly developed biochemical techniques for studying caveolae. One major step forward in this regard has been the development of a procedure to isolate caveolae to homogeneity directly from purified plasma membranes with little, if any, contamination from other sources, including GPI-anchored protein microdomains (Schnitzer et al., 1995b).

Studies have demonstrated that caveolae in endothelium are (i) dynamic carriers with the necessary molecular transport machinery for vesicle docking and fusion (Schnitzer et al., 1995a); (ii) capable of budding from the cell surface via a fission process requiring GTP hydrolysis to form free transport vesicles trafficking molecules into and/or across the endothelium (Schnitzer et al., 1996); and (iii) specialized plasmalemmal subcompartments capable of transducing chemical and mechanical information from the extracellular environment into the cells (Lui et al., 1996; Rizzo et al., 1996). Specific signaling pathways in the caveolae may be important in the budding and transport of macromolecules as well as in the regulation of various key cell functions. Triton-insoluble membranes (TIM) contain both caveolin and GPI-anchored proteins (Lisanti et al., 1994) and are rich in signaling molecules (Sargiacomo et al., 1993). GPI-anchored proteins may also be involved in signal transduction at the cell surface (Cinek and Horejsi, 1992). They were thought for many years to reside concentrated in caveolae based primarily on (i) immunolocalization electron microscopy studies and (ii) various subfractionation procedures (Chang et al., 1994; Lisanti et al., 1994; Smart et al.,

CELL BIOLOGY: A LABORATORY HANDBOOK, Second Edition. Vol. 2.

1995) that isolate membrane vesicles enriched in both caveolin and GPI-anchored proteins. More recent work has revealed methodological shortcomings that invalid these conclusions because (i) antibodies used to localize GPI-anchored proteins can bind and artifactually sequester the GPI-anchored proteins into clusters in or near caveolae (Mayor *et al.*, 1994); (ii) the isolated vesicles contain both small caveolar vesicles as well as much larger contaminating vesicles rich in GPI-anchored proteins; and (iii) caveolae actually can be purified separately from plasmalemmal GPI-anchored protein microdomains (Schnitzer *et al.*, 1995b) as described later. It is clear that the research on caveolae is just beginning and that procedures outlined in this article may help the discovery process.

II. MATERIALS AND INSTRUMENTATION

Sodium chloride (Cat. No. BP358), potassium chloride (Cat. No. BP366), magnesium sulfate (Cat. No. BP213), glucose (Cat. No. BP350), EDTA disodium salt (Cat. No. BP120), sodium bicarbonate (Cat. No. BP328), sucrose (Cat. No. BP220), magnesium chloride (Cat. No. BP214), Tris (Cat. No. BP152), methanol (Cat. No. A412), hydrochloric acid (Cat. No. A144), sodium hydroxide (Cat. No. SS255), sodium carbonate (Cat. No. BP357), 2-mercaptoethanol (Cat. No. BP176), Tricine (Cat. No. BP315), Tween 20 (Cat. No. BP337), silver nitrate (Cat. No. BP360), glacial acetic acid (Cat. No. A38), sodium phosphate dibasic (Cat. No. BP332), potassium phosphate monobasic (Cat. No. BP362), potassium phosphate dibasic (Cat. No. BP363), tygon tubing (Cat. No. 14-169-1B), razor blades (Cat. No. 12-640), and 15-ml centrifuge tubes (Cat. No. 05-539-5) are from Fisher. HEPES (Cat. No. 737 151) and Pefabloc SC (Cat. No. 1 585 916) are from Boehringer Mannheim. MES [2-(N morpholine)-ethanesulfonic acid] (Cat. No. M-3032), leupeptin (Cat. No. L-2884), O-phenanthroline (Cat. No. P-9375), pepstatin A (Cat. No. P-4265), sodium nitroprusside (Cat. No. S-0501), E-64 [*trans*-epoxysuccinyl-L-leucylamido-(4-guanidino)butane] (Cat. No. E-3132), glycerol (Cat. No. G-5516), potassium acetate (Cat. No. P-3542), magnesium acetate (Cat. No. M-0631), EGTA (Cat. No. E-4378) are from Sigma. Percoll (Cat. No. 17-0891-01) is from Pharmacia. Triton X-100 (10%) (Cat. No. 28314) is from Pierce. Goat anti-mouse IgG–HRP conjugate (Cat. No. NA931) and goat anti-rabbit IgG–HRP conjugate (Cat. No. NA934) are from Amersham. Goat anti-mouse IgG–bodipy conjugate (Cat. No. B-2752), goat anti-mouse IgG–Texas red conjugate (Cat. No. T-862), goat anti-rabbit IgG–bodipy conjugate (Cat. No. B-2766), and goat anti-rabbit IgG–Texas red conjugate (Cat. No. T-2767) are from Molecular Probes. Polyacrylic acid (PAA) (Cat. No. 00627) is from Polysciences Inc. Nycodenz (Cat. No. AN7050) is from Accurate Chemicals. Sprague–Dawley rats (male) are from Charles River. SW28 tubes (Cat. No. 344058), SW55 tubes (Cat. No. 344057), 60 Ti tubes (Cat. No. 355642), and cap 60 Ti tubes (Cat. No. 338906) are from Beckman. Ketamine is from Parke Davis. Xylazine is from Fermenta Veterinary Products. The 53-μm Nytex filter (Cat. No. 3-53/41) and the 30-μm Nytex filter (Cat. No. 3-30/21) are from Tetko. The type "AA" homogenizer (Cat. No. 3431-E10), type "BB" homogenizer (Cat. No. 3431-E20), and type "C" homogenizer (Cat. No. 3431-E25) are from Thomas Scientific. Silk suture (3-0) is from Ethicon. Stainless-steel tubing (Cat. No. 38-3069) is from Ranin. The three-way stopcock (Cat. No. 732-8103) is from Bio-Rad. Positively charged colloidal silica particles are from TopoGen.

Ultracentrifugation was carried out in either a Beckman L8-80 or a Sorvall Pro 80. Other centrifugation was carried out in a Brinkman Eppendorf 5415C or IEC Centra MP4R.

Perfusion apparatus: On a ring stand, attach five 60-cc syringes to a series of five three-way stopcocks by way of tygon tubing (0.0625 inch ID). The stopcocks are attached to each other in series and to tygon tubing with a stainless-steel loop (0.046 in i.d.) inserted in the middle of the tubing. The stainless-steel loop is used to regulate the temperature of the solutions perfused into the rat vasculature by emersion in temperature-controlled water. The pressure of flow is measured by a sphyngomometer attached to the 60-cc syringes by tygon tubing and a single hole rubber stopper. The flow for perfusion of the rat lungs is normally 8 mm of Hg. The flow rate at 25°C is approximately 5 ml/min and at 10°C is 4 ml/min.

III. PROCEDURES

A. Purification of Silica-Coated Plasma Membrane (P)

Solutions

1. *Mammalian Ringer's solution (without calcium):* 114 mM sodium chloride, 4.5 mM potassium chloride, 1 mM magnesium sulfate, 11 mM glucose, 1.0 mM sodium phosphate (dibasic), 25 mM sodium bicarbonate in double-distilled H_2O, pH to 7.4, with HCl (filter through a 0.45-μm bottle top filter, can be stored at 4°C); for 2 liters, 13.32 g NaCl, 0.67 g KCl, 0.492 g $MgSO_4 \cdot 7H_2O$, 3.96 g glucose, 0.28 g Na_2HPO_4, and 4.20 g $NaHCO_3$.

2. *Nitroprusside:* 20 mg/ml in double-distilled H_2O; this is 200× stock (always make fresh).

3. *MES-buffered saline (MBS):* 20 mM MES, 135 mM NaCl in double-distilled H_2O, pH to 6.0, with NaOH (filter through a 0.45-μm bottle top filter, can be stored at 4°C); for 2 liters, 7.8 g MES and 14.6 g NaCl.

4. *Positively charged colloidal silica:* From 30% stock solution, dilute to 1% with MBS, pH 6.0 (pH to 6.0 with NaOH if necessary and filter through a 0.45-μm bottle top filter, can be stored at 4°C); for 300 ml, 10 ml stock silica in 290 ml MBS, pH 6.0, pH with pH strips (usually you do not have to adjust pH).

5. *Polyacrylic acid (PAA):* From 25% stock solution, dilute to 0.1% with MBS, pH 6.0 (pH to 6.0 with NaOH and filter through a 0.45-μm bottle top filter, can be stored at 4°C); for 500 ml, 2 ml stock 25% polyacrylic acid in 480 ml MBS, pH 6.0, and bring to 500 ml with MBS, pH 6.0.

6. *Sucrose/HEPES:* 250 mM sucrose, 25 mM HEPES, and 20 mM potassium chloride in double-distilled H_2O (pH to 7.4 with NaOH and filter through a 0.45-μm bottle top filter, can be stored at 4°C); for 2 liters, 171.16 g sucrose, 11.92 g HEPES, and 2.98 g KCl.

7. *Protease inhibitors (200×):* 2 mg/ml leupeptin in double-distilled H_2O, 2 mg/ml pepstatin A in double-distilled H_2O, 10 mg/ml *O*-phenanthroline in double-distilled H_2O, 2 mg/ml E-64 in 50% EtOH, and 2 mg/ml Pefabloc sc in double-distilled H_2O.

8. *Phosphate-buffered saline (PBS):* 137 mM sodium chloride, 2.7 mM potassium chloride, 4.3 mM sodium phosphate (dibasic), 1.4 mM potassium phosphate (monobasic) in double-distilled H_2O, pH to 7.4; for 1 liter, 8 g sodium chloride, 0.2 g potassium chloride, 0.61 g sodium phosphate (dibasic), and 0.2 g potassium phosphate (monobasic).

9. *1 M KCl:* 7.46 g in 100 ml double-distilled H_2O.

10. *60% sucrose:* 385.95 g sucrose in 200 ml double-distilled H_2O, add 10 ml 1 M KCl; after the sucrose dissolves, bring to 500 ml with double-distilled H_2O.

11. *250 mM HEPES pH 7.4:* 119.2 g in 450 ml double-distilled H_2O, pH to 7.4 with NaOH, bring to 500 ml with double-distilled H_2O.

12. *Nycodenz, 102% (v/w):* 102 g Nycodenz in 100 ml of double-distilled H_2O.

    a. 70% Nycodenz in sucrose/HEPES: 7 ml 102% Nycodenz, 1 ml 60% sucrose, 1 ml 250 mM HEPES, pH 7.4, 200 μl 1 M KCl; bring to 10 ml with double-distilled H_2O.

    b. 65% Nycodenz in sucrose/HEPES: 6.5 ml 102% Nycodenz, 1 ml 60% sucrose, 1 ml 250 mM HEPES, pH 7.4, 200 μl 1 M KCl; bring to 10 ml with double-distilled H_2O.

    c. 60% Nycodenz in sucrose/HEPES: 6 ml 102% Nycodenz, 1 ml 60% sucrose, 1 ml 250 mM HEPES, pH 7.4, 200 μl 1 M KCl; bring to 10 ml with double-distilled H_2O.

d. 55% Nycodenz in sucrose/HEPES: 5.5 ml 102% Nycodenz, 1 ml 60% sucrose, 1 ml 250 mM HEPES, pH 7.4, 200 μl 1 M KCl; bring to 10 ml with double-distilled H$_2$O.

13. *Nycodenz:*

    a. 80% Nycodenz: 8 ml 102% Nycodenz, 200 μl 1 M KCl; bring to 10 ml with double-distilled H$_2$O.

    b. 75% Nycodenz: 7.5 ml 102% Nycodenz, 200 μl 1 M KCl; bring to 10 ml with double-distilled H$_2$O.

    c. 70% Nycodenz: 7 ml 102% Nycodenz, 200 μl 1 M KCl; bring to 10 ml with double-distilled H$_2$O.

    d. 65% Nycodenz: 6.5 ml 102% Nycodenz, 200 μl 1 M KCl; bring to 10 ml with double-distilled H$_2$O.

    e. 60% Nycodenz: 6 ml 102% Nycodenz, 200 μl 1 M KCl; bring to 10 ml with double-distilled H$_2$O.

1. Perfusion

Steps

1. Perform a tracheotomy to ventilate the lungs using a respirator.

2. Insert the tubing from the perfusion apparatus into the pulmonary artery via a small cut into the right ventricle and tie a 3-0 silk suture around the artery to fasten. Cut the left atrium to allow the flow to exit.

3. Perfuse Ringer's/nitroprusside for 3–5 min. The temperature of the perfusate begins at room temperature and after 1–1.5 minutes is lowered to around 10°C by placing the stainless-steel loop into an ice-cold bath. Gently drip ice-cold PBS over the lung. The temperature for the rest of the perfusion is done at this cool temperature.

4. Perfuse MBS for 1.5 min.

5. Perfuse 1% colloidal silica solution for 1.5 min. At this point we keep the lungs inflated by replacing the tube attached to the respirator with a syringe (inflate between 3 and 5 ml of air).

6. Flush with MBS for 1.5 min.

7. Perfuse 0.1% PAA for 1.5 min.

8. Flush lungs with 8 ml of sucrose/HEPES with protease inhibitors (1×).

9. Excise lungs from the animal and keep cold by emersion in ice-cold sucrose/HEPES with protease inhibitors (1×).

2. Processing of the Lung

Steps

1. Finely mince the excised lung with a "new" razor blade on a cold aluminum block embedded in packed ice.

2. Add 20 ml of sucrose/HEPES with protease inhibitors and place into a type "C" homogenizer vessel.

3. Homogenize with the corresponding grinder for 12 strokes at 1800 rpm.

3. Purification of Silica-Coated Luminal Endothelial Cell Plasma Membrane (P)

Steps

1. Filter the homogenate through a 53-μm Nytex filter, followed by a 30-μm Nytex filter.

2. Remove 200 μl from the filter solution and save and label homogenate. Bring the remaining solution to 20 ml with cold sucrose/HEPES with protease inhibitors.

3. Add an equal volume of 102% Nycodenz and mix (this is enough for two SW28 tubes). Layer onto a 70–55% continuous Nycodenz sucrose/HEPES gradient (form by placing 3 ml of 70, 65, 60, and 55% Nycodenz sucrose/HEPES and carefully swirling the solution holding the tube at a 45° angle about 5–10 times).

4. Top with sucrose/HEPES with protease inhibitors. Spin at 15,000 for 30 min 4°C in a SW28 rotor. Aspirate off the supernatant. Resuspend the pellet in 1 ml MBS. Add an equal volume of 102% Nycodenz and mix.

5. Layer onto a 80–60% continuous Nycodenz gradient (form by placing 350 μl of 80, 75, 70, 65, and 60% Nycodenz and twirling tube about 5–10 times). Top with 20 mM KCl. Spin at 30,000 for 30 min at 4°C in a SW55 rotor. Aspirate off the supernatant. Resuspend the pellet in 1 ml MBS and label P.

4. Isolation of Silica-Coated Plasma Membranes from Cultured Cells (Monolayer)

Steps

For T75 flasks, increase proportionately for larger or smaller dimensions.

1. Wash cell monolayer three times with Ringer's/nitroprusside and then three times with MBS.

2. Incubate with 1% colloidal silica solution for 10 min. Wash cell monolayer three times with MBS.

3. Incubate with 0.1% PAA for 10 min. Wash cell monolayer three times with MBS.

4. Add 5 ml of sucrose/HEPES with protease inhibitors (1×). Scrape cells and place in 15-ml centrifuge tube. Spin at 1000 g for 5 min at 4°C.

5. Bring to 1 ml with cold sucrose/HEPES with protease inhibitors (1×) (you can use up to six T75 flasks per 1 ml of cold sucrose/HEPES with protease inhibitors). Add an equal volume of 102% Nycodenz and mix (SW55 tube). Layer onto a 70–55% continuous Nycodenz sucrose/HEPES gradient (form by placing 350 μl of 70, 65, 60, and 55% Nycodenz sucrose/HEPES and carefully twirling tube as described earlier).

6. Top with sucrose/HEPES with protease inhibitors. Spin at 30,000 for 30 min at 4°C in a SW55 rotor. Aspirate off the supernatant. Resuspend the pellet in 1 ml MBS and label P.

B. Isolation of Caveolae (V)

Use solution 3 "P" isolation plus the following solutions.

Solutions

1. *10% Triton X-100 in PBS.*

2. *Sucrose:*

   a. 60% sucrose 20 mM KCl: 385.95 g sucrose in 200 ml double-distilled H$_2$O, add 10 ml 1 M KCl; after the sucrose dissolves, bring to 500 ml with double-distilled H$_2$O.

   b. 35% sucrose 20 mM KCl: 5.2 ml 60% Sucrose, 200 μl 1 M KCl; bring to 10 ml with double-distilled H$_2$O.

   c. 30% sucrose 20 mM KCl: 4.4 ml 60% sucrose, 200 μl 1 M KCl; bring to 10 ml with double-distilled H$_2$O.

   d. 25% sucrose 20 mM KCl: 3.8 ml 60% sucrose, 200 μl 1 M KCl; bring to 10 ml with double-distilled H$_2$O.

   e. 20% sucrose 20 mM KCl: 2.8 ml 60% sucrose, 200 μl 1 M KCl; bring to 10 ml with double-distilled H$_2$O.

f. 15% sucrose 20 mM KCl: 2.1 ml 60% sucrose, 200 μl 1 M KCl; bring to 10 ml with double-distilled H_2O.

g. 10% sucrose 20 mM KCl: 1.3 ml 60% sucrose, 200 μl 1 M KCl; bring to 10 ml with double-distilled H_2O.

h. 5% sucrose 20 mM KCl: 0.7 ml 60% sucrose, 200 μl 1 M KCl; bring to 10 ml with double-distilled H_2O.

i. 20 mM KCl: 200 μl 1 M KCl; bring to 10 ml with double-distilled H_2O.

1. Isolation of Caveolae (V) (in the Presence of Detergent)

Steps

1. Take 900 μl "P", add 100 μl 10% Triton X-100 (mix). Save 100 μl "P". Nutate for 10 min at 4°C. Place in type "AA" homogenizer vessel. Homogenize with corresponding grinder for 20 strokes at 1800 rpm.

2. Bring to 40% sucrose/20 mM KCl. Layer with discontinuous 35–0% sucrose/20 mM KCl gradient (350 μl of 35, 30, 25, 20, 15, 10, and 5% sucrose/20 mM KCl and 20 mM KCl).

3. Spin at 30,000 overnight at 4°C in a SW55 rotor. Collect the band between 10 and 15% sucrose density. Dilute the band material two to three times with MBS and spin at 15,000 at 4°C for 2 hr. Resuspend the pellet in MBS, pH 6.0, and label V.

4. Collect and resuspend membrane pellet in MBS and label P-V.

2. Isolation of Caveolae (V′) (Detergent Free)

Steps

1. Process as in above "V" isolation except without the Triton step, and homogenize 60 strokes. Resuspend pellet in MBS and label V′.

2. Collect and resuspend membrane pellet in MBS and label P-V′.

C. Isolation of Budded Caveolae (V$_b$) (without Physical Disruption)

Solutions

Use solutions 1, 2, 6, and 7 from "P" isolation, solutions 1 and 2 from "V" isolation, plus the following solution.

1. *Cytosolic buffer:* 25 mM potassium chloride, 2.5 mM magnesium acetate, 5 mM EGTA, 150 mM potassium acetate, 25 mM HEPES in double-distilled H_2O (pH to 7.4 with KOH and filter through a 0.45-μm bottle top filter, can be stored at 4°C); 500 ml, 0.93 g KCl, 0.27 g $Mg(C_2H_3O_2)_2 \cdot 4H_2O$, 0.95 g $C_{14}H_{24}N_2O_{10}$, 7.36 g $KC_2H_3O_2$, and 2.98 g HEPES.

1. Isolation of Rat Lung Cytosol

Steps

1. Perfuse rat lung as in "P" isolation steps 1, 2, 3, 8, and 9.

2. Homogenize lung in cytosolic buffer (two to three times, v/w) for 10 strokes at 1800 rpm.

3. Spin at 35,000 for 1 hr at 4°C in a SW55 rotor. Place supernatant over PD-10 column and collect first peak. Aliquot (1 ml) and store at −80°C (rat lung cytosol).

2. Isolation of "V$_b$"

Steps

1. Incubate "P" or "PM" (20–50 μg/ml) with rat lung cytosol (1–5 mg/ml) for 1 hr at 37°C.

2. Bring to 40% sucrose with 20 mM KCl. Layer with discontinuous 35–0% sucrose/20 mM KCl gradient (350 μl each layer).

3. Spin at 30,000 overnight at 4°C in a SW55 rotor. Collect the band between 10 and 15% sucrose density. Dilute the band material two to three times with MBS and spin at 15,000 at 4°C for 2 hr. Resuspend pellet in MBS and label V_b.

4. Collect and resuspend membrane pellet in MBS and label P-V_b.

D. Isolation of Nonsilica-Coated Plasma Membranes (PM)
Solutions

Use solutions 1, 2, 6, and 7 from "P" isolation plus the following solutions.

1. *Buffer PM:* 0.25 M Sucrose, 1 mM EDTA, 20 mM Tricine in double-distilled H$_2$O, pH to 7.8 with NaOH (filter through a 0.45-μm bottle top filter, can be stored at 4°C); for 500 ml, 42.79 g sucrose, 0.186 g EDTA (disodium salt), and 1.79 g Tricine.

2. *PBS:* 137 mM sodium chloride, 2.7 mM potassium chloride, 4.3 mM sodium phosphate (dibasic), 1.4 mM potassium phosphate (monobasic) in double-distilled H$_2$O, pH to 7.4; for 1 liter, 8 g sodium chloride, 0.2 g potassium chloride, 0.61 g sodium phosphate (dibasic), and 0.2 g potassium phosphate (monobasic).

3. *30% Percoll in buffer PM:* Mix 30 ml of stock Percoll with 70 ml of buffer PM.

1. Perfusion

Follow steps 1, 2, 3, 8, and 9 from "P" isolation perfusion.

2. Processing of the Lung

Follow Steps from "P" isolation processing, except use 3 ml of buffer PM and type "BB" homogenizer vessel for step 2.

3. Isolation of Fraction Containing Plasma Membranes (PM)
Steps

1. Filter the homogenate through a 53-μm Nytex filter and then through a 30-μm Nytex filter.

2. Remove 200 μl from the filter solution and save and label homogenate. Spin the remaining supernatant at 1000 g for 10 min at 4°C and save supernatant. Resuspend pellet in 3 ml buffer PM and repeat spin. Pool the two supernatants.

3. Split to two tubes and layer onto 17 ml of 30% Percoll in buffer PM (mix). Spin at 84,000 g for 45 min at 4°C (no brakes) in a 60 Ti rotor. Collect the band around two-thirds to three-fourths from the bottom of the tube. Dilute the band material two to three times with PBS and spin at 15,000 at 4°C for 2 hr. Resuspend pellet in PBS and label PM.

E. Isolation of Triton-Insoluble Membranes (TIM)
Solutions

Use solutions 1, 2, 6, and 7 from "P" isolation, solution 2 from "PM" isolation, and solution 1 and 2 from "V" isolation.

1. Perfusion

Follow steps 1, 2, 3, 8, and 9 from "P" isolation perfusion.

2. Processing of the Lung

Follow "P" isolation processing, except add 1% Triton X-100 to sucrose/HEPES solution in step 2.

3. Isolation of Triton-Insoluble Membranes

Steps

1. Filter the homogenate through a 53-μm Nytex filter and then through a 30-μm Nytex filter.

2. Remove 200 μl from the filter solution and save and label homogenate. Bring the remaining supernatant to 40% sucrose/20 mM KCl. Layer with discontinuous sucrose/20 mM KCl from 35 to 0% (3 ml each layer as in V isolation).

3. Spin at 15,000 overnight at 4°C in SW28 rotor. Collect the band between 10 and 15% sucrose density. Dilute the band material two to three times with PBS and spin at 15,000 at 4°C for 2 hr. Resuspend pellet in PBS and label TIM.

F. Isolation of GPI-Anchored Protein Microdomains (G)

Solutions

Use solution 3 from "P" isolation and solutions 1 and 2 from "V" isolation plus the following solution.

1. *Potassium phosphate/polyacrylic acid:* 4 M potassium phosphate (dibasic), 2% poly-acrylic acid in double-distilled H$_2$O, pH to 11 with NaOH; for 10 ml, 6.96 g potassium phosphate (dibasic) and 0.8 ml 25% stock polyacrylic acid.

1. Isolation of G

Steps

1. Resuspend membrane pellet (P-V) from V preparation in 1 ml of MBS, pH 6.0. Add an equal volume of potassium phosphate/polyacrylic acid (pH should be greater than 9.5).

2. Mix, then sonicate on maximum output with cooling ten times (each burst 10 sec). Nutate at room temperature for 8 hr and sonicate as just described but only 5 times.

3. Add Triton X-100 to a final concentration of 1%. Nutate for 10 min at 4°C, then place in a type "AA" homogenizer vessel.

4. Homogenize with corresponding grinder for 30 strokes at 1800 rpm. Bring to 40% sucrose and layer with discontinuous 35–0% sucrose with 20 mM KCl.

5. Spin at 30,000 overnight at 4°C in a SW55 rotor. Collect band between 10 and 15% sucrose density. Dilute the band material two to three times with MBS and spin at 15,000 at 4°C for 2 hr. Resuspend pellet in resuspend in MBS and label G.

6. Collect the stripped membrane pellet and resuspend in MBS and label SP.

IV. COMMENTS

The overall goal here is to excise microdomains specifically from cellular membranes, primarily in this case from plasma membranes. In order to isolate any microdomain from a membrane, certain critical criteria seem to be necessary. First is the need for selective excision of the microdomain. In most cases, because this cannot be accomplished in one step, one must start with the appropriate, highly purified membrane containing the desired microdomain and then utilize this material with a process for specifically excising and, if necessary, differentially isolating the desired microdomain away from other similarly excised domains.

Although desirable because of its simplicity, it is nevertheless clear that TIM, isolated from whole cells or tissue or even from purified plasma membranes, can contain caveolae but also contain microdomains rich in GPI-anchored proteins and cytoskeletal elements (Lisanti *et al.*, 1994; Schnitzer *et al.*, 1995b). Triton X-100 even at low temperatures solubilizes most membrane molecules, except apparently for those located in detergent-resistant domains. In essence, the effect of the detergent is to "melt away" most of the membrane, leaving behind the Triton-insoluble microdomains, which can then be isolated by centrifugation as low-density membranes enriched in caveolin, GPI-anchored proteins, and cytoskeletal proteins.

Although caveolin is an excellent marker for labeling caveolae selectively on the cell surface, it can also be amply present in exocytic vesicles of the *trans*-Golgi network. In cultured cells such as MDCK, fibroblast, or endothelial cells, it is actually found predominately in the Golgi, in part because many cells, when isolated and grown in culture, lose most of their caveolae at the cell surface which may result in significant molecular reorganization in the cell. It is important to note that newly synthesized GPI-anchored proteins acquire resistance to detergent extraction upon entering the *trans*-Golgi compartment; in fact, TIM were first isolated from cultured MDCK cells as a tool for studying this maturation process (Brown and Rose, 1992). In cells with caveolae, both caveolar and GPI-anchored proteins (G) domains will be coisolated in TIM. Even in cells that lack both caveolae and caveolin, TIM can be isolated readily and are rich in GPI-anchored proteins. TIM from cells with and without caveolae have very similar buoyant densities. Thus, it is apparent so far that the physical characteristic of vesicle density is unable to discriminate and thereby separate caveolae from G microdomains, which clearly explains their coisolation in TIM and other more recently developed detergent-free procedures (Smart *et al.*, 1995).

It is clearly advantageous to start with purified plasma membranes when trying to purify caveolae or other plasmalemmal microdomains. The method described here is based on first isolating plasma membranes and then subfractionating them to purify their caveolae separately from other plasmalemmal microdomains, including those rich in GPI-anchored proteins and/or cytoskeletal proteins. The luminal endothelial cell plasma membrane, which is exposed directly to the circulating blood and is the critical interface in most organs between tissue cells and the circulation, is purified directly from tissue. First, the tissue is perfused with polycationic colloidal silica particles that selectively coat this luminal surface to increase the membrane density. Then, after cross-linking, the tissue is homogenized and then subjected to a series of centrifugations to sediment the higher density membranes with the silica coating away from the other tissue components that are not coated and have much lower densities. This procedure can also be performed as described on cultured cells (Schnitzer and Oh, 1996), either in suspension or as a cell monolayer where the top membrane surface opposite to the plastic can be coated and purified. It is important to note that the plasma membranes are isolated to a quite high purity at levels greatly exceeding more standard subcellular fractionation techniques using centrifugation through density gradients (i.e., Percoll). Biochemical and morphological tests (Schnitzer *et al.*, 1995b) reveal little, if any, contamination from tissue or intracellular components, including the endoplasmic reticulum and Golgi, which are well-known contaminates of these latter preparations (Chang *et al.*, 1994; Lisanti *et al.*, 1994; Smart *et al.*, 1995). Figure 1 shows significant enrichments for plasma membrane markers in P without contamination from markers of other membranes such as ϵ-Cop, which resides in the *trans*-Golgi and endosomes.

The silica coating forms a very stable and strongly adherent layer over the flat noninvaginated portions of the plasmalemma because a particle size has been chosen that does not enter the desired invaginations, the caveolae. Thus, when the silica-coated membranes are exposed to shearing forces during homogenization, the caveolar invaginations hanging on the side of the membrane opposite to the silica coating are physically stripped from the membrane quite selectively and then these low density buoyant caveolae can be isolated by simple floatation during sucrose gradient centrifugation. More than 95% of the caveolae on the silica-coated plasma membranes are removed and isolated in this manner (Schnitzer *et al.*, 1995b). Transmission electron microscopy of the isolated membrane shows a rather homogeneous population of vesicles of 80 nm in diameter that maintain their distinctive morphological features, including their flask-like shape. Biochemical analysis reveals ample enrichment for known caveolar markers, including caveolin, G_{M1}, IP_3 receptors and Ca^{2+} ATPase. Both analyses detect little contamination. More recent work in the laboratory has utilized these membranes successfully to produce and characterize monoclonal antibodies specific for the caveolae as assessed by both subfractionation and immunomicroscopic analyses (Tan *et al.*, 1996). Figure 1 shows representative immunoblots for various proteins of interest (see also, Volume 2, Celis *et al.*, "Determination of Antibody Specificity by Western Blotting and

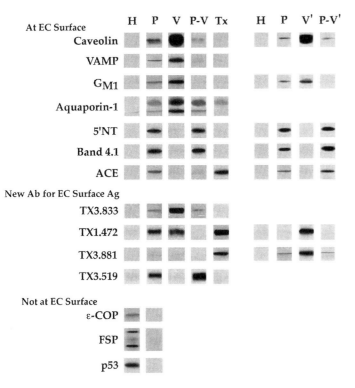

	H	P	V	P-V	Tx	H	P	V'	P-V'
At EC Surface									
Caveolin									
VAMP									
G$_{M1}$									
Aquaporin-1									
5'NT									
Band 4.1									
ACE									
New Ab for EC Surface Ag									
TX3.833									
TX1.472									
TX3.881									
TX3.519									
Not at EC Surface									
ε-COP									
FSP									
p53									

FIGURE 1 Molecular mapping of luminal plasma membrane subfractions. Proteins (5 μg) from the indicated membrane fractions, homogenate (H), silica-coated plasma membrane (P), caveolae [V (in the presence of Triton), V' (in the absence of Triton)], plasma membrane stripped of caveolae [P-V (in the presence of Triton), P-V' (in the absence of Triton)], and Triton-soluble phase (TX) were subjected to SDS–PAGE followed by electrotransfer to filters for immunoblotting with antibodies to the indicated proteins.

Immunoprecipitation," for additional information), including some of the markers discussed earlier and some of our new monoclonal antibodies.

By being stably attached to the plasmalemma proper, the silica coating prevents excision of the flat noninvaginated domains which can be detergent resistant, are rich in GPI-anchored proteins or cytoskeletal elements, and are not by definition the invaginated caveolar microdomains. This purification procedure can be performed in the presence or absence of detergents at low temperatures. The presence of Triton X-100 has the advantage of facilitating the stripping of the caveolae from the plasma membranes and the disadvantage of possibly inducing artifacts, including solubilizing, and thereby removing some caveolar resident proteins such as src-like nonreceptor tyrosine kinases (Schnitzer *et al.*, 1995b). The silica-coated membrane, which has already been stripped of the caveolae and sedimented to form a membrane pellet during the centrifugation, can be resuspended and used to isolate the G domains. Not surprisingly, the silica coating must be removed first before the Triton X-100 treatment, followed by sucrose density centrifugation to isolate the G domains without caveolae, which were removed earlier. Figure 2 shows that each of our tissue subfractions has a distinct protein profile and that caveolae and G domains differ considerably in this regard. Thus, it has been shown that even in highly purified plasma membranes, lacking contamination from Golgi or other intracellular compartments, that caveolae and GPI-anchored proteins exist distinctly at the cell surface and can actually be purified separately (Schnitzer *et al.*, 1995b).

A reconstituted cell-free system has also been developed to study further the dynamics and properties of caveolae (Schnitzer *et al.*, 1996). It shows that caveolae are dynamic structures that can be induced to bud from plasma membrane by GTP. Fission of caveolae from the plasmalemma without any physical disruption requires GTP hydrolysis and provides an alternative, possibly a more physiological source, in isolating caveolae. The comparison of caveolae stimulated by GTP to bud from plasma membrane fractions isolated by Percoll gradients or with the silica-coating technique yield similar protein profiles (Schnitzer *et al.*,

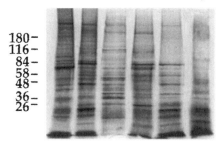

FIGURE 2 One dimensional protein analysis of various membrane subfractions. Proteins (5 µg) from the indicated membrane fractions, silica-coated plasma membrane (P), Triton-soluble phase (TX), caveolae (V), low density detergent-resistant fraction (B), plasma membrane stripped of caveolae (P-V), and GPI-anchored protein microdomain (G) were subjected to SDS–PAGE followed by silver staining.

1996). These budded caveolae are rich in caveolin, G_{M1}, and albondin, but appear to lack GPI-anchored proteins. It is likely that GPI-anchored proteins may be driven into the caveolae through antibody sequestration (Mayor *et al.*, 1994).

V. PITFALLS

1. It is very important that one first check the purity of their starting material, preferably the plasma membrane. One can look for contamination such as the appearance of Golgi elements. One should always be wary of the mass balance of all proteins, which can be accomplished by looking at what is left behind in the P-V fraction and comparing it to both the P fraction and the V fraction to see if the signal for all three fractions could be accounted for by the fractionation.

2. It is very important that one adheres to the temperature presented in the methodology, especially when working with Triton X-100 solubilization. Any increases in temperature may solubilize a greater number of proteins, which may alter the final protein profile.

3. Be careful about the pH of the solutions, especially those associated with the cationic silica particles which bind best at a pH of 6.0.

4. All solutions must be filtered with at least a 0.45-µm filter. This will ensure that your solutions will remain relatively free of contaminants and aggregates.

5. Do not use phosphate solutions because the silica will precipitate.

6. Be careful about any air bubbles in the line of the perfusion apparatus as this will also interfere with your perfusions.

References

Brown, D. A., and Rose, J. K. (1992). Sorting of GPI-anchored proteins to glycolipid-enriched membrane subdomains during transport to the apical cell surface. *Cell* **68**, 533–544.

Chang, W. J., Ying, Y. S, Rothberg, K. G., Hooper, N. M., Turner, A. J., Gambliel, H. A., De, G. J., Mumby, S. M., Gilman, A. G., and Anderson, R. G. (1994). Purification and characterization of smooth muscle cell caveolae. *J. Cell Biol.* **126**, 127–138.

Cinek, T., and Horejsi, V. (1992). The nature of large nonconvalent complexes containing glycosyl-phosphatidylinositol-anchored membrane glycoproteins and protein tyrosine kinases. *J. Immunol.* **149**, 2262–2270.

Lisanti, M., Schrer, P. E., Vidugiriene, J., Tang, Z., Hermanowski-Vosatka, A., Hu, Y., Cook, R. F., and Sargiacomo, M. (1994). Characterization of caveolin-rich membrane domains isolated from an endothelial-rich source: Implications for human disease. *J. Cell Biol.* **126**, 111–126.

Lui, J., Oh, P., Horner, T., Rogers, R., and Schnitzer, J. E. (1997). Organized endothelial cell surface signal transduction in caveolae distinct from glycosylghosphatidylinositol-anchored protein microdomains. *J. Biol. Chem.* **272**, 7211–7222.

Mayor, S., Rothberg, K. G., and Maxfield, F. R. (1994). Sequestration of GPI-anchored proteins in caveolae triggered by cross-linking. *Science* **264**, 1948–1951.

Rizzo, V., Sung, A., Oh, P., and Schnitzer, J. E. (1996). Rapid flow-induced mechanotransduction occurs at the luminal endothelial cell surface in caveolae. Submitted for publication.

Sargiacomo, M., Sudol, M., Tang, Z., and Lisanti, M. P. (1993). The glycosyl-phosphatidylinositol-linked proteins form a caveolin-rich insoluble complex in MDCK cells. *J. Cell Biol.* **122**, 789–807.

Schnitzer, J. E., Liu, J., and Oh, P. (1995a). Endothelial caveolae have the molecular transport machinery for vesicle budding, docking, and fusion including VAMP, NSF, SNAP, annexins, and GTPases. *J. Biol. Chem.* **270**, 14399–14404.

Schnitzer, J. E., McIntosh, D. P., Dvorak, A. M., Liu, J., and Oh, P. (1995b). Separation of caveolae from associated microdomains of GPI-anchored proteins. *Science* **269**, 1435–1439.

Schnitzer, J. E., and Oh, P. (1996). Aquaporin-1 in plasma membrane and caveolae provides mercury-sensitive water channels across lung endothelium. *Am. J. Physiol.* **270**, H416–H422.

Schnitzer, J. E., Oh, P., and McIntosh, D. P. (1996). Role of GTP hydrolysis in the fission of caveolae directly from plasma membranes. *Science* **274**, 239–242.

Smart, E. J., Ying, Y. S., Mineo, C., and Anderson, R. G. (1995). A detergent-free method for purifying caveolae membrane from tissue culture cells. *Proc. Natl. Acad. Sci. USA* **92**, 10104–10108.

Tan, X.-Y., McIntosh, D. P., Oh, P., and Schnitzer, J. E. (1997). A novel strategy of mapping tissue-specific proteins on endothelium and caveolae for vascular targeting. Submitted for publication.

Purification of Rat Liver Golgi Stacks

Norman Hui, Nobuhiro Nakamura, Paul Slusarewicz, and Graham Warren

I. INTRODUCTION

The study of intracellular organelles has been greatly facilitated by their purification from cellular homogenates. Such protocols yield an abundant source of material for both structural and functional studies. This article describes some simple protocols, derived from several earlier methods (Leelavathi *et al.*, 1970; Fleischer and Fleischer, 1970; Hino *et al.*, 1978), for obtaining a highly purified preparation of stacked Golgi apparatus from rat liver and for determining their relative purity over the liver homogenate.

II. MATERIALS AND INSTRUMENTATION

Centrifugation was carried out using an L8-70M preparative ultracentrifuge and either a SW-28 rotor (Cat. No. 342207) containing Ultraclear tubes (Cat. No. 344058) or a SW-40Ti rotor (Cat. No. 331302) containing Ultraclear tubes (Cat. No. 344060) from Beckman Instruments Inc. A 150-μm-mesh stainless-steel laboratory test sieve (Cat. No. SF524-42), a stainless-steel receiver (Cat. No. SF528-08), and a double-ended sieve brush (Cat. No. SF512-10) are from Endecotts Ltd. through Marathon Laboratory Supplies. A 0–50% Delta refractometer (Cat. No. 2-70) is from Bellingham and Stanley Ltd. Dipotassium hydrogen orthophosphate (Cat. No. 10436), potassium dihydrogen orthophosphate (Cat. No. 10203), sucrose (Cat. No. 10274), magnesium chloride (Cat. No. 10149), sodium cacodylate (Cat. No. 30118), manganese chloride (Cat. No. 10152), and dimethyl sulfoxide (DMSO, Cat. No. 103234L) are from BDH Laboratory Supplies. Triton X-100 (Cat. No. T-6878), ovomucoid (Cat. No. T-2011), UDP-galactose (Cat. No. U-4500), ATP (Cat. No. A-5394), aprotinin (Cat. No. A-4529), pepstatin A (Cat. No. P-4265), benzamidine (Cat. No. B-6506), chymostatin (Cat. No. C-7268), and phenylmethylsulfonyl fluoride (PMSF, Cat. No. P-7626) are from Sigma Chemical Company, Ltd. Leupeptin (Cat. No. 1017101) and antipain dihydrochloride (Cat. No. 1004646) are from Boehringer-Mannheim. [^{3}H]UDP-galactose with a specific activity of 30–50 Ci/mmol (Cat. No. NET758) is from NEN Research Products. β-Mercaptoethanol (Cat. No. 361-0710), Tris (Cat. No. 1610719), and SDS (Cat. No. 161-0302) are from Bio-Rad Laboratories Ltd. Ninety-five percent (95%) ethanol (Cat. No. SIN 1170) is from Hayman Ltd. Concentrated hydrochloric acid (Cat. No. H/1200/PB17) is from Fisons Scientific Equipment. The BCA protein assay reagent (Cat. No. 23225) is from Pierce Chemicals. Ready Micro liquid scintillation cocktail (Cat. No. CAS# 68412-54-4) and scintillation counter

LS6000IC are from Beckman Instruments Inc. All reagents were of analytical grade or better and the water used was double distilled and filtered.

III. PROCEDURES

A. Purification of Rat Liver Golgi Stacks

Solutions

1. *0.5 M potassium phosphate, pH 6.7:* Make up 500-ml solutions of 0.5 M anhydrous K_2HPO_4 (43.55 g) and 0.5 M anhydrous KH_2PO_4 (34.02 g). To 400 ml of the latter, gradually add the former until the pH reaches 6.7. Store at 4°C.

2. *2 M sucrose:* Dissolve 342.3 g in water by stirring at 50°C. Make up to a final volume of 500 ml and store at 4°C.

3. *2 M MgCl₂:* Dissolve 40.7 g of $MgCl_2 \cdot 6H_2O$ in water to a final volume of 100 ml. Store at room temperature.

4. *Protease inhibitor cocktail (PIC):* Dissolve 2 mg each of antipain, aprotinin, chymostatin, leupeptin, and pepstatin; 80 mg of PMSF; and 313 mg of benzamidine in 2 ml DMSO as a 1000× stock solution. Store at −20°C.

5. *Gradient buffers:* Buffers A–E can be made up from the preceding three stock solutions and cold water as shown in the following table. The water should be cooled to 4°C overnight to ensure that all the buffers are ice cold.

Buffer	A	B	C	D	E
Sucrose concentration (*M*)	0	0.25	0.5	0.86	1.3
0.5 *M* potassium phosphate, pH 6.7 (ml)	20	40	80	20	12
2 *M* sucrose (ml)	—	25	100	43	39
MgCl₂ (ml)	0.25	0.5	1	0.25	0.15
Water (ml)	79.8	134.5	219	36.8	8.9
Total volume (ml)	100	200	400	100	60
Refractive index (%)		9	16	26.5	38

It is very important to be as accurate as possible when mixing various components and to check the refractive index of the above buffers using a refractometer. The final refractive index should be adjusted to ±0.5% for buffer C and D in particular.

1. Standard Protocol (Medium-Scale Preparation)

This is the protocol used for general biochemical analysis of the Golgi membranes. It suffices for many electron microscopy studies as well.

Steps

1. Starve six female Sprague–Dawley or Wistar rats for 24 hr.

2. Before killing the rats, pour six discontinuous gradients consisting of 13 ml of buffer D underlain with 7.5 ml of buffer E into SW-28 rotor tubes and keep them on ice. Underlaying of buffer E can be performed with a 10-ml syringe connected to a plastic tube.

3. After killing the rats, rapidly remove the livers and place into 200 ml of buffer C. Swirl and squeeze the livers occasionally to expel blood and to speed cooling.

4. Weigh out 48 g of liver into 100 ml of fresh buffer C; cut the liver into several pieces to release as much blood as possible. Pour off excess buffer, leaving a volume of less than

80 ml, and then mince into small pieces (approximately 4–5 mm in diameter) with a clean pair of scissors (Fig. 1A).

5. Homogenize the tissue by gently pressing through a 150-μm-mesh stainless-steel sieve with the bottom of a conical flask in a grinding action (Fig. 1C). Adding a small amount of buffer C to the sieve will help the liver homogenate press through more easily (Fig. 1D). Pour the homogenate into a 100-ml measuring cylinder, adjust to a final volume of 80 ml with buffer C, and mix well.

6. Overlay 13 ml of the homogenate on each of the gradients, top up and balance the tubes with buffer B, and centrifuge in an SW-28 rotor at 28,000 rpm for 1 hr at 4°C. Keep the remaining homogenate on ice for later assay.

7. After centrifugation, aspirate away the lipid layer on the surface of the gradient and collect the Golgi fraction from the interface between buffers C and D using a Pasteur pipette (Fig. 2). Collect approximately 2–3 ml from each gradient.

8. Dilute the pooled Golgi fractions to 0.25 M sucrose using buffer A. Check the concentration with a refractometer (0.25 M reads 9%) and measure the volume of the sample.

9. Aliquot the diluted samples into two fresh SW-28 tubes and keep 100 μl on ice for the enzyme assay. Top up the tubes with buffer B and slowly add 100 μl of buffer E to each, to form a sucrose cushion, by dribbling it down the side of the tube. Centrifuge at 7000 rpm for 30 min at 4°C in the SW-28 rotor.

FIGURE 1 Steps in homogenizing rat liver. (A) Mince rat liver in ice-cold buffer. (B) Gently press the liver through the metal mesh for small-scale preparation. (C) Gently press the minced liver through the metal mesh for medium and large-scale preparations. (D) Add buffer C to dilute liver stuck on the mesh for easier pressing.

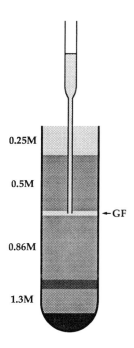

FIGURE 2 Collection of the Golgi fraction from a discontinuous sucrose gradient. The Golgi stacks form a thin band (GF) at the 0.5/0.86 *M* interface and can be collected using a Pasteur pipette.

10. Aspirate and discard the supernatant, resuspend each of the two pellets in 2 ml of buffer B, and pool the suspensions. Make up to 38 ml with buffer B, place into a fresh SW-28 tube, and add 100 μl of buffer E to form another cushion. Balance with a tube containing buffer A alone and spin as in step 9.

11. Aspirate and discard the supernatant and resuspend the final pellet in 4.5 ml of buffer B. Measure the exact volume, aliquot appropriately, flash-freeze in liquid nitrogen, and store at −80°C. Samples can be thawed and refrozen at least twice without significant loss of enzymatic activity or loss of morphology.

2. Small-Scale Preparation

Use this protocol if the morphology of the Golgi membranes is of primary concern, e.g., for electron microscopy studies. It is very important not to overload the gradients. Minimize pipetting wherever possible to preserve the Golgi stacks.

Steps

1. Starve two or three female Sprague–Dawley or Wistar rats for 24 hr.
2. After killing the rats, rapidly remove the livers into ice-cold buffer C.
3. Check that the weight of livers is about 15 g.
4. Transfer the livers into small amounts of fresh buffer C. There is no need to cut liver into small pieces as only two are used. Proceed directly to pressing through a 150-μm-mesh stainless-steel sieve with the bottom of a conical flask using a rolling action (Fig. 1B). The final volume of homogenate for 15 g of liver should be about 25 ml. Keep at least 100 μl for assay.
5. Gradients:
   a. Before killing the rats, place 6.5 ml buffer D in each of the six SW-40 Ultraclear tubes.
   b. Overlay each gradient with 4 ml of the homogenate.
   c. Finally overlay with 2.5 ml buffer B and balance the tubes.

6. Centrifuge at 29,000 rpm in a SW-40Ti rotor for 60 min at 4°C.

7. Aspirate the lipid at the top and collect Golgi fractions that accumulate at the 0.5/0.86 M interface between buffers C and D with a Pasteur pipette (Fig. 2).

8. Pool the fractions and dilute to 0.25 M sucrose using buffer A. Check the refractive index: 0.25 M reads 9%. Record the volume and keep 100 μl for assay. The volume should be about 25 ml.

9. Place 12 ml each into two SW-40 tubes and underlay with 1 ml buffer E as a cushion. Balance with buffer B and spin at 6000 rpm for 20 min at 4°C.

10. Collect the membrane just above buffer E with a plastic pastette and gently resuspend in buffer B with a Gilson P200.

11. Keep 50 μl for assay, and freeze the rest in 50-μl aliquots in liquid nitrogen.

3. Large-Scale Preparation

This protocol ensures large amounts of Golgi membranes with minimal contamination by other membranes. However, stacked cisternal structures are not as well preserved. The addition of protease inhibitors during the procedure is optional (see Section IV). Two consecutive spins at high speed are used to accommodate the large amount of homogenate used.

Steps

1. Use 8–12 female rats depending on size and requirement. Starve for 24 hr. Kill the rats and ensure that all subsequent steps are performed as quickly as possible.

2. Rapidly remove the livers into 100 ml of ice-cold buffer C, while swirling to speed cooling.

3. Weigh the livers and transfer to fresh buffer B. Cut into several pieces to increase surface area and speed cooling. In about 80 ml buffer C mince the tissue into a fine pâté with a pair of scissors (Fig. 1A).

4. Homogenize the tissue by gently pressing through a 150-μm-mesh stainless-steel sieve that has been allowed to cool in the cold room, using the base of a conical flask (Fig. 1C). Take at least 100 μl of the homogenate for assay and record the total volume. The final volume for up to 60 g of liver should be about 120 ml, which is the maximum for six gradients.

5. Gradients:
   a. Before killing the rats, place 16 ml buffer D in each of the six SW-28 Ultraclear tubes.
   b. Overlay each gradient with 18–20 ml of the homogenate.
   c. Finally overlay with buffer B to top up and balance the tubes.

6. Centrifuge at 28,000 rpm in a SW-28 rotor for 60 min at 4°C.

7. Aspirate the lipid and collect Golgi fractions that accumulate at the 0.5/0.86 M interface between buffers C and D. About 3–4 ml should be collected from each gradient.

8. Pool the fractions and dilute to 0.5 M sucrose using buffer A. Check the refractive index: 0.5 M reads 16%. Record the volume and keep 100 μl for assay. The total volume should be about 40 ml and should not exceed 60 ml.

9. Place 10 ml each into SW-40 tubes and underlay with 2 ml buffer D. Top up and balance with buffer C. Spin at 29,000 rpm for 45 min at 4°C.

10. Aspirate and discard supernatant as before. Collect membranes at the 0.5/0.86 M interface. Dilute to 0.25 M sucrose using buffer A. Check refractive index: 0.25 M reads 9%. Adjust volume to 20 ml and keep 100 μl for assay.

11. Put 10 ml each into new SW-40 tubes and underlay with 2 ml buffer E. Top up and balance with buffer B. Spin at 7000 rpm for 30 min.

12. Aspirate supernatant and collect membranes just above buffer E using a plastic pastette. Resuspend gently in buffer B to a final volume of about 5 ml. Freeze samples in appropriate aliquots by immersion in liquid nitrogen and store at −80°C until needed.

B. Determination of β-1,4-Galactosyltransferase Activity

The relative purification of the Golgi stacks can be assessed by measuring the increase in specific activity of a *trans*-Golgi enzyme, β-1,4-galactosyltransferase (GalT), over that of the whole cell homogenate. The enzyme assay used is that of Bretz and Staubli (1977), where the addition of tritiated galactose onto the oligosaccharides of an acceptor protein, ovomucoid, is measured.

Solutions

1. *0.4 M sodium cacodylate, pH 6.6:* Dissolve 17.1 g in 150 ml of water and adjust the pH to 6.6 with HCl. Make up to 200 ml and store at room temperature.

2. *175 mg/ml ovomucoid:* Dissolve 1 g in water and make up to a final volume of 5.7 ml. Filter through a 0.45-μm nitrocellulose filter, aliquot, and store at −20°C.

3. *10 mM UDP-galactose:* Dissolve 25 mg in a final volume of 4.4 ml of water. Aliquot and store at −20°C.

4. *10% (w/v) Triton X-100:* Dissolve 10 g in 80 ml of water and make up to a final volume of 100 ml. Store at 4°C.

5. *0.2 M ATP:* Dissolve 605 mg in 3 ml of water. Adjust the pH to 6.5–7.0 with 1 M NaOH and make up to a final volume of 5 ml with water. Aliquot and store at −20°C.

6. *2 M MnCl$_2$:* Dissolve 9.9 g of MnCl$_2$·4H$_2$O in 15 ml of water and make up to 25 ml. Store at room temperature.

7. *1% phosphotungstic acid/0.5 M HCl (PTA/HCl):* Dissolve 5 g of phosphotungstic acid in 400 ml of water. Add 22 ml of concentrated HCl and make up to 500 ml with water. Store at 4°C.

8. *5% (w/v) SDS:* Dissolve 5 g of SDS in 80 ml of water and make up to 100 ml. Store at room temperature.

9. *2 M Tris:* Dissolve 24.2 g of Tris in 70 ml of water and make up to 100 ml. Store at room temperature.

10. *Assay mixture:* Make up a fresh batch of the assay mixture from the above stocks as follows: 200 μl sodium cacodylate, 200 μl ovomucoid, 6 μl β-mercaptoethanol, 40 μl UDP-galactose, 40 μl Triton X-100, 20 μl ATP, 40 μl MnCl$_2$, 10 μl [^{3}H]UDP-galactose, and 1040 μl of water.

Steps

1. Once the Golgi preparation has been completed, make 1:20 dilutions of the homogenate, intermediate, and Golgi fractions with water.

2. Add 80 μl of assay mixture to screw-capped Eppendorf tubes containing duplicate 20-μl aliquots of the diluted samples and of water (blanks). Vortex and incubate at 37°C for exactly 30 min.

3. Stop the reaction by adding 1 ml of ice-cold PTA/HCl and spin at 14,000 rpm on a bench-top centrifuge for 7 sec.

4. Aspirate and discard the supernatants, and add 1 ml of PTA/HCl. Resuspend the pellets by vortexing or scraping on a rack and spin as in step 3.

5. Aspirate and discard the supernatant, add 1 ml of ice-cold 95% ethanol, and resuspend the pellets as in step 4.

6. Spin as in step 3 and remove the supernatant. Add 50 μl of 2 M Tris followed by 200

μl of 5% SDS and shake or vortex until dissolved. Do not scrape to avoid excessive frothing. Add 10 μl of assay mixture, 40 μl of water, and 200 μl of 5% SDS to a fresh tube to allow determination of the [^{3}H]UDP-galactose specific activity in the mixture.

7. Add 1 ml of scintillation fluid to each sample. Vortex and count in a scintillation counter using the tritium channel.

C. Determination of Protein Concentration

Steps

1. While the GalT assays are incubating, make up the following dilutions of the three samples in water: 1:100 for the homogenate, 1:20 for the intermediate fraction, and 1:5 for the Golgi fraction.

2. Prepare a standard curve by aliquoting 50, 45, 40, 35, and 30 μl of water in duplicate and adding 0, 5, 10, 15, and 20 μl of 2 mg/ml BSA. This gives protein standards containing 0, 10, 20, 30, and 40 μg of BSA.

3. Aliquot 50 μl of each of the diluted samples in duplicate and 50 μl of water as a blank.

4. Add 1 ml of the Pierce protein assay mixture to the standards and the samples, allow to develop, and measure absorbance as described in the manual.

D. Calculation of Purification Tables

Steps

1. Construct a protein standard curve by plotting the absorbance of each standard against the amount of BSA it contains. Calculate the slope (m) and the intercept at the ordinate axis (c). Calculate the protein concentrations of the samples as

$$\text{Protein concentration (mg/ml)} = \frac{(\text{sample absorbance} - c) \times \text{dilution}}{m \times 50}.$$

2. Calculate the specific activity (SA) of [^{3}H]UDP-galactose in the assay mixture as

$$\text{SA [}^3\text{H]UDP-galactose (dpm/nmol)} = \frac{(\text{dpm of standard} - \text{blank})}{2.5}.$$

3. Calculate the concentration of galactosyltransferase in each sample as

$$\text{Concentration GalT (nmol/hr/ml)} = \frac{(\text{average dpm} - \text{blank}) \times 2000}{\text{SA [}^3\text{H]UDP-galactose}}.$$

4. Calculate the SA of GalT by dividing its concentration by the protein concentration of the same sample to give SA in nmol/hr/mg.

5. The yields of Golgi membranes can be calculated from the ratio between the total GalT in the intermediate and Golgi fractions and that of the homogenate.

6. The purification fold is the factor by which the GalT specific activity increases in the intermediate and Golgi fractions over the homogenate.

IV. COMMENTS

This protocol typically yields Golgi membranes that are purified 70- to 90-fold over the homogenate, as depicted in Table I. Purification folds are usually higher when β-1,2-N-acetylglucosaminyltransferase I (NAGT I) is used as the marker, as shown in Table II. This enzyme can be assayed as described by Vischer and Hughes (1981), although this assay is more time-consuming than that for GalT. This difference is probably due to some loss of

TABLE I Enrichment of a *trans*-Golgi Marker, β-1,4-Galactosyltransferase (GalT), over the Homogenate in both the Intermediate Fraction (0.5/0.86 *M* Interface) and the Golgi Preparation[a]

Fraction	Volume (ml)	[Protein] (mg/ml)	[GalT] (nmol/hr/ml)	Specific activity	Yield (%)	Purification (-fold)
Homogenate	78	82.8 ± 3.6	760 ± 57.4	9.4 ± 0.6	100.0	1.0
0.5/0.86 *M* interface	65.6 ± 4.0	5.7 ± 0.4	236.8 ± 16.3	45.0 ± 3.2	25.6 ± 1.4	5.0 ± 0.4
Golgi	4.5 ± 0.1	2.8 ± 0.2	2060.0 ± 228.7	749.0 ± 70.0	15.5 ± 1.5	81.8 ± 6.9

[a] This table was compiled using the results from 24 separate purifications, presented as the mean ± SEM for each parameter. Note that the specific activity, yield, and purification fold are not, therefore, arithmetically related to GalT and protein concentrations.

the *trans*-Golgi network during purification because this organelle contains a significant amount of GalT but little if any NAGT I (see Nilsson *et al.*, 1993).

The Golgi preparations contain very little lysosomal or endoplasmic reticulum contamination as assessed by assay of β-*N*-acetylhexosaminidase (Landegren, 1984) and rotenone-insensitive NADH–cytochrome c reductase (Sottocasa *et al.*, 1967). The stacked nature of these Golgi membranes can be confirmed by examination of preparations by electron microscopy (Fig. 3). Samples are fixed in suspension by the method described by Pypaert *et al.* (1991).

The addition of the protease inhibitors is optional and need only be considered when the Golgi protein of interest is known to be sensitive to proteases. The addition of protease inhibitors can lower the apparent purification factor. This might be caused by the preservation of contaminating proteins that would otherwise be degraded during purification.

V. PITFALLS

1. Like many organelle purification procedures, it is vital to keep all solutions at 4°C during the whole protocol to prevent excessive protease digestion. If possible, steps 3–6 should be performed in a cold room. The addition of a protease inhibitor cocktail is advisable.

2. All steps should be carried out as quickly as possible, and the entire procedure, from the killing of rats to the freezing of final samples, should take approximately 3–3.5 hr.

3. Gradients should not be overloaded by increasing the concentration of the homogenate, as this increases the amount of mitochondrial contamination of the preparation.

4. The final Golgi pellet should be white. A brown pellet indicates the presence of contaminating mitochondria. Such contamination can be reduced by lowering the concentration of the homogenate.

5. The 150-μm sieve will become clogged with connective tissue after excessive use. This can be removed after each preparation by soaking the sieve in 4 *M* NaOH for 2–3 hr followed

TABLE II Relative Purification of a Medial-Golgi Marker, β-1,4-*N*-Acetylglucosaminyltransferase I (NAGT I), over the Homogenate[a]

Fraction	Volume (ml)	[Protein] (mg/ml)	[NAGT I] (nmol/hr/ml)	Specific activity	Yield (%)	Purification (-fold)
Homogenate	78	78.9	45.2	0.6	100	1
0.5/0.86 *M* interface	74.8	5.2	49.6	9.6	88.3	16.6
Golgi	4.9	2.7	370	144	51.0	192

[a] The average value obtained from two experiments.

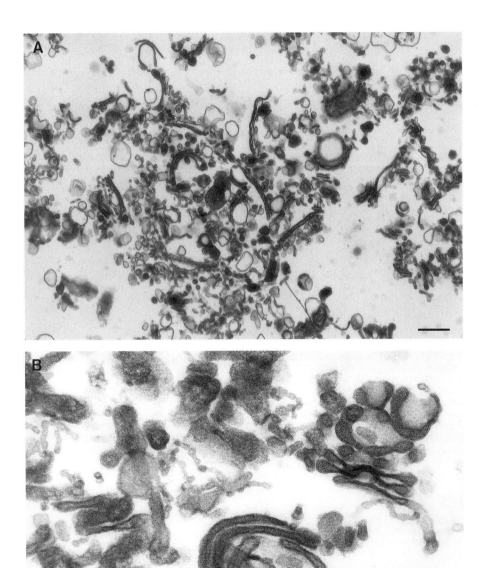

FIGURE 3 Representative micrographs of a typical rat liver Golgi apparatus preparation showing stacked Golgi membranes (arrows) at low (A) and high (B) magnification. Magnification: (A) 22,250 (bar = 0.5 μm) and (B) 97,500 (bar = 0.1 μm).

by washing with copious amounts of water and brushing. Prolonged soaking in 4 *M* NaOH attacks the glue used in binding the mesh to the wall of the sieves and should therefore be avoided.

Acknowledgments

The authors thank Dr. Catherine Rabouille and Dr. Tom Misteli for developing the small-scale preparation and Dr. Francis Barr for providing the photographs in Fig. 1.

References

Bretz, R., and Staubli, W. (1977). Detergent influence on rat-liver galactosyltransferase activities towards different acceptors. *Eur. J. Biochem.* 77, 181–192.

Fleischer, B., and Fleischer, S. (1970). Preparation and characterisation of Golgi membranes from rat liver. *Biochem. Biophys. Acta* **219**, 301–319.

Hino, Y., Asano, A., Sato, R., and Shimizu, S. (1978). Biochemical studies of rat liver Golgi apparatus. I. Isolation and preliminary characterisation. *J. Biochem. Tokyo* **83**, 909–923.

Landegren, U. (1984). Measurement of cell numbers by means of the endogenous enzyme hexosaminidase: Applications to the detection of lymphokines and cell surface antigens. *J. Immunol. Methods* **67**, 379–388.

Leelavathi, D. E., Estes, L. W., Feingold, D. S., and Lombardi, B. (1970). Isolation of a Golgi-rich fraction from rat liver. *Biochem. Biophys. Acta* **211**, 124–138.

Nilsson, T., Pypaert, M., Hoe, M. W., Slusarewicz, P., Berger, E., and Warren, G. (1993). Overlapping distribution of two glycosyltransferases in the Golgi apparatus of Hela cells. *J. Cell Biol.* **120**, 5–13.

Pypaert, M., Mundy, D., Souter, E., Labbe, J.-C., and Warren, G. (1991). Mitotic cytosol inhibits invagination of coated pits in broken mitotic cells. *J. Cell Biol.* **214**, 1159–1166.

Sottocasa, G. L., Kuylenstierna, B., Ernster, L., and Bergstrand, A. (1967). An electron transport system associated with the outer membrane of liver mitochondria. *J. Cell Biol.* **32**, 415–438.

Vischer, P., and Hughes, C. K. (1981). Glycosyl transferases of baby-hamster-kidney (BHK) cells and ricin-resistant mutants. *Eur. J. Biochem.* **117**, 275–284.

Preparation and Purification of Post-Golgi Transport Vesicles from Perforated Madin–Darby Canine Kidney Cells

Lukas A. Huber and Kai Simons

I. INTRODUCTION

An *in vitro* system is described that allows the isolation of transport vesicles derived from the *trans*-Golgi network (TGN) with the aim of identifying the molecular machinery involved in protein sorting and vesicle targeting in polarized epithelia (Bennett *et al.*, 1988). Madin–Darby canine kidney cells (MDCK strain II cells) are grown on permeable filter supports, allowing them to attach tightly to the substrate and to form a fully polarized monolayer. The cells are then infected for short times with influenza or vesicular stomatitis virus (VSV) and incubated at 20°C, causing the transport markers to accumulate in the TGN (Griffiths *et al.*, 1985; Matlin *et al.*, 1983; Hughson *et al.*, 1988). Subsequently, the cells are mechanically perforated with the aid of a nitrocellulose filter. This introduces holes in the plasma membranes but leaves the cells attached to the filter support with their subcellular organization intact (Simons and Virta, 1987). Using this system, we demonstrated previously that HA and VSV G-protein, accumulated in the TGN before perforation, were released from the perforated cells in sealed membrane vesicles (Bennett *et al.*, 1988). The vesicles had the topology expected for authentic transport vesicles and required ATP for their formation and release. The release of membranes from such perforated cells was quite specific: only low levels of resident Golgi markers, endocytic markers, or endoplasmic reticulum (ER)-derived vesicles were recovered in the incubation medium (Bennett *et al.*, 1988). The TGN-derived transport vesicles have been resolved into apical and basolateral fractions and further characterized (Wandinger-Ness *et al.*, 1990; Kurzchalia *et al.*, 1992; Huber *et al.*, 1993).

II. MATERIALS AND INSTRUMENTATION ·

Media and reagents for cell culture are from Gibco Biocult and Biochrom. Growth medium for MDCK strain II cells consists of MEM with Earle's salts (E-MEM) supplemented with 10 mM Hepes, pH 7.3, 10% fetal calf serum (FCS), 100 U/ml penicillin, and 100 μg/ml streptomycin. MDCK II cells are grown and passaged as described previously (see Volume 1, Simmons and Virta, "Growing Madin-Darby Canine Kidney Cells for Studying Epithelial Cell Biology" for additional information). For large-scale isolation of vesicles, cells from a

75-cm^2 flask were seeded on a single 100-mm-diameter, 0.4-μm-pore-size Transwell filter (24 × 10^6 cells) as described (see Volume 1, Simmons and Virta, "Growing Madin-Darbin Canine Kidney Cells for Studying Epithelial Cell Biology" for additional information).

ATP is from Sigma (Cat. No. 5394). Creatine phosphate (Cat. No. 621714, disodium salt) and creatine kinase (CK, Cat. No. 127566, rabbit muscle) are from Boehringer-Mannheim. NC filters (HATF 02500 0.45-μm filter type HA, Triton free) are from Millipore. No. 1 filters are from Whatman Scientific Limited. Cell filters (0.4-μm pore size; No. 3412, 24 mm; No. 3419, 100 mm Transwell) are from Costar.

III. PROCEDURES

A. Generation and Permeabilization of MDCK Cells

Solutions

1. *10× KOAc transport buffer:* To make 200 ml, combine 250 mM HEPES–KOH, pH 7.4 (11.92 g, 1150 mM potassium), acetate (22.58 g), and 25 mM MgCl$_2$ (1.02 g).

2. *10× GGA buffer (transport buffer):* To make 50 ml, prepare three solutions: (a) 1150 mM L-glutamate (MW 185.2, 10.65 g), (b) 1150 mM L-aspartate (MW 171.2, 9.84 g), and (c) 1150 mM potassium gluconate (MW 234.2, 13.47 g). To each solution a–c, add 250 mM HEPES–KOH, pH 7.4, and 25 mM MgCl$_2 \cdot$6H$_2$O. Filter-sterilize the solutions. Mix solutions a–c 1:1:1. Dilute 10× (add 1 mM DTT and 2 mM EGTA). Adjust to pH 7.4 just before use.

3. *ATP regenerating system (ATP mix), 100× stocks, prepare three solutions (10 ml each):*
   a. 100 mM ATP (disodium salt, pH 6–7, neutralized with 2 M NaOH; 0.605 g/10 ml).
   b. 800 mM creatine phosphate (disodium salt, 2620 g/10 ml).
   c. 800 U/mg (at 37°C) creatine kinase (0.5 mg/10 ml in 50% glycerol).

   Store stocks in aliquots at −20°C. Mix solutions a–c 1:1:1 just before use.

Steps

The basic procedure for the generation of perforated cells is quite simple. A monolayer of cells grown on a solid substrate is overlain with a nitrocellulose filter. The filter is allowed to bind under controlled conditions for a specified period and is then peeled away from the cells. The cells remain attached to the substrate but now have holes in their plasma membrane. These holes allow manipulation of the cytoplasmic composition in cells whose overall morphology and organellar organization remain intact.

A number of factors are important in establishing an efficient and reproducible system for the perforation of cells using the nitrocellulose filter procedures. Perhaps the most important consideration is the strength of the cell attachment to the substrate on which they are grown. This will determine the amount of binding between the cells and the nitrocellulose filter that can be allowed to occur. If the cells are weakly adherent, the entire cell layer, instead of small fragments of the plasma membrane, will be removed by the nitrocellulose filter. A second consideration is the method of binding the nitrocellulose filter to the cell layer, in this case, the amount of binding: with insufficient drying, the cells will not be perforated; with excessive drying, the cells will become detached. It is therefore important to establish conditions that allow for reproducible and even drying. A number of variables, including the time, temperature, humidity, and air circulation, should be controlled to obtain the most reproducible results.

The following manipulations are performed in a cold room (4°C) with cold buffers.

MDCK Cells Grown for 3 Days on Transwell Filters (3412 or 3419)

1. Wash filter cells twice with PBS(+) [dip twice in beaker with PBS(+)]. Aspirate excess liquid carefully so as not to destroy filter.

2. Incubate filters for 45–60 min at 20°C (TGN block).

Perforation (in cold room!)

3. Presoak NC filters in KOAc transport buffer.

4. Wash cell filter (Transwell 3412 or 3419) twice in KOAc transport buffer (glass beaker). Cut out cell filter with a sharp scalpel (**Fig. 1a**). Place cell filter in KOAc transport buffer, cell side up.

5. Place predrained NC filter on Whatman No. 1 paper, put another Whatman paper on it, and drain a bit with a L-form Pasteur pipette.

6. Time: 0 sec.

7. Time: 30 sec. Place cell filter in 20°C water bath (on petri dish), cell side up.

8. Time: 60 sec. Place NC filter exactly on cell filter (Fig. 1b) and immediately place a Whatman paper on it; drain it a bit by using a L-form Pasteur pipette (this should take 15 sec) and drain again with a new Whatman paper (Fig. 1c).

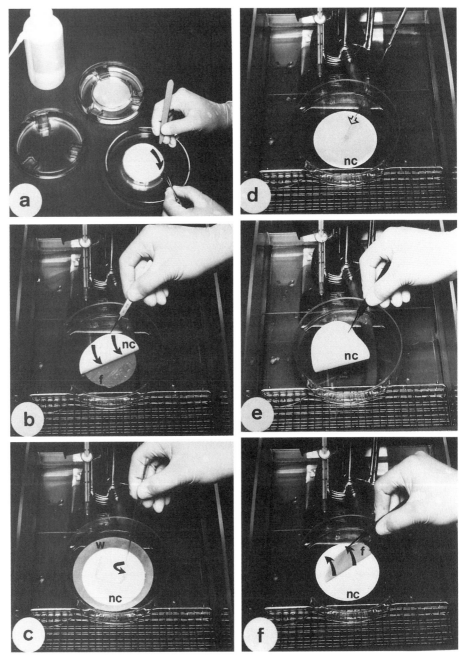

FIGURE 1 Perforation of MDCK cells.

9. Time: 75 sec. During a time course of 90 sec (starting after the first draining) the cells become attached to the NC filter at 20°C.

10. Time: 165 sec (90 sec after the first draining). Add 200 μl of KOAc transport buffer on the cell filter–NC filter sandwich. Aspirate excess moisture (Fig. 1d) and take the petri dish out of the water bath. Flip the filter sandwich around (cell filter–NC filter on top) (Fig. 1e). Peel off the cell filter carefully (still on the petri dish) (Fig. 1f).

11. Place cell filter in 1× GGA buffer + ATP-regenerating system (for Transwell 3412: 1 ml in 3.5-cm dish and 30 μl ATP mix; for Transwell 3419, 10 ml in 10-cm dish and 300 μl ATP mix).

12. Discard NC filter or use it for nuclear control stain (Hoechst). No more than 10% of the cells should be on the NC filter.

13. Incubate cell filter at 37°C for 1 hr in a water bath in the presence of GGA buffer and ATP-generating system (place wet Whatman paper inside the lid to collect condensed water).

B. Preparation of Exocytic Vesicles from Perforated MDCK Cells

Solutions

1. 0.25 M sucrose (8.5 g/100 ml) in 10 mM HEPES (0.238 g/100 ml), 2 mM EGTA (0.076 g/100 ml), and 1 mM dithiothreitol (DTT) (0.015 g/100 ml).

2. 1.5 M sucrose (51.3 g/100 ml) in 10 mM HEPES, 2 mM EGTA, and 1 mM DTT.

3. 1.2 M sucrose (41.04 g/100 ml) in 10 mM HEPES, 2 mM EGTA, and 1 mM DTT.

4. 0.8 M sucrose (27.36 g/100 ml) in 10 mM HEPES, 2 mM EGTA, and 1 mM DTT (filter-sterilize).

Steps

Pelleting vesicles through sucrose cushion

1. Collect GGA buffer with ATP-generating system with pipette (may be kept on ice for up to 30 min). Centrifugation: 1500 rpm, 10 min, 4°C (Falcon 15-ml tube/cell debris in pellet).

2. First ultracentrifugation through the sucrose cushion: SW 40, 36,000–40,000 rpm; 3 hr; 4°C. Overlay 6 ml vesicle solution (from six Transwell 3412 filters) onto 6 ml sucrose solution (0.25 M). For 10 ml vesicle solution (from one Transwell 3419 filter), overlay onto 2 ml sucrose (0.25 M).

3. Aspirate supernatant carefully with a drawn-out Pasteur pipette (pellet is hardly detectable).

4. Resuspend pellet in 100 μl sucrose solution (1.5 M).

Floating of vesicles in discontinous sucrose gradient

1. Bottom: vesicles from the sucrose cushion, resuspended in 1.5 M sucrose in 10 mM HEPES, 2 mM EGTA, 1 mM DTT: 0.4 ml per tube of vesicles from up to 36 ϕ24-mm filters or up to 6 ϕ100-mm filters. Overlay with 1.2 M sucrose in 10 mM HEPES, 2 mM EGTA, 1 mM DTT: 1.7 ml; overlay with 0.8 M sucrose in 10 mM HEPES, 2 mM EGTA, 1 mM DTT: 1.5 ml.

2. Centrifugation: 14 hr, 35,000 rpm, 4°C, SW 60 (no brake).

3. Fractionate gradient into 0.3-ml aliquots from the top to the bottom.

4. Recover the peak fractions containing vesicles at the 0.8/1.2 M sucrose interface and pool for further analysis (immunoisolation).

C. Immunoisolation of Apical and Basolateral Vesicle Fractions

Solutions

1. Cellulose fibers coupled with sheep affinity-purified anti-Fc antibodies.

2. Mouse monoclonal antibodies directed against the cytoplasmic domains of VSV-G protein (Kreis, 1986) and influenza PR8 (Hughson *et al.*, 1988) are used in the form of concentrated hybridoma culture supernatants.

3. Antiprotease mix (CLAP): 10 mg/ml DMSO of each: chymostatin, leupeptin, antipain, pepstatin; use 1:1000. Store stock at $-20°C$.

Steps

Immunoisolation can be used as the final vesicle purification step, allowing for the separation of transport vesicles destined for either the apical or the basolateral plasma membrane domains (Wandinger-Ness *et al.*, 1990). MDCK cells are grown and infected with VSV (for basolateral vesicles) or influenza virus (WSN ts61 influenza strain for apical vesicles) as described previously (Pfeiffer *et al.*, 1985; Wandinger-Ness *et al.*, 1990). We use antibodies directed against the cytoplasmic domains of either VSV G-protein or influenza HA for these studies, but antibodies directed against other markers with cytoplasmically exposed epitopes (e.g., mannose 6-phosphate receptor, clathrin, synaptophysin) could be used to isolate different types of vesicles. The following immunoisolation methods represent a modification of previously published procedures.

1. Divide the vesicle pool from the flotation gradient into three equal aliquots (usually 300 μl each).

2. Dilute the first aliquot fourfold with 10 mM HEPES–KOH, pH 7.4, 1 mM EGTA (to reduce the sucrose concentration to 0.25 M) and centrifuge at 55,000 rpm (200,000 g) in a TLS55 rotor in a Beckman TL-100 centrifuge. One resulting pellet serves as a control for the total particulate material in the starting vesicle fraction. Use the two remaining pellets for immunoisolation using either specific or control monoclonal antibodies. Antibody P5D4 directed against the cytoplasmic domain of VSV G-protein serves as the specific antibody for vesicles obtained from VSV-infected cells and as the control antibody for vesicles obtained from WSN-infected cells (Wandinger-Ness *et al.*, 1990). Antibody 2D1 directed against the cytoplasmic domain of WSN HA serves as the specific antibody for vesicles obtained from WSN-infected cells and as the control antibody for vesicles obtained from VSV-infected cells (Wandinger-Ness *et al.*, 1990).

3. Add 0.5 vol (150 μl) of phosphate-buffered saline (PBS) containing 0.1% (w/v) gelatin and a protease inhibitor cocktail (10 μg/ml each of chymostatin, leupeptin, antipain, and pepstatin) to each aliquot.

4. Add monoclonal antibody (20 μl of a 10-fold concentrated hybridoma culture supernatant) and incubate the sample at 4°C with rotation either for 3 hr for VSV samples or overnight for WSN samples. Cellulose fibers to which an affinity-purified sheep antibody generated against the Fc domain of mouse IgG is covalently coupled are used as the solid support to recover the antibody–vesicle complexes.

5. Add 1 mg of the cellulose fibers [500 μl of 2 mg/ml fibers in PBS containing 0.1% (w/v) gelatin] to each sample and continue the incubation for 1.5 hr at 4°C with rotation.

6. Pellet the fibers in the microfuge at 1500 rpm for 10 min and wash three times with PBS containing 0.1% (w/v) gelatin and once with PBS. Under these conditions, greater than 50% of the viral glycoproteins are recovered on the specific fibers, whereas less than 10% are recovered on the control fibers.

The immunoisolation conditions used are optimized to obtain maximal specific recovery of the desired marker with minimal nonspecific binding. One factor important for maximal recovery is to carry out antibody binding in solution (i.e., in the absence of the solid support) to increase the possibility of antibody–antigen interaction. The HA-containing vesicles proved to be more difficult to recover quantitatively than the VSV G-containing vesicles, perhaps because of the short cytoplasmic domain of WSN HA (only 11 amino acids) or because of differences in antibody affinity for the respective antigens. Two modifications of the VSV G protocol were used that increased the recovery of HA-containing vesicles. The first was to increase the antibody binding time from 3 hr to overnight. The second was the use of a temperature-sensitive mutant of WSN (ts 61). The mutant HA fails to be transported out of the ER at the nonpermissive temperature and therefore accumulates in the ER when the infection is carried out at 39°C. When the cells are subsequently shifted to 20°C the HA is transported out of the ER and accumulates in the TGN at higher concentrations than would be obtained with wild-type virus. This presumably results in a higher concentration of the antigen in the transport vesicles; therefore, a more efficient recovery by immunoisolation is achieved.

Nonspecific binding can be a serious problem when dealing with small amounts of material and with the mild washing conditions required to maintain membrane integrity. The minimization of nonspecific binding required screening of a number of different solid supports (cellulose fibers, magnetic beads, fixed *Staphylococcus aureus* cells) and the use of gelatin as a blocking reagent. In our experience the cellulose fibers gave the lowest nonspecific binding.

Steps

1. Pool vesicle fractions from flotation gradient.

Sample	Volume (μl)	Antibody (20 μl)	Temp/time	PBS/0.1% gelatin (CLAP) 1:1000	Cellulose fibers/anti-Fc
VSV	200	Anti-G	4°C/3 hr	150 μl	500 μl
Influenza	300	Anti-HA	4°C/12 hr	150 μl	500 μl
Influenza	300	Anti-G	4°C/12 hr	150 μl	500 μl

2. Incubate sample (flotation gradient pool from influenza virus) on VSV-infected MDCK cells (perforated, etc.) with appropriate antibody and PBS/0.1% gelatin (CLAP) for the indicated time. Incubate in an Eppendorf tube under rotation in a cold room.

3. Add 1 mg (500 μl) of cellulose fibers containing anti-Fc antibodies, prewashed. To wash fibers, place them in a Falcon tube and fill tube with PBS/0.1% gelatin. Centrifuge at 2000 rpm for 10 min. Aspirate supernatant and repeat. Resuspend fibers into original volume with PBS/0.1% gelatin.

4. Incubate for 1.5–2 hr at 4°C under rotation.

5. Wash fibers three times in 1 ml PBS/0.1% gelatin and once in 1 ml PBS (Biofuge at 2500 rpm and 4°C for 8–10 min).

6. After the last wash, aspirate supernatant and centrifuge for 30 sec in a microfuge to remove all excess liquid. Add 50 μl two-dimensional gel lysis buffer and a few grains of solid urea. Mix well and store −20°C until loading.

7. Concentrate control vesicle samples (sucrose concentration $<0.3\ M$) that have not been immunoisolated in small polyallomer tubes (rotor TLS 55) for 3 hr at 5500 rpm in a Beckman table-top ultracentrifuge. Take off supernatant with a drawn-out Pasteur pipette and process for further analysis (SDS–PAGE, etc.).

IV. COMMENTS AND PITFALLS

The vesicles become sticky and adsorb to the walls of nonpolyallomer tubes. When large polyallomer tubes were used for concentration, vesicles were lost as well.

References

Bennett, M. K., Wandinger-Ness, A., and Simons, K. (1988). Release of putative exocytic transport vesicles from perforated MDCK cells. *EMBO J.* 7, 4075–4085.

Griffiths, G., Pfeiffer, S., Simons, K., and Matlin, K. (1985). Exit of newly synthesized membrane proteins from the trans cisternae of the Golgi complex to the plasma membrane. *J. Biol. Chem.* **101**, 949–964.

Huber, L. A., Pimplikar, S. W., Virta, H., Parton, R. G., Zerial, M., and Simons, K. (1993). rab8, a small GTPase involved in vesicular traffic between the TGN and the basolateral plasma membrane. *J. Cell Biol.* **123**, 35–45.

Hughson, E., Wandinger-Ness, A., Gausepohl, H., Griffiths, G., and Simons, K. (1988). The cell biology of enveloped virus infection of epithelial tissues. *In* "The Molecular Biology of Infectious Diseases: Centenary Symposium of the Pasteur Institute" (M. Schwartz, ed.), pp. 75–89. Elsevier, Paris.

Kreis, T. E. (1986). Microinjected antibodies against the cytoplasmic domain of vesicular stomatitis virus glycoprotein block its transport to the cell surface. *EMBO J.* **5**, 931–941.

Kurzchalia, T. V., Dupree, P., Parton, R. G., Kellner, R., Virta, H., Lehnert, M., and Simons, K. (1992). VIP21, a 21-kD membrane protein, is an integral component of *trans*-Golgi-network-derived transport vesicles. *J. Cell Biol.* **118**, 1003–1014.

Matlin, K., Bainton, D. F., Pesonen, M., Louvard, D., Genty, N., and Simons, K. (1983). Transepithelial transport of a viral membrane glycoprotein implanted into the apical plasma membrane of Madin–Darby canine kidney cells. I. Morphological evidence. *J. Cell Biol.* **97**, 627–637.

Pfeiffer, S., Fuller, S. D., and Simons, K. (1985). Intracellular sorting and basolateral appearance of the G protein of vesicular stomatitis virus in MDCK cells. *J. Cell Biol.* **101**, 470–476.

Simons, K., and Virta, H. (1987). Perforated MDCK cells support intracellular transport. *EMBO J.* **6**, 2241–2247.

Wandinger-Ness, A., Bennett, M. K., Antony, C., and Simons, K. (1990). Distinct transport vesicles mediate the delivery of plasma membrane proteins to the apical and basolateral domains of MDCK cells. *J. Cell Biol.* **111**, 987–1000.

Fluorescence-Activated Sorting of Endocytic Organelles

Guenther Boeck, Peter Steinlein, Michaela Haberfellner, Jean Gruenberg, and Lukas A. Huber

I. INTRODUCTION

Progress in cell biology has been closely associated with the development of quantitative analytical methods applicable to individual cells or cell organelles. In particular, subcellular fractionation has provided the means required to analyze the molecular composition and regulatory properties of purified cellular elements. Most fractionation protocols take advantage of physical properties of intracellular membranes, e.g., density gradient centrifugation (Gruenberg and Howell, 1989; Gruenberg and Gorvel, 1992; Desjardins *et al.*, 1994; Huber and Simons, 1994). Because several intracellular compartments share similar physical properties and cofractionate at least to some extent in conventional gradients, the achievement of a complete purification is a barely accessible goal. It is therefore desirable to have a subcellular fractionation technique that relies on biological properties rather than on physical densities to overcome these technical limitations. A good example for organelles where this could be achieved are endosomes, as they can be accessed from outside the cell and loaded temporally with fluorescent membrane dyes or fluorescently labeled ligands under different conditions. Endosomes containing fluorescent signals can then be analyzed and purified based on their contents using flow cytometry (Siminoski *et al.*, 1986; Wilson and Murphy, 1989; Murphy, 1985, 1988; Murphy and Roederer, 1986; Murphy *et al.*, 1989).

This article describes a protocol for the purification of endosomes from baby hamster kidney cells (BHK-21) that combines established subcellular fractionation techniques (Gruenberg and Gorvel, 1992; Gruenberg *et al.*, 1989) with organelle sorting in a flow cytometer.

II. MATERIALS AND INSTRUMENTATION

A. Reagents

Glasgow MEM (G-MEM) is from GIBCO BRL as a powder (Cat. No. 22100-028), mixed with Milli-Q-filtered H_2O, and sterile filtered. Penicillin–streptomycin solution (5000 IU/ml–5000 µg/ml; Cat. No. 15070-022), trypsin–EDTA solution (10×; Cat. No. 35400-027), and HEPES (Cat. No. 11344-025) are from GIBCO-BRL. Fetal calf serum (FCS) is from Boehringer-Mannheim (Cat. No. 210471). Tryptose phosphate is from GIBCO-BRL (Cat. No. 18050-039). Tissue culture 100-mm plastic dishes (Cat. No. 150350) and four-well multidishes (Cat. No. 176740) are from Nunc. Sterile 15-ml conical tubes are from Falcon (Cat. No. 2095). 1-(4-Trimethylammoniumphenyl)-6-phenyl-1,3,5-hexatriene-*p*-toluenesulfonate

(TMA-DPH) is from Molecular Probes (Cat.No. T-204). Protease inhibitors aprotinin (Cat. No. 236624), pepstatin (Cat. No. 253286), and antipain (Cat. No. 1004646) are from Boehringer Mannheim. Imidazole (1,3-diaza-2,4-cyclopentadiene) and dimethyl sulfoxide (DMSO) are from Sigma (Cat. No. I-0250 and Cat. No. D2650).

B. Cells

Monolayers of the baby hamster kidney (BHK-21) cell line are grown in G-MEM supplemented with 5% FCS, 10% tryptose phosphate broth, and 2 mM glutamine in a 5% CO_2 atmosphere. We find that cells grow as an even monolayer, a critical requirement in these experiments, when seeded 14–16 hrs before use at a high density (4×10^4 cells/cm^2 of culture dish) in 100-mm tissue culture dishes. These conditions yield $\approx 1.3 \times 10^7$ cells per dish (≈ 2.5–3.0 mg total protein) at the time of the experiment. Typically, six to eight dishes (10 cm) are used for one experiment.

C. Dye

TMA-DPH interacts with living cells by instantaneous incorporation into the plasma membrane, following a water (probe not fluorescent)/membrane (probe highly fluorescent) partition equilibrium. TMA-DPH can be used as a fluorescence anisotropy probe for plasma membrane fluidity determinations (Huber *et al.*, 1991) and as a quantitative tracer for endocytosis and intracellular membrane traffic (Illinger *et al.*, 1991a,b, 1995). TMA-DPH can be excited with UV (excitation 355 nm, emission 450 nm); the blue fluorescence is bright and easily detectable with photo detection systems but the dye bleaches rapidly.

D. Flow Cytometer

A fluorescence-activated cell sorter (FACS III, Becton Dickinson), equipped with an UV argon laser (20–100 mW) and a BP 460/60 nm for fluorescence emission, is used for analysis and sorting.

III. PROCEDURES

A. Binding and Internalization of TMA-DPH

Solutions

1. *Phosphate-buffered saline (PBS):* 137 mM NaCl, 2.7 mM KCl, 1.5 mM KH_2PO_4, and 6.5 mM Na_2HPO_4. To prepare 1 liter of PBS, dissolve 0.2 g KCl, 0.2 g KH_2PO_4, 1.15 g Na_2HPO_4, and 8 g NaCl in 1 liter H_2O bidest.

2. *Lyophilized TMA-DPH:* Reconstitute in DMSO and keep light protected at $-20°C$. To make a 20 mM stock solution, dissolve 25 mg of TMA-DPH in 2.71 ml of DMSO.

3. *TMA-DPH/PBS:* Dilute the 20 mM stock solution of TMA-DPH to 50 μM in PBS; 1.0 ml is needed per 100-mm tissue culture dish.

4. *PBS/bovine serum albumin (BSA):* 5 mg BSA/ml PBS, use ice-cold solutions only for washing.

Steps

1. Place six confluent tissue culture dishes of BHK-21 cells on ice plates (fit a wet metal plate into an ice bucket to guarantee good contact with the dish). The dish must lie perfectly flat on the plate.

2. Remove medium with an aspirator connected to a water pump; add 5 ml PBS and leave for 2–3 min on a rocking platform. Repeat this step twice.

3. Continuous internalization: Aspirate the solution and prewarm each dish for 1–2 sec in a water bath at 37°C. Add 1 ml of the 50 μM TMA-DPH/PBS solution to each tissue culture dish.

4. Place dishes on a metal plate in a water bath at 37°C for 5–10 min.

5. Return the dishes to the ice plate on the rocking platform.

6. Wash three times for 10 min with 10 ml of ice-cold 5 mg/ml BSA in PBS.

7. Wash twice with ice-cold 5 ml PBS.

8. Cells are ready for subcellular fractionation.

B. Homogenization of BHK-21 Cells

Solutions

1. *PBS:* 137 mM NaCl, 2.7 mM KCl, 1.5 mM KH$_2$PO$_4$, and 6.5 mM Na$_2$HPO$_4$. To prepare 1 liter of PBS, dissolve 0.2 g KCl, 0.2 g KH$_2$PO$_4$, 1.15 g Na$_2$HPO$_4$, and 8 g NaCl in 1 liter H$_2$O bidest.

2. *Homogenization buffer (HB):* 250 mM sucrose (17.1 g) and 3 mM imidazole, pH 7.4 (0.408 g), bidist. Add H$_2$O to 200 ml.

3. *HB+ with protease inhibitors and EDTA:* 50 mM sucrose (17.1 g) and 3 mM imidazole pH 7.4 (0.408 g), bidist. Add H$_2$O to 200 ml. Add cocktail of protease inhibitors (10 μg/ml aprotinin, 1 μg/ml pepstatin, and 1 μg/ml antipain) and 1 mM EDTA.

Steps

1. Remove the cells from dishes by scraping with a rubber policeman in 2.5 ml PBS on the ice plates. Perform first a circular movement then three parallel ones. Try to get large sheets of cells.

2. Pool scraped cells from three dishes each and transfer with a plastic Pasteur pipette to a 15-ml tube and centrifuge cells at 1000 rpm for 5 min at 4°C.

3. Remove supernatant with a Pasteur pipette and resuspend the cell pellet very gently in 5 ml HB using a pipette with a wide tip.

4. Centrifuge at 2500 rpm for 10 min at 4°C.

5. Remove supernatant and add 0.5 ml HB+, supplemented with 1 mM EDTA and protease inhibitors, to the cell pellet. EDTA has to be included at this step to inhibit Ca^{2+}-dependent proteases that will be released by the homogenization. The cells should not break at this step.

6. Resuspend cell pellet first with a blue tip. Prepare a 1-ml hypodermic syringe with a 22-gauge needle fitted to it. Wet the syringe with HB+, avoiding air bubbles in the needle. Homogenize the cells with the syringe by applying five or more strokes. Pull up the suspension gently and push it down hard.

7. After the first three strokes, apply 20 μl of HB to a microscope slide, push out the rest of the homogenate remaining in the syringe (typically a drop of 20–40 μl) into the drop of HB, mix gently, and mount a coverslip. Look at the homogenate in the microscope. Ideally, all nuclei should be round and dark and no whole cells (shiny and round cells) should be visible. Stop the homogenization when the first broken nuclei are seen.

8. Centrifuge the homogenate at 2500 rpm for 15 min and collect the postnuclear supernatant (PNS).

9. The PNS is ready for separation on the flotation gradient.

C. Flotation Gradient

With this gradient, rab5-positive early endosomes can be separated from late endosomes, which contain rab7 and the cation-independent mannose 6-phosphate receptor(Gorvel *et al.*,

1991). Early endosomal fractions are collected at the 35/25% interface (see also article by Jean Gruenberg, this volume).

Solutions

1. *25% (w/w) sucrose solution in H₂O:* 0.806 M sucrose (27.6 g), 3 mM imidazole, pH 7.4 1 ml (0.0204 g), and 1 mM EDTA bidist H₂O to 100 ml (the refractive index of this solution should be 1.3723 at 20°C).

2. *35% (w/w) sucrose solution in H₂O:* 1.177 M sucrose (40.3 g), 3 mM imidazole, pH 7.4 1 ml (0.0204 g), and 1 mM EDTA bidist H₂O to 100 ml (the refractive index of this solution should be 1.3904 at 20°C).

3. *62% (w/w) sucrose solution in H₂O:* 2.351 M sucrose (80.5 g), 3 mM imidazole, pH 7.4 1 ml (0.0204 g), and 1 mM EDTA bidist H₂O to 100 ml (the refractive index of this solution should be 1.4463 at 20°C).

4. *HB+ with protease inhibitors and EDTA:* 50 mM sucrose (17.1 g) and 3 mM imidazole, pH 7.4 (0.408 g), bidist. Add H₂O to 200 ml. Add cocktail of protease inhibitors (10 μg/ml aprotinin, 1 μg/ml pepstatin, and 1 μg/ml antipain) and 1 mM EDTA.

Steps

1. Adjust the sucrose concentration of the PNS to 40.6% by adding 62% sucrose (start with 1.2× the volume of PNS). Mix carefully with a cut blue tip (endosomes are very fragile under these conditions) and measure the sucrose concentration with a refractometer.

2. Load the diluted PNS in the bottom of a SW 60 centrifuge tube by letting it run down along the wall, leaving a thin layer of solution on which overlaying of the following solutions is easier.

3. Sequentially overlay with 1.5 ml of 35% sucrose and then 1 ml of 25% sucrose.

4. Add HB+ to fill the tube.

5. Mount the tubes in a SW 60 rotor and centrifuge for 1 hr at 14,000 g (35,000 rpm) at 4°C.

6. Collect endosomes at the 35/25% interface. The fractions are now ready for flow cytometry.

D. Flow Cytometry

Analysis should be done with logarithmic amplification for forward scatter (FSC) and side scatter (SSC) light and with logarithmic amplification for fluorescence detection. In some experiments, linear amplifications for the scatter signals may be preferable to discriminate between real particle detection and background, sheath fluid fluctuation, etc. The flow rate for analysis is less than 500 events/sec and for sorting is approximately 3000 events/sec. Sorting has to be done normally at 4°C, as membranes will be more stable and clumping will be reduced. The sort has to be done into glass tubes or into Eppendorf tubes with a platinum wire reaching the suspension connected to ground of the machine. Thresholds on FSC or SSC should be set using sucrose solutions, and sort windows should be defined with unlabeled endosome fractions.

IV. COMMENTS

The method described here (see outline in Fig. 1) is based on the combination of established subcellular fractionation techniques with sorting of organelles in a flow cytometer. Sufficient amounts can be obtained for subsequent biochemical analysis of the sorted fractions (Fig. 1). Sorted(+) fractions show a significant enrichment of the endosomal marker proteins rab5 (Chavrier *et al.*, 1990; Gorvel *et al.*, 1991), transferrin receptor (Gruenberg and Maxfield, 1995), and annexin II (Emans *et al.*, 1993).

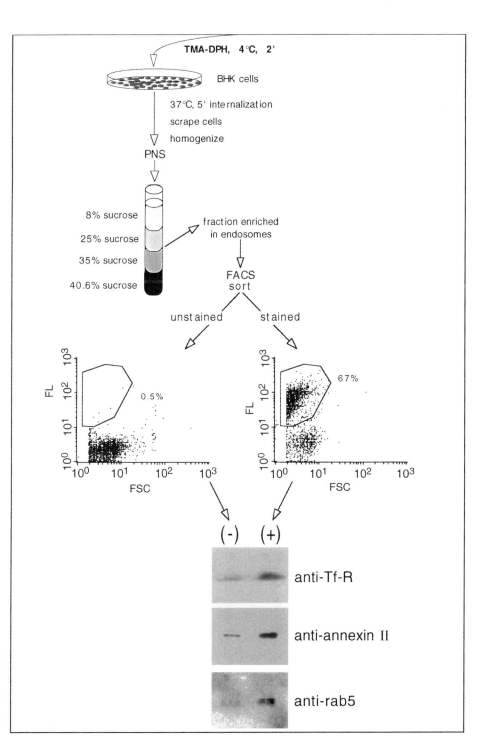

FIGURE 1

The incorporation of TMA-DPH into membranes, monitored by UV absorption measurements, remains proportional to the probe concentration over a wide range (5×10^{-7}M–2.5×10^{-5}M) (Illinger *et al.*, 1995). However, concerning fluorescence, quenching effects in the membranes above certain critical concentrations (≥ 2–$5 \ \mu M$) have been reported (Illinger *et al.*, 1995). In order to reach this critical concentration in endosomes, the amount of TMA-DPH within the lumen of sealed vesicles can be considerably higher, taking into account the water (probe not fluorescent)/membrane (probe highly fluorescent) partition equilibrium. Therefore it is important to apply gentle conditions for homogenization to limit possible damage to endosomal elements, particularly when using TMA-DPH. Broken endosomes would lose TMA-DPH by back-exchange in aqueous solution, resulting in a significant loss of signal. Harsh conditions should, however, always be avoided in order to limit the breakage of lysosomes and consequent proteolysis due to released hydrolases. Because BHK-21 cells are easily homogenized, it is wise to monitor each step of the homogenization process under phase-contrast microscopy.

V. PITFALLS

1. For small vesicles discrimination between real particle detection and background in a flow cytometer can be somehow difficult and some of the following modifications may help:
- use a 0.22 μ pre-filter for sheath flow, filter all diluting buffers
- degas sheath flow (set a vacuum to the sheath container for half an hour)
- use the SSC or fluorescence signal for instrument threshold, for sorting with others than small nozzles this may cause problems because of a "heavy drop drive signal" causing stream fluctuations and therefore wrong events.

2. Using others than the mentioned flow cytometers may need special attention, e.g., the scattered light of the blue excitation of a FACS Vantage can overlap the fluorescence intensity of stained endosomes. Using a BP 450/20 nm instead of a BP 460/60 nm for fluorescence emission can compensate this.

3. To sort into a proteinaceous solution keeps the endosomal membranes intact but the protein may interfere with subsequent procedures (e.g. protein staining after SDS-PAGE). Sorting into some neutral buffer (PBS) can cause membranes to break. All material will still be recovered, but sort purity controls will be misleading. Leaky endosomes will loose the TMA-DPH and a considerable fraction of nonstained particles will be found in the (+) sorted fractions.

4. Rinse cytometer tube system properly, endosomal vesicles can stick and later on contaminate your sort control.

Acknowledgments

This work was supported by grants from the Johnson & Johnson Focused Giving Program and the Swiss Cancer Foundation to LAH and JG and by grants from the FWF/Austria (Grant P-11446-MED) and FFF/Austria (Grant 3/11504) to LAH.

References

Chavrier, P., Parton, R. G., Hauri, H. P., Simons, K., and Zerial, M. (1990). Localization of low molecular weight GTP binding proteins to exocytic and endocytic compartments. *Cell* **62**, 317–329.

Desjardins, M., Celis, J. E., Vanmeer, G., Dieplinger, H., Jahraus, A., Griffiths, G., and Huber, L. A. (1994). Molecular characterization of phagosomes. *J. Biol. Chem.* **269**, 32194–32200.

Emans, N., Gorvel, J.-P., Walter, C., Gerke, V., Kellner, R., Griffiths, G., and Gruenberg, J. (1993). Annexin II is a major component of fusogenic endosomal vesicles. *J. Cell Biol.* **120**, 1357–1369.

Gorvel, J.-P., Chavrier, P., Zerial, M., and Gruenberg, J. (1991). rab5 controls early endosome fusion in vitro. *Cell* **64**, 915–925.

Gruenberg, J., and Gorvel, J.-P. (1992). In vitro reconstitution of endocytic vesicle fusion. *In* "Protein Targetting, a Practical Approach" (A. I. Magee and T. Wileman, eds.), pp. 187–216. Oxford University Press, Oxford.

Gruenberg, J., Griffiths, G., and Howell, K. E. (1989). Characterisation of the early endosome and putative endocytic carrier vesicles in vivo and with an assay of vesicle fusion in vitro. **108**, 1301–1316.

Gruenberg, J., and Howell, K. E. (1989). Membrane traffic in endocytosis: Insights from cell-free assays. *Annu. Rev. Cell Biol.* **5**, 453–481.

Gruenberg, J., and Maxfield, F. R. (1995). Membrane transport in the endocytic pathway. *Curr. Opin. Cell Biol.* **7**, 552–563.

Huber, L. A., Xu, Q.-B., Jürgens, G., Böck, G., Bühler, E., Gey, F., Schönitzer, D., Traill, K, N., and Wick, G. (1991). Correlation of lymphocyte lipid composition, membrane microviscosity and mitogen response in the aged. *Eur. J. Immunol.* **21**, 2761–2765.

Huber, L. A., and Simons, K. (1994). Preparation and purification of post-Golgi transport vesicles from perforated Madin–Darby canine kidney cells. *In* "Cell Biology: A Laboratory Handbook" (J. E. Celis, ed.), Vol. 1, pp. 517–524. Academic Press, San Diego.

Illinger, D., Duportail, G., Mely, Y., Poirel Morales, N., Gerard, D., and Kuhry, J. G. (1995). A comparison of the fluorescence properties of TMA-DPH as a probe for plasma membrane and for endocytic membrane. *Biochim. Biophys. Acta* **1239**, 58–66.

Illinger, D., Poindron, P., and Kuhry, J. G. (1991a). Fluid phase endocytosis investigated by fluorescence with trimethylamino-diphenylhexatriene in L929 cells; the influence of temperature and of cytoskeleton depolymerizing drugs. *Biol. Cell* **73**, 131–138.

Illinger, D., Poindron, P., and Kuhry, J. G. (1991b). Membrane fluidity aspects in endocytosis; a study with the fluorescent probe trimethylamino-diphenylhexatriene in L929 cells. *Biol. Cell* **71**, 293–296.

Murphy, R. F. (1985). Analysis and isolation of endocytic vesicles by flow cytometry and sorting: Demonstration of three kinetically distinct compartments involved in fluid-phase endocytosis. *Proc. Natl. Acad. Sci. USA* **82**, 8523–8526.

Murphy, R. F. (1988). Processing of endocytosed material. *Adv. Cell Biol.* **2**, 159–180.

Murphy, R. F., and Roederer, M. (1986). Flow cytometric analysis of endocytic pathways. *In* "Applications of Fluorescence in the Biomedical Sciences" (D. L. Taylor, A. S. Waggoner, R. F. Murphy, F. Lanni, and R. Birge, eds.), pp. 545–566. A. R. Liss, New York.

Murphy, R. F., Roederer, M., Sipe, D. M., Cain, C. C., and Bowser, R. (1989). Determination of the biochemical characteristics of endocytic compartments by flow cytometric and fluorimetric analysis of cells and organelles. *In* "Flow Cytometry: Advanced Research and Clinical Applications" (A. Yen, ed.), pp. 221–254. CRC Press, Boca Raton, FL.

Siminoski, K., Gonella, P., Bernake, J., Owen, L., Neutra, M., and Murphy, R. A. (1986). Uptake and transepithelial transport of nerve growth factor in suckling rat ileum. *J. Cell Biol.* **103**, 1779–1990.

Wilson, R. B., and Murphy, R. F. (1989). Flow-cytometric analysis of endocytic compartments. *Methods Cell Biol.* **31**, 293–317.

Purification of Clathrin-Coated Vesicles from Bovine Brain, Liver, and Adrenal Gland

Robert Lindner

I. INTRODUCTION

Clathrin-coated vesicles are intermediates in selective membrane transport processes in eukaryotic cells (Smythe and Warren, 1991). Since the first report on the purification of these organelles, a variety of protocols have been published (for review see Pearse, 1989). This article describes a differential centrifugation protocol that has been adapted from Campbell *et al.* (1984) for bovine brain tissue as a source. It rapidly provides crude clathrin-coated vesicles that are ideally suited for the preparation of various coat proteins (see in Keen *et al.*, 1979; Ahle *et al.*, 1988; Ahle and Ungewickell, 1990; Lindner and Ungewickell, 1991, 1992). Because of the existence of tissue-specific isoforms of several coat components, remarks on modifications in the procedure for other bovine organs (adrenal gland, liver) are included.

II. MATERIALS AND INSTRUMENTATION

Ficoll 400 (Cat. No. 17-0400-02) is from Pharmacia. Sucrose (Cat. No. BP 220-1) is from Fisher Scientific. EGTA (Cat. No. E-3889), MES (Cat. No. M-3023), and phenylmethylsulfonyl fluoride (PMSF, Cat. No. P-7626) are from Sigma. All other reagents are from Sigma or Fisher in analytical grade.

The biological material was obtained from an local abattoir within 1 hr of slaughter and was kept on ice until further processing (1–2 hr). Fresh and cleaned material can be frozen in liquid nitrogen and stored at −80°C for several months. It is helpful to cut the tissue into small pieces before freezing, as this supports a rapid drop of temperature in the tissue and minimizes ice crystal formation. The latter process reduces yield and purity of the following preparation, probably due to the destruction of membrane vesicles and the liberation of proteolytic activities.

For homogenization of the tissue, a Waring commercial blender (VWR International, Cat. No. 58977-169) was used. Membrane pellets were resuspended with Potter–Elvehjem homogenizers of various sizes (10–55 ml) obtained from Fisher Scientific (Cat. No. 08414-14 A to D). The metal shaft of the larger homogenizers was attached to a variable-speed overhead drive so that the pestle could be rotated.

Low-speed centrifugations were done in a Sorvall RC-5B centrifuge using GS-3, GSA, or SS-34 heads. High-speed centrifugations were performed with Beckman Ti 45 or Ti 35 rotors in conventional ultracentrifuges.

III. PROCEDURES

A. Cleaning of Bovine Brain Cortices
Solution

1. *Phosphate-buffered saline (PBS):* 137 mM NaCl, 2.7 mM KCl, 8.2 mM Na_2HPO_4, 1.9 mM KH_2PO_4, and 0.02% NaN_3, pH 7.0. To make 1 liter of 10× PBS (stock solution), dissolve 80 g NaCl, 2 g KCl, 2.58 g KH_2PO_4, 11.64 g Na_2HPO_4, and 0.2 g NaN_3 in double-distilled water. Adjust the pH to 7.0 (if necessary) and bring the volume to 1 liter. Dilute this stock solution 1:10 with double-distilled water and chill to 4°C prior to use. Approximately 2–3 liters of 1× PBS are needed per kilogram of tissue.

Steps

1. Separate the cerebellum and the lower part of the brain from the cortex.

2. Take a hemisphere of the cortex, place it in an ice bucket covered with plastic wrap, and remove the meninges along with blood vessels contained therein using forceps.

3. Collect the cleaned cortex hemispheres in a preweighed beaker on ice. Determine the weight of the tissue and estimate the volume of homogenization buffer needed in Section B (approximately 1 liter/kg of tissue).

4. Wash the cortex hemispheres with cold PBS several times to remove the remaining blood. To do this fill the beaker containing the hemispheres with PBS, mix gently, and pour the liquid off. It is helpful to use a household sieve at this step. Repeat the washes until the blood is removed.

5. If you want to store the brain tissue for later processing, cut it into small pieces after the washing step and freeze in liquid nitrogen. Store frozen material at −80°C. Otherwise continue with Section B.

B. Homogenization
Solutions

1. *Buffer A:* 0.1 M MES, 1.0 mM EGTA, 0.5 mM $MgCl_2$, and 0.02% NaN_3, pH 6.5. To make 1 liter of 10× buffer A (stock solution), dissolve 195.2 g MES, 3.8 g EGTA, and 1.0 g $MgCl_2$ in double-distilled water. Adjust the pH with 10 N NaOH to 6.5, add 2 g NaN_3, and bring the volume to 1 liter. Do not add NaN_3 prior to the adjustment of the pH, as MES is acidic and may release HN_3. To prepare 1× buffer A for the homogenization, dilute the stock solution 1:10 with double-distilled water, chill to 4°C, and supplement with 0.1 mM PMSF prior to use. Prepare about 3 liters of buffer A/kg tissue.

2. *Protease inhibitor (PMSF stock solution, 1000× = 0.1 M):* To make 10 ml, dissolve 174 mg PMSF in 10 ml pure methanol, aliquot, and store at −20°C. Dilute 1:1000 to get a working concentration of 0.1 mM. Note that PMSF will hydolyze in water and that it is not soluble in this solvent at high concentrations.

Steps

1. Fill the cup of a Waring commercial blender with 300–400 g of washed tissue and an equivalent amount of cold buffer A containing 0.1 mM freshly added PMSF. Do not fill the cup up to the top, but leave space below the rim.

2. Homogenize the tissue by three to six bursts of 10–15 sec duration with the setting on maximum speed. The number of bursts required to give a good homogenization varies: for brain, usually three bursts are sufficient; other organs, especially the adrenal glands, require more (see comments). Do not increase the length of the bursts, only their number, to prevent heating of the homogenate.

C. Differential Centrifugations

1. Preparation of Postmitochondrial Supernatants

Steps

1. Pour the homogenate into the buckets of a Sorvall GS-3 or GSA rotor, balance the buckets, and centrifuge them in a precooled rotor at 7000 rpm (about 8000 g in both types of rotors) for 50 min at 4°C.

2. Pour the turbid supernatants through a funnel with several layers of gauze to separate floating lipids from the supernatant.

3. The bulky pellets (up to a third of the total volume) should be resuspended with at least an equivalent volume of cold homogenization buffer and recentrifuged under the above conditions.

4. Discard the pellets after the second centrifugation and keep the combined supernatants on ice.

2. Preparation of Microsomal Pellets

Steps

1. Pour the postmitochondrial supernatant into the tubes of a Beckman Ti 35 or Ti 45 rotor, balance the tubes, and ultracentrifuge them at 32,000 rpm and 4°C for 1.6 hr (Ti 35 rotor) or at 40,000 rpm for 1 hr (Ti 45 rotor).

2. After the centrifugation, discard the clear supernatant and either refill the tubes with postmitochondrial supernatant for the next spin (leave the first pellet in the tube without resuspending it to save time) or carefully remove the microsomal pellets when all the postmitochondrial supernatant has been centrifuged or pellets from three to four ultracentrifugations have been collected in one tube.

3. Resuspend the pellets with 1–2 vol of buffer A supplemented with 0.1 mM PMSF using a 10-ml glass pipette in the reverse orientation (the wide top end down) and then thoroughly homogenize by 10–15 strokes in a Potter–Elvehjem device.

3. Purification of Coated Vesicles from Microsomal Pellets

Solution

1. *12.5% Ficoll–sucrose:* To make 1 liter, dissolve 125 g sucrose and 125 g Ficoll 400 in buffer A, pH 6.5, and stir overnight in the cold room. Ficoll 400 dissolves only slowly. Keep at 4°C.

Steps

1. Mix the well-homogenized microsomes with the same volume of Ficoll–sucrose solution and pour into open tubes suitable for an SS-34 Sorvall rotor. Centrifuge the tubes at 43,000 g (19,000 rpm) and 4°C for 40 min in a precooled rotor.

2. After centrifugation, pour the supernatants containing the clathrin-coated vesicles through a funnel with gauze to remove floating lipids. The compact sediment (about a fifth of the total volume) will stay in the tube and can be discarded.

3. Dilute the combined supernatants with 3–4 vol of cold buffer A containing 0.1 mM PMSF.

4. Concentration of Coated Vesicles by Pelleting

Steps

1. Ultracentrifuge the coated vesicles in the diluted Ficoll–sucrose solution as described in Section C2. The pellets obtained in this step are considerably smaller than the microsomal pellets. The color usually varies from yellowish-white to brown and is most likely due to contaminating ferritin (Kedersha and Rome, 1986).

2. Pour off the supernatants and carefully remove the pellets from the tubes with a spatula. Use a small volume of buffer A + PMSF to wash off material still attached to the walls of the tube.

3. Homogenize the combined pellets in about the same volume of buffer A + PMSF using a suitable Potter–Elvjehem homogenizer (5–15 ml volume, 10 strokes).

4. The material obtained after this step can either be directly used for extraction of the coat proteins (for a typical electrophoresis pattern of a Tris extract of bovine brain clathrin-coated vesicles, see Fig. 1) or be frozen in liquid nitrogen and stored at −80°C.

IV. COMMENTS

Following this protocol (for summary, see Fig. 2) 80–90% pure clathrin-coated vesicles are obtained from bovine brain. Major contaminants are smooth vesicles, ferritin, and some filamentous material (Pearse, 1989). The yields usually range from 100 to 150 mg clathrin-coated vesicles/kg brain cortex.

This basic protocol can also be used with other organs rich in clathrin-coated membranes, such as adrenal glands, liver or placenta. For adrenal glands, a more thorough homogenization (usually six or more 10-sec bursts in a Waring commercial blender) is required to break up the very resistant capsule material.

Clathrin-coated vesicle preparations from liver usually contain a considerable amount of ribonucleoprotein complexes termed vaults (Kedersha and Rome, 1986). It is advisable to remove these structures by velocity centrifugation on 5–40% sucrose gradients as an additional step after the preparation described here (for details see Kedersha and Rome, 1986). Vaults are at least partially dissociated by the conventional extraction methods for clathrin coat structures and thus contaminate coat protein extracts.

Although the main intention of the protocol described in this article is to provide a rapid access to various coat proteins, it might serve as a basis for the further purification of clathrin-coated vesicles as well (see Morris *et al.*, 1988 and references therein). A more elaborate protocol for the isolation of fusion-competent clathrin-coated vesicles from tissue culture cells has been described (Woodman and Warren, 1992).

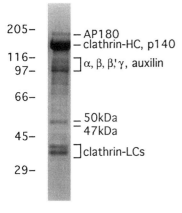

FIGURE 1 Coat proteins obtained by Tris extraction of a crude clathrin-coated vesicle preparation from bovine brain. Coated vesicles were incubated in 0.5 M Tris, pH 7.0, to liberate the peripheral membrane proteins and ultracentrifuged to remove the membranes. Approximately 12 μg Tris extract was electrophoresed in a low bis-SDS–polyacrylamide gel (for details see Lindner and Ungewickell, 1992). Note that in this gel system the brain-specific clathrin-associated protein AP180 is well resolved from the clathrin heavy chain. Both AP180 and another clathrin-associated protein, auxilin, are not detectable in coated vesicles from adrenal gland and liver (data not shown).

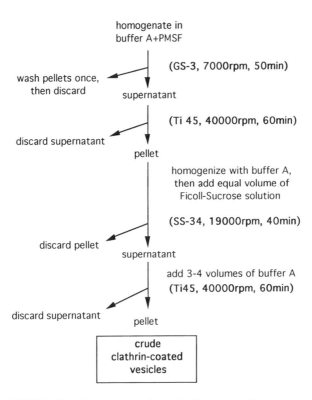

FIGURE 2 Purification scheme for crude clathrin-coated vesicles.

V. PITFALLS

1. Do not mix Ficoll–sucrose solution with microsomal pellets obtained in Section C3 before thorough homogenization of the pellets. The high viscosity of the Ficoll–sucrose solution prevents proper homogenization.

2. In order to quantitatively pellet the clathrin-coated vesicles from the supernatant after the centrifugation in Ficoll–sucrose, dilute with a *minimum* of 3 vol of buffer A + PMSF to decrease both the density and the viscosity of the supernatant.

References

Ahle, S., Mann, A., Eichelsbacher, U., and Ungewickell, E. (1988). Structural relationships between clathrin assembly proteins from the Golgi and the plasma membrane. *EMBO J.* **7**, 919–929.

Ahle, S., and Ungewickell, E. (1990). Auxilin, a newly identified clathrin-associated protein in coated vesicles from bovine brain. *J. Cell Biol.* **111**, 19–29.

Campbell, C., Squicciarini, J., Shia, M., Pilch, P. F., and Fine, R. E. (1984). Identification of a protein kinase as an intrinsic component of rat liver coated vesicles. *Biochemistry* **23**, 4420–4426.

Kedersha, N. L., and Rome, L. H. (1986). Isolation and characterization of a novel ribonucleoprotein particle: Large structures contain a single species of small RNA. *J. Cell Biol.* **103**, 699–709.

Keen, J. H., Willingham, M. C., and Pastan, I. H. (1979). Clathrin-coated vesicles: Isolation, dissociation and factor-dependent reassociation of clathrin baskets. *Cell* **16**, 303–312.

Lindner, R., and Ungewickell, E. (1991). Light-chain-independent binding of adaptors, AP180 and auxilin to clathrin. *Biochemistry* **30**, 9097–9101.

Lindner, R., and Ungewickell, E. (1992). Clathrin-associated proteins from bovine brain coated vesicles: An analysis of their number and assembly-promoting activity. *J. Biol. Chem.* **267**, 16567–16573.

Morris, S. A., Hannig, K., and Ungewickell, E. (1988). Rapid purification of clathrin-coated vesicles by free-flow electrophoresis. *Eur. J. Cell Biol.* **47**, 251–258.

Pearse, B. M. F. (1989). Characterization of coated-vesicle adapters: Their reassembly with clathrin and with recycling receptors. *Meth. Cell Biol.* **31**, 229–243.

Smythe, E., and Warren, G. (1991). The mechanism of receptor-mediated endocytosis. *Eur. J. Biochem.* **202**, 689–699.

Woodman, P. G., and Warren, G. (1992). Isolation and characterization of functional clathrin-coated vesicles. *In* "Methods in Enzymology" (J. E. Rothman, ed.), Vol. 219, pp. 251–260. Academic Press, San Diego.

Isolation of Phagosomes from Professional and Nonprofessional Phagocytes

Michel Desjardins and Sandra Scianimanico

I. INTRODUCTION

Phagosomes are intracellular organelles formed in a variety of cells by the internalization of large particulate materials such as dust or microorganisms. These can be the causal agents of important diseases such as salmonellosis or tuberculosis. Newly formed phagosomes are extensively modified with time to allow the killing and degradation of their content by complex mechanisms (see Desjardins, 1995). Briefly, phagosomes are transformed into phagolysosomes through a series of interactions with endocytic and biosynthetic organelles (Desjardins *et al.*, 1994b; Desjardins, 1995). This complex process is regulated by a variety of molecules allowing the binding of phagosomes to cytoskeletal elements, their movement inside the cell, and their sequential interaction and fusion with subsets of endosomes. Interestingly, some microorganisms have developed ways to escape normal degradation in phagosomes by inhibiting this normal set of interactions. The isolation of phagosomes from professional and nonprofessional phagocytes at precise intervals after their formation enables their fine biochemical analysis and the development of *in vitro* assays to study the role of various molecules in phagolysosome biogenesis. These studies are likely to allow the understanding of the way by which pathogenic microorganisms are able to invade cells and evade normal degradation in phagolysosomes.

II. MATERIALS AND INSTRUMENTATION

Blue dyed latex beads of 0.8 μm diameter (Cat. No. L-1398), carboxylated latex beads of 0.9 μm diameter (Cat. No. CLB-9), imidazole (Cat. No. I-0125), bovine serum albumin (BSA, Cat. No. A-6793), horseradish peroxydase (Cat. No. P-8250), Triton X-100 (Cat. No. T-6878), o-dianisidine (Cat. No. D-3252) are from Sigma. Sodium azide (Cat. No. S-227) and hydrogen peroxide 30% (Cat. No. H325-100) are from Fisher. $Na_2HPO_4,7H_2O$ (Cat. No. 3824-01) and NaH_2PO_4,H_2O (Cat. No. 3818-01) are from Baker Analyzed. Sucrose (Cat. No. 100168) and protease inhibitors leupeptin (Cat. No. 1017 101), aprotinin (Cat. No. 236 624), pepstatin (Cat. No. 253 286), and phenylmethylsulfonyl fluoride (PMSF, Cat. No. 236 608) are from Boehringer-Mannheim. Methionine-free culture medium (Cat. No. TZ 0023) is from Seromed. Promix [^{35}S]methionine–cysteine (Cat. No. SJQ-0079) is from Amersham.

Falcon bottle-top filtration membranes of 0.45 μm (Cat. No. 7109) are from VWR. Syringe filtration membranes of 0.2 μm (Cat. No. 190-2520) are from Nalgene. The rubber policeman to scrape the cell was homemade. Kontes 2-ml Dounce homogenizer (Cat. No. 885303-0002) with a tight pestle (Cat. No. 885302-0002) is from Fisher. The peristaltic pump P-1 (Cat. No. 19-4611-02) is from Pharmacia LKB. SW41 tubes (Cat. No. 344059) are from Beckman.

III. PROCEDURES

A. Phagosome Formation and Isolation

An easy way to isolate phagosomes has been developed by Wetzel and Korn (1969). They took advantage of the ability to form phagosomes by feeding cells with latex particles and the low buoyant density of this material to separate latex bead-containing phagosomes on sucrose step gradients. Latex beads are available from various suppliers and come in different sizes from 30 nm to over 10 μm. They are available in different colors, facilitating their localization throughout phagosome formation and isolation procedure; with modified surfaces such as carboxylated beads, allowing to attach a variety of proteins to their surface; or as iron-containing beads to isolate them on magnets. Latex beads can be phagocytosed by a variety of cells simply by adding these to the culture medium. The following procedure has been used in the past to isolate phagosomes from J774 mouse macrophages (Fig. 1), U937 human monocytes–macrophages, and BHK or NRK cells (Desjardins *et al.*, 1994a).

Solutions

1. *300 mM imidazole stock solution, pH 7.4:* To make 100 ml, dissolve 2.04 g imidazole in distilled water and bring to pH 7.4. Complete to 100 ml with distilled water.

2. *Homogenization buffer and sucrose solutions, pH 7.4, 3 mM imidazole:* To make 100 g of 8.55% (w/w) sucrose solution (homogenization buffer), weigh 8.55 g of sucrose, bring to 99 g with distilled water, and add 1 ml of imidazole stock solution. To make 100 g of 10% (w/w) sucrose solution, weigh 10 g of sucrose, bring to 99 g with distilled water, and add 1

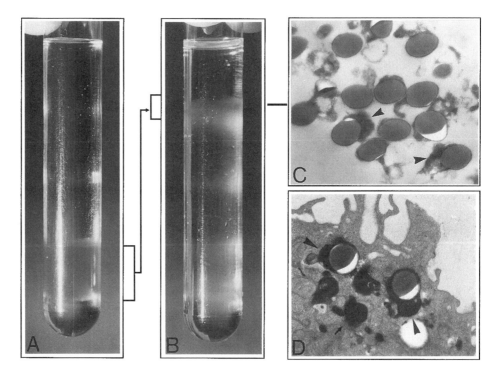

FIGURE 1 Procedure used to isolated phagosomes from J774 macrophages. Cell are grown in petri dishes and fed latex beads diluted in culture medium (D). Broken cells are then centrifuged to remove unbroken cells and nuclei, and the supernatant containing the phagosomes is loaded on a step sucrose gradient (A) and centrifuged for 60 min at 24,000 rpm in a SW40 rotor. The phagosomes collected at the 10 and 25% sucrose interface (B) are devoid of major cellular contaminants (C).

ml of imidazole stock solution. To make 100 g of 25% (w/w) sucrose solution, weigh 25 g of sucrose, bring to 99 g with distilled water, and add 1 ml of imidazole stock solution. To make 100 g of 35% (w/w) sucrose solution, weigh 35 g of sucrose, bring to 99 g with distilled water, and add 1 ml of imidazole stock solution. To make 100 g of 62% (w/w) sucrose solution, weigh 62 g of sucrose, bring to 99 g with distilled water, and add 1 ml of imidazole stock solution. Adjust all solutions at pH 7.4, filter on a 0.45-μm membrane, and keep at 4°C.

3. *Protease inhibitors stock solutions:* To make 1 ml of aprotinin at 10 mg/ml, add 1 ml of distilled water to the 10-mg content tube of aprotinin. To make 0.5 ml of leupeptin at 10 mg/ml, add 0.5 ml of distilled water to the 5-mg content tube of leupeptin. To make 2 ml of pepstatin at 1 mg/ml, add 2 ml of methanol to the 2-mg content tube of pepstatin. To make 10 ml of PMSF at 10 mg/ml, dissolve 100 mg of PMSF in 10 ml of isopropanol. Aliquot and store all solutions at −20°C.

4. *Sucrose solutions and phosphate-buffered saline (PBS) with protease inhibitors:* Add aprotinin, leupeptin, and pepstatin at 1 μg/ml each and PMSF at 100 μg/ml. Add 1 μl of each aprotinin and leupeptin stock solution, 10 μl of pepstatin stock solution, and 100 μl of PMSF stock solution to 10 ml of either sucrose solution or PBS.

Steps

1. Dilute latex beads 1:200 in culture medium and incubate cells in this medium for the desired time at 37°C in a CO_2 incubator. Because the internalization of latex beads is a slow process, allow cells to ingest beads for 20 to 60 min. Alternatively, the time length of bead internalization can be shortened if beads are spun on top of the cells. In this case, fix the dishes of cells in medium containing beads at 4°C on top of a swinging rotor and spin for 5 min at 700 rpm in a table-top refrigerate centrifuge. Then put cells at 37°C for 5 to 15 min to allow synchronous internalization.

2. After bead internalization, wash cells three times for 5 min with cold PBS containing 0.5% BSA. This gets rid of noninternalized particles. However, in most cases, latex beads are still attached to the surface of the culture dish, causing certain problems when phagosomes of a precise age are needed for kinetic studies. Accordingly, when late phagosomes are required, cells can be scraped after the internalization process, washed by centrifugation in PBS (see below), replated in culture dishes, and incubated in medium without latex beads for the desired chase time.

3. After washes, cells can be further incubated in culture medium without latex beads (allowing aging of the phagosomes) or scraped with a rubber policeman in cold PBS. All the following steps are done at 4°C.

4. After scraping, spin cells down at low speed (3 min at 1500 rpm in a 15-ml Falcon tube in a table-top refrigerated centrifuge) in cold PBS. Resuspend the pellet of cells in 1 ml of homogenization buffer by gently knocking the side of the tube to avoid major breakage of the cell pellet. Spin the cells again, resuspend in 1 ml of fresh homogenization buffer, and break the pellet by pipeting it through a blue tip 10 to 12 times.

5. Transfer cells to a Dounce homogenizer and break with a tight pestle. The number of stroke needed depends on the cell type and their conditions of growth. Typically, cells are disrupted after 5 to 25 strokes. To ensure the best results, observe a sample of the broken cells under the light microscope after each set of 5 strokes until most of the cells are broken without major breakage of the nucleus. The phagosomes containing latex beads are highly refringent and can be easily observed floating under the light microscope.

6. Centrifuge the homogenate in a 15-ml Falcon tube at 1500 rpm for 7 min to get rid of the unbroken cells and nucleus. After this centrifugation, three distinct phases are observed: The unbroken cells are at the bottom, nucleus are in the middle phase and the postnuclear supernatant containing the phagosomes is present in the upper phase. Recover the postnuclear supernatant (typically 750 μl to 1 ml volume) and mix with a 62% sucrose solution to bring it to a final concentration of 42% sucrose. This is done by adding a volume of 62% sucrose equal to the sample volume plus 10% of its volume. For example, if you have a sample of 750 μl, one would add $750 + 75 = 825$ μl of 62% sucrose in a 2-ml Eppendorf tube. The phagosomes are then ready to be isolated on the step sucrose gradient.

7. Pour a sucrose gradient consisting of the following steps in a SW41 Beckman tube (all the sucrose solutions contain the protease inhibitor cocktail): a cushion of 1 ml of 62% sucrose, the phagosome sample at 42% sucrose (about 2 ml), 2 ml of 35% sucrose, 2 ml of 25% sucrose, and 2 ml of 10% sucrose. Centrifuge this gradient at 100,000 g (about 24,000 rpm) for 60 min at 4°C.

8. After the centrifugation, phagosomes are present in a single band at the 10–25% sucrose interface. Recover the phagosomes at 4°C with a capillary plugged to a peristaltic pump set to low speed (final volume equal about 1 ml). Resuspend these phagosomes in cold PBS containing protease inhibitors to fill up a clean SW41 tube and pellet by centrifugation at 15,000 rpm for 15 min at 4°C. This step allows a further purification of the phagosome preparations.

9. Resuspend the phagosome pellet in an appropriate buffer for further analyses.

B. Isolation of Phagosomes from Metabolically Labeled Cells

The fine analyses of proteins and polypeptides associated to phagosomes can be made by forming phagosomes in cells that have been previously labeled with radioactive amino acids such as [^{35}S]methionine (see Volume 4, Celis *et al.*, "High Resolution Two-Dimensional Gel Electrophoresis of Proteins: Isoelectric Focusing (IEF) and Nonequilibrium pH Gradient Electrophoresis (NEPHGE)" for additional information).

Solution

1. *Low methionine culture medium:* To make 1 liter, add 100 ml of normal medium containing 10% fetal bovine serum to 900 ml of methionine-free medium containing the usual L-glutamine and antibiotic concentrations.

Steps

1. For J774 macrophages, incubate subconfluent cells (50% confluency) twice for 10 min in a low methionine culture medium at 37°C in the CO$_2$ incubator to allow methionine depletion.

2. Incubate the cells overnight in low methionine medium supplemented with [^{35}S]-methionine–cysteine at a concentration of 50 μCi/ml.

3. The next day, recover the radioactive medium in a Falcon tube and dilute latex beads in this medium (we found out that adding the concentrated latex bead directly to the cell dishes was, in many cases, accompanied by clogging of the beads, resulting in an inhibition of their internalization). Add the beads back to the cells for internalization.

4. Follow steps 2 to 9 (Procedure A).

C. Measurement of Phagosome Latency

The fractionation and isolation procedures can induce breakage of phagosomes. The state of integrity of the phagosome preparations can be assessed by several measurements. To ensure that the isolated phagosomes do not break open, one can measure the latency of the preparation. This measurement determines the proportion of phagosomes present in the preparation that are broken and evaluates the leakage of their lumenal content. An easy way to measure phagosome latency is to form phagosomes as described earlier and then feed cells horseradish peroxidase (HRP) to fill the endocytic and phagocytic organelles.

Solutions

1. *0.5 M phosphate buffer, pH 5:* To make 100 ml, dissolve 1.13 g of NaH$_2$PO$_4$, H$_2$O and 10.71 g of Na$_2$HPO$_4$, 7H$_2$O in distilled water and bring to pH 5. Complete to 100 ml with distilled water.

2. *HRP measurement buffer:* 0.342 mM o-dianisidine and 0.003% H_2O_2 as substrates in 0.05 M phosphate buffer containing 0.3% Triton X-100. To prepare 120 ml, using sterilized glassware, mix 12 ml of 0.5 M phosphate buffer, pH 5, 3.6 ml of 10% Triton X-100, and 12 μl of 30% H_2O_2. Then add 13 mg o-dianisidine and dissolve gently. Filter this on a 0.22-μm membrane in 104 ml of sterilized distilled water. Keep in the dark at 4°C, as long as it does not develop a straw color.

3. *2% NaN$_3$ stock solution:* Dissolve 1 g of NaN_3 in 50 ml of distilled water. Keep at room temperature.

Steps

1. Form phagosomes as described in step 1 of Procedure A. Incubate cells in culture medium containing 10 mg/ml HRP for 30 min at 37°C to allow internalization of the fluid-phase marker to endosomes and its further transfer to phagosome through endosome-phagosome fusions.

2. Wash cells as described in step 2 (Procedure A), and isolate phagosomes as described in steps 4 to 8 (Procedure A).

3. Resuspend the phagosomes gently in 1 ml cold PBS. Centrifuge again as in step 8 (Procedure A).

4. The amount of HRP activity in the pellet and in the supernatant is determined as follows: mix 100 μl of each sample with 900 μl of HRP measurement buffer. Put the samples in the dark and allow the oxidative reaction to proceed for a length of time sufficient to produce a visible brownish reaction (typically 1 to 10 min). Stop the HRP activity by adding sodium azide to a final concentration of 0.01%, and read the samples in a spectrophotometer at 455 nm. Compare the total OD of both samples. The latency is expressed as the percentage of total HRP activity present in the pellet.

D. Isolation of Phagosomes Containing Microorganisms

The protocols just described are simple and efficient, allowing to yield extremely pure preparations of phagosome-containing latex beads. These phagosomes have been shown to be fusogenic both *in vivo* and *in vitro* (Desjardins *et al.*, 1994b; G. Griffiths, personal communication) and to bind and move along microtubules (Burkhardt *et al.*, 1995), as expected from normal phagosomes. These preparations are probably well suited for a variety of analyses and cell biology assays of membrane organelle functions. However, one might need to study phagosomes induced by the internalization of various microorganisms. The density of these organelles does not allow their easy isolation on sucrose gradients. Instead, sedimentation procedures have been used and described in detail elsewhere (Alvarez-Dominguez *et al.*, 1996).

E. Derivation of Latex Beads via Electrostatic Interactions

Latex microspheres can be coated with a variety of proteins via electrostatic interactions. The coating of latex beads with a given protein is done by incubating beads in a solution of the protein of interest. This procedure allows the deposition of proteins at the surface of the bead through hydrophobic interactions. This binding is relatively stable at neutral pH. The stability of the protein–bead interaction is greatly enhanced when covalent amine bindings are induced. This is done by incubating carboxylated latex beads with the protein solution of interest in the presence of 1-ethyl-3-(3-dimethylaminopropyl)carbodiimide)hydrochloride. This approach was used to tag the enzyme invertase to beads (Jahraus *et al.*, 1994).

Steps

1. In siliconed Eppendorf tubes, put 100 μl of latex beads and add 1 ml of distilled water. Mix and sonicate briefly to get rid of aggregates.

2. Wash beads three times with distilled water by centrifugation of 5 min at 9000 rpm in a table-top centrifuge.

3. Add 1 ml of the coating protein solution at 2 (IgG) to 10 mg/ml (BSA). Mix and sonicate briefly. Beads aggregates can be checked under the light microscope. Incubate for 2 hr at 37°C under rotation.

4. Spin the solution. The efficiency of the coating procedure can be assessed by measuring the amount of protein present in the solution before and after coating.

5. Wash the pellet three times in distilled water. Incubate protein-coated beads in BSA 0.5% aqueous solution to block the uncoated parts. Resuspend beads in 100 μl of distilled water and add NaN$_3$ to 0.02%. Beads can be store at 4°C for 1 week. Before use, check the absence of aggregates under the light microscope. Resuspend and sonicate gently if necessary.

IV. COMMENTS

The ease of forming and isolating phagosomes from a variety of cell lines at precise time points after their formation, together with the availability of a wide range of *in vitro* assays, provides an ideal system for testing some of the phagosome functions and interactions with other organelles. Increasing evidence that phagosome composition and the maturation process resemble that of endosomes argues for the use of this organelle as a working system for elucidating mechanisms associated with endovacuolar membrane–membrane interactions.

References

Alvarez-Dominguez, C., Barbieri, A. M., Berón, W., Wandiger-Ness, A., and Stahl, P. D. (1996). Phagocytosed live *Listeria monocytogenes* influences Rab5-regulated *in vitro* phagosome-endosome fusion. *J. Biol. Chem.* **271**, 13834–13843.

Burkhardt, J. K., Blocker, A., Jahraus, A., and Griffiths, G. (1995). Microtubule dependent transport and fusion of phagosomes with endocytic pathway. *In* "Trafficking of Intracellular Membranes," (M. C. Pedroso de Lima, N. Düzgünes, and D. Hoekstra, eds.), pp. 211–222. Springer-Verlag, Berlin.

Desjardins, M. (1995). Biogenesis of phagolysosomes: The "kiss and run" hypothesis. *Trends Cell Biol.* **5**, 183–186.

Desjardins, M., Celis, J. E., van Meer, G., Dieplinger, H., Jahraus, A., Griffiths, G., and Huber, L. A. (1994a). Molecular characterization of phagosomes. *J. Biol. Chem.* **269**, 32194–32200.

Desjardins, M., Huber, L. A., Parton, R. G., and Griffiths, G. (1994b). Biogenesis of phagolysosomes proceeds through a sequential series of interactions with the endocytic apparatus. *J. Cell Biol.* **124**, 677–688.

Jahraus, A., Storrie, B., Griffiths, G., and Desjardins, M. (1994). Evidence for retrograde traffic between terminal lysosomes and the prelysosomal/late endosome compartment. *J. Cell Sci.* **107**, 145–157.

Wetzel, M. G., and Korn, E. D. (1969). Phagocytosis of latex beads by *Acanthamoeba castellanii* (NEFF). III. Isolation of the phagocytic vesicles and their membranes. *J. Cell Biol.* **43**, 90–104.

Preparation of Chloroplasts for Protein Synthesis and Protein Import

Ruth M. Mould and John C. Gray

I. INTRODUCTION

The biogenesis of higher plant chloroplasts involves input from two genomes. The chloroplast genome encodes approximately 10% of chloroplast proteins, which are synthesized within the organelle. The majority of chloroplast proteins are nuclear encoded, synthesized in the cytosol, and imported by the organelle posttranslationally.

This article describes the preparation of chloroplasts suitable for the study of protein synthesis by the organelle and for studying the import of nuclear-encoded proteins *in vitro*. Section IIIA describes a relatively simple method that yields a mixture of broken and intact chloroplasts. This quick method is suitable for protein synthesis studies (described in Section IIIB), as the efficiency of translation by chloroplasts declines rapidly after isolation and protein synthesis can cease within 15 to 30 min (Nivison and Jagendorf, 1984). Translation by isolated chloroplasts can be prolonged using the more complex protocols described by Mullet *et al.* (1986). In Section IIIB, radiolabeled methionine is incorporated into nascent polypeptides within isolated chloroplasts. Intact chloroplasts are subsequently isolated using a Percoll step gradient and are fractionated into stroma and thylakoids. Section IIIC describes the preparation of intact chloroplasts suitable for protein import studies (see Volume 2, Mould and Gray, "Import of Nuclear-Encoded Proteins by Isolated Chloroplasts and Thykaloids" for additional information).

II. MATERIALS AND INSTRUMENTATION

Pea seeds were obtained from a local supplier. Several dwarf varieties are suitable, including Feltham First and Progress No 9. Two plastic $60 \times 32 \times 9$-cm seedling trays with drainage holes were used to grow 1 kg of pea seeds. Levington M3 high nutrient compost is produced by Fisons.

The Polytron PT.K and accessories were made by Kinematica AG and supplied by Philip Harris Scientific. The Polytron PT.K (Cat. No. H62-300) was used with a 20-mm head with knives (Cat. No. H62-334) and a 500-ml borosilicate homogenizing vessel (H62-406) or equivalent. The Polytron PT.K is soon to be superceded; the Polytron PT 3100 (Cat. No. H62-601) used with the appropriate 20-mm head with knives (Cat. No. H62-404) will give the same result. Muslin is unbleached. Acetone (AnalaR grade, Cat. No. 10003) and Tris (Cat. No. 10315) are from Merck. Adenosine 5′-triphosphate (magnesium salt, Cat. No. A-9187), bromophenol blue (Cat. No. B7021), EDTA (disodium salt, Cat. No. ED2SS), HEPES (free acid, Cat. No. H3375), 2-mercaptoethanol (Cat. No. M7154), L-methionine (Cat. No. M9625), Percoll (Cat. No. P1644), sorbitol (Cat. No. S1876), sucrose (Cat. No. S9378), thermolysin (protease type X, Cat. No. P1512), and tricine (Cat. No. T0377) are from Sigma. [^{35}S]Methionine (15 mCi/ml, >1000 Ci/mmol, Cat. No. SJ204) is from Amersham Life Science. Glycerol (Cat. No. G/P450) and sodium dodecyl sulfate (SDS, Cat. No. S/P530) are from Fisher Scientific.

III. PROCEDURES

A. Preparation of Chloroplast Suspension

Growth of Pea Seedlings

Soak 1 kg pea seeds for 4–6 hr in tap water, then sow on approximately 3 cm compost in seedling trays and cover with approximately 2 cm compost. Water with tap water immediately and thereafter every second day. Grow at 18 to 25°C in a greenhouse with supplementary light providing approximately 125 μmol photons m^{-2}sec^{-1} for 16 hr per day. Gently brush compost off the tops of the shoots as they emerge. Seedlings are ready to harvest after 7–9 days when the first two leaves are starting to expand (see Section IV).

Solutions

1. *Sucrose isolation medium (SIM):* To make 1 liter, dissolve 120 g sucrose, 6 g HEPES, and 0.8 g EDTA in 900 ml distilled water. Adjust the pH to 7.6 with NaOH and bring the volume up to 1 liter with distilled water. Store frozen in 200-ml aliquots. To thaw to an icy slush, place at 4°C the evening before the day of use.

2. *2× sorbitol/tricine (2 × ST):* To make 250 ml, dissolve 30.1 g sorbitol and 44.8 g tricine in 200 ml distilled water. Adjust the pH to 8.4 with KOH and bring the volume up to 250 ml with distilled water. Autoclave and store at room temperature.

3. *1× ST:* Dilute 2× ST with an equal volume of sterile distilled water.

4. *80% acetone:* Dilute 100% acetone to 80% with distilled water on the day of use.

Steps

1. Harvest approximately 30 g of pea shoots (cut with scissors) and place in a homogenization chamber. All subsequent steps are performed at 0–4°C unless stated otherwise.

2. Add 200 ml SIM (as an icy slush) and blend for 6–8 sec with a Polytron PT.K homogenizer at setting 7 (or with a Polytron PT 3100 homogenizer for 8–10 sec at a setting of 20,000 rpm).

3. Filter the homogenate through eight layers of muslin into a 500-ml beaker on ice. Gently squeezing the homogenate through the muslin increases yield and saves time.

4. Divide the filtrate into two 250-ml centrifuge pots and centrifuge at 2000 g for 3 min.

5. Carefully decant the supernatant and resuspend the pellets in approximately 5 ml SIM using a cotton swab until the suspension looks uniform. Bring to a total volume of approximately 100 ml with SIM.

6. Centrifuge at 2000 g for 3 min.

7. Gently resuspend the pellets using a cotton swab in a minimal volume of 1× ST (e.g., 500 μl per pellet) and pool aliquots.

8. Determine the chlorophyll concentration of the chloroplast suspension by taking 10 μl and mixing with 990 μl 80% acetone in a microcentrifuge tube. Vortex the sample and then microcentrifuge for 3 min at full speed. The absorbance at 652 nm is determined using 80% acetone as a blank:

$$2.9 \times A_{652} = \text{chlorophyll concentration (mg/ml).}$$

9. Dilute chloroplast suspension to a chlorophyll concentration of 0.3 to 0.5 mg/ml with 1× ST.

B. Radiolabeling of Chloroplast Proteins

Solutions

1. *100 mM MgATP:* Dissolve 50.8 mg of MgATP in 0.5 ml 1× ST, adjust the pH to neutral with 1 M NaOH, testing the pH by spotting a few microliters onto pH indicator paper, and make the volume up to 1 ml. Store at −20°C in 100-μl aliquots.

2. *100 mM methionine:* Dissolve 14.9 mg of L-methionine in 1 ml 1X ST. Store at −20°C in 100-μl aliquots.

3. *40% Percoll/1× ST:* To make 10 ml, mix 4 ml Percoll, 5 ml 2× ST, and 1 ml distilled water on the day of use. Store at 4°C.

4. *100 mM HEPES pH 8:* To make 500 ml, dissolve 11.9 g of HEPES in 400 ml distilled water. Adjust the pH to 8.0 with KOH and make the volume up to 500 ml with distilled water. Autoclave and store at room temperature.

5. *10 mM HEPES pH 8:* To make 20 ml, mix 2 ml 100 mM HEPES, pH 8, with 18 ml sterile distilled water.

6. *Thermolysin:* Dissolve 4 mg thermolysin in 1 ml 10 mM HEPES, pH 8. Store in 50-μl aliquots at −20°C. Thaw aliquots (by hand) just prior to use and discard any unused solution as the protease activity is reduced when thawed, even at 4°C, due to autoproteolysis.

7. *1 M EDTA, pH 8:* Dissolve 74.5 g disodium EDTA in 150 ml distilled water. Adjust the pH to 8.0 with NaOH and make up to 200 ml. Autoclave and store at room temperature.

8. *500 mM EDTA, pH 8:* To make 10 ml, mix 5 ml 1 M EDTA, pH 8, with 0.5 ml 100 mM HEPES, pH 8, and 4.5 ml distilled water. Autoclave and store at room temperature.

9. *SDS sample buffer (SB):* To make 10 ml, mix 2.5 ml 0.5 M Tris–HCl, pH 6.8, 2 ml glycerol, 4 ml 10% (w/v) SDS, and 1.5 ml distilled water. Add a grain of bromophenol blue and store at room temperature. Add 50 μl 2-mercaptoethanol per 1-ml aliquot prior to use.

10. *1× ST:* See Section A.

Steps

1. Isolate chloroplasts as described in Section A and resuspend to 0.3–0.5 mg/ml chlorophyll with 1× ST (Fig. 1).

2. In a 15-ml glass tube, gently mix 400 μl chloroplast suspension, 50 μl MgATP (100 mM stock), 2 μl [^{35}S]methionine (30 μCi), and 48 μl 1× ST.

3. Incubate at 22–26°C for 15 min.

4. Add 110 μl 100 mM methionine (see Section IV).

5. Incubate at 22–26°C for 15 min.

6. Using a wide-bore pipette tip, pipette the sample on top of 3 ml 40% Percoll/1× ST in a 15-ml glass centrifuge tube at 4°C.

7. Centrifuge at 2000 g for 10 min at 4°C in a swing-out rotor with the brake off.

8. Discard the upper layer (broken chloroplasts) and the Percoll layer from the tube and reserve the pellet.

9. Wash the pellet by gently resuspending in 5 ml 1× ST using a cotton swab and centrifuge at 2000 g for 5 min at 4°C.

10. Lyse the chloroplasts by resuspending the pellet to 300 μl 10 mM HEPES, pH 8. (The volume to be added is dependent on the yield of intact chloroplasts after isolation through 40% Percoll/1× ST.) Pipette until the sample is uniform.

11. Place 100 μl in a clean microcentrifuge tube, add 100 μl SB, and store at −20°C (washed chloroplast fraction).

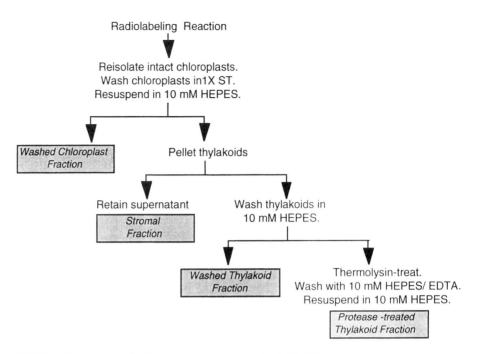

FIGURE 1 Fractionation of chloroplasts after synthesis of radiolabeled proteins.

12. Microcentrifuge the remaining 200 μl at full speed for 5 min at 4°C.

13. Transfer the supernatant (stromal fraction) to a clean microcentrifuge tube and reserve the pellet (thylakoid fraction).

14. Microcentrifuge the stromal fraction at full speed for 5 min at 4°C to pellet any contaminating thylakoids. Transfer the supernatant to a clean microcentrifuge tube and make the volume up to 400 μl with SB. Store at −20°C (stromal fraction).

15. Wash the thylakoid fraction by resuspending the pellet in 1 ml 10 mM HEPES, pH 8.

16. Microcentrifuge at full speed for 5 min at 4°C.

17. Discard the supernatant and resuspend the pellet in 200 μl 10 mM HEPES, pH 8.

18. Divide the sample into two aliquots of 100 μl.

19. To one aliquot add 100 μl SB and store at −20°C (thylakoid fraction).

20. To the second aliquot add 100 μl 10 mM HEPES, pH 8, and 10 μl thermolysin (see Section IV). Incubate on ice for 40 min.

21. Add 900 μl 10 mM HEPES, pH 8, and 130 μl 500 mM EDTA, pH 8.

22. Microcentrifuge at full speed for 5 min at 4°C.

23. Discard the supernatant and resuspend the pellet to 100 μl with 10 mM HEPES, pH 8. Add 100 μl SB and store at −20°C (protease-treated thylakoid fraction).

24. Heat samples to 80°C for 2 min before analyzing by electrophoresis on an SDS–17% polyacrylamide gel followed by fluorography (see Volume 4, Celis and Olsen, "One-Dimensional Sodium Dodecyl Sulfate-Polyacrylamide Gel Electrophoresis" for additional information).

C. Isolation of Intact Chloroplasts

Solutions

1. *5× sorbitol resuspension medium (5× SRM):* To prepare 1 liter, dissolve 300.6 g sorbitol and 59.6 g HEPES in approximately 600 ml of distilled water. Adjust to pH 8.0 using 10 M KOH and then bring to a total volume of 1 liter. Store frozen in 20-ml aliquots.

2. *1× SRM:* To prepare 100 ml, mix 20 ml 5× SRM with 80 ml distilled water and store frozen. Thaw to 4°C before use.

3. *Percoll step gradients:* To prepare two 9-ml gradients, make 10 ml of 80% Percoll/1× SRM by mixing 8 ml Percoll with 2 ml 5× SRM, and 10 ml of 40% Percoll/1× SRM by mixing 4 ml Percoll with 2 ml 5X SRM and 4 ml distilled water. For each gradient, layer 4.5 ml of 40% Percoll/1× SRM in the bottom of a 30-ml glass centrifuge tube, then slowly pipette 4.5 ml of 80% Percoll/1× SRM *underneath* the 40% layer, keeping the pipette tip at the bottom of the tube and taking care not to mix the two layers. If done correctly, an interface should be clearly visible. Store at 4°C. Do not make up the gradients more than a couple of hours before use as the layers start to mix.

4. *40% Percoll/1× SRM:* On the day of use, mix 2 ml Percoll, 1 ml 5X SRM, and 2 ml distilled water. Store at 4°C.

5. *1X SIM:* See Section A.

Steps

1. Follow Section IIIA up to the end of step 6.
2. Gently resuspend pellets in 1 ml of 1X SRM using a cotton swab until the suspension looks uniform and then make up to 4 ml with 1X SRM.
3. Carefully layer 2 ml of the chloroplast suspension on top of each Percoll step gradient (Fig. 2).
4. Centrifuge at 3000 g in a swing-out rotor for 15 min with the brake off.
5. Pipette off and discard the top green layer, which contains lysed chloroplasts, and the 40% Percoll/1× SRM layer.
6. Transfer the green layer, which formed at the interface between the 80 and the 40% Percoll/1× SRM layers, to a clean 30-ml centrifuge tube and add 15 ml 1X SRM.
7. Centrifuge at 2000 g for 3 min at 4°C.
8. Resuspend the pellet gently with a cotton swab in 500 μl 1X SRM.
9. Determine the chlorophyll concentration of the chloroplast suspension (see Section IIIA, step 8).
10. Dilute the chloroplast suspension to a chlorophyll concentration of 1 mg/ml with 1X SRM. These chloroplasts are suitable for import of nuclear-encoded proteins (see Section IV).

IV. COMMENTS

Some chloroplast-encoded proteins are synthesized at easily detectable levels only during early leaf development, i.e., before the leaves have started to expand. Therefore the age of shoots used to prepare chloroplasts for protein synthesis studies depends on the protein(s)

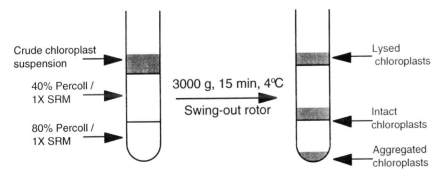

FIGURE 2 Isolation of intact chloroplasts by centrifugation through Percoll step gradients.

of interest to the investigator. The age at which pea shoots are harvested to prepare chloroplasts for protein import studies is less critical (Dahlin and Cline, 1991).

The radiolabeling reaction in Section IIIB is followed by the addition of an excess of unlabeled methionine. The incubation is continued so that the synthesis of most labeled proteins is completed, thereby reducing the number of labeled truncated polypeptides that would be visualized as extra bands when the samples are analyzed by SDS–PAGE followed by fluorography (Mullet *et al.*, 1986).

The addition of 2.5 mM calcium chloride during treatment with thermolysin may increase the efficiency of protein degradation; however, we do not find this necessary.

For protein import assays using isolated thylakoids (see Volume 2, Mold and Gray, "Import of Nuclear-Encoded Proteins by Isolated Chloroplasts and Thykaloids" for additional information), aggregated chloroplasts that form a pellet when centrifuged through 40% Percoll/1× SRM may be used to prepare thylakoids instead of preparing isolated chloroplasts using a Percoll step gradient (see Section IIIC).

V. PITFALLS

1. Poor radiolabeling of chloroplast-encoded proteins (Section IIIB) is usually due to taking too long to prepare the chloroplast suspension or not maintaining the chloroplasts at 0 to 4°C during preparation.

2. High levels of chloroplast lysis can occur if the chloroplasts are broken by pipetting or if the buffer is not isotonic.

References

Blair, G. E., and Ellis, R. J. (1973). Protein synthesis in chloroplasts. I. Light-driven synthesis of the large subunit of Fraction I protein by isolated pea chloroplasts. *Biochim. Biophys. Acta* **319**, 223–234.

Dahlin, C. D., and Cline, K. (1991). Developmental regulation of the plastid protein import apparatus. *Plant Cell* **3**, 1131–1140.

Mullet, J. E., Klein, R. R., and Grossman, A. R. (1986). Optimization of protein synthesis in isolated higher plant chloroplasts: Identification of paused translation intermediates. *Eur. J. Biochem.* **155**, 331–338.

Nivison, H. T., and Jagendorf, A. T. (1984). Factors permitting prolonged translation by isolated pea chloroplasts. *Plant Physiol.* **75**, 1001–1008.

Isolation of Peroxisomes

Alfred Völkl and H. Dariush Fahimi

I. INTRODUCTION

The investigation of unique functional and structural aspects of peroxisomes (PO) requires the preparation of highly purified fractions of this organelle. This is, however, hampered by two serious problems: (1) the relative paucity of PO (2% of total liver protein) and (2) their considerable fragility. Thus, mild homogenization conditions minimizing mechanical, hydrostatic, and osmotic stress have to be sustained.

In general, the isolation of PO is accomplished in three steps: (a) homogenization of the tissue or disruption of the cells; (b) subfractionation of the homogenate by differential centrifugation, usually according to the classical scheme of de Duve *et al.* (1955); and (c) isolation of purified peroxisomes by density gradient centrifugation of the so-called light mitochondrial (λ) fraction.

Homogenization is carried out in an isotonic medium (e.g., see Section IIIA1). High salt concentrations should be avoided as they cause aggregation. However, the addition of a chelator (e.g., EDTA) is feasible as it prevents the aggregation of microsomes that may contaminate PO.

For purification of PO by density gradient centrifugation, three approaches have been developed. In the classic procedure (Leighton *et al.*, 1968), sucrose gradients and the specialized type Beaufay rotor were employed. A self-generating Percoll gradient in conjunction with a vertical rotor is used for the isolation of PO under isotonic conditions (Neat *et al.*, 1980). The most straightforward approach in obtaining highly purified PO makes use of iodinated gradient media such as metrizamide and nycodenz in conjunction with a vertical rotor (Völkl and Fahimi, 1985; Hartl *et al.*, 1985).

PO band because of their permeability to low molecular weight compounds (van Veldhoven *et al.*, 1983), at the high density of 1.24 g/cm^3, well separated from lysosomes as well as from mitochondria and microsomes.

The method described in this article is a modification of an approach developed for isolation of highly purified (>98%) PO from normal rat liver (Völkl and Fahimi, 1985). In the mean time it has been applied to the livers and kidneys of several other mammalian species (Fahimi *et al.*, 1993; Zaar, 1992) and for the isolation of PO from cell cultures (Schrader *et al.*, 1994). Moreover, by adding a differential centrifugation step, even peroxisomal subpopulations may be isolated using this protocol (Lüers *et al.*, 1993).

II. INSTRUMENTATION AND MATERIALS

A. Instrumentation

1. Perfusion device (self made).

2. A 30-ml Potter–Elvehjem tissue grinder (Cat. No. 9651 P25) with loose-fitting pestle (Cat. No. 9657 925 clearance 0.10–0.15 mm) and a motor-driven homogenizer (Cat. No. 5308 50001) was obtained from Bayer and Migge.

3. Refrigerated low- and high-speed centrifuge (e.g., Beckman TJ-6 and J 2-21); ultracentrifuge (e.g., Beckman L90) with corresponding rotors (e.g., Beckman JA-20; VTi 50).

4. Refractometer.

B. Chemicals

Morpholinopropane sulfonic acid (MOPS, Cat. No. 29 836), phenylmethylsulfonyl fluoride (PMSF, Cat. No. 32 395), and dithiothreitol (DTT, Cat. No. 20 710) are from Serva. Ethanol (Cat. No. 100 986), NaCl (Cat. No. 106 404), Ethylenediaminetetraacetic acid (EDTA, 814 696), and ϵ-Aminocaproic acid (Cat. No. 817 010) are from Merck. Sucrose (Cat. No. 4621.1) is from Roth. Metrizamide (Cat. No. 10o 1983N) is from Life Technologies.

C. Animals

Female rats of 220–250 g body weight, starved overnight.

III. PROCEDURES

(Fig. 1)

A. Perfusion and Homogenization
Solutions

1. *Homogenization medium (HM):* To make 1 liter, dissolve 85.55 g of sucrose, 1.406 g of MOPS, 0.292 g of EDTA, and 1 ml of ethanol in distilled water, adjust pH to 7.2 with

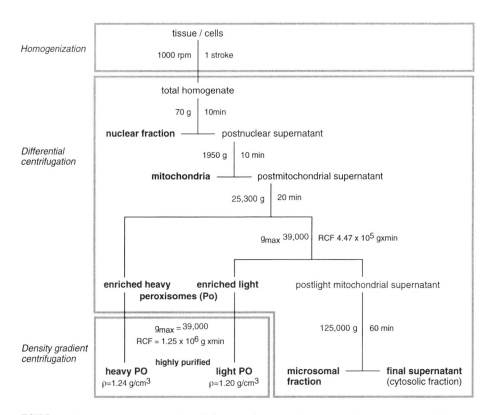

FIGURE 1 Diagrammatic representation of the procedure used for isolating highly purified (>98%) peroxisomes from rat liver.

NaOH, and add water up to 1 liter. Store at 4°C. Prior to use, add per 100 ml: 0.2 ml of 0.1 M PMSF, 0.1 ml of 1 M ϵ-aminocaproic acid, and 20 μl of 1 M DTT.

2. *Saline (0.9%):* To make 1 liter, dissolve 9 g of NaCl in distilled water and adjust to a total volume of 1 liter.

Steps

1. Anesthesize the animal (e.g., by ip injection of chloralhydrate).

2. Weigh the animal, open abdominal cavity, and perfuse liver with 0.9% saline via the portal vein until all blood is drained away.

3. Remove liver, dissect connective tissue, and weigh and cut liver in small pieces into a Potter tissue grinder held in an ice bath containing 3 ml/g (wet liver weight) of ice-cold HM.

4. Homogenize tissue at 1000 rpm with a single down and up stroke using a loose-fitting pestle.

5. Pour homogenate into a 50-ml centrifuge tube.

B. Subcellular Fractionation

Solution

HM: See Section A.

Steps

1. To remove debris, unbroken hepatocytes, blood cells, and concomitantly most of the nuclei, centrifuge the total homogenate at 70 g for 10 min in a refrigerated low-speed centrifuge.

2. Carefully pour off the supernatant (loose pellet), resuspend pellet in 2 ml/g of ice-cold HM, and rehomogenize and spin again under the same conditions.

3. Pour off the second supernatant and combine it with the first one (postnuclear supernatant).

4. Centrifuge postnuclear supernatant at 1950 g for 10 min in a refrigerated high-speed centrifuge.

5. Decant supernatant (firm pellet), resuspend pellet manually in 1 ml/g of ice-cold HM using an appropriate pestle, and spin again at 1950 g. The final pellet contains the majority of *mitochondria,* large microsomal sheets, and some remaining nuclei. The combined supernatants represent the postmitochondrial supernatant.

6. Subject the latter to 25,300 g for 20 min; remove supernatant, including the reddish fluffy layer by suction; resuspend pellet in about 10 ml of ice-cold HM using a glass rod and recentrifuge at 25,300 g for 15 min. Resuspend the final pellet in 5 ml of ice-cold HM by means of a glass rod; this pellet comprises *enriched heavy peroxisomes* [light mitochondrial (λ) fraction]. The corresponding supernatant may be used to prepare a microsomal fraction and a final supernatant (soluble proteins, mostly of cytosolic origin) or further processed for the isolation of "light peroxisomes."

7. To this extent it is centrifuged at an integrated relative centrifugal force (RCF) of 4.47 $\times 10^5$ $g\times$min (g_{max} = 39,000). Resuspend the pellet thus obtained in 5 ml of ice-cold HM using a glass rod; this pellet represents *enriched light peroxisomes.*

C. Metrizamide Density Gradient Centrifugation

Solutions

1. *Gradient buffer (GB):* To make 1 liter, dissolve 1.426 g of MOPS, 0.292 g of EDTA, and 1 ml of ethanol in distilled water, adjust pH to 7.2 with NaOH, and add water up to 1 liter. Store at 4°C. Prior to use, add per 100 ml: 0.2 ml of 0.1 M PMSF, 0.1 ml of 1 M ϵ-aminocaproic acid, and 20 μl of 1 M DTT.

2. *Metrizamide solutions (MS):*

  a. Stock solution, 60% (w/v): Dissolve 60 g of metrizamide in GB by stirring. Add GB up to 100 ml. Store at 4°C.

  b. Gradient solutions: To prepare one gradient, take 3.78, 3.38, 3.53, 2.06, and 3.2 ml of 60% stock solution. To adjust densities to 1.12, 1.155, 1.19, 1.225, and 1.26 g/ml, respectively, add GB up to 10, 7, 6, 3, and 4 ml.

Steps

Preparation of a Metrizamide Gradient

1. Layer sequentially 4, 3, 6, 7, and 10 ml respectively, of MS (1.26–1.12 g/ml) in a 40-ml centrifuge tube (e.g., Quick-seal polyallomer, Beckman) to form a discontiuous gradient.

2. Immediately freeze the gradient in liquid nitrogen and store at −20°C.

3. Thaw the gradient quickly at room temperature using a metallic stand, thus transforming the step gradient into one with an exponential profile.

Gradient Centrifugation

1. Layer 5 ml of the appropriate enriched peroxisomal fraction (corresponding to one liver of approximately 5–6 g) on top of the thawed gradient and seal it.

2. Centrifuge gradients in a vertical-type rotor (e.g., Beckman VTi50) at an integrated force of $1.256 \times 10^6 g \times min$ ($g_{max} = 39{,}000$) using slow acceleration/deceleration. Under the conditions employed, **highly purified heavy peroxisomes** band at 1.23–1.24 g/ml whereas **light peroxisomes** band at 1.20–1.21 g/ml.

3. Recover peroxisomal fraction by means of a fraction collector or by puncturing the gradient tube and aspiration using a syringe. Store fractions at −80°C.

4. To remove metrizamide, which interferes with the determination of some peroxisomal enzymes (e.g., urate oxidase) or of protein (Lowry method), dilute the peroxisome fraction about 10-fold with HM followed by centrifugation at 25,000 and 39,000 g, respectively, to pellet the organelles.

IV. OBSERVATIONS AND COMMENTS

The properties of the heavy peroxisomal fraction are listed in Table I. Estimated by the specific peroxisomal reference enzymes, Table I shows a purification rate of about 38-fold

TABLE I   Properties of Purified Heavy Peroxisomal Fractions from Normal Rat Liver[a]

Enzyme	mU/mg homogenate (mg/g liver)	Rate of recovery (%)	RSA
Protein	256.36 ± 82.94	0.28 ± 0.08	—
Catalase	203 ± 42.7	9.96 ± 1.92	37.67 ± 4.28
HA-oxidase	3.3 ± 0.7	9.46 ± 1.06	40.25 ± 5.83
FA β-oxidation	3.6 ± 0.96	11.18 ± 3.25	36.32 ± 5.09
β-Glucuronidase	38 ± 5.88	0.01 ± 0.01	0.07 ± 0.06
Acid phosphatase	31	0.15	0.07
Esterase	1229 ± 246.7	0.01 ± 0.01	0.09 ± 0.07
NADPH-CcR	18 ± 4.6	0.18	0.702
Cc-oxidase	127 ± 35.6	0.02 ± 0.02	0.09 ± 0.05
L-GlDH	57 ± 16.5	0.08 ± 0.02	0.679

[a] Values given are means ± SD. HA-oxidase, α-hydroxyacid oxidase; FA β-oxidation, fatty acid β-oxidation; NADPH-CcR, NADPH–cytochrome c reductase; Cc-oxidase, cytochrome c oxidase; L-GlDH, L-glutamate dehydrogenase; and RSA, relative specific activity.

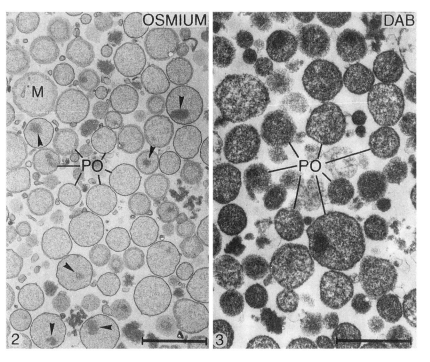

FIGURE 2 Electron microscopic appearance of isolated heavy peroxisomes after fixation in glutaraldehyde and osmium. The fraction consists almost exclusively of peroxisomes (PO) with only a rare mitochondrion (M). Many peroxisomes contain urate–oxidase cores (arrowheads). Bar = 1 μm.

FIGURE 3 A preparation comparable to that in Fig. 2 but incubated in alkaline 3,3'-diaminobenzidine for the localization of catalase (Fahimi, 1969). Note the electron-dense reaction product of catalase over the matrix of most peroxisomes. This illustrates the absence of catalase leakage, confirming their integrity. Reproduced from *J. Cell Biol.*, (1969), **43**, 275–288 by copyright permission of The Rockefeller University Press.

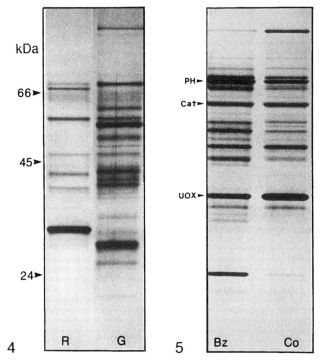

FIGURE 4 SDS–PAGE of highly purified heavy peroxisomes from rat (R) and guinea pig (G) liver. A 10–12.5% resolving gel was used, and the amounts of protein loaded per lane were 2.4 μg (R) and 5.4 μg (G). Silver staining of polypeptide bands. M_r standards: BSA, 66 kDa; ovalbumin, 45 kDa; and trypsinogen, 24 kDa. Note the distinct differences in the polypeptide patterns between rat and guinea pig peroxisomes.

FIGURE 5 SDS–PAGE of highly purified peroxisomes from control (Co) and bezafibrate-treated (Bz) rat liver. A 10–15% resolving gel was used, and 5.0 μg of protein was loaded per lane. Silver staining of bands. Peroxisomal polypeptides indicated by an arrowhead are PH, trifunctional protein; Cat, catalase; and UOX, urate oxidase. Note the induction of PH and the concomitant reduction of Cat and UOX in peroxisomes of the Bz-treated rat.

over the original homogenate. More than 95% of the total protein content of this fraction is contributed by peroxisomes (Völkl and Fahimi, 1985), with mitochondria and microsomes accounting for about 2% each and lysosomes for less than 1%. This is also confirmed by electron microscopy which shows that peroxisomes make up 98–99% of the fraction (Figs. 2 and 3). Many peroxisomes contain the typical inclusions of urate oxidase in matrix, but some extruded free cores are also found between them. The electron-dense cytochemical reaction product of catalase after the incubation of filter preparations in the alkaline 3,3′-diaminobenzidine medium (Fahimi, 1969) is seen over the matrix of the majority of peroxisomes, demonstrating their integrity and the absence of leakage of catalase.

The polypeptide pattern (SDS–PAGE) of rat liver heavy peroxisomes is shown in Fig. 4, confirming their high degree of purity because of the absence of bands typical for mitochondria and microsomes. Figure 4 also shows distinct differences in the protein composition of hepatic peroxisomes between rat and guinea pig. The selective induction of specific peroxisomal proteins such as the trifunctional protein with concomitant reductions of catalase and urate oxidase in rats treated with hypolipidemic fibrates is shown in Fig. 5. The extended procedure outlined in this article has been successfully employed to isolate heavy and light peroxisome subpopulations from normal and regenerating rat liver that differ in density, size, shape, and enzymatic composition (Lüers et al., 1993). Peroxisome subpopulations as divergent as the former have also been obtained from the human heptoma cell line HepG2 (Schrader et al., 1994). In total, these observations are consistent with the concept of heterogeneity of PO and may have some bearing on the biogenesis of this organelle.

Acknowledgments

The original work in the laboratory of the authors has been supported by grants of the Deutsche Forschungsgemeinschaft, Bonn, FRG (Fa 146/1-3; Vo 317/3-1; SFB 352 and Vo 317/4-1) and Landesforschungsschwerpunkt-Programm of the State of Baden-Württemberg, FRG.

References

de Duve, C., Pressman, B. C., Gianetto, R., Wattiaux, R., and Appelmans, F. (1955). Intracellular distribution patterns of enzymes in rat liver tissue. *Biochem. J.* **60**, 604–617.

Fahimi, H. D. (1969). Cytochemical localization of peroxidatic activity of catalase in rat hepatic microbodies (peroxisomes). *J. Cell Biol.* **43**, 275–288.

Fahimi, H. D., Baumgart, E., Beier, K., Pill, J., Hartig, F., and Völkl, A. (1993). Ultrastructural and biochemical aspects of peroxisome proliferation and biogenesis in different mammalian species. *In* "Peroxisomes: Biology and Importance in Toxicology and Medicine" (G. G. Gibson and B. Lake, eds), pp. 395–424. Taylor and Francis Ltd., London.

Hartl, F. U., Just, W. W., Köster, A., and Schimassek, H. (1985). Improved isolation and purification of rat liver peroxisomes by combined rate zonal and equilibrium density centrifugation. *Arch. Biochem. Biophys.* **237**, 124–134.

Leighton, F., Poole, B., Beaufay, H., Baudhuin, P., Coffey, J. W., Fowler, S., and De Duve, C. (1968). The large-scale preparation of peroxisomes, mitochondria and lysosomes from the livers of rats injected with Triton WR-1339. *J. Cell Biol.* **37**, 482–513.

Lüers, G., Hashimoto, T., Fahimi, H. D., and Völkl, A. (1993). Biogenesis of peroxisomes : Isolation and characterization of two distinct peroxisomal populations from normal and regenerating rat liver. *J. Cell Biol.* **121**, 1271–1280.

Neat, C. E., Thomassen, M. S., and Osmundsen, H. (1980). Induction of peroxisomal β-oxidation in rat liver by high-fat diets. *Biochem. J.* **186**, 369–371.

Schrader, M., Baumgart, E., Völkl, A., and Fahimi, H. D. (1994). Heterogeneity of peroxisomes in human hepatoblastoma cell line HepG2: Evidence of distinct subpopulations. *Eur. Cell Biol.* **64**, 281–294.

van Veldhoven, P., Debeer, L. J., and Mannaerts, G. P. (1983). Water-and solute-accessible spaces of purified peroxisomes. *Biochem. J.* **210**, 685–693.

Völkl, A., and Fahimi, H. D. (1985). Isolation and characterization of peroxisomes from the liver of normal untreated rats. *Eur. J. Biochem.* **149**, 257–265.

Zaar, K. (1992). Structure and function of peroxisomes in the mammalian kidney. *Eur. J. Cell Biol.* **59**, 233–254.

Purification of Secretory Granules from PC12 Cells

Jane C. Stinchcombe and Wieland B. Huttner

I. INTRODUCTION

Specialized cells that secrete certain proteins in response to external signals store these secretory products in specific organelles called *secretory granules* (Burgess and Kelly, 1987). The ability to purify secretory granules is an important prerequisite for their molecular characterization, which in turn is required for a full understanding of their function. Most studies dealing with the purification and characterization of secretory granules have been performed with endocrine (e.g., adrenal medulla; reviewed by Winkler *et al.*, 1986) or exocrine (reviewed by Castle *et al.*, 1987) tissue rather than secretory granule-containing cell lines. This is because secretory granules are denser and more abundant in tissue than in cells in culture and, thus, much more easily purified from the former. However, cell lines with secretory granules offer many advantages over tissue as model systems for investigating functional aspects of the secretory process. Ideally, one would therefore want to purify and characterize secretory granules from such cell lines.

The neuroendocrine cell line PC12, derived from a rat pheochromocytoma (Greene and Tischler, 1976), has extensively been used as a model system to study various aspects of secretory granule function (see references in Bauerfeind and Huttner, 1993; Kelly, 1993; Tooze *et al.*, 1993). Several procedures for the preparation of secretory granules from PC12 cells have been reported, which vary in the degree of purity of the final material obtained (Roda *et al.*, 1980; Wagner, 1985; Cutler and Cramer, 1990). This article describes the method developed in our laboratory which yields PC12 cell secretory granules of near-morphological purity (see also Stinchcombe, 1992; Stinchcombe and Huttner, manuscript in preparation). To determine the purification factor and yield, we have used the tyrosine-sulfated secretory granule-specific proteins chromogranin B (CgB, secretogranin I) and secretogranin II (SgII) (Huttner *et al.*, 1991a,b) as markers. CgB and SgII are easily detected either by long-term [^{35}S]sulfate labeling followed by SDS–PAGE (Rosa *et al.*, 1985) or by SDS–PAGE followed by immunoblotting using, in the case of CgB, a commercially available monoclonal antibody (Rosa *et al.*, 1989) (Boehringer-Mannheim, Cat. No. 1112-490).

II. MATERIALS AND INSTRUMENTATION

In addition to standard chemicals and instruments, the following materials are required. Trizma base (Cat. No. T-1503) and leupeptin (Cat. No. L-2884) can be bought from Sigma Chemicals, aprotinin (Cat. No. 236624) from Boehringer-Mannheim, phenylmethylsulfonyl fluoride (PMSF, Cat. No. 32395) from Serva, and HEPES (Cat. No. 9105) and EDTA (Cat. No. 8043) from Roth. Carrier-free [^{35}S]sulfate (25–40 Ci/mg, Cat. No. SJS 1) is from Amersham Buchler. Dulbecco's modified Eagle's medium powder (DMEM, Cat. No. 07401600N) and horse serum (HS, Cat. No. 03406050M) are from GIBCO Laboratories. Fetal calf serum (FCS, Cat. No. S0115) is from Seromed (Berlin).

Cell culture plates (24 × 24 cm, Cat. No. 1-66508A) can be obtained from Nunc, and disposable plastic pipettes and 15-ml tubes (Cat. No. 2095) from Falcon Labware. Ultraclear centrifuge tubes (14 × 95 mm, for an SW40 rotor) are from Beckman (Cat. No. 344060), a 20-ml linear gradient mixing chamber (Cat. No. 42020) from Hoelzel, and an Auto Densiflow IIC gradient maker from Buchler Instruments. The cell cracker is that described by Balch *et al.* (1984) (prepared by the EMBL workshop, Heidelberg) and is used with a ball with an 18-μm clearance. Plastic reaction tubes (1.5 ml, Cat. No. 3810) are from Eppendorf. These should be prepared for collecting 1.1-ml gradient fractions by marking the 1.1-ml level by comparison with a standard obtained by filling one tube with 1.1 ml of distilled H$_2$O. Twelve or 13 Eppendorf tubes are required for each gradient used (48–52 tubes per secretory granule preparation).

In addition, the following equipment is required: 22-gauge needles, 1-ml plastic syringes, 15-ml Corex tubes, and a cell scraper, which can be prepared from a slice of a silicone stopper (approximately 2 cm in diameter) attached to a plastic pipette.

III. PROCEDURE

A. Cell Culture

For one secretory granule preparation, six to eight plates (24 × 24 cm) of subconfluent PC12 cells (clone 251) (Heumann *et al.*, 1983; see Volume 1, Teng *et al.*, "Cultured PC12 Cells: A Model for Neuronal Function, Differentiation, and Survival" for additional information) are required.

Solution

1. *Growth medium:* DMEM supplemented with 5% FCS and 10% HS.

Steps

1. Culture cells in growth medium at 37°C in 10% CO$_2$.

2. Passage cells at dilutions of 1:5 or 1:6 and plate them 5–6 days in advance of the experiment.

B. Long-Term [^{35}S]Sulfate Labeling of PC12 Cells

For monitoring the distribution of the secretory granules during the purification, at least one dish of cells should be incubated with [^{35}S]sulfate to label the sulfated secretory granule content proteins, CgB and SgII.

Solution

1. *Labeling medium:* Supplement DMEM that lacks cysteine and methionine and in which MgSO$_4$ is replaced with MgCl$_2$, with 1% of the normal DMEM concentration of cysteine, 1% of the normal DMEM concentration of methionine, 0.05% FCS and 0.1% HS. The FCS and HS should both be dialyzed against phosphate-buffered saline (PBS).

Steps

1. Remove growth medium from the cells and replace with 20 ml per plate of labeling medium.

2. Add 0.15–0.25 mCi/ml carrier-free [^{35}S]sulfate to each plate.

3. Incubate at 37°C, 10% CO_2 for 12–16 hr.

C. Gradient Preparation

Gradients are prepared immediately before cell homogenization and stored at 4°C until used. Four first gradients (three for samples, one as balance) and two second gradients (one for the sample, one as a balance) are required for one secretory granule preparation.

Solutions

1. *2.0 M sucrose stock solution:* Dissolve 68.46 g of sucrose in a small amount of double-distilled (dd) H_2O, then adjust to a final volume of 100 ml. The solution can be aliquoted and stored at −20°C. The molarity of the sucrose solution can be verified by refractometry.

2. *Sucrose solutions for density gradients:* 0.3, 1.1, 1.2, and 2.0 M sucrose solutions, containing 1 μg/ml each of aprotinin and leupeptin (from a 1 mg/ml stock in ddH_2O) and 0.5 mM PMSF (from a 250 mM stock in 100% ethanol) are prepared as described in Table I.

Steps

First (0.3–1.2 *M* Continuous Sucrose) Gradient

1. Connect a Buchler Auto Densi-flow gradient maker IIC to a 20-ml gradient mixing chamber placed on a magnetic stirrer containing a stirring bar in the output chamber.

2. Place an SW40 tube in the gradient maker, lower the probe to the bottom of the tube, and put the probe in the "up" setting. Set the pump of the gradient maker to "4".

3. Place 6.0 ml of 1.2 *M* sucrose in the output chamber and 5.5 ml of 0.3 *M* sucrose in the other chamber, then switch on the stirrer for gentle mixing.

4. Switch the gradient maker to "deposit." Allow the 1.2 *M* sucrose to exit for about 5–10 sec, then open the channel connecting the two chambers of the gradient mixer, allowing the 0.3 *M* sucrose to enter the output chamber.

5. Check that the sucrose solutions are mixing evenly while being deposited into the tube, until all the sucrose has been used.

Second (1.1–2.0 *M* Continuous Sucrose) Gradient

1. Prepare two continuous sucrose second gradients as described earlier, using 5.5 ml of 1.1 *M* and 5.0 ml of 2.0 *M* sucrose.

TABLE I Sucrose Solutions for Density Gradients[a]

	2.0 *M* sucrose	Distilled H_2O[b]	1 mg/ml aprotinin/leupeptin	250 m*M* PMSF
First gradient				
0.3 *M* sucrose	6 ml	34 ml	40 μl	80 μl
1.2 *M* sucrose	24 ml	16 ml	40 μl	80 μl
Second gradient				
1.1 *M* sucrose	11 ml	9 ml	20 μl	40 μl
2.0 *M* sucrose	20 ml	0 ml	20 μl	40 μl

[a] Solutions are made fresh.
[b] 10 m*M* HEPES-KOH, pH 7.4, may be used instead of ddH_2O.

D. Preparation of the Postnuclear Supernatant

See also Fig. 1. All subsequent steps should be performed at 4°C, and all solutions, centrifuge tubes, and the cell cracker should be precooled to 0°C.

Solutions

1. *Tris-buffered saline (TBS):* 25 mM Tris–HCl, pH 7.4, 137 mM KCl, 0.6 mM Na$_2$HPO$_4$. For 1 liter, dissolve 0.4 g KCl, 3.0 g Trizma base, 8.0 g NaCl, and 0.1 g Na$_2$HPO$_4$ in ≈900 ml ddH$_2$O, adjust the pH to 7.4 with 1 N and then 0.1 N HCl, and adjust the volume to 1 liter with ddH$_2$O. The solution can be aliquoted and stored at −20°C.

2. *Homogenization buffer (HB):* 0.25 M sucrose, 10 mM HEPES, pH 7.4, 1 mM EDTA, containing 1 μg/ml each of aprotinin and leupeptin, and 0.5 mM PMSF. Mix 6.25 ml 2.0 M sucrose stock solution, 0.5 ml 1.0 M HEPES–KOH, pH 7.4, 0.5 ml 100 mM EDTA–NaOH, pH 7.4, and 42.75 ml ddH$_2$O. To the final solution, add 50 μl of 1 mg/ml aprotinin and 1 mg/ml leupeptin in ddH$_2$O and 100 μl of 250 mM PMSF in 100% ethanol. The buffer is made fresh.

Steps

1. Place the plates on ice, remove the growth or labeling medium, and wash each plate three times with 15–20 ml of TBS.

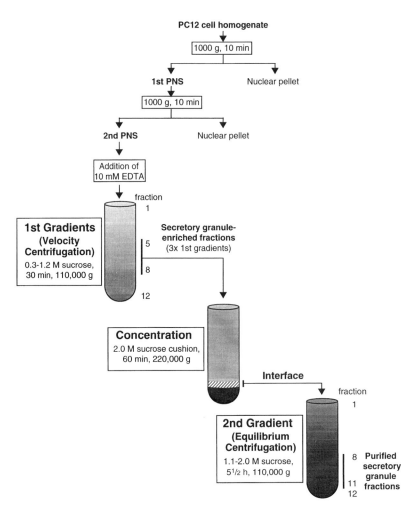

FIGURE 1 Flow scheme illustrating the secretory granule purification procedure.

2. Remove the cells in 10 ml TBS per plate, using the cell scraper. Place the cell suspension from one plate into one 15-ml Falcon tube and pellet the cells by centrifuging for 5 min at 700 g (800 rpm in a Heraeus Minifuge RF).

3. Resuspend each pellet in 1 ml of HB, divide the total suspension between three preweighed (see below) 15-ml Falcon tubes, and adjust each volume to 6 ml. Centrifuge for 5 min at 1700 g (1800 rpm in a Heraeus Minifuge RF).

4. Remove the supernatant. Weigh the tube with the pellet and determine the weight of the pellet by subtracting the weight of the empty tube (see earlier). Resuspend each pellet in 1.5–2 vol of HB. Pool the resuspended pellets. Typically, this yields $\approx$6.0 ml of total cell suspension when starting with six to eight subconfluent 24 $\times$ 24-cm plates of PC12 cells.

5. Pass the cell suspension, in 1-ml aliquots, six times through a 22-gauge needle attached to a disposable 1-ml plastic syringe to disperse cell clumps. Homogenize the cells by passing 1-ml aliquots through the cell cracker until the maximum number of cells is disrupted and the minimum number of nuclei is destroyed. This is typically the case after four to six single passes, and should be determined by phase-contrast microscopy of a small aliquot of the homogenate after staining with trypan blue. After homogenization of each 1-ml aliquot, remove any material remaining within the cell cracker by suction with a syringe and add to the homogenate. In addition, wash out the cell cracker with a minimal volume of HB (about 0.5–0.8 ml) between every two to three 1-ml aliquots, and add the wash to the homogenate.

6. Place the final homogenate (typically 7.0–7.5 ml) in one 15-ml Corex centrifuge tube and centrifuge for 10 min at 1000 g (3000 rpm in a precooled Sorvall SS34 rotor).

7. Carefully remove the resulting postnuclear supernatant (PNS) from the nuclear pellet, transfer it into one fresh 15-ml Corex tube, and recentrifuge for 10 min at 1000 g. Remove the second PNS ($\approx$4 ml, $\approx$14-mg/ml protein) carefully and place in a 15-ml Falcon tube.

8. To prevent contamination of the final secretory granule preparation with rough endoplasmic reticulum (RER) (see Section IV) raise the EDTA concentration of the second PNS to 11 mM EDTA by adding 20 μl of a 500 mM EDTA–NaOH, pH 7.4 solution, for every ml of PNS to the side of the tube, then gently mixing the sample by inversion. Count a 10- to 30-μl aliquot to determine the total amount of [^{35}S]sulfate-labeled material, and take a 1- to 5-μl aliquot for protein analysis by SDS–PAGE.

E. First Gradient (Velocity Centrifugation)

See also Fig. 1.

Steps

1. Divide the EDTA-treated PNS into three 1.2- to 1.4-ml aliquots, and load each onto the top of a first (0.3–1.2 M sucrose) gradient.

2. Centrifuge the gradients at 110,000 g_{max} (25,000 rpm for the Beckman SW40 rotor) with the centrifuge set at maximum acceleration and with the brake on, for 30 min timed from when the rotor reaches full speed.

3. Collect 1.1-ml aliquots from the top of the gradient into premarked Eppendorf tubes using the Buchler Auto Densi-flow set at "4–6." Use the most dense fraction (No. 12) to resuspend the pellet. Count an aliquot (50–150 μl) of each fraction to identify the fractions containing [^{35}S]sulfate-labeled CgB and SgII, i.e., secretory granules. The secretory granule peak is typically found in fractions 3–7, corresponding to a sucrose concentration of 0.4–0.9 M (Figs. 2A and 2B). Remove a 5- to 10-μl aliquot from each fraction for SDS–PAGE minigel analysis of the protein distribution across the gradient.

F. Concentration of Secretory Granules for the Second Gradient

See also Fig. 1.

Solution

1. *0.7 M sucrose solution:* For 10 ml of 0.7 M sucrose solution, dilute 3.5 ml of 2.0 M sucrose stock with 6.5 ml of ddH$_2$O.

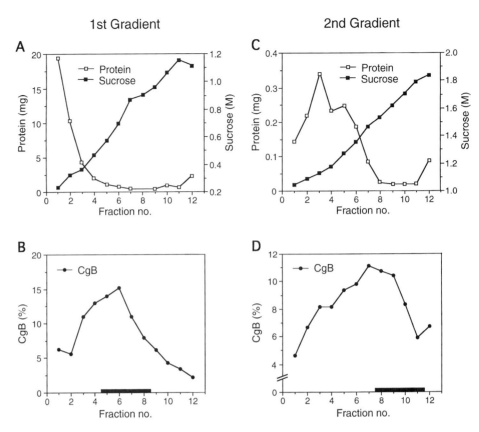

1st Gradient / 2nd Gradient

FIGURE 2 Distribution of secretory granules across the first and second gradients, as indicated by CgB quantitated after immunoblotting (Stinchcombe and Huttner, in preparation). The amount of CgB in each fraction (1 = top) is expressed as a percentage of total per gradient. The bar in B indicates the fractions typically taken for further purification; the bar in D indicates the purified secretory granule fractions.

Steps

1. From the three first gradients, pool fractions 5–8, which are relatively enriched in secretory granules [compare the total protein (Fig. 2A) with CgB (Fig. 2B). Gently mix the pooled fractions ($\approx$13 ml, $\approx$0.7 M sucrose average).

2. Load the pooled material onto a single 700-μl cushion of 2.0 M sucrose at the bottom of an SW40 centrifuge tube.

3. Centrifuge for 60 min at 220,000 g_{max} (36,000 rpm for the Beckman SW40 rotor). After this a thick band should be visible at the cushion interface.

4. Remove most of the material above the interface band, leaving $\approx$0.5 ml of fluid above the interface band. Carefully collect the interface band plus the residual fluid above it, but minimize the amount of 2.0 M sucrose cushion collected.

5. Check that the density of the collected material is less than 1.1 M sucrose by placing a drop of the material onto the top of the second (1.1–2.0 M sucrose) gradient. If the material enters the gradient, carefully reduce the density of the sample by adding small aliquots of fresh 0.7 M sucrose until the sample no longer sinks into the second gradient.

G. Second Gradient (Equilibrium Centrifugation)

See also Fig. 1.

Steps

1. Load the sample onto the top of the second (1.1–2.0 M sucrose) gradient. Centrifuge the gradient for 5.5 hr at 110,000 g_{max} (25,000 rpm for the Beckman SW40 rotor), with the centrifuge set at maximum acceleration and with the brake on.

TABLE II Recovery and Enrichment of CgB in the Purified Granule Preparation

	Protein		CgB			
	Total (mg)	%	Total (arbitrary units)	%	CgB/protein (arbitrary units/mg)	Purification (fold)
PNS	46.500	100.0	139,793	100.0	3,006	1
First gradient						
Total	42.670	91.8	134,390	96.1	3,150	1
Fractions 5–8	2.770	6.0	64,490	46.1	23,282	8
Interface	1.490	3.2	52,715	37.7	35,379	12
Second gradient						
Total	1.613	3.5	47,868	34.2	29,676	10
Fractions 8–11	0.084	0.2	16,933	12.1	201,583	67

2. Collect 1.1-ml aliquots into premarked Eppendorf tubes as before. Take 5- to 30-μl aliquots of each gradient fraction for SDS–PAGE analysis of the distribution and purity of secretory granules (see Section IV). Store the remainder of the fractions at 4°C or frozen, as appropriate.

3. Analyze the gradient fractions by SDS–PAGE using minigels. After electrophoresis, the gel can be (i) stained with silver (Heukeshoven and Dernick, 1988) (CgB and SgII appear blue, other proteins brown) or Coomassie blue, to analyze the total protein profile including CgB and SgII; (ii) subjected to fluorography to detect [^{35}S]sulfate-labeled CgB and SgII; or (iii) processed for Western blotting to detect CgB and SgII immunologically. The purest secretory granule material is typically found in fractions 8–11, corresponding to a sucrose concentration of 1.55–1.8 M (Figs. 2C and 2D).

IV. COMMENTS

The characteristics of the secretory granule preparation obtained by the procedure described here are shown in Table II and Fig. 3. The purified secretory granule preparation has a

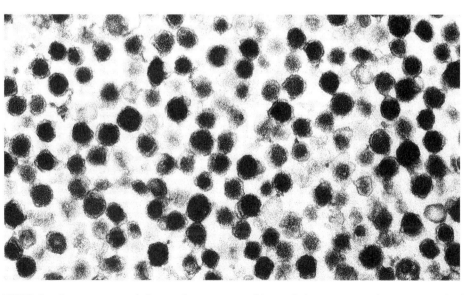

FIGURE 3 Electron micrograph showing the appearance of the purified secretory granule preparation. The diameter of the dense cores is ≈80–120 nm.

protein concentration of ≈ 20 μg/ml. Using CgB as marker, $\approx 12\%$ of the secretory granules present in the EDTA-treated PNS are recovered in the purified secretory granule preparation (Table II). The purification factor, calculated from the ratio of CgB to total protein, is ≈ 70-fold as compared with the EDTA-treated PNS (Table II). By electron microscopy, at least 90% of the structures are indistinguishable from dense-cored secretory granules of intact PC12 cells (Fig. 3), and thus the secretory granule preparation approaches morphological homogeneity by this criterion. The yield of secretory granules can be increased, although at the expense of purity, by taking the entire secretory granule peak from the first and second gradients (see Fig. 2).

Raising the EDTA concentration in the PNS before velocity centrifugation is crucial. This treatment reduces the buoyant density of the RER by dissociating the membrane-bound polysomes. Without this treatment, significant amounts of RER are recovered in the secretory granule preparation (Stinchcombe, 1992). EDTA treatment cannot be carried out with the homogenate as nuclei will lyse and the released DNA will aggregate organelles.

The procedure described in this article has been developed for PC12 cells. The principles of the procedure should, however, be applicable to the purification of secretory granules from other cell lines if the specific parameters are adjusted to the cell and secretory granule type under study.

V. PITFALLS

1. Too stringent homogenization of the cells will result in lysis of nuclei, release of DNA, and aggregation of membranes.

2. Sudden reduction of the sucrose concentration by dilution, and freezing and thawing, may result in lysis of secretory granules.

References

Balch, W. E., Dunphy, W. G., Braell, W. A., and Rothman, J. E. (1984). Reconstitution of the transport of protein between successive compartments of the Golgi measured by the coupled incorporation of N-acetylglucosamine. *Cell* 39, 405–416.

Bauerfeind, R., and Huttner, W. B. (1993). Biogenesis of constitutive secretory vesicles, secretory granules, and synaptic vesicles. *Curr. Opin. Cell Biol.* 5, 628–635.

Burgess, T. L., and Kelly, R. B. (1987). Constitutive and regulated secretion of proteins. *Annu. Rev. Cell. Biol.* 3, 243–293.

Castle, J. D., Cameron, R. S., Arvan, P., von Zastrow, M., and Rudnick, G. (1987). Similarities and differences among neuroendocrine, exocrine and endocytic vesicles. *Ann. N.Y. Acad. Sci.* 493, 448–459.

Cutler, D. F., and Cramer, L. P. (1990). Sorting during transport to the surface of PC12 cells: Divergence of synaptic vesicle and secretory granule proteins. *J. Cell Biol.* 110, 721–730.

Greene, L. A., and Tischler, A. S. (1976). Establishment of a noradrenergic clonal line of rat adrenal pheochromocytoma cells which respond to nerve growth factor. *Proc. Natl. Acad. Sci. USA* 73, 2424–2428.

Heukeshoven, J., and Dernick, R. (1988). Improved silver staining procedure for fast staining in Phast System Development Unit. I. Staining of sodium dodecyl sulfate gels. *Electrophoresis* 9, 28–32.

Heumann, R., Kachel, V., and Thoenen, H. (1983). Relationship between NGF-mediated volume increase and "priming effect" in fast and slow reacting clones of PC12 pheochromocytoma cells. *Exp. Cell Res.* 145, 179–190.

Huttner, W. B., Gerdes, H.-H., and Rosa, P. (1991a). Chromogranins/secretogranins: Widespread constituents of the secretory granule matrix in endocrine cells and neurons. *In* "Markers for Neural and Endocrine Cells: Molecular and Cell Biology, Diagnostic applications" (M. Gratzl and K. Langley, eds.), pp. 93–131. VCH, Weinheim.

Huttner, W. B., Gerdes, H.-H., and Rosa, P. (1991b). The granin (chromogranin/secretogranin) family. *Trends Biochem. Sci.* 16, 27–30.

Kelly, R. B. (1993). Storage and release of neurotransmitters. *Cell* 72(Suppl.), 43–53.

Roda, L. G., Nolan, J. A., Kim, S. V., and Hogue-Angeletti, R. A. (1980). Isolation and characterisation of chromaffin granules from a pheochromocytoma (PC12) cell line. *Exp. Cell Res.* 128, 103–109.

Rosa, P., Hille, A., Lee, R. W. H., Zanini, A., De Camilli, P., and Huttner, W. B. (1985). Secretogranins I and II: Two tyrosine-sulfated secretory proteins common to a variety of cells secreting peptides by the regulated pathway. *J. Cell Biol.* 101, 1999–2011.

Rosa, P., Weiss, U., Pepperkok, R., Ansorge, W., Niehrs, C., Stelzer, E. H. K., and Huttner, W. B. (1989). An antibody against secretogranin I (chromogranin B) is packaged into secretory granules. *J. Cell Biol.* **109**, 17–34.

Stinchcombe, J. C. (1992). "The Purification and Characterisation of Secretory Storage Granules from PC12 Cells." Ph.D. thesis, Council of National Academic Awards (U.K.).

Tooze, S. A., Chanat, E., Tooze, J., and Huttner, W. B. (1993). Secretory granule formation. *In* "Mechanisms of Intracellular Trafficking and Processing of Proproteins" (Y. Peng Loh, ed.), pp. 157–177. CRC Press, Boca Raton, FL.

Wagner, J. A. (1985). Structure of catecholamine secretory vesicles from PC12 cells. *J. Neurochem.* **45**, 1244–1253.

Winkler, H., Apps, D. K., and Fischer-Colbrie, R. (1986). The molecular function of adrenal chromaffin granules: Established facts and unresolved topics. *Neuroscience* **18**, 261–290.

Preparation of Synaptic Vesicles from Mammalian Brain

Johannes W. Hell and Reinhard Jahn

I. INTRODUCTION

Synaptic vesicles are secretory organelles that store the neurotransmitter in presynaptic nerve endings. When an action potential arrives in the nerve terminal, the plasma membrane is depolarized, leading to the opening of voltage-gated Ca^{2+} channels. The rise in intracellular Ca^{2+} concentration leads to the exocytosis of synaptic vesicles within a time interval that can be as short as 200 μsec (reviewed by Südhof, 1995).

Synaptic vesicles possess several remarkable properties that distinguish them from most other organelles involved in membrane traffic. First, they are very abundant in brain tissue. Model calculations show that an average neuron contains approximately 10^6 synaptic vesicles, with a total of around 10^{17} in the human central nervous system (CNS) (Jahn and Südhof, 1993). Approximately 5% of the protein of CNS tissue is contributed by synaptic vesicles; thus, about a 20-fold enrichment from the homogenate is sufficient to obtain a pure preparation. Second, synaptic vesicles are highly homogenous in size and shape and, in addition, are smaller than most other organelles, with an average diameter of only 50 nm. Therefore, size-fractionation techniques can be applied for the isolation of synaptic vesicles. Third, synaptic vesicles do not contain a matrix of soluble proteins (Jahn and Südhof, 1993) as they recycle many times in the nerve terminal and thus can only be reloaded with nonpeptide transmitters by means of specific transport systems.

The study of synaptic vesicles has been facilitated by recent advances in understanding their protein composition. To date, more than half a dozen protein families have been shown to be localized specifically on the membrane of synaptic vesicles. With the exception of some variation due to isoforms, most of these proteins are residents of all synaptic vesicles irrespective of their neurotransmitter content or of the location of the neuron. These include the synapsins, synaptophysins, synaptotagmins, SV2s, synaptobrevins/VAMPs, rabs, cysteine string proteins, synaptogyrin, and the subunits of the vacuolar proton pump (for a review see Südhof, 1995). Rapid progress during the last few years has clarified the function of many of these proteins in membrane traffic. Synaptic vesicles are thus presently regarded as the best characterized "model" trafficking organelle of eukaryotic cells (Südhof, 1995). Antibodies against synaptophysin and synaptobrevin/VAMP are commercially available from several sources and may be used as probes to assess synaptic vesicle purity. In addition, synaptic vesicles contain neurotransmitter transporters that are specific for neurons exhibiting the corresponding neurotransmitter phenotype. Two of these transporters (the monoamine transporter and the acetylcholine transporter) have been cloned and sequenced (Peter *et al.*, 1995; Usdin *et al.*, 1995).

Purification protocols for synaptic vesicles can be divided into three groups. The first group involves the preparation of isolated nerve terminals (synaptosomes) by differential centrifugation that are subsequently lysed in order to release the synaptic vesicles. An advantage of this procedure is that small membrane fragments generated during homogenization are removed prior to vesicle extraction as synaptosomes sediment at lower centrifugal forces than these fragments. The first protocol described here belongs in this category. These protocols are laborious and result in relatively low yields, but the resulting synaptic vesicle preparations are of exceptionally high purity. In the second group of protocols, synaptic vesicles are directly purified from homogenate, without prior isolation of synaptosomes. In order to obtain high yields, initial homogenization is harsh in order to break up as many nerve terminals as possible, e.g., by freeze-powder homogenization as described in the second procedure. In both cases, a combination of differential centrifugation, rate-zonal density gradient centrifugation or isopycnic density gradient centrifugation, and size exclusion chromatography is employed for purification (Fig. 1). The third group involves immunoisolation using antibodies specific for synaptic vesicle proteins (see, e.g., Burger *et al.*, 1989; Walch-Solimena *et al.*, 1993). These procedures allow for the rapid isolation of small quantities of highly pure organelles from brain homogenates. However, they require access to large amounts of specific antibodies (preferably monoclonals) and will therefore not be further discussed here.

II. MATERIALS AND INSTRUMENTATION

The following chemicals are used: HEPES (American Bioanalytical, Cat. No. AB892), sucrose (J.T. Baker, Cat. No. 4072-05), glycine (Bio-Rad, Cat. No. 161-0718), phenylmethylsulfonyl fluoride (PMSF, Pierce, Cat. No. 36978), pepstatin A (ICN, Cat. No. 0219536805), and dimethyl sulfoxide (American Type Culture Collection, Cat. No. 442608). Controlled pore glass beads are from CPG Inc. (see Appendix). Note that the standard reagents can also be obtained from other sources.

The following instrumentation is required: a loose-fitting, motor-driven glass–Teflon homogenizer (Braun), a cooled centrifuge [Sorvall RC5 (DuPont) or comparable, SS34 rotor], an ultracentrifuge with fixed angle and swing-out rotors [Beckman L80 (Beckman Instruments) or comparable, Ti70 or Ti50.2 rotor, SW-28 rotor, Ti 45 rotor] and corresponding tubes, equipment for column chromatography (peristaltic pump, UV monitor, fraction collector), a gradient mixer for forming continuous sucrose gradients, a filtration device for the filtration of buffers using 0.45-μm membranes (Millipore), and glass columns (see Appendix).

III. PROCEDURES

A. Preparation of Synaptic Vesicles from Synaptosomes

In this protocol (Nagy *et al.*, 1976; Huttner *et al.*, 1983), a crude synaptosomal fraction (P2) is first isolated by differential centrifugation. The synaptosomes are then lysed by osmotic shock and synaptic vesicles are released into the medium. After the removal of synaptosomal fragments and large membranes, synaptic vesicles are sedimented by high-speed centrifugation. The resulting pellet, already five- to sixfold enriched in synaptic vesicles, is then further purified by sucrose velocity-density gradient centrifugation and size-exclusion chromatography on controlled-pore glass beads (CPG). This procedure is the standard method for obtaining synaptic vesicles of the highest purity, with less than 5% contamination as judged by electron microscopy and biochemical analysis. This preparation does contain, however, decoated coated vesicles that lost their clathrin but retained adaptors and endosomes derived from nerve terminals. The degree of this contamination is probably minor but cannot be easily quantified (Fig. 1).

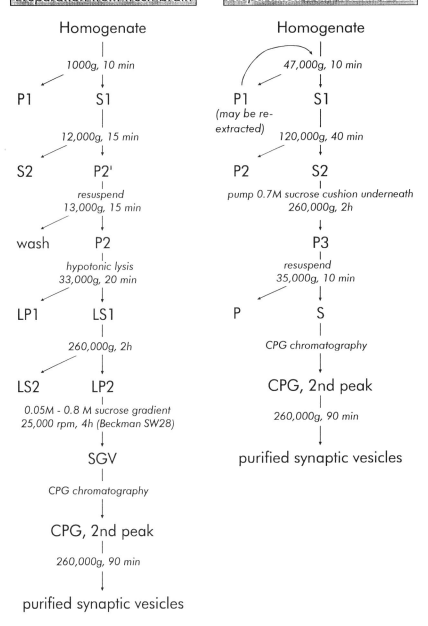

Preparation from fresh brain

Homogenate
|
1000g, 10 min
|
P1   S1
|
12,000g, 15 min
|
S2   P2'
|
resuspend
13,000g, 15 min
|
wash   P2
|
hypotonic lysis
33,000g, 20 min
|
LP1   LS1
|
260,000g, 2h
|
LS2   LP2
|
0.05M - 0.8 M sucrose gradient
25,000 rpm, 4h (Beckman SW28)
|
SGV
|
CPG chromatography
|
CPG, 2nd peak
|
260,000g, 90 min
|
purified synaptic vesicles

Preparation from frozen brain

Homogenate
|
47,000g, 10 min
|
P1   S1
(may be re-
extracted)
|
120,000g, 40 min
|
P2   S2
|
pump 0.7M sucrose cushion underneath
260,000g, 2h
|
P3
|
resuspend
35,000g, 10 min
|
P   S
|
CPG chromatography
|
CPG, 2nd peak
|
260,000g, 90 min
|
purified synaptic vesicles

FIGURE 1 Flow chart depicting the main steps of the two preparation methods for synaptic vesicles described in the text.

Solutions

1. *Homogenization buffer*: 320 mM sucrose, 4 mM HEPES–NaOH, pH 7.3 (HEPES is optional; we found no difference when the buffer is omitted. Other buffers such as MES and slightly lower pH are also acceptable).

2. *1 M HEPES–NaOH, pH 7.4.*

3. *40 mM sucrose.*

4. *50 mM sucrose.*

5. *800 mM sucrose.*

6. *Glycine buffer*: 300 mM glycine, 5 mM HEPES–KOH, pH 7.4, degassed and filtered through a 0.45-μm filter.

7. *Protease inhibitors:* 1 mg/ml pepstatin A in dimethyl sulfoxide and 200 mM phenyl-methylsulfonyl fluoride in dry ethanol. Keep stocks at room temperature. Add 1/1000 vol where indicated. Note that PMSF is unstable in aqueous solutions and should not be added to buffers prior to use.

Steps

After collecting the brains, all steps are carried out on ice or at 4°C.

1. Decapitate 20 rats (180–200 g); remove the brains, avoiding myelin-rich areas such as corpus callosum or medulla oblongata; place into 180 ml ice-cold homogenization buffer; and homogenize in several aliquots with a loose-fitting glass–Teflon homogenizer (nine strokes, 900 rpm). Add protease inhibitors.

2. Centrifuge the homogenate for 10 min at 1000 g_{max} (2700 rpm in a Sorval SS34 rotor); discard the resulting pellet (P1) containing large cell fragments and nuclei, and collect the supernatant (S1).

3. Centrifuge S1 for 15 min at 12,000 g_{max} (SS34 rotor; 10,000 rpm); remove the supernatant (S2) containing small cell fragments such as microsomes or small myelin fragments and soluble proteins. Wash the P2 by carefully resuspending in 120 ml homogenization buffer (pipette, avoiding the dark brown bottom part of the pellet that consists mainly of mitochondria) and recentrifugation at 11,000 rpm (SS34 rotor; 13,000 g_{max}); discard the supernatant (S2′). The resulting pellet (P2′) represents a crude synaptosomal fraction.

4. To release synaptic vesicles from the synaptosomes, resuspend P2′ in homogenization buffer to yield a final volume of 12 ml. Transfer this fraction into a glass–Teflon homogenizer, add 9 vol (108 ml) ice-cold water and perform three up-and-down strokes at 2000 rpm. Add 1 ml of 1 M HEPES–NaOH, pH 7.4, and protease inhibitors.

5. Centrifuge the suspension for 20 min at 33,000 g_{max} (16,500 rpm, SS34 rotor) to yield the lysate pellet (LP1) and the lysate supernatant (LS1). Using an electric pipetter, carefully remove LS1 immediately after the end of the run without disturbing LP1. It is crucial that LS1 does not get contaminated even with traces of membrane fragments from LP1 (rather, leave 1–2 ml behind in the tube): this is the most common error which significantly reduces the purity of the final vesicle fraction.

6. Centrifuge LS1 for 2 hr at 260,000 g_{max} (50,000 rpm in a Beckman 60Ti or comparable rotor). Discard the supernatant (LS2) and resuspend the pellet (LP2) in 6 ml of 40 mM sucrose, utilizing a small, tight-fitting glass–Teflon homogenizer, followed by extruding the sample consecutively through a 23-gauge and a 27-gauge hypodermic needle attached to a 10 ml syringe (avoid air bubbles).

7. Layer the suspension (3-ml aliquots) on top of a linear sucrose gradients, formed from 18.5 ml of 800 mM sucrose and 18.5 ml of 50 mM sucrose (prepare two tubes containing identical gradients in advance), and centrifuge for 4 hr at 25,000 rpm in a Beckman SW-28 rotor (65,000 g_{av}). After the run a turbid (white-opaque) zone is visible in the middle of the gradient (in the range of 200 to 400 mM sucrose, best visible when viewed against a black background with light from the top). These bands are collected with the aid of a glass capillary connected to a peristaltic pump, yielding a combined volume of 25–30 ml. This fraction represents synaptic vesicles that are 8- to 10-fold enriched over the homogenate (Jahn *et al.*, 1985). Note that synaptic vesicles do not reach isopycnic equilibrium during centrifugation (velocity-gradient centrifugation); changes of angular velocity or of the run time will therefore affect the result.

8. Equilibrate a CPG-3000 column (180 × 2 cm, see Appendix) with 10 column volumes of glycine buffer (optimally done overnight before the preparation). Load the sample on top of the resin and overlay it carefully with glycine buffer without diluting the sample. Elute the column with glycine buffer at a flow rate of 40 ml/hr, collecting 6- to 8-ml fractions. Monitor protein efflux at 280 nm. The first peak, which contains plasma membranes and some microsomes, is usually smaller than the second peak containing synaptic vesicles. If no separation into two clearly distinguishable peaks is obtained, the column may need to be repacked. Fractions of the second peak are pooled and centrifuged for 90 min at 260,000 g_{max} (50,000 rpm, Beckman 60Ti rotor). The synaptic vesicle pellet should have a glassy appearance, being completely transparent and colorless. It is resuspended in the desired buffer as in step 6. The suspension is frozen rapidly (e.g., in liquid nitrogen) and stored at $-70°C$. Yields are typically between 2 and 3 mg of protein, based on one of the commercially available Coomassie blue protein determination kits.

Note: Size-exclusion chromatography on glyceryl-coated CPG beads or on Sephacryl S-1000 is omitted in many protocols and, if applied, is the last step of the procedure. Both resins have a relatively low capacity, do not tolerate overloading, and require some experience in their use. Sephacryl S-1000 has a higher separation capacity than CPG per gel volume, but the columns have low flow rates, do not tolerate increased pressure, and have a tendency to adsorb proteins and membrane particles, particularly during the first few separation runs in the life of the column. CPG columns are more difficult to set up and tolerate less material. However, glass beads are noncompressible and allow high flow rates, substantially shortening separation times. The experimenter who does not shy away from the effort to set up a large CPG column is rewarded with highly reliable results for many runs and exceptionally clean synaptic vesicles preparations. In our laboratory, we utilized a CPG-3000 column (3×180 cm) continuously for 10 years for more than 200 synaptic vesicles preparations, with only a few repackings required, usually caused by experimental error (running dry). Column profiles and synaptic vesicle purity were highly reproducible.

B. Preparation of Synaptic Vesicles from Frozen Brain

This procedure starts with a harsh homogenization of frozen brains to efficiently break up the nerve terminals, thus releasing synaptic vesicles. Frozen brains are ground in a precooled mortar to yield a fine powder. This treatment does not affect the function or integrity of the small synaptic vesicles, but larger membrane structures are ruptured. After resuspending the tissue powder in sucrose solution, most of the cell fragments are removed by centrifugation with low and intermediate angular velocities, leaving synaptic vesicles in the supernatant. Synaptic vesicles are then sedimented at a high speed through a cushion of 0.7 M sucrose, removing soluble proteins and membrane contaminants of lower buoyant density (mostly myelin). Synaptic vesicles are five- to sixfold enriched in the pellet and can be purified further by CPG chromatography (Fig. 1).

The final enrichment factor for synaptic vesicles purified by this protocol is 15–20 (Hell *et al.*, 1988), which is somewhat lower than in the previous method. However, there are several advantages: First, the tissue can be collected before the experiment and can be stored in liquid nitrogen for more than 1 year, allowing a more efficient use of experimental animals. Second, the yield is severalfold higher under optimal conditions than the yield of the preparation from synaptosomes. Third, the procedure is faster, requiring only 12 hr for completion.

Solutions

1. *Homogenization buffer:* 320 mM sucrose, degassed.
2. *700 mM sucrose:* 700 mM sucrose and 10 mM HEPES–KOH, pH 7.3.
3. *Resuspension buffer:* 320 mM sucrose and 10 mM HEPES–KOH, pH 7.3.
4. *Glycine buffer:* 300 mM glycine and 5 mM HEPES–KOH, pH 7.3, degassed.
5. *Protease inhibitors:* 1 mg/ml pepstatin A dissolved in dimethyl sulfoxide and 200 mM PMSF in dry ethanol; add 1/1000 vol where indicated.

Steps

After powdering the frozen brains, all steps are carried out on ice or at 4°C.

1. Decapitate 40 rats (180–200 g body weight); remove the brains, avoiding myelin-rich areas such as corpus callosum or medulla oblongata; and freeze immediately in liquid nitrogen.

Immediate shock freezing is essential. According to our experience, frozen brains available from commercial sources are usually not satisfactory for this reason.

2. To create a tissue powder, place the frozen brains into a porcelain mortar precooled with liquid nitrogen, cover them with cheesecloth, and break them carefully using a porcelain pestle. Grind to a fine powder. This step is crucial for obtaining high yields. After evaporation of the liquid N_2, suspend the powder in 320 ml ice-cold homogenization buffer (magnetic stirrer) and homogenize with a glass–Teflon homogenizer (eight strokes, 1000 rpm).

3. Centrifuge the homogenate for 10 min at 47,000 g_{max} (20,000 rpm in a Sorval SS34 rotor). Collect the S1. Pellet P1 contains large cell fragments and nuclei, but also some entrapped synaptic vesicles. To increase the yield, the pellet may therefore be reextracted with 160 ml homogenization buffer by means of one slow stroke in the glass–Teflon homogenizer followed by centrifugation as described earlier. The resulting supernatant (S1′) is combined with S1.

4. Centrifuge S1 for 40 min at 120,000 g_{max} (32,000 rpm in a Beckman 45Ti rotor). Using an electric pipetter, collect S2 carefully without disturbing P2. It is crucial that S2 is not contaminated with membrane fragments from soft pellet P2. S2 should be clear with a reddish color. If it is turbid, it should be recentrifuged using the same conditions to remove contaminating membrane fragments.

5. To sediment synaptic vesicles through a sucrose cushion, 25-ml centrifuge tubes fitting into a Beckman 60Ti rotor are filled with 20 ml S2. The sucrose cushion is formed by pumping 5.5 ml of 700 mM sucrose underneath S2 using a peristaltic pump and a glass capillary. Centrifuge for 2 hr at 260,000 g_{max} (50,000 rpm, Beckman 60Ti rotor). Remove supernatant S3 and resuspend pellet P3 in 6–10 ml resuspension buffer. This sample represents a crude synaptic vesicles fraction. Clear the suspension by a short spin (17,000 rpm for 10 min, SS34, 35,000 g_{max}) before loading onto the CPG column.

6. Equilibrate a CPG column (see Appendix) with 10 column volumes of glycine buffer. Load the sample on top of the resin and overlay carefully with glycine buffer without disturbing the sample. A column size of 85 × 1.6 cm has a maximal capacity of 15 mg of protein, requiring several consecutive runs if all material is to be chromatographed. Elute the column with glycine buffer at a flow rate of 80 ml/hr collecting 2-ml fractions. Follow the elution of protein with a UV detector at 280 nm. The first peak, containing plasma membranes and microsomes, is usually larger than the second peak containing synaptic vesicles. The two peaks are not completely separated. The shoulder frequently observed at the end of the second peak represents soluble protein. Pool the fractions of the second peak, centrifuge for 2 hr at 260,000 g_{max} (60Ti rotor; 50,000 rpm), and resuspend the synaptic vesicle pellet in the desired buffer.

IV. COMMENTS

Scaling up or down is acceptable, but it should be kept in mind that changing rotors or using half-filled centrifuge tubes may significantly and adversely affect yields and purity. Contamination by other subcellular compartments can be conveniently monitored by assaying for marker enzymes, particularly for plasma membranes, mitochondria, or endoplasmic reticulum (Hell *et al.*, 1988). In parallel to a decrease of these marker enzymes, proteins specific for synaptic vesicles, namely synaptophysin (p38), for which antibodies are commercially available (e.g., from Boehringer-Mannheim), should be enriched about 20- to 25-fold over the homogenate (Jahn *et al.*, 1985), best quantitated by immunoblotting. Synaptobrevin is less reliable for quantitation by SDS–PAGE/immunoblotting as histones present in the homogenate migrate alongside synaptobrevin in nuclei-containing fractions (homogenate) and interfere with the signal, resulting in overestimation of the enrichment factor. The protein profile of the synaptic vesicle preparation, as observed after SDS–PAGE, exhibits a characteristic pattern (Huttner *et al.*, 1983; Hell *et al.*, 1988), with the prominent membrane proteins synaptobrevin/VAMP, synaptophysin (p38), synaptotagmin (p65), and synapsin I being clearly visible. Synaptic vesicle preparations contain various amounts of soluble proteins with affinity for membranes such as glyceraldehyde phosphate dehydrogenase, aldolase, actin, and tubulin. These proteins may be partially removed by a salt wash (resuspend synaptic vesicles in 160 mM KCl, 10 mM HEPES–KOH, pH 7.4, and centrifuge them for 2 hr at 50 000 rpm 260,000 g_{max}). However, this treatment also removes the synaptic vesicle protein synapsin I.

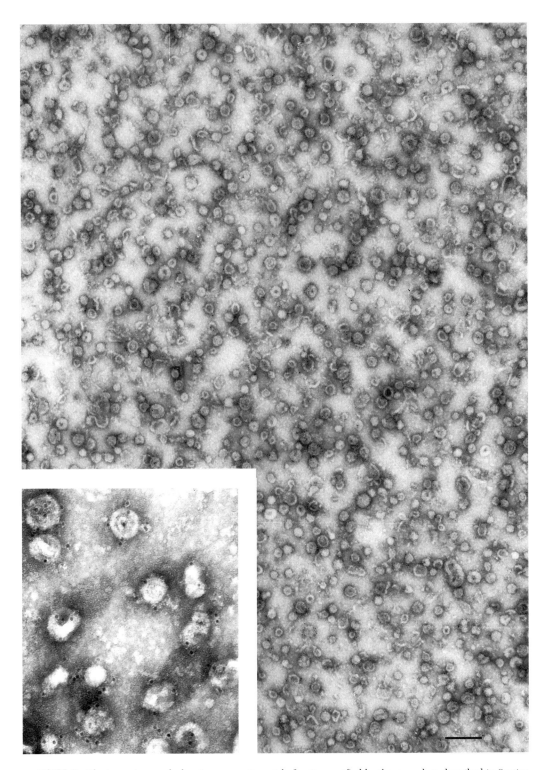

FIGURE 2 Electron micrograph showing a synaptic vesicle fraction purified by the procedure described in Section IIIB (negative staining). The inset shows a magnification of a field following immunogold labeling for the synaptic vesicle protein synaptophysin. For methods, see Jahn and Maycox (1988). Bar: 200 nm. (Photograph courtesy of Dr. Peter R. Maycox, London, UK.)

The morphology of the synaptic vesicles fraction can be studied using electron microscopy (Fig. 2). Membranes can be visualized easily, e.g., by negative staining (Hell *et al.*, 1988). Synaptic vesicle membranes are identified by their very uniform appearance (small vesicular profiles of approximately 50 nm in diameter). Confirmation can also be obtained by immunogold labeling for the vesicle protein synaptophysin, which can be carried out conveniently on a single day when combined with negative staining (Jahn and Maycox, 1988).

V. APPENDIX: PREPARATION AND MAINTENANCE OF CPG COLUMNS

Controlled pore glass beads (CPG 3000, glycerol coated) are obtained from CPG Inc. Column dimensions may vary, but the diameter to length ratio should be at least 1:20–1:50. Use a sturdy, tension-free glass column that withstands mechanical stress during packing.

Steps

1. Resuspend CPG in distilled water and degas thoroughly.

2. Connect the column to a vibration device (e.g., an immobilized vortex apparatus) using a stiff, nonbreakable connection (e.g., a plastic cylinder). Attach the column with several clamps to a strong support such as a heavy stand or wall-mounted rack, with uneven spacing between the clamps to avoid the generation of waves with large amplitudes.

3. Fill the column with the CPG slurry. Vibrate the column at a high speed in the vortex apparatus and keep adding slurry. Avoid the generation of settled zones between additions. After filling, keep vibrating until the resin does not settle for at least 30 min. Packing a small column (85×1.6 cm) requires around 4 hr, whereas a large column (180×2 cm) requires approximately 10 hr.

4. Equilibrate the column with 10 column volumes of column buffer.

5. For maintenance, the column should be stored in buffer containing 0.02% sodium azide to prevent microbial growth. To avoid accumulation of debris, all loaded samples should be cleared by a short centrifugation and all buffers filtered by ultrafiltration. The column should not be allowed to run dry. If this occurs, the air may be removed by pumping large amounts of extensively degassed and temperature-equilibrated column buffer from the bottom through the column. However, if air bubbles remain, repacking is unavoidable. If contaminants accumulate (after 10–20 runs), the column can be cleaned by washing with 1–2 bed volumes of 4 M urea, buffered to pH 7.0, followed by extensive washing. The first run after urea cleaning results in lower yields.

Acknowledgment

The authors thank Dr. Julia Avery (New Haven) for critically reading the manuscript.

References

Burger, P. M., Mehl, E., Cameron, P., Maycox, P. R., Baumert, M., Lottspeich, F., De Camilli, P., and Jahn, R. (1989). Synaptic vesicles immunoisolated from rat cerebral cortex contain high levels of glutamate. *Neuron* **3**, 715–720.

Hell, J. W., Maycox, P. R., Stadler, H., and Jahn., R. (1988). Uptake of GABA by rat brain synaptic vesicles isolated by a new procedure. *EMBO J.* **7**, 3023–3029.

Huttner, W. B., Schiebler, W., Greengard, P., and De Camilli, P. (1983). Synapsin I (protein I), a nerve terminal-specific phosphoprotein. III. Its association with synaptic vesicles studied in a highly purified synaptic vesicle preparation. *J. Cell Biol.* **96**, 1374–1388.

Jahn, R., and Maycox, P. R. (1988). Protein components and neurotransmitter uptake in brain synaptic vesicles. *In* "Molecular Mechanisms in Secretion" (N. A. Thorn, M. Treiman, and O. H. Peterson, eds.), pp. 411–424. Munksgaard, Copenhagen.

Jahn, R., Schiebler, W., Ouimet, C., and Greengard, P. (1985). A 38,000 dalton membrane protein (p38) present in synaptic vesicles. *Proc. Natl. Acad. Sci. USA* **82**, 4137–4141.

Jahn, R., and Südhof, T. C. (1993). Synaptic vesicle traffic: Rush hour in the nerve terminal. *J. Neurochem.* **61,** 12–21.

Nagy, A., Baker, R. R., Morris, S. J., and Whittaker, V. P. (1976). The preparation and characterization of synaptic vesicles of high purity. *Brain Res.* **109,** 285–309.

Peter, D., Liu, Y., Brecha, N., and Edwards, R. H. (1995). The transport of neurotransmitters into synaptic vesicles. *Prog. Brain Res.* **105,** 273–281.

Südhof, T. C. (1995). The synaptic vesicle cycle: A cascade or protein-protein interactions. *Nature* **375,** 645–653.

Usdin, T. B., Eiden, L. E., Bonner, T. I., and Erickson, J. D. (1995). Molecular biology of the vesicular ACh transporter. *Trends Neurosci.* **18,** 218–224.

Walch-Solimena, C., Takei, K., Marek, K., Midyett, K., Südhof, T. C., De Camilli, P., and Jahn, R. (1993). Synaptotagmin: A membrane constituent of neuropeptide-containing large dense-core vesicles. *J. Neurosci.* **13,** 3895–3903.

Method of Centrosome Isolation from Cultured Animal Cells

Mohammed Moudjou and Michel Bornens

I. INTRODUCTION

The centrosome, a central body playing a key role in the temporal and spatial distribution of the interphasic and mitotic microtubule network, could be considered a major determinant of the overall organization of the cytoplasm and of the fidelity of cell division (Bornens, 1992; Kalt and Schliwa, 1993). In animal cells, the centrosome is composed of two centrioles surrounded by the so-called pericentriolar material (PCM), which consists of a complex thin filament network and two sets of appendages (Paintrand et al., 1992). Contrasting wih the importance given to this organelle at early days of cell biology, little is known about the molecular mechanisms that govern its main functions: the nucleation of microtubules and the controlled cycle of its duplication, the two duplicated entities functioning as mitotic spindle poles during subsequent cell division. The biochemical and functional identification of centrosomal components has been difficult for several reasons. Genetic approaches were rare and only recently have a few systems been found to be useful for this strategy (for review see Huang, 1990). Preparative isolation of this organelle was also limited as the centrosome exists as one copy per interphasic cell, and good assays for obtaining enriched fractions of centrosomal proteins were lacking. In the recent years, a number of laboratories have developed protocols for centrosome isolation from cultured cells, fungi, or tissues (Mitchison and Kirschner, 1986; Bornens et al., 1987; Rout and Kilmartin, 1990; Komesli et al., 1989). Indirect strategies were also developed to identify centrosomal components from *Drosophila* embryos (Raff et al., 1993). This article describes a protocol of centrosome isolation from cultured human lymphoblastic cells. This method is largely based on an earlier one (Bornens et al., 1987). Several modifications are introduced to refine both quantitative and qualitative aspects of the method. A recent report using such modified protocole has been published showing both the two-dimentional pattern of centrosomal proteins and the cellular distribution of γ-tubulin (Moudjou et al., 1996). This includes a biochemical and ultrastructural analysis of the centrosomal form of γ-tubulin and further demonstrates that a vast majority of this protein is not centrosome associated.

II. MATERIALS AND INSTRUMENTATION

Culture medium RPMI 1640 is from Eurobio Laboratories. Fetal calf serum (FCS) is from Jacques Boy. Cytochalasin D (Cat. No. C-8273), HEPES (Cat. No. H-3375), Nonidet P-40 (NP-40, Cat. No. N-3516), Tween 20 (Cat. No. P-5927), phenylmethylsulfonyl fluoride

(PMSF, Cat. No. P-7626), EGTA (Cat. No. E-4378), and Triton X-100 are from Sigma. Nocodazole (Cat. No. 152 405) is from ICN. DNase I (Cat. No. 104 159), PIPES (Cat. No. 1359 045), GTP (Cat. No. 414 581), and protease inhibitors aprotinin (Cat. No. 236 624), leupeptin (Cat. No. 1017 101), and pepstatin (Cat. No. 258 286) are from Boehringer. Sucrose (Cat. No. 35579) is from Serva. Tris (Cat. No. 8382) and β-mercaptoethanol (Cat. No. 805740) are from Merck. Glutaraldehyde EM-25% (Cat. No. G004) is from TAAB. Glycerol redistilled (RP, Normapur 99.5%, Cat. No. 24388.295) and $MgCl_2$ (Cat. No. 25 108 295) are from Prolabo. Bovine serum albumin (BSA, Cat. No. 82.045.1, fraction V protease free) is obtained from Miles.

The density gradient fractionator (Model 183) is from Instrumentation Specialities Company (ISCO). Corex tubes (15 ml) were equipped with a plastic adapter (or plug) (homemade) to support a 12-mm round glass coverslip for sedimentation of isolated centrosomes (see Evans *et al.*, 1985). Prior to use, coverslips were washed with a mixture of methanol and hydrochloric acid at a 2:1 ratio for 2 hr and rinsed for 1 hr with distilled water and finally with pure ethanol for at least 2 hr. They were transferred one by one to filter paper to dry and were conserved in a closed petri box until use. The immunofluorescence box was homemade.

The monoclonal anti-α tubulin antibody (Cat. No. N356) is from Amersham. Fluorescein (DTAF)-conjugated affini-Pure donkey anti-rabbit IgGs (Cat. No. 711-015-132) and lissamine rhodamine (LRSC)-conjugated affini-Pure goat anti-mouse IgGs (Cat. No. 115-085-068) are from Jackson ImmunoResearch Laboratories and are used as secondary antibodies. Mounting solution (Citifluor Cat. No. AF1) used to prevent fading of chromophore-labeled materials is from Citifluor Limited.

III. PROCEDURES

A. Centrosome Isolation

Solutions

1. *Cytochalasin D at 5 mg/ml:* To make 10 ml, solubilize 10 mg of cytochalasin D in 2 ml of pure dimethyl sulfoxide (DMSO). Store at 4°C.

2. *$10^{-3}M$ Nocodazole:* To make 10 ml, solubilize 3 mg of Nocodazole in 10 ml of pure DMSO. Store as aliquots at −20°C.

3. *DNase I (1 mg/ml):* To make 10 ml, solubilize 10 mg of DNase I in 10 ml of distilled water. Store as aliquots at 4°C.

4. *TBS buffer:* 10 mM Tris–HCl, pH 7.4, and 150 mM NaCl. To make 1 liter, add 1.2 g of Tris and 8.7 g of NaCl to distilled water, adjust pH to 7.4, and complete the volume to 1 liter. Store at 4°C.

5. *TBS 1/10–8% (w/v) sucrose buffer:* To make 1 liter, add 100 ml of TBS buffer, pH 7.4, and 80 g of sucrose to distilled water and bring to a total volume of 1 liter. It can be freshly prepared or frozen at −20°C.

6. *1 M HEPES pH 7.2, stock solution:* To make 200 ml, add 47.6 g of HEPES to distilled water, adjust pH to 7.2 with 10 N NaOH, and complete the volume to 200 ml. Store at 4°C.

7. *1 M PIPES, pH 7.2, stock solution:* To make 200 ml, add 60.5 g of PIPES to distilled water, adjust pH to 7.2 with 10 N KOH, and complete the volume to 200 ml. Store at 4°C.

8. *1 M $MgCl_2$ stock solution:* To make 100 ml, add 5.84 g of $MgCl_2$ to distilled water. Store at 4°C.

9. *Lysis buffer:* 1 mM HEPES, pH 7.2, 0.5% NP-40, 0.5 mM $MgCl_2$, 0.1% β-mercaptoethanol, protease inhibitors (leupeptin, pepstatin, and aprotinin) at 1 μg/ml each, and 1 mM of PMSF. To make 1 liter, add 1 ml of 1 M HEPES pH 7.2 stock solution, 5 ml of pure NP-40, 0.5 ml of 1 M $MgCl_2$ stock solution, 1 ml of pure β-mercaptoethanol, 1 mg of aprotinin, leupeptin, and pepstatin, and 174.2 mg of PMSF

to distilled water and complete to 1 liter. It can be freshly prepared or conserved as aliquots at −20°C.

10. *Gradient buffer:* 10 mM PIPES pH 7.2, 0.1% Triton X-100, and 0.1% β-mercaptoethanol. To make 1 liter, add 10 ml of 1 M PIPES pH 7.2 stock solution, 1 ml of pure Triton X-100, and 1 ml of pure β-mercaptoethanol to distilled water and complete to 1 liter. Store at 4°C.

11. *Sucrose solutions:*

   a. To make 200 g of sucrose 70% (w/w) solution, weigh 140 g of sucrose and adjust to 200 g with gradient buffer. This gives a 150 ml final volume. Store at −20°C.

   b. To make 400 g of sucrose 60% (w/w) solution, weigh 120 g of sucrose and adjust to 400 g with gradient buffer. This gives a 300 ml final volume. Store at −20°C.

   c. To make 150 g of sucrose 50% (w/w) solution, weigh 75 g of sucrose and adjust to 150 g with gradient buffer. This gives a 120 ml final volume. Store at −20°C.

   d. To make 150 g of sucrose 40% (w/w) solution, weigh 60 g of sucrose and adjust to 150 g with gradient buffer. This gives a 130 ml final volume. Store at −20°C.

12. *10 mM PIPES, pH 7.2:* This is used for the sedimentation of isolated centrosomes on glass coverslips. To make 500 ml, add 5 ml of 1 M PIPES pH 7.2 stock solution to distilled water and complete to 500 ml. Store at 4°C.

Steps

1. Culture the human lymphoblastic KE37 cell line in suspension in RPMI 1640 medium supplemented with 10% FCS at 37°C and 5% CO_2. To 1000 ml of cell suspension (8×10^5 to 1×10^6 cells/ml), add 200 μl of Nocodazole 10^{-3} M and 200 μl of cytochalasin D at 5 mg/ml, and incubate cells for 1 hr at 37°C. All the next steps are done at 4°C except where indicated.

2. Sediment cells by centrifugation at 280 g (1200 rpm) for 8 min. Wash cells with half of the initial cell suspension volume of TBS by gentle resuspension with a 10-ml pipette. Repeat this step once with TBS 1/10–8% sucrose buffer (half volume of the previous step).

3. Before lysis, resuspend cells in 20 ml of TBS 1/10–8% sucrose buffer, and then add the lysis buffer to obtain a concentration of 1×10^7 cells/ml. Shake cells slowly and resuspend them four to five times with a 10-ml pipette until chromatin aggregates become visible.

4. Remove an aliquot of 200 μl to control the quality of the lysis step by immunofluorescence microscopy, and centrifuge down the swollen nuclei, chromatin agregates, and unlysed cells at 2500 g (3500 rpm) for 10 min.

5. Filter the lysis supernatant through a nylon mesh (Crin Polyamide 125 μm Fyltis Motte) into a 250-ml Nalgene tube and add concentrated solutions of 1 M HEPES and DNase I at 1 mg/ml to make a final concentration of 10 mM and 2 units/ml (1 μg/ml) of each component, respectively.

6. Let the suspension sit for 30 min and remove a 200-μl aliquot for immunofluorescence observation of the lysis supernatant.

7. Using a 50-ml syringue and a long needle with a flat end, place 10 ml of the 60% sucrose solution at the bottom of the Nalgene tube and sediment centrosomes onto this cushion by centrifugation at 10,000 g (7500 rpm) for 30 min in a J2-21M/E Beckman centrifuge equiped with a JS 7.5 rotor.

8. During the preceding centrifugation, prepare a discontinuous sucrose gradient in a 38-ml SW28 Beckman ultraclear tube containing, from the bottom, 5, 3, and 3 ml of 70, 50, and 40% sucrose solutions, respectively. After centrosomes are concentrated on the 60% sucrose cushion by the previous centrifugation (step 7), remove the supernatant until only about 25–30 ml remain at the bottom of the 250-ml Nalgene tube.

9. Vortex vigorously the obtained centrosomal suspension (which is now about 20–25% sucrose) and fill the SW28 Beckman tube with this sample. Centrifuge the gradient at 40,000 g (25,000 rpm) for 1 hr in a Beckman L8.50B ultracentrifuge equipped with a SW28 rotor.

10. Place the SW28 tube on the density gradient fractionator (ISCO): optimally, the tube is clamped in the ISCO apparatus, perforated at the bottom with a two holes needle, and 14 fractions of 0.5 ml each are manually recovered from the bottom into 1.5-ml Eppendorf tubes.

11. Remove 10 μl from each fraction for monitoring the centrosomes concentration (see below). Freeze the fractions in liquid nitrogen and conserve at −80°C.

B. Immunofluorescence Analysis of Sucrose Gradient Fractions

Solutions

1. *10 mM PIPES, pH 7.2:* To make 500 ml, add 5 ml of 1 M PIPES pH 7.2 stock solution to distilled water and complete to 500 ml. Store at 4°C.

2. *Phosphate-buffered saline (PBS) 10× stock buffer:* 100 mM phosphate and 1.5 M NaCl, pH 7.4.

   a. Solution A: 0.2 M Na_2HPO_4; to make 500 ml, add 14.2 g of Na_2HPO_4 to distilled water.

   b. Solution B: 0.2 M NaH_2PO_4; to make 100 ml, add 2.76 g of Na_2HPO_4 to distilled water.

   Mix solution A with solution B, add 87.66 g of NaCl, and complete to 1 liter with distilled water by checking the pH. Store at room temperature.

3. *PBS–0.1% Tween 20 buffer:* To make 100 ml, add 10 ml of PBS 10× stock buffer and 100 μl of pure Tween 20 to distilled water and complete to 100 ml. Store at 4°C.

4. *PBS–3% BSA buffer:* To make 50 ml, add 5 ml of PBS 10x stock buffer and 1.5 g of BSA to distilled water. Stir and complete to 50 ml. Store at 4°C.

5. *Secondary labeled antibodies:* Rhodamine-conjugated goat anti-mouse and fluorescein labeled goat anti-rabbit are used as secondary antibodies.

Steps

1. Prepare 12 15-ml Corex tubes containing special adapters and 12-mm-round coverslips as described by Evans *et al.* (1985) and fill them with 3 ml of 10 mM PIPES, pH 7.2.

2. Disperse 10 μl of fractions 4 to 13 in the corresponding tubes and centrifuge them at 20,000 g (10,000 rpm) with a JS 13.1 rotor in a Beckman J2-21M/E centrifuge. At the same time, sediment 150 μl from aliquots of the lysis step and the lysis supernatant on glass coverslips.

3. After centrifugation, remove the PIPES buffer from the tubes, place the coverslips in a coverslip-holder box, and immerse them in methanol at −20°C for 6 min. All the following steps take place at room temperature.

4. Rinse the coverslips three times with PBS–0.1% Tween 20 buffer and place them in a chamber suitable for immunofluorescence experiments with coverslips. Generally, a double immunofluorescence staining with a polyclonal anti-pericentriolar material and a monoclonal anti-α tubulin antibody (the latter stains centrioles) is performed (an example is shown in Fig. 1). The primary antibodies are diluted in PBS–BSA 3%. Incubate sedimented centrosomes with the primary antibodies for at least 30 min.

5. Rinse the coverslips three times with PBS–0.1% Tween 20 buffer and add mixed fluorescein-labeled anti-rabbit IgGs and rhodamine-conjugated anti-mouse IgGs (both diluted at 1/750 in PBS–BSA 3%) as secondary antibodies. Let incubate for at least 30 min.

6. Rinse coverslips with PBS–Tween 20 buffer and place them in pure ethanol for 2 min. When coverslips are dry, mount them on a drop of Citifluor placed on microscope slides.

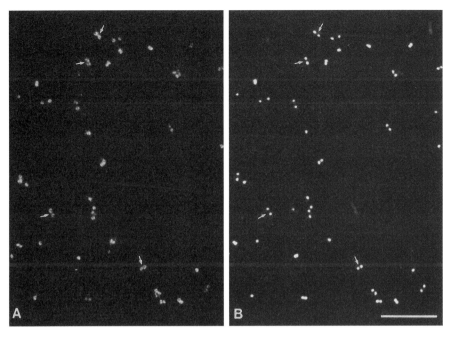

FIGURE 1 An example of double immunofluoresence staining of centrosomes present in fraction 5 of the sucrose gradient. Centrosomes were stained with an antipericentriolar material antibody (A) and an anti-α tubulin antibody that stains centrioles (B). Centrosomes were homogeneously distributed on the glass coverslip. They were mainly recovered as a pair of centrioles, their native configuration (arrows in A and B). Bar = 10 μm.

C. Quantification of Centrosomes per Sucrose Gradient Fraction

Steps

1. Observe the coverslips with an epifluorescence microscope (Leitz Dialux 20), and count centrosomes on a defined field outlined by the camera.

2. Measure the area of this field using a slide graduated with a micrometer, and the ratio to the total area on which centrosomes have been sedimented (i.e., the transverse section of the Corex tube) is used to calculate the number of centrosomes in the aliquot. From that, the concentration of each fraction is obtained (Fig. 2A).

D. Biochemical Analysis of Sucrose Gradient Fractions

Solution

1. *10 mM PIPES, pH 7.2:* To make 500 ml, add 5 ml of 1 *M* PIPES pH 7.2 stock solution to distilled water and complete to 500 ml. Store at 4°C.

Steps

1. Dilute $1-2 \times 10^7$ centrosomes from each fraction with 1 ml of 10 m*M* PIPES buffer, pH 7.2, and sediment them by centrifugation with a 2-MK Sigma centrifuge at 15,000 rpm for 10 min.

2. Remove all the supernatant, resuspend the centrosomal pellets in a known volume of the Laemmli sample buffer (Laemmli, 1970), and heat each tube for 5 min in boiling water. A short centrifugation is sometimes necessary to remove residual insoluble materials.

3. Analyze the fractions by conventional electrophoresis technique on a 6–15% SDS–PAGE gradient (Fig. 2B).

4. Stain the gel by a classical silver nitrate staining method.

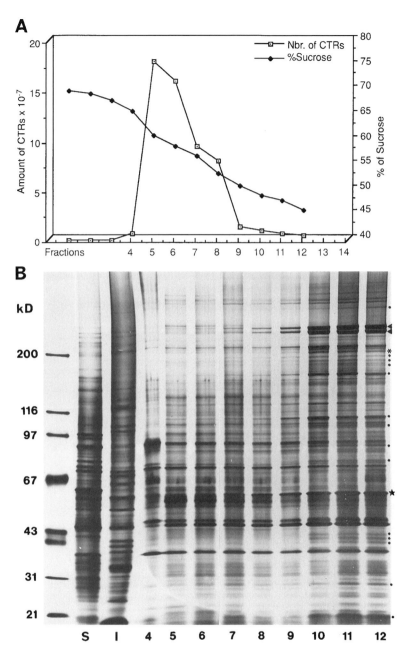

FIGURE 2 (A) Quantification of centrosomes recovered in each fraction of the sucrose gradient used as the late stage of centrosome isolation. The bottom of the gradient is on the left of the figure. This preparation used 3 × 10⁹ cells. Note that fractions 5 and 6 are more enriched. (B) Electrophoretic analysis of the same fractions represented in A. A 6–15% SDS–PAGE gradient was used to detect simultaneously the high and low molecular weight polypeptides present in each fraction. The gel was stained by the silver nitrate technique. Lanes S and I correspond to Triton X-100 soluble (S) and insoluble (I) proteins of KE 37 lymphoblastic cells from which the centrosomes were isolated. Only fractions 5 and 6 (also 7) are constant in their protein composition. Many additional bands appear at the top of the sucrose gradient from fractions 8 to 9 and are prominent in fractions 10, 11, and 12. Among these additional contaminating bands (dots), the major ones have been identified as fodrin (arrowheads), myosin heavy chain (asterisk), and vimentin (star).

E. Functional Assay of Isolated Centrosomes

Isolated centrosomes are usually tested for their capacity to promote the nucleation of microtubules. α/β tubulin dimers were purified from bovine or porcine brain using temperature-dependent polymerization/depolymerization cycles, followed by a phosphocellulose column.

Solutions

1. *Purified tubulin solution at 2.5 mg/ml.*

2. *1 M PIPES, pH 6.8, stock solution:* To make 200 ml, add 60.5 g of PIPES to distilled water, adjust pH to 6.8 with 10 N KOH, and complete the volume to 200 ml. Store at 4°C.

3. *1 M MgCl$_2$ stock solution:* To make 100 ml, add 5.84 g of MgCl$_2$ to distilled water. Store at 4°C.

4. *0.2 M EGTA stock solution:* To make 200 ml, solubilize 15.2 g of EGTA in distilled water, adjust the pH to 7 with NaOH, and complete the volume to 200 ml.

5. *RG2 buffer:* 80 mM PIPES pH 6.8, 1 mM MgCl$_2$, and 1 mM EGTA. To make 1 liter, add 80 ml of 1 M PIPES pH 7.2 stock solution, 1 ml of 1 M MgCl$_2$ stock solution, and 5 ml of 0.2 M EGTA stock solution to distilled water, check the pH, and adjust the volume to 1 liter. Store at 4°C.

6. *0.2 M GTP stock solution:* Solubilize 1 g of GTP in 9.5 ml of distilled water. Adjust the pH to 7 with NaOH. Store as aliquots at −20°C.

7. *RG1 buffer:* RG2 buffer + 1 mM GTP. To make 2 ml, add 10 μl of 0.2 M GTP stock solution to 1.990 ml of RG2 buffer. Make fresh and maintain at 4°C.

8. *Glutaraldehyde 1% (v/v) solution:* To make 1.2 ml, add 48 μl of glutaraldehyde 25% to 1.152 ml of RG1. Make fresh and maintain at room temperature.

9. *Glycerol 25% (v/v) solution:* To make 40 ml, add 10 ml of pure glycerol solution to 30 ml of RG2 buffer. Make fresh and maintain at 4°C.

10. *Triton X-100 1% (v/v) solution:* To make 10 ml, add 0.1 ml of pure Triton X-100 to 9.9 ml of RG2 buffer. Make fresh and maintain at 4°C.

Steps

1. Add 50 μl of tubulin solution, 5 μl of centrosomes (containing 2 to 5 $\times$ 10^5 centrosomes), and 10 μl of RG1 buffer to an Eppendorf tube. Conserve the Eppendorf tube at 4°C.

2. Place the mixture at 37°C for 8 min.

3. Add 200 μl of the 1% glutaraldehyde solution, and place the tube at 25°C for 3 min.

4. Place the tube on ice and add 1 ml of RG2 cold buffer.

5. Prepare 15-ml Corex tubes with corresponding adapters and glass coverslips (Evans *et al.*, 1985). Fill them with 5 ml of the glycerol 25% solution. Overlay this solution with the nucleated microtubule sample and sediment the asters at 20,000 g (10,000 rpm) for 10 min with the JS 13.1 rotor in a Beckman J2-21M/E centrifuge at 4°C.

6. Aspire 1 ml from the top of the Corex tube and replace it with 1 ml of the 1% Triton X-100 solution.

7. Remove all of the 5 ml of glycerol solution leaving just the 1 ml of the Triton X-100 solution over the coverslip, which is recovered from the Corex tube. Classically, a double immunofluorescence experiment (see Section B) is run to detect the centrosomes and microtubule asters (Fig. 3).

IV. COMMENTS

The method described here is based on lysis of cells in a very low ionic strength to separate the centrosome from the nucleus after the disassembly of microfilament and microtubule cytoskeletal networks. EDTA has been eliminated from all steps of centrosome isolation, as the structure of centrosomes is affected by this agent (see Paintrand *et al.*, 1992; Moudjou and Bornens, 1992). This method may be modified for each cell type, and preliminary tests of cell lysis in various ionic strength buffers are warranted before undertaking centrosome isolation. As seen in Fig. 1B, the electrophoretic profile of sucrose gradient fractions vary

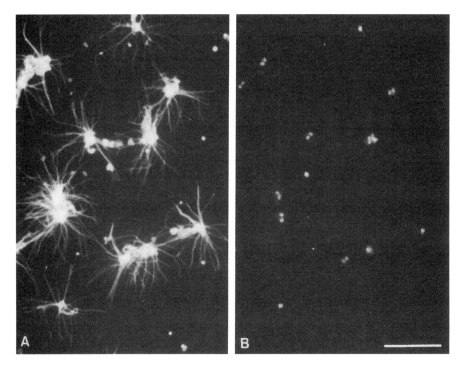

FIGURE 3 Functional assay of isolated centrosomes. The microtubule nucleating activity of isolated centrosomes was tested as described in Section V. After sedimentation on the glass coverslip, aster microtubules were labeled with the anti-α tubulin monoclonal antibody (A) and the centrosomes were detected with an anticentrosome antibody (B). Note that only one centrosome was usually present at the center of each aster. The bright isolated dots in A correspond to tubulin aggregates. Bar = 10 μm.

from bottom to top. Some proteins appear, especially in the top of the gradient (from fraction 8 and 9). The presence of these proteins is correlated with the appearance of centrosome aggregates associated in a large network. The biochemical nature of this network is not understood and could correspond to fodrin aggregates, intermediate, or acto-myosin filaments. Indeed, vimentin, fodrin, and myosin heavy chain (and other polypeptides) begin to be detected from fractions 8 and 9. Fractions 5, 6, and 7 contain at least 70% of the total centrosomes recovered in the sucrose gradient. Immunofluorescence, ultrastructural, and biochemical analyses have led us to conclude that these fractions are the most representative of isolated centrosomes from KE37 human lymphoblastic cells. Transmission electron microscopic analysis of fractions 5 and 6 showed that their main constituent is centrosomes distributed in a homogenous fashion (see Paintrand *et al.*, 1992). For any stringent biochemical study of a new centrosomal component, it is necessary to work with only these fractions. The isolated centrosomes can be stored at $-80°C$ for many months. They conserve their capacity to nucleate microtubules (Fig. 3) and to promote a parthenogenetic development after injection into *Xenopus* eggs (Tournier *et al.*, 1991). Finally, the convergence between biochemical studies of isolated centrosomes, or its related structures, and genetics approaches in appropriate systems will be key to unraveling the interesting aspects of centrosomal functions.

V. PITFALLS

1. The state of cells before centrosome isolation is critical for the quality of centrosomal fractions. Cells must be in an exponential phase of growth (8×10^5 to 1×10^6 cells/ml).

2. To avoid contamination of solution conserved at 4°C or at room temperature, 1 mM of sodium azide should be added.

3. All solutions must be filtered through a 0.22-μm filter, except for sucrose solutions, which must be sterilized with an autoclave at 105°C.

4. It is important to have a conventional method to recognize the side on which the material was sedimented on the coverslip for immunoflurescence experiments. Make sure that the coverslips never dry.

Acknowledgments

We thank Dr. Spencer Brown for checking the English version. The excellent technical assistance of C. Celati is gratefully acknowledged. This work was supported by CNRS, Ligue Nationale de Lutte contre le Cancer, and Fondation pour la Recherche Medicale (FRM). M. Moudjou was a recipient of a postdoctoral fellowship from "Association pour la Recherche sur le Cancer" (ARC).

References

Bornens, M. (1992). Structure and functions of isolated centrosomes. *In* "The Centrosome" (V.I. Kalnins, ed.), pp 1–43. Academic Press, New York.

Bornens, M., Paintrand, M., Marty, M. C., and Karsenti, E. (1987). Structural and chemical characterization of isolated centrosomes. *Cell Motil and Cytosk.* 8, 238–249.

Evans, L., Mitchison, T. J., and Kirschner, M. W. (1985). Influence of centrosome on the structure of nucleated microtubules. *J. Cell Biol.* 100, 1185–1191.

Huang, B. (1990). Genetics and biochemistry of centrosomes and spindle poles. *Curr. Opin. Cell Biol.* 2, 28–32.

Kalt, A., and Schliwa, M. (1993). Molecular components of the centrosome. *Trends Cell Biol.* 3, 118–128.

Komesli, S., Tournier, F., Paintrand, M., Margolis, R. L., Job, D., and Bornens, M. (1989). Mass isolation of calf thymus centrosomes: Identification of a specific configuration. *J. Cell Biol.* 109, 2869–2878.

Laemmli, U. K. (1970). Cleavage of structural proteins during the assembly of the head of bacteriophage T4. *Nature (London)* 227, 680–685.

Mitchison, T. J., and Kirschner, M. W. (1986). Isolation of mammalian centrosomes. *In* "Methods in Enzymology" (R. B. Vallee, ed.), Vol. 134, pp. 261–268. Academic Press, San Diego.

Moudjou, M., and Bornens, M. (1992). Is the centrosome a dynamic structure? *Compt. Rend. Acad. Sci. (Paris)* 315, 527–534.

Moudjou, M., Bordes, N., Paintrand, M., and Bornens, M. (1996). γ-Tubulin in mammalian cells: The centrosomal and the cytosolic forms. *J. Cell Sci.* 109, 875–887.

Paintrand, M., Moudjou, M., Delacroix, H., and Bornens, M. (1992). Centrosome organization and centriole architecture: Their sensitivity to divalent cations. *J. Struct. Biol.* 108, 107–128.

Raff, J. W., Kellogg, D. R., and Alberts, B. A. (1993). Drosophila γ-tubulin is part of a complex containing two previously identified centrosomal MAPs. *J. Cell Biol.* 121, 823–835.

Rout, M. P., and Kilmartin, J. V. (1990). Components of the yeast spindle and spindle pole body. *J. Cell Biol.* 11, 1913–1927.

Tournier, F., Komesli, S., Paintrand, M., Job, D., and Bornens, M. (1991). The intercentriolar linkage is critical for the ability of heterologous centrosomes to induce parthenogenesis in Xenopus. *J. Cell Biol.* 113, 1361–1369.

Preparation of Yeast Spindle Pole Bodies

Michael P. Rout and John V. Kilmartin

I. INTRODUCTION

Spindle pole bodies (SPBs) are the sole microtubule organizing centers of budding yeast cells. SPBs are embedded in the nuclear envelope, which remains intact during mitosis, thus spindles are intranuclear (Fig. 1). SPBs can be enriched 600-fold and in high yield from *Saccharomyces uvarum* (Rout and Kilmartin, 1990). The procedure involves the preparation of nuclei by a modification of an existing method (Rozijn and Tonino, 1964). The nuclei are then lysed and extracted to free the SPBs from the nuclear envelope, followed by two gradient steps to separate the SPBs from other nuclear components. These SPBs, which are about 10% pure (Fig. 2), have been used to prepare monoclonal antibodies (mAbs) and thereby identify components of the SPB and spindle (Rout and Kilmartin, 1990, 1991).

II. MATERIALS AND INSTRUMENTATION

All catalog numbers are in parentheses. Anti-foam B (A-5757), sorbitol (S-1876), PVP-40 (PVP-40), pepstatin (P-4265), phenylmethylsulfonyl fluoride (PMSF, P-7626), digitonin (D-1407), DNase I (DN-EP), RNase A (R-5503), GTP (G-5884), bis–Tris (B-9754), and EGTA (E 4378) are from Sigma. Ficoll-400 (17-0400-01) and Percoll (17-0891-01) are from Pharmacia. Triton X-100 (30632), glucose (10117), and dimethyl sulfoxide (DMSO, 10323) are from BDH. Sucrose (5503UA) is from GIBCO-BRL. The yeast extract (1896) is from Beta Lab. Bactopeptone (0118-08-1) is from Difco. Glusulase (NEE-154) is from Du Pont. Zymolyase 20T (120491) is from Seikagaku. SP 299 is from Novo (Mutanase is no longer commercially available). Liquid malt extract is from a local health food store.

Foam stoppers (FPP6) to seal the neck of the fermentor flask are from Scientific Instruments. *S. uvarum* (NCYC 74) is from The National Collection of Yeast Cultures. Centrifuge tubes, SW28 Ultraclear (344058) and 70 TI (355654), are from Beckman. The hemocytometer (AC-1) is from Weber Scientific. The continuous flow centrifuge (KA 1-06-525) is from Westfalia. The Polytron (PT 10/35 with PTA 10S probe) is from Kinematica. The refractometer (144974) is from Zeiss.

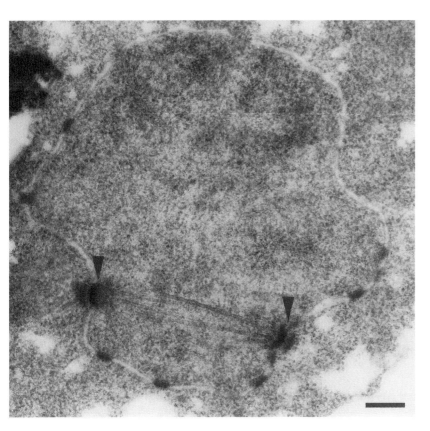

FIGURE 1 Electron micrograph of a thin section of a diploid yeast intranuclear mitotic spindle. Arrowheads show the two SPBs. Bar: 0.2 μm.

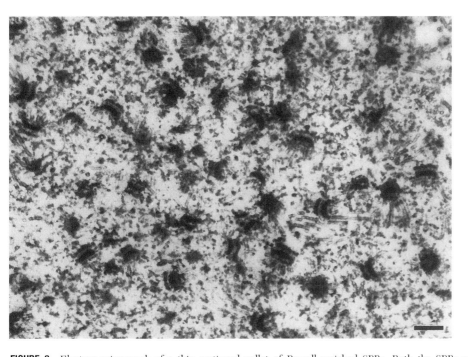

FIGURE 2 Electron micrograph of a thin sectioned pellet of Percoll-enriched SPBs. Both the SPBs and their attached microtubules are clearly visible. Bar: 0.2 μm.

III. PROCEDURES

A. Preparation of Nuclei from *Saccharomyces uvarum*

The following solutions are required for processing 36 liters of yeast cells.

Solutions

1. *5 liters of cold distilled water.*

2. *1.1 M sorbitol:* To make 2 liters, dissolve 400 g sorbitol in distilled water and make up to 2 liters. Store at 4°C.

3. *7.5% Ficoll-400 in 1.1 M sorbitol:* To make 400 ml, dissolve 30 g Ficoll-400 in 1.1 *M* sorbitol, it dissolves very slowly, make up to 400 ml. Store at 4°C.

4. *8% PVP:* To make 1 liter, dissolve 80 g PVP-40, 1.57 g KH_2PO_4, 1.46 g K_2HPO_4, and 152 mg $MgCl_2 \cdot 6H_2O$ in distilled water, check the pH, and adjust to 6.53; it usually needs 15 μl of concentrated H_3PO_4, make up to 1 liter. Store at 4°C. Prepare 1.5 liters if also making up sucrose solutions.

5. *Solution P:* Dissolve 2 mg pepstatin A and 90 mg PMSF in 5 ml absolute ethanol. Store at 4°C and discard after 3 weeks. Note that PMSF is highly poisonous.

6. *20% Triton X-100:* Dissolve 20 g Triton X-100 in distilled water and make up to 100 ml. Store at −20°C; retain a small aliquot at 4°C.

7. *Sucrose solutions for the gradient:* Prepare the sucrose solutions by weighing out sucrose and 8% PVP solution in a beaker and dissolve by stirring the beaker in a large dish of hot water. When the sucrose has dissolved, remove from the hot water and measure the refractive index (RI). Continue stirring and adding 8% PVP until the correct refractive index is obtained (within 0.0003). Store in 50-ml sterile tubes at −20°C. For 2.01 *M* sucrose, weigh out 183.3 g sucrose and add 8% PVP solution to 338 g in total, then adjust to a final RI of 1.4370. For 2.10 *M* sucrose, weigh out 193 g sucrose and add 8% PVP to a total of 340 g (final RI 1.4420). For 2.30 *M* sucrose, weigh out 216 g sucrose and add 8% PVP to a total of 340 g (final RI 1.4540). Note that this nomenclature differs from that used in Rout and Kilmartin (1990), which used apparent (rather than actual) sucrose molarities.

8. *Wickerham's medium:* Prepare Wickerman's medium 10× concentrated. Dissolve 108 g malt extract, 108 g yeast extract, 180 g bactopeptone, and 360 g glucose in 3.6 liters distilled water and remove turbidity by centrifuging at 10,000 rpm for 10 min. Dilute the supernatant 10-fold.

Steps

1. Two days before the nuclei preparation, prepare the concentrated Wickerham's medium to grow the yeast cells. The cells can be grown in a commercial fermentor or the simple system described below can be used. Dispense 9 liters of diluted medium into four 10-liter glass Pyrex bottles containing a heavy magnetic stirrer. Add 2 ml of anti-foam B (use a disposable plastic pipette). Prepare a glass aeration tube by attaching a small plastic vial with four or five needle-sized holes in the bottom of one end of the glass tube via a rubber stopper or thick-walled rubber tubing. Attach a glass fiber filter on the inlet of the aeration tube using rubber tubing. Cut halfway into a foam stopper slightly larger than the diameter of the neck of the flask and use this to position the aeration tube and to seal the flask. Seal the stopper and the filter at the end of the aeration tube with aluminium foil and cover both the aeration tube and the top of the flask with a foil cap. Alternatively, simply cover the top of the flask with three or more layers of foil and tape it in place with autoclave tape; the aeration, in the form of a sterile disposable 25-ml pipette attached to the bench air supply by sterilized plastic tubing, can be added after flask autoclaving. Autoclave for 1 hr and 1 hr exhaust or for 90 min with faster exhaust. It is important to autoclave properly, especially if the medium was not ultrafiltered, as there are spores of a remarkably heat-resistant bacterium in the medium. Allow the flasks to cool at room temperature overnight.

2. Set up a starter culture from a single yeast colony in 10–20 ml medium and grow at 25°C in a rotating incubator or small bubbler for up to 24 hr until the stationary phase is reached (1.0×10^8 cells/ml). Use a hemocytometer to count the cells, use shearing or sonication to break up any clumps of cells.

3. In the afternoon of the day before the nuclei preparation, set up the 10-liter flasks to grow overnight. Aim for 1×10^7 cells/ml in the morning; it is essential to have midlog-phase cells to prepare good spheroplasts. It is probably pointless to proceed with the spheroplasting if the cells have overgrown beyond 50% above the optimal count. To calculate the number of starter cells to add, first work out the number of hours the cells will grow at log phase. If the starter cells are stationary phase (1×10^8 cells/ml), subtract one doubling time (1.75 hr for *S. uvarum*) to allow for the lag period of growth. Divide the number of hours of log-phase growth by the doubling time to give the number of divisions (d). The number of cells needed is the final number of cells divided by 2^d (2^d = anti-log$_{10}d/3.32$).

4. Transfer these cells by sterile technique to the 10-liter flasks and place in a water bath set at 25°C (the depth of water need be no more than 25 cm). If room temperature is 23°C or above, the water bath is not necessary, but an appropriate correction in the cell doubling time calculations is required. Place magnetic stirrers under the bath to stir the flasks during growth, connect the aeration tube to a compressed air source and water bubbler to humidify the air, and position the aeration tube so that the bubbles of air are blown at the rotating stirring bar, thereby dispersing throughout the flask. If just 25-ml pipettes, and no stirring bars are used, then bubble air through the medium at a rate sufficient to keep the cells from settling but not too violent (as this will cool the medium). The quality of aeration can be checked by growing one culture to stationary phase, it should reach a cell count of 1×10^8 cells/ml.

5. Harvest the cells using a Westfalia continuous flow centrifuge or in individual centrifuge pots and spin at 4000 g for 5 min (all centrifugation steps are at 4°C).

6. Combine the pellets in one 500-ml centrifuge pot, wash twice with cold distilled water, resuspend the pellet in 1.1 *M* sorbitol using a thick glass rod, and spin at 5000 g for 5 min. Repeat the resuspension and pelleting in 1.1 *M* sorbitol and finally resuspend in about one pellet volume of sorbitol for spheroplasting. Dilute this suspension 1:100, count the cells, and then adjust the suspension to 1.5×10^9 cells/ml.

7. Add 50 μl glusulase, 20 μl 1% zymolyase 20T, and 3 μl 5% mutanase or SP 299 per milliliter of suspension. The enzyme solutions should be spun or filtered beforehand to remove insoluble material. Incubate at 30°C, leaving the top of the centrifuge pot a little loose as some gas evolves, and mix carefully from time to time. This can be done either in a water bath, manually stirring the pot every 10–15 min, or in an air incubator with a moderate speed of stirring. The digestion, which takes 2–4 hr, can be monitored using the hemocytometer by diluting 10 μl of digesting cells in 0.2 ml of sorbitol to determine the proportion of spheroplasts and another 10 μl in 0.2 ml of water to determine the number of completely intact cells. After 1 hr in sorbitol there should be a few spheroplasts and less than 1% live intact cells in water. After about 1.5 hr the spheroplasts become very clumpy and irregular in shape. After further digestion at 2–3 hr they separate into round spheroplasts with some clumps remaining. In water there should be complete lysis with no undigested cells and less than 5×10^4 black cells or ghosts/ml.

8. While the cells are digesting, make up the gradient tubes: 12 Beckman Ultraclear SW28 tubes or equivalent for a 36-liter preparation, *provided not more than 230 ml of cells has been digested,* otherwise increase the gradient tubes in proportion. Add solution P at 1:1000 to the three sucrose solutions and prepare a step gradient in each tube of 8 ml of 2.30 *M*, 8 ml of 2.10 *M*, and 8 ml of 2.01 *M* sucrose–PVP. Leave the tubes in an ice bucket in the cold room covered with foil.

9. While the cells are digesting, calculate the approximate amount of 8% PVP needed to lyse the spheroplasts (see below), then prepare the same volume of 0.6 *M* sucrose–PVP (205.2 g sucrose/liter 8% PVP) and enough 1.7 *M* sucrose–PVP (581.4 g sucrose/liter 8% PVP) to resuspend the crude nuclei for the gradient (about 12 ml per tube); cool these solutions in ice. Also prepare two to four 100-ml Ficoll cushions in 250-ml clear Sorvall pots. Note that the maximum volume that can be conveniently lysed in a 250-ml pot is 150 ml, so have no more than 1.5×10^{11} diploid spheroplasts in each pot.

10. At the end of the digest add an equal volume of 1.1 M sorbitol and pellet the spheroplasts in a swinging bucket centrifuge at 5000 g for 20 min. Aspirate off the supernatant, immediately wash the sides of the tube with sorbitol, and aspirate off again. Remove as much enzyme as possible as proteases are present. Resuspend the pellet in the same volume of sorbitol. The spheroplasts are *very fragile* so be very careful to avoid lysis: use a thick glass rod to partially resuspend as lumps and then gently shake in ice on a rotating shaker. Use the glass rod occasionally to disperse the lumps.

11. When the cells are resuspended, overlay 0.2 ml onto 0.4 ml 7.5% Ficoll in 1.1 M sorbitol in a small glass test tube to check that the cells are not too dense to overlay; if the cells fall through the Ficoll, dilute them further with 1.1 M sorbitol until they do not. Then overlay the cells onto the 100-ml cushions of 7.5% Ficoll in 1.1 M sorbitol in 250-ml clear centrifuge pots. Spin in a swinging bucket centrifuge at 5000 g for 20 min.

12. Aspirate the supernatant down to the Ficoll layer and then wash the sides of the pot several times with sorbitol, aspirating each time to remove as much of the enzyme-containing supernatant as possible. Finally, aspirate down through the Ficoll layer to the pellet and wash the sides of the tube with 8% PVP and aspirate that.

13. Add 10 ml 8% PVP, 60 μl solution P, and 12 μl 20% Triton X-100 per 10^{10} spheroplasts. This should prove to be about a total of 300 ml of 8% PVP plus 1.8 ml solution P and 0.36 ml 20% Triton X-100. Use the Polytron probe at a low speed first to resuspend the pellet, then increase the speed setting to 3.5 to 5 (almost always keep at 3.5) and Polytron throughout the solution for 1 min; drag the probe across the bottom of the pot during this time to try to resuspend as much of the pellet as possible. Place the pot in ice and remove a 5-μl sample to observe by phase-contrast microscopy at a magnification of about 1000$\times$. There should be *complete* cell lysis; plenty of nuclei, which look like small perfectly round black balls, often with a slightly darker crescent region; and vacuoles, which look like white balls the same size as nuclei. The presence of vacuoles is a reassuring sign that the correct quantity of Triton has been added, i.e., enough to lyse the spheroplasts without disrupting too many other membranes. Unlysed spheroplasts are highly swollen with a prominent vacuole. Check to see what proportion of nuclei is trapped in lysed but undispersed cells. Repeat the Polytron step for another minute. If there is a high proportion of trapped nuclei or unlysed cells, then extend the time of the Polytron treatment. Decant the lysed cells into clean centrifuge pots or a cold conical flask and add an equal volume of cold 0.6 M sucrose–PVP, mix carefully, and spin in a 6 × 250-ml anglehead rotor for 15 min at 10,000 g. Note that this Polytron and spin step must be completed *as fast as possible*, as after lysis nuclei slowly clump, trapping cytoplasmic debris and thereby contaminating the gradient.

14. Retain the supernatant from the spin to check that nuclei have pelleted and then resuspend nuclei in cold 1.7 M sucrose–PVP (12 ml per SW28 tube) using the Polytron at low speed (speed 2–3; try also to resuspend any nuclei that have been daubed up the side of the tube by the spin) and load the gradient tubes with about 15 ml using either a measuring cylinder or a 10-ml pipette. Weigh the tubes and balance by overlaying with 1.0 M sucrose–PVP and spin at 28,000 rpm for 4 hr (Beckman SW28 rotor).

15. Nuclei mainly band at the 2.10/2.30 M layer. Unload by aspirating down from the top, wiping the top part of the tube with tissue as you go down. If the wad of material at the top is particularly thick, scoop it out with the end of a Pasteur pipette. Stop at the 2.01/2.10 M interface for the first tube only and remove 5 μl (wipe the sides of the tip to remove sucrose) to check by phase-contrast microscopy if there are significant quantities of nuclei there (usually there are a few but they are too contaminated with debris to be of much use); also check the 2.10 M layer. Aspirate down to just above the nuclei band in the 2.10/2.30 M layer, then quickly wipe the sides of the tube with a rolled up tissue, paying particular attention to that part where the main band was at the sample to the 2.01 M interface. Try to get as close to the nuclei as possible as osmiophilic membranes (probably from the plasma membrane) are concentrated in the 2.10 M layer. Ring the nuclei layer with a sealed-off

Pasteur pipette to separate it from the walls of the tube and then suck off with a 10-ml pipette, going down partially into the 2.30 M layer to remove as much nuclei as possible. Do not get too close to the pellet containing empty cell walls. Store nuclei at $-70°C$; they seem to be very stable to storage at this temperature.

B. Preparation of Spindle Pole Bodies

Solutions

1. *0.9 M sucrose–PVP:* Dissolve 30.8 g sucrose in 8% PVP and make up to 100 ml. Store at $-20°C$.

2. *0.01 M bis–Tris–HCl pH 6.5, 0.1 mM MgCl$_2$ (bt buffer):* Prepare a 0.1 M stock of bis–Tris, dissolve 20.9 g bis–Tris in distilled water, adjust the pH to 6.5 with concentrated HCl, and make up to 100 ml. Store at $-20°C$. To prepare 100 ml bt buffer, add 10 ml 0.1 M bis–Tris–HCl, pH 6.5, and 10 μl 1 M MgCl$_2$ (dissolve 0.2 g MgCl$_2$ in a total of 1 ml water) to 90 ml of distilled water.

3. *Sucrose solutions in bt buffer:* Prepare stock 2.50 M sucrose in bt buffer (214 g/250 ml) as described for the sucrose–PVP solutions; the refractive index should be 1.4533. Prepare 1.75, 2.00, and 2.25 M sucrose–bt solutions by diluting the stock 2.50 M solution with bt buffer; the final refractive indices should be 1.4174, 1.4295, and 1.4414, respectively. Add a 1:1000 dilution of solution P just before pouring the gradient. Store at $-20°C$.

4. *Nuclear lysis solution:* For 10 ml, add 0.2 g digitonin and 0.15 ml Triton X-100 to 2 ml DMSO and about 1 ml distilled water (wear gloves as digitonin is toxic). Microwave to almost boiling and then stir to dissolve the digitonin. Make up to 10 ml in a measuring cylinder by slowly adding distilled water down the side of the cylinder without mixing; add 1 μl 1 M MgCl$_2$. Place the solution in the cold room and mix immediately before use, it should be a clear solution.

5. *2% DNase I:* Dissolve 5 mg of DNase I in 0.25 ml 0.25 M sucrose, 0.05 M triethanolamine, pH 7.4, 0.25 M KCl, and 5 mM MgCl$_2$. Store at $-20°C$.

6. *RNase A:* Dissolve 2 mg RNase A in 1 ml of distilled water.

7. *DMSO buffer:* Prepare this immediately before use. Prepare stocks of 0.1 M GTP (dissolve 55 mg GTP in 1 ml water; store at $-20°C$), 0.1 M EGTA (dissolve 3.8 g EGTA in 100 ml water and adjust the pH to 7.0 with 10 N NaOH), 1.0 M DTT (dissolve 154 mg DTT in 1 ml water; store at $-20°C$). For 300 ml of DMSO buffer, use 30 ml 0.1 M bis–Tris–HCl, pH 6.5, 30 μl 1.0 M MgCl$_2$, 300 μl 0.1 M GTP, 300 μl 0.1 M EGTA, pH 7.0, 30 μl 1.0 M DTT, 300 μl solution P, and 60 ml DMSO; make up to 300 ml with distilled water and cool in ice.

8. *bt–DMSO buffer.* For 100 ml, use 10 ml 0.1 M bis–Tris–HCl, pH 6.5, 10 μl 1.0 M MgCl$_2$, 100 μl solution P, and 20 ml DMSO and make up to 100 ml with distilled water. Cool in ice.

Steps

1. Pellet the nuclei by first decreasing the sucrose concentration from about 2.4 to 2.1 M (refractive index 1.434) by adding 0.9 M sucrose–PVP. Mix very thoroughly by shaking. Calculate the total number of ODs of nuclei by measuring the OD at 260 nm of 10 μl nuclei in 1 ml of 1% SDS (10^{10} nuclei is about 100 OD$_{260\text{ nm}}$). Decant the nuclei into Beckman 70 TI tubes or equivalent. Six tubes are usually enough for nuclei from 3×10^{11} spheroplasts. Mix the tubes thoroughly again; any unmixed sucrose will prevent proper pelleting of the nuclei. Spin the 70 TI tubes in a 70 TI rotor at 40,000 rpm for 1 hr. After spinning, immediately remove the supernatant by aspiration, being careful not to disturb the pellet. These pellets can be stored in the tubes at $-70°C$ until needed.

2. Prepare the gradients in Beckman SW28 Ultraclear tubes. Four tubes are usually enough. Each tube contains 2.5 ml 2.50 M sucrose–bt, 7.5 ml 2.25 M sucrose–bt, 5.0 ml 2.00 M sucrose–bt, and 5.0 ml 1.75 M sucrose–bt. Place tubes in ice.

3. To lyse the nuclei, thaw the 70 TI tubes containing the nuclei pellets (if frozen), aspirate any residual sucrose–PVP, and place in ice. Add 1.0 ml nuclear lysis solution, 10 μl solution P and 1 μl 2% DNase I for each 100 $OD_{260\,nm}$ in the nuclear pellets and resuspend at 4°C by mixing vigorously until about a minute after the last traces of pellet have disappeared. The suspension will froth a lot.

4. After resuspension, warm the tubes in your hand, let them stand for 5 min at room temperature, and then add 2.50 M sucrose–bt buffer (at room temperature) equal to the volume of nuclear lysate to each tube. The nuclear lysate must be properly warmed up before incubation at room temperature. If the lysate has not reached room temperature, then the DNA digestion may not be complete at the end of the incubation. Shake well and spin in a 70 TI rotor at 6000 rpm for 6 min. Do not resuspend any pellet that may form at this stage. Distribute the supernatant equally on top of the SW28 gradients (check on a balance); each gradient tube should be loaded to within 5 mm of its top and overlaid if necessary with 1.0 M sucrose–bt. Spin tubes in an SW28 rotor at 28,000 rpm for 6 hr.

5. Remove the gradient layers from the top of the tube using a Pasteur pipette, taking special care when removing material from around the sides of the tubes or at the gradient interfaces. Take off the top layer (sample layer) to within 5 mm of the first interface (sample/ 1.75 M) and then save each interface fraction separately, taking from 5 mm above one interface through to 5 mm above the next. Resuspend the final interface (2.25/2.50 M) and pellet with a sealed Pasteur pipette and whirl mix before removing. The spindle pole bodies should be mainly (~70%) in the 2.00/2.25 M fraction. It is usual to assay the SPB count and protein concentration protein assay (Bradford's) of all the fractions taken. Store these fractions at −20°C.

6. SPBs can be further enriched on a Percoll gradient. For each SW28 tube, add 20 μl of 2 mg/ml RNase A to 2 ml of the 2.00/2.25 M sucrose SPBs and incubate for 15 min at room temperature. Then add, in turn, mixing at each step, 7 ml 2.50 M sucrose–bt, 1 ml Percoll, and 4 ml cold DMSO buffer. Cool to 4°C and place in the SW28 tube, gently overlaying with cold DMSO buffer. Spin at 28,000 rpm for 6 hr.

7. After the spin the SPBs should be visible as a faint band (viewed against a black background) about 1 cm into the gradient. Mark its position on each tube and then collect the gradient solution above it to within 2–3 mm of this band (~3.5 ml) using a Pasteur pipette. This is fraction 1. Collect the next layer (fraction 2), right through the SPB band to about 17 mm from the bottom of the tube (~4.5 ml). This fraction contains 60–70% of the SPBs originally loaded on the gradient. Collect the next 3.5 ml (fraction 3), then whirl mix the final 2.0 ml and save (fraction 4). Store the fractions at −20°C.

8. To pellet the SPB fractions, add 3.5 vol of cold bt–DMSO to each, mix well by shaking, and portion between Beckman 70 TI tubes (about three-quarters full in each). Centrifuge in a 70 TI rotor at 40,000 rpm for 1 hr. The SPBs pellet is visible as a faint translucent layer on top of a transparent Percoll pellet, and after 5 min on ice this layer slides off the Percoll pellet to the bottom of the tube. Carefully aspirate off the supernatant and recover the delicate SPB layer in about 0.2 ml of liquid per tube. The presence of SPBs can be assayed by Coomassie staining of SDS gradient gels. A comparison of fractions 1–4 should show the enrichment of the tubulin (55 kD) and 110-kD bands associated with the SPBs in fraction 2. Alternatively, immunoblotting with anti-SPB mAbs (Rout and Kilmartin 1990, 1991) could also be used to detect the presence of SPBs. A fast and quantitative assay for SPBs is by dark-field microscopy (Rout and Kilmartin, 1990).

IV. COMMENTS AND MODIFICATIONS

This procedure for spindle pole enrichment has also been applied to *Saccharomyces cerevisiae* strains. The extent of enrichment is not as good because these strains do not spheroplast as well as *S. uvarum*, leading to contamination of the nuclei band with cells. In addition, these strains do not appear to disperse their cellular contents during lysis as well as *S. uvarum*, leading to further contamination of the nuclei layer with large aggregated cytoplasmic masses.

The lower quality of the nuclei leads to a corresponding decrease in the quality of the SPBs. However, a number of modifications can be incorporated so that this procedure can be applied successfully to *S. cerevisiae* strains if necessary on a much smaller (and so more convenient) scale, allowing different strains to be processed in parallel. The quality of these nuclei can be nearly as high as those from *S. uvarum*. They may be prepared on such a *small scale* by the following modifications to the numbered steps in the original protocol. A *large-scale S. cerevisiae* nuclei preparation can also be made by scaling up these same modifications as appropriate.

Steps

1–4. Set up an overnight starter culture of the *S. cerevisiae* strain(s) you wish to use in 5 ml of autoclaved YPD medium (1% yeast extract, 2% bactopeptone, 2% glucose) in a rotating incubator at 30°C. Prepare 1–2 liters of autoclaved YPD on the same day in 1-liter culture flasks. On the following day, pour the starter culture into 50 ml of YPD and grow during the day in a rotating incubator at 30°C. In the afternoon/evening, set up the 1-liter cultures to grow overnight at 30°C in a rotating incubator, aiming for between 2 (if diploid) and 5×10^7 cells/ml (if haploid). The only additional sucrose solution to be prepared is 0.3 *M* sucrose–PVP, made by mixing equal volumes of the stock PVP solution and 0.6 *M* sucrose–PVP.

5. Harvest the cells in the appropriate number of 500-ml centrifuge pots (3000 *g* for 5 min).

5a. Resuspend the cells in distilled water to an approximate volume of 50 ml. Place the suspension in a 50-ml graduated Falcon tube and centrifuge in a swing-out rotor at ~1500 *g* for 5 min.

5b. Resuspend the cell pellet in ~50 ml 100 m*M* Tris–HCl and 10 m*M* DTT, pH 9.4, and incubate with occasional swirling for 10 min at 30°C. Then centrifuge again as in step 5a.

5c. Repeat step 5a.

6. Resuspend the cell pellet with ~40 ml of 1.1 *M* sorbitol and centrifuge again as in step 5a. Note the volume of the packed cell pellet and resuspend it (by vigorous shaking of the sealed tube) with an equal volume of 1.1 *M* sorbitol.

7. Add 0.1 vol glusulase and 0.01 vol 1% zymolyase 20T/1% mutanase. Incubate at 30°C as normal.

8. Usually one gradient tube is adequate for 2×10^{10} diploid cells or 5×10^{10} haploid cells.

9. For each strain, prepare one 15-ml Ficoll cushion in a 50-ml round-bottomed polypropylene centrifuge tube and chill on ice.

10. At the end of the digest, add an equal volume of ice cold 1.1 *M* sorbitol and pellet the spheroplasts in a swing-out rotor at 1500 *g* for 10 min at 4°C. Using a Pasteur pipette, gently resuspend the spheroplasts with 20 ml of ice-cold 1.1 *M* sorbitol and pellet again in a swing-out rotor at 1500 *g* for 5 min at 4°C.

11. Gently resuspend the spheroplasts to a total of 20 ml with ice-cold 1.1 *M* sorbitol, overlay onto the Ficoll cushion, and centrifuge in a swing-out rotor at 10,500 *g* for 15 min at 4°C. At this stage, prepare 20 ml 8% PVP solution, containing 100 μl 1 *M* DTT, 200 μl solution P, 50 μl 10% Triton X-100; and 10 ml 0.3 *M* sucrose–PVP containing 100 μl solution P for each spheroplast pellet and cool on ice.

12. Aspirate and rinse the tube and pellet as in the original protocol.

13. To each spheroplast pellet add the 20 ml of ice-cold PVP solution and Polytron to resuspend at a setting of 3.5 for 1 min. Allow the suspension to sit on ice for 1 min, during which time the degree of spheroplast lysis should be monitored as usual. Repeat the Polytron and monitoring steps until lysis is complete. Then, gently underlay each tube with 10 ml of the 0.3 *M* sucrose–PVP and centrifuge in a swing-out rotor at 16,500 *g* for 20 min at 4°C.

14. Resuspend each crude nuclei pellet with 6 ml of cold 1.7 M sucrose–PVP containing 60 μl solution P, using the Polytron on setting 2.5 for two 30-sec bursts separated by 1 min on ice. Any nuclei adhering to the tube bottom can be dislodged by scraping with a sealed Pasteur pipette during the 1-min rest. The nuclei resuspension is then adjusted to a refractive index of 1.425 with 2.30 M sucrose–PVP, making a total volume of ~12 ml; this is usually sufficient to fill one gradient tube, but if necessary the overlayer can be added. Centrifuge the gradient tubes as usual, except in two cases: (i) if the strain is haploid or (ii) if nuclear fragmentation is observed during spheroplast lysis, in which case the centrifuge run time should be increased to 8 hr.

15. Unloading and storage of the nuclei are as previously described.

V. PITFALLS

1. The quality of the spheroplast preparation largely determines the quality of the subsequent nuclei preparation. The amount of glusulase added seems to be crucial: the amount given for this protocol is a minimum for reasons of expense; thus any problems with spheroplast quality can probably be cured by adding more glusulase.

2. The nuclei sucrose gradient is very sensitive to the presence of excess amounts of empty partially digested cell walls. These appear to aggregate at the nuclei band (2.1/2.3 M sucrose–PVP), forming a solid mass, making it impossible to unload the nuclei. It is important to adhere to the loading limits suggested in the nuclei protocol.

3. The type of digitonin seems to be important in successful nuclear extraction; always use a water-soluble type.

4. One should note that the procedures for preparing the SPB fraction described in this article do not work well when nuclei prepared using techniques other than the one described are used as a starting material. This may be related to the fact that the conditions used to pack the nuclear pellet cannot be varied much. If the pellet is too loose, nuclei resuspend without lysing properly, whereas if too firm, the vortexing will not resuspend them at all. Also, variations in available equipment may necessitate some experimentation to determine the correct lysis conditions for each laboratory. These points cannot be stressed too hard because they have been the source of much frustration for people trying to reproduce this technique.

References

Rout, M. P., and Kilmartin, J. V. (1990). Components of the yeast spindle and spindle pole body. *J. Cell Biol.* **111**, 1913–1927.

Rout, M. P., and Kilmartin, J. V. (1991). Yeast spindle pole body components. *Cold Spring Harbor Symp. Quant. Biol.* **56**, 687–691.

Rozijn, Th. H., and Tonino, G. J. M. (1964). Studies on the yeast nucleus. I. The isolation of nuclei. *Biochim. Biophys. Acta* **91**, 105–112.

Preparation of Proteasomes

Keiji Tanaka and Nobuyuki Tanahashi

I. INTRODUCTION

Proteasomes catalyze the nonlysosomal proteolytic pathway in eukaryotic cells. They are involved in a variety of important biological processes, including selective removal of proteins with aberrant structures and naturally occurring short-lived proteins, generation of class I MHC-associated antigenic peptides, and processing of NF-κB (Coux et al., 1996). Proteasomes consist of the common catalytic core (called 20S proteasome) and two distinct associated complexes that have regulatory roles, called PA700 and PA28 (DeMartino and Slaughter, 1993), to form dumbbell-shaped or football-shaped proteasomes, respectively (see EM structures in Fig. 1). The 20S proteasome (also called a multicatalytic proteinase) catalyzes the endoproteolytic cleavage of peptide bonds on the carboxyl side of basic, neutral, and acidic amino acid residues of proteins (Rivett, 1993) and belongs to a novel class of a proteolytic enzyme identified for the first time as threonine protease (Coux et al., 1996). The dumbbell-shaped proteasome (called 26S proteasome) is an eukaryotic ATP-dependent protease responsible for the degradation of proteins with covalently attached ubiquitin as a signal for their selective breakdown or short-lived proteins, such as ornithine decarboxylase, without ubiquitination (Peters, 1994; Hochstrasser, 1995). The football-shaped proteasome has been implicated to be involved in antigen processing (Lupas et al., 1993; DeMartino and Slaughter, 1993; Tanaka et al., 1997). Thus proteasomes appear to be organized into a major proteolytic system as supramolecular multisubunit complexes for rapid and highly regulated degradations of numerous proteins. Therefore, they are assumed to be "a protein death machinery." This article briefly reviews procedures for isolating not only 20S proteasome, but also dumbbell-shaped and football-shaped proteasomes from mammalian tissues and indicates some important aspects of the procedures.

II. MATERIALS AND INSTRUMENTATION

Q-Sepharose (Cat. No. 17-1014-03) and heparin–Sepharose CL-6B (Cat. No. 17-0467-09) can be purchased from Pharmacia LKB Biotechnology. Bio-Gel A-1.5m (Cat. No. 151-0449) and hydroxylapatite Bio-Gel HTP (Cat. No. 130-0420) are from Bio-Rad. Polyethylene glycol 6000 (Cat. No. P-2139), ubiquitin (Ub, Cat. No. U-6253), and succinyl-Leu-Leu-Val-Tyr-4-methylcourmaryl-7-amide (Suc-LLVY-MCA, Cat. No. S-6510) are from Sigma. Amicon PM-10 and PM-30 membranes (Cat. No. 13132) can be obtained from Amicon.

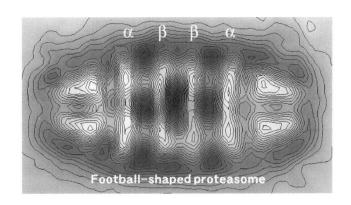

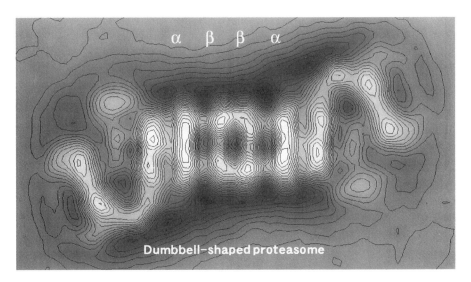

FIGURE 1 Averaged image, based on electron micrographs, of the complex of football-shaped (top) and dumbbell-shaped (bottom) proteasomes from rat. The α and β rings of the 20S proteasome are indicated. (Photograph courtesy of W. Baumeister.)

III. PROCEDURES

Solutions

1. *Buffer A:* 25 mM Tris–HCl (pH 7.5) containing 1 mM diothreitol (DTT) (or 10 mM 2-mercaptoethanol) and 20% glycerol.
2. *Buffer B:* 10 mM phosphate buffer (pH 6.8) containing 1 mM DTT and 20% glycerol.
3. *Buffer C:* Buffer A containing 0.5 mM ATP.
4. *Buffer D:* Buffer B containing 5 mM ATP.

A. Preparation of Catalytic 20S Proteasome

Steps

1. Homogenize 200- to 400-g samples of animal tissues in 3 vol of 25 mM Tris–HCl buffer (pH 7.5) containing 1 mM DTT and 0.25 M sucrose in a Potter–Elvehjem homogenizer. Centrifuge the homogenate for 1 hr at 70,100 g and use the resulting supernatant as the crude extract.

2. Add glycerol at a final concentration of 20% to the crude extract. Then mix the extract with 500 g of Q-Sepharose that has been equilibrated with buffer A. Wash the Q-Sepharose with the same buffer on a Büchner funnel and transfer to a column (5 × 60 cm). Wash the column with buffer A, elute the material with 2 liters of a linear gradient of 0–0.8 M NaCl in the same buffer, and measure the activity of proteasomes using Suc-LLVY-MCA as a substrate (for details, see Section IV).

3. Pool fractions containing 20S proteasomes from the Q-Sepharose column and add 50% polyethylene glycol 6000 (adjust to pH 7.4) at a final concentration of 15% with gentle stirring. After 15 min, centrifuge the mixture at 10,000 g for 20 min, dissolve the resulting precipitate in a minimum volume (approximately 50 ml) of buffer A, and centrifuge at 20,000 g for 10 min to remove insoluble material.

4. Fractionate the material precipitated with polyethylene glycol on a Bio-Gel A-1.5m column (5 × 90 cm) in buffer A. Collect 10-ml fractions and assay their proteasome activity. Pool fractions of 20S proteasomes.

5. Apply the active fractions from the Bio-Gel A-1.5m column directly to a column of hydroxylapatite equilibrated with buffer B. Wash the column with the same buffer and elute the material with 400 ml of a linear gradient of 10–300 mM phosphate. Collect 4-ml fractions. Elute the 20S proteasomes with about 150 mM phosphate.

6. Combine the active fractions from the hydroxylapatite, dialyze against buffer A, and apply to a coulmn of heparin–Sepharose CL-6B equilibrated with buffer A. Wash the column with the same buffer until the absorbance of the eluate at 280 nm returns to baseline. Then eluate with 200 ml of a linear gradient of 0–0.4 M NaCl in the same buffer and collect 2-ml fractions. Eluate the 20S proteasomes with approximately 75 mM NaCl.

7. Pool the fractions with high proteasomal activity, dialyze against buffer A, and concentrate to about 5 mg/ml protein by ultrafiltration in an Amicon cell with a PM-10 membrane. The enzyme can be stored at −80°C for at least 2 to 3 years.

B. Preparation of Dumbbell-Shaped 26S Proteasome

Steps

1. Homogenize 200- to 400-g samples of animal tissues in 3 vol of 25 mM Tris–HCl buffer (pH 7.5) containing 1 mM DTT, 2 mM ATP, and 0.25 M sucrose in a Potter–Elvehjem homogenizer. Centrifuge the homogenate for 1 hr at 70,100 g, and use the resulting supernatant as the starting material.

2. Recentrifugation of the crude supernatant for 5 hr at 70,100 g precipitates 26S proteasomes almost completely. Dissolve the precipitate in a suitable volume (40–50 ml) of buffer A and centrifuge at 20,000 g for 30 min to remove insoluble material.

3. Apply samples of the preparation from step 2 to a Bio-Gel A-1.5m column (5 × 90 cm) in buffer C. Collect 10-ml fractions and assay the 26S proteasome activity in the fractions (for the assay, see Section IV). Pool fractions of 26S proteasomes.

4. Add ATP at a final concentration of 5 mM to the pooled fractions of 26S proteasomes from the Bio-Ggel A-1.5m column. Apply a sample directly to a hydroxylapatite column with a 50-ml bed volume that has been equilibrated with buffer D. Recover the 26S proteasomes in the flow-through fraction, as they do not associate with this column in the presence of 5 mM ATP. Approximately 70% of the proteins, including free 20S proteasomes, bind to the hydroxylapatite resin.

5. Apply the flow-through fraction from the hydroxylapatite column to a Q-Sepharose column that has been equilibrated with buffer C without ATP and washed with 1 bed volume of buffer C. Wash the column with 5 bed volumes of buffe C and elute the adsorbed materials with 300 ml of a linear gradient of 0–0.8 M NaCl in the same buffer. Collect 3.0-ml fractions of eluate. Proteins with the ability to degrade Suc-LLVY-MCA with or without 0.05% SDS are eluted with about 0.4 M NaCl as a single symmetrical peak. ATPase activity and the ATP-dependent activity to degrade ^{125}I-lysozyme-Ub conjugates are observed at the same position as the peptidase activity and are eluted as superimposable symmetrical peaks, suggesting a specific association of ATPase with the 26S proteasome complex. Collect protein in the fractions exhibiting high activity.

6. Concentrate the 26S proteasome fraction obtained by Q-Sepharose chromatography to 2.0 mg/ml by ultrafiltration with an Amicon PM-30 membrane and subject 2.0-mg samples of protein to 10–40% glycerol density-gradient centrifugation (30 ml in buffer C containing 2 mM ATP). Centrifuge for 22 hr at 82,200 g in a SW rotor, and collect fractions of 1 ml from the bottom of the centrifuged tube. A single major peak of peptidase activity in the absence of SDS is eluted around fraction 15, but when the activity is assayed with 0.05% SDS, another small peak is observed around fraction 20. The latter peak corresponds to the elution position of 20S proteasomes. ATPase activity is observed at the same position as peptidase activity. Activity for the ATP-dependent degradation of ^{125}I-lysozyme-Ub conjugates is also observed as a single symmetrical peak, coinciding in position with the ATPase and peptidase activities in the absence of SDS. No significant ^{125}I-lysozyme-Ub conjugate-degrading activity is detected in fractions of 20S proteasomes. Pool fractions 12–16 and store at −80°C.

Preparation of the Regulatory Complex (PA700) of 26S Proteasomes. Steps for homogenization, ultracentrifugation, and Bio-Gel A-1.5m gel filtration are the same for the preparation of 26S proteasomes. Apply the pooled fractions of 26S proteasomes from the Bio-Gel A-1.5m column directly to a hydroxylapatite column with a 50-ml bed volume that has been equilibrated with buffer C. Wash the column with the same buffer, and elute the adsorbed materials with 300 ml of a linear gradient of 10–300 mM phosphate. Collect 3.0-ml fractions of eluate. Note that the 26S proteasome can be adsorbed in the hydroxylapatite column under a low concentration of ATP and that the 20S proteasome and PA700 regulatory complex can be eluted separately at different phosphate concentrations of approximately 150 and 50 mM, respectively. The 20S proteasome is detected by measuring the Suc-LLVY-MCA-degrading activity with 0.05% SDS as described earlier, whereas the PA700 complex is monitored by immunoblotting with antibodies against their subunits reported so far (Tanaka, 1995). Collect protein in fractions containing the PA700 complex from hydroxylapatite chromatography, concentrate it to 2.0 mg/ml by ultrafiltration with an Amicon PM-30 membrane, and subject 1.0- and 2.0-mg samples of protein to 10–30% glycerol density-gradient centrifugation (see step 6 in the preparation of the 26S proteasome). Collect 1-ml fractions from the bottom of the centrifuged tube. The PA700 complex can be monitored by ATPase activity and/or an immunoblotting analysis mentioned earlier. Pool fractions 14–18 containing the PA700 complex and store at −80°C.

C. Preparation of Football-Shaped Proteasome

Steps

1. Perfuse animal tissues with 25 mM Tris–HCl buffer (pH 7.5) containing 1 mM DTT, 1 mM phenylmethylsulfonyl fluoride, 20 μg/ml E64, and 0.25 M sucrose, and then homogenize in 3 vol of the same buffer in a Potter–Elvehjem homogenizer. Centrifuge the homogenate for 1 hr at 70,100 g, and use the resulting supernatant as the crude extract.

2. Apply the crude extract directly to a Q-Sepharose column that has been equilibrated with buffer A. Wash the column with 5 bed volumes of buffer A and elute adsorbed materials with a linear gradient of 0–0.8 M NaCl in the same buffer. For detection of the 20S proteasome activator protein termed PA28, the hydrolysis of Suc-LLVY-MCA is assayed after preincubation for 10 min at 4°C with approximately 0.5 μg of the latent 20S proteasome purified as described in Section A. Eluate the PA28 activator with about 0.3 M NaCl. The endogenous activities of 20S and 26S proteasomes are monitored by assaying Suc-LLVY-MCA degradation with or without 0.05% SDS, respectively, which are coeluted at approximately 0.45 M NaCl.

3. Combine the PA28 fractions from the Q-Sepharose column, dialyze against buffer A, and apply it to a heparin–Sepharose CL-6B column that has been equilibrated in the same buffer. Recover PA28 in the flow-through fraction, as it does not bind to this resin.

4. Apply the flow-through fraction from the heparin–Sepharose CL-6B column directly to the hydroxylapatite column that has been equilibrated with buffer B. Wash the column with 5 bed volumes of the same buffer and elute the adsorbed material with a linear gradient of 10–200 mM phosphate. Pool fractions of PA28.

5. Concentrate the PA28 activator from the hydroxylapatite column to 2.0 mg/ml by ultrafiltration with an Amicon PM-10 membrane. Incubate the PA28 with the purified 20S proteasome (about 0.5 mg) for 30 min at 4°C to form the PA28–20S proteasome complex, and subject samples of 2.0 mg of protein to 10–40% glycerol density-gradient centrifugation (see step 6 in the preparation of the 26S proteasome). Note that excess PA28 should be used for association with the 20S proteasome because the PA28–20S proteasome complex is hardly separated from the 20S proteasome, unlike the PA28 complex, by the density-gradient centrifugation analysis. Collect 1-ml fractions from the bottom of the centrifuged tube. The PA28–20S proteasome complex is monitored by assaying Suc-LLVY-MCA-degrading activity. Pool fractions 14–18 and store at −80°C.

Preparation of the 20S Proteasomal Activator (PA28). Apply the PA28–20S proteasome complex from the glycerol density-gradient centrifugation, a final material for preparation of the football-shaped proteasome, directly to a Q-Sepharose column that has been equilibrated in buffer A and wash extensively with the same buffer. Elute the adsorbed materials with a linear gradient of 0–0.8 M NaCl in the same buffer, as 20S proteasome and PA28 are separated by this column operation. Eluate the PA28 activator with about 0.3 M NaCl (for details, see step 2). Pool fractions containing PA28 and store at −80°C.

D. Properties of Isolated Proteasomes

The protein and gene structures of all proteasomes are described in various reviews (Rivett, 1993; Peters, 1994; Coux *et al.*, 1996; Tanaka *et al.*, 1997). On electron microscopy, the 26S proteasome appears dumbbell-shaped, consisting of two rectangular domains attached to a thinner central structure with four protein layers that we assume is the 20S proteasome (Fig. 1). 20S and 26S proteasomes from other eukaryotic cells have essentially the same molecular properties, suggesting that their gross sizes and shapes have been highly conserved during evolution (Lupas *et al.*, 1993). The 26S proteasome with approximately 2000 kDa is composed of components of both 21–31 and 25–110 kDa, the former and latter components are the 20S proteasome and PA700 regulatory complex, respectively (Tanaka, 1995). The proteasome activator PA28 is composed of three family of subunits called PA28α, PA28β, and PA28γ with 28–32 kDa, but the native molecule has a molecular mass of about 170–180 kDa, suggesting that it is an oligomeric complex consisting of at least six copies of the PA28 subunit (DeMartino and Slaughter, 1993). PA28 forms a ring-shaped particle, and caps that are associated with the 20S proteasome at both or either end give a football-like appearance (Fig. 1). The overall structural organization of the football-like proteasome somewhat resembles that of the 26S dumbbell-shaped proteasome, indicating that PA28 occupies the same site on the 20S core particle as the 26S proteasomal regulator does on the 26S proteasome complex (Coux *et al.*, 1996).

IV. COMMENTS

For measuring 20S proteasomal activity, various fluorogenic peptides are useful because proteasomes show broad substrate specificity. However, Suc-LLVY-MCA is recommended as a sensitive substrate. Latent 20S proteasomes can be activated in various ways (Rivett, 1993). We recommend using SDS at low concentrations of 0.02 to 0.08% for the activation of Suc-LLVY-MCA breakdown; the optimal concentration depends on the enzyme source and the protein concentration used. The fluorogenic peptide (Suc-LLVY-MCA) can be used for assaying 26S proteasomes and the PA28–20S proteasome complex, but they are active without any treatment, unlike latent 20S proteasomes. For specific assays, ATP-dependent degradation of ubiquitinated ^{125}I-lysozyme should be measured, although this assay is not easy, as multiple Ub-ligated enzymes must be purified for the *in vitro* preparation of the ubiquitinated substrate (for procedures, see Tamura *et al.*, 1991). Because 26S proteasomes and the PA700 regulator complex have intrinsic ATPase activity, their purification can be monitored by measuring ATPase activity at later steps of their purifications.

Proteasomes have been purified from a variety of eukaryotic cells by many investigators. Many purification methods have been reported, but special techniques are not necessary because 20S proteasomes are very stable and abundant in cells, consituting 0.5–1.0% of the total cellular proteins. Thus, other procedures described here, such as cellulose phosphate gel or phenyl–Sepharose chromatography, are also useful for their purification. Procedures for the purification of 20S proteasomes obviously differ, depending on whether they are small or large operations. For their isolation from small amounts of starting materials, such as cultured cells, 10–40% glycerol density-gradient centrifugation analysis is very effective. 20S proteasomes are present in a latent form in cells and can be isolated in this form in the presence of 20% glycerol. For their isolation in high yield, a key point is to keep them in their latent form because their activation results in an autolytic loss of a certain subunit(s) and in a marked reduction of enzymatic activities, particularly their hydrolyses of various proteins. Accordingly, all buffers used contain 20% glycerol as a stabilizer. A reducing agent is required because 20S proteasomes precipitate in its absence. All purification procedures are performed at 4°C, but operations in an HPLC apparatus can be carried out within a few hours at room temperature.

For purification of the 26S proteasome, ATP (0.5 or 2 mM), together with 20% glycerol and 1 mM DTT, should be added to all solutions used because they strongly stabilize the 26S proteasome complex: the purified enzyme is stable during storage at −70°C for at least 6 months in the presence of 2 mM ATP and 20% glycerol. Various drastic chromatographies should be avoided because they may result in dissociation of the 26S complex into its constituents. Alternative methods of purification of the PA28 and PA700 regulatory complexes are reviewed by DeMartino and Slaughter (1993).

References

Coux, O., Tanaka, K., and Goldberg, A. L. (1996). Structure and functions of the 20S and 26S proteasomes. *Annu. Rev. Biochem.* **65**, 801–847.

DeMartino, G. N., and Slaughter, C. A. (1993). Regulatory proteins of the proteasome. *Enzyme Protein* **47**, 314–324.

Hochstrasser, M. (1995). Ubiquitin, proteasomes, and the regulation of intracellular protein degradation. *Curr. Opin. Cell Biol.* **7**, 215–223.

Lupas, A., Koster, A. J., and Baumeister, W. (1993). Structural features of 26S and 20S proteasomes. *Enzyme Protein* **47**, 252–273.

Peters, J.-M. (1994). Proteasomes: Protein degradation machines of the cells. *Trends Biochem. Sci.* **19**, 377–382.

Rivett, A. J. (1993). Proteasomes; multicatalytic proteinase complexes. *Biochem. J.* **291**, 1–10.

Tamura, T., Tanaka, K., Tanahashi, N., and Ichihara, A. (1991). Improved method for preparation of ubiquitin-ligated lysozyme as substrates of ATP-dependent proteolysis. *FEBS Lett.* **292**, 154–158.

Tanaka, K., and Tsurumi, C. (1997). The 26 Sproteasome: subunits and functions. *Mol. Biol. Rep.* **24**, 3–11.

Tanaka, K., Tanahasi, N., Tsurumi, C., Yokota, K., and Shimbara, N. (1997). Proteasomes and antigen processing. *Adv. Immunol.*, in press.

Preparation of Ribosomes and Ribosomal Proteins from Cultured Cells

Anna Greco and Jean-Jacques Madjar

I. INTRODUCTION

For many studies concerning cell metabolism or gene expression, including ribosomal protein gene expression under normal or pathological conditions, it may be useful to prepare ribosomes and ribosomal proteins from eukaryotic cells in culture. Moreover, synthesis of ribosomal proteins being under translational control (Amaldi and Pierandrei-Amaldi, 1996; Meyuhas et al., 1996), preparation of a mixture of pure ribosomal proteins may also be required for the study of ribosome assembling. This can be easily achieved by simple and reproducible procedures of cell fractionation that do not require specific equipment. Ribosomes consist of about 80 basic proteins that interact with each other and with four different ribosomal RNAs in the ribosomal particle made up of two subunits (Wool et al., 1996). In an active ribosome, they also interact with many other proteins involved in the complicated process of protein synthesis and with other RNA molecules such as tRNA and mRNA. Preparing a mixture of pure total ribosomal proteins (TP80S) implies getting rid of contaminating nonribosomal proteins and of virtually all RNA molecules. The procedure described in this article allows us to reach this goal in a way that permits the subsequent separation and analysis of all ribosomal proteins by two-dimensional polyacrylamide gel electrophoresis (Kaltschmidt and Wittmann, 1970; Madjar et al., 1979; Sherton and Wool, 1972).

II. MATERIALS AND INSTRUMENTATION

Acetic acid (Art. 100063), acetone (Art. 100014), fuming hydrochloric acid (Art. 100317), magnesium acetate tetrahydrate (Art. 105819), magnesium chloride hexahydrate (Art. 105833), potassium chloride (Art. 104936), disodium hydrogen phosphate (Art. 106580), and potassium dihydrogen phosphate (Art. 104873) are from Merck. Guanidine hydrochloride (Cat. No. G-4505), iodoacetamide (Cat. No. I-6125), sodium chloride (Cat. No. S-5886), sucrose (Cat. No. S-0389), and 1,4-dithioerythritol (DTE, Cat. No. D-8255) are from Sigma. Nonidet P-40 (NP-40, Cat. No. 1 332 473) and 2-amino-2-(hydroxymethyl)-1,3-propanediol (Tris, Cat. No. 708 976) are from Boehringer Mannheim.

Cell tissue culture flasks of 162 cm^2 (Cat. No. 003 150) and 50-ml conical polypropylene tubes (Cat. No. 016 751) are from Costar. Dialysis tubes (1-8/32 in.) are from Medicell International Ltd. Microtubes of 1.5 ml (Cat. No. 33605) are from Eppendorf. Tubes of 15 ml (Corex tube, Cat. No. 03286) are from Sorvall. A TL-100 table-top ultracentrifuge with a TLA-100.3 fixed-angle rotor and thick-wall polycarbonate tubes (Part No. 349622) are from Beckman.

III. PROCEDURES

A. Cell Fractionation for Preparation of Ribosomes

Solutions

1. *Phosphate-buffered saline (PBS):* To make 2 liters, dissolve 15.3 g of NaCl, 1.45 g of $Na_2HPO_4 \cdot 2H_2O$, and 0.42 g of KH_2PO_4 in distilled water and adjust the volume to 2 liters with distilled water. Store at 4°C.

2. *Potassium chloride stock solution:* 2 M KCl. To make 1 liter, dissolve 149.12 g of KCl in distilled water and adjust the volume to 1 liter with distilled water. Store at 4°C.

3. *Magnesium chloride stock solution:* 1 M $MgCl_2$. To make 100 ml, dissolve 20.33 g of $MgCl_2 \cdot 6H_2O$ in distilled water and adjust the volume to 100 ml with distilled water. Store at 4°C.

4. *Tris–HCl, pH 7.4, stock solution:* 1 M Tris–HCl, pH 7.4. To make 100 ml, dissolve 12.11 g of Tris in 80 ml of distilled water. Adjust the pH to 7.4 with fuming HCl at room temperature. Adjust the volume to 100 ml with distilled water. Store at 4°C.

5. *Buffer A:* 0.25 M sucrose, 25 mM KCl, 5 mM $MgCl_2$, and 50 mM Tris–HCl, pH 7.4. To make 100 ml of the buffer, add 8.55 g of sucrose, 1.25 ml of a 2 M stock solution of KCl, 0.5 ml of a 1 M stock solution of $MgCl_2$, and 5 ml of a 1 M stock solution of Tris–HCl, pH 7.4. After dissolving, complete to 100 ml with distilled water. Store at 4°C.

6. *Buffer B:* 0.25 M sucrose, 0.5 M KCl, 5 mM $MgCl_2$, and 50 mM Tris–HCl, pH 7.4. To make 100 ml of the buffer, add 8.55 g of sucrose, 25 ml of a 2 M stock solution of KCl, 0.5 ml of a 1 M stock solution of $MgCl_2$, and 5 ml of a 1 M stock solution of Tris–HCl, pH 7.4. After dissolving, complete to 100 ml with distilled water. Store at 4°C.

7. *Buffer C:* 0.25 M sucrose, 2 M KCl, 5 mM $MgCl_2$, and 50 mM Tris–HCl, pH 7.4. To make 10 ml of the buffer, add 0.85 g of sucrose, 1.49 g of KCl, 50 μl of a 1 M stock solution of $MgCl_2$, and 0.5 ml of a 1 M stock solution of Tris–HCl, pH 7.4. After dissolving, complete to 10 ml with distilled water. Store at 4°C.

8. *Sucrose cushion:* 1 M sucrose, 0.5 M KCl, 5 mM $MgCl_2$, and 50 mM Tris–HCl, pH 7.4. To make 50 ml of the buffer, add 17.1 g of sucrose, 12.5 ml of a 2 M stock solution of KCl, 0.25 ml of a 1 M stock solution of $MgCl_2$, and 2.5 ml of a 1 M stock solution of Tris–HCl, pH 7.4. After dissolving, complete to 50 ml with distilled water. Store at 4°C.

9. *Nonidet P-40:* 20% (w/v) NP-40 in H_2O. To make 10 ml of the solution, weigh 2 g of NP-40 and complete to 10 ml with distilled water. Store at room temperature.

Steps

1. Wash the cells in the 162-cm^2 flask (about 15×10^6 HeLa cells) three times with cold PBS kept at 4°C. Scrape the cells off the flask with a cell lifter in 10 ml of PBS and pour into a 50-ml conical polypropylene tube kept on ice. Rinse the flask twice with 5 ml of PBS, and add the buffer containing the residual cells to the same conical tube.

2. Centrifuge cell suspension at 500 g for 5 min at 4°C. Remove the supernatant and resuspend the cells in 0.3 ml of cold buffer A (about three times the volume of the cell pellet). Transfer the cell suspension to a 1.5-ml microtube.

3. Stir slowly with a Vortex while adding 14 μl of the 20% NP-40 solution to make it 0.7% and keep on ice for 10 min.

4. Centrifuge the cell lysate at 750 g for 10 min at 4°C to spin down the nuclei. Carefully remove the postnuclear supernatant and put it in another microtube. If necessary, save the pellet for other studies.

5. Centrifuge the postnuclear supernatant at 12,500 g for 10 min at 4°C to spin down the mitochondria. Carefully decant the postmitochondrial supernatant into another microtube. If necessary, save the pellet for other studies.

6. Measure precisely the volume (which should be about 0.4 ml) and add 0.32 vol of buffer C (128 μl for 0.4 ml) to adjust the final concentration of KCl to 0.5 M.

7. Layer the postmitochondrial supernatant on top of 1 ml of the 1 M sucrose cushion in a thick-wall polycarbonate centrifuge tube. Equilibrate the tubes precisely by adding a few drops of buffer B. Place the tubes in the TLA-100.3 fixed-angle rotor. It is not necessary to fill the tubes.

8. After centrifugation in the TL-100 table-top Beckman ultracentrifuge for 2 hr at 75,000 rpm (245,000 g) at 4°C, with the brake on for maximum deceleration rate, remove the supernatant on top of the translucent pellet of ribosomes. If necessary, save the supernatant for other studies. Quickly rinse the pellet twice with distilled water.

9. Resuspend the ribosome pellet in 100 μl of buffer A. Wash the tube twice with 100 μl of buffer A and pool the 300 μl into an Eppendorf microtube.

10. Measure the optical density at 260 nm. Estimate the amount of ribosomes assuming that 14 A_{260} units correspond roughly to 1 mg of ribosomes and to about 0.5 mg of total ribosomal proteins. Using this procedure, 15×10^6 HeLa cells (grown in one 162-cm^2 flask) give about 2.8 A_{260} units, i.e., about 200 μg of ribosomes and about 100 μg of total ribosomal proteins (TP80S).

B. Extraction, Alkylation, and Lyophilization of Ribosomal Proteins

Solutions

1. *Magnesium acetate:* 1 M magnesium acetate. To make 100 ml of the solution, weigh 21.45 g of magnesium acetate. After dissolving in distilled water, complete to 100 ml with distilled water. Store at 4°C.

2. *Acetic acid:* 1 M acetic acid. To make 1 liter of the solution, measure 57 ml of glacial acetic acid and complete to 1 liter with distilled water. Store at 4°C.

3. *Tris base:* 1 M Tris base. To make 100 ml of the solution, weigh 12.14 g of Tris base. After dissolving in distilled water, complete to 100 ml with distilled water. Do not adjust the pH. Store at 4°C.

4. *1,4-Dithioerythritol:* 0.5 M DTE. To make 10 ml of the solution, weigh 0.77 g of DTE. After dissolving in distilled water, complete to 10 ml with distilled water. Make 0.5-ml aliquots in microtubes and store at −20°C.

5. *Guanidine hydrochloride–Tris buffer*: 6 M guanidine hydrochloride, and 0.5 M Tris–HCl, pH 8.5. To make 10 ml, dissolve 5.7 g of guanidine hydrochloride in 5 ml of the 1 M Tris base solution. Adjust the pH to 8.5 with fuming HCl (37%). The final volume should then be 10 ml. Prepare just before use and filter if necessary.

6. *Reducing buffer:* 6 M guanidine hydrochloride, 0.5 M Tris–HCl, pH 8.5, and 10 mM dithioerythritol. Add 82 μl of the 0.5 M DTE solution to 4 ml of guanidine hydrochloride–Tris buffer. Prepare just before use.

7. *Alkylating buffer:* 6 M guanidine hydrochloride, 0.5 M Tris–HCl, pH 8.5, and 40 mM iodoacetamide. Weigh about 200 mg of iodoacetamide and dissolve in the guanidine hydrochloride–Tris buffer at the rate of 92.5 mg/ml. Prepare just before use.

Steps

1. Adjust the magnesium acetate concentration to 0.2 M by adding one-fourth of its volume of the 1 M magnesium acetate solution to the ribosome suspension.

2. Add 2 vol of glacial acetic acid, i.e., 2.5 vol of the initial ribosome solution. Stir with a Vortex and keep on ice for 1 hr.

3. Centrifuge at 12,000 rpm for 10 min at 4°C to pellet the RNA which appears as white.

Remove the supernatant containing the ribosomal proteins and transfer it in a 15-ml Corex tube.

4. Reextract the ribosomal proteins from the RNA pellet by resuspending the pellet in 200 μl of 0.1 M magesium acetate (20 μl of 1 M magnesium acetate plus 180 μl of distilled water) and 400 μl of glacial acetic acid. Stir with a Vortex and keep on ice for 10 min. Centrifuge again at 12,000 rpm for 10 min at 4°C. Remove the supernatant containing the residual ribosomal proteins and pool with the first ribosomal protein extract in the 15-ml Corex tube.

5. Precipitate the ribosomal proteins by adding at least 5 vol of cold acetone. Let the proteins precipitate at −80°C for 2 hr or at 4°C overnight.

6. Centrifuge at 10,000 rpm. Remove the liquid carefully and dry the pellet in a Speed-Vac Savant concentrator.

7. For reducing the proteins, solubilize the dried TP80S in the reducing buffer (1 to 10 mg/ml or less) in the Corex tube. Let stand at room temperature under nitrogen for 30 min. For this, flow nitrogen over the reducing buffer containing the solubilized TP80S and close the tube with a piece of Parafilm.

8. For alkylating the proteins, add 8.7 μl of alkylating buffer per 100 μl of solubilized proteins. Let stand at room temperature under nitrogen for 1.5 hr, as above.

9. Transfer the protein solution to a dialysis tube and rinse the Corex tube with the 1 M acetic acid solution. Dialyze at least three times against 100 vol of 1 M acetic acid at 4°C.

10. After dialysis, TP80S can be stored in 1 M acetic acid at −20°C for years. For analysis by two-dimensional gel electrophoresis (2D PAGE), TP80S are first lyophilized in a microtube in a Speed-Vac concentrator.

IV. COMMENTS

By using the protocol described in this article it is possible to prepare crude ribosomes from cell tissue culture with a good yield. They are salt washed by 0.5 M KCl and are prepared from a mixture of both free and membrane-bound ribosomes after lysis of the cell membranes with Nonidet P-40; however, this treatment does not disrupt the inner nuclear membrane, allowing easy disposition of the nuclei during the cell fractionation procedure. TP80S can be isotopically labeled. TP80S extracted by the acetic acid procedure are extremely stable in 1 M acetic acid, in which they can be stored frozen at −20°C or less. The alkylation step by iodoacetamide after reduction by dithioerythritol allows the blocking of all the SH groups without change of the net charge of the protein (Madjar and Traut, 1980). Rinsing the tube with 1 M acetic acid and dialyzing against the same solution permit lowering the pH to stop the alkylation reaction. After lyophilization, TP80S are free of contaminating salts and are very easily solubilized in the sample buffer containing at least 8 M urea. Under these conditions, TP80S can be separated by 2D PAGE, according to charge in the presence of at least 8 M urea in the first dimension, according to mass in the second dimension, in the presence of either SDS or urea in a highly reticulated polyacrylamide gel. As most ribosomal proteins are very basic, they are separated by electrophoresis by migration from the anode to the cathode (except if the second dimension is run in the presence of SDS). Under these conditions, alkylation of the proteins avoids possible reoxidation of the proteins which are overflowed by the ammonium persulfate migrating in the opposite direction. Moreover, alkylation of the SH groups does not prevent microsequencing of the proteins (Diaz *et al.*, 1993).

V. PITFALLS

1. To avoid breaking the cells, do not spin them down at more than 500 g.

2. To obtain the postmitochondrial supernatant, first spin down the nuclei after the cell lysis. The risks involved in omitting this step are disruption of the nuclei during centrifugation of the mitochondria at 12,500 g and contamination (with fragments of DNA and nucleosomes) of ribosomes obtained through the 1 M sucrose cushion. This contamination can be visualized by the presence of histones among ribosomal proteins after separation by 2D PAGE.

3. The ribosome pellet must be resuspended very carefully to avoid the loss of ribosomes, to measure the optical density with precision, and to obtain a good yield of protein extracted by the acetic acid procedure.

References

Amaldi, F., and Pierandrei-Amaldi, P. (1996). Top genes: A translationally controlled class of genes including those coding for ribosomal proteins. *In* "Cytoplasmic Fate of Eukaryotic mRNA" (P. Jeanteur, ed.), *Prog. Mol. Subcell. Biol.*, Vol. 18, pp. 1–17. Springer-Verlag, Heidelberg.

Diaz, J.-J., Simonin, D., Massé, T., Deviller, P., Kindbeiter, K., Denoroy, L., and Madjar, J.-J. (1993). The herpes simplex virus type 1 US11 gene product is a phosphorylated protein found to be nonspecifically associated with both ribosomal subunits. *J. Gen. Virol.* **74**, 397–406.

Kaltschmidt, E., and Wittmann, H. G. (1970). Two-dimensional polyacrylamide gel electrophoresis for fingerprinting of ribosomal proteins. *Anal. Biochem.* **36**, 401–412.

Madjar, J.-J., Arpin, M., Buisson, M., and Reboud, J.-P. (1979). Spot position of rat liver ribosomal proteins by four different two-dimensional electrophoreses in polyacrylamide gel. *Mol. Gen. Genet.* **171**, 121–134.

Madjar, J.-J., and Traut, R. R. (1980). Differences in electrophoretic behaviour of eight ribosomal proteins from rat and rabbit tissues and evidence for proteolytic action on liver proteins. *Mol. Gen. Genet.* **179**, 89–101.

Meyuhas, O., Avni, D., and Shama, S. (1996). Translational control of ribosomal protein mRNA in eukaryotes. *In* "Translational Control", (J. W. B. Hershey, M. B. Mathews, and N. Sonenberg, eds.), pp. 363–388. Cold Spring Harbor Laboratory Press, Cold Spring Harbor, NY.

Sherton, C. C., and Wool, I. G. (1972). Determination of the number of proteins in liver ribosomes and ribosomal subunits by two-dimensional polyacrylamide gel electrophoresis. *J. Biol. Chem.* **247**, 4460–4467.

Wool, I. G., Chan, Y-L., and Glück, A. (1996). Mammalian ribosomes: The structure and the evolution of the proteins. *In* "Translational Control", (J. W. B. Hershey, M. B. Mathews, and N. Sonenberg, eds.), pp. 685–718. Cold Spring Harbor Laboratory Press, Cold Spring Harbor, NY.

Nucleus and Nuclear Structures

Isolation of Yeast Nuclear Pore Complexes and Nuclear Envelopes

Michael P. Rout and Caterina Strambio-de-Castillia

I. INTRODUCTION

The yeast *Saccharomyces*, a mainstay system for the geneticist and molecular biologist, has become increasingly amenable to cell biological and biochemical techniques. This, plus the completion of the *Saccharomyces* genome sequence project, has considerably increased the potential of this organism as a model system for the cell biologist. A method for the production of a highly enriched fraction of yeast nuclei has been described previously (see Volume 2, Rout and Kilmartin, "Preparation of Yeast Spindle Pole Bodies" for additional information), as well as a procedure for the isolation of spindle pole bodies from this fraction. This article describes two more isolation procedures for yeast, one producing nuclear pore complexes (NPCs) and the other producing nuclear envelopes (NEs) and nuclear membranes. Both result in material of sufficiently high yield and degree of enrichment to be potentially useful in a variety of preparative and analytical studies, including the identification of NPC components and their localization in the NE. The relationship between these procedures is diagrammed in Fig. 1.

II. MATERIALS AND INSTRUMENTATION

All catalog numbers are indicated in parentheses. Nycodenz (Accudenz, AN 7050) is from Accurate Chemical and Scientific Corporation, Bradford's assay stock solution (Bio-Rad protein assay, 500-0006) is from Bio-Rad, heparin (sodium salt, grade 1-A; H-3393) and sodium taurodeoxycholate (T-0875) are from Sigma, dithiothreitol (DTT, Cleland's reagent, 233155) is from Calbiochem, formaldehyde (~37% stock solution in 10% methanol, 47629) is from Fluka, glutaraldehyde (~25% stock solution, 360802F) is from BDH, Tween 20 (Surfact-Amps-20, 10% stock solution, 28320) is from Pierce, pancreatic RNase (RNase A, 109169) is from Boehringer, and centrifuge tubes are from Beckman (SW55 Ultraclear, 344057) or from Sorvall (10 ml Ultra bottles, 03020, and screw caps, 03613). Carbon/formvar-coated 300 mesh copper electron microscope grids (01821) and uranyl acetate (19481) are from Ted Pella, Inc. All other materials were obtained as described (see article by Rout and Kilmartin, this volume).

III. PROCEDURES

As discussed, both procedures require the same starting material, highly enriched yeast nuclei (Rozijn and Tonino, 1964; Kilmartin and Fogg, 1982; Rout and Kilmartin, 1990). The method

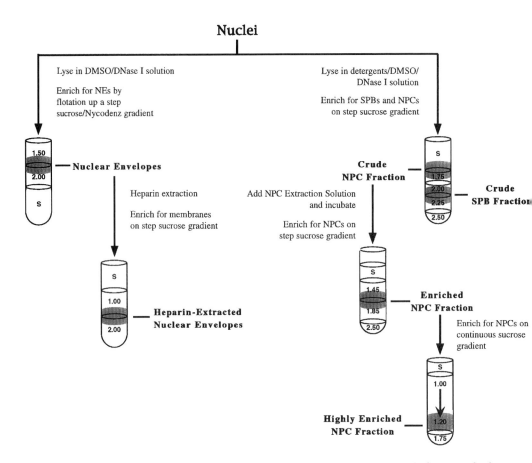

Nuclei

Lyse in DMSO/DNase I solution

Enrich for NEs by
flotation up a step
sucrose/Nycodenz gradient

1.50
2.00
S

Nuclear Envelopes

Heparin extraction

Enrich for membranes
on step sucrose gradient

S
1.00
2.00

**Heparin-Extracted
Nuclear Envelopes**

Lyse in detergents/DMSO/
DNase I solution

Enrich for SPBs and NPCs
on step sucrose gradient

S
1.75
2.00
2.25
2.50

**Crude
NPC Fraction**

**Crude
SPB Fraction**

Add NPC Extraction Solution
and incubate

Enrich for NPCs on
step sucrose gradient

S
1.45
1.85
2.50

**Enriched
NPC Fraction**

Enrich for NPCs on
continuous sucrose
gradient

S
1.00
1.20
1.75

**Highly Enriched
NPC Fraction**

FIGURE 1 Flow diagram of the various enrichment procedures using the highly enriched yeast nuclei fraction a
their starting material.

for the production of this nuclear fraction has been described previously (see article by Rout
and Kilmartin, this volume). The best results are obtained with nuclei from S. *uvarum*, but
both procedures have been successfully applied to S. *cerevisiae* strains with the inclusion of
the appropriate modifications where necessary.

A. Preparation of a Highly Enriched Nuclear Pore Complex Fraction

Solutions

All solutions other than those listed below were prepared as described previously (see Volume
2, Rout and Kilmartin, "Preparation of Yeast Spindle Pole Bodies" for additional information).
The following solutions can all be stored for long periods at either 4°C or −20°C.

1. *NPC extraction solution:* 10 mM bis–Tris–HCl, pH 6.50, 0.1 mM MgCl$_2$, 1.0% sodium
taurodeoxycholate, 10 μg / ml RNase A, and 0.5 mM DTT.

2. *Sucrose solutions in bt buffer:* Prepare the 1.45 and 1.85 M sucrose–bt solutions
(refractive indexes of 1.4032 and 1.4225, respectively) by diluting the stock 2.50 M sucrose–
bt solution with bt buffer plus 0.02 vol of 10% Triton X-100. It is very important that the
refractive index (measured at room temperature), rather than weight or volume, is used to
measure the molarity of the sucrose solutions.

3. *Sucrose solutions in bt–DMSO buffer:* Prepare 1.75 M sucrose–bt–DMSO solution
by dissolving 120 g sucrose in 20 ml of 0.1 M bis–Tris–HCl, pH 6.50, 20 μl 1 M MgCl$_2$,
0.2 ml 10% Tween 20, and distilled water to a total volume of 150 ml. When the sucrose
has dissolved, add 40 ml of DMSO and make up to a total volume of 200 ml with distilled
water. Make sure that the refractive index is close to 1.450. Prepare 1.00 sucrose–bt–DMSO
solution by dissolving 342.3 g sucrose in 100 ml of 0.1 M bis–Tris–HCl, pH 6.50, 100 μl 1

M MgCl$_2$, 1 ml 10% Tween 20, and distilled water to a total volume of 700 ml. When the sucrose has dissolved, add 200 ml of DMSO and make up to a total volume of ~1000 ml with distilled water to a refractive index of 1.4120. Prepare 1.20 M sucrose–bt–DMSO solution similarly, but by starting with 408.9 g sucrose and making to a refractive index of 1.4220.

4. *Solution P2:* Dissolve 90 mg of 4-(2-aminoethyl)benzenesulfonyl fluoride and 2 mg pepstatin A in 5 ml dry absolute ethanol.

Steps

1. All solutions contain a 1:1000 dilution of solution P (freshly added) and are kept on ice, unless otherwise stated. Solution P can also be substituted with the more stable (and less toxic) solution P2. Nuclear lysis and the separation of the lysate over a stepped sucrose gradient are exactly as described, with the crude NPC fraction being recovered from the S/1.75 M interface. Usually, ~5 ml of the fraction is recovered per Beckman SW28 tube, resulting in a total of 15–20 ml per 36-liter culture (Rout and Kilmartin, 1990; see Volume 2, Rout and Kilmartin, "Preparation of Yeast Spindle Pole Bodies" for additional information).

2. The S/1.75 M fraction must be assayed in order to determine the correct amount of heparin to add for the next enrichment step. Three assays can be used to arrive at a consensus.

Assay 1: To 100-μl aliquots of the S/1.75 M fraction, add 20 μl of NPC extraction solution and various amounts of 10- and 100-mg/ml heparin stock solutions, covering a range of 0–0.5 mg/ml final heparin concentrations. Incubate for 1 hr on ice and then measure and compare their absorbance ($A_{600 \text{ nm}}$) with a 100-μl aliquot of the S/1.75 M fraction containing no additions. The correct amount of heparin (together with the NPC extraction solution) usually results in a reduction of turbidity to 40% of the value of the untreated fraction.

Assay 2: Place 5-μl aliquots of Assay 1 incubations onto glow-discharged electron microscope grids and allow to sit for 5 min at room temperature. Rinse each grid with two drops of bt–DMSO and then invert (sample side down) onto a drop of bt–DMSO containing 4% formaldehyde and 0.5% glutaraldehyde sitting on a sheet of Parafilm. After 10 min at room temperature, rinse the grid once more with several drops of bt–DMSO and then stain with 6–10 drops of 4% uranyl acetate in distilled water. Wick away the excess staining solution with a sliver of filter paper; after air drying the grid is ready to examine in the electron microscope. NPCs appear as donuts containing a plug in the middle; too much heparin causes eventual disintegration of the plug and the round donuts become distorted and oval. If there is too little heparin, however, contaminating material appears as aggregates of the same size or larger than the NPCs; enough heparin must be added to reduce the particulate contaminants to a size significantly smaller than that of the NPCs.

Assay 3: Determine the protein concentration of the S/1.75 M crude NPC fraction using the Bradford protein assay from Bio-Rad following the manufacturer's instructions. The correct amount of heparin (from a 100-mg/ml stock) to add to the NPC extraction reaction (step 4) is then 0.045 mg per 1.0 mg of the S/1.75 M fraction protein. This assay is the most reliable and the one used routinely.

3. Before starting, make up the required number of gradient tubes: one Beckman Ultraclear SW55 tube per 1 ml of S/1.75 M fraction to be processed, each containing 0.5 ml of the 2.50 M sucrose–bt solution, 1.5 ml of the 1.85 M sucrose–bt solution, and 1.5 ml of the 1.45 M sucrose–bt solution (layered from bottom to top in that order). Also prepare a beaker full of distilled water prechilled by the addition of ice to a temperature of 10°C.

4. To each 1 ml of the S/1.75 M fraction in a 50-ml Falcon tube, quickly add 0.2 ml of NPC extraction solution, 5 μl Solution P, and the correct amount of heparin (usually 0.045 mg heparin per milligram of fraction protein, as in Assay 3). Start a timer immediately and thoroughly vortex the mixture for 30 sec before placing the tube in the chilled beaker of water.

5. Incubate for a total of 15 min, vortexing the tube for 5 sec every 3 min.

6. Centrifuge the tube to remove the froth (700 g, 4 min, 4°C) and then load the resulting extracted fraction onto the gradient tubes, 1.2 ml per tube. Overlay this sample layer with

~0.4 ml of bt–DMSO (for balance) and centrifuge the tubes in a Beckman SW55 Ti rotor at 50,000 rpm for 5 hr at 4°C.

7. After the spin is complete, hold the tubes against a black background with a strong light illuminating from above. A strong sharp white band should be visible at the S/1.45 M interface whereas there should be a weaker diffuse white band at the 1.45/1.85 M interface. Occasionally, a faint sharp white band can be found immediately above the diffuse band, but this has not seemed to be a cause for concern. A very faint band is also often discernible at the 1.85/2.50 M interface, particularly if the tubes are allowed to warm slightly. Unload the tubes by volume from the top, at the walls of the tube. Collect the top fraction from *every* tube and pool before starting to collect the second fraction from any tube, and so on for each fraction. Collect the first 1.5 ml from each tube; the resulting pool from all tubes is termed the S fraction. The second 1.5-ml pool (containing the strong sharp white band) is called the S/1.45 fraction. The enriched NPC fraction is recovered in 1.3 ml from the 1.45/1.85 M interface (containing the weaker diffuse white band) and is termed the 1.45/1.85 fraction. The final 0.8 ml containing the 1.85/2.50 M interface (installed for diagnostic purposes; insufficient extraction will cause NPC material to begin to accumulate in this fraction) is called the 1.85/2.50 fraction. The fractions can be stored at this stage by freezing at −70°C. For most analytical purposes the enriched NPC fraction (in which the NPCs represent 20–30% of the total protein) is clean enough and has the advantage of being significantly more concentrated than the highly enriched NPC fraction. The appearance of this fraction as prepared by the method in Assay 2 is shown in Fig. 2. Note the presence of a significant amount of particulate contaminants.

8. If a highly enriched NPC fraction is required, then immediately beforehand make up one gradient tube per 2 ml of the 1.45/1.85-enriched NPC fraction to be processed, each being a Beckman Ultraclear SW28 tube containing a 5-ml cushion of the 1.75 M sucrose–bt–DMSO solution overlaid with a continuous linear gradient formed from 14.5 ml of the 1.20 M sucrose–bt–DMSO and 14.5 ml of the 1.00 M sucrose–bt–DMSO solution. These gradients are delicate and so must be handled carefully.

9. Dilute the enriched NPC fraction with an equal volume of bt–DMSO buffer containing a 1:500 dilution of solution P.

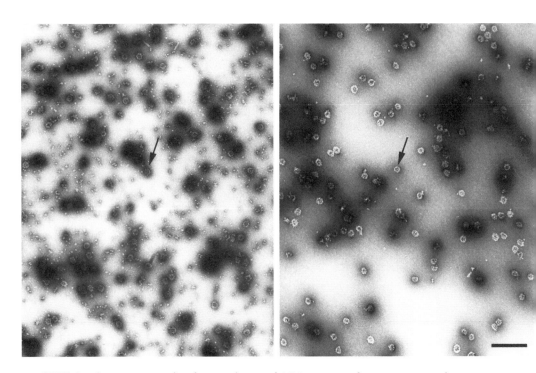

FIGURE 2 Electron micrographs of negatively stained NPC-containing fractions. Arrows indicate representative NPCs. Bar = 0.5 μm. (Left) Enriched NPC fraction (1.45/1.85) showing NPCs and contaminating particulate material. (Right) Highly enriched NPC fraction (No. 4) showing NPCs and much less contaminating material.

10. Overlay 4-ml aliquots of the diluted, enriched NPC fractions *gently* onto each gradient tube, and then overlay this sample layer *gently* with a little bt–DMSO for balance if necessary. Centrifuge all the tubes in a Beckman SW28 rotor at 28,000 rpm for 24 hr at 4°C.

11. After centrifugation, remove each tube from the rotor and hold it against a black background with a strong light illuminating it from above. There should be two diffuse bands visible: an extremely faint white band about one-third of the way down the tube, and a stronger band just above the 1.20/1.75 M interface; the latter band contains the NPCs.

12. Because the DMSO also contributes to the refractive indexes, it is difficult to completely prevent batch-to-batch variations in the sucrose–bt–DMSO solutions. Therefore, the desired separation is sometimes not seen. The 24-hr centrifuge run seems to be a minimum, so any such variation can usually be dealt with by centrifuging the tubes for longer times. Normally the tubes should be replaced in the rotor, spun again at 28,000 rpm for 2 hr at 4°C, and the position of the lower band noted again at the end of the run. Repeat this step until the required band separation is achieved.

13. Unload each gradient tube from the top, at the tube wall. Usually, a total of 5 fractions are collected from each gradient tube. The first 9 ml is termed fraction No. 1 and the second 9 ml, fraction No. 2. The third 9 ml, fraction No. 3, is collected to within about 5 mm above the lower diffuse white band. Fraction No. 4 is the 10 ml above and containing the 1.20/1.75 M interface. This fraction contains the NPCs and is referred to as the highly enriched NPC fraction. Fraction No. 5 is the final 1 ml or so collected after vortexing to resuspend any pellet. A more cautious approach is to collect more fractions, each containing less volume; for example, if the position of the NPCs is not certain, then 13 fractions may be collected, representing 3 ml per fraction from each gradient tube. Store the fractions at −70°C. Assay 2, or SDS–polyacrylamide gel analysis (below), can be used to determine the position of the NPCs, and the NPC-containing fractions are then pooled. The appearance of this final fraction, as prepared by the method in Assay 2, is shown in Fig. 2. Note the removal of almost all of the particulate contaminants.

An important method of quality control is to run SDS–polyacrylamide gels of aliquots from each fraction. An aliquot of each fraction proportional in volume to the total volume of that fraction should be taken. Hence, if 4.0 ml of the S/1.75 M fraction was processed to yield 6.0 ml of S, 6.0 ml of S/1.45, 5.2 ml of 1.45/1.85, and 0.8 ml of 1.85/2.50 fractions, then aliquots of (say) 12 μl of the S, 12 μl of the S/1.45, 10.4 μl of the 1.45/1.85, and 1.6 μl of the 1.85/2.50 fractions should be taken. An example of such an SDS–PAGE analysis is shown in Fig. 3, in which characteristic Coomassie-stained bands from known NPC proteins can be seen to coenrich with the NPC fractions. Failure of the preparation is immediately apparent after such an analysis, as the NPC protein bands cannot be seen or are found in the wrong fractions.

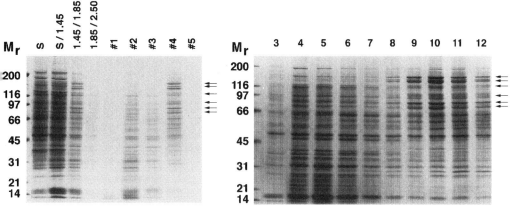

FIGURE 3 SDS–PAGE analysis of fractions from the NPC fractionation procedure, loaded as described in step 13. The fractions are indicated above each lane. Arrows indicate some groups of known coenriching NPC proteins. (Left) Profile for a typical preparation. (Right) Fractions from the last gradient in which 13 × 3-ml fractions were collected; only the protein-containing fractions (3–12) are shown.

B. Preparation of a Highly Enriched Nuclear Envelope Fraction

Solutions

All solutions other than those listed below were prepared as described for the NPC preparation (Section A) or in the article by Rout and Kilmartin (this volume).

1. *Sucrose/Nycodenz solution:* Warm 15 ml of H_2O in the microwave until it reaches close to boiling temperature. Add 20 g of Nycodenz and dissolve it by vigorous stirring on a heating plate. When the Nycodenz is dissolved, add 10.0 ml of 100 mM bis–Tris–HCl, pH 6.50, 10 μl of 1.0 M $MgCl_2$, and 71.88 g of sucrose and H_2O to just below 100 ml. After the sucrose has dissolved, let the solution cool before bringing the volume to 100 ml.

2. *Sucrose solutions in bt buffer:* Prepare 1.50 M (refractive index 1.4057), 2.00 M (refractive index 1.4295), and 2.25 M (refractive index 1.4414) sucrose–bt solutions by diluting the stock 2.50 M solution with bt buffer. It is very important that the refractive index (measured at room temperature), rather than weight or volume, is used to measure the molarity of the sucrose solutions.

Steps

1. Prepare nuclei pellets as described previously (see article by Rout and Kilmartin, this volume). Place the tubes containing the nuclei pellets on ice.

2. Carry out the following procedure in the cold room using one tube at a time. Add to the first tube the following precooled solutions: 1.0 ml bt–DMSO, 1.0 μl 2.0% DNase I, and 10.0 μl solution P per 100 OD_{260} units. Immediately resuspend the pellet by vortexing at the maximum setting. Note that it is extremely important that the solution swirls around the sides of the tube, such that the shear force needed to lyse the nuclei efficiently is developed. This step can take up to 10 min. It is important to start vortexing the tubes immediately after adding the nuclear lysis solution to the nuclei pellet. Typically all solutions should be prepared in advance, precooled on ice, and transported to the cold room where a vortexer is already in place. Add the solutions at the last minute, start the vortexing and continue with the minimum number of interruptions until the pellet has dissolved. When the pellet is completely resuspended, transfer the tube to room temperature, warm it to approximately 25°C by holding it in your hand, and then incubate it for 5–10 min at room temperature. The nuclear lysate must be properly warmed up before incubation at room temperature. If the lysate has not reached room temperature, then the DNA digestion may not be complete at the end of the incubation, and the nuclear envelopes will not float up properly in the ensuing gradient. After this incubation, replace the tube on ice. Repeat the procedure described in this step for the other tubes. Successful nuclear lysis may be checked by light microscopy. Under a 100× phase-contrast oil immersion objective the NEs appear as short black lines, many in the shape of a "C" (retaining the original curvature of the NE); there should be little other large particulate debris in the lysate, although there should be many small particles.

3. Add 1.0 vol of sucrose/Nycodenz solution to each tube, thoroughly mix the samples and centrifuge the tubes in a Beckman 70 Ti rotor or equivalent (i.e., Beckman 50.2 Ti or Sorvall T865 rotor) at 6000 rpm for 6 min at 4°C. Remove the supernatants without disturbing the pellets and pool in a fresh tube. Add 2.0 times the *initial* volume of sucrose/Nycodenz solution to the pooled supernatants and mix the samples thoroughly.

4. Divide the sample into the appropriate number of Beckman SW28 Ultraclear tubes by adding ~13 ml of it to each tube. A 36-liter nuclei preparation usually yields enough nuclear lysate to fill six Beckman SW28 Ultraclear tubes. Overlay each tube with 12 ml 2.00 M sucrose–bt and 12 ml 1.50 M sucrose–bt, and centrifuge in a Beckman SW28 rotor at 28,000 rpm for 24 hr at 4°C.

5. Unload the tubes from the top. A faint white band at the top of the tube contains a few vesicular remnants and should be completely removed (collect ~6.0 ml per tube). NEs are found at the 1.50/2.00 M interface, appearing as a broad, white, slightly flocculent band (collect ~12.0 ml per tube). The protein concentration of this fraction is typically ~0.5–1.0

ng/ml, i.e., ~35–70 mg total is obtained from a 36-liter nuclei preparation. The next band is a dense, sharp yellowish/white band containing a few nuclear envelopes and dead cell remnants (collect ~12.0 ml per tube). The final ~6.0 ml, including a dense brownish/white pellet, contains soluble and particulate matter mainly derived from chromatin and cell wall remnants that contaminated the nuclei preparation. The NE-containing fraction may be stored at −70°C. When examined by electron microscopy using the procedure described earlier (Section A, step 2, Assay 2), NEs appear as sheets studded with small white circles (the ribosomes on the outer nuclear membrane) and dark structured larger circles (NPCs). Microtubules from SPBs can sometimes be seen projecting from the NEs, as can occasional isolated SPBs torn out of their NEs (Fig. 4).

C. Preparation of an Enriched Nuclear Membranes Fraction

Solutions

1. *Heparin extraction solution (HES):* Prepare a 100-mg/ml heparin stock by dissolving 5 g of heparin in distilled water and making up to 50 ml (store at −20°C). To prepare 50 ml of HES (prepare fresh every time), add 5 ml 100 mg/ml heparin, 5 ml 100 mM bis–Tris–HCl, pH 6.50, 5 μl 1 M MgCl$_2$, 5 μl 1 M DTT, and 0.25 μl solution P to 39.74 ml of distilled water.

2. *RNase stock:* Dissolve 100 mg of RNase A in 10 ml 0.01 M Na acetate, pH 5.2. Heat to 100°C for 15 min. After cooling slowly to room temperature, adjust the pH by adding 0.1 vol of 1 M Tris–HCl, pH 7.40. Store aliquots at −20°C.

3. *Sucrose solutions in bt buffer:* Prepare 1.00 M (refractive index 1.3815) and 2.00 M (refractive index 1.4295) sucrose–bt solutions by diluting the stock 2.50 M solution with bt buffer, as described earlier.

4. *KCl extraction solution (KES):* To prepare 50 ml of KES (prepare fresh every time),

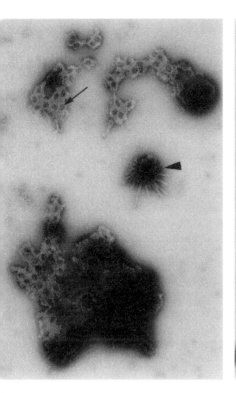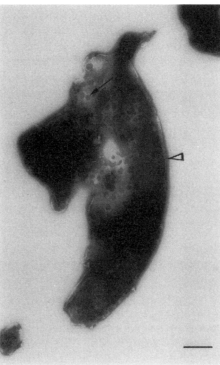

FIGURE 4 Electron micrographs of negatively stained NE fractions. Bar = 0.5 μm. (Left) NE fraction showing NPCs (arrows) and SPBs (solid arrowhead). (Right) H-NE fraction, showing NPC pores (arrows) and curve of envelope (open arrowhead).

mix 12.5 ml 2.0 M KCl, 5 μl 1 M DTT, and 0.25 μl solution P with 37.25 ml of distilled water.

Steps

1. Mix the 1.50/2.00 nuclear envelope fraction with 5 vol of HES on ice. Typically 0.2 vol (i.e., ~14 ml) of the total amount obtained from 3.6×10^{11} spheroplasts are used to prepare nuclear membranes and mixed with 70 ml of HES. Vortex the mixture twice for 10 sec each time at maximum setting and incubate for 1 hr on ice. At the end of the incubation add the RNase stock solution to a final concentration of 10 μg/ml and continue the incubation for 15 min at 10°C in a plastic beaker containing water and ice (as with Section A, step 5).

2. Prepare the sucrose step gradients in the appropriate number of Beckman Ultraclear SW28 tubes, keeping in mind that ~20 ml of extracted nuclear envelope sample is loaded over each gradient. To prepare the gradients, overlay 9 ml of 1.00 M sucrose–bt onto 9 ml of 2.00 M sucrose–bt in each centrifuge tube. Load the heparin-extracted nuclear envelope sample over the sucrose–bt step gradients (~20 ml per gradient). Centrifuge the tubes in a Beckman SW28 rotor at 28,000 rpm for 1 hr and 15 min at 4°C.

3. Collect the fractions from the top of the tube using a hand-held pipette. Collect the first fraction (~18 ml per tube) by aspirating until just above the 1.00 M interface. This fraction is clear and contains the bulk of the soluble proteins. Collect the next fraction (~9 ml per tube) from just above the first interface to just above the 1.00/2.00 M interface. This fraction is also clear and contains some of the soluble proteins together with a few of the nuclear envelope membranes. The bulk of heparin-extracted nuclear membranes is recovered at the 1.00/2.00 M interface and appears as a rather tight white band (~5 ml per tube). The last fraction (~6 ml per tube) consists of the remainder of the gradient, including the pellet. This fraction is clear and sometimes contains small amounts of nuclear membranes. The enriched nuclear membrane fraction can be stored at −70°C.

4. If necessary, traces of heparin can be removed from the enriched nuclear membrane fraction by extracting it with 0.5 M KCl. Thoroughly mix the nuclear membrane fraction mixed with 3 vol of KES and incubate for 1 hr on ice. After the incubation, load the mixture over a 5-ml 1.00 M sucrose–bt cushion in the appropriate number of Beckman SW28 Ultraclear tubes, and decant the nuclear membranes by centrifugation at 28,000 rpm for 1 hr and 15 min at 4°C. Resuspend the pellet of nuclear membranes in 6.5 ml of 1.50 M sucrose–bt per tube. The KCl-washed nuclear membranes fraction can be stored at −70°C. When examined by electron microscopy using the procedure described earlier (Section A, step 2, Assay 2), the H-NEs have lost the ribosomes, giving them a smooth appearance, and the exposed NPC proteins have also been removed, leaving only the empty NPC pores. Often the H-NEs have a smooth curvature, likely retaining the original curvature of the NEs (Fig. 4).

IV. MODIFICATIONS

When NEs are required from *S. cerevisiae* strains they may be prepared by the following modifications to the numbered steps in the original NE preparation protocol (Section B). This procedure has been found to produce more reproducible results and can be scaled up appropriately.

1. Load ~50 OD_{260} units of nuclei in a Sorvall 10-ml Ultra bottle, add 0.2 vol of 8% PVP solution and 0.01 vol of solution P, and mix well by vortexing. Allow to settle on ice, overlay to the top of the tube with 8% PVP solution, and centrifuge in a Beckman 70.1 Ti rotor or Sorvall T-875 rotor at 40,000 rpm for 1 hr at 4°C.

2. Aspirate the supernatant thoroughly and resuspend by vortexing the nuclei into 0.5 ml of bt buffer (*not* bt–DMSO buffer) plus 10 μM CaCl$_2$ and 10 μM ZnCl$_2$, 0.5 μl 2.0% DNase I, and 5 μl solution P. Allow to stand for 5 min at room temperature.

3. Centrifuge for 1 min at 2000 g. Add 2.0 ml of sucrose/Nycodenz solution to the resulting supernatant and vortex for 1 min.

4. Place in a Beckman SW55 centrifuge tube. Overlay this with 1.5 ml of 2.25 M sucrose–bt containing 0.005 vol of solution P and then overlay *this* to within 2 mm of the top of the tube with 1.50 M sucrose–bt plus 0.005 vol of solution P. Centrifuge in a Beckman SW55 rotor at 50,000 rpm for 24 hr at 4°C.

5. After the spin, a dense flocculent white band should be visible at the 1.50/2.25 M interface; this is where the NEs may be found. The gradient can then be unloaded from the top, in whatever volume fractions seem appropriate (depending on the size of the NE band).

V. PITFALLS

One should note that the procedures used for preparing the highly enriched NPC fraction and NE fraction described in this article do not work well when nuclei prepared using techniques other than the one described (see Volume 2, Rout and Kilmartin, "Preparation of Yeast Spindle Pole Bodies" for additional information) are used as a starting material. This may be related to the fact that the conditions used to pack the nuclear pellet cannot be varied much. If the pellet is too loose, the nuclei resuspend without lysing properly, whereas if it is too firm, the vortexing will not resuspend them at all. Also, variations in available equipment may necessitate some experimentation to determine the correct lysis conditions for each laboratory. These points cannot be stressed enough because they have been the source of much frustration for people trying to reproduce these techniques.

Acknowledgment

We are extremely indebted to Gunter Blobel for providing us with the support necessary to develop the techniques described in this article and for allowing us the use of his laboratory to do so.

References

Kilmartin, J. V., and Fogg, J. (1982). Partial purification of yeast spindle pole bodies. *In* "Microtubules in Microorganisms" (P. Cappucinelli and N. R. Morris, eds.), pp. 157–170. Dekker, New York.
Rout, M. P., and Blobel, G. (1993). Isolation of the yeast nuclear pore complex. *J. Cell Biol.* **123**, 771–783.
Rout, M. P., and Kilmartin, J. V. (1990). Components of the yeast spindle and spindle pole body. *J. Cell Biol.* **111**, 1913–1927.
Rozijn, Th. H., and Tonino, G. J. M. (1964). Studies on the yeast nucleus. I. The isolation of nuclei. *Biochim. Biophys. Acta* **91**, 105–112.
Strambio-de-Castillia, C., Blobel, G., and Rout, M. P. (1995). Isolation and characterization of nuclear envelopes from the yeast *Saccharomyces*. *J. Cell Biol.* **131**, 19–31.

Preparation of Nuclei and Nuclear Envelopes

Madeleine Kihlmark and Einar Hallberg

I. INTRODUCTION

The chromatin of the eukaryotic cell nucleus is surrounded by a nuclear envelope (NE) (Fig. 1). The NE consists of three morphologically and biochemically distinct domains. The outer nuclear membrane (ONM), with its attached ribosomes, is continuous with the rough endoplasmic reticulum (RER). The inner nuclear membrane (INM) aligns the inner surface of the nucleoplasm and is attached to a network of intermediate filament proteins called the nuclear lamina. At numerous circumscribed points, referred to as the pore membrane domain (PMD), the ONM and INM are connected with each other, forming circular nuclear pores. The nuclear pores harbor the large multiprotein structures called the nuclear pore complexes (NPCs). One of the principal functions of the nuclear envelope is to serve as ports of entry of soluble and integral membrane proteins for their destinations in the nucleoplasm or nuclear membranes, respectively.

The uniformity in size and density makes it possible to isolate pure intact nuclei at high yields from tissue homogenates by centrifugation through a sucrose cushion (Blobel and Potter, 1966). Nuclei prepared in such a manner can be used as a source for isolation of nuclear components or in various *in vitro* studies of nuclear transport or mitotic disassembly. Nuclei can also be purified from cells in tissue culture (see Wood and Earnshaw, 1990). Monolayer COS cells are frequently used for transient overexpression of proteins in order to study, e.g., intracellular protein trafficking. After sorting to its proper intracellular location, biochemical extraction might give additional information about to what extent the overexpressed protein becomes integrated in biological membranes or multiprotein complexes. A convenient method for the isolation of nuclei from tissue culture cells considerably facilitates further biochemical fractionation and analysis.

NEs can be prepared from isolated nuclei by enzymatic degradation of their nucleic acid content (Dwyer and Blobel, 1976). The resulting NEs are structurally well conserved with attached ribosomes, NPCs, and nuclear lamina. Further fractionation of the NEs have been very useful for the identification and characterization of NE proteins (see Hallberg *et al.*, 1993; Radu *et al.*, 1993).

II. MATERIALS AND INSTRUMENTATION

Filters 0.45 μm, (Cat. No. HAWP04700) are from Millipore. SW-28 tubes are from Beckman Instruments, Inc. HB-4 tubes are obtained from (Sorvall) Du Pont Co. Chemicals should be

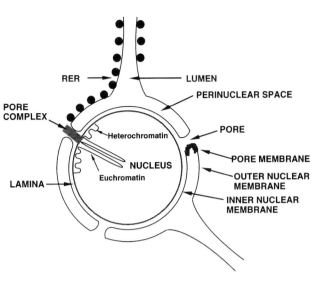

FIGURE 1 Schematic representation of the interface nucleus.

of analytical grade. Phenylmethylsulfonyl fluoride (PMSF, Cat. No. P-7626), DNase I (Cat. No. DN-25), and RNase (Cat. No. R4875) are from Sigma Chemical Co. Dithiothreitol (DTT, Cat. No. 709000), cytochalasin B (Cat. No. 18015), trypsin inhibitor (Cat. No. 109886), pepstatin (Cat. No. 253286), and leupeptin (Cat. No. 1017101) are from Boehringer Mannheim Biochemicals.

III. PROCEDURES

A. Preparation of Nuclei from Rat Liver

Stocks

1. *2.5 M sucrose (85%, w/v):* To make 1 liter, dissolve 850 g of sucrose in distilled water and adjust to a total volume of 1 liter. Allow a day for solving (at 60°C) and filtration of the sucrose. Store at 4°C.

2. *2 M KCl (filter):* To make 1 liter, dissolve 149.1 g of KCl in distilled water and adjust to a total volume of 1 liter. Filter and store at 4°C.

3. *1 M triethanolamine–HCl, pH 7.5 (filter):* To make 1 liter, dissolve 185 g of triethanolamine–HCl in distilled water, adjust pH with HCl, and dilute to a total volume of 1 liter. Filter and store at 4°C.

4. *1 M MgCl$_2$ (filter):* To make 1 liter, dissolve 95.2 g of MgCl$_2$ in distilled water and adjust to a total volume of 1 liter. Filter and store at 4°C.

5. *0.1 M PMSF in ethanol:* To make 10 ml, dissolve 174 mg of PMSF in 10 ml of ethanol. Keep on ice. Prepare fresh.

6. *1 M DTT:* To make 10 ml, dissolve 1.54 g of DTT in distilled water. Keep on ice. The DTT will form a precipitate, which has to be dissolved by moderate heating just before use. Prepare fresh.

Solutions

7. *0.25 M STEAKM:* 0.25 M sucrose, 25 mM KCl; 50 mM triethanolamine–HCl, pH 7.5, 5 mM MgCl, 0.5 mM PMSF, and 1 mM DTT. To make 1 liter, mix 100 ml of 2.5 M sucrose, 12.5 ml 2 M KCl, 50 ml 1 M triethanolamine–HCl, pH 7.5, 5 ml of 1 M MgCl$_2$, 1 ml 1 M DTT, and 5 ml 0.1 M PMSF. Adjust to 1 liter and keep on ice.

8. *2.3 M STEAKM:* 2.3 M sucrose, 25 mM KCl, 50 mM triethanolamine–HCl, pH 7.5,

5 mM MgCl, 0.5 mM PMSF, and 1 mM DTT. To make 250 ml, mix 230 ml 2.5 M sucrose, 3.125 ml 2 M KCl, 12.5 ml 1 M triethanolamine–HCl, pH 7.5, 1.25 ml MgCl$_2$, 250 μl 1 M DTT, and 1.25 ml 0.1 M PMSF. Adjust to 250 ml and keep on ice.

Steps

1. Trim livers from six 150- to 200-g starved Sprague–Dawley rats from fat and chop into small (~1–2 mm) pieces using two razor blades.

2. Make a 30% homogenate (15 strokes) in solution 7 using a tight-fitting Potter–Elvhjelm homogenizer.

3. Filter the homogenate through four layers of cheesecloth and add 5 μl of solution 6 per milliliter of homogenate and spin in a HB-4 rotor for 15 min at 2300 rpm (800 g).

4. Remove the supernatant and homogenize the large and loose pellet in a clean Potter–Elvhjelm homogenizer and dilute to 40 ml in solution 7.

5. Add exactly 2 vol (80 ml) of solution 8, mix well, and layer on top of 5 ml solution 8 in SW28 tubes. Disrupt the interface a little bit with a Pasteur pipette, fill up the tubes completely, and balance. Spin at 27,000 rpm (124,000 g) in a SW28 rotor for 1 hr at 4°C.

6. Discard the supernatant by inverting the tubes. Cut the tubes right above the pellet, which should appear white. Harvest the pellet with a spatula and resuspend in ~20 ml of solution 7 and homogenize (only a few strokes are necessary).

7. Spin the suspension in a HB-4 rotor at 15 min at 2300 rpm (800 g). Resuspend the nuclei again in ~10 ml of solution 7 and measure the OD at 260 nm (1 OD$_{260}$ represents 3 × 10^6 isolated rat liver nuclei). The yield is typically between 200 and 300 OD$_{260}$ per rat liver. The purified nuclei can be used immediately or stored at −70°C as a pellet.

B. Preparation of Nuclear Envelopes from Rat Liver Nuclei

Stocks

1. *2.5 M sucrose (85%, w/v):* To make 1 liter, dissolve 850 g of sucrose in distilled water and adjust to a total volume of 1 liter. Allow a day for solving (at 60°C) and filtration of the sucrose. Store at 4°C.

2. *1 M MgCl$_2$ (filter):* To make 1 liter, dissolve 95.2 g of MgCl$_2$ in distilled water and adjust to a total volume of 1 liter. Filter and store at 4°C.

3. *1 M triethanolamine–HCl, pH 7.5 (filter):* To make 1 liter, dissolve 185 g of triethanolamine–HCl in distilled water, adjust pH with HCl, and dilute to a total volume of 1 liter. Filter and store at 4°C.

4. *1 M triethanolamine–HCl, pH 8.5 (filter):* To make 1 liter, dissolve 185 g of triethanolamine–HCl in distilled water, adjust pH with HCl, and dilute to a total volume of 1 liter. Filter and store at 4°C.

5. *0.1 M PMSF in ethanol:* To make 10 ml, dissolve 174 mg of PMSF in 10 ml of ethanol. Keep on ice. Prepare fresh.

6. *1 M DTT:* To make 10 ml, dissolve 1.54 g of DTT in distilled water. Keep on ice. The DTT will form a precipitate, which has to be dissolved by moderate heating just before use. Prepare fresh.

Solutions

7. *0.1 mM MgCl$_2$, 1 mM DTT, and 0.1 mM PMSF:* To make 100 ml, mix 10 μl 1 M MgCl$_2$, 100 μl 1 M DTT, and 100 μl 0.1 M PMSF and dilute to 100 ml with distilled water. Keep on ice.

8. *10% sucrose:* 20 mM triethanolamine, pH 7.5, 0.1 mM MgCl$_2$, 1 mM DTT, and 0.1 mM PMSF. To make 250 ml, mix 29.4 ml 2.5 M sucrose, 5 ml 1 M triethanolamine–HCl, pH 7.5, 25 μl 1 M MgCl$_2$, 250 μl 1 M DTT, and 250 μl 0.1 M PMSF. Adjust the volume to 250 ml with distilled water. Make up just prior to use.

9. *10% sucrose, 20 mM triethanolamine, pH 8.5, 0.1 mM MgCl$_2$, 1 mM DTT, and 0.1 mM PMSF:* To make 250 ml, mix 29.4 ml 2.5 M sucrose, 5 ml 1 M triethanolamine–HCl, pH 8.5, 25 μl 1 M MgCl$_2$, 250 μl 1 M DTT, and 250 μl 0.1 M PMSF. Adjust the volume to 250 ml with distilled water. Make up just prior to use.

10. *30% sucrose, 20 mM triethanolamine, pH 7.5, 0.1 mM MgCl$_2$, 1 mM DTT, and 0.1 mM PMSF:* To make 250 ml, mix 88.2 ml 2.5 M sucrose, 5 ml 1 M triethanolamine–HCl, pH 7.5, 25 μl 1 M MgCl$_2$, 250 μl 1 M DTT, and 250 μl 0.1 M PMSF. Adjust the volume to 250 ml with distilled water. Make up just prior to use.

11. *2 mg/ml DNase I:* Dissolve 1 mg DNase I in 500 μl distilled water.

12. *10 μg/ml RNase.* Dissolve 1 mg RNase in 1 ml of distilled water. Dilute 10 μl of this solution to 1 ml.

Steps

1. Suspend 500 OD$_{260}$ of pelleted rat liver nuclei (procedure A) in 5 ml of solution 7 by gentle suction up and down in a 10-ml pipette using a pipette aid.

2. Add 20 ml of solution 9 and 25 μl each of solutions 11 and 12. Mix and incubate for 15 min at room temperature.

3. Underlay with 5 ml of solution 10 and spin for 20 min at 10,000 rpm (16,000 g) in a HB-4 rotor at 4°C.

4. Resuspend the pellet in 5 ml of solution 8 as in step 1. Add 25 μl each of solutions 11 and 12. Mix and incubate for 15 min at room temperature.

5. Repeat step 3. The pelleted NEs can be suspended or stored frozen as a pellet. One OD$_{260}$ of rat liver NEs, i.e., the amount of rat liver NEs isolated from 1 OD$_{260}$ of isolated rat liver nuclei, is equivalent to 10 μg of protein.

C. Preparation of Nuclei from COS Cells

Stocks

1. *0.5 M N-2-hydroxyethylpiperazine-N′-2-ethanesulfonic acid (HEPES) pH 7.5:* To make 50 ml, dissolve 5.96 g of HEPES in 40 ml of distilled water and adjust pH to 7.5. Add distilled water to a total volume of 50 ml.

2. *2.5 M sucrose (85%, w/v):* To make 10 ml, dissolve 8.50 g of sucrose in distilled water and adjust to a total volume of 10 ml. Allow a day for solving (at 60°C) and filtration of the sucrose. Store at 4°C.

3. *1 M KCl (filter):* To make 10 ml, dissolve 746 mg of KCl in distilled water and adjust to a total volume of 10 ml. Filter and store at 4°C.

4. *1 M MgCl$_2$ (filter):* To make 10 ml, dissolve 952 mg of MgCl$_2$ in distilled water and adjust to a total volume of 10 ml. Filter and store at 4°C.

5. *0.1 M PMSF in ethanol:* To make 1 ml, dissolve 17.4 mg of PMSF in 1 ml of ethanol. Keep on ice. Prepare fresh.

6. *1 M DTT:* To make 1 ml, dissolve 154 mg of DTT in 1 ml distilled water. Keep on ice. The DTT will form a precipitate, which has to be dissolved by moderate heating just before use. Prepare fresh.

7. *1 mg/ml pepstatin in ethanol:* To make 1 ml, dissolve 1 mg of pepstatin in 1 ml of ethanol. Keep on ice. Prepare fresh.

8. *1 mg/ml leupeptin:* To make 1 ml, dissolve 1 mg of leupeptin in 1 ml of distilled water. Keep on ice. Prepare fresh.

9. *1 mg/ml trypsin inhibitor:* To make 1 ml, dissolve 1 mg of trypsin inhibitor in 1 ml of distilled water. Keep on ice at 4°C. Prepare fresh.

10. *2 mM cytochalasin B in DMSO:* To make 2 ml, dissolve 1.92 mg of cytochalasin B in 2 ml of DMSO. Keep on ice. Prepare fresh.

11. *Phosphate-buffered saline (PBS), pH 7.4:* To make 1 liter, dissolve 8 g NaCl, 0.2 g KCl, 1.44 g Na_2HPO_4, and 0.24 g of KH_2PO_4 in 800 ml of distilled water. Adjust the pH to 7.4 and add H_2O to 1 liter. Store at 4°C.

12. *Dulbecco's modified Eagle medium (DMEM):* Purchase from Life technologies. Store at 4°C.

Solutions

13. *DMEM containing 10 μM cytochalasin B:* To make 10 ml, add 50 μl of solution 10 to 9.95 ml of solution 12.

14. *1× NB:* 10 mM HEPES, pH 7.5, 10 mM KCl, 1.5 mM $MgCl_2$; 1 mM DTT, 0.1 mM PMSF, 1 μg/ml pepstatin, 1 μg/ml leupeptin, and 1 μg/ml trypsin inhibitor: To make 10 ml, mix 200 μl 0.5 M HEPES, 100 μl 1 M KCl, 15 μl 1 M $MgCl_2$, 10 μl 1 M DTT, 10 μl 0.1 M PMSF, 10 μl 1 mg/ml pepstatin, 10 μl 1 mg/ml leupeptin, and 10 μl 1 mg/ml trypsin inhibitor and adjust to a total volume of 10 ml.

15. *Cytochalasin B in 1× NB:* 10 mM HEPES, pH 7.5, 10 mM KCl, 1.5 mM $MgCl_2$, 1 mM DTT, 0.1 mM PMSF, 1 μg/ml pepstatin, 1 μg/ml leupeptin, 01 μg/ml trypsin inhibitor, and 10 μM cytochalasin B. To make 10 ml, add 50 μl of solution 10 to 9.95 ml of solution 14.

16. *Underlay:* 10 mM HEPES, pH 7.5, 10 mM KCl, 1.5 mM $MgCl_2$, 1 mM DTT, 0.1 mM PMSF, 1 μg/ml pepstatin, 1 μg/ml leupeptin, 1 μg/ml trypsin inhibitor, and 40% sucrose. To make 1 ml, mix 20 μl 0.5 M HEPES, 10 μl 1 M KCl, 1.5 μl 1 M $MgCl_2$, 1 μl 1 M DTT, 1 μl 0.1 M PMSF, 1 μl 1 mg/ml pepstatin, 1 μl 1 mg/ml leupeptin, 1 μl 1 mg/ml trypsin inhibitor, and 470 μl solution 2 and adjust to a total volume of 1 ml.

Steps

1. Aspire off the medium from one 100-mm petri dish of untransfected or transfected monolayer COS-1 cell cultures (cf. Söderqvist *et al.*, 1996) and wash twice with solution 11 (PBS).

2. Add 5 ml of solution 13 (DMEM containing cytochalasin B) and let the cells incubate for 40 min at 37°C.

3. Wash the petri dish twice with solution 11 (PBS). Suspend the cells by scraping and pipetting up and down through a 10-ml pipette and transfer to a 15-ml Falcon tube. Collect cells by centrifuging at 800 *g* for 5 min.

4. Wash the cells in 2 ml of solution 14 (1× NB).

5. Place the cell pellet on ice and resuspend in 1 ml of solution 15. Let the cells incubate on ice for an additional 20 min and transfer to a chilled 1-ml Weaton glass homogenizer.

6. Disrupt the plasma membranes by 20 gentle strokes. The efficiency of cell breakage may be monitored by phase-contrast microscopy. Transfer the homogenate to a 1-ml Eppendorf tube.

7. Underlay the homogenate with 200 μl of solution 16, place the Eppendorf tube inside a 50-ml Falcon tube, and centrifuge at 800 *g* for 15 min in a bench-top centrifuge with a swinging bucket rotor. Aspire off the supernatant and the sucrose cushion and resuspend the pellet in 200 μl of solution 11 (PBS).

8. This procedure yields apparently intact nuclei (Fig. 2), which are less than 10% contaminated, as judged by LDH activity.

IV. COMMENTS

The pellet of isolated nuclei (Section IIIA) should be completely white and the nuclei should look intact when observed in the phase-contrast microscope. On the ultrastructural level, the

Phase	DNA	GFP

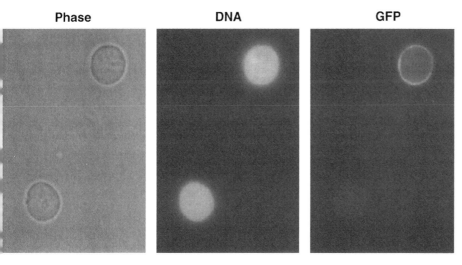

FIGURE 2 Purified nuclei from transfected COS cells. The three panels show the same field of nuclei purified from COS-1 cells transfected with cDNA encoding a chimerical fusion between the pore membrane protein POM121 (Hallberg *et al.*, 1993; Šoderqvist *et al.*, 1996) and GFP (green fluorescent protein). Nuclei appear intact and free of cellular contamination as judged by phase contrast (left) and DNA staining using Hoechst 33258 (middle). In the FITC channel (right) the overexpressed POM121–GFP fusion protein gives rise to a continuous rim staining of the nuclear envelope, indicating that the nuclear periphery is intact. The lower nucleus comes from a nonexpressing cell.

INM, pores, NPCs, and the nuclear lamina appear intact. Holes are produced in the ONM, however, as a result of tearing off the RER during homogenization, exposing the perinuclear space to the surrounding medium. Nuclei with intact NEs can be obtained by "sealing" the ONM using *Xenopus* egg extracts.

Nuclei isolated from transfected COS cells (Section IIIC) display an intact appearance in phase-contrast microscopy, a normal DNA staining and a normal distribution of an overexpressed chimerical POM121-GFP fusion protein delineating an intact nuclear periphery (Fig. 2). Nuclei isolated in this way may be used for further fractionation and analysis.

NEs isolated as described in Section IIIB are freed of chromatin, nucleoplasmic, and nucleolar material but show a well-conserved ultrastructure with clearly identifiable ONM, INM, pores, NPCs, and lamina (Dwyer and Blobel, 1976).

Salt extraction of NEs can be used to eliminate less well-attached NE proteins, whereas extractions with Triton X-100/high salt produce a structurally well-conserved pore complex lamina fraction free of membranes (Dwyer and Blobel, 1976). Integral and nonintegral membrane proteins can be separated by extractions in 7 *M* urea (Hallberg *et al.*, 1993).

V. PITFALLS

1. In order to avoid the accumulation (via repeated pelletation) of dust particles present in bulk chemicals, it is important to filter the stock solutions through a 0.45-μm Millipore filter.

2. It is necessary to starve the rats for 24 hr before sacrifice in order to reduce the liver glycogen content.

3. All steps should be performed on ice in the cold room with chilled solutions. The speed of the preparation is essential for the yield.

4. It is absolutely essential for the outcome of the preparation of nuclei (Section IIIA) that the ratio of solution 8 to the first pellet (suspended in solution 7) is exactly 2:1.

References

Blobel, G., and Potter, R. V. (1966). Nuclei from rat liver: Isolation method that combines purity with high yield. *Science* **154**, 1662–1665.

Dwyer, N., and Blobel, G. (1976). A modified procedure for the isolation of a pore complex-lamina fraction from rat liver nuclei. *J. Cell Biol.* **70**, 581–591.

Hallberg, E., Wozniak, R. W., and Blobel, G. (1993). An integral membrane protein of the pore membrane domain of the nuclear envelope contains a nucleoporin-like region. *J. Cell Biol.* **122**, 513–521.

Radu, A., Blobel, G., and Wozniak, R. W. (1993). Nup155 is a novel nuclear pore complex protein that contains neither repetitive sequence motifs nor reacts with WGA. *J. Cell Biol.* **121**, 1–9.

Söderqvist, H., Jiang, W.-Q., Ringertz, N. R., and Hallberg, E. (1996). Formation of nuclear bodies in cells overexpressing the nuclear pore protein POM121. *Exp. Cell Res.* **225**, 75–84.

Wood, E. R., and Earnshaw, W. C. (1990). Mitotic chromatin condensation in vitro using somatic cell extracts and nuclei with variable levels of endogeneous topoisomerase II. *J. Cell Biol.* **111**, 2839–2850.

Affinity Purification of Protein A-Tagged Nuclear Pore Proteins from Yeast

Symeon Siniossoglou, Paola Grandi, and Eduard C. Hurt

I. INTRODUCTION

Epitope tagging is a powerful approach that allows subsequent purification of proteins from whole cell extracts. Using standard DNA recombinant techniques, it is possible to fuse the coding sequence of a given protein of interest to another protein domain (epitope tag) that exhibits a high affinity to a ligand. Total extracts can be prepared from cells expressing such fusion proteins and purified by passing them over a column bearing the immobilized ligand. Given the high affinity and specificity of the epitope tag/ligand interaction, one-step purification can result in a high enrichment of the fusion protein. Tags that are often used include glutathione S transferase (GST), dihydrofolate reductase (DHFR), (histidine)$_6$ tag, and small peptide epitopes for which monoclonal antibodies are available, such as the influenza hemagglutinin (HA) tag.

Protein A is a cell wall component of *Staphylococcus aureus* that binds to the constant region (Fc) of IgG. It has a molecular mass of 42kDa and consists of four N-terminal IgG-binding domains along with a C-terminal domain that links the protein to the cell wall. The affinity constants for the binding of protein A to human polyclonal IgG or rabbit IgG are 4.8 $\times$ 10^7 liter/mol and 4.1 $\times$ 10^8 liter/mol, respectively (Lindmark *et al.*, 1981).

We exploited this high binding affinity of protein A to IgG by expressing and purifying under native conditions protein A-tagged nuclear pore proteins from the budding yeast *Saccharomyces cerevisiae* (Grandi *et al.*, 1993). This allowed the identification of copurifying components that physically interact with the tagged proteins. Reverse genetics can then be used to clone the genes of the interacting proteins. Protein A tagging can also be used in order to immunolocalize the tagged protein. Protein A fusions have been previously used to purify bacterially expressed insulin (Moks *et al.*, 1987) and to isolate calmodulin-binding proteins in yeast (Stirling *et al.*, 1992).

The protein A tag can be inserted in frame, either at the amino or at the carboxy terminus of the protein of interest. A unique restriction site can be introduced, via polymerase chain reaction (PCR), just after the start codon in the first case or before the stop codon, in the second case. The protein A tag, encoding usually two IgG-binding domains (14 kDa), can then be introduced to this site (Fig. 1B). Expression can be driven either from the authentic promoter of the tagged gene or from another strong yeast promoter, in case high expression levels are desired. If the fusion protein is functional, it should rescue the lethal or sick phenotype of a strain lacking the authentic protein (Fig. 1A). The tagged protein can then be immunopurified under native conditions on IgG–Sepharose columns or localized in the cell by indirect immunofluorescence.

A

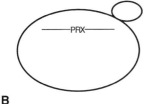

wild-type haploid yeast strain carrying one chromosomal copy of the *PRX* gene

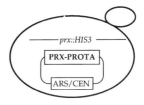

haploid yeast strain disrupted for the *prx* gene (*prx::HIS3*) and complemented with a single copy plasmid carrying the *PRX* gene fused to two IgG-binding domains from protein A (PRX-PROTA)

B

a
5' .....ATT.ACT.GCC.CAA.CAA.TAA.ATA.AGA..... 3'
H₂N..... Ile . Thr . Ala . Gln . Gln . **Stop.**

b
5' .....ATT.ACT.GCC.CAA.CAA.**GGA.TCC**.TAA.ATA.AGA..... 3'
H₂N..... Ile . Thr . Ala . Gln . Gln . **Gly . Ser** . Stop.

c
5'.....ATT.ACT.GCG.CAA.CAA. **GGA.TCC** . CTT.GCG.........AAT.TCA **GGA.TCC**.TAA.ATA.AGA..... 3'
H₂N..... Ile . Thr . Ala . Gln . Gln . **Gly . Ser** . Leu . Ala .... Asn . Ser **Gly . Ser** .Stop.

Protein A

FIGURE 1   (A) Construction of a yeast strain expressing a protein A-tagged version of the protein X of interest (PRX). If the fusion protein (PRX-PROTA) is functional *in vivo* and the tag does not interfere with the properties of the authentic protein, PRX-PROTA will complement the growth defect of a yeast strain for which the *prx* gene has been disrupted, e.g., by the *HIS3* marker. (B) Construction of the protein A fusion protein. In this example, the protein A tag (for most applications, consisting of two IgG-binding domains) is inserted at the carboxy-terminal end of the protein. (a) Sequence of the wild-type gene around the stop codon. (b) Introduction by PCR of a unique restriction site (in this example, a *Bam*HI site; GGATCC) before the stop codon. (c) In-frame fusion of the protein A-coding sequence inserted as a *Bam*HI–DNA restriction fragment at the restriction site generated in step b.

II. MATERIALS AND INSTRUMENTATION

Yeast extract (Cat. No. 0127-08-0), bactopeptone (Cat. No. 0118-08-1), and yeast nitrogen base without amino acids (Cat. No. 0919-07-3) for preparing the culture media are all from Difco. Complete supplement mixture minus adenine (Cat. No. 4510-022), histidine (Cat. No. 4510-322), leucine (Cat. No. 4510-522), uracil (Cat. No. 4511-222), or tryptophan (Cat. No. 4511-022) are from BIO101. Glucose monohydrate (Cat. No.108346) is from Merck, sorbitol (Cat. No. 6313.2) and Tris (Cat. No. 4855.2) are from Roth, Triton X-100 (Cat. No. X-100) and Tween 20 (Cat. No. P-1379) are from Sigma, and Zymolyase 20T (Cat. No. 120491) or 100T (Cat. No. 120493) is from Seikagaku Corporation. IgG–Sepharose 6 fast flow (Cat. No. 17-0969-01) is from Pharmacia. Antipain (Cat. No. A-61919), leupeptin (Cat. No. L2884), trypsin inhibitor (Cat. No. T-2011), chymostatin (Cat. No C-7268), and pepstatin (Cat. No. P4265) are from Sigma. Aprotinin (Cat. No. A162.3) is from Roth. Chromatography columns (Cat. No. 731-1550) are from Bio-Rad. Rabbit antibodies coupled to horseradish peroxidase (Cat. No. Z0113) are from DAKO, primary rabbit anti-protein A antibodies for immunofluorescence are from Sigma (Cat. No. P-3775), and secondary donkey anti-rabbit antibodies coupled to Cy3 (Cat. No. 711-165-152) are from Jackson. Bisbenzimide H 33342 (Cat. No. 14533) is from Fluka.

III. PROCEDURES

A. Preparation of Yeast Spheroplasts for IgG Affinity Purification

Solutions

1. *Alkaline dithiothreitol (DTT) buffer:* 1 M Tris–HCl, pH 9.4, and 10 mM DTT. To make 200 ml, dissolve 24.2 g of Tris in distilled water, adjust to pH 9.4, add 310 mg of DTT, and complete the volume to 200 ml. Prepare fresh.

2. *1 M K$_2$HPO$_4$/KH$_2$PO$_4$ buffer, pH 7.4:* Dissolve 22.8 g of K$_2$HPO$_4\cdot$3H$_2$O in 100 ml of distilled water. Dissolve 13.6 g of KH$_2$PO$_4$ in 100 ml of distilled water. Mix the two buffers and adjust the pH to 7.4. Store at 4°C.

3. *2.4 M sorbitol stock solution:* To make 1 liter, dissolve 437 g of sorbitol in distilled water. Heat up to dissolve and bring to a total volume of 1 liter.

4. *Spheroplasting buffer:* 1.2 M sorbitol, 20 mM K$_2$HPO$_4$/KH$_2$PO$_4$, pH 7.4. To make 1 liter, add 500 ml of 2.4 M sorbitol stock solution and 20 ml of K$_2$HPO$_4$/KH$_2$PO$_4$ buffer, pH 7.4, to 480 ml of distilled water. Store at 4°C.

Steps

1. Grow yeast cells in YPD liquid medium (1% yeast extract, 2% peptone, 2% D-glucose). If the tagged protein is carried on a plasmid and is not essential for growth of the yeast strain, grow cells in selective medium. Dilute a 100-ml preculture of OD$_{600}$ 0.4 per milliliter of culture to 1 liter liquid medium and let grow at 30°C to OD$_{600}$ 2.0 to 2.5 per milliliter of culture.

2. Spin at 3000 rpm for 5 min in a Sorvall GSA rotor. Pour off supernatant, wash cells with 100 ml water, and centrifuge at 3000 rpm for 5 min in the same rotor.

3. Resuspend cells in alkaline DTT buffer and incubate for 10 min at room temperature under moderate shaking. Use 100 ml of buffer per 1000 ODs of cells. Centrifuge as above.

4. Weigh cell pellet and resuspend in 100 ml of spheroplasting buffer per 1000 ODs of cells. Add 5 mg of Zymolyase 20T per gram of wet cells. Incubate under gentle shaking at 30°C for 20–30 min and monitor spheroplasting by diluting 50 μl of the yeast suspension in 1 ml of water. Loss of turbidity indicates cell lysis. Centrifuge spheroplasted cells at 2000 rpm for 5 min in a Sorvall SS34 rotor.

5. Carefully wash spheroplasts with spheroplasting buffer. Avoid any mechanical stress. Centrifuge at 2000 rpm for 5 min as above. Pour off supernatant and freeze spheroplasts at −20°C or proceed as described below.

B. Lysis of Spheroplasts and Affinity Isolation on IgG–Sepharose Column

Solutions

1. *Lysis buffer:* 150 mM KCl, 20 mM Tris–HCl, pH 8.0, 5 mM MgCl$_2$, and 1% Triton X-100. To make 500 ml, dissolve 1.2 g of Tris, bring pH to 8.0, and add 5.6 g of KCl and 0.51 g of MgCl$_2\cdot$6H$_2$O. Add 50 ml of Triton X-100 (10% solution) and bring volume to 500 ml with distilled water. Store at 4°C.

2. *TST buffer:* 150 mM NaCl, 50 mM Tris–HCl, pH 7.5, and 0.05% Tween 20. To make 500 ml, disolve 3 g of Tris, adjust pH to 7.5, and add 4.4 g of NaCl and 2.5 ml of Tween 20 from a 10% stock. Bring volume to 500 ml. Store at 4°C.

3. *Alati stock solution:* Mix protease inhibitors antipain, leupeptin, aprotinin, and trypsin inhibitor. To make 1 ml stock solution, dissolve 2.4 mg antipain, 1.72 mg leupeptin, 1.8 mg aprotinin, and 2 mg trypsin inhibitor in 1 ml distilled water. Freeze immediatly and keep at −20°C. This is stable for at least 2 months at −20°C.

4. *PC stock solution:* Mix the protease inhibitors pepstatin A and chymostatin. To make 1 ml stock solution, dissolve 5 mg pepstatin A and 6 mg chymostatin in 1 ml dimethyl sulfoxide. Freeze immediatly and store at −20°C. This is stable for several months at −20°C.

5. *Wash buffer:* 5 mM ammonium acetate, pH 5. To make 100 ml, solubilize 38.5 mg of ammonium acetate and adjust the pH to 5.0. Bring the volume to 100 ml with distilled water. Keep at 4°C.

6. *0.5 M acetic acid, pH 3.4:* To make 100 ml, mix 3.1 ml of a 100% stock solution of acetic acid with distilled water and bring pH to 3.4 with ammonium acetate. Bring volume to 100 ml. Keep at 4°C.

Steps

All steps are at 4°C.

1. Pack the column with 100 μl bed volume of IgG–Sepharose beads per gram of spheroplasts to be lysed. Wash with 2 ml TST buffer, 1 ml 0.5 M HAc, pH 3.4, and 1 ml TST. Again apply 1 ml 0.5 M HAc, pH 3.4, and then TST buffer until the pH of the flow through is back to 7.4.

2. To lyse the spheroplasts, use 15 ml of lysis buffer per gram of pelleted cells. Prior to lysis, add Alati inhibitors in a 1:400 dilution and PC inhibitors in a 1:1000 dilution. Dounce with 10 strokes using a Dounce homogenizer. Centrifuge the lysate at 15,000 rpm for 10 min at 4°C in a Sorvall SS34 rotor.

3. Load supernatant on the column and let it flow by gravity feed. Wash with 10 ml lysis buffer, followed by 1 ml 5 mM ammonium acetate, pH 5.0. Elute the bound proteins with 1 ml 0.5 M acetic acid, pH 3.4, freeze it in liquid N_2, and lyophylize to completion.

4. Resuspend the lyophilized probe in SDS sample buffer (80 to 150 μl of sample buffer per 3–4 g of wet cells) and analyze it by SDS–polyacrylamide gel electrophoresis followed by Coomassie staining or Western blotting.

NOTE

Alternatively, the fusion protein can be eluted by adding SDS sample buffer (80 to 150 μl of sample buffer per 3–4 g of wet cells) to the beads after the ammonium acetate wash. Incubate for 2 min at room temperature, spin in a microfuge, and check the supernatant by SDS–polyacrylamide gel electrophoresis.

C. Step Elution of Components Bound to Protein A Fusion Proteins

In case that interacting components copurify with the protein A fusion protein, it is possible to elute them from the column using a gradient of increasing salt concentrations. If $MgCl_2$ is used, the fusion protein is eluted at 4.5 M $MgCl_2$, whereas the interacting proteins are usually eluted at lower $MgCl_2$ concentrations (Fig. 2). Thus, the salt elution can be used to demonstrate different binding affinities among the components of a purified complex (Siniossoglou et al., 1996).

Solutions

1. *Dilution buffer:* 50 mM Tris–HCl, pH 7.4, and 0.05% Triton X-100. To make 100 ml, solubilize 0.6 g of Tris, adjust the pH to 7.4, and add 500 ml of a 10% stock solution of Triton X-100. Fill up the volume to 100 ml with distilled water. Store at 4°C.

2. *$MgCl_2$ step gradient:* 1, 2, 3, and 4.5 M solutions of $MgCl_2$ in dilution buffer. To make 10 ml of each, dissolve 2, 4, 6.1, and 9.1 g of $MgCl_2 \cdot 6H_2O$ in dilution buffer, then fill up the volume to 10 ml with dilution buffer. Store at 4°C.

3. *Dialysis buffer:* 50 mM Tris–HCl, pH 7.4, and 150 mM NaCl. To make 1 liter, solubilize 6 g of Tris, adjust the pH to 7.4, and add 8.7 g of NaCl. Adjust the volume to 1 liter with distilled water. Store at 4°C.

Steps

1. Proceed for the purification of the tagged protein as described in Section B, up to step 2. Load supernatant on IgG–Sepharose beads and wash with lysis buffer. Divide beads in two aliquots. Treat one aliquot as described earlier, i.e., elute the fusion protein with acetic acid (steps 2 and 3, Section B).

2. Transfer the second aliquot of beads in a 1.5-ml Eppendorf microtube and incubate with 3 bed volumes of 1 M $MgCl_2$ in dilution buffer for 5 min at room temperature on a turning wheel. Spin at 3000 rpm, collect the supernatant, and repeat the incubation with the 2 M $MgCl_2$ solution in dilution buffer.

3. Repeat incubations as decribed in step 2, with the 3 and 4.5 M $MgCl_2$ solutions in dilution buffer.

4. Dialyze the supernatants from steps 2 and 3 against 2 liters dialysis buffer for at least 5 hr at 4°C.

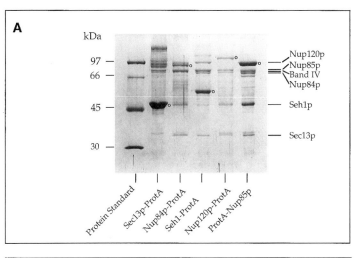

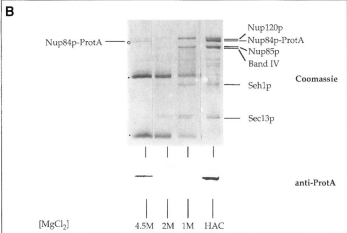

FIGURE 2 (A) SDS–PAGE of acetic acid eluates from affinity purification of five different yeast proteins (Sec13p, Nup84p, Seh1p, Nup120p, and Nup85p) tagged with protein A and purified on IgG–Sepharose columns as described in this protocol. Because all five proteins are subunits of a complex that is localized at the nuclear pore, the pattern of copurifying components is complementary (Siniossoglou *et al.*, 1996). The positions of the corresponding protein A fusion proteins are marked by open circles, the names of the associated proteins are indicated on the right, and the molecular masses of a protein standard are shown on the left. (B) SDS–PAGE of MgCl₂ eluates from affinity purification of the Nup84p-ProtA nuclear pore protein. The Nup84-ProtA fusion was absorbed on IgG–Sepharose beads that contained the other five associated nuclear pore proteins. Half of the beads were then eluted with acetic acid ("HAc"). The other half of the beads were treated with increasing MgCl₂ concentration (steps of 1, 2, and 4.5 *M*) as described in the text. The interacting proteins are mainly released from the beads at 1 *M* MgCl₂, whereas the fusion protein (marked by an open circle) is predominantly released at 4.5 *M*. The heavy and light chains of IgGs are marked with squares.

5. Concentrate the dialyzed fractions with Centricon ultrafiltration units. Spin at 3000 *g* for 2 hr to get roughly a 20× concentration of the dialyzed eluates.

6. Analyze the samples by SDS–polyacrylamide gel electrophoresis followed by Coomassie staining or Western blotting (Fig. 2).

D. Localization of Protein A-Tagged Proteins by Indirect Immunofluorescence

Solutions

1. For solutions 1, 2, 3, and 4, see Section A.

2. *Poly-L-lysine stock solution (1 mg/ml):* To make 1 ml, dissolve 1 mg of poly-L-lysine in 1 ml of distilled water. Store at 4°C.

---

3. *Phosphate-buffered saline (PBS buffer):* To make 1 liter, dissolve 8 g of NaCl, 0.2 g of KCl, 1.44 g of Na_2HPO_4, and 0.24 g of KH_2PO_4 in distilled water. Adjust the pH to 7.4 and add distilled water to 1 liter. Store at room temperature.

4. *Bovine serum albumin (BSA) 20% in PBS:* To make 1 ml, dissolve 200 mg of BSA in 1 ml of PBS buffer. Store at 4°C.

Steps

1. Grow yeast cells in 50 ml YPD or synthetic liquid medium to an OD_{600} 0.5 to 1.0. Add 3.7% formaldehyde (final concentration) directly to the medium and fix the cells for 45 min at room temperature with gentle shaking.

2. Spin down the cells at 3000 rpm for 4 min in a SS34 rotor. Resuspend cells in 25 ml alkaline DTT buffer and incubate for 10 min at room temperature under gentle shaking.

3. Spin down cells as above. Wash the cell pellet with 1 ml spheroplasting buffer. Spin 10 sec in a table-top microfuge and resuspend cells in 1 ml spheroplasting buffer with 0.2 mg/ml Zymolyase 100T. Incubate suspension at 30°C and check for spheroplasting under the microscope. The degree of spheroplasting can vary from strain to strain, but in most cases requires 10 to 30 min.

4. Spin down cells at 3000 rpm for 4 min in a table-top microfuge. Remove supernatant and resuspend cells carefully in 1 ml spheroplasting buffer. Avoid mechanical stress. Spin as before, and resuspend cells in 50–300 ml spheroplasting buffer. Spheroplasts can be stored at −20°C and used in several experiments.

5. Coat 11-mm-round coverslips with 1 mg/ml poly-L-lysine using a brush and let dry for 5 min.

6. Put 20 μl of spheroplasts onto the poly-L-lysine-coated coverslip, spread them with a pipette, and incubate for 5 min. Put the coverslip upside up into a well of a six multiwell dish containing 5 ml of PBS and wash three times with PBS.

7. Add 20 μl of the primary antibody (rabbit anti-protein A, 1:10,000 diluted in PBS/ 0.2% BSA) on Parafilm and put coverslip upside down onto the antibody drop. Incubate for 30 min at room temperature.

8. Wash as in step 6 and add 20 μl of the secondary antibody (donkey anti-rabbit Cy3 conjugated, 1:600 diluted in PBS/0.2% BSA) as in step 7. Incubate as in step 7.

9. Wash as in step 6. Add 200 μl of a saturated bisbenzimide solution in 10 ml PBS and incubate the coverslip for 2 min. Wash as in step 6.

10. Immerse coverslip into ethanol (100%), let dry for 10 min at room temperature, and mount on a slide with 2 μl of Moviol. Let dry for 20 min and observe cells in the rhodamine channel.

References

Grandi, P., Doye, V., and Hurt, E. C. (1993). *EMBO J.* **12**, 3061–3071.
Lindmark, R., Biriell, C., and Sjöquist, J. (1981). *Scand. J. Immunol.* **14**, 409.
Moks, T., Abrahmsen, L., Österlöf, B., Josephson, S., Östling, M., Enfors, S., Persson, I., Nilsson, B., and Uhlen, M. (1987). *Bio/Technology* **5**, 379–382.
Siniossoglou, S., Wimmer, C., Rieger, M., Doye, V., Tekotte, H., Weise, C., Emig, S., Segref, A., and Hurt, E. C. (1996). *Cell* **84**, 265–275.
Stirling, D. A., Petrie, A., Pulford, D. J., Patterson, D. T. W., and Stark, M. J. R. (1992). *Mol. Microbiol.* **6**, 703–713.

Isolation of 40S Monomer HnRNP Core Particles, Polyparticle Complexes, and the Assembly of HnRNP Particles *in Vitro*

James G. McAfee, Sunita Iyengar, and Wallace M. LeStourgeon

I. INTRODUCTION

In the presence of nuclease inhibitors, the great majority of the pre-mRNA molecules released from isolated HeLa S3 nuclei are recovered as poly hnRNP complexes that sediment from about 30–300S in density gradients (reviewed in McAfee *et al.*, 1997; LeStourgeon *et al.*, 1990; Beyer and Osheim, 1990). Upon brief exposure to nuclease, these complexes are converted to 30–40S monoparticles that possess the same general protein composition as polyparticle complexes. The most abundant proteins (the "core" proteins A1, A2, B1, B2, C1, and C2) possess the intrinsic ability to spontaneously form 40S monoparticles on approximately 700 nucleotide (nt) lengths of RNA, and each multiple of this length supports the assembly of an additional particle (Conway *et al.*, 1988; Huang *et al.*, 1994). Based on hydrodynamic and protein cross-linking studies it was initially suggested that most of the core proteins exist as tetramers of $(A1)_3B2$, $(A2)_3B1$, and $(C1)_3C2$ (Barnett *et al.*, 1988, 1991). However, when expressed separately in bacterial cells, proteins C1 and C2 spontaneously form homo $C1_4$ and $C2_4$ tetramers (McAfee *et al.*, 1996), and it is likely that homotypic oligomers of the core proteins also exist *in vivo*. Individually, the nuclear concentration of the core hnRNP proteins A1, A2, and C protein is about one-third to one-half that of the core histones. Together these proteins are the most abundant RNA-binding proteins in the nucleus of actively growing HeLa cells (LeStourgeon *et al.*, 1990; Kiledjian *et al.*, 1994). This article describes refinements in the methods for purification of 40S core hnRNP particles, polyparticle complexes, and individual core particle proteins. Also described are methods for the assembly and purification of various hnRNP assembly intermediates (see Huang *et al.*, 1994; Rech *et al.*, 1995).

II. MATERIALS AND INSTRUMENTATION

HeLa S3 cells (ATCC line CCL2.2) are cultured in minimal essential medium (S-MEM, with L-glutamine, without sodium bicarbonate, Cat. No. 410-1400EH) from GIBCO Laboratory. Pluronic (Cat. No. P1300), sodium bicarbonate (Cat. No. S5761), penicillin-G (Cat. No. P7794), streptomycin sulfate (Cat. No. S9137), Tris (Cat. No. T6791), and ribonuclease A (5× crystallized, protease-free, Cat. No. R-4875) are from Sigma. Bovine calf serum (Cat. No. A-2151-L) is from HyClone Laboratory.

The reagents used in these procedures are obtained as follows. Sodium Chloride (Cat. No. S671-3), Ethylenediaminetetracetic acid disodium salt (EDTA, Cat. No. 0311-100), magnesium chloride (Cat. No. M33-500), glycerin (Cat. No. G33-500), and calcium chloride (Cat. No. C79-79197) are from Fisher Scientific. Sodium dodecyl sulfate (SDS, Cat. No. 1680024), dithiothreitol (DTT, Cat. No. 1346782), acrylamide (Cat. No. 1812270), and bisacrylamide (Cat. No. 1198373) are from Eastman Kodak Company. Micrococcal nuclease (Nuclease S7, Cat. No. 107-921) is from Boehringer Mannheim Biochemical. Triton X-100 (Cat. No. 23472-9) is from Aldrich Chemical Company. Phenylmethylsulfonyl fluoride (PMSF, Cat. No. 20203) is from United States Biochemical Corporation. GTP (Cat. No. 27-2000-02), CTP (Cat. No. 27-1200-02), UTP (Cat. No. 28-0700-01), and Sephadex G-50 (medium, Cat. No. 11-0043-02) are from Pharmacia LKB Biotech Inc.

Equipment used in these procedures include the sonifier cell disrupter Model W140 from Heat-Systems Ultrasonics Inc. The Centriprep membrane concentrator (Centriprep-30, Cat. No. 4306) is from Amicon Corporation and the glycerol gradient maker (Cat. No. SG30) is from Hoffer Scientific Instruments. The HP8452 diode array spectrophotometer is from Hewlett Packard. The Mono-Q anion-exchange column (Model HR5/5, Cat. No. 17-0546-01) is from Pharmacia LKB Biotech Inc. The Sorvall RC2-B centrifuge, HB4 rotor, and HS-4 rotor are from the Dupont Company. The Beckman L3-50 preparative ultracentrifuge, SW27.1 rotor, and SW28 rotor are from Beckman. The Model 4451 lyophilizer is from Labconco and the Model SE660 24-cm vertical slab gel unit and glass plates for electrophoresis are from Hoffer Scientific Instruments. The dialysis tubing (with a molecular weight cutoff of 14,000, Cat. No. 08-667A) is from Spectrum Medical Industries, Inc. The Model 203 fraction collector is from Gilson Medical Electronics, Inc.

III. PROCEDURES

A. HeLa Cell Growth, Labeling, Nuclear Isolation, and 40S hnRNP Monoparticle Extraction

Adapted from Beyer *et al.* (1977), Lothstein *et al.* (1985), and Huang *et al.* (1994).

Solutions

1. *1.0 M MgCl$_2$ stock solution:* To make 500 ml of solution, dissolve 101.7g of MgCl$_2 \cdot$6H$_2$O in distilled water and adjust the volume to 500 ml. Autoclave and store the solution at 4°C.

2. *10% (v/v) Triton X-100 solution:* For a 100-ml stock solution, bring 10 ml of Triton X-100 to a final volume of 100 ml with sterile distilled water. Mix well and store the solution at 4°C.

3. *10× HeLa nuclear isolation buffer (HNIB stock solution):* To make 1 liter, dissolve 2.42 g Tris base in about 950 ml distilled water and add 2.0 ml 10% Triton X-100 stock solution and 1.0 ml of 1.0 M MgCl$_2$ stock solution. Adjust the pH to 7.5 at room temperature with 6 N HCl then qs to 1000 ml. Autoclave and store the solution at 4°C or freeze 100-ml aliquots and dilute them to 1000 ml of 1× HNIB before use.

4. *1× HeLa nuclear isolation buffer (1× HNIB solution):* Prepare 1 liter of fresh solution for each nuclear isolation by diluting 100 ml 10× HNIB with distilled water to 1000 ml. Place the solution on ice and readjust the pH of the cold solution to 7.5 before use.

5. *10× salt−Tris−magnesium buffer (STM), pH 8.0:* To make a liter of 10× stock, dissolve 52.6 g NaCl and 24.21 g Tris base in about 900 ml distilled water. Add 10.0 ml of 1.0 M MgCl$_2$ stock solution and then pH the solution to 8.0 with 6 N HCl. Adjust the final volume to 1000 ml. Autoclave and store the solution at 4°C.

6. *1× STM, pH 7.2:* Make 1000 ml of fresh solution for each nuclear isolation preparation. Dissolve 5.26 g NaCl and 2.42 g Tris base in about 980 ml distilled water. Add 1.0 ml of 1.0 M MgCl$_2$ and then pH the solution to 7.2 with 6 N HCl. Adjust the volume to 1000 ml.

7. *15% glycerol STM solution:* Make the gradient solutions using sterile, DEPC-treated water and sterile glassware. For 500 ml, add 75 ml glycerol and 50 ml of 10× STM (pH 8.0) stock solution to 250 ml of distilled water. Stir the solution aggressively until homogeneous and adjust to 500 ml. Autoclave for 15 min and store at −20°C.

8. *30% glycerol STM solution:* To make a 500 ml solution, add 150 ml glycerol and 50 ml of 10× STM (pH 8.0) stock solution to 200 ml distilled water. Stir until glycerol dissolves and adjust to 500 ml with DEPC-treated distilled water. Autoclave for 15 min and store at −20°C.

Steps

1. Grow HeLa cells in suspension culture in modified Eagle's minimal essential medium (Eagle's salt base without $CaCl_2$, with 9.61 mM NaH_2PO_4 and 10 mg/ml phenol red) supplemented with 5% calf serum, 0.09% pluronic acid, streptomycin (6 mg/liter), penicillin (6000 U/liter). Twenty-four hours before labeling, add [^{35}S]methionine to fresh medium to a final concentration of 1 μCi/ml.

2. Harvest the cells at a density of 2.0–6.0 × 10^5 cells/ml and store the cells at −70°C in 50% glycerol.

3. Perform the following nuclear isolation procedures at 0°C on ice using either fresh or frozen cells. Wash the cells twice in hypotonic 1× HNIB buffer. If using large numbers of cells (between 3 to 9 billion), wash the cells with 1× HNIB for two more times so that enough hypotonic exposure will lyse the cells. It may be necessary to aid cell lysis with several strokes of a tight-fitting Dounce homogenizer. Wash the nuclei three times with STM buffer (pH 7.2) and resuspend them in STM buffer (pH 8.0) at a final concentration of 1.3 × 10^9 nuclei/ml.

4. Transfer the tube containing the nuclear pellet to a beaker of ice at 0°C. Disrupt nuclei by sonication at 50–100 W. Apply three to four 15-sec bursts with a 15-sec pause. After sonication examine the mitigated nuclear preparation under a phase-contrast microscope for complete disruption. A few intact nuclei per field are acceptable.

5. Incubate the nuclear sonicate at 37°C for 15 min in a water bath with occasional gentle agitation. This step allows the endogenous nuclease to digest the polyparticle complexes into monoparticles.

6. Centrifuge the sonicate at 8000 g (7000 rpm, HB4 rotor, Sorvall RC2B centrifuge) for 10 min at 4°C to remove chromatin, nucleoli, and nuclear membranes.

7. Load the supernatant containing the RNP complexes onto a 15–30% (v/v) linear glycerol gradient and centrifuge at 25,000 rpm in a Beckman SW28 rotor for 16 hr at 4°C.

8. Fractionate the gradients from the bottom by pumping with a peristaltic pump. Monitor the distribution of optical density at 256–266 nm, using the HP8452 diode array spectrophotometer.

9. Collect gradient fractions using a Gilson Model 203 fraction collector. Take 50- to 100-μl aliquots from each fraction for SDS–PAGE in a 0.75-mm Laemmli gel (see Fig. 1). Pool 4.0–5.0 ml of peak fractions that correspond to the 40S zone in glycerol gradients.

10. Estimate the 40S monoparticle concentration by calculating the product of its optical density at 260 nm and its volume. Under optimal conditions, each 1.0 billion cell preparation yields 4.0–5.0 ml of 40S monoparticles with an $OD_{260\,nm}$ of 1.5–2.0. Store the sample at −70°C. No differences have been observed between fresh samples and those stored in this condition for months.

B. Isolation of hnRNP Polyparticle Complexes

The procedures described earlier are designed to provide high yields of 40S core hnRNP monoparticles. If the following modifications are followed (designed mostly to prevent nuclease action), one can obtain a heterodisperse population of polyparticle complexes that sediment in gradients from the position of 40S monoparticles to complexes composed of 6–10 particle arrays sedimenting near the bottom of 15–30% glycerol or sucrose gradients.

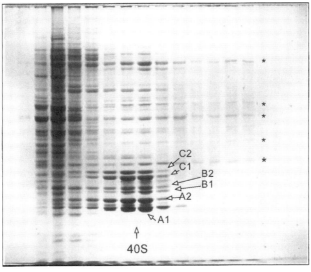

Sedimentation $\longrightarrow$

FIGURE 1 The sedimentation and protein composition of intact 40S hnRNP particles extracted from HeLa cell nuclei. The extracted monoparticles were sedimented in 15–30% glycerol gradients. This Coomassie-stained SDS–PAGE gel shows the protein present in successive gradient fractions. Note that the protein bands denoted with an asterisk do not sediment specifically with the 40S hnRNP monoparticles but overlap their distribution in gradients.

Steps

1. Suspend $1–2 \times 10^9$ fresh or frozen HeLa cells in 1.0 liters of 1× HNIB buffer (pH 7.5) containing 5.0 μg/ml yeast tRNA. Let stand on ice for 10 min to facilitate cell lysis. Dounce cells a few times if necessary to facilitate cell lysis and to remove cytoplasmic tags.

2. Centrifuge the lysed cell preparation for 8 min at 1600 rpm in an IEC refrigerated centrifuge (do not use electronic brake). Discard the supernatant and resuspend each nuclear pellet in 15 ml of STM, pH 7.2, containing 5.0 μg/ml yeast tRNA. Combine the pellets and centrifuge for 8 min at 1800 rpm in the IEC centrifuge.

3. Resuspend the nuclear pellet in 5.0 ml of STM, pH 8.0, containing tRNA as just described along with 5000 units of RNasin. Sonicate as described earlier until most of the nuclei are fragmented. Add 30 units of RQ1DNase I to the suspension and incubate the preparation in a 37°C water bath for 10 min. This step frees nascent transcripts from the DNA template. Terminate the digestion by adding 40 μl of 0.02 mM EDTA.

4. Stir the preparation at 10°C for 30 min using a "flea" (small Teflon-coated stirring bar) and then centrifuge the sample for 5 min at 2500 rpm to pellet large aggregates of nuclear material.

5. Load the supernatant on a prechilled 15–30% linear glycerol gradient containing 60 μl of 0.03 mM EDTA/ml of gradient solution (see solutions 3 and 4 in Section E).

6. Centrifuge the sample for 16 hr at 25,000 rpm at 4°C. The polyparticle complexes will sediment throughout the gradient regions from the 40S monoparticle peak to the bottom of the gradient (see Figs. 2A and 2C).

C. Purification of the C Protein Tetramer

Modified from Barnett *et al.*, (1988).

Solutions

1. *1.0 M DTT solution:* Dissolve 15.43 g DTT in distilled water to a final volume of 100 ml and divide the solution into 1.0-ml aliquots. Freeze at −20°C.

NOTE

The C protein tetramer is the first core particle protein to bind pre-mRNA during hnRNP assembly. Three tetramers bind approximately 700 nt through a highly cooperative binding mode to form a 19S triangular structure that nucleates correct binding of the basic A and B proteins (see Huang *et al.*, 1994; McAfee *et al.*, 1996).

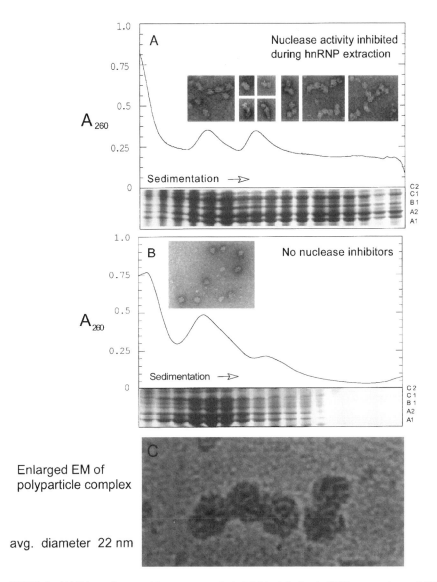

FIGURE 2 (A) Ribonuclease activity was aggressively inhibited during hnRNP preparation and DNase I was used to free nascent hnRNP complexes from chromatin. hnRNP polyparticle complexes sediment throughout the gradient with the largest arrays of particles recovered from bottom gradient fractions. The solid line tracing shows the distribution of RNA (monitored as OD_{260}) throughout the gradient, and the inset electron micrographs reveal the general morphology of the complexes in respective gradient regions. The lower inset in A is a Coomassie-stained gel showing the distribution of the major hnRNP core proteins coincident with RNA (compare with Fig. 1). Note that protein composition and stoichiometry are the same throughout the gradient. (B) No inhibitors of nuclease activity were used during hnRNP extraction, and a brief incubation step was employed to facilitate endogenous nuclease cleavage at labile sites between particles. Note that most of the RNA (solid tracing) sediments at 40S with monoparticles shown in the inset electron micrograph. Note also that the protein composition is the same as for the polyparticle complexes shown in the upper panel. (C) An enlarged electron micrograph of a polyparticle array taken from the bottom third of the gradient shown in A. The particles have an average diameter of 22 nm and are generally regular.

2. *200 mM buffer B:* To prepare 1000 ml, dissolve 11.6 g of NaCl, 2.42 g of Tris base, and 0.203 g of $MgCl_2$ in about 900 ml of water, then add 1.0 ml of 1.0 M DTT solution. Adjust the pH to 8.0 and the volume to 1000 ml. Filter the solution with a 500-ml 0.22 μM Fisher bottle top filter.

3. *600 mM buffer B:* To prepare 1000 ml, dissolve 35.0 g of NaCl, 2.42 g of Tris base, and 0.203 g of $MgCl_2$ in about 900 ml of water, then add 1.0 ml of 1.0 M DTT solution. Adjust the pH to 8.0 and the volume to 1000 ml. Filter the solution with a 500-ml 0.22 μM Fisher bottle top filter.

4. *460 mM buffer B:* To prepare 1000 ml, mix 350 ml of 200 mM buffer B with 650 ml of 600 mM buffer B. Filter the solution as described earlier.

Steps

1. Equilibrate the Mono-Q HR5/5 anion-exchange column with 200 mM buffer B for 30 min at a flow rate of 0.5 ml/min. Back wash the column (according to Pharmacia protocol) if it has been used repeatedly and/or if the back pressure is greater the 3.0 mPa.

2. Before loading the gradient-isolated 40S monoparticles (in 22–25% glycerol STM, pH 8.0) on the strong anion-exchange column, remove any particulates by centrifugation at 8000 g (7000 rpm in the HB-4 rotor) in the Sorvall RC-2B centrifuge for 10 min.

3. Load the sample (up to 40 ml or 60 OD) at a flow rate of 0.5 ml/min.

4. Elute the column with a 27.0-ml salt gradient (from 200 to 600 mM NaCl, buffer B) at a flow rate of 0.5 ml/min and monitor the optical density of the column effluent at 214 nm using a Pharmacia UV-M monitor. The C protein tetramer elutes at a salt concentration near 460 mM.

5. Collect and pool the 2.0- to 3.0-ml peak C protein fractions. Take a 30-μl aliquot for SDS–PAGE and another aliquot for quantitation; freeze the rest of the sample at $-70°C$.

6. Quantify the C protein by reading the absorbance at 214 nm and by BCA assay as described in the next four steps.

7. Dilute the C protein sample 10 times by mixing 50 μl of C protein with 450 μl of 460 mM column elution buffer and estimate the amount of C protein using a molar extinction coefficient "E" of 1.6×10^{-6}. This is valid only if the absorbance is below 0.4. If it is above 0.4 then dilute the sample further for the BCA assay below.

8. Prepare BSA standards with a range of several concentrations that cover the C protein concentration determined from the previously described assay and preform the BCA assay using the Pharmacia protocol.

9. Read the absorbance of each sample at 562 nm.

10. Plot the BSA standard curve and extrapolate the C protein concentration. Use a correction factor of 0.845 to obtain the true C protein concentration. In such a preparation, the protein concentration ranges from 50 to 400 μg/ml.

D. Purification of the (A2)₃B1 Tetramers

Modified from Lothstein *et al.,* (1985) and Barnett *et al.,* (1991).

Solutions

1. *RNase A stock solution (1 mg/ml):* To make a 50-ml solution, weigh out 50 mg RNase A on an analytical balance and add STM (pH 8.0) buffer to make a 1-mg/ml solution. Divide into 1-ml aliquots and boil for 3 min. Cap tightly and store at $-20°C$.

2. *Micrococcal nuclease stock solution (10 units/μl):* Dissolve the lyophilized powder in STM (pH 8) to a concentration of 10 units/μl. Divide the solution into small aliquots and store at $-70°C$.

3. *100 mM DTT:* Dissolve 1.513 g of DTT in 100 ml distilled water and immediately divide the solution into small aliquots and freeze at $-20°C$.

4. *0.1 M CaCl₂:* Dissolve 14.7 g of $CaCl_2 \cdot 2H_2O$ in distilled water and adjust the volume to 100 ml. Autoclave the solution.

5. *0.5 M EDTA (pH 8.0):* To make 100 ml, dissolve 18.61 g EDTA to 80 ml water, stir, and adjust the pH to 8.0 with 6 N NaOH. Adjust the volume to 100 ml. Autoclave the solution for 15 min.

Steps

1. Prepare nuclei and the nuclear sonicate as described in procedure A (steps 1–4). Add RNase A (5 μg/10^8 nuclei) to crude nuclei sonic extracts. Incubate the preparation at 37°C for 10 to 15 min.

2. Centrifuge this sonicate at 8000 g (7000 rpm, HB4 rotor, Sorvall RC2B centrifuge) for 10 min at 4°C to remove chromatin, nucleoli, and nuclear membranes.

3. Load the supernatant containing RNP complexes onto a 15–30% linear glycerol gradient and centrifuge at 25,000 rpm in a Beckman SW28 rotor for 16 hr at 4°C.

4. Fractionate the gradients and collect fractions using the instruments and procedures described in procedure A (steps 7–9). Analyze the protein composition by taking 50- to 100-μl aliquots from each fraction for SDS–PAGE in 0.75-mm Laemmli gels. Pool 3.0–4.0 ml of peak fractions that correspond to the 43S zone in glycerol gradients.

5. Estimate the 43S monoparticle concentration by calculating the product of its optical density at 260 nm and its volume. Under optimal conditions, each 2.0–3.0 billion cell aliquot will yield 3.0–4.0 ml of sample with an $OD_{260\,nm}$ of 0.4–0.6.

6. Dialyze the 43S complex for 4 hr against STM (pH 8.0) buffer to remove the glycerol. Change the buffer after 2 hr.

7. Add $CaCl_2$ and DTT to a final concentration of 1 mM and add micrococcal nuclease to a concentration of 300 U/ml to digest the endogenous RNA. Perform the digestion for 1 hr at 0°C. After complete digestion of the endogenous RNA, inactivate the enzyme by adding 0.5 M EDTA (pH 8.0) to a final concentration of 10 mM. It is very important to keep this preparation at 0°C because insoluble helical fibers spontaneously form through the association of $(A2)_3B1$ tetramers at temperatures above 10°C. Fiber formation can also be slowed with EDTA and DTT.

E. *In Vitro* Assembly of the 19S C Protein–RNA Complex

Solutions

1. *10× salt–Tris–EDTA buffer (STE):* To prepare 1 liter of 10× stock, dissolve 8.0 g of NaOH in about 900 ml of water, dissolve 37.22 g of disodium dihydrogen edetate dihydrate, and dissolve 29.22 g of NaCl and 12.12 g of Tris base. Add HCl to adjust the pH to 8.0 and adjust the volume to 1000 ml.

2. *1% PMSF in propanol:* Dissolve 1.0 g of PMSF in 100 ml of propanol, divide the solution into small aliquots, and freeze at −20°C.

3. *15% glycerol STE solution:* Make the gradient solutions using sterile, DEPC-treated water and sterile glassware. For 500 ml, add 75 ml glycerol and 50 ml of 10× STE (pH 8.0) stock solution to 250 ml of distilled water. Stir the solution aggressively until all the glycerol dissolve and adjust to 500 ml. Autoclave for 15 min and store in a −20°C freezer.

4. *30% glycerol STE solution:* To make 500 ml of solution, add 150 ml glycerol and 50 ml of 10× STE (pH 8.0) stock solution to 200 ml of distilled water. Stir until glycerol dissolves in solution and adjust to 500 ml with DEPC-treated distilled water. Autoclave for 15 min and store in a −20°C freezer.

Steps

1. Prepare the C protein samples for assembly by dialyzing the purified protein 1× STE (pH 8.0) buffer for 4 hr at 4°C in the presence of 0.001% PMSF. Change the buffer after 2 hr of dialysis. The C protein concentration ranges from 50 to 400 μg/ml.

2. Prepare the RNA substrates at a concentration of 0.5–1.0 μg/μl in TE buffer. Check each RNA sample in 8% urea denaturing gel for homogeneity.

3. Mix the dialyzed C protein with the *in vitro*-synthesized RNA at the specified ratios and concentration. For a comparison of sedimentation rates among a series of RNP assembled

on different species of RNA, use the same amount of purified C protein tetramers and RNA in each reaction.

4. Adjust the above reaction mixes to the same volume, in general, 1.0–2.0 ml with STE buffer. The mixture may be allowed to stand for 1 hr on ice.

5. Prepare 15.0 ml 15–30% glycerol gradients (in STE buffer) for sedimentation analysis using a Hoffer SG30 gradient maker. For a comparison of sedimentation of assembled RNP complexes from different experiments, it is important to prepare parallel gradients by carefully controlling the density, volume, and flow rate of the glycerol solutions.

6. For sedimentation analysis, load each sample on the above gradient and centrifuge at 25,000 rpm in a Beckman SW28 rotor at 4°C. For different separation purposes, sedimentation times may range from 16 to 30 hr; longer spin times give slightly better resolution.

7. Fractionate the gradients by pumping from the bottom of the gradient tubes at a flow rate of 16.7 μl/sec using a 20-cm-long 23-gauge needle inserted to the bottom of the gradients.

8. Monitor the distribution of optical density throughout the gradient for each nanometer from 256 to 266 nm and at 300–310 nm. Correct for OD due to light scatter by subtracting the 300- to 310-nm values using the HP8452 diode array spectrophotometer. For the comparison of RNP from a series of gradients, it is important to control the pump speed, as each collected data point represents a spectroscopic reading monitored at 1-sec intervals on a 100-μl sample window (Fig. 2).

9. Collect gradient fractions using a Gilson Model 203 fraction collector. Pool 3.0–4.0 ml of peak fractions that correspond to the RNP complex zone in glycerol gradients.

10. Analyze the assembled RNP products by electrophoresis of the protein and RNA components in each fraction in 7.75% Laemmli gels.

11. Determine the yield of the various C protein–RNA complexes by calculating the product of its optical density at 260 nm and its volume. Store the sample at −70°C.

IV. COMMENTS

In order to isolate 40S hnRNP core particles or polyparticle complexes that possess correct protein composition and stoichiometry, it is very important to minimize both protease and excessive nuclease activity. In the presence of protease activity, isolated monoparticles will be deficient in proteins A1 and C as these are the most protease labile of the core proteins (Lothstein *et al.*, 1985). Clearly, polyparticle complexes cannot be isolated in the presence of ribonuclease activity, but ribonuclease-free DNase I is critically necessary to free nascent hnRNP complexes from chromatin. It is especially helpful if a cell line such as HeLa cells (or other cells in tissue culture) can be used because, in comparison to solid tissues such as liver, far fewer nucleases and proteases are present and nuclei can be isolated in a matter of minutes. It is also important not to overload glycerol or sucrose gradients with nuclear extract as gradient resolution will be impaired. Additionally, if SDS–PAGE gels are overloaded (i.e., the core proteins are present at more than 3 μg/band), one will underestimate their contribution to the total protein present in each fraction. In numerous places above the amount of cells used in a particular procedure reflects the authors access to an efficient tissue culture laboratory. All of the procedures described in this article can be performed successfully with much less starting material. When characterizing the RNA-binding properties of a given hnRNP protein, it is also very important to know the RNA-binding site size of the protein and its binding mode (cooperative or noncooperative). Nonequilibrium-binding studies in the absence of this information have led to incorrect assumptions regarding protein–RNA interactions (see McAfee *et al.*, 1996a,b).

V. PITFALLS

1. In addition to the nuclease and protease problems mentioned earlier, it is of significant value in the various *in vitro* RNP assembly studies to cosediment the various purified RNAs (used as binding substrates) in parallel gradients. This routine procedure provides a second check (in addition to electrophoresis) for RNA homogeneity, it provides an important standard for quantification following centrifugation, and it allows one to readily distinguish the protein-bound from the protein-free peaks (especially under assembly conditions where the RNA is in excess).

2. Under conditions of extreme protein excess ($20–50\ M$), about one-third of the purified C in a binding reaction bind RNA to form a fast-sedimenting, artifactual complex which, based on its density in CsCl, is probably composed of two tetramers and eight RNA molecules. This is not inconsistent with the presence of four separate RNA-binding sites in each tetramer. It also suggests that tetramer–tetramer interactions may exist at high protein concentrations prior to RNA addition. Under these assembly conditions, the efficiency of the 19S complex is decreased. Therefore we recommend conducting the protein–RNA-binding experiments using RNA excess or slight protein excess. The artifactual complex is either absent or present only in trace amounts when RNA is mixed with protein in a ratio greater than 25:100 μg (molar ratio of 1:7.46).

Acknowledgment

The work leading to the development of these methods was supported by NSF Grant MCB 8819051 and NIH Grant GM48567 to W.M.L.

References

Barnett, S. F., LeStourgeon, W. M., and Friedman, D. L. (1988). Rapid purification of native C protein from nuclear ribonucleoprotein particles. *J. Biochem. Biophys. Meth.* **16**, 87–98.

Barnett, S. F., Northington, S. J., and LeStourgeon, W. M. (1990). Isolation and in vitro assembly of nuclear ribonucleoprotein particles and purification of core particle proteins. In "Methods in Enzymology" (J. Dahlberg and J. N. Abelson, eds.), Vol. 181, pp. 293–307. Academic Press, San Diego.

Barnett, S. F., Theiry, T. A., and LeStourgeon, W. M. (1991). The core proteins A2 and B1 exist as (A2)3B1 tetramers in 40S nuclear ribonucleoprotein particles. *Mol. Cell. Biol.* **11**, 864–871.

Beyer, A. L., Christensen, M. E., Walker, B. W., and LeStourgeon, W. M. (1977). Identification and characterization of the packaging proteins of core 40S hnRNP Particles. *Cell* **11**, 127–138.

Beyer, A. L., and Osheim, Y. N. (1990). Ultrastructural analysis of the ribonucleoprotein substrate for pre-mRNA processing. In "The Eukaryotic Nucleus: Molecular Biochemistry and Macromolecular Assemblies" (P. R. Strauss and S. H. Wilson, eds.), Vol. 2, pp. 477–502. Telford Press, Caldwell, NJ.

Conway, G., Wooley, J., Bibring, T., and LeStourgeon, W. M. (1988). Ribonucleoproteins package 700 nucleotides of pre-mRNA into a repeating array of regular particles. *Mol. Cell Biol.* **8**, 2884–2895.

Huang, M., Rech, J. E., Northington, S. J., Flicker, P., Mayeda, A., Krainer, A. R., and LeStourgeon, W. M. (1994). The C protein tetramer binds 230-240 nucleotides of pre-mRNA and nucleates the assembly of 40S heterogeneous nuclear ribonucleoprotein particles. *Mol. Cell. Biol.* **14**, 518–533.

Kiledjian, M., Burd, C. G., Gorlach, M., Portman, D. S., and Dreyfuss, G. (1994). Structure and function of hnRNP proteins in RNA protein interactions. In "Frontiers in Molecular Biology" (I. Mattaj and K. Nagai, eds.), pp. 127–149. Oxford University Press, Oxford.

LeStourgeon, W. M., Barnett, S. F., and Northington, S. J. (1990). Tetramers of the core proteins of 40S nuclear ribonucleoprotein particles assemble to package nascent transcripts into a repeating array of regular particles. In "The Eukaryotic Nucleus: Molecular Biochemistry and Macromolecular Assemblies" (P. R. Strauss and S. H. Wilson, eds.), Vol. 2, pp. 477–502. Telford Press, Caldwell, NJ.

Lothstein, L., Arenstorf, H. P., Chung, S., Walker, B. W., Wooley, J. C., and LeStourgeon, W. M. (1985). General organization of protein in HeLa 40S nuclear ribonucleoprotein particles. *J. Cell Biol.* **100**, 1570–1581.

Maniatis, T., Fritsh, E. F., and Sambrook, J. (1982). "Molecular Cloning: A Laboratory Manual." Cold Spring Harbor. Laboratory Press, Cold Spring Harbor, NY.

McAfee, J. G., Huang, M., Soltaninassab, S., Rech, J. E., Iyengar, S., and LeStourgeon, W. M. (1997). The packaging of pre-mRNA. In "Frontiers of Molecular Biology" (A. Krainer, ed.). Oxford University Press, Oxford, in press.

McAfee, J. G., Soltaninassab, S. R., Lindsay, M. E., and LeStourgeon, W. M. (1996). Proteins C1 and C2 of heterogeneous nuclear ribonucleoprotein complexes bind RNA in a highly cooperative fashion: Support for their contiguous deposition on pre-mRNA during transcription. *Biochemistry* **35**, 1212–1222.

Rech, J. E., Huang, M. H., LeStourgeon, W. M., and Flicker, P. F. (1995). An ultrastructural characterization of in vitro-assembled hnRNP C protein-RNA complexes. *J. Structural Biol.* **114**, 84–92.

Preparation of U Small Nuclear Ribonucleoprotein Particles

Sven-Erik Behrens, Berthold Kastner, and Reinhard Lührmann

I. INTRODUCTION

Small nuclear ribonucleoprotein particles (snRNPs) occupy a central position in the expression of proteins in eukaryotic cells, as it is their task to remove the introns from newly transcribed messenger RNA. This RNA splicing reaction is reviewed by Green (1991) and Moore *et al.* (1993). In addition, it has been shown that a number of autoimmune diseases are associated with the production of autoantibodies against proteins of the snRNPs (reviewed by Tan, 1989; van Venrooij and Sillekens, 1989). For both of these reasons, snRNPs have come to occupy an increasingly important place in molecular and cell biology, and thus much attention has been paid to questions of their mechanism of action and their biogenesis. A thorough investigation of these issues naturally presupposes the purification of snRNPs in an as nearly as possible native form and their complete characterization, and much effort has been devoted to meeting this challenge.

Four basic snRNP particles—U1, U2, U5, and U4/U6—are essential for pre-mRNA splicing. Each contains one or two characteristic RNA molecules and a set of "common" proteins along with a variable number of "specific" proteins. The snRNPs assemble together with additional non-snRNP splicing factors in an ordered pathway onto an intron of the pre-mRNA and form the spliceosome, a 50–60 S RNP complex that facilitates the catalysis of the splicing reaction (Steitz *et al.*, 1988; Moore *et al.*, 1993).

Under conditions of an *in vitro* splicing reaction in Hela cell nuclear extracts (i.e., at about 100 mM salt concentration), the spliceosomal snRNPs are organized in three RNP forms: 12 S U1, 17 S U2, and 25 S [U4/U6.U5] tri-snRNP. These snRNP complexes may be considered as functional subunits of the spliceosome. The composition and structures of the snRNPs are reviewed elsewhere (see Lührmann *et al.*, 1990).

The procedures given here for the preparation of snRNPs have been in use in our laboratory for some years. These protocols were developed with a view to obtaining the purest possible snRNPs and snRNP components, a task that has proved very difficult on account of the ease with which the snRNPs lose their proteins, on the one hand, and their higher-order structure, on the other. This has given rise to problems of characterization, and many of the methods described here were developed to overcome these problems. Essential aspects of the methods to be described include the following. (1) The salt concentration must always be exactly right because the binding of the specific proteins is frequently salt labile, as is the 17 S U2 snRNP and the [U4/U6.U5] tri-snRNP complex. (2) Fortunately, all snRNPs possess a characteristic structural feature, an m$_3$G group ("cap") at the 5' end of the RNA. This has allowed the development of chromatographic procedures employing immobilized antibodies against the m_3G cap; these bind the snRNPs selectively, which are subsequently eluted by displacement with free m_3G or cross-reacting m^7G. (3) Despite their large size and very similar structures, the snRNPs can be separated from each other by HPLC using the anion exchanger Mono Q.

II. MATERIALS AND INSTRUMENTATION

S-MEM (Cat. No. 072-01400) and RPMI1640 (Cat. No. 041-02400) cell media, newborn calf serum, (NCS, Cat. No. 021-06010), and fetal calf serum (FCS, Cat. No. 011-06290) are from GIBCO-BRL. All inorganic salts are from Merck AG, as are all organic solvents: acetic acid, 98%, p.a.; methanol, 100%, p.a.; ethanol, 100%, p.a.; acetone, 100%, p.a.; formamide, p.a.; dimethysulfoxide (DMSO); glycerol, 87%, p.a.; chloroform, 100%, p.a. Penicillin (Cat. No. 31749) and streptomycin (Cat. No. 35500) are from Serva. Urea (Cat. No. U-5378), N,N,N',N'-tetramethylethylendiamine (TEMED, Cat. No. T-9281), SDS (Cat. No. L-3771), HEPES (Cat. No. H-7523), phenylmethylsulfoxide fluoride (PMSF, Cat. No. P-7626), dimethyl pimelimidate (DMP, Cat. No. D-8388), dithioerythritol (DTE, Cat. No. D-8255), β-mercaptoethanol (Cat. No. M-3148), Coomassie brilliant blue G and R (Cat. Nos. B-2025 and B-0149), bromophenol blue (Cat. No. B-5525), xylene cyanol blue (Cat. No. X-4126), and N^7-methylguanosine (Cat. No. M-0627) are from Sigma. Gel solutions are from Roth AG, either as Rotiphorese Gel 30 (30% w/v acrylamide, ratio of acrylamide to bisacrylamide = 30:0.8, Cat. No. 3029.1) or as Rotiphorese Gel 40 (40% w/v acrylamide, ratio = 19:1, Cat. No. 3030.1). Phenol (Cat. No. 0038), Tris (Cat. No. 4855), and glycine (Cat. No. 3908) are also from Roth. Antibodies such as the anti-m_3G antibody and H386 were raised in our laboratory (see Bochnig et al., 1987; Reuter and Lührmann, 1986). The anti-mouse IgM antibody used for immunoprecipitation and immunoaffinity purification with H386 is a goat anti-mouse IgM (Cat. No. M-8644) from Sigma. Marker proteins for gel electrophoresis (Cat. No. 161-0303 and 161-0304) are from Bio-Rad. T4 RNA ligase (Cat. No. 1449478) is from Boehringer-Mannheim. [^{32}P]pCp (Cat. No. PB10208) is from Amersham/Buchler.

Cells are broken with a "douncer" (Cat. No. K-88530) from Kontes Glass. Dialysis tubing with a 3.5-kDa cutoff (Spektrapor 3, Cat. No. 132725) is from Spektrum. Protein A–Sepharose CL-4B (Cat. No. 17-0780-01), protein G–Sepharose (Cat. No. 17-0618-01), and cyanogen bromide-activated Sepharose 4D (Cat. No. 17-0430-01) are from Pharmacia, as is the whole FPLC system (which includes a Mono Q column type HR 5/5). Minichromatographic columns (Cat. No. 731-1550) are from Bio-Rad. Glycerol density gradients are poured with the Gradient Master Model 106 from BioComp Instruments. Where necessary, gentle agitation is provided by an end-over-end rotor (type 7637-01, Cole–Parmer). Centrifugation is performed in Heraeus Biofuge A, Heraeus Cryofuge 6000, Sorvall RC-5D centrifuges (rotor types GS3 and SS34), and the Beckman ultracentrifuge type L-60 (rotor types SW 28 and SW 40). Peptides were synthesized on an Applied Biosystems Synthesizer 430A and purified on an PD 10 (G-25) Sephadex column (Cat. No. 17-0851-01) from Pharmacia.

III. PROCEDURES

All buffers are prepared by mixing sterilized (either autoclaved or sterile-filtered) stock solutions and adding autoclaved, doubly distilled water.

Stock Solutions

1. *3 M NaCl:* To make 1 liter, add 175.3 g of NaCl to water and bring to a total volume of 1 liter.
2. *1 M K_2HPO_4:* To make 1 liter, add 174.2 g of K_2HPO_4 to water and bring to a total volume of 1 liter.

3. *1 M KH₂PO₄:* To make 1 liter, add 136.1 g of KH_2PO_4 to water and bring to a total volume of 1 liter.

4. *1 M Na₂HPO₄:* To make 1 liter, add 142.0 g of Na_2HPO_4 to water and bring to a total volume of 1 liter.

5. *1 M NaH₂PO₄:* To make 1 liter, add 138.0 g of $NaH_2PO_4 \cdot H_2O$ to water and bring to a total volume of 1 liter.

6. *7.5% (w/v) NaHCO₃:* To make 1 liter, add 75 g of $NaHCO_3$ to water and bring to a total volume of 1 liter.

7. *1 M HEPES, pH 8:* To make 1 liter, add 238.3 g of HEPES to water, adjust pH to 8.0 with 5 M KOH, and bring to a total volume of 1 liter.

8. *1 M MgCl₂:* To make 1 liter, add 203.3 g of $MgCl_2 \cdot 6H_2O$ to water and bring to a total volume of 1 liter.

9. *5 M MgCl₂:* To make 1 liter, add 1016.5 g of $MgCl_2 \cdot 6H_2O$ to water and bring to a total volume of 1 liter.

10. *0.5 M EDTA, pH 8:* To make 1 liter, add 146.1 g of EDTA to water, adjust pH to 8.0 with 5 M NaOH, and bring to a total volume of 1 liter.

11. *3 M KCl:* To make 1 liter, add 223.7 g of KCl to water and bring to a total volume of 1 liter.

12. *1 M Tris, pH 7:* To make 1 liter, add 121.1 g of Tris to water, adjust pH to 7.0 with 1 M HCl, and bring to a total volume of 1 liter.

13. *250 mM DTE:* To make 0.1 liter, add 3.86 g of DTE to water and bring to a total volume of 0.1 liter.

14. *50% (v/v) glycerol:* To make 1 liter, add 575 ml of 87% glycerol to water and bring to a total volume of 1 liter.

15. *20% (w/v) NaN₃:* To make 0.1 liter, add 20.0 g of NaN_3 to water and bring to a total volume of 0.1 liter.

16. *100 mM sodium pyruvate:* To make 1 liter, add 11.0 g of sodium pyruvate to water and bring to a total volume of 1 liter.

17. *100 mM PMSF:* To make 0.1 liter, add 1.74 g of PMSF to ethanol p.a. and bring to a total volume of 0.1 liter with ethanol p.a.

Apart from pyruvate, DTE, and PMSF, which are stored at −20°C, all stock solutions and readymade buffer solutions are kept at 4°C. All steps are carried out at 4°C unless otherwise specified.

A. Preparation of HeLa Nuclear Extracts

Solutions

1. *PBS Earle:* 130 mM NaCl and 20 mM K_2HPO_4/KH_2PO_4, pH 7.4. To make 1 liter, add 50 ml of 3 M NaCl and 16.2 ml of 1 M K_2HPO_4, adjust pH to 7.4 with approximately 3.8 ml of 1 M KH_2PO_4, and bring to a total volume of 1 liter.

2. *Buffer A:* 10 mM HEPES–KOH, pH 8, 10 mM KCl, 1.5 mM $MgCl_2$, 0.5 mM DTE. To make 1 liter, add 10 ml of 1 M HEPES, pH 8, 3.3 ml of 3 M KCl, 1.5 ml 1 M $MgCl_2$, and 2 ml of 250 mM DTE and bring to a total volume of 1 liter.

3. *Buffer C:* 20 mM HEPES–KOH, pH 8, 420 mM NaCl, 1.5 mM $MgCl_2$, 0.5 mM DTE, 0.5 mM PMSF, 0.2 mM EDTA, pH 8, and 25% (v/v) glycerol. To make 1 liter, add 20 ml of 1 M HEPES, pH 8, 140 ml of 3 M NaCl, 1.5 ml of 1 M $MgCl_2$, 2 ml of 250 mM DTE, 5 ml of 100 mM PMSF (add PMSF very slowly while the solution is stirring), 0.4 ml of 0.5 M EDTA, and 500 ml of 50% glycerol and bring to a total volume of 1 liter.

4. *Buffer G:* 20 mM HEPES–KOH, pH 8, 150 mM KCl, 1.5 mM $MgCl_2$, 0.5 mM DTE, 0.5 mM PMSF, or 5% (v/v) glycerol. To make 1 liter, add 20 ml of 1 M HEPES, pH 8, 50

ml of 3 M KCl, 1.5 ml of 1 M MgCl$_2$, 2 ml of 250 mM DTE, 5 ml of 100 mM PMSF (see buffer C), and 100 ml of 50% glycerol and bring to a total volume of 1 liter.

Steps

1. Allow HeLa S3 cells to grow in suspension culture in S-MEM supplemented with 5% (v/v) NCS, 50 μg/ml penicillin, and 100 μg/ml streptomycin at 37°C, keeping the cells at a density between 2.5 and 5 × 10^5/ml medium at logarithmic growth rate. To harvest sufficient amounts of U snRNPs, the number of cells has to be increased to 5 × 10^9 (corresponding to about 8–10 liters of medium).

2. Harvest the cells using a Heraeus Cryofuge 6000 with swinging bucket rotor for 10 min at 1000 g.

3. Resuspend cells in PBS–Earle (20 ml/10^9 cells) and pellet again in a Sorvall HB4 rotor for 10 min at 1000 g.

4. Determine the volume of the cell pellet and resuspend in 5 vol of buffer A.

5. Leave the cells to swell for 10 min, pellet again (see step 3), and resuspend in 2 vol of buffer A.

6. Break the cells by 10 strokes in a 40-ml douncer.

7. Remove the cytoplasm by two successive centrifugations in a Sorvall SS 34 rotor for 10 min at 1000 g and then 20 min at 25,000 g.

8. Resuspend the pellet (cell nuclei) in buffer C (3 ml/10^9 cells).

9. Break the nuclei by 10 strokes in a 40-ml douncer.

10. Transfer the resulting suspension into a beaker and stir carefully on ice for 30 min.

11. Remove the nuclear membrane by centrifugation in an SS 34 rotor at 25,000 g for 30 min.

12. The resultant supernatant is the nuclear extract obtained under high-salt conditions (420 mM).

13. To obtain an extract active in splicing, dialyze to a lower salt concentration (150 mM) with 100 vol of buffer G for 4–5 hr. Nuclear extracts prepared in this way are active when they make up ca. 40–60% of the final volume in the usual splicing assay (Krainer *et al.*, 1984); 5 × 10^9 cells yield about 18 ml of low-salt nuclear extract.

B. Glycerol Gradient Centrifugation

Solutions

1. *Buffer C containing 10% (v/v) glycerol:* Make buffer C as described in Section A, but take 200 ml of 50% glycerol.

2. *Buffer C containing 30% (v/v) glycerol:* Make buffer C as described in Section A, but take 600 ml of 50% glycerol.

3. *Buffer G containing 10% (v/v) glycerol:* Make buffer G as described in Section A, but take 200 ml of 50% glycerol.

4. *Buffer G containing 30% (v/v) glycerol:* Make Buffer G as described in Section A, but take 600 ml of 50% glycerol.

Steps

1. Pour 10–30% glycerol gradients with either buffer C ("high-salt gradient") or buffer G ("low-salt gradient") using sterilized SW 28 tubes (see also the manual of the BioComp Gradient Master Model 106). This can be done at room temperature.

2. Store the gradients to even out any irregularities for at least 1 hr (maximum overnight) at 4°C.

3. Load the nuclear extract onto the gradient carefully (use an Eppendorf pipette). It is possible to load up to 8 ml extract onto a 30-ml SW 28 gradient.

4. Start the centrifugation using a low acceleration rate and run it for 17 hr at 27,000 rpm (10^5 g). Stop the centrifugation without the brake.

5. Harvest the gradients manually in 1.5-ml fractions from top to bottom, using an Eppendorf pipette.

6. To analyze the U snRNP content, take one-tenth of each fraction, extract with phenol/chloroform (1/1), and precipitate the RNA of the aqueous phase by adding 0.1 vol sodium acetate (3 M, pH 4.8) and 3 vol ethanol. Check the RNA by gel electrophoresis followed by silver staining using standard procedures (Fig. 1).

C. Immunoprecipitation of U Small Nuclear Ribonucleoproteins

Solutions

1. *PBS:* 130 mM NaCl and 20 mM Na_2HPO_4/NaH_2PO_4, pH 8. To make 1 liter, add 50 ml of 3 M NaCl and 18.92 ml of 1 M Na_2HPO_4, adjust pH to 7.4 with approximately 1.08 ml of 1 M NaH_2PO_4, and bring to a total volume of 1 liter.

2. *Buffer G:* See Section A.

3. *Antibodies:* Antibodies are used either directly as hybridoma supernatants [RPMI 1640, 10 mM HEPES, 50 μg/ml penicillin, 100 μg/ml streptomycin, 0.2% (w/v) $NaHCO_3$, 10% (v/v) FCS, 1 mM sodium pyruvate] or as purified IgG fractions (1 mg/ml, dissolved in buffer G) using a standard procedure that employs protein A–Sepharose (Harlowe and Lane, 1988).

Steps

1. Hydrate protein A–Sepharose CL-4B or protein G–Sepharose overnight in PBS (1 ml per 100 mg Sepharose) using an end-over-end rotor. For long-term storage, add NaN_3 to a final concentration of 0.02% (w/v).

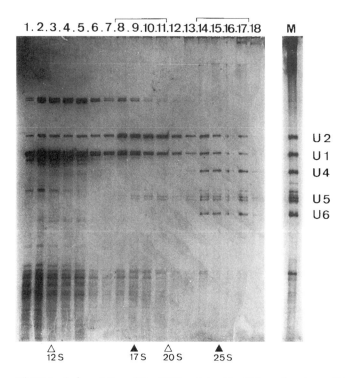

FIGURE 1 Sedimentation of U snRNP particles. The HeLa cell nuclear splicing extract was sedimented in a linear, 10–30% glycerol gradient at a low salt concentration from left to right. Small nuclear RNAs in each fraction were isolated as described in the text and analyzed on a 10% polyacrylamide–urea gel. Positions of some sedimentation coefficients are indicated. M is a U snRNA marker.

2. Wash the Sepharose twice, each time by adding PBS and then centrifuging (1 min, 10,000 *g*) in a Sorvall SS 34 tube.

3. Determine the approximate volume of the Sepharose pellet and resuspend it in the same volume of PBS.

4. For analytical immunoprecipitation assays (i.e., to identify or assay the precipitated U snRNP species) use 20 μl; for preparative assays (to determine the protein composition of the precipitated U snRNPs), take 100 μl of the 1/1 (v/v Sepharose/buffer) mixture.

5. For binding of the antibody in an analytical assay, incubate the Sepharose either with 100–200 μl hybridoma supernatant or with 5–10 μg purified antibodies in a total volume of 500 μl PBS overnight at 4°C with end-over-end rotation. For a preparative assay, incubate under the same conditions with 1–2 ml supernatant per 50- to 100-μg antibody. If the precipitations have to be performed with an IgM antibody, an IgG against the IgM must first be coupled to the Sepharose.

6. Wash the Sepharose pellet four times with 1 ml ice-cold buffer G to remove unbound antibodies. Transfer the pellet with the last wash into a new reaction tube and wash again.

7. As antigenic material use either purified U snRNPs (5–10 μg per analytical assay, 50–100 μg per preparative assay, prepared as in Section D) or glycerol gradient fractions (200 μl or 2 ml, respectively) of density gradient-fractionated nuclear extract (see Fig. 1). For precipitations with IgM antibody, first preincubate the U snRNPs/gradient fractions with the antibody for at least 2 hr on ice, with occasional shaking.

8. For binding of the antigen, incubate the Sepharose-coupled antibodies with the respective antigen in a total volume of 500 μl buffer G for at least 2 hr at 4°C with end-over-end rotation.

9. Remove the unbound antigen by washing five times with buffer G (see step 2). Again, change the reaction tube between the fourth and fifth wash.

10. For analytical assays, resuspend the Sepharose pellet in 300 μl buffer G, extract the U snRNAs with phenol (without chloroform), precipitate with ethanol (see Section B, step 6), and analyze the precipitate by 3′-end labeling with [^{32}P]pCp using T4 RNA ligase (England and Uhlenbeck, 1978). For preparative assays, resuspend the Sepharose pellet directly in protein sample buffer (Laemmli, 1970), boil for 5 min at 90°C, centrifuge briefly, and analyze the proteins of the supernatant by SDS–PAGE (see Volume 4, Celis and Olsen, "One-Dimensional Sodium Dodecyl Sulfate–Polyacrylamide Gel Eletrophoresis" for additional information).

D. Purification of U Small Nuclear Ribonucleoproteins by Anti-m₃G Immunoaffinity Chromatography

Solutions

1. *Buffer C low:* Same composition as buffer C (see Section A), but only 5% glycerol. Make buffer C as in Section A, but take 100 ml of 50% glycerol.

2. *Buffer F:* Same composition as buffer C low, except that the concentration of NaCl is 250 m*M*. Make buffer C as in Section A, but take 83.3 ml of 3 *M* NaCl and 100 ml of 50% glycerol.

Steps

1. Equilibrate an anti-m₃G immunoaffinity column (anti-m₃G IgG is coupled to CNBr-activated Sepharose by a standard procedure, see Bochnig *et al.*, 1987; Bringmann *et al.*, 1983) by washing with about 5 column volumes of buffer C low. For a 15-ml nuclear extract (corresponding to about 5×10^9 HeLa cells), a column with a bed volume of at least 5 ml must be used to achieve sufficient retardation of U snRNPs.

2. Apply nuclear extract prepared under high-salt conditions (see Section B, step 12) to the affinity column at about 1.5 ml/hr.

3. Elute nonspecifically bound components of the extract with about 6 column volumes of buffer C low.

4. Elute the specifically bound U snRNPs using 15 mM m⁷G nucleoside dissolved either in high-salt buffer (buffer C low) or at more moderate salt concentrations (buffer F). If buffer F is used, the U4, U5, and U6 snRNAs can be eluted as complete [U4/U6.U5] tri-snRNP complexes (see also Section F). From a 15-ml nuclear extract, about 4 mg U snRNPs (determined by the method of Bearden, 1978) can be retarded and eluted.

5. Remove the antibody-bound m⁷G by passing a solution of 6 M urea in buffer C low down the column.

6. Regenerate the affinity column by washing with 20 column volumes of buffer C low. For long-term storage, add NaN_3 to give a final concentration of 0.2% (v/v).

E. Purification of U1 Small Nuclear Ribonucleoproteins by Mono Q Chromatography

Solutions

1. *Buffer Q-0:* 20 mM Tris–HCl, pH 7.0, 1.5 mM $MgCl_2$, 0.5 mM DTE, and 0.5 mM PMSF. To make 1 liter, add 20 ml of 1 M Tris, pH 7, 1.5 ml of 1 M $MgCl_2$, 2 ml of 250 mM DTE, and 5 ml of 100 mM PMSF (see buffer C, procedure A) and bring to a total volume of 1 liter.

2. *Buffer Q-50:* Buffer Q-0 plus 50 mM KCl. Make buffer Q-0 as before, but add also 16.7 ml of 3 M KCl.

3. *Buffer Q-1000:* Buffer Q-0 plus 1000 mM KCl. Make buffer Q-0 as before, but add 333.3 ml of 3 M KCl.

Steps

1. Wash the Pharmacia FPLC System, which includes a 50-ml "superloop" and a Mono Q HR 5/5 column (1-ml bed volume), with a volume of buffer Q-1000 equal to 20 times the total volume of the system.

2. Wash and equilibrate the column with buffer Q-50 (same volume as in step 1). Determine the absorbance at 280 nm. The value obtained is the zero point for subsequent absorbance measurements.

3. Dilute the U snRNPs obtained by anti-m₃G chromatography with buffer Q-0 so as to bring the concentration of univalent ions below 200 mM.

4. Load U snRNPs (1–40 mg) onto the Mono Q column, using the superloop, with a flow rate of 2 ml/min. The pressure should not exceed 3.0 MPa.

5. Wash with buffer Q-50 until the absorbance reaches zero.

6. Elute the snRNPs at a flow rate of 1 ml/min using the following KCl gradient. Start with 50 mM KCl (buffer Q-50). Increase the amount of buffer Q-1000 by 5.4% per minute for 4 min, 1% per minute for 30 min, and then 4.2% per minute for 10 min. Finally, elute the column with pure buffer Q-1000 for 4 min. During the elution, collect 1-ml fractions. The U1 snRNPs elute in the first main peak at 350 to 370 mM KCl.

7. Determine the concentration of U1 snRNPs in the fractions by measurement of their absorbance at 280 nm (1 A_{280} unit is about 0.35 mg/ml) and analyze the RNA and protein content by standard PAGE procedures.

8. Contaminating 20 S U5 snRNPs can be separated from the 12 S U1 snRNPs by glycerol density gradient centrifugation (see Section B) immediately after the Mono Q chromatography (see also Bach *et al.*, 1989).

F. PURIFICATION OF THE [U4/U6.U5] tri-snRNP COMPLEX

Solutions

1. *Buffer G:* See Section A.
2. *Phosphate buffer:* 10 mM phosphate, pH 7.2. To make 1 liter, add 7.2 ml of 1 M
   Na_2HPO_4, adjust pH to 7.2 with approximately 2.8 ml of 1 M NaH_2PO_4, and bring to
   a total volume of 1 liter.

NOTE

This procedure employs the antibody H386, which is a monoclonal antibody of the IgM subclass and can be coupled via an anti-IgM antibody to protein A–Sepharose with DMP in a standard procedure. It was originally raised against the U1 snRNP-specific 70K protein (Reuter and Lührmann, 1986), but it also exhibits cross-reactivity with a 100-kDa protein that is a component of the [U4/U6.U5] tri-snRNP complex (Behrens and Lührmann, 1991).

Steps

1. Pool the 25 S fractions from a nuclear extract fractionated on a glycerol gradient (see Section B and Fig. 1), dilute with buffer G and pour slowly (1 ml/min) onto an H386 immunoaffinity column. Up to 7 ml, corresponding to four fractions, containing about 100–150 μg U snRNPs, can be loaded onto a column of 2-ml bed volume.

2. Elute nonspecifically bound components of the extract with about 20 column volumes of buffer G.

3. Elute the specifically bound U snRNPs with 5 column volumes of a 0.01 mM solution in buffer G of a competing 32-mer peptide, which corresponds to the primary epitope of H386 (see Behrens and Lührmann, 1991). Collect in 500-μl fractions and analyze one-tenth of each fraction for RNA and protein content using standard procedures.

4. If necessary, separate minor amounts of coretarded and coeluted 12 S U1 snRNPs (see Fig. 1) from the 25 S [U4/U6.U5] tri-snRNP complexes by glycerol gradient centrifugation (see Section B).

5. Elute the antibody-bound peptide with 5 column volumes of phosphate buffer and then with 5 column volumes of 3.5 M $MgCl_2$ in the same buffer.

6. Regenerate the affinity column by washing with 10 column volumes of buffer G. For long-term storage, add NaN_3 to give a final concentration of 0.02% (w/v).

G. Purification of 17 S U2 Small Nuclear Ribonucleoproteins

Solution

1. *Buffer G:* See Section A.

Steps

See also Behrens *et al.* (1992).

1. Pool the 17 S fractions from a glycerol gradient-fractionated nuclear extract (see Fig. 1). Remove small quantities of contaminating 12 S U1 snRNPs, U5 snRNPs, and [U4/U6.U5] tri-snRNP complexes by passing twice over an H386 immunoaffinity column (see Section F).

2. Apply the H386 flow-through, which contains the 17 S U2 snRNPs, to an anti-m_3G immunoaffinity column (see Section D).

3. Elute nonspecifically bound components with 20 column volumes of buffer G.

4. Elute the 17 S U2 snRNPs with m^7G and analyze as described in Section A, step 13.

5. Regenerate the affinity columns (see Section F, step 6, or Section D, step 6).

IV. COMMENTS

With these methods it is possible to purify U snRNPs from a HeLa nuclear extract under conditions close to those that allow the splicing reaction *in vitro*. Figure 2 shows various U snRNP species and those of their protein components that have been identified so far by one-dimensional SDS–PAGE.

NAME	appM$_R$ kDa	Presence in snRNP particles		
		12S U1	17S U2	25S U4/U6.U5
G	9	●	●	●
F	11	●	●	●
E	12	●	●	●
D1	16	●	●	●
D2	16,5	●	●	●
D3	18	●	●	●
B	28	●	●	●
B′	29	●	●	●
C	22	◐		
A	34	◐		
70K	70	◐		
B″	28,5		◍	
A′	31		◍	
	33		◍	
	35		◍	
	53		◍	
	60		◍	
	66		◍	
	92		◍	
	110		◍	
	120		◍	
	150		◍	
	160		◍	
	15			◫
	40			◫
	52			◫
	65			◫
	100			◫
	101			◫
	102			◫
	110			◫
	116			◫
	200			◫
	220			◫
	60			⊕
	90			⊕
	15,5			⊞
	20			⊞
	27			⊞
	61			⊞
	63			⊞

FIGURE 2 Protein composition of the 12 S U1 snRNP, the 17 S U2 snRNP, and the 25 S [U4/U6.U5] tri-snRNP complex. The apparent molecular weights of the common proteins (black dots) and the specific proteins (light dots) are indicated.

V. PITFALLS

1. For the preparation of nuclear extracts and the purification of U snRNPs, the following precautions are essential to keep all solutions and glassware free of RNase contamination and activity:

   a. All glassware must be washed thoroughly.

   b. Glassware should be rinsed extensively with distilled water.

   c. Glassware should be dried thoroughly and then heated for at least 1 hr at 250°C.

   d. All buffers and solutions should be autoclaved and then stored, if appropriate, at 4°C.

   e. Sterile gloves should be worn during all steps.

   f. All steps should be performed at 4°C unless otherwise specified.

2. Immunoaffinity purification protocols are very sensitive to any kind of variations in the

pH and salt concentration. For this reason, the pH of all buffers should be monitored very carefully, and buffer solutions should be checked regularly for absence of any precipitate.

3. For reproducible results using immunoaffinity columns, the following precautions should be observed in connection with the bound antibodies. For long-term storage, NaN$_3$ must be added to the storage buffer. Antibodies must never be incubated with denaturing solutions such as urea or 3.5 M MgCl$_2$ for longer than 4–5 hr; shorter periods are preferable. Affinity columns should always be kept at 4°C.

REFERENCES

Bach, M., Winkelmann, G., and Lührmann, R. (1989). 20 S small nuclear ribonucleoprotein U5 shows a surprisingly complex protein composition. *Proc. Natl. Acad. Sci. USA* **86**, 6038–6042.

Bearden, J. C. (1978). Quantitation of submicrogram quantities of proteins by an improved protein dye binding assay. *Biochim. Biophys. Acta* **553**, 525–529.

Behrens, S. E., and Lührmann, R. (1991). Immunoaffinity purification of a [U4/U6.U5] tri-snRNP complex from human cells. *Genes Dev.* **5**, 1429–1452.

Behrens, S. E., Tyc, K., Kastner, B., Reichelt, J., and Lührmann, R. (1992). Small nuclear ribonucleoprotein (RNP) U2 contains numerous additional proteins and has a bipartite RNP structure under splicing conditions. *Mol. Cell. Biol.* **13**, 307–309.

Bochnig, P., Reuter, R., Bringmann, P., and Lührmann, R. (1987). A monoclonal antibody against 2,2,7-trimethyl-guanosine that reacts with intact U snRNPs as well as with 7-methylguanosine capped RNAs. *Eur. J. Biochem.* **168**, 461–467.

Bringmann, P., Rinke, J., Appel, B., Reuter, R., and Lührmann R. (1983). Purification of snRNPs U1, U2, U4, U5 and U6 with 2,2,7-trimethylguanosine-specific antibody and definition of their constituent proteins reacting with anti-Sm and anti-(U1)RNP antisera. *EMBO J.* **2**, 1129–1135.

England, T. E., and Uhlenbeck, O. (1978). 3′-Terminal labelling of RNA with T4 RNA ligase. *Nature* **275**, 560–561.

Green, M. R. (1991). Biochemical mechanisms of constitutive and regulated pre-mRNA splicing. *Annu. Rev. Cell. Biol.* **7**, 559–599.

Harlowe, E., and Lane, D. (1988). "Antibodies: A Laboratory Manual." Cold Spring Harbor Laboratory, Cold Spring Harbor, NY.

Krainer, A. R., Maniatis, T., Ruskin, B., and Green, M. (1984). Normal and mutant human β-globin pre-mRNAs are faithfully and efficiently spliced in vitro. *Cell* **36**, 993–1005.

Laemmli, U. K. (1970). Cleavage of structural proteins during the assembly of the head of bacteriophage T4. *Nature* **227**, 680–685.

Lührmann, R., Kastner, B., and Bach, M. (1990). Structure of spliceosomal snRNPs and their role in pre-mRNA splicing. *Biochim. Biophys. Acta Gene Struct. Express.* **1087**, 265–292.

Moore, M. J., Query, C. C., and Sharp, P. A. (1993). Splicing of precursors to messenger RNAs by the spliceosome. *In* "RNA World" (R. F. Gesteland and J. F. Atkins, eds.), pp. 303–358. Cold Spring Harbor Press, NY.

Reuter, R., and Lührmann, R. (1986). Immunization of mice with purified U1 small nuclear ribonucleoprotein (RNP) induces a pattern of antibody specificities characteristic of the anti-Sm and anti-RNP autoimmune response of patients with lupus erythematosus, as measured by monoclonal antibodies. *Proc. Natl. Acad. Sci. USA* **83**, 8689–8693.

Steitz, J. A., Black, D. L., Gerke, V., Parker, K. A., Krämer, A., Frendeway, D., and Keller, W. (1988). Functions of the abundant U-snRNPs. *In* "Structure and Function of Major and Minor Small Nuclear Ribonucleoprotein Particles" (M. L. Birnstiel, ed.), pp. 115–154. Springer-Verlag, Berlin/New York.

Tan, E. M. (1989). Antinuclear antibodies: Diagnostic markers for autoimmune disease and probes for cell biology. *Adv. Immunol.* **44**, 93–152.

Van Venrooij, W. J., and Sillekens, P. T. G. (1989). Small nuclear RNA associated proteins: Autoantigens in connective tissue disease. *Clin. Exp. Rheumatol.* **7**, 1–11.

Isolation and Visualization of the Nuclear Matrix, the Nonchromatin Structure of the Nucleus

Jeffrey A. Nickerson, Gabriela Krockmalnic, and Sheldon Penman

I. INTRODUCTION

When the light microscope was the only tool available to examine the cell nucleus, most of the nuclear interior, except for the nucleolus and a few larger patches of heterochromatin, appeared to be clear. This transparency suggested a structureless space that was called the *nuclear sap* or the *karyolymph*. Nuclear metabolism was conceptualized as a soluble process occurring in this nuclear solution.

The electron microscope subsequently showed the nuclear interior to be much more highly structured than previously imagined. What had been termed *sap* was actually highly structured. The architecture of the nuclear interior was composed of two nucleic acid-containing structures: a DNA-containing structure called the chromatin and a RNA-containing structure that could be selectively stained (Bernard, 1969). The RNA-containing structure consisted of granules and fibers and is the only portion of the nuclear matrix that can be seen without first removing chromatin by a suitable extraction protocol.

Beginning with the pioneering efforts of Berezney and Coffey (1974), many procedures have been used to separate the nuclear matrix from the much larger mass of chromatin. The success of these protocols should be judged by how well they preserve nuclear matrix ultrastructure and composition. Studies in this laboratory developed gentler nuclear matrix isolation procedures that provided a greater preservation of matrix structure while effectively removing chromatin (Capco *et al.*, 1982; Fey *et al.*, 1986; He *et al.*, 1990).

Development of these procedures required a better electron microscopic technique for visualizing filamentous cell structures. The conventional resin-embedded thin section can only visualize those filaments that happen to lie at the section surface. To overcome this limitation, we adopted resinless section electron microscopy in which the sample is embedded and sectioned, as in conventional thin section microscopy, but in which the embedding material is removed before observation. Once the embedding resin is removed, it is easy to obtain high contrast, three-dimensional images of the nuclear filaments without staining (Capco *et al.*, 1984; Nickerson *et al.*, 1990). The entire contents of the section are imaged, not just the stained surface. The resinless section technique combines the image clarity of whole mount electron microscopy with the ability to section through the nucleus. This article presents protocols for isolating the nuclear matrix, for imaging its fine structure, and for localizing specific molecules in that structure using gold-conjugated antibodies (Fig. 1).

CELL BIOLOGY: A LABORATORY HANDBOOK, Second Edition. Vol. 2. Copyright © 1998 by Academic Press. <span></span>

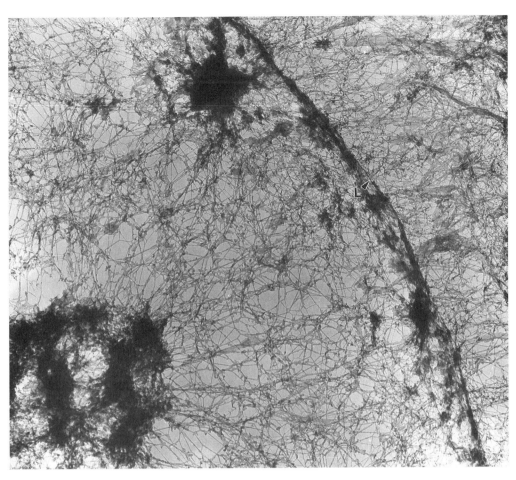

FIGURE 1 Core filaments of the nuclear matrix. This HeLa cell was extracted to reveal the core filaments of the nuclear matrix. Soluble proteins were removed by extraction with 0.5% Triton X-100. Chromatin was removed by DNase I digestion followed by extraction first with 0.25 M ammonium sulfate and then with 2 M NaCl. The network of highly branched, 10-nm nuclear core filaments is connected to the inside of the nuclear lamina (L). Some masses of material remain enmeshed in the core filament web. The intermediate filaments of the cytoskeleton are connected to the outside of the nuclear lamina. The lamina integrates the nuclear matrix structure and the cytoskeleton into a single cell structure, the fundamental organizing scaffold of the cell.

II. MATERIALS AND INSTRUMENTATION

A. Sequential Fractionation

Vanadyl riboside complex (VRC) is available from 5 Prime to 3 Prime (Cat. No. 5302-752369) and other suppliers. RNase-free DNase I is from Boehringer Mannheim (Cat. No. 776 785). The serine protease inhibitor 4-(2-aminoethyl)benzenesulfonyl fluoride (AEBSF) is available from Boehringer Mannheim under the trade name Pefabloc SC. Triton X-100, available as a 10% solution, and all other extraction chemicals are also available from Boehringer Mannheim.

B. Resinless Section Electron Microscopy

An electron microscopy, an ultramicrotome, and an oven are essential. A carbon evaporator, BEEM-embedding capsules, and an infrared lamp are optional, but may be required for certain experiments. Either diamond or fresh glass knives are required. Thermanox coverslips can be purchased from Nunc, Inc. Diethylene glycol distearate (DGD) is available from Polysciences Inc. A critical point drier or hexamethyldisilazane (HMDS), available from Sigma, is required. EM grids with carbon-coated plastic support film, forceps, razor blades, distilled water, Pasteur pipettes, ethanol, butanol, and toluene are required.

C. Antibody Staining of Resinless Sections

Colloidal gold-conjugated second antibodies are available from several suppliers. We use electron microscopy grade gold-conjugated antibodies from Amersham. These are available with gold bead sizes of 5–20 nm with different binding specificities. This allows colocalizations to be performed with first antibodies of different types and corresponding second antibodies with different sized beads.

III. PROCEDURES

A. Sequential Fractionation

Solutions

1. *Cytoskeletal buffer:* 10 mM PIPES, pH 6.8, 300 mM sucrose, 100 mM NaCl, 3 mM MgCl$_2$, and 1 mM EGTA. Freeze in aliquots at $-20°$C. Before use, add Triton X-100 to a final concentration of 0.5% from the 20× stock solution, add VRC to a final concentration of 20 mM from the 100× stock solution, and add AEBSF to a final concentration of 1 mM from the 100× stock solution. To make 1 liter of this stock solution, use 3.024 g of PIPES, 102.69 g of sucrose, 5.844 g of NaCl, 0.6099 g of MgCl$_2$·6H$_2$O, and 0.3804 g of EGTA. Titrate to pH 6.8 with 1 M NaOH.

2. *Extraction buffer:* 10 mM PIPES, pH 6.8, 250 mM ammonium sulfate, 300 mM sucrose, 3 mM MgCl$_2$, and 1 mM EGTA. Freeze in aliquots at $-20°$C. Before use, add Triton X-100 to a final concentration of 0.5% from the 20× stock solution, add VRC to a final concentration of 20 mM from the 100× stock solution, and add AEBSF to a final concentration of 1 mM from the 100× stock solution. To make 1 liter of this stock solution, use 3.024 g of PIPES, 102.69 g of sucrose, 33.035 g of ammonium sulfate, 0.6099 g of MgCl$_2$·6H$_2$O, and 0.3804 g of EGTA. Titrate to pH 6.8 with 1 M NaOH.

3. *Digestion buffer:* 10 mM PIPES, pH 6.8, 300 mM sucrose, 50 mM NaCl, 3 mM MgCl$_2$, and 1 mM EGTA. Freeze in aliquots at $-20°$C. Before use, add Triton X-100 to a final concentration of 0.5% from the 20× stock solution, add VRC to a final concentration of 20 mM from the 100× stock solution, and add AEBSF to a final concentration of 1 mM from the 100× stock solution. To make 1 liter of this stock solution, use 3.024 g of PIPES, 102.69 g of sucrose, 2.922 g of NaCl, 0.6099 g of MgCl$_2$·6H$_2$O, and 0.3804 g of EGTA. Titrate to pH 6.8 with 1 M NaOH.

4. *2 M NaCl buffer:* 10 mM PIPES, pH 6.8, 300 mM sucrose, 2 M NaCl, 3 mM MgCl$_2$, and 1 mM EGTA. Freeze in aliquots at $-20°$C. Before use, add VRC to a final concentration of 20 mM from the 100× stock solution and add AEBSF to a final concentration of 1 mM from the 100× stock solution. To make 1 liter of this stock solution, use 3.024 g of PIPES, 102.69 g of sucrose, 116.88 g of NaCl, 0.6099 g of MgCl$_2$·6H$_2$O and 0.3804 g of EGTA. Titrate to pH 6.8 with 1 M NaOH.

5. *Triton stock:* 10% (w/v) Triton X-100 frozen in aliquots at $-20°$C. This is a 20× stock solution.

6. *AEBSF stock:* 100 mM AEBSF, hydrochloride frozen in aliquots at $-20°$C. This is a 100× stock solution. Alternatively, phenylmethylsulfonyl fluoride (PMSF) can be used. The PMSF stock solution is 20 mg/ml in isopropanol, stored at room temperature, and is used at a final concentration of 0.2 mg/ml. Other protease inhibitors can be added if proteolysis is suspected, but do not use EDTA as divalent ions are necessary for structural integrity.

7. *Phosphate-buffered saline (PBS):* 10 mM Na$_2$HPO$_4$, 1 mM KH$_2$PO$_4$, 137 mM NaCl, and 2.7 mM KCl. To make 1 liter, use 1.420 g of Na$_2$HPO$_4$, 0.136 g of KH$_2$PO$_4$, 8.006 g of NaCl, and 0.2013 g of KCl. Autoclave and store in aliquots at room temperature.

Steps

1. Cells are most conveniently processed for biochemical experiments in suspension following trypsinization or scraping. For microscopy, cells can be extracted in monolayers and grown on glass coverslips for light microscopy and on Thermanox coverslips for electron microcopy. Suspension processing can be done by sequentially resuspending cell pellets in the extraction solutions and centrifuging at 1000 g for 3 min at 4°C between steps. The supernatant fractions can be saved for biochemical analysis. The extracted cell structure at each step is in the pellet. Coverslips are moved between the different extraction solutions when cells are processed in monolayer.

2. Wash cells once with PBS at 4°C. Cells to be processed in monolayer on glass or Thermanox are washed by immersing in PBS. Subsequent extraction steps are done by immersing the monolayer in the extraction solution. Cells to be processed in suspension are centrifuged at 1000 g for 3 min at 4°C between steps and are resuspended in the next wash or extraction solution.

3. Extract cells in cytoskeletal buffer containing 0.5% Triton X-100 at 4°C for 3–5 min. We use about 1 ml for each 10^7 cells until the digestion step and then halve the volume. This step will remove soluble proteins, both cytoplasmic and nucleoplasmic.

4. Extract cells with extraction buffer at 4°C for 3–5 min. This step will remove histone H1 and will strip the cytoskeleton except for the intermediate filaments that remain tightly anchored to the nuclear lamina.

5. Remove chromatin by digestion with RNase-free DNase I in digestion buffer containing 0.5% Triton X-100. Digestion is with 200–400 units of DNase I for 30–50 min at 32°C. It is important to optimize the removal of chromatin for the cell type selected. This should be done by labeling cell DNA with [^{3}H]thymidine (5 μCi/ml of culture medium) overnight and then quantitating the release of DNA by scintillation counting. This step will remove DNA and the remaining histones. What is left with the structure is the complete nuclear matrix.

6. Extract the structure in 2 M NaCl buffer at 4°C for 3–5 min. This step strips some proteins from the nuclear matrix and either uncovers or stabilizes a highly branched network of 10-nm filaments that form the core structure of the nuclear matrix. We believe that the nuclear matrix core filament network is the basic organizing scaffolding of the cell nucleus. Nuclear matrix proteins that are not part of this core filament network should be in the supernatant fraction.

It is important to increase the ionic strength gradually or in steps. Direct application of 2 M NaCl causes some collapse of the structure; much better preservation is achieved by the two-step 0.25 M ammonium sulfate–2 M NaCl procedure.

B. Resinless Section Microscopy

Solutions

Buffers

1. *Digestion buffer:* See Section A.
2. *Cacodylate buffer:* The stock solution is 0.2 M sodium cacodylate, pH 7.2–7.4, and is prepared according to the following method.

   a. Solution A: 42.8 g sodium cacodylate [$Na(CH_3)_2AsO_2 \cdot 3H_2O$] and 1000 ml distilled water.

   b. Solution B (0.2 M HCl): 10 ml concentrated HCl (36–38%) and 603 ml distilled water.

The stock solution of the desired pH can be obtained by adding solution B as shown below to 50 ml of solution A and diluting to a total volume of 200 ml.

Solution B (ml)	pH of buffer
6.3	7.0
4.2	7.2
2.7	7.4

The 0.2 M sodium cacodylate solution is stable for few months and should be kept at 4°C. The washing buffer is 0.1 M sodium cacodylate and is prepared by mixing together 1:1 (v/v) 0.2 M sodium cacodylate and distilled water.

Fixatives

1. *Glutaraldehyde solution:* 2.5% glutaraldehyde in digestion buffer is used to fix the samples. The glutaraldehyde concentration in the fixative may vary from 1–3%. Only EM grade glutaraldehyde should be used, which is available from any electron microscopy supplier. Glutaraldehyde is packaged in 1-ml ampules of 8 or 25% aqueous solution. The fixative has to be prepared freshly, within 3 hr prior to fixation, and stored at 4°C.

2. *Osmium solution:* A solution of 1% osmium tetroxide in 0.1 M sodium cacodylate, pH 7.2–7.4, is the optional second fixative. Osmium tetroxide can be purchased as a stock solution in distilled water from most electron microscopy suppliers. The fixative is prepared by mixing the osmium stock solution with 0.2 M sodium cacodylate buffer and distilled water. The fixative is stable for 1–2 months at 4°C.

Stains

1. *Hematoxylin:* Delafield's hematoxylin was used and purchased as a solution from Rowley Biochemical Institute.

2. *Eosin:* A saturated solution of eosin Y (Sigma) in 70% ethanol was used. The solution can be stored at room temperature. The pipetting of the stain should be done with care so as to not disturb the eosin deposit. Alternatively, a small quantity of stain can be centrifuged before use.

Steps

The following procedures are suitable for cells processed in suspension or preserved as monolayers.

1. *Cell culture and processing:* Fractionate, fix, and process cells grown on Thermanox coverslips through all steps *in situ* while still attached to the coverslip. Harvest, extract, and fix suspended cells as a small pellet. After fixation, place the pellet into a BEEM capsule for further processing. The height of the pellet in the BEEM capsule should be less than 2 mm; if larger, then divide the sample or allow more time for individual steps.

2. *Fixation:* The nuclear matrices should be fixed immediately after fractionation. Fix the nuclei in 2.5% glutaraldehyde in digestion buffer for 40 min at 4°C. Wash the fixed nuclei in digestion buffer for 5 min at 4°C.

   Optional postfixation. Wash the samples in 0.1 M sodium cacodylate buffer for 5 min at 4°C. Fix the samples in 1–2% osmium tetroxide in 0.1 M sodium cacodylate buffer for 30 min at 4°C. Wash the samples again in 0.1 M sodium cacodylate for 5 min. The fixed nuclear matrices can be stored overnight at 4°C in 0.1 M sodium cacodylate buffer.

3. *Dehydration:* Dehydration is performed at room temperature. Transfer the samples from digestion buffer (or 0.1 M cacodylate buffer) to 50% ethanol for 5 min. Dehydrate the samples in increasing ethanol concentrations ending with three changes of 100% ethanol for 5 min each. The dehydrated matrices can be stored overnight at 4°C.

4. *Block staining (optional):* The following staining procedure is not necessary for the ultrastructural visualization of the sample but may facilitate the localization of nuclei in the DGD block prior to sectioning. There are two alternative methods for staining; both are performed at room temperature.

   a. Staining with eosin Y: After the first dehydration step, transfer the samples to a freshly prepared, saturated solution of eosin Y in 70% ethanol for 20 min. Briefly wash the samples in 70% ethanol and then dehydrate as in step 3.

   b. Staining with hematoxylin: Prior to the first dehydration change, incubate in hematoxylin for 20 min. Continue the dehydration as in step 3.

5. *Transition fluid:* Ethanol and DGD are not miscible so a transition fluid, butanol or toluene, is used. Transfer the samples to a 1:1 (v/v) mixture of ethanol:butanol (or

toluene) for 5 min and through two changes of 100% butanol (or toluene) for 5 min each. Put the samples in a 56–60°C oven.

6. *DGD infiltration:* Infiltration with DGD is performed at 56–60°C. DGD is melted to this temperature in an oven and samples can be handled in the oven or on the lab bench with an infrared lamp.

   Prepare a mixture of 1:1 (v/v) DGD:butanol (or toluene) and pour it over the samples (use prewarmed transition fluid and molten DGD.) Leave the samples in the oven, without a cover, for 30 min to allow the transition solvent to evaporate. Replace the mixture with two changes of pure molten DGD for 30–60 min each to ensure proper infiltration.

7. *Mounting, trimming and sectioning:* Allow the samples to cool and solidify at room temperature. Peel the Thermanox coverslips from DGD. Cut squares (about 16 mm²) from the area that contains nuclei and mount them on DGD blocks with a drop of molten DGD.

   For samples in BEEM capsules, cut away the capsule plastic and the trim the remaining block with razor blades. For thin sections, 0.15 μm or less, trim the block face to as small a size possible. For thicker sections, up to 1 μm, the block face can be as large as 5 mm².

   Sectioning is done with glass or diamond knives with troughs filled with distilled water. Collect the sections on plastic-covered, carbon-coated grids. The section thickness is estimated by the continuous interference color (Peachy, 1958), which is essentially the same as in epon sections.

8. *DGD removal:* Immerse the grids in toluene or butanol and incubate them at room temperature. Wax removal is very rapid with toluene; incubation for an hour is sufficient. Butanol affects a slower extraction and samples are best left for a few hours or overnight.

9. *Final sample preparation:* There are two different methods for transferring the samples to air which allow a good preservation of the three-dimensional architecture. Critical point drying is the standard method. The second method, HMDS drying, although less completely tested, is far simpler, quicker and does not require a special apparatus. The technique uses HMDS, which is a solvent with a very low surface tension. All of the following steps are performed at room temperature.

   a. Critical point drying: Transfer the grids with sections to a 1:1 (v/v) mixture of ethanol with the dewaxing solvent (butanol or toluene) for 5 min, and then to three changes of 10 min each of 100% ethanol. Place the samples in 100% ethanol in the critical point dryer and process according to the apparatus instructions.

   b. HMDS drying: Transfer the grids to 100% ethanol following the same procedure as in a. After the last change of 100% ethanol is completed immerse the grids for 5 min in a 1:1 (v/v) mixture of ethanol and HMDS. Then do three changes of pure HMDS, for 10 min each. Place the grids on filter paper to air-dry.

10. *Carbon coating (optional):* When dry, the grids are ready to be placed in the electron microscope. However, many samples are quite delicate and vulnerable to beam damage. Many electron microscopes, especially older ones, require a high minimum beam current for adequate viewing and focusing. The specimens can be stabilized by a light coating with carbon in a standard carbon coating apparatus.

C. Staining of Resinless Sections with Antibodies

The location of specific proteins in the nuclear matrix or cytoskeleton can be determined by resinless section electron microscopy with the use of specific antibodies and colloidal gold-conjugated second antibodies. This requires a few modifications of and additions to the basic resinless section protocol.

Solutions

Buffers

1. *Digestion buffer:* See Section A.

2. *Cacodylate buffer:* See Section B.

3. *TBS-1:* 10 mM Tris–HCl, pH 7.7, 150 mM NaCl, 3 mM KCl, 1.5 mM MgCl$_2$, 0.05% (v/v) Tween 20, 0.1% (w/v) bovine serum albumin, and 0.2% (w/v) glycine. For 1 liter of buffer use 10 ml Tris–HCl from a 1 M Tris–HCl stock solution, 8.766 g NaCl, 0.224 g KCl, 0.352 g MgCl$_2$, 1 g bovine serum albumin, 2 g glycine, and 500 μl Tween 20.

4. *TBS-2:* 20 mM Tris–HCl, pH 8.2, 140 mM NaCl, and 0.1% (w/v) bovine serum albumin. For 1 liter of buffer use 20 ml Tris–HCl from a 1 M Tris–HCl stock solution, 8.176 g NaCl, and 1 g bovine serum albumin.

Fixatives

1. *Formaldehyde solution:* To preserve the antigenicity, paraformaldehyde is used in concentrations varying from 3 to 4%. The fixative is prepared fresh, just before use, from a stock solution of 16% of paraformaldehyde EM grade. Sometimes, glutaralehyde in concentrations varying from 0.5 to 1% is added, which slightly improves the ultrastructural preservation. The buffer for this fixative can be cacodylate buffer prepared as described in Section B, cytoskeletal buffer prepared as described in Section A, or digestion buffer prepared as described in Section A.

2. *Glutaraldehyde solution:* See Section B.

3. *Osmium solution:* See Section B.

Blocking Reagent

1. *5% (w/v) nonfat dry milk in TBS-1:* Prepare by adding 5 g of dry milk to 100 ml of TBS-1.

Quenching Reagent

1. *Sodium borohydride solution:* 0.5 mg/ml NaBH$_4$ in TBS-1.

Stains

1. *Hematoxylin:* See Section B.

2. *Eosin:* See Section B.

Steps

The following procedure is suitable for cells processed as monolayers.

1. *Cell culture and processing:* Fractionate, fix, and process the cells grown on Thermanox or glass coverslips through all steps *in situ* while still attached to the coverslip.

2. *Fixation:* The nuclear matrices should be fixed immediately after fractionation. Fix the nuclei in the paraformaldehyde fixative for 30 min at 4°C. Wash the fixed nuclei in the fixative buffer for 5 min at 4°C. The fixed nuclear matrices can be stored overnight at 4°C in 0.1 M sodium cacodylate buffer.

3. *Quenching:* If the initial fixative contains glutaralehyde, the free aldehyde groups may be quenched with sodium borohydrate solution (0.5 mg/ml NaBH$_4$ in TBS-1). This is in addition to the quenching provided by the glycine present in TBS-1. The quenching should be performed at room temperature for 15 min and followed by three quick rinses in TBS-1.

4. *Blocking:* In order to prevent nonspecific binding of antibodies, incubate the samples at room temperature for 15–60 min with 5% (v/v) normal goat serum in TBS-1.

5. *Antibody staining:*

   a. First antibody incubation: The first antibody incubation can be done either for 1–3 hr at room temperature or overnight at 4°C. Dilute the first antibody in TBS-1 containing1% (v/v) normal goat serum. The optimal antibody concentration should be determined empirically. We usually determine the antibody concentration giving the maximal signal and minimal background by immunofluorescence microscopy. The optimal concentration for staining resinless sections may be slightly higher.

b. Washing: Rinse the samples three to five times for 5–10 min in TBS-1.

c. Blocking:

    1. Incubate in TBS-1 containing 5% (w/v) nonfat dry milk for 10 min at room temperature, followed by three quick rinses in TBS-1. This step is optional.

    2. Incubate in 5% (v/v) normal goat serum in TBS-1 for 15–30 min at room temperature.

d. Second antibody incubation: Without rinsing, incubate the samples in the appropriate gold-conjugated second antibody diluted 1:3 to 1:10 in TBS-2. Incubate for 1–3 hr at either room temperature or 37°C.

e. Washing: After incubation, rinse the samples three times, for 10 min each, in TBS-1 and then rinse in 0.1 M sodium cacodylate, pH 7.4.

6. *Postfixation:*

a. Wash the samples in 0.1 M sodium cacodylate buffer for 5 min at 4°C.

b. Fix the sample in 2–2.5% (v/v) glutaraldehyde in 0.1 M sodium cacodylate buffer for 30–60 min at 4°C.

c. Fix the samples in 1–2% osmium tetroxide in 0.1 M sodium cacodylate buffer for 30 min at 4°C. Wash the samples again in 0.1 M sodium cacodylate for 5 min. This step is optional.

7. *Dehydration:* Dehydration is performed at room temperature. Transfer the samples from 0.1 M cacodylate buffer to 50% ethanol for 5 min. Dehydrate the samples in increasing ethanol concentrations, ending with three changes of 100% ethanol for 5 min each. The dehydrated matrices can be stored overnight at 4°C.

8. *Block staining (optional):* The following staining procedure is not necessary for the ultrastructural visualization of the sample, but may facilitate the localization of nuclei in the DGD block prior to sectioning. There are two alternative methods for staining; both are performed at room temperature.

a. Staining with eosin Y: After the first dehydration step, transfer the samples to a freshly prepared, saturated solution of eosin Y in 70% ethanol for 20 min. Briefly wash the samples in 70% ethanol and then dehydrate as in step 3.

b. Staining with hematoxylin: Prior to the first dehydration change, incubate in hematoxylin for 20 min. Continue the dehydration as in step 3.

9. *Transition fluid:* Ethanol and DGD are not miscible so a transition fluid, butanol or toluene, is used. Transfer the samples to a 1:1 (v/v) mixture of ethanol:butanol (or toluene) for 5 min and through two changes of 100% butanol (or toluene) for 5 min each. Put the samples in a 56–60°C oven.

10. *DGD infiltration:* Perform the infiltration with DGD at 56–60°C. DGD is melted to this temperature in an oven and samples can be handled in the oven or on the lab bench with an infrared lamp.

    Prepare a mixture of 1:1 (v/v) DGD:butanol (or toluene) and pour it over the samples (use prewarmed transition fluid and molten DGD.) Leave the samples in the oven, without a cover, for 30 min to allow the transition solvent to evaporate. Replace the mixture with two changes of pure molten DGD for 30–60 min each to ensure proper infiltration.

11. *Mounting, trimming and sectioning:* Allow the samples to cool and solidify at room temperature. Peel the Thermanox coverslips from DGD. Cut squares (about 16 mm^2) from the area that contains nuclei and mount them on DGD blocks with a drop of molten DGD.

The sectioning is done with glass or diamond knives with troughs filled with distilled water. Collect the sections on plastic-covered, carbon-coated grids. The section thickness is estimated by the continuous interference color (Peachy, 1958), which is essentially the same as in resin sections.

12. *DGD removal:* Immerse the grids in toluene or butanol and incubate them at room temperature. Wax removal is very rapid with toluene; incubation for an hour is sufficient. Butanol affects a slower extraction and samples are best left for a few hours or overnight.

13. *Final sample preparation:* There are two different methods for transferring the samples to air that allow a good preservation of the three-dimensional architecture. Critical point drying is the standard method. The second method, HMDS drying, although less completely tested, is far simpler, quicker, and does not require a special apparatus. The technique uses HMDS, which is a solvent with a very low surface tension. All of the following steps are performed at room temperature.

   a. Critical point drying: Transfer the grids with sections to a 1:1 (v/v) mixture of ethanol with the dewaxing solvent (butanol or toluene) for 5 min, and then to three changes of 10 min each of 100% ethanol. Place the samples in 100% ethanol in the critical point dryer and process according to the apparatus instructions.

   b. HMDS drying: Transfer the grids to 100% ethanol following the same procedure as in a. After the last change of 100% ethanol is completed, immerse the grids for 5 min in a 1:1 (v/v) mixture of ethanol and HMDS. Then do three changes of pure HMDS, for 10 min each. Place the grids on filter paper to air-dry.

14. *Carbon coating (optional):* When dry, the grids are ready to be placed in the electron microscope. However, many samples are quite delicate and vulnerable to beam damage. Many electron microscopes, especially older ones, require a high minimum beam current for adequate viewing and focusing. The specimens can be stabilized by a light coating with carbon in a standard carbon coating apparatus.

References

Bernhard, W. (1969). A new procedure for electron microscopical cytology. *J. Ultrastruct. Res.* **27**, 250–265.

Capco, D. G., Krochmalnic, G., and Penman, S. (1984). A new method of preparing embedment-free sections for TEM: Application to the cytoskeletal framework and other three-dimensional networks. *J. Cell Biol.* **98**, 1878–1885.

Capco, D., Wan, K., and Penman, S. (1982). The nuclear matrix: Three-dimensional architecture and protein composition. *Cell* **29**, 847–858.

Fey, E. G., Krochmalnic, G., and Penman, S. (1986). The non-chromatin substructures of the nucleus: The RNP-containing and RNP-depleted matrix analyzed by sequential fractionation and resinless section electron microscopy. *J. Cell Biol.* **102**, 1653–1665.

He, D. C., Nickerson, J., and Penman, S. (1990). RNA containing core filaments of the nuclear matrix. *J. Cell Biol.* **110**, 569–580.

Nickerson, J. A., Krockmalnic, G., He, D., and Penman, S. (1990). Immuno-localization in three dimensions; immuno-gold staining of cytoskeletal and nuclear matrix proteins in resinless EM sections. *Proc. Natl. Acad. Sci. USA.* **87**, 2259–2263.

Peachy, L. D. (1958). Thin sections. I. A study of section thickness and physical distortion produced during microtomy. *J. Biophys. Biochem. Cytol.* **4**, 233–242.

Micromanipulation of Chromosomes Using Laser Microsurgery (Optical Scissors) and Laser-Induced Optical Forces (Optical Tweezers)

Michael W. Berns, Hong Liang, Gregory J. Sonek, and Yagang Liu

I. INTRODUCTION

Individual chromosomes in living cells and/or suspensions can be manipulated by optical scissors and optical tweezers. The first experiments (Berns *et al.*, 1969) demonstrated that a pulsed argon laser focused onto chromosomes of living mitotic salamander cells resulted in the production of a 0.5-μm lesion in the irradiation region of the chromosome. Subsequent studies indicated that the laser microbeam could be used to selectively inactivate the nucleolar genes (Berns, 1978). By use of the 266-nm wavelength of a Nd:YAG laser, not only could the nucleolar genes be selectively deleted, causing a loss of nucleoli in subsequent cell generations, but also a corresponding lack of one light staining Giemsa band in the nucleolar organizers region of the chromosome could be demonstrated in cells proliferated from the single irradiated cell (Berns *et al.*, 1979). With the further development of cloning techniques specific for single irradiated cells, cellular sublines with deleted ribosomal genes resulting from laser microbeam irradiation of the rDNA on the mitotic chromosome were established (Liang and Berns, 1983).

In 1987, Ashkin and Dziedzic first used a tightly focused laser beam to generate optical trapping forces to move biological objects. The manipulation of chromosomes in living cells and in isolation buffer was reported by Berns and colleagues in 1989. In an extended study, an optical force applied to a late moving metaphase chromosome caused it to accelerate toward the metaphase plate. In addition, anaphase chromosomes could be held motionless by optical trapping forces (Liang *et al.*, 1991). A further study (Liang *et al.*, 1993) indicated that it was possible to combine the optical tweezers to grasp and pull with the cutting and ablation capacity of the optical scissors. A wavelength dependence of induced chromosome bridges and cell cloning efficiency have been reported (Vorobjev *et al.*, 1993; Liang *et al.*, 1996). It has been demonstrated that a specific region of a single human chromosome can be dissected and manipulated by the cutting beam and trapping beam simultaneously (He, 1995). This improves on the current needle collection methods used for generating region-specific DNA libraries. Cell biologists now have a complete set of optical tools to manipulate chromosomes for the study of chromosome movements, spindle function, and cell genetics.

II. MATERIALS AND INSTRUMENTATION
A. Cells

1. Male rat kangaroo (*Potorous tridactylis*) kidney cells PTK$_2$.
2. Human–rodent hybrid cell line containing the human chromosome of interest.

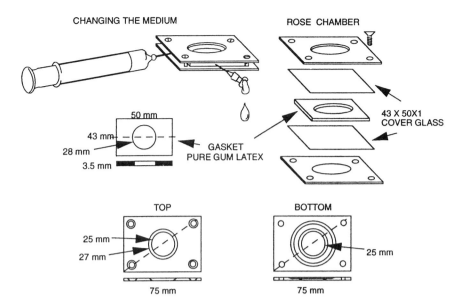

FIGURE 1 Rose multipurpose chamber with its component parts.

B. Media, Chemicals, and Supplies for Cell Culture and Isolation of Chromosomes

Modified Eagle's medium (Cat. No. 410-1500 ED), penicillin-G (Cat. No. 600-5140AG), 100 units/ml (working concentration for experiment, same as follows); streptomycin sulfate (Cat. No. 600-5140AG), 100 μg/ml; trypsin (Cat. No. 610-5050AJ), 0.25%; L-glutamine (Cat. No. 320-5030AJ), 2 mM; pancreatin (Cat. No. 610-5720AG), phenol red (Cat. No. 15-100-019), phosphate-buffered saline (PBS) without Ca^{2+} and Mg^{2+} (Cat. No. 310-4200), and fetal bovine serum (Cat. No. 230-6140AJ) are from GIBCO. EDTA (Cat. No. 34103) is from Calbiochem. $CaCl_2$ (Cat. No. O-7902), colcemid solution (Cat. No. D-1925), glutaraldehyde solution (Cat. No.G-7526), and potassium chloride (Cat. No. P-4504) are from Sigma. Triton X-100 (Cat. No. BP151-100) is from Fisher Scientific. Culture flasks (T_{25}, Cat. No. 08-772-4E; T_{75}, Cat. No. 08-772) and centrifuge tubes (15 ml, Cat. No. 2097; 50 ml, Cat. No. 2098) are from Falcon. The hemacytometer "Bright Line" (Cat. No. B3180-2) is from Baxter. Centrifuge Dynac II is provided by Clay Adams.

C. Rose Multipurpose Chamber

Chamber tops and bottoms, screws, sterile gaskets, sterile needles, and sterile syringes are components of the Rose multipurpose chamber (see Fig. 1).

D. Laser and Observation Instrumentation

1. A pulsed laser for laser surgery, with wavelength $\lambda < 550$ nm and output energy adjustable up to 10 mJ per pulse (e.g., Continuum SLI-10 Laser System, $\lambda = 532$ nm).

2. A continuous wavelength (CW) laser for laser trapping, with wavelength $\lambda > 650$ nm and output power P adjustable up to 1 W (e.g., a Coherent Model 889 titanium–sapphire laser, $\lambda = 700$–1000 nm, or, for an economic choice, a Quantronix 116 YAG laser system, $\lambda = 1.06$ μm).

3. A circular variable attenuator A1 (Newport, 50G00AV.2). A high-power variable attenuator A2 (Newport, M-935-5-OPT).[1] Three electronic shutters (Melles Griot, 04 IES 001, and 04 ISC 001). A lens L1 with focal length $f = 150$ mm (Newport, KPX100AR.16).[1] Three to six pieces of aluminum-coated mirror (Newport, 10D20ER.1).[1]

---

[1] For the wavelength of CW trapping laser $\lambda = 1.06$ μm and pulsed surgery laser $\lambda = 0.532$ μm only. Titanium–sapphire laser at 700- to 1000-μm wavelengths, the polarized beam splitter (Newport Corp 05FC16PB5) can be used; other alternatives can be done in consultation with researchers in this field.

4. A dichroic beam splitter DBS1 (CVI, BSR-51-2025).[1] A chromatic beam splitter DBS2 (CVI, HM-0803-45).[1] Two polarized beam splitters PBS (Newport Corp. 05FC16PB7).[1]

5. A Zeiss Universal M microscope (an inverted microscope is preferred). A motor stage control system with joystick (Zeiss, MSP65). A Ph3 Neofluar 100X, 1.3 N.A. oil immersion microscope objective lens (Zeiss, 440481).

6. A power/energy meter (Sciencetech, 362002), an infrared sensor card (Newport, F-IRC4), and an infrared viewer (FJW Optical systems, Inc., 58100) for detecting and monitoring the laser.

7. A CCD camera (Panasonic, GP-MF 502), a VCR (Panasonic, AG-6030), and a TV monitor (Mitsubishi, CS-20EX1) for monitoring and recording the image from the microscope.

8. A scanning mirror system: three precision gimbal optical mounts (Newport Corp., 605-4-OMA), six close-loop micrometer actuators (Newport Corp., 850B-05), and a motion controller (Newport Corp., MM2000RX-8).

9. Two polarization beam splitter cubes used for splitting one beam into two beams without power loss and, more importantly, without interference between these two beams with perpendicular polarizations (Fig. 2).

III. PROCEDURES

A. Preparation of PTK2 Dividing Cells in a Rose Chamber

Solution

1. *0.125% viokase solution:* To make 100 ml, dissolve 5 ml of pancreatin, 0.1 g of EDTA, and 0.25 ml of phenol red into 95 ml of PBS without Ca^{2+} and Mg^{2+} and adjust pH to 7.4.

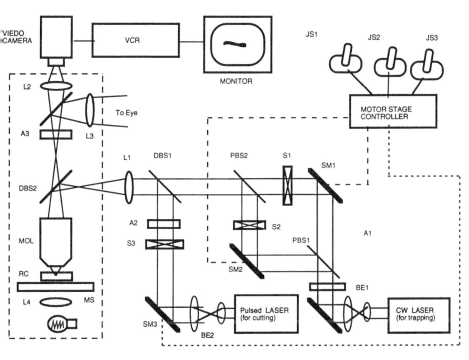

FIGURE 2 Schematic diagram for the combined use of optical scissors and optical tweezers. A1, A2, and A3, attenuators; DBS1 and DBS2, dichroic beam splitters; PBS1 and PBS2, polarized beam splitters; BE1 and BE2, beam expanders; SM1, SM2, and SM3, scanning mirrors; JS1, JS2, and JS3, joysticks; L1, L2, L3, and L4, lenses; S1, S2, and S3, shutters; MOL, microscope objective lens; MS, motor stage; and RC, Rose chamber.

Steps

1. Select a healthy, confluent or nearly confluent flask of cells.

2. Remove the old medium from the flask of cells using an unplugged sterile pipette attached to a vacuum flask.

3. Add 1.0 ml of viokase solution to the flask of cells.

4. Place the flask of cells with viokase in the 37°C incubator for 7–10 min. When the cells begin to lift free from the flask, rap the flask sharply two or three times to dislodge the cells completely.

5. Add 5 ml MEM to inactive the viokase and wash any adhering cells free.

6. Transfer the medium, viokase, and cell mixture to a sterile centrifuge tube.

7. Centrifuge the cell suspension for 4–5 min at 800–1000 rpm.

8. After centrifugation, carefully remove the stopper from the tube and very carefully aspirate the supernatant from the tube.

9. Resuspend the cell pellet in the drop remaining in the tube bottom.

10. Add 5 ml of MEM to the resuspended pellet and take a sample to count on a hemacytometer. Count all four corners (i.e., four groups of 16 squares each), divide the result by 4, and multiply by 10^4. This gives the concentration of cells per milliliter of resuspended material.

11. Adjust the cell concentration to give 2.5 to 3.5×10^4 cells/ml and inject these into the Rose chambers using a sterile syringe and 23-gauge needle.

12. Incubate the chambers (cell side down) at 37°C in a 5–7.5% CO_2 incubator. After 36–60 hr, nonconfluent chambers with dividing cells are desirable for experimentation.

B. Preparation of Chromosome in Suspension

Solutions

1. *Colcemid solution:* Make 50 ml with concentration of 0.1 μg/ml and solubilize 500-μl stock solution (concentration of 10 μg/ml) in 50 ml media. Store at −20°C.

2. *40 mM potassium chloride solution:* Dissolve 0.298 g potassium chloride in 100 ml sterilized double-distilled water.

3. *0.5 mM propedium iodide solution.*

4. *1% Triton X-100:* Mix 1 ml of Triton X-100 in 99 ml of water. Store at room temperature.

5. *8% glutaraldehyde solution.*

6. *1× PBS.*

Steps

1. Culture the human–rodent hybrid cell line containing the human chromosome of interest (e.g., GM10611 for human chromosome 9) in four T_{75} tissue culture flasks up to 80% confluent.

2. Remove the medium, add 10 ml medium containing colcemid solution, and incubate at 37°C for 2 hr.

3. Shake off the mitotic cells and collect the midium into a 50-ml tube.

4. Rinse the flasks with 2 ml 1X PBS each and collect them into the same 50-ml tube.

5. Centrifuge at 1000 rpm for 10 min to collect the mitotic cells.

6. Resuspend cells in 800 μl of 40 mM KCl solution and transfer to a 1.7-ml microtube. Let set at room temperature for 15 min.

7. Add 400 μl of 1% Triton X-100 and 40 μl of propedium iodide solution to the tube, leave it at room temperature for another 3 min.

8. Push cell suspension gently through a 23-gauge needle to release the chromosomes.

9. Add 1 μl of 8% glutaradehyde solution and leave the sample at room temperature for 20 min.

10. Store at 4°C overnight.

11. Remove the top two-thirds of solution, resuspend chromosomes in 1.5 ml 1× PBS, and store at 4°C.

C. Alignment of Laser Microbeam System

Steps

1. Turn on the CW trapping laser; adjust output power of the laser to 10–200 mW.

2. Send the laser beam onto the PBS1. The beam is split into two beams with perpendicular polarization after the PBS1. Make sure the beams are at the center of mirrors SM1, SM2 checking with infrared sensor card (IRC) or infrared viewer.

3. Send the beam (from SM1) to the center of DBS2 by adjusting SM1.

4. Put the IRC under the microscope objective lens; finely adjust SM1 until the beam shape shown on the IRC is symmetrically round.

5. Insert the lens L1 in optical path, with the distance from the objective lens at 310 mm. Finely adjust the position of L1 until the spot on the IRC is brightest and still symmetrically round.

6. Insert PBS2 in between SM1 and L1.

7. Place an IRC after PBS2 (as close as possible). Adjust SM2 to overlap the beam 2 with beam on the IRC.

8. Place an IRC under objective lens. Block beam 1. Then adjust PBS2 until the beam shape shown on the IRC is symmetrically round.

9. Place S1 and S2 after SM1 and SM2, respectively. Turn the shutters off.

10. Put the Rose chamber with test sample under the microscope and bring into focus.

11. Adjust the video camera to see the image of the sample.

12. Put a high optical density of IR-absorbing filter in front of the video camera. Turn on the shutters. If you cannot see any laser spots on the monitor screen, replace the IR filter to lower optical density which transmits a small portion of scattering light to the video monitor for tracing the laser beams.

13. Turn on the pulsed surgery laser and adjust its output energy to between 10 and 50 μJ per pulse, with a repetition rate greater than 20 Hz.

14. Using the infrared viewer (IRV) to see the spot of the trapping beam on the surface of DBS1, adjust SM3 until the surgery beam overlaps with the trapping beam on the surface of BS1.

15. Adjust DBS1 to let the surgery beam hit the center of DBS2.

16. Remove the test Rose chamber. Put a white paper card under the objective lens and finely adjust DBS1 until the beam is brightest and symmetrically round on the card.

17. Place the sample under the microscope and send in trapping and cutting laser beams. If the cutting beams is not focused at the same plane of the trapping beam, slightly adjust BE2 to bring them together.

> **NOTE**
>
> The preceding steps complete the alignment of the trapping beam. The following steps are for alignment of the surgery beam.

D. Microsurgery of Chromosomes in Living Cells

Steps

1. After the alignment of the laser microbeam, place a dried smear of red blood cells under the microscope objective. Fire a few pulses of the surgery laser beam on the red blood cells to produce a small hole (<1 μm) to verify that the cross hair on the TV screen is directly over the lesion. If the hole is too large, attenuate or reduce the laser output until a small threshold lesion is produced.

2. Remove the red blood cell slide, and place the experimental sample under the microscope.

3. Select dividing cells that appear healthy and flat (they should have very few vacuoles and the cytoplasm should be free of small dark granules).

4. The mitotic stage of dividing cells should be determined by the specific needs of the experiment.

5. Move the microscope stage so the specific target site of the selected chromosome is under the cross hair on the monitor screen.

6. Fire the laser on the selected chromosome site. Gradually increase the laser power until the desired lesion appears (see Fig. 3).

7. Videotape the entire experiment or take photographs with a 35-mm camera. Record the image before and after irradiation.

8. Under sterile condition, remove unirradiated cells from near the target cell using a micromanipulator. Close the Rose chamber.

9. Check the chamber at 12 hr, and use the 532-nm laser beam to kill cells migrating into the area of the cell being followed.

10. Monitor the proliferation of the target cell using the VCR or simply by observation.

11. Collect descendant "clonal" cells by 0.125% viokase solution and then transfer into one well of a 12-well culture cluster containing normal medium.

12. Collect "clonal" cells with the 0.125% viokase solution until they are confluent and transfer them into T_{25} culture flask.

13. When the proliferation of descendant "clonal" cells reaches a sufficient number, they can be subjected to standard karyotypic and/or biochemical analysis.

E. Optical Trapping of Chromosomes in Living Cells

Steps

1. Place a Rose chamber under the microscope that contains dividing cells.

2. Select a specific chromosome under the microscope for experimentation.

> **NOTE**
>
> In the case of genetic studies, the irradiated cell may be isolated and cloned into a viable population. Follow steps 8 to 12 (see Fig. 4).

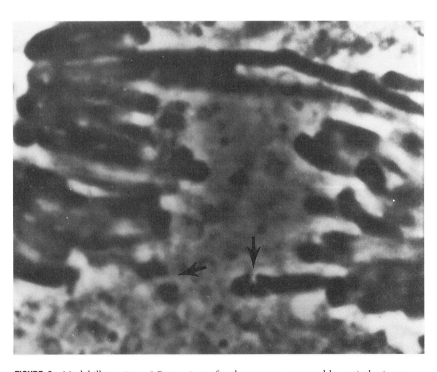

FIGURE 3 Model illustrating a 0.5-μm piece of a chromosome removed by optical scissors.

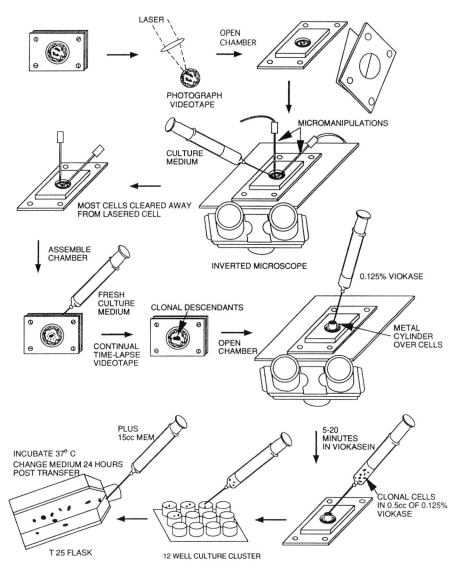

FIGURE 4 Diagrammatic representation of the procedure for cloning cells that have been irradiated at a specific chromosomal site.

3. Flat and large cells are especially good for micromanipulation. The selection of the mitotic stage depends on the specific goal of the experiment.

4. Locate the specific site of the selected chromosome at the cross hair on the monitor screen.

5. Open the shutter allowing the trapping beam to enter the microscope. The trapping laser is focused at the prealigned site which is located in the image plane of the microscope objective.

6. The chromosome near the focal point of the trapping beam will be drawn into the focal point.

7. Move the specimen stage at a speed less than 25 μm/sec in the desired direction. The chromosome will be held at the trapping position. (Usually the sites on which the largest trapping force can be applied are at either ends of the chromosome.)

8. If the trapping force is not large enough to hold the chromosome, increase the power of the trapping beam by adjusting either the beam attenuator or the output of the laser. (The trapping force is linearly proportional to the incident power of trapping beam.)

9. Videotape the entire experiment and take photographs with a 35-mm camera. Record data before, during, and after the manipulation by the optical trapping force (see Fig. 5).

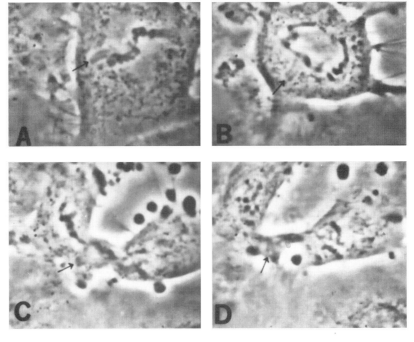

FIGURE 5 Model illustrating the holding of chromosomes in living cell with optical tweezers. (A) A live rat kangaroo cell (PTK2) in the metaphase of cell division is having a pair of large chromatids (arrow) held with the optical tweezers (wavelength is 1.06 mm and 60 mW power). (B) In anaphase the same two chromatids are still being held by the laser tweezers (arrow). (C) The cell is now in late telophase and undergoing cytokineses. The two chromatids (arrow) are trapped in the midbody between the two daughter nuclei. (D) The two chromatids are trapped in the interzone between the two daughter cells and are eventually lost from both.

F. Trapping and Cutting Chromosomes in Suspension

1. Dilute the chromosome suspension (see Procedure B, step 11) 1:1 with 1× PBS and inject them into the Rose chamber.

2. Align the trapping and cutting beam following the instructions described in Procedure C.

3. Adjust the cutting laser power using red blood smear sample as in Procedure D1.

4. Set the sample on the microscope stage. Adjust the focus until the clear chromosome images are shown.

5. Use one of the trapping beams to grab the chromosome of interest and trap it close to the second trapping beam; turn on the second beam and trap the other end of the chromosome.

6. Immobilize the chromosome in the center of the field with two trapping beams, then turn on the cutting beam, and dissect the chromosome by moving the cutting beam slowly over the chromosome.

7. Turn off the cutting beam.

8. The dissected chromosome pieces can be individually manipulated with two trapping beams (Fig. 6).

IV. COMMENTS

Newt lung cells, another desirable cell type, can be used for chromosome studies by optical scissors or tweezers. They are large and flat.

Laser light at 532-, 355-, and 266-nm wavelengths are most often used in chromosome microsurgery. If a 266-nm UV laser light is used, either a quartz-ultrafluar objective or a reflective objective must be employed. Quartz Rose chamber windows must also be used.

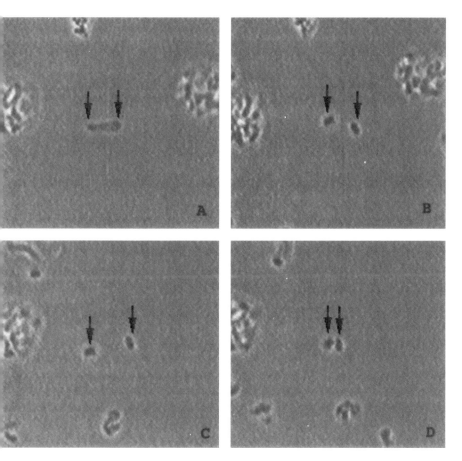

FIGURE 6 Model illustrating laser trapping and cutting the human chromosomes in suspension. (A) A single chromosome was held in position using two trapping beams which hold the ends of the chromosomes as indicated by arrows. (B) A cutting laser was used to dissect this chromosome into two pieces which were individually held by each trapping beam. (C) The individual trapping beams were used to move the chromosome fragments away from each other. (D) The individual trapping beams were used to move the chromosome fragments toward each other.

In case more trapping laser power and only one tweezers are needed, just simply remove PBS1 and PBS2. The beam is, therefore, controlled by SM1.

V. PITFALLS

1. Variation in laser output power will cause inaccurate experimental results. Turn on the lasers 30 min to 1 hr before the experiment and keep monitoring the output powerof the trapping laser beam and output energy of the surgery laser beam using a power/energy detector.

2. Living cells will be damaged if exposed to the laser beam for a long period of time or at high intensities. A power $P < 100$ mW for the trapping beam and an energy $E < 10$ mJ per pulse for the surgery beam are recommended. Generally, only about one-third of the near-infrared trapping beam power and two-thirds of the visible surgery laser beam power are transmitted through the objective lens.

3. Photodamage to cells should also be considered. Choose a laser with the appropriate wavelength to avoid regions of strong photon absorption. The near-infrared wavelength (e.g., 1.06 mm from a Nd:YAG laser) is a relatively safe wavelength for most cells.

4. Microscope optics can be damaged if the laser beam is too intense. In order to prevent damage to the objective lens, do not adjust optical components (e.g., mirrors, beam splitters) while being exposed to the intense laser beam.

5. Laser exposure may produce eye injury and physical burns. Never view a laser beam directly or by specular reflection. Use laser safety glasses whenever possible. Make sure that the laser microscope has appropriate filtration or beam blocks so that the laser beam does not go directly through the oculars into the eyes.

References

Ashkin, A., and Dziedzic, J. M. (1987). Optical trapping and manipulation of viruses and bacteria. *Science* **235**, 1517–1519.

Berns, M. W. (1978). The laser microbeam as a probe for chromatin structure and function. *In* "Methods in Cell Biology" (G. Stein, J. Stein, and L. Kleinsmith, eds.), Vol. 18, pp. 277–294. Academic Press, New York.

Berns, M. W., Aist, J. R., Wright, W. H., and Liang, H. (1992). Optical trapping in animal and plant cells using a tunable near-infrared titanium-sapphire laser. *Exp. Cell Res.* **198**, 375–378.

Berns, M. W., Chong, L. K., Hammer-Wilson, M., Miller, K., and Siemens, A. (1979). Genetic microsurgery by laser: Establishment of a clonal population of rat kangaroo cells (PTK$_2$) with a directed deficiency in a chromosomal nucleolar organizer. *Chromosoma* **73**, 1–8.

Berns, M. W., Olson, R. S., and Rounds, D. E. (1969). *In vitro* production of chromosomal lesion using an argon laser microbeam. *Nature* **221**, 74–75.

Berns, M. W., Wright, W. H., Tromberg, B. J., Profeta, G. A., Andrews, J. J., and Walter, R. J. (1989). Use of a laser-induced optical force trap to study chromosome movement on the mitotic spindle. *Proc. Natl. Acad. Sci. USA* **86**, 4539–4543.

Berns, M. W., Wright, W. H., and Wiegand Steubing, R. (1991). Laser microbeam as a tool in cell biology. *Int. Rev. Cytol.* **129**, 1–44.

He, W. (1995). "Laser Microdissection and Its Application to the Human Tuberous Sclerosis 1 Gene region on Chromosome 9q34." Ph.D. thesis, University of California, Irvine.

Liang, H., and Berns, M. W. (1983). Establishment of nucleolar deficient sublines of PTK$_2$ (*Potorous tridactylis*) by ultraviolet laser microirradiation. *Exp. Cell Res.* **144**, 234–240.

Liang, H., Vu, K. T., Krishnan, P., Trang, T. C., Shin, D., Kimel, S., and Berns, M. W. (1996). Wavelength dependence of cell cloning efficiency after optical trapping. *Biophys. J.* **70**, 1529–1533.

Liang, H., Wright, W. H., Cheng, S., He, W., and Berns, M. W. (1993). Micromanipulation of chromosomes in PTK$_2$ cells using laser microsurgery (optical scalpel) in combination with laser induced optical force (optical tweezers). *Exp. Cell Res.* **204**, 110–120.

Liang, H., Wright, W. H., He, W., and Berns, M. W. (1991). Micromanipulation of mitotic chromosomes in PTK$_2$ cells using laser induced optical forces ("optical tweezers"). *Exp. Cell Res.* **197**, 21–35.

Liang H., Wright, W. H., Reider, C. L., Salmon, E. D., Profeta, G., Andrews, J., Liu, Y., Sonek, G. J., and Berns, M. W. (1994). Directed movement of chromosome arms and fragments in mitotic newt lung cells using optical scissors and optical tweezers. *Exp. Cell Res.* **213**, 308–312.

Vorobjev, I. A., Liang H., Wright, W. H., and Berns, M. W. (1993). Optical trapping for chromosome manipulation: A wavelength dependence of induced chromosome bridges. *Biophys. J.* **64**, 533–538.

Proteins

Preparation of Tubulin from Bovine Brain

Anthony J. Ashford, Søren S. L. Andersen, and Anthony A. Hyman

I. INTRODUCTION

Recent research on microtubules has given much insight into many fundamental problems in cell biology, such as cell structure and polarity, vesicular transport, cell division, and chromosomal segregation. Furthermore, the ability to prepare polarity-marked microtubules (Howard and Hyman 1993) has contributed much to our understanding of the behavior of microtubule motors.

Microtubules are polymers of the heterodimeric protein tubulin. Because these heterodimers have the same orientation in the microtubule lattice and the dimers themselves are not symmetric, it follows that the microtubule itself must have an inherent polarity. Polymerization can take place from either end of the growing microtubule, although at different rates, with a slow growing (−) and more rapidly growing (+) end. The tubulin heterodimer has binding sites for two GTP molecules in its unpolymerized state, only one of which is hydrolyzable (GTP that is bound to the E site of the tubulin β subunit). The current model, first proposed by Mitchison and Kirschner (1984a) and further developed since then (reviewed by Hyman and Karsenti 1996), suggests that GTP-unpolymerized tubulin is incorporated at the + growing end of the microtubule, and that after a delay, during which time more GTP-bound tubulin is bound at the growing end, the GTP on the β subunit is then hydrolyzed to GDP, possibly destabilizing the microtubule lattice. Because GTP-bound tubulin dissociates from the microtubules only slowly and the dissociation rate of exposed bound GDP–tubulin two or three orders of magnitude higher, it follows that the presence of a so-called GTP–tubulin "cap" will protect the polymerized microtubule from depolymerization. Alternatively, the GTP cap structure may be inherently more stable, effectively preventing the GDP (destabilized)-polymerized tubulin from dissociating from the lattice. Thus the GTP cap hypothesis states that if the microtubule loses this GTP cap structure because of slow growth (e.g., due to lack of free tubulin, cold temperature, presence of Ca^{2+}), the GDP–tubulin dissociates from the lattice. Microtubules growing more slowly are more likely to have GDP–tubulin exposed and thus rapidly dissociate, causing microtubule shrinkage. Although it is not within the scope of this article to examine all the properties of microtubules and tubulin, some of these are of direct relevance to their isolation.

The purification of tubulin described in this article utilizes the mechanisms just described. Through a series of polymerizations and depolymerizations (see Fig. 1), microtubules and associated proteins (MAPs) are obtained. This is followed by ion-exchange chromatography to separate the tubulin from the MAPs. This protocol is based on that originally described by Shalanski *et al.* (1973) and modified by Weingarten *et al.* (1974) and subsequently described by Mitchison and Kirschner (1984b). Pure tubulin can be further modified for other uses (Hyman *et al.,* 1991).

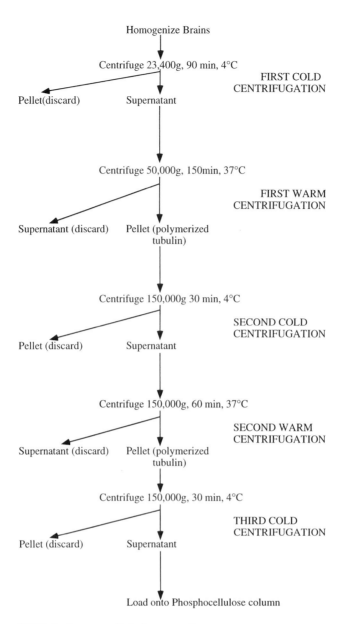

Homogenize Brains

Centrifuge 23,400g, 90 min, 4°C — FIRST COLD CENTRIFUGATION

Pellet(discard)   Supernatant

Centrifuge 50,000g, 150min, 37°C — FIRST WARM CENTRIFUGATION

Supernatant (discard)   Pellet (polymerized tubulin)

Centrifuge 150,000g 30 min, 4°C — SECOND COLD CENTRIFUGATION

Pellet (discard)   Supernatant

Centrifuge 150,000g, 60 min, 37°C — SECOND WARM CENTRIFUGATION

Supernatant (discard)   Pellet (polymerized tubulin)

Centrifuge 150,000g, 30 min, 4°C — THIRD COLD CENTRIFUGATION

Pellet (discard)   Supernatant

Load onto Phosphocellulose column

FIGURE 1 Summary of tubulin preparation.

This article describes how to make pure tubulin with an emphasis on large (gram quantities)-scale purification. This, as made clear in later sections, is quite an exercise in both organization and logistics. However, the method can easily be scaled downward if large amounts of tubulin are not required.

Because the principle of tubulin preparation relies on successive rounds of polymerization and depolymerization steps based on temperature shifts, delays and not having the proper equipment at the right temperature will drastically reduce the yield. Consequently, after one centrifuge run is completed, it is a good practice to reset the temperature for the next centrifugation and to switch on the vacuum, checking the centrifuge regularly for faults. This also applies to the rotors, and it is a good idea to have some large water baths set at 37°C and some ice troughs in which to either raise or lower the rotor temperature accordingly. For the polymerization of tubulin, a good supply of hot running water is essential as slow polymerization will reduce the overall yield.

II. MATERIALS AND INSTRUMENTATION

The protocol, as stated previously, is suitable for the preparation of large (2- to 3-g) quantities of pure tubulin. To obtain this quantity it will be necessary to procure approximately 10 to

12 bovine brains. It follows from this that the volumes involved in processing this number of brains, at least in the initial stages of preparation, will be quite large and many centrifuges will be necessary.

A. Centrifuges, Rotors, and Bottles

At least six Sorval RC5s or equivalent centrifuges with GSA-type rotors or, better still, Super Lite SLA 1500 (since these can be run at a higher speed) are needed together with 36 GSA rotor bottles (250-ml capacity) and their adapters.

Four ultracentrifuges with three Beckman-type 19 rotors and 24 corresponding bottles are needed along with four Beckman-type 45Ti rotors and 24 (70-ml capacity) polycarbonate bottles. Also, for the cycling of tubulin, a Beckman Ti50 rotor and/or Beckman TLA100 and 100.2 table-top centrifuge rotors are needed.

As can be seen from this list, the rotors required are quite large and heavy, and it is therefore prudent to check the vacuum and refrigeration efficiency of all the centrifuges beforehand. Regarding the rotors themselves, they should be inspected beforehand, and any seals and "O" rings should be checked for signs of perishing, replacing where necessary. It is also important that these "O" rings are greased lightly with vacuum grease and that the screw threads for the rotor lids are lightly smeared with Beckman "Spinkote" to ensure that the samples are sealed and that the vacuum is maintained during the run. Finally, the centrifuge rotors should be checked for the condition of the overspeed decals to prevent premature termination of the run. This is particularly true of the Beckman-type 19 rotors, which seem especially sensitive on this point.

The centrifuge bottles should also be inspected for cracks and warping (particularly true of the polycarbonate Beckman-type 45 bottles), and the lids and caps should be checked for condition and integrity of the "O" ring.

B. Other Equipment

A motor-driven homogenizer such as that made by Waring along with a 4.5-liter capacity homogenizing beaker, a continuous flow homogenizer such as that made by Yamato for resuspending tubulin pellets or a very large Dounce hand homogenizer, and a 1-liter phospho-cellulose (PC) column are needed. Details of how to prepare such a column are described in detail later in this article.

C. Chemicals

All chemicals should be of analytical grade and can be obtained from a variety of suppliers. PIPES (Cat. No. P-6757), EGTA (Cat. No. E-4378), $MgCl_2$ (Cat. No. 104-20), GTP (Cat. No. G-8877), ATP (Cat. No. A-1388) and β-mercaptoethanol (Cat. No. M-6250) are from Sigma-Aldrich. EDTA (Cat. No. 1.08418.1000), anhydrous glycerol (Cat. No. 1.04093.2500), NaCl (Cat. No. 1.06404.1000), NaOH (Cat. No. 1.06498.1000), HCl (Cat. No. 1.00319.2500), KH_2PO_4 (Cat. No. 4873.1000), and K_2HPO_4 (Cat. No. 5099.1000) are from Merck.

III. PROCEDURES

A. Preparation of Tublin

Solutions

The buffer system of choice in the preparation of tubulin is 1,4-piperazinediethanesulfonic acid (PIPES). When making up K-PIPES buffer solutions, it is first necessary to raise the pH of the solution to about pH 6.0, using KOH in order to get the PIPES to go into solution. This initial pH adjustment must be done *without the use of a pH meter* (i.e., using pH indicator paper), as any undissolved PIPES can damage the electrode.

1. *Assembly buffer (1 liter):* 30.2 g of 0.1 *M* PIPES, pH 6.8, 0.5 ml of 1 *M* stock of 0.5 m*M* MgCl$_2$, 0.761 g of 2.0 m*M* EGTA, and 37 mg of 0.5 m*M* EDTA. About 6 liters of AB is needed for a large tubulin preparation; if preferred, it can be made as a 5× stock solution and then diluted with water at 4°C before use.

2. *Column buffer (CB) (liter):* 151.2 g of 50 m*M* PIPES, pH 6.8, 3.8 g of 1 m*M* EGTA, and 2 ml of 1 *M* stock of 0.2 m*M* MgCl$_2$. Weigh out 151.2 g of PIPES, 3.8 g of EGTA, and 2 ml of a 1 *M* stock solution of magnesium chloride and add 45.6 g of potassium hydroxide. This will adjust the pH to about pH 6.7 and further adjustments can be made using a 10 *M* solution of potassium hydroxide. Approximately 5 liters of CB will be needed for such a preparation; make up as a 10× solution and dilute with water at 4°C prior to use.

3. *BRB80 conversion buffer:* To make up 150 ml, mix 47.6 g PIPES, 4.2 ml MgCl$_2$ (1 *M*), and 1.25 ml EGTA (0.2 *M*). Adjust the pH to 6.8 with potassium hydroxide. Having the correct pH is very important. Add this buffer to the purified tubulin that has been eluted from the phosphocellulose column to convert the buffer from CB to BRB80 prior to storage of the tubulin. Add the BRB80 conversion buffer in a ratio of one part conversion buffer to 20 parts tubulin in column buffer.

4. *BRB80 Buffer (5× stock solution):* Add 12.0 g of 80 m*M* K-PIPES (pH 6.8), 0.5 ml of 1 *M* stock of 1 m*M* MgCl$_2$, and 2.5 ml of 0.2 *M* stock of 1 m*M* EGTA. Adjust the pH to 6.8 with KOH and add water to 100 ml.

5. *Glycerol PB:* For 100 ml, add 2.4 g of 80 m*M* K-PIPES (pH 6.8), 0.5 ml of 1 *M* stock of 5 m*M* MgCl$_2$, 0.5 ml of 0.2 *M* stock of 1 m*M* EGTA, 52.3 mg of 1 m*M* GTP, and 33 ml anhydrous 33% (v/v) glycerol. Adjust pH to 6.8 with KOH and add water to 100 ml.

Nucleotides

Nucleotides are made as stock solutions as 100 m*M* ATP and 200 m*M* GTP. It is important that the pH of these nucleotides does not become acidic as hydrolysis will result. Consequently, they are made up in water and adjusted to pH 7.5 with sodium hydroxide. The ATP solution is made 200 m*M* with respect to MgCl$_2$. This is *not* the case with the GTP solution as this will cause precipitation. Store these solutions at −80°C until needed.

For 20 ml ATP, dissolve 1.102 g ATP with 0.813 g MgCl$_2$ and make up to 19 ml with water. Adjust pH to 7.5 with sodium hydroxide and make up the volume to 20 ml.

For 20 ml GTP, take 2.37 g Na$_3$GTP and make up to 18 ml with water, adjust the pH to 7.5 with sodium hydroxide, and make up to 20 ml with water.

Protein Concentration Determination Solutions

Protein concentration determinations are made using the Bio-Rad version of the Bradford protein assay, reading the absorption at 595 nm using bovine serum albumin as a reference standard.

Steps

Because brain tissue has a high density of microtubules in the dendrites and axons of its nerve cells, this is the tissue of choice for tubulin preparation. Bovine brain is readily available from slaughterhouses. It is essential, however, to get the brains as fresh as possible, as protein degradation will begin to take place soon after death. The brains should be warm when they are received and should be immediately plunged into a mixture of ice and saline solution for transport back to the laboratory. Do not use cold brains.

1. Preparation of Brain Tissue for the First Cold Spin

1. At the laboratory cold room, strip the brains of brain stems, blood clots, and meninges (kitchen paper tissue is very good for this purpose). Weigh the brains, transfer them to a Waring blender, and add cold AB containing 0.1% β-mercaptoethanol and 1 m*M* ATP, but without any protease inhibitors, in a ratio of 1 liter of buffer per kilogram of brain tissue.

Homogenize the brains twice for approximately 30 sec each, and then pour directly into the centrifuge bottles. Typically, approximately 5 liters of AB from 12 brains gives about 10 liters of total homogenate. Centrifuge these homogenates at 23,400 g for 90 min (GSA rotor) or at 34,200 g for 60 min (SLA1500) at 4°C.

2. First Warm Centrifugation

1. Pool and pour the supernatants (approximately 5 liters for a 12 brain preparation) into two 5-liter Erlenmeyer flasks.

2. Add a half volume of anhydrous glycerol that has been prewarmed to 37°C and adjust the GTP, ATP, and MgCl$_2$ concentrations to 0.5, 1.5, and 3 mM, respectively. It is essential to bring the tubulin solution above 30°C as quickly as possible.

3. Warm the tubulin to 30°C under a continuously flowing hot tap while swirling the flask. At some point after the tubulin has reached 30°C, the solution becomes noticeably more viscous, indicating that polymerization is beginning to take place.

4. Incubate the tubulin for another 60 min at 37°C in a water bath before recovery by centrifugation at 50,000 g for 150 min at 37°C in a Beckman-type 19 rotor. If difficulties are experienced with the centrifuges at this stage, the centrifugation time can be shortened, since the bulk of the tubulin is pelleted after 90 min.

3. Second Cold Centrifugation

1. Resuspend the pellets in approximately 700 ml of AB (with 0.1% β-mercaptoethanol) at 4°C using either a Dounce or a continuous flow homogenizer (such as the Yamato). At this point, determine the protein concentration of the resuspended pellet solution and lower the concentration to 25 mg/ml by adding cold AB.

2. Leave the solution on ice for 40 min to allow for complete depolymerization before centrifuging at 150,000 g for 30 min in a Beckman-type 45Ti rotor at 4°C.

4. Second Warm Centrifugation

1. Pool supernatants from this centrifugation and adjust the concentrations of ATP, GTP, and MgCl$_2$ to 1, 0.5, and 4 mM, respectively.

2. As before, add a half volume of anhydrous glycerol prewarmed to 37°C and then warm the entire mixture to 37°C; allow the tubulin to polymerize at this temperature for another 40 min prior to centrifugation at 150,000 g at 37°C for 1 hr in a Beckman-type 45Ti rotor. If time is short, freeze the polymerized tubulin pellets in liquid nitrogen and store at −80°C.

5. Third Cold Centrifugation

1. Collect and resuspend the polymerized tubulin pellets in a total volume of 200 ml of AB with 0.1% β-mercaptoethanol using either a continuous flow or a Dounce homogenizer.

2. Determine the protein concentration and dilute the solution with AB as necessary to 35 mg/ml.

3. Allow the tubulin to depolymerize on ice for another 40 min and then centrifuge at 150,000 g for 30 min in a Beckman-type 45Ti rotor.

4. The supernatant is now ready for loading onto the PC column to separate the tubulin from the MAPs.

6. Purification of Tubulin over a Phosphocellulose Column

1. Load the depolymerized tubulin onto a 1-liter sized column (at 4°C) that has been previously equilibrated in CB. Tubulin should be loaded onto the column at a flow rate of approximately 1.5 ml/min. Once loaded, however, the flow rate can be increased to 6 ml/

min or even faster if the phosphocellulose shows little sign of compressing, although some compression is inevitable.

2. Elute purified tubulin in the flow through, pool the protein peak, and determine the protein concentration. The purified tubulin in CB can then be snap frozen in liquid nitrogen or converted to tubulin in BRB80 using BRB80 conversion buffer prior to freezing and subsequent storage at −80°C. Figure 2 shows samples of the purification steps and purified tubulin separated on a 12% SDS–polyacrylamide gel stained with Coomassie. Purified tubulin runs as a single band with no trace of contaminating proteins. MAPs are retained on the column, and these proteins can be eluted using CB containing 1 *M* NaCl.

B. Cycling Tubulin

It is inevitable, however, that some tubulin will become inactivated or denatured during passage through the phosphocellulose column, as tubulin is an unstable entity. It is therefore recommended to "cycle" the tubulin after it passes through the column. Cycling tubulin means that it is polymerized at 37°C, reisolated by centrifugation through a cushion, and depolymerized at 4°C prior to freezing. This can be performed either immediately after elution from the column or after it has been stored at −80°C in CB. In both cases the tubulin buffer should be converted to BRB80 as described earlier. There are several advantages to cycling the tubulin: it enriches for active tubulin dimers, removes free nucleotides, removes denatured tubulin and other impurities, and concentrates the tubulin to approximately 10– 20 mg/ml (cf. 5–10 mg/ml as it comes through the column). The resulting cycled tubulin is suitable for use in *in vitro* assays, such as video microscopy (e.g., Andersen *et al.*, 1994). It is also recommended to use cycled tubulin for biochemical studies (e.g., purification of microtubule-associated proteins) because it is much easier to control the amount of microtubules formed when this highly active tubulin is used compared to that obtained directly from the phosphocellulose column.

Steps

To obtain a large stock of identical cycled tubulin, it is recommended to cycle approximately 30 ml of tubulin obtained directly from the phosphocellulose column.

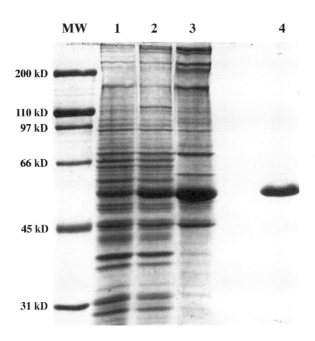

FIGURE 2 Purification steps: Lane 1, second cold centrifugation pellet; lane 2, second warm centrifugation supernatant; lane 3, third cold centrifugation supernatant (column load); and lane 4, purified tubulin.

1. Prewarm the rotor (a Beckman Ti50 rotor is suitable) in a water bath at 37°C. Fill the rotor tubes to half their volume with the glycerol cushion consisting of 60% glycerol in BRB80 (no GTP) and place these in the rotor and allow them to equilibrate to 37°C. While this is happening, proceed with the polymerization of the tubulin.

2. Thaw the tubulin rapidly using a 37°C water bath until the tube is half full of ice and then continue to thaw the remainder on ice. Adjust the solutes to glycerol PB using anhydrous (100%) glycerol. Allow polymerization to occur for 40 min at 37°C.

3. Layer the polymerized tubulin onto the prewarmed cushions using tips with large openings to avoid depolymerizing the microtubules. Centrifuge at 226,240 g (50,000 rpm in the Beckman Ti50 rotor) for 60 min or, alternatively, at 70,000 rpm for 30 min in the TLA100 rotor at 37°C.

4. Aspirate away the supernatant above the cushion. Rinse the cushion interface twice with water. Aspirate away the cushion. Resuspend the pellet in 0.25× BRB80 + 0.1% β-mercaptoethanol on ice, using an homogenizer to depolymerize the microtubules. The volume of buffer used for the resuspension is chosen so that the final tubulin concentration is between 10 and 20 mg/ml (based on the assumption that approximately half the tubulin from the PC column will polymerize). Incubate on ice for 15 min and then add 5× BRB80 to adjust the buffer to 1× BRB80. (*Note:* The volume of 5× BRB80 to add is 3/16ths of the volume of 0.25× BRB80–tubulin.)

5. Sediment the undepolymerized microtubules by centrifuging the sample at 213,483 g (70,000 rpm in the Beckman TLA100.2 rotor) for 15 min at 4°C.

6. Aliquots (10–200 μl) of the concentrated tubulin can be made, which should then be snap frozen in liquid nitrogen. The tubulin can then be stored either in liquid nitrogen (indefinitely) or at −80°C (for at least 12 months).

C. Preparation of a Phosphocellulose Column

For large-scale preparations of tubulin (finally giving 1 g or more of purified tubulin), a PC column of approximately 1 liter volume will be necessary. For successful preparation of PC, it has to be equilibrated *for short periods* first in base and then in acid, interspersed with water washes. This is normally achieved by suspension of the PC in either the acid or the base solution and then rapid filtration over a sintered glass funnel where the PC can be washed with large volumes of water. Large volumes of PC are, however, quite cumbersome to handle, and the importance of setting up an efficient filtration system *before* commencing the PC preparation cannot be overemphasized. With volumes as large as 1 liter it may be unwise to rely on running water aspirators as the vacuum produced may be insufficient (but for small volumes of PC these may be suitable). It is better to use a membrane-type pump and, if possible, to connect this to a wide, sintered glass funnel over which Whatman 3MM filter paper has been placed. Alternatively, and we have used this quite successfully, use some nylon or polypropylene meshing (available from SpectraMesh) as this allows rapid filtration rates with little risk of tearing.

When considering what volume of phosphocellulose is needed, our laboratory uses 200 ml phosphocellulose to obtain 300–600 mg of purified tubulin. When equilibrated in buffer, 1 g of PC powder will give between 5 and 6 ml of PC column matrix (depending on the salt concentration), but do allow for some loss due to the removal of fines when calculating the amount of PC to prepare.

Phosphocellulose is from Whatman Scientific Ltd. Column and adaptors are from Kontes, but any column with similar dimensions (10 × 30 × 4.8 cm) should be suitable.

Steps

All stages are performed at 4°C.

1. Having determined the amount of phosphocellulose that is needed using the information just given, slowly hydrate the dry powder by washing twice in 95% ethanol, once in 50% ethanol, and once in water.

2. Resuspend the hydrated phosphocellulose in 25 vol (liquid volume per original dry weight of phosphocellulose) of 0.5 M NaOH and stir *gently* (stirring too fast produces fines that will result in slow column flow rates if not removed later) for 5 min. Filter the phosphocellulose rapidly to remove the NaOH solution, and continue to rinse with water until the pH of the washings is lower than 10 (usually at least three times the volume of the NaOH solution used).

3. As soon as the pH is sufficiently low, resuspend the phosphocellulose in 25 vol of 0.5 M HCl (i.e., the same volume as the NaOH solution as used before), again stir gently for 5 min, and quickly filter. Continue to wash with water until the pH is no longer acid, usually around 10 times the volume of the HCl solution used.

4. If the column is not poured at this stage, resuspend the phosphocellulose and stir for 5 min in a 2 M solution of potassium phosphate, pH 7.0. The phosphocellulose can then be resuspended and stored in a solution of 0.5 M potassium phosphate, pH 7.0, containing 20 mM sodium azide as a preservative. When the phosphocellulose is removed from storage, it should be resuspended and washed in at least 3 vol of water.

5. At this stage, whether the column was stored or not, resuspend the phosphocellulose in at least 3 vol of water and transfer to a large measuring cylinder. Stir the phosphocellulose and then allow it to settle naturally. After the bulk of the phoshocellulose has settled, there will be a cloudy layer just above it. These are the fines and should be removed by aspiration to ensure fast column flow rates. Repeat this cycle of resuspension, stirring, and aspiration of the fines until no more fines are visible.

6. Resuspend the phosphocellulose in CB and then degas for 30 min prior to pouring the column.

7. After use, wash the phosphocellulose column extensively and replace the buffer with 50 mM potassium phosphate buffer containing 20 mM sodium azide.

References

Andersen, S., Buendia, B., Domínguez, J., Sawyer, A., and Karsenti, E. (1994). Effect on microtubule dynamics of XMAP230, a microtubule-associated protein present in *Xenopus laevis* eggs and dividing cells. *Cell Biol.* **127**, 1289–1299.

Howard, J., and Hyman, A. (1993). Preparation of marked microtubules for the assay of microtubule-based motors by fluorescence microscopy. *In* "Methods in Cell Biology," Vol. 39, pp. 105–113.

Hyman, A., Drechsel, D., Kellogg, D., Salser, S., Sawin, K., Steffen, P., Wordeman, L., and Mitchison, T. (1991). Preparation of modified tubulins. *In* "Methods in Enzymology" (R. B. Vallee, ed.), Vol. 196, pp. 478–485. Academic Press, San Diego.

Hyman, A., and Karsenti, E. (1996). Morphogenetic properties of microtubules and mitotic spindle assembly. *Cell* **84**, 401–410.

Mitchison, T., and Kirschner, M. (1984a). Dynamic instability of microtubule growth. *Nature* **312**, 237–242.

Mitchison, T., and Kirschner, M. (1984b). Microtubule assembly nucleated by isolated centrosomes. *Nature* **312**, 232–237.

Shalanski, M., Gaskin, F., and Cantor, C. (1973). Microtubule assembly in the absence of added nucleotides. *Proc. Natl. Acad. Sci. USA* **70**(3), 765–768.

Weingarten, M., Suter, M., Littman, D., and Kirschner, M. (1974). Properties of the depolymerization products of microtubules from mammalian brain. *Biochemistry* **13**(27), 5529–5537.

Purification of Human Erythrocyte Spectrin and Actin

David M. Shotton

I. INTRODUCTION

Evenly covering and tightly laminated to the cytoplasmic surface of the phospholipid bilayer of the erythrocyte membrane is the membrane skeleton, an isotropic two-dimensional meshwork of protein that is elastic and slightly contractile, composed primarily of the proteins spectrin, ankyrin, actin, tropomyosin, and "Band 4.1." Ankyrin, a 200-kDa monomeric linker protein, serves to laminate the membrane skeleton to the phospholipid bilayer by binding both to β-spectrin and to the cytoplasmic domain of the integral anion transporter "Band 3." It is this laminated complex of the phospholipid bilayer with its integral proteins together with the membrane skeleton that constitute the erythrocyte membrane. The preparation of the proteins of the membrane skeleton has been described by Horne *et al.* (1989). The properties of the most abundant proteins of the erythrocyte membrane are summarized in Table I.

A. Actin

Actin is present as the β-actin isoform, organized into short protofilaments, averaging ~13 monomers in length, which act as multiple binding sites for the distal ends of spectrin tetramers, enabling the spectrin molecules to be linked into an isotropic two-dimensional meshwork. A pair of tropomyosin molecules are thought to complex with and stabilize each of these protofilaments. Myosin II is also present as minor components of the membrane skeleton, although there is little evidence that active acto–myosin contraction contributes to the functional state of the membrane skeleton. The interactions between actin and the distal ends of the spectrin tetramers are stabilized by the monomeric protein Band 4.1, together with the smaller protein adducin. Band 4.1 is also reported to have binding affinities for glycophorin C and the anion transporter, further laminating the membrane skeleton to the phospholipid bilayer.

B. Spectrin

The spectrin molecule is an $\alpha_2\beta_2$ heterotetramer of two large elongated polypeptide chains that show structural homologies with one another (Fig. 1). The spectrin molecule has a high α-helical content (70–75%), and the conformations of the tetramers revealed by low angle rotary shadowing show the extended tetramer to be about 200 nm long; the individual

TABLE I   Major Proteins of the Erythrocyte Membrane[a]

Band number	Name	Monomer Mr (kDa)	Oligomeric form	% Total protein	Copies per cell
1	α-Spectrin	240,000	$\alpha_2\beta_2$ tetramer	25	100,000 tetramers
2	β-Spectrin	220,000			
2.1	Ankyrin	200,000	Monomer	5	100,000
3	Anion transporter	93,000	Dimer	25	600,000 dimers
4.1		78,000	Monomer	5	200,000
4.2		72,000	Tetramer	5	50,000 tetramers
5	β-Actin	43,000	α_{13}–α_{14} protofilament	5	30,000 protofilaments
Sialoglycoproteins					
PAS-1	Glycophorin A	31,000 (60% sugars)	Dimer, or heterodimer with glycophorin B	2	210,000 dimers
PAS-2	Glycophorin B	~30,000 (60% sugars)	Dimer, or heterodimer with glycophorin A	Trace	35,000 dimers
PAS-3	Glycophorin C	~30,000 (60% sugars)	Monomer	Trace	35,000 dimers

[a] All numerical data are approximate. Adapted from Shotton (1983).

polypeptide chains are approximately 100 nm in length (Shotton *et al.*, 1979). This is approximately one-third of the length expected if the chains existed as single, fully extended α-helices, and is slightly wider than the width of the tail of a myosin II molecule, which is composed of two supercoiled α-helices. From this, the individual chains were predicted to have a triple-helical conformation. This prediction was borne out when peptide sequence studies showed that both spectrin polypeptides were largely composed of repeating homologous domains, 106 amino acid residues in length, in which helix predictions suggested three α-helices of almost equal length folding back on one another to form a sausage-shaped domain connected to the next domain by a short stretch of extended chain to form an elongated flexible molecule (Speicher and Marchesi, 1984).

Subsequent spectrin gene sequences confirmed these findings. The spectrin α chain was shown to be composed of 20 such repeats, of which numbers 10 and 20 were irregular; the whole chain showed evidence of gene duplication to double the length of an initial 10-domain polypeptide. Similarly, the β chain had 18 such domains. Interestingly, the polarity of the chains in the spectrin dimers is antiparallel, with the N terminus of the α chain and the phosphorylated C terminus of the β chain being at the dimer–dimer-binding site of the tetramer, whereas the C-terminal domain of the α chain and the N terminus of the β

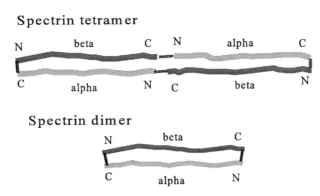

FIGURE 1   Schematic diagram of the arrangements of the polypeptides in extended spectrin tetramers and dimers.

chain form an actin- and calcium-binding domain at the distal ends of the tetramer (Fig. 1). The binding site for ankyrin is on the β chain, quite close to the C terminus.

Heterotetramers formed by head-to-head dimerization of spectrin $\alpha\beta$ dimers are unable to polymerize further, and thus require the accessory proteins actin and Band 4.1 to permit cross-linking into the isotropic molecular network that forms the erythrocyte membrane skeleton. However, evidence shows that, to a limited extent, spectrin dimers also assemble by quasi-equivalent binding into head-to-head-to-head hexamers and higher oligomers, thus providing further branch points in the spectrin lattice, the extent of such polymerization being limited only by steric constraints. Substantial evidence shows that the spectrin molecules in the native erythrocyte membrane exist not as fully extended chains, but in a more compact folded quaternary structure, about 70 nm in length, from which they can be distorted easily and to which they return by elastic recoil, explaining the extraordinary viscoelastic properties of the membrane itself.

Spectrin homologs, also known as fodrin, are widely distributed throughout almost every tissue and cell type of the vertebrate body, forming actin-binding membrane skeletons associated with anion transporter molecules homologous to Band 3 of the erythrocyte membrane, or with a variety of other functional integral membrane proteins such as the Na^+/K^+-ATPase of kidney epithelial cells (Bennet, 1990). Genes encoding spectrin isoforms have been identified, cloned, and sequenced from a wide variety of tissues and species, and structural studies using NMR and X-ray crystallography have been undertaken on domains of both α and β chains of *Drosophila* spectrin expressed in transfected cells, confirming and extending the structural predictions made from earlier electron microscopic and protein sequence studies. The normal expression of these various spectrin genes is developmentally regulated not only in terms of their tissue distribution, but also as a function of the developmental stage. Sequence comparisons have revealed important wider homologies between spectrin and other actin-binding proteins, particularly α-actinin and dystrophin.

C. Spectrin and Actin Extraction from Erythrocyte Ghosts

Isolated erythrocyte membrane are called erythrocyte ghosts. Spectrin and actin can be easily removed from erythrocyte ghosts by low ionic strength extraction at slightly alkaline pH. Extraction at 4°C by dialysis for 48 hr against such buffers retains the native tetrameric structure of spectrin, while brief extraction at 37°C leads to breakdown of the tetramer into $\alpha\beta$ dimers (Ralston *et al.*, 1977). Quasi-equivalent binding between the ends of the dimers (Fig. 1) permits these alternative quaternary conformations to exist.

The following method is that used by Shotton *et al.* (1979) and uses as starting material human erythrocyte ghosts prepared as described previously (see volume 2, David M. Shotton, "Preparation of Human Erythrocyte Ghosts" for additional information). If the extraction is to be performed immediately after ghost preparation, the initial wash may be performed in the same centrifuge tubes in which the ghosts were prepared.

II. MATERIALS AND INSTRUMENTATION

A. Buffers and Reagents

Extraction buffer: 1 mM EDTA and 10 μM dithiothreitol, pH 9.5. Mix 20.17 mg of anhydrous (MW 336.2 g) EDTA or 22.33 mg of dihydrate (MW 372.2 g) EDTA with 9.23 mg of dithiothreitol (MW 154.2 g). Add ice-cold distilled water to 6 liters. Titrate dropwise with stirring to pH 9.5 with 1 M ammonium hydroxide.

B. Equipment Requirements

1. Laboratory Equipment

Magnetic stirrer, vortex mixer, water pump aspirator with Büchner flask trap, top pan balance weighing to 0.1 g (for balancing centrifuge tubes), and a spectrophotometer with quartz cuvettes.

2. Refrigerated Ultracentrifuge (up to at least 50,000 rpm) and Suitable Rotor and Tubes

For example, using the Beckman L8 ultracentrifuge, one requires a Beckman 70 Ti fixed angle rotor (23° to vertical; 8 × 38.5 ml): 50,000 rpm = 257,000 g_{max}. Use with Beckman 28-ml capacity polycarbonate bottles (Cat. No. 340382) and aluminum sealing cap assemblies (Cat. No. 341279). Alternatively, one may also use a Beckman 70.1 Ti fixed angle rotor (24° to vertical; 12 × 13.5 ml): 50,000 rpm = 230,000 g_{max}. Use with Beckman 10-ml capacity polycarbonate bottles (Cat. No. 339573) and noryl sealing cap assemblies (Cat. No. 339574).

III. PROCEDURES

Steps

1. Using the same techniques and equipment described for the preparation of the erythrocyte ghosts (see Volume 2, David M. Shotton, "Preparation of Human Erythrocyte Ghosts" for additional information), wash each milliliter of ghosts in a 35-ml Oak Ridge style screwtop centrifuge tube, using ~30 ml of ice-cold extraction buffer, spin at 30,000 g and 4°C for 10 min to resediment the ghosts, and remove the supernatant completely. This will lower the ionic strength to which the ghosts are exposed, enabling 37°C extraction to be accomplished by simple warming, and promoting the rate of 4°C extraction.

2. Pool all the washed ghosts, using a little extraction buffer to sequentially wash out each centrifuge tube, pooling the washings with the ghosts. The total volume of washings should not exceed the volume of the ghosts. Thus eight 30-ml tubes should yield 8 ml of ghosts diluted by washings to no more than 16 ml.

3. Wearing ear protectors, disrupt the ghosts to reduce the ghosts to small vesicles, permitting easier release of spectrin and actin, by sonicating for 30 sec at full power using the sonicator probe of an ultrasonic disintegrator. This step is not essential, but increases the efficiency of the subsequent spectrin extraction. If an ultrasonic disintegrator is not available, alternative means of disruption may be employed, such as shearing in a glass Potter homogenizer.

4. For extraction of spectrin as $\alpha\beta$ dimers, seal the ghost container and incubate at 37°C in a water bath, with occasional mixing, for 60 min. Alternatively, for extraction of spectrin as $\alpha_2\beta_2$ tetramers, transfer the sonicated erythrocytes to a suitable length of dialysis tubing, previously boiled in a dilute sodium bicarbonate solution and rinsed in distilled water. Taking care not to overfill the tubing, seal the ends securely and dialyze for 48 hr against approximately 100 vol of ice-cold extraction buffer, ideally changing the buffer after 4, 16, and 28 hr.

5. Transfer the incubated or dialyzed ghost preparation to suitably sized ultracentrifuge tubes, balance pairs of tubes to within 0.1 g by weighing, seal according to the manufacturer's instructions, and centrifuge at 230,000 g and 4°C in an ultracentrifuge for 60 min (or for proportionately shorter times at higher speeds).

6. Remove the tubes onto ice. With a Pasteur pipette, taking great care not to disturb the translucent pellet of extracted erythrocyte membranes, carefully remove the top ~90% of the supernatant and save it on ice. This will contain the extracted spectrin and actin. Discard the remaining supernatant and any of the pellet accidentally drawn off with it. Save the remaining pellet on ice for further extraction or analysis.

7. Measure the absorbance of the spectrin/actin supernatant at 280 nm relative to the extraction buffer, using quartz cuvettes, and calculate the approximate yield of extracted protein, assuming that a 1-mg/ml solution has an absorbance at 280 nm of 1.0. Typically, 20 ml blood will yield 10 to 12 ml of supernatant with a protein concentration of between 0.7 and 1.2 mg/ml.

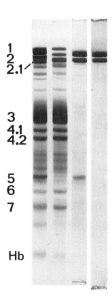

a b c d

1
2
2.1
3
4.1
4.2
5
6
7
Hb

FIGURE 2 SDS–polyacrylamide gel fractionation of erythrocyte membrane proteins, stained with Coomassie brilliant blue, showing a 0°C extraction of spectrin (bands 1 and 2) and actin (band 5) from fresh human erythrocyte ghosts, and the column-purified spectrin preparation from such an extract. The tube gels were loaded with (a) 7.5 μl of fresh erythrocyte ghosts, (b) 7.5 μl of extracted ghost membranes resuspended to their original volume in water, (c) 10 μl of the low ionic strength spectrin/actin extract, and (d) 5 μg of the spectrin sample chromatographically purified from c. [Reproduced from Shotton *et al.* (1979) by permission of Academic Press.]

IV. EVALUATION OF ERYTHROCYTE MEMBRANE PROTEINS

The quality of erythrocyte ghost preparation and the efficiency of spectrin and actin extraction may be conveniently analyzed by fractionating the erythrocyte membrane proteins by SDS–polyacrylamide gel electrophoresis (Shotton *et al.*, 1979). Because this method is described later in Volume 4 by Celis and Olsen ("One-Dimensional Sodium Dodecyl Sulfate-Polyacrylamide Gel Electrophoresis"), it will not be detailed here. Suitable loadings are 40 μg of total ghost protein per tube gel or 10 μg of total protein per slab gel well; 10 μl of packed ghosts contains approximately 40 μg of total protein.

From the gel results, estimate the proportion of spectrin and actin extracted from the ghosts and note any proteolytic degradation of the spectrin. Results should resemble those shown in Fig. 2, from which it can be seen that actin has been entirely extracted from the ghosts, that about 90% of the spectrin has been extracted, and that the spectrin shows no signs of proteolytic degradation.

Spectrin and actin may be separated from one another by passing the ultracentrifugation supernatant down a Sepharose 4B (Pharmacia) gel filtration column equilibrated with 100 μM EDTA, 10 μM dithiothreitol, 1 mM Tris–HCl, pH 8.0, at 4°C, as described by Shotton *et al.* (1979). Spectrin molecules appear in the void volume (V_0), whereas β G-actin elutes at a position $1.5 \times V_0$. Spectrin tetramers and dimers can then be fractionated from one another by rerunning on the same material in nonvolatile or volatile buffers of higher ionic strength, under the conditions described by Shotton *et al.* (1979).

Purified erythrocyte spectrin and actin preparations should ideally be used soon after preparation. Brief storage at 4°C is permissable, and more prolonged storage at −20°C is possible but not recommended if spectrin precipitation is to be totally avoided.

References

Bennet, V. (1990). Spectrin: A structural mediator between diverse plasma membrane proteins and the cytoplasm. *Curr. Opin. Cell Biol.* **2**, 51–56.

Horne, W. C., Leto, T. L., and Anderson, R. A. (1989). Preparation of red cell membrane skeleton proteins. *In* "Methods in Enzymology" (S. Fleischer and B. Fleischer, eds.), Vol. 173, pp. 380–391. Academic Press, San Diego.

Ralston, G. B., Dunbar, J., and White, M. (1977). Temperature dependent dissociation of spectrin. *Biochim. Biophys. Acta* **491**, 345–348.

Shotton, D. M. (1983). The proteins of the erythrocyte membrane. *In* "Electron Microscopy of Proteins" (J. R. Harris, ed.), Vol. 4, pp. 205–330. Academic Press, London.

Shotton, D. M., Burke, B. E., and Branton, D. (1979). The molecular structure of human erythrocyte spectrin: Biophysical and electron microscopic studies. *J. Mol. Biol.* **131**, 303–329.

Speicher, D. W., and Marchesi, V. T. (1984). Erythrocyte spectrin is composed of many homologous triple helical segments. *Nature* **311**, 177–180.

RNA

Single-Step Method of Total RNA Isolation by Acid Guanidine–Phenol Extraction

Piotr Chomczynski and Karol Mackey

I. INTRODUCTION

The single-step method (Chomczynski and Sacchi, 1987) has become a method of choice for the isolation of total RNA from a variety of sources. As compared with other RNA isolation methods (Chomczynski and Sacchi, 1990; Sambrook *et al.*, 1989), it substantially reduces the amount of time required to isolate RNA, without sacrificing its quality. The single-step method performs well with small and large quantities of tissues or cells and allows simultaneous isolation of a large number of samples. A biological sample is homogenized in 4 *M* guanidine thiocyanate solution and mixed with phenol to form a monophase solution. After the addition of chloroform or bromochloropropane and phase separation, RNA remains exclusively in the aqueous phase, whereas DNA and proteins are allocated to the interphase and organic phase. RNA is precipitated from the aqueous phase with isopropanol, reprecipitated, and washed with 75% ethanol. The entire procedure can be completed in less than 4 hr.

The use of a single-step method for commercial purposes is restricted by U.S. Patent No. 4,843,155.

II. MATERIALS AND INSTRUMENTATION

Guanidine thiocyanate is from Fluka Chemie AG (Cat. No. 50990) and from Amresco, Inc. (Cat. No. 380). Phenol is obtained from Life Technologies, Inc. (Cat. No. 5509). Stabilized formamide (Cat. No. FM 121) and bromochloropropane (Cat. No. BP 151) are from Molecular Research Center, Inc. All other reagents are from Fisher Scientific: ethanol (Cat. No. A407), chloroform (Cat. No. C297), glacial acetic acid (Cat. No. A35), isoamyl alcohol (Cat. No. A393), isopropanol (Cat. No. A416), lithium chloride (Cat. No. L119), 2-mercaptoethanol (Cat. No. O34461), sarcosyl (Cat. No. BP234), sodium acetate (Cat. No. S210), and sodium citrate (Cat. No. S279).

Centrifugation was performed in a Sorvall RC-5B centrifuge (Du Pont Instruments) in a SS-34 rotor with rubber adapters.

III. PROCEDURES

A. Homogenization

Solutions

1. *Denaturing solution:* 4 *M* guanidine thiocyanate, 25 m*M* sodium citrate, pH 7.0, 0.5% sarcosyl, and 0.1 *M* 2-mercaptoethanol. Prepare a stock solution by dissolving the following ingredients in 319 ml of water (at 50–60°C): 250 g guanidine thiocyanate, 2.6 g sarcosyl, and 17.6 ml 0.75 M sodium citrate, pH 7.0. The stock solution can be stored at least 3 months at room temperature. The denaturing solution is prepared by adding 0.36 ml of 2-mercaptoethanol per 50 ml of stock solution. The denaturing solution can be stored for 1 month at room temperature.

2. *2 M sodium acetate, pH 4:* The solution is 2 M with respect to the sodium ions. Dissolve 16.42 g of sodium acetate (anhydrous) in 40 ml water and 35 ml glacial acetic acid. Adjust the solution to pH 4.0 with glacial acetic acid and the final volume to 100 ml with water.

3. *Water-saturated phenol:* Dissolve 100 g of phenol by incubating with 100 ml of water at 50–60°C. Aspirate the upper aqueous phase and store the liquified phenol at 4°C for up to 1 month. Use distilled water without buffer or other additives.

Steps

1. Immediately after removal from the animal, mince the tissue on ice and homogenize with 1 ml of denaturing solution per 100 mg of tissue. Perform homogenization at room temperature using a glass–Teflon or Polytron homogenizer. Tissues with a very high RNase content, such as pancreas, should be frozen and powderized in liquid nitrogen before mixing with the denaturing solution.

2. Lyse cells grown in a monolayer directly in a culture dish by adding 1 ml of denaturing solution per 10 cm² of the dish area. Cells grown in suspension are sedimented first and then lysed by the addition of denaturing solution (1 ml per 10^7 cells). Pass the cell lysate several times through a pipette.

B. Phase Separation

Solution

Chloroform–isoamyl alcohol (49:1, v/v) or bromochloropropane.

Steps

1. Transfer the homogenate (or lysate) into a polypropylene or glass tube and add sequentially per 1 ml of denaturing solution used for the homogenization: 0.1 ml of 2 M sodium acetate, pH 4, 1 ml of water-saturated phenol, and 0.2 ml of chloroform–isoamyl alcohol mixture or 0.1 ml of bromochloropropane.

2. Thoroughly shake the suspension for 10–20 sec and incubate it at 0–4°C for 15 min. During this incubation, the suspension separates into the top aqueous phase and the bottom organic phase.

3. Complete the phase separation by centrifuging the suspension for 20 min at 10,000 g at 4°C. Following centrifugation, total RNA remains exclusively in the top aqueous phase, whereas DNA and proteins remain in the interphase and organic phase. The volume of the aqueous phase is approximately equal to the initial volume of the denaturing solution.

C. Precipitation and Wash

Solutions

1. *Isopropanol (100%).*

2. *75% ethanol:* Prepare 75% ethanol by mixing ethanol (100%) with diethyl pyrocarbonate (DEPC)-treated water (Sambrook *et al.*, 1989).

Steps

1. Transfer the aqueous phase to a fresh tube and precipitate RNA by mixing with an equal volume of isopropanol. Place the mixture at −20°C for 30 min.

2. Sediment the RNA precipitate by centrifugation at 10,000 g for 20 min at 4°C.

3. Dissolve the RNA pellet in 0.3–0.5 ml of denaturing solution.

4. Reprecipitate RNA by mixing with 1 vol of isopropanol, storing the mixture at −20°C for 30 min, and centrifuging at 10,000 g for 10 min at 4°C.

5. Remove the supernatant and suspend the RNA pellet in 75% ethanol by vortexing.

6. Centrifuge the suspension at 10,000 g for 10 min at 4°C.

D. Solubilization

Solutions

1. *Water:* Water used for RNA solubilization should be made RNase-free by DEPC treatment.

2. *Formamide:* Prepare freshly deionized formamide by stirring with ion-exchange resin (10 ml/g of Bio-Rad AG 501-X8 resin, Cat. No. 143-7425) for 30 min and filtering at room temperature. Alternatively, use a commercially available, stabilized formamide (Molecular Research Center, Inc., Cat. No. FO-121).

Steps

1. Dry briefly the RNA pellet under vacuum for 5 min. Drying is not necessary for solubilization of RNA in formamide.

2. Dissolve the RNA pellet in water or formamide by passing the solution through a pipette tip a few times and incubate for 10–15 min at 55–60°C.

IV. COMMENTS

Formamide is a more convenient solubilization agent for RNA than water. RNA solubilized in formamide is protected from degradation by RNase and can be stored at −20°C for months (Chomczynski, 1992).

An additional step can be employed for isolating RNA from tissues with a high glycogen content such as liver (Puissant and Houdebine, 1990). Following precipitation of RNA with isopropanol (step 3), resuspend the RNA pellet with 4 M LiCl by vortexing and sediment the insoluble RNA at 5000 g for 10 min. Dissolve the pellet in denaturing solution and proceed with the RNA precipitation step as described earlier.

The single-step method isolates a whole spectrum of RNA molecules. As shown in Fig 1,

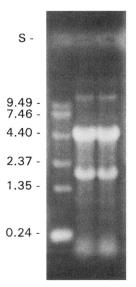

FIGURE 1 Electrophoretic separation of total RNA isolated by the single-step method from rat liver. RNA (5 μg per lane) and molecular weight markers (1.5 μg) were electrophoresed in 1% agarose–formaldehyde gel. Following electrophoresis, the gel was washed with water (3 × 20 min) and stained with ethidium bromide.

the ethidium bromide staining of the RNA isolated from mammalian cells and separated in agarose gel visualizes two predominant bands of small (about 2 kb) and large ribosomal RNA (4–5 kb), low molecular weight RNA (0.1–0.3 kb), and discrete bands of high molecular weight (7–15 kb) RNA.

V. PITFALLS

1. At the end of the procedure, do not let the RNA pellet dry completely as this greatly decreases its solubility. Avoid drying the pellet by centrifugation under vacuum.

2. Use gloves when working with RNA, as hands are a likely source of ribonuclease contamination.

3. Always store tissue or cell samples and aqueous RNA solutions at −70°C. Even an overnight storage at −20°C may result in RNA degradation.

References

Chomczynski, P. (1992). Solubilization in formamide protects RNA from degradation. *Nucleic Acid Res.* **20**, 3791–3792.

Chomczynski, P., and Mackey, K. (1995). Substitution of chloroform by bromochloropropane in the single-step method of RNA isolation. *Anal. Biochem.* **225**, 163–164.

Chomczynski, P., and Sacchi, N. (1987). Single-step method of RNA isolation by acid guanidinium thiocyanate-phenol-chloroform extraction. *Anal. Biochem.* **162**, 156–159.

Chomczynski, P., and Sacchi, N. (1990). Single-Step RNA isolation from cultured cells and tissues. *In* "Current Protocols in Molecular Biology (F. M. Ausbel, *et al.*, eds.), Vol. 1, pp. 4.2.4–4.2.8. Greene Publ. Assoc. and Wiley-Interscience, New York.

Puissant, C., and Houdebine, L. M. (1990). An improvement of the single-step method of RNA isolation by acid guanidinium thiocyanate-phenol-chloroform extraction. *BioTechniques* **8**, 148–149.

Sambrook, J., Fritsch E. F., and Maniatis, T. (1989). Extraction and purification of RNA. *In* "Molecular Cloning," Vol. 1, pp. 7.3–7.23. Cold Spring Harbor Laboratory Press, Cold Spring Harbor, NY.

PART 4

Assays

Endocytic and Exocytic Pathways

Permeabilized Epithelial Cells to Study Exocytic Membrane Transport

Frank Lafont, Elina Ikonen, and Kai Simons

I. INTRODUCTION

Polarized exocytic transport in epithelial cells can be measured by performing *in vitro* assays on filter-grown Madin–Darby canine kidney cells (MDCK strain II cells). Techniques have been developed to establish conditions where the cytosol-dependent transfer of a viral marker protein is monitored in MDCK cells permeabilized with the cholesterol-binding and pore-forming toxin *Streptococcus* streptolysin-O (SLO) (Gravotta *et al.*, 1990; Kobayashi *et al.*, 1992; Pimplikar and Simons, 1993). The transport is assayed early in the biosynthetic pathway, from the endoplasmic reticulum (ER) to the Golgi complex, from the *trans*-Golgi network (TGN) to the apical, or from the TGN to the basolateral plasma membrane [see Lafont *et al.* (1995) for a detailed characterization of the assays and their properties].

The assays provide the possibility to analyze the effects of exogenous molecules added to the cytosol. Cytosolic molecules can be inactivated prior to their addition to the permeabilized cells or depleted using antibodies (Lafont *et al.*, 1994; Ikonen *et al.*, 1995, 1996). Membrane-impermeable molecules can gain access to lipids or membrane proteins facing the cytosol. We have used successfully fluorescent lipid analogs (Kobayashi *et al.*, 1992), chemicals (Pimplikar and Simons, 1993; Lafont *et al.*, 1994), antibodies (Pimplikar and Simons, 1993; Lafont *et al.*, 1994; Fiedler *et al.*, 1995; Ikonen *et al.*, 1995, 1996), peptides (Huber *et al.*, 1993; Pimplikar and Simons, 1993; Pimplikar *et al.*, 1994; Fiedler *et al.*, 1995), purified proteins (Lafont *et al.*, 1994; Ikonen *et al.*, 1995), and toxins (Pimplikar and Simons, 1993; Ikonen *et al.*, 1995). The specificity of the inhibition of transport may be tested by rescuing transport when the purified molecule is added back to the transport reaction (Lafont *et al.*, 1994; Ikonen *et al.*, 1995).

II. MATERIALS AND INSTRUMENTATION

The recombinant SLO fused to a maltose-binding protein is from Dr. S. Bhakdi (University of Mainz, Mainz, Germany). PhosphorImager and ImageQuant software are from Molecular Dynamics. Media and reagents for cell culture are from GIBCO Biocult and Biochrom.; chemicals are from Merck; ATP (Cat. No. A-5394), antipain (Cat. No. A-6271), CaCO3 (Cat. No. C-5273), chymostatin (Cat. No. C-7268), cytochalasin D (Cat. No. C-8273), dimethylsulf-oxide (DMSO, Cat. No. D-5879), leupeptin (Cat. No. L-2884), pepstatin (Cat. No. P-4265), and pyruvate (Cat. No. P-2256) are from Sigma; creatine kinase (Cat, No. 127566), creatine phosphate (Cat. No. 621714), endoglycosidase H (Cat. No. 1088726), and NADH (Cat No.

107727) are from Boehringer Mannheim; the trypsin and soybean trypsin inhibitor are from Worthington Biochemical Corporation; protein A–Sepharose (Cat. No. 17-0780-01) is from Pharmacia; and cell filters (0.4-μm pore size; No. 3401, 12-mm-diameter Transwell polycarbonate filters) are from Costar.

III. PROCEDURES

A. SLO Standardization

Each batch of toxin is standardized for the amount of lactate dehydrogenase (LDH) released from the filter-grown MDCK cells, and the amount of LDH released is determined according to a previously described protocol (Bergmeyer and Bent, 1984). Alternatively, it is possible to use antibody accessibility to monitor SLO permeabilization (Pimplikar, 1994).

Solutions

1. *Growth medium (GM):* Eagle's minimal essential medium with Earle's salts (E-MEM) supplemented with 10 mM HEPES, pH 7.3, 10% (v/v) fetal calf serum (FCS), 2 mM glutamine, 100 U/ml penicillin, and 100 μg/ml streptomycin.

2. Either purified SLO (Bhakdi *et al.*, 1984, 1993) or recombinant SLO fused to a maltose binding protein is obtained from Dr. S. Bhakdi (University of Mainz, Mainz, Germany). The recombinant SLO preparation is stored as a lyophilisate at −80°C.

3. *10× KOAc transport buffer:* To prepare 1 liter, dissolve 250 mM HEPES (59.6 g), 1150 mM KOAc (112.9 g), and 50 mM MgCl$_2$ (25 ml from 1 M stock) in 800 ml water. Adjust the pH to 7.4 with ~50 ml 1 M KOH and make it 1 liter with water.

4. *KOAc+ buffer:* KOAc buffer containing 0.9 mM CaCO$_3$ (90 mg/liter) and 0.5 mM MgCl$_2$ (0.5 ml/liter from 1 M stock).

5. *Assay buffer:* KOAc plus 0.011% Triton X-100 for media and KOAc for filter.

6. *NADH 14 mg/ml KOAc.*

7. *Pyruvate 60 mM:* 30 mg in 50 ml water.

Steps

1. Grow MDCK cells strain II on 1.2-mm-diameter filters for 3.5 days *in vitro* and change the medium every 24 hr.

2. Wash cells by dipping in KOAc.

3. Prepare a range of SLO, e.g., from 0.5 to 10 μg of recombinant SLO per filter. Activate the SLO for 30 min at 37°C in 50 μl KOAc plus 10 mM dithiothreitol (DTT) per filter. Place the filters on a Parafilm sheet and put on a metal plate on an ice bucket.

4. Add 50 μl of activated SLO on the apical side for 15 min at 4°C.

5. Wash excess SLO by dipping the filters twice in KOAc+.

6. Transfer the filters to a 12-well dish containing 0.75 ml TM at 19.5°C. Add 0.75 ml TM to the apical side and incubate at 19.5°C for exactly 30 min in a water bath.

7. Collect apical and basolateral media, cut the filters, and shake the filters with 1 ml KOAc containing 0.1% Trifon X-100.

8. Mix in a disposable cuvette:

    a. 800 μl of assay buffer (KOAc or KOAc/Triton X-100)

    b. 200 μl of sample

    c. 10 μl of NADH

    d. 10 μl of pyruvate and turn the cuvette upside down twice using a Parafilm to close it.

9. Read immediately the OD at 340 nm for 60 sec (the OD should be around 1.5).

10. Calculate the LDH activity as the change in OD in 10 sec, taking the average of the first 30 sec.

11. Dissolve the recombinant SLO in KOAc buffer and store in aliquots at −80°C with the amount necessary for one set of assays per aliquot.

B. Preparation of Cytosols

Solutions

1. *Growth medium:* For HeLa cells use Joklik's medium supplemented with 50 ml/liter newborn calf serum (heat inactivated for 30 min at 56°C), 2 mM glutamine, 100 U/ml penicillin, 100 μg/ml streptomycin, and 10 M 150 μl NaOH. For MDCK cells use E-MEM supplemented with 10 mM HEPES, pH 7.3, 10% (v/v) fetal calf serum, 2 mM glutamine, 100 U/ml penicillin, and 100 μg/ml streptomycin.

2. *10× KOAc transport buffer:* To prepare 1 liter, dissolve 250 mM HEPES (59.6 g), 1150 mM KOAc (112.9 g), and 50 mM $MgCl_2$ (25 ml from 1 M stock) in 800 ml water. Adjust the pH to 7.4 with ~50 ml 1 M KOH and make it 1 liter with water.

3. *Phosphate-buffered saline (PBS)+:* PBS containing 0.9 mM $CaCl_2$ and 0.5 mM $MgCl_2$.

4. *Swelling buffer (SB):* For 100 ml prepare 1 mM EGTA (200 μl of 0.5 M stock), 1 mM $MgCl_2$ (100 μl of 1 M stock), 1 mM DTT (100 μl of 1 M stock), and 1 μM cytochalasin D (100 μl of 1 mM stock).

5. *Protease inhibitor cocktail, CLAP:* To prepare a 1000× stock, dissolve antipain, chymostatin, leupeptin, and pepstatin each at 25 μg/ml DMSO and combine.

Steps

HeLa Cytosol

1. To grow 20 liters of HeLa cells in suspension to a density of 6×10^5 cells/ml:
   a. Inoculate 250 ml cell suspension (6×10^5 cells/ml) in 1 liter of medium and leave stirring in a 37°C room.
   b. On the third day, split 1:4 by adding 250 to 750 ml of fresh medium. Leave stirring in the room. This will give 4 liters of cell suspension.
   c. On the fifth day, again split 1:4 by adding 1 liter of the cell suspension to 3 liters of fresh medium in a 6-liter round-bottom flask. If the cells seem overgrown, add some fresh medium. Thus, between 16 and 20 liters of cell suspension is obtained.

2. Concentrate the cell suspension to 2 liters either by centrifugation or with any cell-concentrating system.

3. Centrifuge the cells at 5000 rpm in a Sorvall GS-3 rotor (400-ml buckets) at 4°C for 20 min.

4. Discard the supernatant and wash the cells by resuspending the pellet with 10-ml of ice-cold PBS- with a sterile 10-ml pipette (10-ml PBS/bucket). Pool the suspension in 250-ml Falcon tubes.

5. Spin at 3000 rpm for 10 min at 4°C in a minifuge. The pellet will be loose. Discard the supernatant.

6. Fill the tube with cold SB (about 35 ml). Resuspend the cells carefully with a sterile 10-ml pipette in SB. Let the pellet swell for 5 min on ice.

7. Spin in the minifuge at 2000 rpm for 10 min at 4°C. Remove as much as possible of the supernatant. The pellet should be loose.

8. Transfer the pellet to a 30-ml Dounce homogenizer (use a spatula) and perform 5 strokes. Add 0.1 vol of KOAc 10× and homogenize further by 15–20 strokes.

9. Spin the homogenate at 10,000 rpm for 25 min at 4°C in SS-34 tubes (Sorvall). Collect the supernatant.

10. Spin at 50,000 rpm in Ti70 tubes (Beckman) for 90 min at 4°C. Aliquot the supernatant in screw-top 1.5-ml tubes. Avoid including lipids that might be on top of the supernatant. Freeze the aliquots in liquid N_2 and store at −70°C.

11. Measure the protein concentration; it should be around 5 mg/ml. Lower concentrations do not work.

12. When needed, thaw the aliquots quickly and keep them on ice (up to 6 hr) until use. Aliquots can be refrozen at least twice.

MDCK Cytosol

1. Trypsinize 30 24 × 24-cm dishes of confluent MDCK cells. Resuspend trypsinized cells in cold growth medium (containing 5% FCS to inactivate trypsin) and leave on ice until all dishes are trypsinized and cells are pooled. After this, all handling should be done on ice using ice-cold solutions.

2. Centrifugate at 4°C for 10 min at 2000 rpm in a RF Heraeus centrifuge. Wash medium away with PBS.

3. Wash in PBS containing 2 mg/ml STI. Wash out STI with PBS, resuspend the cell pellet in SB, and keep on ice for 10 min.

4. Centrifugate at 2000 rpm for 10 min at 4°C in a RF Heraus centrifuge.

5. Sonicate the loose cell pellet (∼10 ml) (power 6, 0.5-sec pulse, Sonifier cell disruptor B 15, Branson) until cells are broken as judged by light microscopy. Add 0.1 vol of 10× KOAc to the sample and spin for 20 min at 3000 rpm.

6. Spin the supernatant again at 75,000 rpm for 1 hr in a TLA 100.2 rotor.

This procedure routinely yields ∼6 ml of cytosol at 14 mg/ml. Other cell types, e.g., NIH 3T3 fibroblasts, can also be used as a starting material for cytosol preparation using the protocol just described. We have not observed significant differences between the efficiencies of the different cytosols to support transport. However, HeLa cytosol is used routinely because of the ease of preparing large quantities.

C. Transport Assays

Solutions

All solutions are sterilized and kept at 4°C unless indicated.

1. *Growth medium:* E-MEM supplemented with 10 mM HEPES (10 ml/liter), pH 7.3, 10% (v/v) fetal calf serum, 2 mM glutamine (10 ml/liter of 200 mM stock), 100 U/ml penicillin (10 ml/liter of 10^4 U/ml stock), and 100 μg/ml streptomycin (10 ml/liter of 10^4 μg/ml stock).

2. *Infection medium (IM):* E-MEM supplemented with 10 mM HEPES, pH 7.3, 0.2% (w/v) bovine serum albumin (BSA), 100 U/ml penicillin, and 100 μg/ml streptomycin.

3. *Labeling medium (LM):* Methionine-free E-MEM containing 0.35 g/liter sodium bicarbonate instead of the usual 2.2 g/liter, 10 mM HEPES, pH 7.3, and 0.2% (w/v) BSA supplemented with 16.5 μCi of [^{35}S]methionine/filter.

4. *Chase medium (CM):* Labeling medium without [^{35}S]methionine and containing 20 μg/ml cycloheximide and 150 μg/ml cold methionine.

5. *10× KOAc transport buffer:* To prepare 1 liter, dissolve 59.6 g HEPES (250 mM) and 112.9 g KOAc (1150 mM) in 800 ml water and add 25 ml from 1 M stock MgCl$_2$ (50 mM). Adjust the pH to 7.4 with ∼50 ml 1 M KOH and make it 1 liter with water.

6. *KOAc+ buffer:* KOAC buffer containing 0.9 mM CaCO$_3$ (90 mg/liter) and 0.5 mM MgCl$_2$ (0.5 ml/liter from 1 M stock).

7. *PBS+:* PBS containing 0.9 mM CaCl$_2$ (0.13 g/liter CaCl$_2$·2H$_2$O) and 0.5 mM MgCl$_2$ (0.5 ml/liter from 1 M stock).

8. *0.5 M EGTA:* To prepare 100 ml, dissolve 19 g in 60 ml water. The solution should be turbid (pH 3.5). Add slowly ~10 ml 10 M KOH to the suspension. When the pH of the solution starts to clear, adjust to 7.4. Make it 100 ml with water.

9. *0.1 M Ca and 0.5 M EGTA:* To prepare 100 ml, stir 1 g $CaCO_3$ and 3.8 g EGTA in ~70 ml water for at least 45 min (degas). After adding 2 ml of 10 M KOH, the pH should be about 6. Finally, add a few drops of 1 M KOH to make the pH 7.4. Make it 1 liter with water.

10. *Transport medium (TM):* To prepare 50 ml, combine 50 μl 1 M DTT, 200 μl 0.5 M EGTA, and 1 ml 0.5 M EGTA/0.1 M $CaCO_3$ in 1× KOAc.

11. *ATP-regenerating system (ARS):* To make 100× stock, prepare three solutions (10 ml each):

    a. 100 mM ATP (disodium salt, pH 6–7, neutralized with 2 M NaOH; 0.605 g/10 ml).

    b. 800 mM creatine phosphate (disodium salt, 2.620 g/10 ml).

    c. 800 U/mg (at 37°C) creatine kinase (0.5 mg/10 ml in 50% glycerol).

    Store stocks in aliquots at −20°C. Mix solutions a–c 1:1:1 just before use.

12. *Lysis buffer (LB):* PBS+ containing 2% NP-40 and 0.2% SDS.

13. *Protease inhibitor cocktail, CLAP:* To prepare a 1000× stock, combine 25 μg/ml DMSO of antipain, chymostatin, leupeptin, and pepstatin.

Steps

The basic steps of the assays can be summarized as follows. Grow MDCK cells on a permeable filter support until a tight monolayer is formed (see Volume 1, Virta and Simons, "Growing Madin-Darby Canine Kidney Cells for Studying Epithelial Cell Biology" for additional information). Grow cells on 12-mm-diameter filters and use when they display a transmonolayer electrical resistance of at least 50 $\Omega \times cm^2$. Infect cells layers with either vesicular stomatitis (VSV) or influenza virus using the G glycoprotein of VSV (VSV G) or the hemagglutinin (HA) of influenza virus N as basolateral or apical markers, respectively (Simons and Fuller, 1985). After a short pulse of radioactive methionine, block the newly synthesized viral proteins either in the ER at 4°C or in the TGN after a 20°C incubation. At this stage, one cell surface is permeabilized with SLO which allows leakage of cytosolic proteins whereas membrane constituents, including transport vesicles, are retained inside the cells. After removal of the endogenous cytosol by washing, add cytosol and ATP and raise the temperature to 37°C. Transport of the viral proteins from the ER to the Golgi complex or from the TGN to the apical or the basolateral plasma membrane is reconstituted in a cytosol-, energy-, and temperature-dependent manner. Measure the amount of viral proteins reaching the acceptor compartment by endoglycosidase H treatment (ER to Golgi transport), trypsinization of HA on the apical surface (TGN to apical transport), or immunoprecipitation of the VSV G at the basolateral surface (TGN to basolateral transport). Obtain quantitations of viral polypeptides resolved on SDS–PAGE by PhosphorImager analysis.

Due to handling and the various steps, the entire procedure for running one assay requires about 9 hr for the apical assay and 11 hr for the basolateral assay. The immunoprecipitation for the basolateral assay is carried out the following day (takes 5 hr). The ER to Golgi transport takes about 6 hr before an overnight enzymatic treatment step. Samples can be frozen at −20°C before running the SDS–PAGEs (routinely performed the following day). It is worth noting that for the reproducibility of the results the assays should be performed strictly obeying the schedule indicated in the protocol.

Grow the N strain influenza virus (A/chick/Germany/49/Hav2Neq1) in 11-day embryonated chick eggs as described (Matlin and Simons, 1983). Prepare a stock of phenotypically mixed VSV (Indiana strain) grown in Chinese hamster ovary C15.CF1 cells which express HA on their plasma membrane as described (Bennett *et al.*, 1988). Prepare the affinity-purified antibody raised against the luminal domain of the VSV G as described (Pfeiffer *et al.*, 1985).

Apical Transport Assay

1. *Cell culture:* Seed 1:72 of the MDCK cells from one confluent 75-cm^2 flask per filter (6×10^4 cells/filter) with 0.75 ml of growth medium apically and 2 ml basolaterally. Change the growth medium every 24 hr and use the cells 3.5 days after plating.

2. *Infection:* Prior to viral infection, wash the monolayers in warm PBS+ then IM and infect with the influenza virus in 50 μl IM (20 PFU/cell; enough to obtain 100% infection as judged by immunofluorescence) on the apical side. After allowing adsorption of the virus to the cells for 1 hr at 37°C, remove the inoculum and continue the infection for an additional 3 hr after adding 0.75 ml IM on the apical side and 2 ml IM on the basolateral side of the filter.

3. *Metabolic labeling:* Rinse the cell monolayers by dipping in beakers containing warm PBS+ and LM at 37°C. Place a 25-μl drop of the LM containing 12.5 μCi of [^{35}S]methionine on a Parafilm sheet in a wet chamber at 37°C in a water bath. Add 100 μl of only LM to the apical side of the filters and place them basal side down on the drop. Incubate for 6 min at 37°C.

4. *TGN block:* Terminate all the pulses by moving the filters to a new 12-well plate containing 1.5 ml CM already at 20°C. Add 0.75 ml of CM (20°C) on the apical side and incubate for 75 min in a 19.5°C water bath.

5. *SLO permeabilization:* Activate the SLO in KOAc buffer containing 10 mM DTT at 37°C for 30 min and keep at 4°C until used for permeabilization. Use the toxin within 1 hr of activation. Wash the filters twice in ice-cold KOAc+ by dipping. Carefully remove the excess buffer from the apical side and blot the basolateral side with a Kleenex. Place 25-μl drops of the activated SLO (enough to release 60% LDH in 30 min, about 20 mg/ml) on a Parafilm sheet placed on a metal plate on an ice bucket. Leave the apical side without buffer. Place the filters on the drop for 15 min. Wash the basolateral surface twice by dipping in ice-cold KOAc+ buffer. Transfer the filters to a 12-well dish containing 0.75 ml TM at 19.5°C. Add 0.75 ml TM to the apical side and incubate at 19.5°C for exactly 30 min in water bath.

6. *Transport:* Remove the filters from the water bath, rinse them once with cold TM, blot the basolateral side, and place a 35-μl drop of either TM (control) or HeLa cytosol ($\pm$treatment or molecule to be tested) supplemented each with ARS (3 μl/100 μl TM or HeLa cytosol). Add the ARS to TM or cytosol immediately before dispensing the drops onto which the filters are placed. Layer 100 μl of TM on the apical side and incubate at 4°C for 15 min in a moist chamber. Transfer the chamber to a warm water bath for 60 min at 37°C. Terminate the transport by transferring on ice and washing the filters with cold PBS+ three times.

7. *Trypsinization:* Add to the apical surface 250 μl of 100 μg/ml trypsin freshly prepared in PBS+ and then add 2 ml of PBS+ to the basolateral chamber. Keep on ice for 30 min and stop the reaction by adding 3 μl of soybean trypsin inhibitor (STI; 10 mg/ml) to the apical side and then wash the apical surface three times with PBS+ containing 100 μg/ml STI.

8. *Cell lysis:* Solubilize the monolayers in 100 μl LB containing freshly added CLAP. Scrape the cells and spin for 5 min in microfuge; discard the pellet. Analyze 20 μl per sample by SDS–PAGE on a 10% acrylamide gel, fix, and dry the gel.

9. *Quantitation:* Scan the gels with a PhophorImager and measure the band intensities using the ImageQuant software. Calculate the amount of HA transported as the percentage of HA (68 kDa) transported to the cell surface = $(2 \times HA2/[HA + 2 \times HA2]) \times 100$ with HA2 (32 kDA) being the small trypsin cleavage product of HA.

Basolateral Transport Assay

1. *Cell culture:* Seed 1:72 of the MDCK cells from one confluent 75-cm^2 flask per filter (6×10^4 cells/filter) with 0.75 ml of growth medium apically and 2 ml basolaterally. Change the growth medium every 24 hr and use cells 3.5 days after plating.

2. *Infection:* Prior to viral infection, wash the monolayers in warm PBS+ and IM and infect with the vesicular stomatitis virus in 50 μl IM (20 PFU/cell; enough to obtain 100%

infection as judged by immunofluorescence) on the apical side. After allowing adsorption of the virus to the cells for 1 hr at 37°C, remove the inoculum and continue the infection for an additional 3 hr after adding 0.75 ml IM on the apical side and 2 ml IM on the basolateral side of the filter.

3. *Metabolic labeling and chase:* Rinse the cell monolayers by dipping in beakers containing warm PBS+ and LM at 37°C. Place a 25-μl drop of the LM containing 12.5 μCi of [^{35}S]methionine on a Parafilm sheet in a wet chamber at 37°C in a water bath. Add 100 μl of LM to the apical side of the filter and place the filter on the drop. Incubate for 6 min at 37°C and incubate for an additional 6 min at 37°C in CM with 0.75 ml on the apical side and 2 ml on the basolateral side before the 19.5°C block.

4. *TGN block:* Terminate the pulse by moving the filters to a new 12-well plate containing 1.5 ml CM already at 20°C. Add 0.75 ml of CM (20°C) on the apical side and incubate for 60 min in a 19.5°C water bath.

5. *SLO permeabilization:* Activate the SLO in KOAc buffer containing 10 mM DTT at 37°C for 30 min and keep at 4°C until used for permeabilization. Use the toxin within 1 hr of activation. Wash the filters twice in ice-cold KOAc+ by dipping. Carefully remove excess buffer from the apical side and blot the basolateral side with a Kleenex. Add 50-μl drops of the activated SLO (enough to release 60% LDH in 30 min, about 40 mg/ml) on a Parafilm sheet placed on a metal plate layered on an ice bucket. Leave the basolateral side without buffer. Incubate for 15 min. Wash the apical surface twice with 0.75 ml of ice-cold KOAc+ buffer. Transfer the filters to a 12-well dish containing 0.75 ml TM at 19.5°C. Add 0.75 ml TM to the apical side and incubate at 19.5°C for exactly 30 min in water bath.

6. *Transport:* Remove the filters from the water bath, rinse them once with cold TM, blot the basolateral side and add 100 μl of either TM (control) or HeLa cytosol (±treatment or moleucule to be tested) supplemented with ARS each (3 μl/100 μl TM or HeLa cytosol) on the apical side. Add the ARS to TM or cytosol prior to dispension onto drops. Incubate at 4°C for 15 min. Put the filters on already dispensed 35-μl drops of TM supplemented with ARS on a Parafilm in a moist chamber in a water bath at 37°C. Incubate for 60 min. Terminate the transport by transferring on ice and washing the filters with cold PBS+ three times.

7. *Anti-VSV G binding:* Wash twice (2–5 min) with 2 ml of CM containing 10% FCS (CM-FCS) on the basolateral side and once with 0.75 ml on the apical side. Dilute anti VSV-G antibodies in CM-FCS (1:18, i.e., 50 μg/ml; 300 μl) and place the filters on 25-μl drops of PBS+/ab (1.5 μl ab /filter). Add nothing to the apical side and incubate in a cold room on a metal plate placed on an ice bucket for 90 min. Remove the filters and wash (3×) on the basolateral side with CM-FCS (2 ml with constant rocking) and once on the apical side. Shake the filters gently for the efficient removal of unbound antibodies. Wash the filters once with PBS+ and place each on a 25-μl drop of PBS+ supplemented with cold virus (1:25; 300 μl). Incubate for 10 min in the cold.

8. *Cell lysis:* Solubilize the monolayers in 200 μl LB containing freshly added CLAP and cold virus (1:166). Scrape the cells, spin for 5 min, and discard the pellet. Remove a 10-μl aliquot from each sample (total).

9. *Immunoprecipitation:* Wash protein A–Sepharose powder with PBS (3×), let it swell for 10 min in PBS, wash it with LB, and store as a 1:1 slurry in LB at 4°C for 3 weeks maximum. Add 30 μl of the 1:1 slurry of protein A–Sepharose to the lysate. Rotate in cold room for 60 min. Spin the resin down and wash (3×) with 500 μl LB and elute the bound sample with 35 μl 2× Laemmli buffer and boil for 5 min at 95°C. Load 20 μl of bound material and 10 μl of total (after boiling with 10 μl of 2× Laemmli buffer) on SDS–PAGE gels (10%). Run SDS–PAGE on a 10% acrylamide gels, fix, and dry the gels (see Volume 4, Celis and Olsen, "One-Dimensional Sodium Dodecyl Sulfate-Polyacrylamide Gel Electrophoresis" for additional information).

10. *Quantitation:* Scan the gels with a PhophorImager and measure the band intensities using the ImageQuant software. Calculate the amount of VSV G (67 kDa) transported as follows: percentage of VSV G on the cell surface = (surface immunoprecipitated VSV G/total VSV G) × 100.

ER to Golgi Transport Assay

In the ER to Golgi transport assay, influenza N- or VSV-infected MDCK monolayers can be used. An infection with the influenza virus is used here as an example.

1. *Cell culture:* Seed 1:72 of the MDCK cells from one confluent 75-cm² flask per filter with 0.75 ml of growth medium apically and 2 ml basolaterally. Change the growth medium every 24 hr and use cells 3.5 days after plating.

2. *Infection:* Prior to viral infection, wash the monolayers in warm IM and then infect with the influenza virus in 50 µl IM (20 PFU/cell; enough to obtain 100% infection as judged by immunofluorescence) on the apical side. After allowing adsorption of the virus to the cells for 1 hr at 37°C, remove the inoculum and continue the infection for an additional 3 hr after adding 0.75 ml IM on the apical side and 2 ml IM on the basolateral side of the filter.

3. *Metabolic labeling:* Rinse the cell monolayers by dipping in beakers containing warm PBS+ and then LM at 37°C. Place a 25-µl drop of the LM containing 12.5 µCi of [³⁵S]-methionine on a Parafilm sheet in a wet chamber at 37°C in a water bath. Add 100 µl of LM to the apical side of the filter and place the filter on the drop. Incubate for 6 min at 37°C.

4. *Termination:* Terminate the pulse by moving the filters to a new 12-well plate containing 1.5 ml CM already at 4°C. Add 0.75 ml of CM (4°C) on the apical side and incubate for 30 min at 4°C.

5. *SLO permeabilization:* Activate the SLO in KOAc buffer containing 10 mM DTT at 37°C for 30 min and keep at 4°C until used for permeabilization. Use the toxin within 1 hr of activation. Wash the filters twice in ice-cold KOAc+ by dipping. Carefully remove the excess buffer from the apical side and blot the basolateral side with a Kleenex. Place 25-µl drops of the activated SLO (enough to release 60% LDH in 30 min, about 20 mg/ml) on a Parafilm sheet placed on a metal plate layered on an ice bucket. Leave the apical side without buffer. Place the filters on the drops for 15 min. Wash the basolateral surface twice by dipping in ice-cold KOAc+ buffer. Transfer the filters to a 12-well dish containing 0.75 ml TM at 4°C. Add 0.75 ml TM to the apical side and incubate at 37°C for 3 min and add 1.5 ml on the basolateral side of the filter (formation of pores). This is followed by an incubation at 4°C for 20 min in fresh TM on both sides (cytosol depletion).

6. *Transport:* Rinse the filters once with cold TM, blot on the basolateral side, and put on a 35-µl drop of either TM (control) or HeLa cytosol (±treatment or molecule to be tested) supplemented with ARS each (3 µl/100 µl TM or HeLa cytosol). Add the ARS to TM or cytosol prior to dispension onto drops. Layer the apical side with 100 µl of TM and incubate at 4°C for 15 min in a moist chamber. Transfer the chamber to a water bath at 37°C for 45 min. Terminate the transport by transferring on ice and washing the filters three times with cold PBS+.

7. *Cell lysis:* Solubilize the monolayers in 100 µl LB containing freshly added CLAP. Scrape the cells and spin for 5 min in microfuge, discard the pellet.

8. *Endoglycosidase H treatment:* Remove a 75-µl aliquot and add to 25 µl of 0.2 M sodium citrate buffer, pH 5.0. The resulting 100-µl mixture has a pH of 5.3. Divide it into two 50-µl aliquots: One receives 5 µl of 1 U/ml endoglycosidase H and the other receives only the citrate buffer. After 20 hr at 37°C, terminate the reaction by boiling in Laemmli buffer. Analyze the samples by running SDS–PAGE on a 10% acrylamide gel, fix, and dry the gel.

9. *Quantitation:* Scan the gels with a PhophorImager and measure the band intensities using the ImageQuant software. Calculate the amount of HA transported as the percentage of HA acquiring endoglycosidase H resistance with the following formula: percentage of HA reaching the Golgi complex = (endo H-resistant HA / total HA) × 100.

IV. COMMENTS AND PITFALLS

In all cases the values are expressed as the control cytosol-dependent transport being 100% (transport in the presence of cytosol minus transport in the absence of added cytosol). For ach manipulation a matched control is used (e.g., antibody or peptide tested vs control antibody or peptide, respectively). Assays are carried out routinely on duplicate filters, and quantifications represent the mean ± SEM obtained in several experiments.

A critical parameter for the successful performance of these transport assays is the quality of SLO. MDCK cells are difficult to permeabilize compared to several other cell types (e.g., BHK, CHO, and L-cells) (Miller and Moore, 1991; Krijnse-Locker et al., 1994; Yoshimori t al., 1996), and with the available commercial sources of SLO, the degree of permeabilization, as measured by LDH release, has not been satisfactory. The wild-type toxin purified rom *Streptococci* (Bhakdi et al., 1984, 1993) and the recombinant toxin produced in *Escherichia coli* have both worked equally well.

The available amount of reagent often determines whether the preincubation during cytosol depletion is possible as cytosol depletion must be carried out in an excess volume of buffer. The routinely used volume (750 μl per filter) can be, however, somewhat reduced. By using 500 μl per filter there is not yet a significant effect on transport efficiency, whereas using 200 μl per filter cytosol depletion is moderately compromised, which increases the background and cytosol-independent transport, resulting in a transport efficiency of about three-fourths of normal.

Because the transport is carried out in a leaky cellular microenvironment with diluted cytosolic components, the increase in transport obtained with exogenous cytosol is usually two- to threefold (three- to fourfold in the ER to Golgi assay). This is the window in which the differences in cytosol-dependent transport are measured. In assays measuring transport in the late secretory pathway, part of the efficiency is lost due to the retention of some viral marker early in the exocytic route. However, because both markers, HA and VSV G, are glycoproteins whose mobility on SDS–PAGE shifts according to the degree of glycosylation, careful examination of their mobilities will reveal, in the apical and basolateral assays, if the test condition retarded significantly the processing of the marker to the terminally glycosylated form. This may therefore serve as an internal control for the specificity of inhibition. A more accurate way to test the effect of a reagent in the early secretory pathway is to assay the ER to Golgi transport. The real advantage of having established similar procedures for three different transport assays is the possibility of using them as internal controls for each other. This enables the identification of molecules that are specifically involved in either apical or basolateral transport routes and allows the discrimination between compounds that are needed only in the polarized routes versus those that are common to all three transport processes.

References

Bennett, M. K., Wandinger-Ness, A., and Simons, K. (1988). Release of putative exocytic transport vesicles from perforated MDCK cells. *EMBO J.* **7**, 4075–4085.

Bergmeyer, H. U., and Bent, E. (eds.) (1984). "Methods of Enzymatic Analysis," pp. 574–579. Verlag Chemie, Weinheim and Academic Press, New York.

Bhakdi, S., Roth, M., Sziegoleit, A., and Tranum-Jensen, J. (1984). Isolation and identification of two hemolytic forms of steptolysin-O. *Infect. Immun.* **46**, 394–400.

Bhakdi, S., Weller, U., Walev, I., Martin, E., Jonas, D., and Palmer, M. (1993). A guide to the use of pore-forming toxins for controlled permeabilization of cell membranes. *Med. Microbiol. Immunol.* **182**, 167–175.

Fiedler, K., Lafont, F., Parton, R. G., and Simons, K. (1995). Annexin XIIIb: A novel epithelial specific annexin is implicated in vesicular traffic to the apical plasma membrane. *J. Cell Biol.* **128**, 1043–1053.

Gravotta, D., Adenisk, M., and Sabatini, D. (1990). Transport of influenza HA from the trans-Golgi network to the apical surface of MDCK cells permeabilized in their basolateral plasma membranes: Energy dependence and involvement of GTP-binding proteins. *J. Cell Biol.* **111**, 2893–2908.

Huber, L. A., Pimplikar, S., Parton, R. G., Virta, H., Zerial, M., and Simons, K. (1993). Rab8, a small GTPase involved in vesicular traffic between the TGN and the basolateral plasma membrane. *J. Cell Biol.* **123**, 35–45.

Ikonen, E., Parton, R., Lafont, F., and Simons, K. (1996). Analysis of the role of p200-containing vesicles in post-Golgi traffic. *Mol. Cell Biol.* **7**, 961–974.

Ikonen, E., Tagaya, M., Ullrich, O., Montecucco, C., and Simons, K. (1995). Different requirements for NSF, SNAP and Rab proteins in apical and basolateral transport in MDCK cells. *Cell* **81**, 571–580.

Kobayashi, T., Pimplikar, S. W., Parton, R. G., Bhakdi, S., and Simons, K. (1992). Sphingolipid transport from the trans-Golgi network to the apical surface in permeabilized MDCK cells. *FEBS Lett.* **300**, 227–231.

Krijnse-Locker, J., Ericsson, M., Rottier, P. J. M., and Griffith, G. (1994). Characterization of the budding compartment of mouse hepatitis virus: Evidence that transport from the RER to the Golgi complex requires only one vesicular transport step. *J. Cell Biol.* **124**, 55–70.

Lafont, F., Burkhardt, J. K., and Simons, K. (1994). Involvement of microtubule motors in basolateral and apical transport in kidney cells. *Nature* **372**, 801–803.

Lafont, F., Simons, K., and Ikonen, E. (1995). Dissecting the molecular mechanisms of polarized membrane traffic. Reconstitution of three transport steps in epithelial cells using Streptolysin-O permeabilization. *Cold Spring Harb. Symp. Quant. Biol.* **LX**, 753–762.

Matlin, K. S., and Simons, K. (1983). Reduced temperature prevents transfer of a membrane glycoprotein to the cell surface but does not prevent terminal glycosylation. *Cell* **34**, 233–243.

Miller, S., G., and Moore, H.-P. H. (1991). Reconstitution of constitutive secretion using semi-intact cells: Regulation by GTP but not calcium. *J. Cell Biol.* **112**, 39–54.

Pfeiffer, S., Fuller, S. D., and Simons, K. (1985). Intracellular sorting and basolateral appearance of the G protein of vesicular stomatitis virus in MDCK cells. *J. Cell Biol.* **101**, 470–476.

Pimplikar, S. (1994). Use of toxins in studying epithelial membrane transport. *In* "Bacterial Protein Toxins, Zbl. Bakt. Suppl. 24, (Freer et al., editors) pp. 277–284. Gustav Fischer, Stuttgart, Jena, New York.

Pimplikar, S., and Simons, K. (1993). Regulation of apical transport in epithelial cells by a Gs class of heterotrimeric G protein. *Nature* **362**, 456–458.

Pimplikar, S. W., Ikonen, E., and Simons, K. (1994). Basolateral protein transport in streptolysin O permeabilized MDCK cell. *J. Cell Biol.* **125**, 1025–1035.

Simons, K., and Fuller, S. D. (1985). Cell surface polarity in epithelia. *Annu. Rev. Cell Biol.* **1**, 243–288.

Yoshimori, T., Keller, P., Roth, M. G., and Simons, K. (1996). Different biosynthetic transport routes to the plasma membrane in BHK and CHO cells. *J. Cell Biol.* **133**, 247–256.

Studying Protein Sorting and Transport Vesicle Assembly from the *trans*-Golgi Network in Intact and Semi-intact Epithelial and Neuronal Cells following RNA Viral Infection or Adenovirus-Mediated Gene Transfer

Charles Yeaman, Debra Burdick, Anne Muesch, and Enrique Rodriguez-Boulan

INTRODUCTION

Study of the mechanisms of polarized protein sorting in epithelial and neuronal cells has been greatly facilitated by the use of enveloped RNA viruses, such as vesicular stomatitis virus (VSV) and influenza virus, which bud from the basolateral and apical plasma membranes, respectively, in MDCK cells (Rodriguez-Boulan and Sabatini, 1978). Following infection, a rapid onset of viral protein synthesis occurs, leading to the vectorial transport of envelope glycoproteins to either the apical or the basolateral surface. This model continues to provide information on the mechanisms of protein sorting (Muesch *et al.*, 1996) and the basic protocols will be included here. However, because cells cannot be coinfected efficiently with both types of viruses due to the reciprocal inhibition of protein synthesis, a major drawback of this paradigm is the inability to study the segregation of apical and basolateral proteins from one another in the same cell.

An ideal system for studying the molecular process by which polarized sorting occurs in the *trans*-Golgi network (TGN) should meet several basic criteria: (i) provide high-level expression of both an apical and a basolateral marker *simultaneously* by all of the cells in the culture; (ii) preserve the ability of the cell to synthesize and transport endogenous apical and basolateral proteins; (iii) provide the ability to metabolically label and accumulate these proteins in the TGN; and (iv) allow for comparison of sorting pathways and mechanisms in different cell types both *in vitro* and *in vivo*.

The use of recombinant adenovirus vectors can provide these advantages. We are currently developing a novel strategy using these vectors for introducing foreign cDNAs into polarized epithelial cells and neurons. Perhaps most importantly, multiple recombinant viruses may be effectively introduced into cells, permitting coexpression of apical and basolateral protein markers in the same cell. Because the adenovirus vectors carry mutations in the dominant regulatory E1 region, cytopathic effects encountered with live viruses, such as repression of host cell mRNA and protein synthesis, are not observed. This can allow the study of marker proteins in a more normal cellular environment than that produced by the enveloped RNA viruses. These vectors also produce the high level of marker protein expression that is essential in obtaining suffient incorporation of radioactive precursors for pulse–chase studies and for immunoisolation of transport vesicles. Endogenous proteins are generally synthesized at rates too low for these analyses. Furthermore, the levels of expression may be manipulated in this sytem to allow study of the ability to saturate the various sorting pathways available to the cell. Finally, these particles are efficiently taken up into a wide variety of cells that contain the appropriate receptors, making this technology readily applicable to many cell lines. Superinfection is possible with these vectors and in all cell lines tested by this laboratory, dual infection with adenovirus vectors was efficiently achieved. The ability to transduce two or more gene products with this procedure can make it exceedingly useful for many applications. Because procedures for transduction in different cell lines vary, MDCK cells and PC12 cells will be discussed for comparison.

In addition, the adenovirus system has several advantages over transfection protocols. The first advantage is the lack of clonal variation, which sometimes confounds the interpretation of data achieved with stably transfected cells. The level of protein synthesis can be easily and reproducibly controlled by varying either the multiplicity of infection (m.o.i.) or the time in culture following the infection. Once titrated and optimized, 100% of the cells express the desired protein(s), and pools of cells are produced that are remarkably consistent from experiment to experiment. Second, the tedious process of producing clonal populations of stably transfected cells is eliminated, although the production of the viral vectors is itself non trivial (see Volume 1, Hitt *et al.*, "Construction and Propagation of Human Adenovirus Vectors" for additional information). Finally, it is relatively simple to introduce exogenous cDNAs into cells *in situ*, allowing for the first time a comparison of sorting pathways available to native cells with those described in cultured cell lines.

This article describes an assay that monitors post-Golgi vesicle budding from semi-intact MDCK cells following infection either with enveloped RNA viruses or with recombinant adenovirus vectors. A similar assay using recombinant adenoviruses is presented for PC12 cells. The adenoviruses we have found most useful for these applications encode receptor for neurotrophins (p75^{NTR}) and low density lipoprotein (LDLR), which were previously shown to be sorted to the apical and basolateral surfaces of polarized MDCK cells, respectively (Le Bivic *et al.*, 1991; Hunziker *et al.*, 1991). Each of these proteins, when expressed in MDCK cells, incorporates radiosulfate into carbohydrate moieties during posttranslational processing late in the secretory pathway, providing a convenient method to label markers in the sorting compartment. This method, as well as methods for labeling these proteins in PC12 cells and labeling RNA virus reporter proteins, will be presented.

II. MATERIALS

Reagents for cell culture, including Dulbecco's modified Eagle's medium (DMEM, Cat. No. 12800), MEM SelectAmine kits (Cat. No. 19050-012), MEM nonessential amino acids (Cat. No. 11140), MEM vitamins (Cat. No. 11120), penicillin–streptomycin (Cat. No. 15140), L-glutamine (Cat. No. 25030), 1 M HEPES (Cat. No. 15630), 7.5% bovine serum albumin (BSA, Cat. No. 15260), and donor horse serum (Cat. No. 16050) can be purchased from GIBCO-BRL. Heat inactivated fetal bovine serum (FBS) is from Gemini Bioproducts. Polycarbonate filter (Transwells; 0.4-mm pore size, Cat. No. 3412 for 24-mm filters) can be purchased from Corning Costar Corp. Tissue culture grade plasticware is from Corning Plasticware. Tran^{35}S-label (Cat. No. 51006), [^{35}S]cysteine (Cat. No. 51002), and H$_2$^{35}SO$_4$ (Cat. No. 64040) can be purchased from ICN. Reagents for production of viruses are described elsewhere (Rodriguez-Boulan and Sabatini, 1978; see Volume 1, Hitt *et al.*, "Construction and Propagation of Human Adenovirus Vectors" for additional information). All other reagents are standard reagent grade and available from several sources. Sterile solutions were autoclaved or sterilized by ultrafiltration (0.2 μm).

III. PROCEDURES

A. Adenovirus Transduction

1. Coinfection of MDCK Cells with Recombinant Adenovirus Vectors

Solutions

1. *Dulbecco's modified Eagle's medium:* Dissolve DMEM powder in 1 liter H$_2$O. Sterilize by filtering through 0.2-μm pore filter.

2. *Complete DMEM (cDMEM):* Add 50 ml heat-inactivated FBS, 10 ml 0.2 *M* ʟ-glutamine, 5 ml penicillin–streptomycin (10,000 U/ml and 10 mg/ml, respectively) and ̣ ml MEM nonessential amino acids to 430 ml DMEM. Store at 4°C for up to 2 weeks.

Steps

1. For infection with replication-defective viruses, grow cells on semipermeable polycarbonate filter supports. Seed cells at confluency on day 1 and culture for 4 days with medium changed daily. MDCK strain II cells are confluent at a density of ca. 7×10^5 cells/cm^2, so approximately 3.3×10^6 cells should be seeded on each 24-mm filter.

2. Before applying adenovirus vectors, rinse cultures twice with serum-free DMEM (2 ml for the apical chamber and 2.5 ml for the basolateral chamber). Removal of serum proteins from the culture medium results in enhanced adsorption of adenovirus particles to the cell surface and improves infection efficiency. MDCK cells do not appear to be adversely affected by serum deprivation during the infection period, but this should be checked with other cell types before attempting infection.

3. Add recombinant adenoviruses, diluted in serum-free DMEM, at m.o.i. ranging from to 1000 PFU/cell. Incubate at 37°C for 1 hr, gently tilting the plates every 15 min to mix. Two or more adenovirus vectors can be mixed together and applied simultaneously to cells. It is recommended that serial threefold dilutions of viruses be tested in order to determine the optimal m.o.i. required for quantitative expression of the marker proteins.

4. Vectors must be applied to the *apical* domain of epithelial cells, as infection is markedly more efficient from this surface. The reason for this is unclear, but a preference for adenovirus entry through the apical surface has been observed in every polarized epithelial cell type we have examined, including canine (MDCK), bovine (MDBK), and porcine (LLC-PK$_1$) kidney, rat thyroid (FRT) and retinal pigment epithelia (RPE-J), and human intestine (Caco-2). Serum-free DMEM, without virus, should be applied to the basolateral chamber. Infection should be performed in a minimum volume of DMEM required to keep the filter submerged. A recommended volume is 0.25 ml apical/0.5 ml basolateral for a 24-mm filter.

5. Following the infection period, add 2 ml cDMEM to both apical and basolateral chambers. It is not necessary to aspirate the virus because the addition of serum effectively terminates the infection. Culture the cells at 37°C for the desired incubation time. Expression of adenovirus-encoded proteins should be detectable by 4–6 hr following infection, will rise gradually over the next 18 hr, and should reach a plateau by 24 hr. This level of expression will be maintained for at least 1 week, provided the cells are fed daily. We routinely use cultures of polarized MDCK cells between 20 and 24 hr postinfection.

6. Monitor infection after 24 hr as follows:

a. Indirect immunofluorescent staining is used to determine (i) the percentage of cells in the culture expressing the transfected gene products and (ii) the intracellular distribution of the gene products. Under optimum infection conditions, at least 95% of the cells will express each adenovirus-encoded protein. If the ultimate goal of the experiment is to study the molecular trafficking of two or more proteins in the same cell, it is essential that all of the cells that express one marker also express the second marker. Methods for fixation, permeabilization, and staining of epithelial cells grown on polycarbonate filters are described in Volume 3, Reinsch *et al.*, "Confocal Microscopy of Polarized Epithelial Cells."

b. Immunoprecipitation or Western blotting is used to determine (i) the molecular weight of the adenovirus-encoded proteins and (ii) the level of expression of the proteins in the culture. Ideally, both adenovirus-encoded proteins will be expressed at comparable levels.

2. Coinfection of PC12 Cells with Recombinant Adenovirus Vectors

The adenovirus-mediated dual infection technique is being employed in PC12 cells to study the sorting of proteins exiting TGN. Dual infection may allow elucidation of the various transport vesicle populations and their mechanisms of formation or delivery. Thus, proteins that are destined for the plasma membrane or those in constitutive or regulated secretory granules may be compared and contrasted.

Solutions

1. *PC12 growth medium:* DMEM with 10% fetal bovine serum and 5% horse serum. Medium also includes a 1:100 penicillin–streptomycin solution, 1:100 nonessential amino acid solution, 1:100 L-glutamine solution, and 1:100 HEPES buffer (all are GIBCO reagents listed in Section II).

2. *Dilute collagen solution:* Prepare rat tail collagen in 0.1% acetic acid as previously discribed (Greene *et al.*, 1991). Store aliquots at $-20°C$. Once thawed, it may be stored for several months at 4°C. Dilute the stock in sterile water just before use and determine an optimal final dilution empirically. The ideal concentration of collagen will promote PC12 cell adhesion, yet not impede neurite extension.

3. *Infection medium:* DMEM with 1.5% fetal bovine serum. Reduced serum is needed to improve virus adherence to the cell surface. The labeling medium for PC12 differs from that for MDCK cells due to PC12 cell increased sensitivity to serum deprivation.

4. *Serum mix:* Two-thirds fetal bovine serum and one-third horse serum.

Steps

1. Plate PC12 cells onto the collagen-coated dishes in growth medium. The collagen coat promotes cell adhesion and allows for multiple manipulations of the culture without fear of cell loss. Add dilute collagen solution to tissue culture dishes and spread evenly with a blunt cut rubber policeman to coat the surface of the plastic. Allow to dry in a tissue culture hood.

Passage PC12 cells when ca. 80% confluent at a 1:5 dilution. They are passaged by trituration of the conditioned medium to dislodge the cells. Maintain cells in a mixture of one-fifth conditioned medium and four-fifths new growth medium. (See Volume 1, Teng *et al.*, "Cultured PC12 Cells: A Model for Neuronal Function, Differentiation, and Survival" for additional information).

2. Cells are generally infected 3 days after plating to ensure that they are 70–80% confluent and well adhered. Remove the growth medium and replace with a small volume of infection medium. Because PC12 cells are weakly adherent, add the medium slowly with the pipette against the side of the dish to avoid disturbing the cell layer. Include a rinse step to help remove serum proteins from the cell surface. This may not be feasible if the cells are weakly adherent, in which case this step should be deleted.

3. Adenovirus vectors are diluted into infection medium to m.o.i. of 1–3000 PFU/cell. The optimal concentration for 100% transduction is determined empirically as monitored by immunofluorescence and biochemical assays as described for MDCK cells. The optimum m.o.i. for PC12 cells may be higher than that determined for MDCK cells due to the inclusion of serum in the infection medium. Because PC12 cells are much less adherent than MDCK cells, do not rock the plates during the infection. Instead, use a volume of infection medium that completely covers the cell layer. Return the cells to the incubator and allow the infection to proceed for 1.5 hr. Add serum mix to the plates to adjust the serum concentration to 15% and add additional growth medium.

4. Cells are used for experimental procedures 20–30 hr after infection. Expression of the exogenous protein may not have reached a maximum at shorter time points. Unlike MDCK cells, at time points beyond 30 hr cytopathic effects may be seen at m.o.i. which are sufficient to produce protein expression in 100% of the cells.

Comments

A similar protocol may be used with cultures of hippocampal neurons, except that the adenovirus may be added directly to the normal medium in which these neurons are grown because it is serum free. The ability to remove the medium replacement step reduces the disturbance to the cell layer in these weakly adherent cells.

Pitfalls

Each batch of adenovirus virus vector must be tested for optimum tranduction. We find a batch-to-batch variability in the correspondence of PFU obtained from plaque assays to the m.o.i. needed.

B. Infection of MDCK Cells with Enveloped RNA Viruses

Solutions

1. *DEAE-Dextran (100× stock):* Dissolve 100 mg DEAE-dextran (Sigma Cat. No. D-162) in 10 ml H_2O. Filter sterilize and store 1-ml aliquots at $-20°C$.

2. *Infection medium:* Add 13 ml 7.5% BSA and 5 ml 1 M HEPES, pH 7.4, to 482 ml DMEM. Filter sterilize and store at 4°C. Immediately before use, add 0.1 ml DEAE-dextran stock to 10 ml medium. DEAE-dextran is only necessary for infection with VSV.

3. *Virus stocks:* Vesicular stomatitis virus, Indiana strain (VSV), and influenza virus A (WSN strain) are grown in MDCK strain II cells, harvested, and plaque assayed as described (Rodriguez-Boulan and Sabatini, 1978).

Steps

1. Set up 10-cm dishes of MDCK strain II cells, passages 6–20, and allow them to reach confluency. Cultures are infected with VSV or influenza virus 3 days after becoming confluent.

2. Before infecting cells, rinse cultures twice with infection medium. For viral infection, inoculate MDCK cells with 50 PFU/cell VSV or influenza WSN in 3.5 ml infection medium containing 0.1 mg/ml DEAE-dextran.

3. Incubate cultures for 1 hr at 37°C.

4. Aspirate viral medium and rinse cultures twice with fresh infection medium.

5. Return VSV-infected cultures to 37°C and incubate a further 3.5 hr before metabolic radiolabeling (see later). Incubate influenza WSN-infected cultures for 4.5 hr at 37°C before labeling. Cultures should be examined hourly to monitor cytopathic effects.

C. Metabolic Radiolabeling and Accumulation of Marker Proteins in the *trans*-Golgi Network

1. Radiosulfate Labeling of Glycoproteins at 20°C

Both p75^{NTR} and LDLR are sulfated when expressed in MDCK cells. Sulfation occurs largely on asparagine-linked carbohydrate moieties on both proteins. Because this posttranslational modification occurs late in the secretory pathway, likely in the *trans*-Golgi or TGN, it provides a convenient method to label markers in the sorting compartment. When labeling is performed at the reduced temperature of 20°C, labeled markers accumulate in the TGN because post-Golgi vesicular transport is inhibited.

Solutions

1. *Sulfate-free labeling medium:* Essentially, labeling medium is DMEM in which the MgSO$_4$ is replaced by MgCl$_2$. Combine 100 ml 10× DME salts (Ca^{2+}, Mg^{2+} free), 10 ml 100× Ca^{2+}, Mg^{2+} stock, 10 ml MEM vitamins, 10 ml each of 100× stock MEM amino acid solutions (arginine, glutamine, histidine, isoleucine, leucine, lysine, phenylalanine, threonine, tryptophan, tyrosine, glycine, serine, and valine), 1 ml each of 100× stock MEM solutions of methionine and cysteine, 10 ml MEM nonessential amino acid solution, 20 ml 1 M HEPES,

pH 7.4, 27 ml 7.5% BSA stock, and H_2O to a final volume of 1000 ml. Filter sterilize and store at 4°C.

2. *10× DME salts (Ca^{2+}, Mg^{2+} free):* Combine 50 ml 100 × $Fe(NO_3)_3$, 50 ml 100× NaH_2PO_4, 50 ml 100X KCl, 22.5 g dextrose, 32 g NaCl, and 15 ml phenol red solution. Adjust volume to 500 ml with H_2O. Filter sterilize and store at 4°C.

3. *$Fe(NO_3)_3$ stocks:* Add 0.05 g $Fe(NO_3)_3$ to 50 ml H_2O to prepare a 100,000× stock solution. Dilute 100 μl into 100 ml H_2O to prepare a 100× stock solution. Filter sterilize and store at 4°C.

4. *100× NaH_2PO_4:* Add 1.25 g NaH_2PO_4 to 100 ml H_2O. Filter sterilize and store at 4°C.

5. *100× KCl:* Add 4.0 g KCl to 100 ml H_2O. Filter sterilize and store at 4°C.

6. *100× Ca^{2+}, Mg^{2+}:* Add 2.96 g $CaCl_2 \cdot 2H_2O$ and 3.02 g $MgCl_2 \cdot 6H_2O$ to 100 ml H_2O. Filter sterilize and store at 4°C.

Steps

1. Twenty-four to 48 hr following adenovirus infection, aspirate culture medium and rinse filters three times with sulfate-free labeling medium. Sulfate starve cells for 30 min at 37°C in this medium.

2. Label cells for 1 hr at 20°C in sulfate-free labeling medium containing $H_2^{35}SO_4$. To label cells on one 75-mm Transwell filter, we use 0.5 mCi $H_2^{35}SO_4$ in 500 μl sulfate-free labeling medium. Place medium, preequilibrated at 20°C, on a sheet of Parafilm in a humid chamber and place the Transwell filter on this so that the label is exposed to the basolateral surface. Apply 2.5 ml sulfate-free labeling medium, without label, to the apical chamber to prevent drying.

2. Pulse–Chase Labeling with [^{35}S]Methionine/Cysteine
Solutions

1. *Methionine/cysteine-free labeling medium:* Prepare 1000 ml of medium following the product specification insert (GIBCO SelectAmine kit, Cat. No. 19050), excluding the methionine and cysteine in the kit. Add 10 ml 1 *M* HEPES and 27 ml 7.5% BSA stock solution. Filter sterilize and store at 4°C.

For PC12 cells, replace the BSA with 20 ml dialyzed serum. To prepare dialyzed serum, dialyze a mixture of two-thirds fetal bovine serum and one-third horse serum under sterile conditions against PBS in 12,000 molecular weight cutoff dialysis tubing for 12–20 hr. Dialyze serum to remove small molecules that may be used by the cells to scavenge sulfate.

2. *Chase medium:* Add 5 ml of 100× MEM methionine and cysteine (left over from the Selectamine kit) solutions to 40 ml complete DMEM (or PC12 growth medium for PC12 cells). Immediately before use, add cycloheximide to a concentration of 20 μg/ml.

Steps

1. Rinse MDCK cultures three times in methionine/cysteine-free labeling medium before incubation at 37°C for the final 30 min of incubation following viral infection. For PC12 cells, extensive rinsing may not be possible, so rinse once before the 30-min incubation.

2. Label VSV-infected MDCK cells with [^{35}S]methionine/cysteine (Tran^{35}S-label, ICN Cat. No. 51006). Label influenza WSN-infected cells with [^{35}S]cysteine (ICN Cat. No. 51002). Use 0.5 mCi, in a total volume of 3.5 ml labeling medium, to label each 10-cm plate. Pulse label cells for 10 min at 37°C. Medium can be recycled twice if multiple dishes are labeled.

Label LDLR- and p75-infected PC12 cells with [^{35}S]cysteine because these proteins are not sulfated as they are in MDCK cells. Use 0.5 mCi, in a total volume of 1 ml labeling medium, to label each 10-cm plate. Pulse label cells for 15 min at 37°C, with gentle rocking every 5 min.

3. For MDCK cells, aspirate labeling medium and rinse plates three times with chase medium. For PC12 cultures, just aspirate medium.

4. Chase cultures at 20°C for 2 hr in chase medium. During the chase at 20°C, roughly 60% of the labeled VSV G/influenza HA proteins are accumulated in the TGN (Muesch *et al.*, 1996).

D. Vesicle Budding from the TGN in Semi-intact Cells

Semi-intact MDCK cells are prepared after accumulating marker proteins in the TGN (Muesch *et al.*, 1996). Cells are first swollen in a low salt buffer and subsequently scraped from the substratum, which produces large tears in the plasma membrane. Endogenous cytosol and peripheral membrane proteins are removed by washing with a high salt buffer. Addition of an exogenous source of cytosol, an energy-regenerating system, and incubation at 37°C typically results in the release of 25–65% of the total marker accumulated in the TGN into sealed vesicles. Budded vesicles are separated from the material that remains by a brief, low-speed centrifugation step.

Solutions

1. *Swelling buffer:* Add 7.5 ml 1 M HEPES/KOH, pH 7.2, and 7.5 ml 1 M KCl to 485 ml H$_2$O. Store at 4°C.

2. *10× transport buffer:* Add 20 ml 1 M HEPES/KOH, pH 7.2, 2 ml 1 M Mg(OAc)$_2$, and 18 ml 5 M KOAc to 60 ml H$_2$O. Store at 4°C.

3. *1× transport buffer:* Add 1 ml 10× transport buffer to 9 ml H$_2$O. Immediately before use, bring to 1 mM DTT and add protease inhibitors to 1× concentration.

4. *High salt buffer:* Add 10 ml 1 M HEPES/KOH, pH 7.2, 50 ml 5 M KOAc, and 1 ml 1 M Mg(OAc)$_2$ to 439 ml H$_2$O. Store at 4°C. Immediately before use, bring to 1 mM DTT and add protease inhibitors to 1× concentration.

5. *1 M dithiothreitol stock*

6. *500× protease inhibitor stock:* Dissolve 5 mg of each of the following inhibitors individually in 330 μl dimethyl sulfoxide (DMSO) and combine the three solutions: pepstatin A (Sigma Cat. No. P-4265), leupeptin (Sigma Cat. No. L-8511), and antipain (Sigma Cat. No. A-6191).

7. *100 mM PMSF stock*

8. *Energy mix:* In the following order pipette 3 μl ATP, 2 μl GTP, 4 μl creatine phosphate, and 3 μl creatine kinase

   a. 0.1 M ATP (Boehringer Mannheim Cat. No. 519 987)

   b. 0.2 M GTP (Boehringer Mannheim Cat. No. 106 399)

   c. 0.6 M creatine phosphate (Boehringer Mannheim Cat. No. 621 722)

   d. 8 mg/ml creatine kinase (Boehringer Mannheim Cat. No. 127 566)

9. *Bovine brain cytosol:* Prepare gel-filtered bovine brain cytosol in batches exactly as described previously (Malhotra *et al.*, 1989). The protein concentration should be 10–20 mg/ml. Snap freeze 50-μl aliquots in liquid nitrogen and store at −80°C.

Steps

All steps are performed on ice, unless otherwise specified.

1. Following the 20°C incubation, wash the monolayer twice briefly with ice-cold swelling buffer and incubate in the same for 15 min.

2. Scrape cells from the filter (or plastic dish) with a rubber policeman into 2.5 ml transport buffer. DiSPo scrapers (Baxter Scientific) work well for this purpose. The best way to scrape cells from the Transwell filter is to place the filter inside the lid of the dish so that the bottom lies flat against the plastic. This prevents the scraper from

poking through the filter, but care must be exercised to prevent tearing the filter. Scraping does not have to be vigorous, but should be done with long, gentle strokes. Transfer cells to 1.5-ml microfuge tubes and rinse the filter (or dish) with 2.5 ml fresh transport buffer. Combine with cells from the first scraping. Discard the filter as radioactive waste.

3. Pellet cells by centrifugation at 800 g for 5 min in a refrigerated microfuge.

4. Pool semi-intact cells into one tube and wash with 1.5 ml high salt buffer on ice for 10 min. Pellet cells by centrifugation at 800 g for 5 min in a refrigerated microfuge.

5. Resuspend cells in transport buffer. A volume of 250 μl is used to resuspend cells from one filter (or dish).

6. Set up vesicle budding assay.

   a. In standard assays, suspend semi-intact cells (ca. 10 μl = 20–25 μg protein) in an assay volume of 50 μl transport buffer, supplemented with 50 μg gel-filtered bovine brain cytosol and an energy-regenerating system (1 mM ATP, 1 mM GTP, 5 mM creatine phosphate, and 0.2 IU creatine kinase). Combine components in the following order: mix 26 μl H$_2$O, 1 μl 100 mM PMSF, 4 μl 10$\times$ transport buffer, 4 μl energy mix, 5 μl cytosol (final concentration = 1 mg/ml), and 10 μl semi-intact cells.

   b. Important controls include assays in which either the cytosol or the energy mix or both components are omitted. For the complete depletion of energy from the system, cytosol and semi-intact cells must be preincubated for 10 min on ice with 0.6 U/ml apyrase before assembling the assay. In the absence of either cytosol or energy, vesicular release from semi-intact cells should be negligible. We suggest that serial dilutions of cytosol (i.e., 0.1–10 mg/ml) be tested in order to determine the optimal range of cytosol-dependent vesicle budding in the assay.

7. Incubate assays at 37°C for desired time. In standard assays, we incubate for 30–45 min. However, we suggest that a time course of vesicle budding be performed to optimize the assay for different marker proteins. As additional controls, two complete assays should be assembled and incubated at 0 and 20°C. At these reduced temperatures, vesicular release from semi-intact cells should be insignificant.

8. Pellet semi-intact cells by centrifugation at 800 g for 5 min in a refrigerated microfuge. Transfer supernatant fractions to clean microfuge tubes. The pellets (containing nonbudded material remaining in the TGN) and supernatants (containing the vesicles released during the 37°C incubation) can be analyzed further.

   a. To quantify the efficiency of vesicular release of each marker under different conditions, samples can be lysed in SDS–PAGE sample buffer and analyzed directly by PAGE. In cells infected with VSV or influenza WSN, the viral proteins should be the only labeled proteins in the lysates. Alternatively, following the adenovirus-mediated transfer of cDNAs encoding p75^{NTR} and LDLR into MDCK cells, these proteins are by far the most heavily labeled proteins when radiosulfate is used as a precursor.

   b. To confirm that markers are present inside sealed vesicles, the supernatant fraction should be treated with either proteinase K or trypsin. In the absence of Triton X-100, only the cytoplasmic domains of the proteins will be cleaved and this can be detected as a relatively small mobility increase during SDS–PAGE. In contrast, protease treatment in the presence of 1% Triton X-100 will result in complete digestion of markers. Use three 50-μl assay samples for this analysis. To tube 1, add nothing. To tube 2, add 2.5 μl protease (10 mg/ml stock $\rightarrow$ 0.5 mg/ml final). To tube 3, add 2.5 μl protease and 2.5 μl 20% Triton X-100. Incubate on ice for 30 min. Inactivate protease with 1 mM PMSF or 1 mg/ml soybean trypsin inhibitor before lysing samples in SDS–PAGE sample buffer. Analyze products by SDS–PAGE.

   c. Immunoisolation of specific classes of transport vesicles is performed using antibodies against the cytoplasmic portions of cargo proteins as well as appropriate negative controls.

i. Use 5 mg protein A–Sepharose (Pharmacia, Cat. No. 17-0780-01) for each immunoisolation. Swell in transport buffer for 10 min.

ii. If using a murine monoclonal primary antibody, use a bridge. Incubate 5 mg protein A–Sepharose with 50 μg rabbit anti-mouse IgG (Rockland Labs, supplied by VWR, Cat. No. 610-4102) in 1 ml transport buffer for 60 min at room temperature. Wash twice with transport buffer.

iii. Block nonspecific binding sites in 1 ml transport buffer containing 0.2% BSA for 60 min at room temperature.

iv. Couple primary antibody for 2 hr at room temperature in transport buffer. Wash twice with transport buffer. Titrate each antibody to determine the amount needed for the quantitative recovery of vesicles.

v. Incubate immunoadsorbant with the supernatant fraction from the vesicle budding assay in a total volume of 1 ml transport buffer for 2–18 hr at 4°C with end-over-end rotation. In some cases, vesicle coat proteins may mask epitopes on the cytoplasmic tails of cargo. Therefore, it may be necessary to wash the vesicles in high salt buffer prior to immunoisolation to strip coat proteins. Add 33 μl of 1 M KOAc to 50 μl vesicles. Incubate on ice for 10 min. Add 333 μl salt-free transport buffer (20 mM HEPES/KOH, 2 mM Mg(OAc)$_2$).

vi. Wash immunoprecipitates six times with transport buffer. Elute bound markers by boiling 5 min in SDS–PAGE sample buffer. Analyze by SDS–PAGE.

E. Vesicle Budding from TGN-Enriched Membranes

A TGN-enriched membrane fraction is prepared from metabolically labeled PC12 cells after marker proteins have been accumulated in the TGN. Initially, a postnuclear supernatant is prepared following the method of Tooze and Huttner (1992). From the postnuclear supernatant, a TGN-enriched membrane fraction is prepared (Xu *et al.*, 1995). As with semi-intact MDCK cells, the addition of an exogenous source of cytosol and an energy-regenerating system leads to the release of accumulated marker protein from the TGN in sealed vesicles. All steps for the vesicle budding assay are identical to those for intact MDCK cells, except that 50-μg aliquots of TGN-enriched membranes are used for each individual assay condition.

Acknowledgments

This work was supported by NIH grants to ERB, an NRSA award to DB, and a National Kidney Foundation fellowship to CY. ERB holds a Jules and Doris Stein Professorship from the Research to Prevent Blindness Foundation.

References

Greene, L. A., Sobeih, M. M., and Teng, K. K. (1991). Methodologies for the culture and experimental use of the PC12 rat pheochrmocytoma cell line. *In* "Culturing Nerve Cells" (G. Banker and K. Goslin, eds.), pp. 207–226. MIT Press, Cambridge, MA.

Hunziker, W., Harter, C., Matter, K., and Mellman, I. (1991). Basolateral sorting in MDCK cells requires a distinct cytoplasmic domain determinant. *Cell* **66**, 907–920.

Le Bivic, A., Sambuy, Y., Patzak, A., Patil, N., Chao, M., and Rodriquez-Boulan, E. (1991). An internal deletion in the cytoplasmic tail reverses the apical localization of human NGF receptor in transfected MDCK cells. *J. Cell Biol.* **115**, 607–618.

Muesch, A., Xu, H., Sheilds, D., and Rodriguez-Boulan, E. (1996). Transport of vesicular stomatitis virus G protein to the cell surface is signal mediated in polarized and non-polarized cells. *J. Cell Biol.* **133**, 543–558.

Rodriguez-Boulan, E., and Sabatini, D. D. (1978). Asymmetric budding of viruses in epithelial monolayers: A model system for study of epithelial cell polarity. *Proc. Natl. Acad. Sci. USA* **75**, 5071–5075.

Tooze, S. A., and Huttner, W. B. (1992). Cell-free formation of immature secretory granules and constitutive secretory vesicles from trans-Golgi network. *In* "Methods in Enzymolgy."

Xu, H., Greengard, P., and Gandy, S. (1995). Regulated formation of Golgi secretory vesicles containing Alzheimer b-amyloid precursor protein. *J. Biol. Chem.* **270**, 23243–23245.

Assays Measuring Membrane Transport in the Endocytic Pathway

Linda J. Robinson and Jean Gruenberg

I. INTRODUCTION

Significant progress has been made in understanding mechanisms regulating endocytic membrane traffic using cell-free assays (Braell, 1987; Davey et al., 1985; Diaz et al., 1988; Gruenberg and Howell, 1986; Woodman and Warren, 1988) (see Fig. 1). Both early and late endosomes exhibit homotypic fusion properties in vitro, as in vivo, yet they do not fuse with each other (Aniento et al., 1993). Transport from early to late endosomes is achieved by multivesicular intermediates termed endosomal carrier vesicles (ECV/MVB), which are presumably translocated on microtubules between the two compartments (Aniento et al., 1996; Bomsel et al., 1990; Gruenberg et al., 1989). The vectorial or heterotypic interactions of ECV/MVBs with late endosomes have also been reconstituted in vitro, as has the involvement of microtubules and motor proteins in this process (Aniento et al., 1993; Bomsel et al., 1990). By reducing the components in these in vitro assays to a cytosol source, an ATP-regenerating system, salts, and the purified endosomal membranes, the specificity of endosomal fusion events has been addressed, and the molecules and mechanisms involved have been studied. In fact, a number of conserved molecules, as well as molecules specific for different steps of the endocytic pathway, have been identified and/or characterized using cell-free assays such as those described in this protocol (for review, see Gruenberg and Maxfield, 1995).

This assay for endocytic vesicle fusion is based on the formation of a complex resulting from a reaction between two products present in separate populations of endosomes: avidin and biotinylated HRP (bHRP). These reaction products can be internalized into endosomes by fluid phase or receptor mediated endocytosis in vivo. Avidin and a biotinylated compound are used to provide a fusion-specific reaction because of the high binding affinity and low dissociation constant of avidin for biotin . Following internalization, cells are homogenized and purified endosomal fractions are prepared, which are combined in the assay together with cytosol and ATP. If fusion occurs, a complex is formed between avidin and bHRP. At the end of the assay, the reaction mixture is extracted in detergents in the presence of excess biotinylated insulin, as a quenching agent. The avidin–bHRP complex is then detected by immunoprecipitation with antiavidin antibodies, and the enzymatic activity of the bHRP associated with the immunoprecipitate is quantified. This article describes techniques for the preparation and partial purification of three different loaded endosomal fractions from BHK cells: early endosomes, endosomal carrier vesicles, and late endosomes. In addition, this article describes the preparation of the cytosol source used, as well as the techniques for the fusion assays themselves.

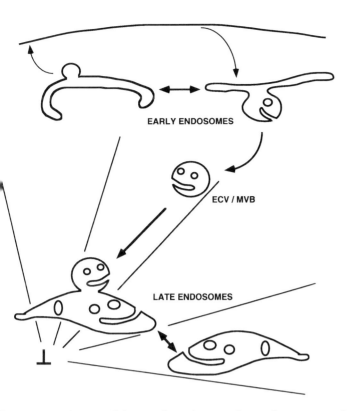

EARLY ENDOSOMES

ECV / MVB

LATE ENDOSOMES

FIGURE 1 Membrane trafficking in the endocytic pathway. The reconstituted steps of the endocytic pathway described in this protocol are: (a) the fusion of early endosomes with each other, (b) the fusion of ECV/MVBs with late endosomes, and (c) the fusion of late endosomes with each other. An *in vitro* budding assay for the formation of ECV/MVBs from early endosomes, which are competent to fuse with late endosomes, is described in Aniento *et al.* (1996). As shown, ECV/MVBs are transported along microtubules from early to late endosomes. If microtubules are depolymerized *in vivo*, prior to the loading of cells with an endocytic tracer, this tracer will accumulate in ECV/MVBs. These vesicles will then fuse with late endosomes, loaded with a different marker, *in vitro*.

I. MATERIALS AND INSTRUMENTATION

Standard laboratory rockers for washing cells and a large 37°C water bath, which can fit a metal plate of dimensions of 20 × 33 cm, were used. Large rectangular ice buckets (Cat. No. 1-6030), from NeoLab GmbH, can also accommodate metal plates of the same dimensions. Cell scrapers (flexible rubber policemen) with a silicone rubber piece of about 2 cm, cut at a sharp angle, and attached to a metal bar, were made. A standard low-speed cell centrifuge and Beckman ultracentrifuges and rotors were used. The refractometer (Cat. No. 79729) was from Carl Zeiss Inc., and the pump for collecting sucrose gradients (peristaltic pump P-1) was from Pharmacia Fine Chemicals. A rotating wheel (such as Snijders Model 34528) with a speed of about 10 rotations per minute should be used. All tissue culture reagents, including modified Eagle's medium (MEM), were from either Sigma Chemical Company or GIBCO-BRL/Life Technologies. Peroxidase from horseradish (HRP) (Cat. No. P-8250), ATP (disodium salt, Cat. No. A-5394), and deuterium oxide (D_2O, Cat. No. D-4501) were from Sigma Chemical Company, Ltd. Biotinyl-ε-aminocaproic acid N-hydroxysuccinimide ester (biotin-X-NHS, Cat. No. 203188) was from Calbiochem. Avidin (egg white, Cat. No. A-887) was from Molecular Probes. Creatine phosphate (Cat. No. 621714), creatine phosphokinase (Cat. No. 127566), and hexokinase [$(NH_4)_2SO_4$ precipitate of yeast hexokinase, 1400 U/ml, Cat. No. 1426362] were from Boehringer-Mannheim GmbH. Protein A–Sepharose beads (Cat. No. CL-4B) were from Pharmacia. Antiavidin antibodies were generated by injecting purified avidin into rabbits, and were affinity purified prior to use. Antiavidin antibodies are also commercially available from several companies. BCA protein assay reagents (Cat. No. 23223) were from Pierce, and Bio-Rad protein assay reagents (Cat. No. 500-0006) were from Bio-Rad Laboratories GmbH.

III. PROCEDURES

A. Internalization of Endocytic Markers into Early Endosomes (EE) from BHK Cells

Solutions

1. *Internalization media (IM):* MEM containing 10 mM HEPES and 5 mM D-glucose pH 7.4. Filter sterilize and store at 4°C.

2. *Phosphate-buffered saline (PBS):* 137 mM NaCl, 2.7 mM KCL, 1.5 mM KH$_2$PO$_4$, and 6.5 mM Na$_2$HPO$_4$; should be pH 7.4. Filter sterilize and store at 4°C.

3. *Biotinylated horseradish peroxidase (bHRP):* Dissolve 20 mg of HRP in 9.5 ml of 0.1 M NaHCO$_3$/Na$_2$CO$_3$ pH 9.0 buffer (make fresh and check pH carefully) in a small glass Erlenmeyer flask. Dissolve 20 mg of biotin-X-NHS in 0.5 ml dimethylformamide. Mix by adding the biotin dropwise to the HRP mixture while gently stirring or shaking the Erlenmeyer and incubate at room temperature with gentle stirring for at least 45 min (a 50:1 molar excess of biotin is important). Quench unreacted active groups with 1 ml of 0.2 M glycine, pH 8.0 (use KOH to pH), by adding dropwise while mixing, and mix for an additional 15 min at room temperature. Transfer to 4°C. Dialyze the mixture extensively against PBS− or IM at 4°C (at least four changes of 200 ml each time). The final dialysis should be in IM. Measure protein concentration (should be about 2 mg/ml) and HRP enzymatic activity (should be unchanged). Aliquot in sterile tubes, freeze in liquid N$_2$, and store at −20°C until use. Immediately before use, thaw quickly and warm to 37°C.

4. *Avidin:* Avidin powder dissolved in IM at 3 mg/ml. Make fresh immediately before use and warm to 37°C.

5. *PBS/BSA:* 5 mg/ml BSA in PBS−. Make fresh before use and cool to 4°C.

Steps

1. *Cell culture:* Maintain monolayers of baby hamster kidney (BHK-21) cells as described in Gruenberg *et al.* (1989). For a fusion assay of 5–10 points, eight petri dishes (10 cm diameter) should be prepared 16 hr before the experiment: four for preparing bHRP-labeled EEs and four for preparing avidin-labeled EEs.

2. *Fluid-phase internalization:* Wash each 10-cm dish of cells 2× with 5 ml ice-cold PBS− on ice. This and other washes on ice to follow are most easily performed by placing four dishes onto a metal plate in a large ice bucket on a rocker. After the last wash, remove PBS− and place the dish on a metal plate in a 37°C water bath. Add at least 3 ml/dish bHRP or avidin solution, prewarmed to 37°C. Incubate for 5 min.

3. *Washes:* From now on, all work should be done at 4°C or on ice. Return the dishes to the metal plate in the ice bucket. Remove the avidin or bHRP solution, and wash dishes 3× for 5 min with 5 ml ice-cold PBS/BSA followed by 2× 5 min with 5 ml ice-cold PBS.

4. *Homogenization and fractionation:* Go directly to Procedure C of this protocol.

B. Internalization of Endocytic Markers into Endosomal Carrier Vesicles (ECV) and Late Endosomes (LE)

Solutions

1. *Nocodazole stock:* 10 mM in dimethyl sulfoxide (DMSO), aliquoted and stored at −20°C.

2. *IM/BSA:* IM containing 2 mg/ml BSA. Make fresh before use and warm to 37°C.

3. All solutions for Procedure A.

Steps

1. *Cell culture:* For a fusion assay of 5–10 points, 10 dishes (10 cm) of BHK cells should be prepared as described in Procedure A, step 1. For ECV–LE fusion assays, use 5 dishes

for bHRP-labeled ECVs and 5 dishes for avidin-labeled LEs. For LE–LE fusion assays, use 5 dishes for bHRP-labeled LEs and 5 dishes for avidin-labeled LEs.

2. *Nocodazole pretreatment for ECV preparation:* Intact microtubules are required for the delivery of endocytosed markers to the LE. Therefore, markers accumulate in transport intermediates (ECVs) in the absence of microtubules. Whereas stable microtubules are cold sensitive, dynamic microtubules are easily depolymerized in the presence of nocodazole (Aniento *et al.*, 1993; Bomsel *et al.*, 1990). In BHK cells, microtubules can be efficiently depolymerized in the presence of nocodazole, whereas cold treatment is without effect. For ECV preparation, depolymerize the microtubules immediately before the experiment with 10 μM nocodazole at 37°C for 1–2 hr in media used to grow cells in a 5% CO_2 incubator. Following this step, nocodazole (10 μM) should remain present in all solutions up to the homogenization step. For LE preparation, do not treat with nocodazole or include nocodazole in any solutions.

3. *Fluid-phase internalization:* Wash each 10-cm dish of cells 2× with 5 ml ice-cold PBS+/− 10 μM nocodazole on ice, as in Procedure A, step 2. After the last wash, remove the PBS− and place the dish on a metal plate in a 37°C water bath. Add at least 3 ml bHRP or avidin solution for making LEs or bHRP + 10 μM nocodazole for making ECVs. Incubate for 10 min.

4. *Chase:* Remove bHRP or avidin, and wash 2× quickly at 37°C with 10 ml PBS/BSA +/− 10 μM nocodazole, prewarmed to 37°C. Remove last wash, and add 8 ml IM/BSA+/− 10 μM nocodazole, prewarmed to 37°C. Incubate at 37°C (in water bath or in a 37°C incubator without CO_2) for 45 min.

5. *Washes:* Remove IM/BSA, move dishes to ice bucket, and wash 2× 5 min with 5 ml cold PBS/BSA followed by 1× 5 min with 5 ml cold PBS on ice.

C. Homogenization and Fractionation of Cells

Solutions

1. *PBS−:* See Procedure A.

2. *300 mM imidazole stock:* Dissolve imidazole in H_2O and adjust pH to 7.4 with NaOH. Filter sterilize and store at 4°C.

3. *Homogenization buffer (HB):* Add imidazole from 300 mM stock to H_2O, and dissolve sucrose such that the final concentrations are 250 mM sucrose and 3 mM imidazole. Filter sterilize and store at 4°C.

4. *62% sucrose solution:* For 100 ml, add 1 ml of imidazole from 300 mM stock to 15 ml H_2O. Add 80.4 g sucrose and dissolve by stirring at 37°C. Add H_2O and mix until the refractive index is 1.4464.

5. *10 and 16% sucrose solutions in D_2O:* For 100 ml, add 1 ml imidazole from 300 mM stock to 50 ml D_2O. For 10% solution, add 10.4 g sucrose, and for 16% solution add 17.0 g sucrose. Dissolve sucrose, add D_2O, and mix until the refractive index is 1.3479 for the 10% solution and 1.3573 for the 16% solution.

Steps

1. *Cell scraping:* All of the following steps should be performed on ice or at 4°C. After the last wash, remove all PBS−. Add 2 ml/dish PBS−, and rock the dish so that cells do not dry. Using a flexible rubber policeman, scrape round 10-cm dishes by first scraping in a circular motion around the outside of the dish, followed by a downward motion in the middle of the dish. Scrape gently in order to obtain "sheets" of cells. Using a plastic Pasteur pipette, gently transfer the scraped "sheets" of cells from four or five dishes into a 15-ml tube on ice.

2. Centrifuge at 1200 rpm for 5 min at 4°C. Gently remove supernatant.

3. Add 1 ml HB to pellet, using a plastic Pasteur pipette, gently pipette up and down one time and add an excess of HB (4–5 ml) to change buffer. Centrifuge again at 2500 rpm for 10 min at 4°C. Remove supernatant.

4. *Homogenization:* It is important that cells are homogenized under conditions where endosomes are released from cells, yet where latency is high so that the endosomes are not broken and retain their internalized marker. First add 0.5 ml HB to the cell pellet. Using a 1-ml pipetman, gently pipette up and down until the pellet is resuspended, and particles can no longer be seen by eye. Do not introduce air bubbles. Using a 22-gauge needle connected to a narrow 1-ml Tubercutine syringe, prewet the needle and syringe with HB so that no air is introduced. Insert the needle into the cell homogenate and slowly pull up on the syringe until most of the cell homogenate is in the syringe, and gently expel without bubbles. Repeat this procedure until plasma membranes are broken, yet nuclear membranes are not. Monitor homogenization as follows. Take 3 μl of homogenate and place in a 50-μl drop of HB on a glass slide. Mix and cover with a glass coverslip. Observe by phase-contrast microscopy, using a 20× objective. Homogenize until unbroken cells, are no longer observed, yet nuclei, which appear as dark round or oblong structures, are not broken. Usually between 3 and 10 up-and-down strokes through the needle are necessary. Centrifuge homogenate at 2000 rpm for 10 min at 4°C, and carefully collect the postnuclear supernatant (PNS) and nuclear pellet.

5. Save a 50-μl aliquot of each PNS fraction for measuring latency and for calculating the balance sheet as described in Procedure D. Adjust the sucrose concentration of the remaining PNS to 40.6% by adding about 1.1 vol of 62% sucrose solution per volume of PNS. Mix gently but thoroughly, without bubbles. Check sucrose concentration using a refractometer.

6. Place adjusted PNS in the bottom of a SW60 centrifuge tube. On top of the PNS layer 1.5 ml of 16% sucrose solution in D_2O, followed by 1 ml of 10% sucrose solution in D_2O, and fill tube with HB. Steps should be layered so that interfaces are clearly seen and not disturbed. See Gruenberg and Gorvel (1992) for diagram of gradients.

7. Centrifuge gradients in SW60 rotor at 35K rpm for 1 hr at 4°C.

8. Carefully remove the interfaces from the gradients after centrifugation by first placing gradients in a test tube rack with a black backdrop. The interfaces should appear white. The layer of white lipids on top of the gradient should be carefully removed. Collect fractions at 4°C using a peristaltic pump at speed 2, with capillary tubes connected to each end. Place the outgoing end into a collection tube and carefully collect the top interface (10%/HB interface = LE + ECV fraction) first. Collect by holding the capillary tube directly in the middle of the wide interface, and slowly move in a circular motion until most of the white interface is collected into the smallest possible volume. Wash the pump tubing with water and then collect the EE (16/10%) interface into another tube. Fractions can be frozen and stored in liquid N_2 until use in fusion assays if they are carefully frozen quickly in liquid N_2 and thawed quickly at 37°C, immediately before use.

D. Measurement of Latency and Balance Sheet for Gradients

Solutions

1. *HRP stocks:* 1–10 ng HRP in 0.1 ml HB, for standards.

2. *HB:* See Procedure C.

3. *HRP reagent:* 0.342 mM *o*-dianisidine and 0.003% H_2O_2 in 0.05 *M* Na-phosphate buffer, pH 5.0, containing 0.3% Triton X-100. To prepare, use very clean glassware or plasticware (as in for tissue culture) and mix 12 ml of 0.5 *M* Na-phosphate buffer, pH 5.0 (filter sterilized), and 6 ml of 2% Triton X-100 (filter sterilized) with 111 ml sterile H_2O. Add 13 mg *o*-dianisidine, dissolve gently, and add 1.2 ml 0.3% H_2O_2 (filter sterilized). Avoid magnetic stirring. Solution should be clear. Store at 4°C in the dark.

4. *1 mM KCN in H_2O.*

5. Protein assay system (such as the BCA protein assay reagent, or the Bio-Rad protein assay system).

Steps

1. Load a 20-μl aliquot of bHRP PNS into an airfuge tube or a small table-top ultracentrifuge tube of the Beckman TL-100 type and fill the tube with a known volme of HB. Mix thoroughly by pipetting without air bubbles. Centrifuge at 4°C for 20 min at 20 psi in an airfuge or at 200,000 g for 20 min in a table-top ultracentrifuge rotor (such as Beckman TLA-100.1). Transfer the supernatant to another tube. Resuspend the pellet in 50 μl HB.

2. To measure the latency, adjust samples, blanks, and standards with HB so that the final volume of each is 0.1 ml. Assay both the pellet and the supernatant of the latency measurement. If the supernatant volume is over 0.1 ml, assay only 0.1 ml. Add 0.9 ml of HRP reagent to each tube, mix quickly, and record the time with a stop clock. Allow color to develop in the dark, as this reagent is light sensitive. When a brown color has begun to develop, read the absorbance at 455 nm and record the time (results expressed as OD units/min or ng HRP/min). Stop the reaction with 10 μl of 1.0 m*M* KCN if necessary.

3. Calculate latency by first adding the value (OD/min) for HRP in the pellet to that of HRP in the supernatant (OD/min after correcting for total supernatant volume). The value for the pellet divided by the total value is the percentage latency. Latency should be over 70% in order to measure endosome fusion.

4. The amount of HRP in each gradient fraction collected from the bHRP gradient can be measured by assaying an aliquot (about 50 μl) of each fraction as described in step 2.

5. Measure the amount of protein in each gradient fraction using a standard protein assay system, as described in the manual.

6. Calculate percentage yield (percentage of HRP in each fraction compared to total amount of HRP in PNS), specific activity (SA) (HRP activity per unit protein), and relative specific activity (RSA) (divide specific activity of each fraction by the specific activity of the PNS). See Gruenberg and Gorvel (1992) for an example of a typical balance sheet.

E. Preparation of BHK Cell Cytosol

Solutions

1. *PBS−:* See Procedure A.

2. *HB:* See Procedure C.

3. *HB + protease inhibitors:* HB with the following protease inhibitors added immediately before use: 10 μ*M* leupeptin, 1 μ*M* pepstatin A, 10 ng/ml aprotinin, and, if needed, 1 μ*M* PMSF.

Steps

Two possible cytosol sources for all of the assays described are BHK and rat liver cytosol. For rat liver cytosol preparation, refer to Aniento *et al.* (1993).

1. BHK cells, maintained as described in Procedure A, should be plated approximately 16 hrs before the experiment. Large (245 × 245 × 25 mm) square dishes are convenient for large cytosol preparations.

2. All steps should be performed on ice or at 4°C. Wash dishes 4× with excess PBS− (50 ml per dish for large square dishes).

3. Remove PBS− from the last wash, add 12 ml PBS− per dish, and rock the dish so that cells do not dry. Scrape cells with a rubber policeman using firm, downward motions, going from top to bottom while holding the plate at an angle, as described in Procedure C, step 1.

4. Collect scraped cells into 15-ml tubes (1 tube per dish). Centrifuge at 1200 rpm for 5 min at 4°C.

5. Remove supernatant and gently add 5 ml HB with a plastic Pasteur pipette and pipette up and down one time.

6. Centrifuge at 2500 rpm for 10 min at 4°C. Remove supernatant and resuspend pellet in 1.2 ml HB + protease inhibitors. Separate into two tubes (about 0.7 ml/tube) for homogenization, and homogenize as described in step 4 of Procedure C.

7. Centrifuge at 2500 rpm for 15 min at 4°C. Add supernatant (PNS) to a centrifuge tube for the TLS-55 rotor (for the Beckman TL-100 table-top ultracentrifuge) and centrifuge in TLS-55 for 45 min at 55K rpm at 4°C. Remove fat from the top using an aspirator. Transfer supernatant (cytosol fraction) to a new tube without disturbing the pellet. Determine the protein concentration of supernatant. Cytosol should be at least 15 mg/ml to give a good signal for fusion assays. Aliquot on ice, freeze quickly and store in liquid N_2 until use.

F. Preparation of Antiavidin Beads for the *in Vitro* Fusion Assay Described in Procedure G

Solutions

1. *PBS/BSA:* Dissolve 5 mg/ml BSA in PBS. Filter sterilize and store at 4°C.

2. *Sterile PBS:* PBS as described in Procedure A, filter sterilize or autoclave, and store at 4°C.

3. *Antiavidin antibody:* Affinity purify and store aliquoted in 50% glycerol/PBS at −20°C.

Steps

To determine how many antiavidin beads to prepare, first determine the number of fusion assay points. From a typical gradient (see Gruenberg and Gorvel, 1992) about 150 μg of EE and 70 μg of ECV or LE are obtained. Optimal amounts of endosomes to use for fusion assays are 20 μg of each EE fraction and 10 μg of each ECV or LE fraction. Therefore, a typical experiment (one gradient each of avidin and bHRP-labeled fractions) will provide enough endosomes for about seven fusion assay points.

1. Swell 1.5 g of protein A–Sepharose beads in 10 ml sterile H_2O at room temperature overnight.

2. Wash beads 3× in 10 ml sterile PBS by centrifuging beads in 15-ml tubes at 3000 rpm for 2 min, resuspending in PBS each time.

3. After the final wash, resuspend beads in an equal volume of sterile PBS per volume of packed beads. Store beads this way up to several months at 4°C.

4. One hundred microliters of this 1:1 slurry is required per fusion assay point. Therefore, for 10 assay points, block 1 ml of beads by washing 3× in 10 ml PBS/BSA, as described in step 2.

5. After final wash, resuspend beads in 10 ml PBS/BSA. For 10 assay points, add 50 μg of antiavidin antibody (5 μg per 100-μl beads). Rotate tube for at least 5 hr at 4°C.

6. Wash beads 4× in PBS/BSA. After the last wash, for 10 assay points, resuspend beads in 10 ml PBS/BSA.

7. Aliquot 1 ml to each of 10 labeled Eppendorf tubes. Centrifuge in Eppendorf centrifuge at maximum speed for 2 min. Remove supernatant. Beads are now ready for the immunoprecipitation step of the fusion assay (Procedure G, step 10).

G. *In Vitro* Assay of Endocytic Vesicle Fusion

Solutions

1. *50× salts:* 0.625 M HEPES, 75 mM Mg-acetate, 50 mM dithiothreitol, pH 7, with KOH. Filter sterilize, aliquot, and store at −20°C.

2. *K-acetate (KOAc stock):* 1 M in H_2O. Filter sterilize, aliquot, and store at −20°C. Note: Depending on the counterion requirement of the experiment, KOAc must be replaced by KCl (see Aniento *et al.*, 1993).

3. *Biotinylated insulin:* 1 mg/ml in H_2O. Store at 4°C.

4. *ATP-regenerating system (ATP-RS):* mix 1:1:1 volumes of the following immediately before use:

    a. *100 mM ATP:* dissolve in ice-cold H_2O, titrate to pH 7.0 with 1 *M* NaOH, filter sterilize, aliquot on ice, and store at −20°C.

    b. *800 mM creatine phosphate:* dissolve in ice-cold H_2O, filter sterilize, aliquot on ice, and store at −20°C.

    c. *4 mg/ml creatine phosphokinase:* to make 4 ml, add 80 μl of 0.5 *M* $NaHPO_4$ buffer, pH 7.0, to 1.6 ml H_2O on ice. When cool, add 16 mg creatine phosphokinase. Vortex until dissolved. Add 2.3 ml ice-cold 87% glycerol. Vortex until well mixed. Aliquot on ice and store at −20°C.

5. *Hexokinase:* Vortex the suspension, pipette the desired amount (e.g., 10 μl = 0.1 mg for one assay point), centrifuge for 2 min in Eppendorf at maximum speed, and aspirate supernatant. Dissolve pellet in the same volume of 0.25 *M* D-glucose. Prepare immediately before use.

6. *Tx100 stock:* 10% stock of Triton X-100 in H_2O.

7. *PBS/BSA and sterile PBS:* See Procedure F.

8. *HRP reagent:* See Procedure D.

9. *PBS/BSA/Tx100:* PBS/BSA containing 0.2% Triton X-100, make immediately before use.

Steps

1. For each fusion assay, at least three points should be included: −ATP, +ATP, and the total. To determine fusion efficiency, determine the total (maximal possible fusion value) by mixing 50 μl of each endosomal fraction in an Eppendorf tube on ice. Add 25 μl Tx100 stock and vortex well. Leave on ice at least 30 min, and add PBS/BSA and continue as described in step 9, below.

2. For all other fusion assay points, 3 μl of 50X salts, 8 μl of biotinylated insulin, and 11 μl of KOAc stock are needed for each point. Make a mixture of these three components by multiplying the number of assay points by 3, 8, and 11, and mix the respective amounts of each component together in one tube. Number Eppendorf tubes for the appropriate number of assay points and put them on ice. Add 22 μl of the above mixture to each tube.

3. Add 50 μl (750 μg–1 mg) of cytosol to each tube and mix.

4. Add either 5 μl of ATP-RS or 10 μl of hexokinase to each tube, as appropriate.

5. Add 50 μl (7–25 μg) of bHRP-labeled endosomes and 50 μl (7–25 μg) of avidin-labeled endosomes to each tube. Endosomal fractions from the gradients can be diluted in IB prior to this step, if desired. Mix gently; avoid introducing air bubbles. Leave tubes on ice for 3 min.

6. Transfer tubes to 37°C for 45 min. Avoid agitation during this time.

7. Return tubes to ice. Add 5 μl of biotinylated insulin to each tube and mix.

8. Add 25 μl of Tx100 stock and vortex well. Leave tubes on ice for 30 min.

9. Add 1 ml of sterile PBS/BSA to each tube and mix well.

10. Centrifuge for 2 min at maximum speed in an Eppendorf centrifuge. Transfer supernatants to numbered tubes containing antiavidin beads, prepared as in Procedure F.

11. Rotate beads for at least 5 hr at 4°C.

12. Centrifuge in Eppendorf centrifuge at maximum speed for 2 min. Remove supernatant and wash 4× with PBS/BSA/Tx100. Wash 1× with sterile PBS.

13. Remove final supernatant and add 900 μl of HRP reagent to each tube. Allow color to develop in the dark at room temperature. Vortex periodically for 2–3 hr or put tubes on rotating wheel in the dark at room temperature while color develops.

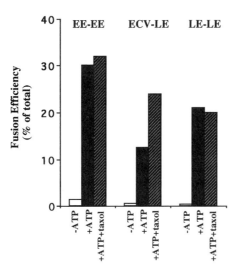

FIGURE 2 Typical fusion assay results. Fusion efficiency is expressed as a percentage of total fusion between each set of endosomal membranes. Total, or maximal, endosome fusion is measured by mixing bHRP and avidin-containing endosomal fractions together in the presence of detergent, followed by immunoprecipitation with antiavidin antibodies and HRP determination. Typical "total" values (measured as absorbance at 455 nm) are in the range of 0.6–1.0 A_{455} units for EE fusion assays and 0.3–0.7 A_{455} units for ECV and LE fusion assays. As a control for nonspecific reactions, the assay is typically carried out without ATP (−ATP). As shown, the polymerization of endogenous tubulin present in the cytosol in the presence of taxol is sufficient to facilitate interactions between ECVs and LE (see Aniento *et al.*, 1993).

14. Centrifuge tubes for 2 min in Eppendorf centrifuge. Measure the absorbance of the supernatants at 455 nm.

IV. COMMENTS

Refer to Gruenberg and Gorvel (1992) for an example of a typical balance sheet for the sucrose gradient fractionation step. Typical results for fusion assays are shown in Fig. 2.

Highly purified loaded endosomes can be prepared by immunoisolation as described in Howell *et al.* (1989). Immunoisolated endosomes can then be used in the fusion assay described in Procedure G. See Gruenberg and Gorvel (1992) and Howell *et al.* (1989) for details.

ECV–LE fusion is stimulated by the addition of polymerized microtubules to the fusion assay. Endogenous microtubules can be polymerized by adding 20 μM taxol to the fusion assay. The preparation of microtubules is described in Volume 2, Ashford *et al.*, "Preparation of Tubulin from Bovine Brain." See Fig. 2 and Aniento *et al.* (1993) for more details on the effects of microtubules and MAPs on ECV–LE fusion.

V. PITFALLS

For ECV and LE preparations, cells should be homogenized until vesicles are no longer seen around the periphery of the nuclei. If nuclei begin to aggregate during homogenization, however, this is a sign that some are broken as free DNA causes aggregation. Freezing and thawing of endosomes may cause a partial loss in latency. Use very clean plasticware or glassware for all fusion assay manipulations as HRP contamination can easily occur. Nocodazole, *o*-dianisidine, and KCN are very toxic.

References

Aniento, F., Emans, N., Griffiths, G., and Gruenberg, J. (1993). Cytoplasmic dynein-dependent vesicular transport from early to late endosomes. *J. Cell Biol.* **123**, 1373–1387.

Aniento, F., Gu, F., Parton, R. G., and Gruenberg, J. (1996). An endosomal βCOP is implicated in the pH-dependent formation of transport vesicles destined for late endosomes. *J. Cell Biol.* **133**, 29–41.

Aniento, F., Roche, E., Cuervo, A., and Knecht, E. (1993). Uptake and degradation of glyceraldehyde-3-phosphate dehydrogenase by rat liver lysosomes. *J. Biol. Chem.* **268**, 10463–10470.

Bomsel, M., Parton, R., Kuznetsov, S. A., Schroer, T. A., and Gruenberg J. (1990). Microtubule- and motor-dependent fusion *in vitro* between apical and basolateral endocytic vesicles from MDCK cells. *Cell* **62**, 719–731.

Braell, W. A. (1987). Fusion between endocytic vesicles in a cell-free system. *Proc. Natl. Acad. Sci. U.S.A.* **84**, 1137–1141.

Davey, J. S., Hurtley, S. M., and Warren, G. (1985). Reconstitution of an endocytic fusion event in a cell-free system. *Cell* **43**, 643–652.

Diaz, R., Mayorga, L., and Stahl, P. D. (1988). *In vitro* fusion of endosomes following receptor-mediated endocytosis. *J. Biol. Chem.* **263**, 6093–6100.

Gruenberg, J., and Gorvel, J.-P. (1992). *In vitro* reconstitution of endocytic vesicle fusion. *In* "Protein Targetting, a Practical Approach" (A. I. Magee, and T. Wileman, eds.), pp. 187–216. Univ. Press, Oxford.

Gruenberg, J., Griffiths, G., and Howell, K. E. (1989). Characterization of the early endosome and putative endocytic carrier vesicles *in vivo* with an assay of vesicle fusion *in vitro*. *J. Cell Biol.* **108**, 1301–1316.

Gruenberg, J., and Howell, K. E. (1986). Reconstitution of vesicle fusions occurring in endocytosis with a cell-free system. *EMBO J.* **5**, 3091–3101.

Gruenberg, J., and Maxfield, F. R. (1995). Membrane transport in the endocytic pathway. *Curr. Opin. Cell Biol.* **7**, 552–563.

Howell, K. E., Schmid, R., Ugelstad, J., and Gruenberg J. (1989). Immuno-isolation using magnetic solid supports: Subcellular fractionation for cell-free functional studies. *Meth. Cell Biol.* **31A**, 264–292.

Woodman, P. G., and Warren, G. (1988). Fusion between vesicles from the pathway of receptor-mediated endocytosis in a cell-free system. *Eur. J. Biochem.* **173**, 101–108.

Microsome-Based Assay of Endoplasmic Reticulum to Golgi Transport in Mammalian Cells

Tony Rowe and William E. Balch

I. INTRODUCTION

The trafficking of proteins along the first stage of the secretory pathway is mediated by small vesicles that bud from the endoplasmic reticulum (ER) and subsequently fuse with the *cis*-Golgi compartment. This article describes a new biochemical assay using mammalian microsomes that can be used to measure these events independently. The microsomes are prepared from cells infected at the restrictive temperature (39.5°C) with the ts045 strain of vesicular stomatitis virus (VSV) (Lafay, 1974). As a reporter molecule the assay utilizes ts045 VSV-glycoprotein (VSV-G), which is retained in the ER during infection due to a thermoreversible folding defect; incubation *in vitro* at the permissive temperature (32°C) results in the synchronous folding and transport of VSV-G to the Golgi complex. To follow vesicle formation, a differential centrifugation procedure is employed to separate the more rapidly sedimenting ER and Golgi membranes from the slowly sedimenting vesicles. Consumption is analyzed using a two-stage assay in which vesicles isolated by differential centrifugation during stage 1 are subsequently added to stage 2 (fusion) reactions containing acceptor Golgi membranes. Transport to the Golgi is measured by following the oligosaccharide processing of VSV-G from the high mannose ER form, which is sensitive to endoglycosidase H (endo H), to the *cis/medial*-Golgi form, which is endo H resistant (Schwaninger *et al.*, 1992). The biochemical characteristics of the overall ER to Golgi transport reaction and the vesicle formation and consumption assays are described elsewhere (Rowe *et al.*, 1996).

II. MATERIALS AND INSTRUMENTATION

Culture medium (α-MEM; Cat. No. 11900-099) is from Life Technologies. The medium is supplemented with penicillin/streptomycin from a 100× stock solution (Cat. No. P0781 Sigma). Fetal bovine serum (FBS; Cat. No. FB-01) is from Omega Scientific. D-Sorbitol (Cat. No. S-1876), leupeptin (Cat. No. L-2884), chymostatin (Cat. No. C-7268), pepstatin A (Cat. No. P-4265), phenylmethylsulfonyl fluoride (PMSF; Cat. No. P-7626), actinomycin D (Cat. No. A-1410), uridine 5'-diphospho-*N*-acetylglucosamine (UDP-GlcNAc; Cat. No. U-4375), and dimethyl sulfoxide (DMSO; Cat. No. D-2650) are from Sigma. The nitrocellulose membrane (Cat. No. 68260) is from Schleicher & Schuell. Horseradish peroxidase-conjugated goat anti-rabbit IgG (Cat. No. 31460) is from Pierce. Chemiluminescence reagent (Cat . No

NEL-101) and autoradiography film ("Reflection") are from NEN. Polyallomer microfuge tubes (Cat. No. 357448) are supplied by Beckman Instruments Inc. A polyclonal antibody to VSV-G was generated in rabbits immunized with the C-terminal 16 amino acids of VSV-G (Indiana serotype) coupled to KLH (Plutner *et al.*, 1991). Centrifugation at 20,000 or 100,000 g was performed using an Optima TL ultracentrifuge (Beckman) equipped with a TLA 100.3 rotor. A laser scanning densitometer (personal densitometer; Molecular Dynamics) was used to quantitate data.

III. PROCEDURES

The following procedures are performed on ice unless otherwise stated.

A. Preparation of Cytosol

The procedure described below for the preparation of rat liver cytosol is based on that described by Davidson *et al.* (1992).

Solutions

1. *Phosphate-buffered saline (PBS) (10× stock):* 90 mM phosphate and 1.5 M NaCl (pH 7.4). To make 1 liter, add 80 g NaCl, 2 g KCl, 2 g KH_2PO_4, and 21.6 g $Na_2HPO_4 \cdot 7H_2O$ to distilled water. Store at room temperature.

2. *25/125:* 0.125 M KOAc, 25 mM HEPES (pH 7.4). To make 100 ml, add 3.125 ml of 4 M KOAc stock and 2.5 ml of 1 M HEPES–KOH (pH 7.4) stock to distilled water. Store at 4°C.

3. *Protease inhibitor cocktail (PIC):* 10 μg/ml leupeptin, 10 μg/ml chymostatin, 0.5 μg/ml pepstatin A, and 0.5 mM PMSF. To supplement 50 ml of buffer (e.g., 25/125) with PIC, add 50 μl of 10 mg/ml leupeptin stock (in H_2O), 50 μl of 10 mg/ml chymostatin stock (in DMSO), 5 μl of 5 mg/ml pepstatin A stock (in DMSO), and 250 μl of 0.1 M PMSF stock (in ethanol). Use PIC buffer immediately or store at −20°C.

Steps

1. Decapitate two anethetized adult Sprague–Dawley rats (~250 g), remove the livers, and place the tissue in a 250-ml glass beaker. Determine the weight of the tissue (typically ~20 g), then wash three times with ~50 ml of 1× PBS and twice with ~50 ml of 25/125.

2. Finely mince the tissue using a pair of scissors and then homogenize in 3 vol (ml per gram of tissue) of 25/125 (supplemented with 1 mM ATP and PIC) with 20 strokes using a 40-ml Dounce (Wheaton). Use a "loose-fitting" Dounce for the first 10 strokes followed by a "tight-fitting" one for the second 10 strokes.

3. Pour the homogenate into a 38-ml polycarbonate tube (Nalgene; Cat. No. 3117-0380) and centrifuge for 10 min at 12,000 g (10,000 rpm) in a Beckman JA20 rotor (Beckman Instruments Inc.). Using a pipette, transfer ~12 ml of the supernatant into each of two 14 × 89-mm Ultra clear centrifuge tubes (Beckman; Cat. No. 344059) and centrifuge at 150,000 g (35,000 rpm) for 90 min in a Beckman SW41 rotor.

4. After centrifugation, remove the overlying lipid layer by aspiration and then withdraw the remaining supernatants (cytosol) from each tube using a pipette.

5. Divide the cytosol into 250-μl aliquots in 0.5-ml microfuge tubes, freeze in liquid N_2, and store at −80°C. The protein concentration of the cytosol is ~25 mg/ml.

B. Preparation of Microsomes

Solutions

1. *Actinomycin D (200× stock):* Add 10 mg actinomycin D to 10 ml of ethanol. Store at −20°C.

2. *Homogenization buffer:* 0.375 M sorbitol and 20 mM HEPES (pH 7.4). To make 500 ml, add 34.2 g sorbitol and 10 ml of 1 M HEPES–KOH (pH 7.4) stock to distilled water. Store at 4°C.

3. *0.21 M KOAc buffer:* 0.21 M KOAc, 3 mM Mg(OAc)$_2$, and 20 mM HEPES (pH 7.4). To make 100 ml, add 5.25 ml of 4 M KOAc stock, 0.3 ml of 1 M Mg(OAc)$_2$ stock, and 2 ml of 1 M HEPES–KOH (pH 7.4) stock to distilled water. Store at 4°C.

4. *Transport buffer:* 0.25 M sorbitol, 70 mM KOAc, 1 mM Mg(OAc)$_2$, and 20 mM HEPES (pH 7.4). To make 100 ml, add 10 ml of 2.5 M sorbitol stock, 1.75 ml of 4 M KOAc stock, 0.1 ml of 1 M Mg(OAc)$_2$ stock, and 2 ml of 1 M HEPES–KOH (pH 7.4) stock to distilled water. Store at −20°C.

Steps

1. Prepare vesicular stomatitis virus (VSV; Indiana serotype) strain ts045 according to Schwaninger *et al.* (1992). Store the virus in 1-ml aliquots in screw-capped tubes at −80°C.

2. Grow normal rat kidney (NRK) cells on 150-mm tissue culture dishes (Cat. No. 3025; Falcon) in α-MEM medium supplemented with 5% FBS at 37°C and 5% CO$_2$. At confluency infect the cells with a 5-ml cocktail (per dish) containing 0.5 ml of ts045 VSV (thawed at 32°C) and 25 μg of actinomycin D in serum-free α-MEM as described by Schwaninger *et al.* (1992). Rock the dishes by hand at 5-min intervals to ensure an even spread of the infection cocktail. After infection, add 20 ml of α-MEM medium supplemented with 5% FBS to each dish and incubate in the presence of 5% CO$_2$ for 3 hr and 40 min to 4 hr at the restrictive temperature (39.5°C) (see Comment 1). The method detailed below is based on a typical 12-dish microsome preparation.

3. Following incubation at 39.5°C, transfer each dish to ice, immediately aspirate the medium, and add 12 ml of ice-cold 1× PBS to cool the cells as quickly as possible.

4. Remove the PBS by aspiration, add 5 ml of homogenization buffer, and scrape the cells from the dishes using a rubber policeman. Use a pipette to transfer the cells to 50-ml plastic tubes (Cat. No. 25325-50: Corning Inc.), then repeat the scraping procedure to ensure that all the cells are collected. Centrifuge at 720 g for 3 min and remove the supernatant by aspiration.

5. Resuspend each cell pellet (from four dishes) in 0.9 ml of homogenization buffer supplemented with PIC and homogenize by three complete passes (three downward strokes with both plungers) through a 1-ml ball-bearing homogenizer (Balch and Rothman, 1985).

6. Combine the cell homogenates and dilute with an equal volume (~3 ml) of homogenization buffer + PIC. Divide the diluted homogenate into six 1.0-ml aliquots in 1.5-ml microfuge tubes and centrifuge at 720 g for 5 min.

7. Carefully remove the postnuclear supernatant (PNS) fractions and combine in a plastic 15-ml tube (Cat. No. 25319-15: Corning Inc.). Add 0.5 vol (~2.5 ml) of 0.21 M KOAc buffer to the PNS and mix. Divide the mixture into 0.8- to 1.0-ml aliquots in 1.5-ml microfuge tubes and centrifuge at 12,000 g (12,200 rpm) for 2 min in an Eppendorf Model 5402 refrigerated microfuge using the soft spin function.

8. Remove the supernatants by aspiration and resuspend the pellets (including any membranes on the sides of the tubes) using a P1000 Gilson tip in a total volume of 1.0 ml of transport buffer + PIC. Dispense 0.5-ml aliquots into two 1.5-ml microfuge tubes and recentrifuge the microsomes at 12,000 g for 2 min as described earlier.

9. Resuspend the membrane pellets by repeated trituration using a P1000 Gilson tip in 6–8 vol (~1 ml per 75 mg membranes) of transport buffer + PIC. In a typical 12 dish preparation, 1.0–1.5 ml of resuspended microsomes (at a protein concentration of 3–4 mg/ml) is obtained depending on the starting cell density. Pool the membranes, divide into 50- or 100-μl aliquots in 0.5-ml microfuge tubes, freeze in liquid N$_2$, and store at −80°C. The microsomes can be stored for several months with no loss of transport activity.

C. Preparation of Acceptor Golgi Membranes

The procedure described below for the preparation of Golgi membranes by flotation on a sucrose density gradient is a modification of that originally described by Balch *et al.* (1984).

Solution

1. *87.5 mM KOAc buffer:* 87.5 mM KOAc, 1.25 mM Mg(OAc)$_2$, and 20 mM HEPES (pH 7.4). To make 100 ml, add 2.18 ml of 4 M KOAc stock, 0.125 ml of 1 M Mg(OAc)$_2$ stock, and 2 ml of 1 M HEPES–KOH (pH 7.4) stock to distilled water. Store at 4°C.

Steps

1. Prepare an enriched Golgi membrane fraction from noninfected wild-type Chinese hamster ovary cells by flotation in sucrose density gradients as described by Beckers and Rothman (1992). Recover the membranes at the 29–35% sucrose interface, mix thoroughly with 4 vol of 87.5 mM KOAc buffer, and divide into 0.5- to 1.0-ml aliquots in microfuge tubes. Centrifuge at 16,000 g (soft spin) for 10 min.

2. Remove the supernatants by aspiration, wash the pellets with 2 ml (final volume) of transport buffer, and combine the membranes into two 1.5-ml microfuge tubes. Centrifuge at 16,000 g (soft spin) for 10 min.

3. Resuspend the membranes using a P1000 Gilson tip in transport buffer (total volume of 1.0 ml per 1×10^6 cells). Divide the Golgi membrane fraction into 50- to 100-μl aliquots in 0.5-ml microfuge tubes, freeze in liquid N$_2$, and store at −80°C.

D. Reconstitution of ER to Golgi Transport

Steps

1. Set up 40-μl transport reactions containing the components indicated in Table I in 1.5-ml microfuge tubes (see Comment 2). A reaction cocktail consisting of the salts, ATP-regenerating system, and water is added first, followed by the cytosol and finally the microsomes. Mix by pipetting up and down four times using a P20 Gilson tip.

2. Transfer the reactions to a 32°C waterbath and incubate for 75–90 min.

3. Terminate the reactions on ice, harvest the membranes by centrifugation at 20,000 g (27,000 rpm) in a Beckman TLA 100.3 rotor, and remove the supernatant by aspiration (see Comment 3).

TABLE I Reaction Cocktail for Standard ER to Golgi Transport Assay

Solution	Volume (μl)	Final concentration
Microsomes	5	~0.5 mg/ml protein [31.25 mM sorbitol, 8.75 mM KOAc, 0.125 mM Mg(OAc)$_2$, 2.5 mM HEPES (pH 7.4)]
Cytosol	8	5 mg/ml protein [25 mM KOAc, 5 mM HEPES (pH 7.4)]
20× ATP-regenerating system	2	1 mM ATP, 5 mM creatine phosphate, and 0.2 IU creatine phosphate kinase
40 mM UDP-GlcNAc	1	1 mM
10× Ca^{2+}/EGTA buffer	4	5 mM EGTA, 1.8 mM Ca(Cl)$_2$ (100 nM free Ca^{2+})
0.1 M Mg(OAc)$_2$	1	2.5 mM
1 M KOAc	1	40 mM
2.5 M sorbitol	3.5	218.75 mM
1 M HEPES (pH 7.4)	1.1	27.5 mM
H$_2$O	To 40 μl (final)	

4. Solubilize the membranes in 0.3% SDS, incubate overnight in the presence of 3 mU of endo H, and terminate the reactions by adding Laemmli sample buffer as described by Schwaninger *et al.* (1992).

5. Separate the endo H-sensitive and -resistant forms of VSV-G on 6.75% (w/v) SDS–polyacrylamide gels as described by Schwaninger *et al.* (1992).

6. Transfer the proteins to a nitrocellulose membrane and perform Western blotting using anti-VSV-G polyclonal primary antibody (1:5000) and peroxidase-conjugated anti-rabbit IgG secondary antibody (1:20,000) according to Rowe *et al.* (1996).

7. Develop the blots using enhanced chemiluminescence and expose to autoradiography film. Quantitate the relative band intensities of the endo H-sensitive and -resistant forms of VSV-G in each lane by densitometry (see Comments 4 and 5).

E. Vesicle Formation Assay

Solution

1. *Resuspension buffer:* 0.25 M sucrose and 20 mM HEPES (pH 7.4). To make 100 ml, add 8.55 g of sucrose and 2 ml of 1 M HEPES–KOH (pH 7.4) stock to distilled water. Store at 4°C.

Steps

1. Set up 40-μl reactions as described in Procedure D, step 1, containing 10 μl of microsomes, 10–12 μl of cytosol, salts, and an ATP-regenerating system (see Comment 6).

2. Incubate at 32°C for 0–60 min and harvest the membranes as described in Procedure D, step 3. The membrane pellets can be stored for several hours at this stage prior to the differential centrifugation procedure described below.

3. Add 50 μl of resuspension buffer and disperse the membrane pellets by pipetting up and down 10 times using a P200 Gilson tip (see Comments 3 and 7). Incubate the membranes for 10 min on ice and repeat the trituration procedure to completely resuspend the membranes. Add 10.6 μl of a salt mix [9.1 μl of 1 M KOAc + 1.5 μl of 0.1 M Mg(OAc)$_2$] to the resupended membranes and mix by pipetting up and down 5 times with a P20 Gilson tip. Perform the differential centrifugation step at 16,000 g (14,000 rpm) for 2 min in an Eppendorf Model 5402 refrigerated microfuge using the soft spin function.

4. Using a p200 Gilson tip, carefully take the top 42.4-μl supernatant fraction from the side of the tube opposite the pellet. Transfer to a 1.5-ml Polyallomer microfuge tube and centrifuge at 100,000 g (60,000 rpm) for 20 min. Carefully aspirate the remaining supernatant fraction from the 16,000 g (medium speed) pellet and the entire supernatant from the 100,000 g (high speed) pellet (see Comment 8).

5. Add 50 and 35 μl of 1× Laemmli sample buffer (Laemmli, 1970) to the medium speed pellet (MSP) and high-speed pellet (HSP) fractions, respectively, and boil at 95°C for 5 min. Determine the relative amounts of VSV-G in the MSP and HSP from each reaction by SDS–PAGE and quantitative immunoblotting as described by Rowe *et al.* (1996).

F. Two-Stage Fusion Assay

Solution

1. *0.25 M sorbitol buffer:* 0.25 M sorbitol and 20 mM HEPES (pH 7.4). To make 100 ml, add 10 ml of 2.5 M sorbitol stock and 2 ml of 1 M HEPES–KOH (pH 7.4) stock to distilled water. Store at 4°C.

Steps

1. For stage 1 incubations, prepare scaled-up (100 μl) vesicle formation reactions containing 25 μl of microsomes and 30 μl of cytosol as described in Procedure E, step 1. Incubate for 10 min at 32°C.

2. Terminate the reactions by transfer to ice and sediment the membranes as described in Procedure D, step 3.

3. Resuspend the membranes in 90 μl of resuspension buffer, add 9.8 μl of salt mix [7.3 μl of 2 M KOAc + 2.5 μl of 0.1 M Mg(OAc)$_2$], and perform the differential centrifugation step as described in Procedure E, step 3.

4. Withdraw the top 75-ml medium-speed supernatant fraction and recover the vesicles by centrifugation at high speed as described in Procedure E, step 4.

5. Resuspend the HSPs in 25 μl of 0.25 M sorbitol buffer by pipetting up and down 10 times with a P200 Gilson tip, then add 1.9 μl of 1 M KOAc (70 mM KOAc final concentration) (see Comment 9).

6. Set up 40-μl stage 2 reactions containing 10 μl of resuspended HSP fraction, 4 μl of Golgi membranes, 8 μl of cytosol, and an ATP-regenerating system and salts at the final concentrations indicated in Table I (see Comment 6).

7. Incubate for 60 min at 32°C and terminate the reactions on ice. Harvest the membranes and quantitate the conversion of VSV-G to the endo H-resistant form as described in Procedure D, steps 3–7.

IV. COMMENTS

1. To reproduce the temperature-sensitive phenotype of ts045 VSV-G transport *in vitro* it is necessary to supplement the postinfection medium and homogenization buffer with dithiothreitol as described by Aridor *et al.* (1996).

2. The preparation of the ATP-regenerating system and Ca^{2+}/EGTA buffer are described by Schwaninger *et al.* (1992). Transport inhibitors such as antibodies are dialyzed against 25/125 prior to addition to the assay, and the volumes of the salts in the reaction cocktail are adjusted to achieve the final concentrations described in Table I. The reactions can be preincubated on ice for up to 45 min with no loss of transport activity.

3. All of the membrane-bound VSV-G is sedimented at 20,000 g following the 32°C incubation. In the vesicle formation assay, membranes are resuspended in sucrose buffer prior to differential centrifugation.

4. The detection system is linear over the range of VSV-G concentrations tested.

5. As an alternative to densitometry, VSV-G bands can be detected by direct fluorescence imaging (e.g., using a Bio-Rad GS-363 molecular imaging system).

6. In the vesicle formation and two-stage assays, the volumes of salts (see Table I) added to the reaction cocktail are adjusted to account for the salts present in the cytosol and membrane preparations in the assay. UDP-GlcNAc is omitted from the vesicle formation assay and from stage 1 of the two-stage assay.

7. Although in the original description of the assay (Rowe *et al.*, 1996) a 0.25 M sorbitol buffer was used to resuspend the membranes prior to differential centrifugation, we have subsequently found that resuspension in 0.25 M sucrose buffer gives higher yields of vesicles at this step.

8. Because the HSP is small and translucent, care should be taken to avoid losing it during aspiration of the high-speed supernatant.

9. In a typical experiment, multiple stage 1 reactions are performed and the HSPs are resuspended successively in the desired final volume of sorbitol buffer, then adjusted to 70 mM KOAc.

Acknowledgments

This work was supported by postdoctoral fellowships from The International Human Frontier Science Program Organization and Muscular Dystrophy Association to TR and grants from the National Institutes of Health (GM 42336; CA586689) to WEB. We thank Helen Plutner for technical assistance.

References

Aridor, M., Rowe, T., Bannykh, S. I., and Balch, W. E. (1996). Vesicular stomatitis virus regulates budding from the endoplasmic reticulum. Submitted for publication.

Balch, W. E., Dunphy, W. G., Braell, W. A., and Rothman, W. E. (1984). Reconstitution of the transport of protein between successive compartments of the Golgi measured by the coupled incorporation of N-acetylglucosamine. *Cell* **39**, 405–416.

Balch, W. E., and Rothman, J. E. (1985). Characterization of protein transport between successive compartments of the Golgi apparatus: Asymmetric properties of donor and acceptor activities in a cell-free system. *Arch. Biochem. Biophys.* **240**, 413–425.

Beckers, C. J. M., and Rothman, J. E. (1992). Transport between Golgi cisternae. *In* "Methods in Enzymology" (J. E. Rothman, ed.), Vol. 219, pp. 5–12. Academic Press, San Diego.

Davidson, H. W., McGowan, C. H., and Balch, W. E. (1992). Evidence for the regulation of exocytic transport by protein phosphorylation. *J. Cell Biol.* **116**, 1343–1355.

Laemmli, U. K. (1970). Cleavage of structural proteins during the assembly of the head of bacteriophage T4. *Nature* **227**, 680–685.

Lafay, F. (1974). Envelope viruses of vesicular stomatitis virus: Effect of temperature-sensitive mutations in complementation groups III and V. *J. Virol.* **14**, 1220–1228.

Plutner, H., Cox, A. D., Pind, S., Khosravi-Far, R., Bourne, J. R., Schwaninger, R., Der, C. J., and Balch, W. E. (1991). Rab1b regulates vesicular transport between the endoplasmic reticulum and successive Golgi compartments. *J. Cell Biol.* **115**, 31–43.

Rowe, T., Aridor, M., McCaffery, J. M., Plutner, H., Nuoffer, C., and Balch, W. E. (1996). COPII vesicles derived from mammalian ER microsomes recruit COPI. *J. Cell Biol.,* **135**, 895–911.

Schwaninger, R., Beckers, C. J. M., and Balch, W. E. (1991). Sequential transport of protein between the endoplasmic reticulum and successive Golgi compartments in semi-intact cells. *J. Biol. Chem.* **266**, 13055–13063.

Schwaninger, R., Plutner, H., Davidson, H. W., Pind, S., and Balch, W. E. (1992). Transport of protein between the endoplasmic reticulum and Golgi compartments in semiintact cells. *In* "Methods in Enzymology" (J. E. Rothman, ed.), Vol. 219, pp. 110–124. Academic Press, San Diego.

Cotranslational Translocation of Proteins into Microsomes Derived from the Rough Endoplasmic Reticulum of Mammalian Cells

Bruno Martoglio, Stefanie Hauser, and Bernhard Dobberstein

I. INTRODUCTION

In mammalian cells, secretory proteins and membrane proteins of the organelles of the secretory pathway are transported across and integrated into the membrane of the rough endoplasmic reticulum (ER), respectively (for review see Rapoport *et al.*, 1996). This process can be studied in a cell-free translation/translocation system in which a newly synthesized protein is translocated across microsomal membranes derived from the rough ER. Components used in such a system are rough microsomes (RM) prepared from dog pancreas, a cytosolic extract supporting protein synthesis, mRNA coding for a secretory or a membrane protein, and [^{35}S]methionine to label the newly synthesized protein.

Purification of components involved in translocation revealed that a cytosolic component, the signal recognition particle (SRP), is required for targeting nascent chains to the ER membrane. SRP is found in cytosolic extracts and is variably found associated with microsomes. It can be purified from a microsomal high salt wash fraction and may be used to optimize the translocation system.

Upon translocation across the ER membrane, proteins become modified, fold with the help of lumenal chaperones, and assemble into oligomeric complexes. These functions are maintained in rough microsomes and can be studied. This article describes the basic *in vitro* translocation system and how to analyze the translocation of proteins across and integration into the membrane of microsomal vesicles as well as their glycosylation.

II. MATERIALS AND INSTRUMENTATION

RNase inhibitor (Cat. No. 2684) is from Ambion. [^{35}S]methionine (Cat. No. AG 1594) is from Amersham. ATP (Cat. No. 1140922), CTP (Cat. No. 1140922), GTP (Cat. NO. 1140957), UTP (Cat. No. 1140949), creatin kinase (Cat. No. 736988), creatine phosphate (Cat. No. 621714), and SP6 RNA polymerase (Cat. No. 810274) are from Boehringer-Mannheim. EDTA (Cat. No. 03610), octaethylene glycol monododecyl ether (Cat. No. 74680), and SDS (Cat. No. 71725) are from Fluka. Acetic acid (Cat. No. 100063), acetone (Cat. No. 100014), calcium chloride (Cat. No. 102389), HCl (Cat. No. 100317), isopropanol (Cat. No. 109634), magnesium acetate (Cat. No. 105819), magnesium chloride (Cat. No. 105833), 2-mercaptoethanol (Cat. No. 805740), potassium acetate (Cat. No. 104820), proteinase K (Cat.

No. 124568), sodium acetate (Cat. No. 106268), sodium carbonate (Cat. No. 106392), sodium citrate (Cat. No. 106448), sucrose (Cat. No. 107654), trichloroacetic acid (Cat. No. 100810), and Triton-X 100 (Cat. No. 9108605) are from Merck. ^{7}mG(5′)ppp(5′)G (Cat. No. 1404) and endoglycosidase H (Cat. No. 702) are from New England Biolabs. DEAE-Sepharose (Cat. No. 17-0709-01), Sephadex G-150 (Cat. No. 17-0070-01), and Sephadex G-25 (Cat. No. 17-0033-01) are from Pharmacia. Amino acid mixture minus methionine (Cat. No. L9961) is from Promega. Dithiothreitol (DTT, Cat. No. D-0632), HEPES (Cat. No. H-3375), phenyl-methylsulfonyl fluoride (PMSF, Cat. No. P-7626), and Tris base (Cat. No. T-1503) are from Sigma. Ultrafiltration unit (Cat. No. 8050) and YM100 membrane (Cat. No. 14422) are from Amicon. Homogenizer POTTER S is from Braun Biotech Intl. Nucleobond Kit PC 2000-1 (Cat. No. 730 570) is from Macherey and Nagel. The GradiFrac chromatography system (Cat. No. 13-2192-01) is from Pharmacia.

III. PROCEDURES

A. Preparation of Components Required for *In Vitro* Protein Translocation Across ER-Derived Rough Microsomes

1. Preparation of Rough Microsomes from Dog Pancreas

In principle, RMs can be prepared from most tissues or cells in culture. Dog pancreas, however, is a very good source for functional RMs as this tissue is specialized on secretion and contains an extended rough ER. Most importantly, pancreas prepared from dog, in contrast to other animals, contains very little ribonuclease that would digest mRNA in the *in vitro* translation system. Dogs may be obtained from an experimental surgery department. If so, one has to take care that the dogs are not operated on for more than 1 hr. In general, the pancreas should be excised less than 1 hr after death.

Solutions

Glassware and plasticware should be autoclaved, and stock solutions and distilled water should be either autoclaved or filtered through 0.45-μm-pore-sized filters.

1. *2 M sucrose:* To make 200 ml of the solution, dissolve 136.9 g sucrose in water and complete to 200 ml.

2. *1 M HEPES–KOH, pH 7.6:* To make 200 ml of the buffer, dissolve 47.7 g HEPES in water, adjust pH to 7.6 with KOH solution, and complete to 200 ml.

3. *4 M KOAc:* To make 200 ml of the solution, dissolve 78.5 g KOAc in water, neutralize to pH 7 with diluted acetic acid, and complete to 100 ml.

4. *1 M Mg(OAc)$_2$:* To make 100 ml of the solution, dissolve 21.5 g Mg(OAc)$_2 \cdot$4H$_2$O in water, neutralize to pH 7 with diluted acetic acid, and complete to 100 ml.

5. *500 mM EDTA:* To make 100 ml of the solution, dissolve 14.7 g EDTA in water, adjust to pH 8.0 with NaOH solution, and complete to 100 ml.

6. *1 M DTT:* To make 10 ml of the solution, dissolve 1.54 g in water and complete to 10 ml. Store 500-μl aliquots at −20°C.

7. *10 mg/ml PMSF:* To make 10 ml of the solution, dissolve 100 mg PMSF in isopropanol and complete to 10 ml. Store at 4°C.

8. *Homogenization buffer:* 250 mM sucrose, 50 mM HEPES–KOH, pH 7.6, 50 mM KOAc, 6 mM Mg(OAc)$_2$, 1 mM EDTA, 1 mM DTT, and 10 μg/ml PMSF. To make 500 ml of the buffer, add 62.5 ml 2 M sucrose, 25 ml 1 M HEPES–KOH, pH 7.6, 6.25 ml 4 M KOAc, 3 ml 1 M Mg(OAc)$_2$, 1 ml 500 mM EDTA, 0.5 ml 1 M DTT, and 0.5 ml 10 mg/ml PMSF; complete to 500 ml with water.

9. *Sucrose cushion:* 1.3 M sucrose, 50 mM HEPES–KOH, pH 7.6, 50 mM KOAc, 6 mM Mg(OAc)$_2$, 1 mM EDTA, 1 mM DTT, and 10 μg/ml PMSF. To make 200 ml of the solution, add 130 ml 2 M sucrose, 10 ml 1 M HEPES–KOH, pH 7.6, 2.5 ml 4 M KOAc, 1.2 ml 1 M

Mg(OAc)$_2$, 0.4 ml 500 mM EDTA, 0.2 ml 1 M DTT, and 0.2 ml 10 mg/ml PMSF; complete to 200 ml with water.

10. *RM buffer:* 250 mM sucrose, 50 mM HEPES–KOH, pH 7.6, 50 mM KOAc, 2 mM Mg(OAc)$_2$, 1 mM DTT, and 10 μg/ml PMSF. To make 100 ml of the buffer, add 12.5 ml 2 M sucrose, 5 ml 1 M HEPES–KOH, pH 7.6, 1.25 ml 4 M KOAc, 0.2 ml 1 M Mg(OAc)$_2$, 0.1 ml 1 M DTT, and 0.1 ml 10 mg/ml PMSF; complete to 100 ml with water.

Steps

The procedure is performed in the cold room. Samples and buffers should be kept on ice.

1. Excise the pancreas (40–50 g) from a dog (e.g., beagle or fox hound) and place into homogenization buffer.

2. Remove connective tissue and fat, cut the pancreas into small pieces, and place them into 120 ml of homogenization buffer.

3. Pass the tissue through a tissue press with a 1-mm mesh steel sieve.

4. Homogenize the pancreas in a 30-ml glass–Teflon potter with five strokes at full speed (1500 rpm).

5. Transfer homogenate into 30-ml polypropylene tubes. Centrifuge at 3000 rpm (1000 g) for 10 min at 4°C in a Sorvall SS34 rotor.

6. Collect the supernatant, avoiding the floating lipid. Extract the pellet once more in 50 ml homogenization buffer as described in step 4 and centrifuge again (see step 5).

7. Transfer the two 1000 g supernatants to 30-ml polypropylene tubes. Centrifuge at 9500 rpm (10,000 g) for 10 min at 4°C in a Sorvall SS34 rotor.

8. In the meantime, prepare four 70-ml polycarbonate tubes for a Ti 45 rotor and add 25 ml sucrose cushion per tube.

9. After centrifugation (step 7), collect the supernatant and apply carefully, without mixing, onto the 25-ml sucrose cushions (see step 8).

10. Centrifuge at 35,000 rpm (142,000 g) for 1 hr at 4°C in a Beckman Ti 45 rotor.

11. Discard the supernatant and resuspend the membrane pellet, the rough microsomes (RM), in 30 ml RM buffer using a Dounce homogenizer.

12. Measure the absorption at 260 and 280 nm of a 1:100 dilution of the RM suspension in 0.5% (w/v) SDS. Usually an absorption of 0.05–0.1 A_{280}/ml and a ratio A_{260}/A_{280} of 1.8 are obtained.

13. Freeze 500-μl aliquots in liquid nitrogen and store at −70°C until use.

NOTE

RMs prepared by this procedure largely retain SRP on the membrane.

2. Preparation of Signal Recognition Particle

SRP is a ribonucleoprotein particle that targets nascent secretory and membrane proteins to the ER membrane. To optimize this process, purified SRP can be added to the translation/translocation system. This is particularly useful when the wheat germ system is used for *in vitro* translation because wheat germ extract contains low amounts of functional SRP. SRP is most conveniently isolated from RMs by treatment with high salt to release SRP. SRP is then purified by gel filtration and anion-exchange chromatography.

Solutions

Glassware and plasticware should be autoclaved, and stock solutions and distilled water should be either autoclaved or filtered through 0.45-μm-pore-sized filters.

1. *1 M Tris–OAc, pH 7.5:* To make 1 liter of the buffer, dissolve 121.1 g Tris base in water, adjust to pH 7.5 with acetic acid, and complete to 1 liter.

2. *10% (w/v) octaethylene glycol monododecyl ether:* To make 10 ml of the solution, dissolve 1 g of octaethylene glycol monododecyl ether in water and complete to 10 ml.

3. *RM buffer:* See Procedure A1.

4. *High salt solution:* 1.5 M KOAc, and 15 mM Mg(OAc)$_2$. To make 50 ml of the solution, add 18.75 ml 4 M KOAc and 0.75 ml 1 M Mg(OAc)$_2$; complete to 50 ml with water.

5. *Sucrose cushion:* 500 mM sucrose, 50 mM Tris–OAc, pH 7.5, 500 mM KOAc, 5 mM Mg(OAc)$_2$, and 1 mM DTT. To make 100 ml of the solution, add 25 ml 2 M sucrose, 5 ml 1 M Tris–OAc, pH 7.5, 12.5 ml 4 M KOAc, 0.5 ml 1 M Mg(OAc)$_2$, and 0.1 ml 1 M DTT; complete to 100 ml with water.

6. *Gel filtration buffer:* 50 mM Tris–OAc, pH 7.5, 250 mM KOAc, 2.5 mM Mg(OAc)$_2$, 1 mM DTT, and 0.01% octaethylene glycol monododecyl ether. To make 1 liter of the buffer, add 50 ml 1 M Tris–OAc, pH 7.5, 62.5 ml 4 M KOAc, 2.5 ml 1 M Mg(OAc)$_2$, 1 ml 1 M DTT, and 1 ml 10% octaethylene glycol monododecyl ether; complete to 1 liter with water.

7. *Washing buffer:* 50 mM Tris–OAc, pH 7.5, 350 mM KOAc, 3.5 mM Mg(OAc)$_2$, 1 mM DTT, and 0.01% octaethylene glycol monododecyl ether. To make 100 ml of the buffer, add 5 ml 1 M Tris–OAc, pH 7.5, 8.75 ml 4 M KOAc, 0.35 ml 1 M Mg(OAc)$_2$, 0.1 ml 1 M DTT, and 0.1 ml 10% octaethylene glycol monododecyl ether; complete to 100 ml with water.

8. *SRP buffer:* 50 mM Tris–OAc, pH 7.5, 650 mM KOAc, 6 mM Mg(OAc)$_2$, 1 mM DTT, and 0.01% octaethylene glycol monododecyl ether. To make 100 ml of the buffer, add 5 ml 1 M Tris–OAc, pH 7.5, 16.25 ml 4 M KOAc, 0.6 ml 1 M Mg(OAc)$_2$, 0.1 ml 1 M DTT, and 0.1 ml 10% octaethylene glycol monododecyl ether; complete to 100 ml with water.

Steps

The procedure is performed in the cold room and samples and buffers should be kept on ice.

1. Prepare ER-derived rough microsomes from one dog pancreas as described earlier and resuspend the final RM pellet (see step 8 in Procedure A1) in 50 ml RM buffer using a Dounce homogenizer.

2. Add 25 ml high salt solution and incubate for 15 min at 4°C on a turning wheel.

3. Distribute the membrane suspension equally to two 70-ml polycarbonate tubes for a Ti 45 rotor onto 25 ml sucrose cushion per tube, avoid mixing.

4. Centrifuge at 32,000 rpm (120,000 g) for 1 hr at 4°C in a Beckman Ti 45 rotor.

5. In the meantime, prepare ten 10.4-ml polycarbonate tubes for a Ti 70.1 rotor and add 1 ml sucrose cushion per tube.

6. After centrifugation (step 4), collect the supernatant and apply carefully, without mixing, onto the 1-ml sucrose cushions (see step 5).

7. Centrifuge at 65,000 rpm (388,000 g) for 1 hr at 4°C in a Beckman Ti 70.1 rotor.

8. Collect the supernatant again.

9. Concentrate the supernatant to approximately 10 ml in a 50-ml Amicon ultrafiltration unit equipped with a YM100 membrane.

10. Load the concentrated sample onto a Sephadex G-150 column (2.6 × 20 cm) equilibrated with gel filtration buffer and elute with gel filtration buffer. The flow rate is 1 ml/min. Follow elution with a UV monitor (λ = 280 nm) using, e.g., the GradiFrac chromatography system from Pharmacia.

11. Collect flow through (20–25 ml) and load immediately onto a DEAE-Sepharose column (1 × 3 cm) equilibrated with gel filtration buffer. The flow rate is again 1 ml/min.

12. Wash column with 20 ml gel filtration buffer and with 20 ml washing buffer.

13. Elute SRP with SRP buffer and collect peak fraction (2–3 ml). The SRP eluate has an absorption of 1–4 A_{260}/ml and a ratio A_{260}/A_{280} of approximately 1:4. Freeze 100-μl aliquots in liquid nitrogen and store at −70°C until use.

B. *In Vitro* Translation and Translocation Assay

Solutions

1. *Wheat germ extract:* Wheat germ extract for cell-free *in vitro* translation is prepared as described by Erickson and Blobel (1983) except that we use a gel filtration buffer with lower potassium and magnesium concentrations [40 mM HEPES–KOH, pH 7.6, 50 mM KOAc, 1 mM Mg(OAc)$_2$, and 0.1% 2-mercaptoethanol]. Fresh wheat germ may be purchased from a local mill or from General Mills California. Wheat germ is stored in a desiccator over silica gel beads at 4°C. Considerable differences in translation efficiency may yield from different batches. Wheat germ extract is stored in 110-μl aliquots at −70°C and is thawed only once.

2. *Capped mRNA:* To obtain mRNA coding for a secretory or a membrane protein, a cDNA encoding the protein of interest is cloned into a suitable expression vector downstream of a T7 or a SP6 promotor (e.g., pGEM from Promega Biotech). We generally prefer the SP6 promotor as the respective transcripts yield more efficient translation. Plasmid DNA is prepared from *Escherichia coli* cultures using the Nucleobond DNA purification kit from Macherey and Nagel. For transcription, the purified plasmid DNA is linearized with a suitable restriction enzyme that cuts downstream of the coding sequence. *In vitro* transcription is subsequently performed according to a standard protocol (see, e.g., Sambrook *et al.*, 1989) or by using a commercially available transcription kit (e.g., mMESSAGAmMACHINE kits from Ambion, Cat. Nos. 1340 and 1344). The resulting capped messenger RNA is dissolved in water after extraction with phenol and chloroform and precipitation with sodium acetate and ethanol. mRNA is stored at −70°C.

3. *Energy mix:* 50 mM HEPES–KOH, pH 7.6, 12.5 mM ATP, 0.25 mM GTP, 110 mM creatine phosphate, 10 mg/ml creatine kinase, and 0.25 mM of each amino acid, *except* methionine. To make 1 ml of the solution, dissolve 41 mg creatine phosphate (disodium salt · 4H$_2$O) and 10 mg creatine kinase in 590 μl water and add 50 μl 1 M HEPES–KOH, pH 7.6, 125 μl 100 mM ATP solution (Boehringer-Mannheim), 2.5 μl 100 mM GTP solution (Boehringer-Mannheim) and 250 μl 19 amino acids mix without methionine (Promega, 1 mM of each amino acid). The energy mix is stored in 22-μl aliquots at −70°C. Do not refreeze!

4. *Salt mix:* 500 mM Hepes-KOH, pH 7.6, 1 M KOAc, 50 mM Mg(OAc)$_2$. To make 1 ml of the solution, add 500 μl 1 M Hepes-KOH pH 7.6, 250 μl 4 M KOAc, 70 μl 1 M Mg(OAc)$_2$ to 180 μl water.

5. *SRP buffer:* See Procedure A2.

6. *20% (w/v) trichloroacetic acid:* To make 100 ml of the solution, dissolve 20 g trichloroacetic acid in water and complete to 100 ml.

Steps

1. Per assay (25 μl), mix on ice 6 μl water, 1 μl salt mix, 2 μl energy mix, 10 μl wheat germ extract, 2 μl SRP buffer or SRP solution (see Procedure A2, step 12), 2 μl RM suspension (see Procedure A1, step 8), 1 μl [^{35}S]methionine (>1000 Ci/mmol), and 1 μl capped mRNA.

2. Incubate for 60 min at 25°C.

3. Add 25 μl (1 vol) 20% trichloroacetic acid to precipitate proteins and centrifuge at 14,000 rpm for 3 min at room temperature in an Eppendorf centrifuge.

4. Discard supernatant. Wash pellet with 1 ml cold acetone and centrifuge as in step 3.

5. Discard supernatant.

6. Add 20–30 μl sample buffer for SDS–polyacrylamide gel electrophoresis (see Volume 4, Celis and Olsen, "One-Dimensional Sodium Dodecyl Sulfate-Polyacrylamide Gel Electrophoresis" for additional information) and heat for 10 min at 65°C.

7. Analyze the sample by SDS–polyacrylamide gel electrophoresis. The radiolabeled proteins can be visualized by autoradiography, fluorography, or on a PhosphorImager.

Controls: Translation without mRNA (add water instead). Translation without membranes (add RM buffer instead, see Procedure A1).

Note: Translation and translocation assays have to be optimized for each mRNA with respect to magnesium and potassium concentrations as well as to the amount of mRNA, SRP, and membranes. Optimal salt concentrations vary from 1 to 3.5 mM magnesium and from 70 to 150 mM potassium and are adjusted by using adopted salt mixes. The amount of SRP and membranes may be varied by diluting SRP and RM solutions with the respective buffers.

C. Assays to Characterize Translocation Products

1. Protease Protection Assay

Proteinase K is used to test the translocation of proteins or parts of proteins across microsomal membranes (Blobel and Dobberstein, 1975). Membranes are impermeable to the protease and therefore only proteins or protein domains exposed on the cytoplasmic side of the microsomes are digested. To demonstrate that only intact microsomal vesicles protect lumenal proteins or protein domains, nonionic detergent (e.g., Triton X-100) is added to open the membrane.

Protease treatment is also used to characterize the topology of membrane proteins. In this case, protease treatment is often followed by immunoprecipitations with antibodies directed against defined regions of the protein investigated. Successful immunoprecipitations indicate that the respective domains are exposed on the lumenal side of the microsomes and protected from the protease.

Solutions

1. *3 mg /ml proteinase K:* To make 1 ml of the solution, dissolve 3 mg proteinase K in 1 ml water.

2. *10 mg/ml PMSF solution:* See Procedure A1.

3. *10% (w/v) Triton X-100:* To make 10 ml of the solution, dissolve 10 g Triton X-100 in water and complete to 10 ml.

4. *20% (w/v) trichloroacetic acid:* See Procedure B.

Steps

1. Perform a translation/translocation assay (25 μl) as described in Procedure B (steps 1 and 2).

2. After translation/translocation, split the sample into three 8-μl aliquots.

    a. Add 2 μl water to the first aliquot (mock treatment).

    b. Add 1 μl proteinase K solution (3 mg/ml) and 1 μl water to the second aliquot.

    c. Add 1 μl proteinase K solution (3 mg/ml) and 1 μl 10% Triton X-100 to the third aliquot.

3. Incubate the samples for 10 min at 25°C.

4. Stop proteolysis by adding 2 μl 10 mg/ml PMSF per sample.

5. Add 50 μl water and 50 μl 20% trichloroacetic acid to each sample to precipitate protein.

6. Centrifuge samples, wash protein pellets with acetone, and analyze samples by SDS–polyacrylamide gel electrophoresis as described in Procedure B, steps 3–7.

2. Sodium Carbonate Extraction

By alkaline treatment with sodium carbonate at pH 11, microsomal membranes are opened, releasing their contents and peripherally associated proteins. The method is used to separate these proteins from integral membrane proteins and proteins bound tightly to the membrane (Fujiki *et al.*, 1982).

Solutions

1. *0.1 M Na₂CO₃:* To make 10 ml of the solution, dissolve 106 mg Na$_2$CO$_3$ in water and complete to 10 ml. Prepare just prior to use.

2. *Alkaline sucrose cushion:* 0.1 M Na$_2$CO$_3$ and 250 mM sucrose. To make 10 ml of the solution, dissolve 106 mg Na$_2$CO$_3$ in water, add 1.25 ml 2 M sucrose (see Procedure A1), and complete to 10 ml with water.

3. *20% (w/v) trichloroacetic acid:* See Procedure B.

Steps

1. Perform a translation/translocation assay (25 μl) as described in Procedure B (steps 1 and 2).

2. Add 25 μl carbonate solution and incubate for 15 min on ice.

3. In the meantime, prepare 200 μl polycarbonate tubes for a TLA 100 rotor and add 100 μl alkaline sucrose cushion.

4. For centrifugation, apply the sample (step 2) carefully onto a sucrose cushion in the polycarbonate tube.

5. Centrifuge at 55,000 rpm (130,000 g) for 10 min at 4°C in a Beckman TLA 100 rotor.

6. Recover the supernatant and pellet.

   a. Add 150 μl 20% trichloroacetic acid to the supernatant to precipitate proteins. Centrifuge the sample, wash the protein pellet with acetone, and prepare the sample for SDS–polyacrylamide gel electrophoresis as described in Procedure B, steps 3–6.

   b. Add 20–30 μl sample buffer directly to the pellet for SDS–polyacrylamide gel electrophoresis (see Procedure B, step 6).

7. Analyze samples by SDS–polyacrylamide gel electrophoresis and autoradiography. (see Volume 4, Celis and Olsen, "One-Dimensional Sodium Dodecyl Sulfate-Polyacrylamide Gel Electrophoresis" for additional information).

3. Inhibition of N-Glycosylation with Glycosylation Acceptor Tripeptide

The recognition sites for N-glycosylation in the ER are Asn-X-Ser and Asn-X-Thr. In the *in vitro* translation/translocation system described herein, the tripeptide N-benzoyl-Asn-Leu-Thr-methylamide efficiently competes with newly synthesized proteins for N-glycosylation (Lau *et al.*, 1983). The translocated protein is therefore not glycosylated in the presence of acceptor tripeptide and the effects of oligosaccharides, e.g., on protein folding and assembly, may be investigated.

Solution

1. *Acceptor tripeptide solution:* Synthesize N-benzoyl-Asn-Leu-Thr-methylamide on a peptide synthesiser. Dissolve the tripeptide in methanol at a concentration of 0.5 mM (0.23 mg/ml) and store the stock solution at −70°C.

Steps

1. Evaporate per translation/translocation assay (25 μl) 1.5 μl acceptor tripeptide solution in a test tube using a Speed-Vac centrifuge.

---

2. Add the components for the translation/translocation assay to the tripeptide, vortex gently, and perform the assay as described in Procedure B (steps 1 and 2).

3. Precipitate the sample with trichloroacetic acid and analyze by SDS–polyacrylamide gel electrophoresis and autoradiography (see Procedure B, steps 3–7).

4. Endoglycosidase H Treatment

Treatment with endoglycosidase H is used to test N-glycosylation of proteins translocated into microsomal membranes. The glycosidase cleaves oligosaccharides of the high mannose type from glycoproteins, leaving an N-acetylglucosamine residue attached to the polypeptide (Tarentino *et al.*, 1978).

Solutions

1. *Denaturing solution:* 5% (w/v) SDS and 10% (w/v) 2-mercaptoethanol. To make 10 ml of the solution, dissolve 0.5 g SDS in water, add 1 ml 2-mercaptoethanol, and complete to 10 ml with water.

2. *0.5 M Na–citrate, pH 5.5:* To make 10 ml of the buffer, dissolve 1.47 g Na_3-citrate $\cdot$ 2 H_2O in water, adjust to pH 5.5 with HCl solution, and complete to 10 ml.

Steps

1. Perform a translation/translocation assay (25 μl) as described in Procedure B (steps 1 and 2).

2. Add 2.5 μl denaturing solution and incubate for 10 min at 95°C.

3. Add 2.8 μl reaction buffer and 1 μl endoglycosidase H (1000 U/μl). Incubate for 1 hr at 37°C.

4. Precipitate the sample with trichloroacetic acid and analyse by SDS–polyacrylamide gel electrophoresis and autoradiography (see Procedure B, steps 3–7).

IV. COMMENTS

Translation can also be done in reticulocyte lysate (Jackson and Hunt, 1983) or lysate from other cells (Garoff *et al.*, 1978). Because reticulocyte lysate contains SRP, no additional SRP is usually required for the optimization of translocation. When reticulocyte lysate is used, RMs should be treated with microccocal nuclease to digest endogenous mRNA (Garoff *et al.*, 1978). Reticulocyte elongation factors will otherwise promote completion of nascent pancreatic secretory proteins.

Translation/translocation systems of usually good quality are also commercially available. We have tested translation systems from Ambion, Amersham, and Promega. RMs of variable quality are available from Amersham.

References

Blobel, G., and Dobberstein, B. (1975). Transfer of proteins across membranes. II. Reconstitution of functional rough microsomes from heterologous components. *J. Cell Biol.* **67**, 852–862.

Erickson, A. H., and Blobel, G. (1983). Cell-free translation of messenger RNA in a wheat germ system. *In* "Methods in Enzymology" (S. Fleischer and B. Fleischer, eds.), Vol. 96, pp. 38–50. Academic Press, San Diego.

Fujiki, Y., Hubbard, A. L., Fowler, S., and Lazarow, P. B. (1982). Isolation of intracellular membranes by means of sodium carbonate treatment: Application to endoplasmic reticulum. *J. Cell Biol.* **93**, 97–102.

Garoff, H., Simons, K., and Dobberstein, B. (1978). Assembly of the semliki forest virus membrane glycoproteins in the membrane of the endoplasmic reticulum *in vitro*. *J. Mol. Biol.* **124**, 587–600.

Jackson, R. J., and Hunt, T. (1983). Preparation and use of nuclease-treated rabbit reticulocyte lysate for the translation of eukaryotic messenger RNA. *In* "Methods in Enzymology" (S. Fleischer and B. Fleischer, eds.), Vol. 96, pp. 50–74. Academic Press, San Diego.

Lau, J. T. Y., Welply, J. K., Shenbagamurthi, P., Naider, F., and Lennarz, W. J. (1983). Substrate recognition by oligosaccharyl transferase: Inhibition of translational glycosylation by acceptor peptides. *J. Biol. Chem.* **258**, 15255–15260.

Rapoport, T. A., Jungnickel, B., and Kutay, U. (1996). Protein transport across the eukaryotic endoplasmic reticulum and bacterial inner membranes. *Ann. Rev. Biochem.* **65**, 271–303.

Sambrook, J., Fritsch, E. F., and Maniatis, T. (1989). "Molecular cloning: A Laboratory manual." Cold Spring Harbor Press, Cold Spring Harbor, NY.

Tarentino, A. L., Trimble, R. B., and Maley, F. (1978). Endo-β-*N*-acetylglucosaminidase from Streptomyces plicatus. *In* "Methods in Enzymology" (V. Ginsberg, ed.), Vol. 50, pp. 574–580. Academic Press, New York.

Walter, P., and Johnson, A. E. (1994). Signal sequence recognition and protein targeting to the endoplasmic reticulum membrane. *Annu. Rev. Cell Biol.* **10**, 87–119.

Warren, G., and Dobberstein, B. (1978). Protein transfer across microsomal membranes reassembled from separated membrane components. *Nature* **273**, 569–571.

Mitochondria and Chloroplasts

Protein Translocation into Mitochondria

Sabine Rospert and Gottfried Schatz

I. INTRODUCTION

Mitochondria from different sources such as rat liver, rabbit brain, or yeast can be isolated as intact organelles. Isolated mitochondria are able to respire, maintain a membrane potential across their inner membrane, possess an active ATP synthase, and shuttle nucleotides across their membranes. In addition, even a process as complicated as import of mitochondrial precursor proteins can be studied outside the living cell. For this purpose, radiolabeled precursor proteins, synthesized in an *in vitro* transcription/translation system, are mixed with isolated mitochondria (Melton *et al.*, 1984; Glick, 1991). In the presence of ATP, precursor proteins will cross the mitochondrial membranes, become processed to their mature form, and fold to their native state. Building on this basic "import assay," sophisticated experiments have been developed and the results of these experiments provide most of what is known about mitochondrial import today (Schatz, 1995).

This article describes a standard protocol for the *in vitro* synthesis of a radiolabeled precursor protein and the subsequent import of this precursor into isolated yeast mitochondria. As examples, we have selected the precursor proteins bovine mitochondrial rhodanese (Rospert *et al.*, 1996) and Su9-DHFR, a fusion protein consisting of the first 69 amino acids of subunit 9 of the ATPase, and mouse dihydrofolate reductase (Pfanner *et al.*, 1987). The N-terminal presequence of the artificial Su9-DHFR precursor, as is of most mitochondrial precursor proteins, is removed by a protease localized in the mitochondrial matrix (Jensen and Yaffe, 1988). In contrast, rhodanese is not processed after import. The mRNA of both precursor proteins is transcribed with SP6 RNA polymerase (Melton *et al.*, 1984).

II. MATERIALS AND INSTRUMENTATION

SP6 RNA polymerase (Cat. No. 810 274); RNase inhibitor from human placenta (Cat. No. 799017); tRNA, from calf liver (Cat. No. 647225); set of ATP, CTP, GTP, UTP, lithium salts, 100 mM solutions (Cat. No. 1277057); creatine kinase from rabbit muscle (Cat. No. 127566); creatine phosphate, disodium salt (Cat. No. 126969); and proteinase K (Cat. No. 1092766) are from Boehringer. Tris(hydroxymethyl)aminomethan (Tris) (Cat. No. 108382); KCl (Cat. No. 104936); KOH (Cat. No. 105021); MgCl$_2$ (Cat. No. 105835); (NH$_4$)$_2$SO$_4$ (Cat. No. 112019); NaCl (Cat. No. 101540); trichloroacetic acid (TCA) (Cat. No. 810100); NaN$_3$ (Cat. No. 822335); CaCl$_2$, dihydrate (Cat. No. 102383); 25% NH$_3$ solution (Cat. No. 105432); ethanol (Cat. No. 100983); and sodium salicylate (Cat. No. 106602) are from Merck. Spermidine (Cat. No. S-0266); bovine serum albumin, essentially fatty acid free (BSA) (Cat. No. A-7511); dithiothreitol (DTT) (Cat. No. D-5545); (*N*-[2-hydroxyethylpiperazine-*N'*-[2-ethanesulfonic acid]) (HEPES) (Cat. No. H-7523); trypsin (Cat. No. T-9003); trypsin inhibitor, from soybean (Cat. No. T-8642); β-nicotinamide adenine dinucleotide disodium salt, reduced form (NADH) (Cat. No. N-8129); ethylenediaminetetraacetic acid (EDTA) (Cat. No. E-9884); valinomycin (Cat. No. V-0627); ATP, disodium salt (Cat No. A-7699); glycerol (Cat. No. G-6279); potassium acetate (Cat. No. P-5708); KH_2PO_4, 1 *M* solution (Cat. No. P-8709); L-methionine (Cat. No. M-6039); and Triton X-100 (Cat. No. T-9284) are from Sigma. $m^7G(5')ppp(5')G$ (G-cap) (Cat. No. 27 4635 02) is from Pharmacia. Rabbit reticulocyte lysate (Cat. No. L 4970) and amino acid mixture, minus methionine (Cat. No. L 9961), are from Promega. Urea (Cat. No. 24524) and phenylmethylsulfonyl fluoride (PMSF) (Cat. No. 32395) are from Serva. Sorbitol (Cat. No. 2039) is from Baker. L-[^{35}S]Methionine, > 1000 Ci/mmol (Cat. No. SJ 235), is from Amersham. Beckman Airfuge, Beckman. Eppendorf Centrifuge 5417 R, "microfuge" Eppendorf. Fifteen-milliliter Falcon tubes (Cat. No. 420955), Falcon. X-ray film X-OMAT AR (Cat. No. 1651454); Kodak X-ray cassettes, CAWO, Zürich, Switzerland. Plasmids are Su9-DHFR [pSu9-DHFR (Pfanner *et al.*, 1987)] and bovine rhodanese [pSR6 (Rospert *et al.*, 1996)].

III. PROCEDURES

Solutions used directly as obtained from the supplier are only listed in Section II. Protocols for the preparation of solutions used throughout the procedure are only given once.

A. Transcription Using SP6 Polymerase

Solutions

1. *1 M Tris–HCl stock solution, pH 7.5:* Dissolve 12.1 g Tris in 80 ml H_2O and adjust pH to 7.5 with 5 *M* HCl. Add H_2O to 100 ml. Autoclave and store at room temperature.

2. *1 M HEPES–KOH, pH 7.4:* Dissolve 23.8 g HEPES in 80 ml H_2O, adjust pH to 7.4 using 4 *M* KOH. Add H_2O to 100 ml. Filter sterilize and store at room temperature.

3. *1 M spermidine:* Dissolve 145 mg spermidine in 1 ml H_2O. Store at −20°C.

4. *100 mg/ml BSA:* Dissolve 500 mg BSA in 5 ml H_2O. Store at −20°C.

5. *2.5 M $MgCl_2$:* Dissolve 50.8 g $MgCl_2$ in 100 ml H_2O. Autoclave and store at 4°C.

6. *2.5 M KCl:* Dissolve 18.6 g KCl in 100 ml H_2O. Autoclave and store at 4°C.

7. *100 mM DTT:* Dissolve 15.4 mg DTT in 1 ml H_2O. Store at −20°C. Make a fresh solution about every 4 weeks.

8. *5× reaction buffer:* 200 mM Tris–HCl, pH 7.5, 30 m*M* $MgCl_2$, 10 mM spermidine, and 0.5 mg/ml BSA. To obtain 10 ml of a 5× reaction buffer, mix 2 ml 1 *M* Tris–HCl, pH 7.5, 120 μl 2.5 *M* $MgCl_2$, 100 μl 1 *M* spermidine, and 50 μl 100 mg/ml BSA. If necessary, readjust the pH to 7.5. Store in 1-ml aliquots at −20°C.

9. *G-cap, ($m^7G(5')ppp(5')G$):* Dissolve 25 A_{250} units in 242 μl H_2O. Freeze 10-μl aliquots in liquid nitrogen. Store at −70°C.

10. *5 mM NTP–GTP:* To make a 500-μl stock, add 25 μl 100 mM ATP, 25 μl 100 mM UTP, and 25 μl 100 mM CTP to 425 μl 20 mM HEPES–KOH, pH 7.4. Store in 100-μl aliquots at −70°C.

11. *5 mM GTP:* Mix 475 μl 20 mM HEPES–KOH, pH 7.4, with 25 μl 100 mM GTP solution.

12. *RNase inhibitor buffer:* 20 mM HEPES–KOH, pH 7.4, 50 m*M* KCl, 10 m*M* DTT, and 50% glycerol. Make 10 ml of the buffer by mixing 200 μl 1 *M* HEPES–KOH, pH 7.4, 200 μl 2.5 *M* KCl, 1 ml 100 mM DTT, and 5 ml glycerol. Add H_2O to 10 ml and store at −20°C.

13. *4 units/μl RNase inhibitor:* Add 500 μl RNase inhibitor buffer to 2000 units of RNase inhibitor. Store at −20°C for up to 6 months.

14. *1 μg/μl linerarized plasmid DNA:* Prepare the linearized plasmid (pSu9-DHFR and pSR6) according to standard molecular biology procedures.

Steps

1. Mix the following solutions carefully, avoiding the formation of air bubbles. Follow the indicated order of addition because the DNA might precipitate in 5× SP6 buffer. Precipitation of DNA can also occur if the mixture is placed on ice. Incubate the mixture at 40°C for 15 min.

H_2O	12 μl
RNase inhibitor	1 μl
1 μg/μl linear plasmid (pSu9-DHFR or pSR6)	5 μl
5 m*M* rNTP's minus GTP	5 μl
5 m*M* G-cap	5 μl
100 m*M* DTT	5 μl
5× SP6 buffer	10 μl
SP6 polymerase	2 μl

2. Start transcription by adding 2 μl 5 mM GTP solution and incubate for 90 min at 40°C.

3. The mRNA obtained by this procedure is used directly in the translation protocol. mRNA can be stored in 10-μl aliquots at −70°C. If frozen mRNA is used for translation, thaw rapidly and keep at room temperature before adding the mRNA to the translation reaction.

B. Translation Using Reticulocyte Lysate

Solutions

1. *1 M DTT:* Dissolve 154 mg DTT in 1 ml H_2O. Store at −20°C. Make a fresh solution about every 4 weeks.

2. *8 mg/ml creatine kinase:* Dissolve 8 mg creatine kinase in 475 μl H_2O. Add 20 μl 1 M HEPES–KOH, pH 7.4, 5 μl 1 M DTT, and 500 μl glycerol. Freeze in 10-μl aliquots in liquid nitrogen and store at −70°C.

3. *5 mg/ml tRNA:* Dissolve 10 mg tRNA in 2 ml H_2O. Store in 100-μl aliquots at −20°C.

4. *400 mM HEPES–KOH, pH 7.4:* Mix 6 ml H_2O with 4 ml 1 M HEPES–KOH, pH 7.4. If necessary, readjust pH.

5. *10 mM GTP:* Mix 450 μl 20 mM HEPES–KOH, pH 7.4, with 50 μl 100 mM GTP. Store at −20°C.

6. *100 mM ATP:* Dissolve 55.1 mg ATP in 900 μl H_2O. Adjust to pH ~7 using 4 M NaOH and pH indicator paper. Adjust volume to 1 ml and store at −20°C.

7. *600 mM creatine phosphate:* Dissolve 153.06 mg creatine phosphate in 1 ml H_2O. Store at −20°C.

8. *4 M potassium acetate:* Dissolve 3.92 g potassium acetate in 10 ml H_2O. Do not adjust the pH. Store at −20°C.

Steps

1. Prepare the reticulocyte lysate mix and the tRNA mix fresh. To obtain two 50-μl translation reactions, mix the following solutions.

Reticulocyte lysate mix		tRNA mix	
0.4 M HEPES–KOH	5 μl	5 mg/ml tRNA	4 μl
10 mM GTP	0.6 μl	4 units/μl RNase inhibitor	4 μl
100 mM ATP	0.5 μl	4 M potassium acetate	2 μl
1 mM amino acid mix	3.75 μl		
0.6 M creatine phosphate	2 μl		
Reticulocyte lysate	50 μl		
8 mg/ml creatine kinase	2 μl		

2. Use the mRNA obtained in Procedure A. It is possible to use mRNA produced in a different transcription system, e.g., with T7 RNA polymerase. For each of the two translation reactions, mix 30 μl reticulocyte lysate mix, 5 μl tRNA mix, 9 μl mRNA, 5 μl [^{35}S]methionine, and 1 μl 1 M DTT.

3. Incubate this mixture for 60 min at 30°C. Shield it from light to prevent heme-induced photooxidation of the precursor proteins. Remove ribosomes after the translation reaction by centrifugation for 15 min at 150,000 g (30 psi in an Airfuge). Remove the supernatants, being careful not to disturb the ribosomal pellet.

C. Denaturation of Radiolabeled Precursor Protein
Solutions

1. *Saturated $(NH_4)_2SO_4$ solution:* Weigh 100 g $(NH_4)_2SO_4$ and add H_2O to a final volume of 100 ml. Stir for 30 min at room temperature. The $(NH_4)_2SO_4$ will not dissolve entirely. Remove the supernatant and keep at room temperature.

2. *8 M urea:* Dissolve 4.85 g urea in a final volume of 10 ml 25 mM Tris–HCl, pH 7.5, containing 25 mM DTT.

Steps

1. Proteins synthesized in reticulocyte lysate are either folded or bound to chaperone proteins present in the lysate (Wachter *et al.*, 1994). In order to unfold the protein prior to import, it can be precipitated by high concentrations of ammonium sulfate and subsequently denatured in 8 M urea.

2. Add 100 μl of the $(NH_4)_2SO_4$ solution to each of the 50-μl translation reactions. Mix well and allow precipitation of the protein for 30 min on ice. Collect precipitates by centrifugation in an Eppendorf centrifuge at 20,000 g for 10 min.

3. Discard the supernatants and dissolve each pellet in 50 μl of 8 M urea solution. Combine the two solutions of denatured precursor proteins and keep at room temperature for 10–30 min. This precursor mix is used for the import reaction (Procedure D) and preparation of the precursor standard (Procedure G).

D. Import of Denatured Radiolabeled Precursor Proteins
Solutions

For additional solutions required, see Procedures A and B.

1. *2.4 M sorbitol:* Dissolve 43.7 g of sorbitol in a final volume of 100 ml H_2O. Autoclave and store at 4°C.

2. *1 M HEPES–KOH, pH 7.0:* Dissolve 23.8 g HEPES in 80 ml H_2O and adjust pH to 7.0 with 4 M KOH. Add H_2O to a final volume of 100 ml. Filter sterilize and store at room temperature.

3. *250 mM EDTA, pH 7.0:* Resuspend 7.3 g of EDTA in 70 ml H_2O. Adjust pH to 7.0

using 5 M NaOH. Add H_2O to a final volume of 100 ml. Filter sterilize and keep at room temperature.

4. *2× import buffer:* 1.2 M sorbitol, 100 mM HEPES–KOH, pH 7.0, 100 mM KCl, 20 mM MgCl$_2$, 5 mM EDTA, pH 7.0, 4 mM KH$_2$PO$_4$, 2 mg/ml BSA, and 1.5 mg/ml methionine. To make 100 ml of 2× import buffer, mix 50 ml 2.4 M sorbitol, 400 μl 1 M KH$_2$PO$_4$ solution, 4 ml 2.5 M KCl, 10 ml 1 M HEPES–KOH, pH 7.0, 0.8 ml 2.5 M MgCl$_2$, 2 ml 250 mM EDTA, pH 7.0, 150 mg methionine, and 200 mg BSA. Adjust pH to 7.0 and add H_2O to 100 ml. Store at −20°C.

5. *500 mM NADH:* Dissolve 35.5 mg of NADH in a final volume of 100 μl 20 mM HEPES–KOH, pH 7.0.

6. *1 mg/ml valinomycin:* Dissolve 2 mg valinomycin in 2 ml ethanol. Store at −20°C.

7. *Purified yeast mitochondria:* 25 mg mitochondrial protein/ml. Store at −70°C in 0.6 M sorbitol, 20 mM HEPES–KOH, pH 7.4, and 10 mg/ml BSA. Thaw rapidly at 25°C immediately before the experiment. Do not refreeze. A detailed protocol of the purification procedure is given in Glick and Pon (1995).

Steps

1. Import into the matrix of mitochondria requires a membrane potential across the inner mitochondrial membrane. Therefore, the most thorough control for the specificity of an import reaction is to determine its dependence on a membrane potential. This potential is generated by adding ATP and the respiratory substrate NADH. (Note that mammalian mitochondria cannot oxidize added NADH.)

2. Perform two import reactions, one in the absence and one in the presence of valinomycin. Preincubate the import reaction in a 15-ml Falcon tube at 25°C for 1–2 min.

Reaction 1 (+ membrane potential)		Reaction 2 (no membrane potential)	
2× import buffer	500 μl	2× import buffer	500 μl
100 mM ATP	20 μl	100 mM ATP	20 μl
500 mM NADH	4 μl	500 mM NADH	4 μl
Yeast mitochondria	20 μl	Yeast mitochondria	20 μl
H_2O	406 μl	H_2O	405 μl
		1 mg/ml valinomycin	1 μl

3. Add 50 μl each of the denatured precursor mixture (see Procedure C) containing denatured rhodanese and denatured Su9-DHFR to reactions 1 and 2. Intact mitochondria should be handled gently. However, it is essential to mix the denatured precursor protein into the import reaction rapidly. Agitate the import reaction gently on a Vortex mixer while dropwise adding the denatured precursor mixture. If mixing is performed only after addition, the precursor proteins tend to aggregate and become import incompetent.

4. Incubate at 25°C for 10 min. Agitate gently every other minute to facilitate gas exchange. Stop the import reaction by transferring the tubes onto ice. Add 1 μl of 1 mg/ml valinomycin to reaction 1.

5. Remove 200 μl each from reactions 1 and 2. These samples represent the *total* of the two import reactions (Fig. 1, lanes 2 and 5). Add 22 μl 50% TCA to each. Keep on ice and process further after all samples have been acid denatured.

E. Protease Treatment of Intact Mitochondria

Solutions

1. *10 mg/ml trypsin:* Dissolve 3 mg of trypsin in 300 μl H_2O. Make fresh.

2. *20 mg/ml trypsin inhibitor:* Dissolve 6 mg of trypsin inhibitor in 300 μl H_2O. Make fresh.

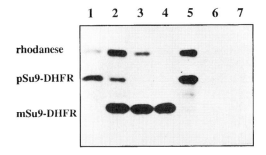

FIGURE 1 Coimport of radiolabeled bovine mitochondrial rhodanese and the fusion protein Su9-DHFR into isolated yeast mitochondria. Lane 1, 10% of the material added to each import reaction; lanes 2–4, import in the presence of ATP and a membrane potential across the inner mitochondrial membrane, and lanes 5–7, import in the absence of a membrane potential across the inner mitochondrial membrane. Lanes 2 and 5: total; material isolated together with the mitochondrial pellet. Lanes 3 and 6: import; material protease protected in intact mitochondria. Lanes 4 and 7: folded; material protease resistant even after solubilization of the mitochondria with Triton X-100. pSu9-DHFR, precursor form of Su9-DHFR; mSu9-DHFR, mature form of Su9-DFHR. Rhodanese does not become processed after import into the matrix. For experimental details see text.

3. *1× import buffer minus BSA:* Prepare 2× import buffer exactly as described in Procedure D, but without BSA. To obtain 1× import buffer minus BSA, mix 2 ml 2× import buffer minus BSA with 2 ml H$_2$O.

Steps

Perform the following steps in parallel with both import reactions.

1. To digest precursor proteins that stick to the surface of the mitochondria, add 8 μl 10 mg/ml trypsin (final concentration 100 μg/ml). Incubate for 30 min on ice.

2. Add 8 μl 20 mg/ml trypsin inhibitor (final concentration 200 μg/ml) and incubate on ice for 5 min.

3. Spin for 3 min in an Eppendorf microfuge at 10,000 g. Remove the supernatant carefully by aspiration.

4. Carefully resuspend the mitochondrial pellet in 800 μl 1× import buffer minus BSA. It is extremely important to resuspend the pellet completely. Add 100 μl of 1× import buffer minus BSA and resuspend mitochondria by pipetting up and down. Add another 700 μl of 1× import buffer minus BSA.

5. Remove 200 μl of each sample and add 22 μl 50% TCA. Keep on ice. These samples represent the material that has crossed the outer membrane completely (import, Fig. 1, lanes 3 and 6).

F. Inherent Protease Resistance of Imported Protein
Solutions

1. *1 M Tris-HCl stock solution, pH 8.0:* Dissolve 12.1 g Tris in 80 ml H$_2$O and adjust pH to 8.0 using 5 M HCl. Add H$_2$O to a final volume of 100 ml. Autoclave and store at room temperature.

2. *2 M CaCl$_2$:* Dissolve 14.7 g CaCl$_2$ in 50 ml H$_2$O. Autoclave and store at room temperature.

3. *10% Triton X-100 (w/v):* Dissolve 10 g of Triton X-100 in a final volume of 100 ml H$_2$O. Store at room temperature in the dark.

4. *10% NaN$_3$:* Dissolve 1 g NaN$_3$ in a final volume of 10 ml H$_2$O. Store at room temperature.

5. *10 mg/ml proteinase K stock:* Dissolve 5 mg of proteinase K in 50 mM Tris–HCl, pH 8.0, 1 mM CaCl$_2$, and 0.02% NaN$_3$. Store at 4°C for up to 1 week without loss of activity.

6. *200 μg/ml proteinase K solution:* Mix 70 μl of the 10-mg/ml proteinase K stock with 290 μl H$_2$O. Add 200 μl 10% Triton X-100. Use fresh.

7. *200 mM PMSF:* Make a fresh solution of PMSF by dissolving 34.85 mg of PMSF in 1 ml of ethanol.

Steps

1. Transfer 200 μl from the remainder of the import reaction into a fresh Eppendorf tube. Add 200 μl of the 200-μg/ml proteinase K solution and mix rapidly. Leave the tube on ice for 15 min.

2. Add 2 μl 200 mM PMSF while agitating on a Vortex mixer. Keep on ice for 5 min. Add 44 μl 50% TCA. Add 300 μl of acetone to dissolve the Triton X-100 that precipitates in the presence of TCA. These samples measure the fraction of the precursor protein that has completely crossed the outer membrane and has reached the folded state (folded, Fig. 1, lanes 4 and 7).

G. Final Processing of Samples and Preparation of a Precursor Standard

Steps

1. To inactivate proteases, incubate the TCA-precipitated samples (total, import, folded) at 65°C for 5 min. Place on ice for 5 min and subsequently collect the TCA precipitate by spinning for 10 min at 20,000 g.

2. Remove supernatant by aspiration and dissolve the pellets in 50 μl 1× sample buffer. If the sample buffer turns yellow, overlay the sample with NH$_3$ gas taken from above a 25% NH$_3$ solution. Agitate to mix the gaseous NH$_3$ into the sample buffer until the color turns blue again.

3. Incubate the samples for 5 min at 95°C.

4. To estimate the efficiency of the import reaction, the amount of precursor protein added to the import reaction has to be determined. The efficiency of import for most precursor proteins is between 5 and 30%. Here we use a 10% standard for both rhodanese and Su9-DHFR (Fig. 1, lane 1).

5. To obtain a 10% standard, mix 4 μl of purified yeast mitochondria (see Procedure D) with 50 μl 1× sample buffer. Incubate at 95°C for 3 min.

6. Add 1 μl of the precursor mix (Procedure C) and incubate for 5 min at 95°C.

H. SDS-Gel Electrophoresis and Processing of the Gel

Solutions

1. *5% TCA:* To make 5 liter, add 250 g of TCA to 5 liters H$_2$O.

2. *1 M Tris base:* Dissolve 121 g of Tris in 1 liter H$_2$O.

3. *1 M sodium salicylate:* Dissolve 160 g in 1 liter of H$_2$O.

Steps

1. Run samples on a 10% Tris–Tricine gel (Schägger and von Jagow, 1987) stabilized by the addition of 0.26% linear polyacrylamide prior to polymerization.

2. To reduce radioactive background, boil 5% TCA in a beaker under the hood. Add the gel to the boiling TCA and incubate for 5 min.

3. Recover the gel and place it into a tray. Wash briefly with water. Neutralize by incubation in 1 M Tris base for 5 min on a shaker.

4. Wash briefly with water. Add 1 M salicylate and incubate for 20 min on a shaker.

5. Dry gel onto a Whatman Filter paper and expose to a Kodak X-OMAT X-ray film for the desired time. Exposure time for the experiment shown in Fig. 1 was 2 hr.

IV. COMMENTS

The method describes a standard experiment to test a precursor protein that has not been used in mitochondrial import before. Most importantly, as demonstrated here for rhodanese and Su9-DHFR, the protocol will reveal if import is dependent on a membrane potential (compare Fig. 1 lanes 3 and 6). This is essential, as sometimes protease-resistant precursor proteins tend to stick to the outside of mitochondria, thereby "mimicking" import.

The efficiency of import can be deduced by a comparison of the amount of imported material with a precursor standard (compare Fig. 1 lanes 1 and 3). In addition, the experiment reveals if a precursor protein folds to a protease-resistant conformation after its import into the mitochondrial matrix. Under the conditions chosen here, complete protease resistance was obtained for Su9-DHFR (Fig. 1, lanes 3 and 4). However, rhodanese did not become resistant to protease during the time course of the experiment (Fig. 1, lanes 3 and 4). This is consistent with the earlier finding that rhodanese interacts with hsp60 in the mitochondrial matrix and becomes protease resistant only after incubation times exceeding 15 min (Rospert *et al.*, 1996). The dramatic difference between Su9-DHFR and rhodanese with respect to import efficiency and folding rate exemplifies the need to determine import and folding kinetics for each precursor protein.

V. PITFALLS

The quality of the DNA used for transcription is essential for efficiency. Use a clean, RNA- and RNase-free plasmid preparation (e.g., purified with a QIAGEN plasmid kit, Qiagen). Linearize plasmid by cutting with a restriction enzyme behind the coding region of the gene of interest. Extract with phenol/chloroform, and then with chloroform, precipitate with 100% ethanol, and wash with 70% ethanol. Resuspend the dried pellet in H_2O at a concentration of 1 $\mu g/\mu l$ and store at 4°C. Never freeze DNA templates used for transcription.

To avoid RNase contamination, solutions used for transcription and translation have to be prepared with special caution. Always wear gloves, even when loading pipette tips into boxes. If initiation at downstream AUG codons is a problem, try diluting the reticulocyte lysate up to fourfold.

It is important to establish that import is linear with time. To establish those conditions it is necessary to perform time course experiments of the import reaction and to try import at different temperatures.

Methods for determining the intramitochondrial localization of an imported precursor protein (Glick, 1991), investigating the energy requirements of mitochondrial import (Glick, 1995), and detecting interactions between imported precursor proteins and matrix chaperones (Rospert and Hallberg, 1995) have been published elsewhere.

Acknowledgments

We thank Carla Köhler and Yves Dubaquié for help with the manuscript. This study was supported by Grant 31-40510.94 from the Swiss National Science Foundation. S.R. was supported by an EMBO long-term fellowship.

References

Glick, B. S. (1991). Protein import into isolated yeast mitochondria. *Methods Cell Biol.* **34,** 389–399.
Glick, B. S. (1995). Pathways and energetics of mitochondrial protein import in Saccharomyces cerevisiae. *Methods Enzymol.* **260,** 224–231.
Glick, B. S., and Pon, L. A. (1995). Isolation of highly purified mitochondria from Saccharomyces cerevisiae. *Methods Enzymol.* **260,** 213–223.
Jensen, R. E., and Yaffe, M. P. (1988). Import of proteins into yeast mitochondria: The nuclear MAS2 gene encodes

a component of the processing protease that is homologous to the MAS1-encoded subunit. *EMBO J.* **7,** 3863–3871.

Melton, D. A., Krieg, P. A., Rebagliati, M. R., Maniatis, T., Zinn, K., and Green, M. R. (1984). Efficient in vitro synthesis of biologically active RNA and RNA hybridization probes from plasmids containing a bacteriophage SP6 promoter. *Nucleic Acids Res.* **18,** 7035–7053.

Pfanner, N., Müller, H. K., Harmey, M. A., and Neupert, W. (1987). Mitochondrial protein import: Involvement of the mature part of a cleavable precursor protein in the binding to receptor sites. *EMBO J.* **6,** 3449–3454.

Rospert, S., and Hallberg, R. L. (1995). Interaction of HSP60 with proteins imported into the mitochondrial matrix. *Methods Enzymol.* **260,** 287–292.

Rospert, S., Looser, R., Dubaquié, Y., Matouschek, A., Glick, B. S., and Schatz, G. (1996). Hsp60-independent protein folding in the matrix of yeast mitochondria. *EMBO J.* **15,** 764–774.

Schägger, H., and von Jagow, G. (1987). Tricine-sodium dodecyl sulfate-polyacrylamide gel electrophoresis for the separation of proteins in the range from 1 to 100 kDa. *Anal. Biochem.* **166,** 368–379.

Schatz, G. (1995). Mitochondria: Beyond oxidative phosphorylation. *Biochim. Biophys. Acta* **1271,** 123–126.

Wachter, C., Schatz, G., and Glick, B. S. (1994). Protein import into mitochondria: The requirement for external ATP is precursor-specific whereas intramitochondrial ATP is universally needed for translocation into the matrix. *Mol. Biol. Cell* **5,** 465–474.

Import of Nuclear-Encoded Proteins by Isolated Chloroplasts and Thylakoids

Ruth M. Mould and John C. Gray

I. INTRODUCTION

The vast majority of chloroplast proteins are nuclear encoded, synthesized in the cytosol and imported by the organelle posttranslationally. Most nuclear-encoded chloroplast proteins are synthesized with N-terminal presequences that contain targeting information for translocation across the envelope membranes. A specific stromal processing peptidase removes the presequences of stromal proteins and most integral thylakoid membrane proteins generating mature-size forms. Targeting of some thylakoid lumen proteins is more complex; these proteins are synthesized in the cytosol with presequences that contain two targeting signals in tandem. The first domain is structurally and functionally equivalent to the presequence of a stromal protein and is usually cleaved off in the stroma, yielding an intermediate-size form of the protein. The second domain contains information required for translocation across the thylakoid membrane and is cleaved off in the thylakoid lumen by a thylakoidal-processing peptidase generating the mature-size protein. Although nuclear-encoded proteins are translocated across the envelope via a common pathway, there are several different pathways for the targeting of proteins from the stroma into or across the thylakoid membrane; the pathway taken depends on the protein in question (reviewed in Cline and Henry, 1996). Protein translocation across the thylakoid membrane can be driven by the ΔpH generated across the membrane in the light (Mould and Robinson, 1991) and require no stromal proteins (Mould et al., 1991) or require ATP and be dependent on the protein CPSecA (Yuan et al., 1994). Integration of a protein into the thylakoid membrane can be spontaneous (Michl et al., 1994) or require the stromal protein 54CP (Li et al., 1995). The translocation pathways outlined earlier were elucidated using assays for protein translocation in which isolated chloroplasts or thylakoids were incubated with radiolabeled precursor proteins synthesized in vitro. This article describes protocols which enable the investigator with a cloned cDNA encoding a chloroplast protein to determine the location of that protein within the chloroplast. The targeting pathways of thylakoid proteins can be studied in isolation from translocation across the envelope using Procedure D (see Comment 6). The ΔpH-dependent pathway for translocation across the thylakoid membrane can be inhibited by the addition of the ionophore nigericin to import assays (Mould and Robinson, 1991) and the CPSecA-dependent pathway can be inhibited by the addition of azide (Knott and Robinson, 1994).

II. MATERIALS AND INSTRUMENTATION

See all materials and instrumentation required in volume 2, Mould and Gray, "Preparation of Chloroplasts for Protein Synthesis and Protein Import."

CAP (m^{7}G(5′)ppp(5′)G) dilithium salt (Cat. No. 904988), ribonucleoside triphosphates (set of 100 mM CTP, GTP, ATP, and UTP lithium salts at pH 7, Cat. No. 1277057), and RNA polymerases (20 U/μl) supplied with 10× transcription buffer (SP6, Cat. No. 810274; T3, Cat. No. 1031163, T7, Cat. No. 881767) are from Boehringer-Mannheim. RNasin (RNase inhibitor at 20–40 U/μl, Cat. No. N2111) and wheat germ extract, minus methionine (supplied with 1 M potassium acetate and 1 mM amino acid mixture minus methionine, Cat. No. L4380), are from Promega.

Import reactions are carried out in an illuminated water bath at approximately 25°C; for small numbers of samples, a glass beaker with one or two desk lamps illuminating the samples from below is adequate, providing the temperature is closely monitored to prevent heating over 27°C.

III. PROCEDURES

A. Transcription of Cloned cDNA *in Vitro*

Solutions

1. *CAP:* Dissolve CAP to 1 mM in sterile water (add 123 μl to 5 A_{250} units) and store at −20°C.

2. *Ribonucleoside triphosphate mixture:* To make 100 μl, mix 5 μl 100 mM CTP, 5 μl 100 mM ATP, 5 μl 100 mM UTP, 1 μl 100 mM GTP, and 84 μl sterile water.

3. *8 mM GTP:* To make 100 μl, mix 8 μl 100 mM GTP with 92 μl sterile water.

Steps

1. The cDNA should be inserted into a plasmid vector suitable for transcription with SP6, T7, or T3 RNA polymerase, and a high-quality plasmid preparation resuspended to 1 μg/μl in sterile water (see Comment 1).

2. Mix at room temperature: 2 μl 1 mM CAP, 4 μl ribonucleoside triphosphate mixture, 2 μl 10× transcription buffer, 1 μl RNasin (40 units), 7 μl sterile distilled water, 2 μl DNA template, and 2 μl RNA polymerase (SP6, T7, or T3 at 20 U/μl). Mixture totals 20 μl.

3. Incubate at 37°C for 30 min.

4. Add 2 μl 8 mM GTP and continue incubation for another 30 min.

5. Store transcription products at −80°C until required (see Comment 2).

B. Production of Radiolabeled Proteins *in Vitro*

Steps

1. Gently mix by pipetting: 5.0 μl wheat germ extract, 0.2 μl RNasin (40 U/μl), 0.4 μl 1 M potassium acetate, 0.8 μl 1 mM amino acid mixture, 0.5 μl [^{35}S]methionine (see Comment 3), 2.3 μl sterile distilled water, and 0.8 μl transcription product. Mixture totals 10.0 μl.

2. Incubate at 25°C for 1 hr.

3. One microliter of translation product analyzed by electrophoresis on an SDS–polyacrylamide gel and detected by fluorography should give an intense band after an overnight exposure (see Comment 4).

C. Import of Proteins by Isolated Chloroplasts

The following procedure enables the investigator to determine whether a radiolabeled protein synthesized *in vitro* is imported by isolated chloroplasts and if it is targeted to the stroma or

NOTE

The optimal translation conditions for each transcription product should be determined. For a 10-μl reaction we generally try 0, 0.4, and 0.8 μl 1 M potassium acetate and 0.4, 0.8, and 1.6 μl of transcription products, altering the volume of sterile water accordingly.

thylakoid network (Fig. 1). For targeting to the chloroplast envelope membranes, chloroplasts may be fractionated by procedures described by Keegstra and Yousif (1986) and Schnell and Blobel (1993).

Solutions

1. *40% Percoll/1× SRM/50 mM EDTA:* To make 5 ml, mix 2 ml Percoll, 1 ml 5× SRM, 0.25 ml 1 *M* EDTA, pH 8, and 1.75 ml distilled water on the day of use. Store at 4°C.

2. *500 mM EDTA/1× SRM:* To make 10 ml, mix 5 ml 1 *M* EDTA, pH 8, 2 ml 5× SRM, and 3 ml distilled water. Make on the day of use and store at 4°C.

3. 5× SRM, 1× SRM, 1 *M* EDTA, pH 8, 10 m*M* HEPES, pH 8, and SB (SDS sample buffer) are prepared as described in Volume 2, Mould and Gray, "Preparation of Chloroplasts for Protein Synthesis and Protein Import."

4. 100 m*M* methionine, 100 m*M* MgATP, and thermolysin are prepared as described in the article by Ruth M. Mould and John C. Gray, this volume, except that compounds are dissolved in 1× SRM instead of in 10 m*M* HEPES, pH 8.

Steps

1. Set up translation reactions as in Procedure B.

2. Prepare isolated chloroplasts as described in Volume 2, Mould and Gray, "Preparation of Chloroplasts for Protein Synthesis and Protein Import," and resuspend to 1 mg/ml chlorophyll with 1× SRM.

NOTE

Retain a small amount of translation product (e.g., 0.5 μl) for electrophoresis alongside import reactions on SDS–polyacrylamide gels (see Figs. 2 and 3).

3. In a microcentrifuge tube, gently mix 100 μl chloroplast suspension, 20 μl MgATP (100 m*M* stock), 10 μl methionine (100 m*M* stock), 10 μl translation product, and 60 μl 1× SRM. Mixture totals 200 μl.

4. Place samples in polystyrene floats with the chloroplast suspension exposed and incubate at 22–26°C for 40 min in an illuminated water bath (100 μmol photons m^{-2} sec^{-1}).

5. Add 600 μl ice-cold 1× SRM (to give a total volume of 800 μl).

6. Using a wide-bore pipette tip, pipette 200 μl (this will be for the washed chloroplast fraction), and the remaining 600 μl, into separate microcentrifuge tubes on ice.

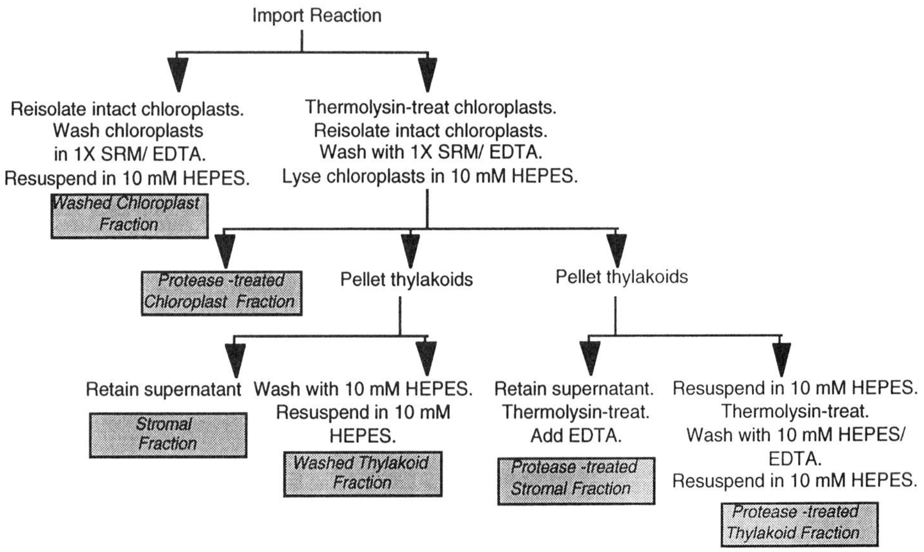

FIGURE 1 Fractionation of chloroplasts after import reaction.

7. Add 30 μl thermolysin (4 mg/ml stock) to the 600-μl portion and incubate on ice for 40 min.

8. Add 70 μl 500 mM EDTA/1× SRM to the thermolysin-treated portion.

9. Layer each portion on top of 800 μl 40% Percoll/ 1× SRM/ 50 mM EDTA at 4°C.

10. Microcentrifuge each portion at full speed for 5 min at 4°C.

11. Discard the upper layer (broken chloroplasts) and Percoll layer from each tube and reserve the pellets.

12. Wash each pellet by gently resuspending in 1 ml 1× SRM (using a wide-bore pipette tip or by inverting the tube several times) and microcentrifuge at half speed (2000–3000 *g*) for 5 min at 4°C.

13. Resuspend the nonthermolysin-treated chloroplast pellet in 50 μl 10 mM HEPES, pH 8, by pipetting and add 50 μl SB. Store at −20°C (washed chloroplast fraction).

14. Lyse the thermolysin-treated chloroplast pellet in 150 μl 10 mM HEPES, pH 8, by pipetting.

15. Divide the sample into three aliquots of 50 μl.

16. Add 50 μl SB to 1 aliquot and store at −20°C (protease-treated chloroplast fraction).

17. Microcentrifuge the remaining two aliquots at full speed for 5 min at 4°C.

18. Transfer the supernatants (stromal fractions) to clean microcentrifuge tubes and reserve the pellets (thylakoid fractions) on ice.

19. Microcentrifuge the stromal fractions at full speed for 5 min at 4°C to pellet any contaminating thylakoids. Transfer the supernatants to clean microcentrifuge tubes.

20. Make the volume of one stromal fraction up to 100 μl with SB. Store at −20°C (stromal fraction).

21. Add thermolysin to the remaining stromal fraction to a final concentration of 0.2 mg/ml and incubate on ice for 40 min. Add 500 mM EDTA stock to give a final concentration of 50 mM and make the volume up to 100 μl with SB (protease-treated stromal fraction).

22. Resuspend one thylakoid pellet in 1 ml 10 mM HEPES, pH 8, by pipetting. Microcentrifuge at full speed for 5 min at 4°C. Discard the supernatant and resuspend the pellet in 50 μl 10 mM HEPES, pH 8. Add 50 μl SB and store at −20°C (washed thylakoid fraction).

23. Resuspend the remaining thylakoid pellet to 200 μl 10 mM HEPES, pH 8. Add 10 μl thermolysin (4 mg/ml stock) and incubate on ice for 40 min.

24. Add 1 ml 10 mM HEPES, pH 8, and 130 μl 500 mM EDTA.

25. Microcentrifuge at full speed for 5 min at 4°C.

26. Discard the supernatant and resuspend the pellet to 50 μl with 10 mM HEPES, pH 8. Add 50 μl SB and store at −20°C (protease-treated thylakoid fraction).

27. Heat samples to 80°C for 2 min before analyzing by SDS–polyacrylamide gel electrophoresis followed by fluorography. Figure 2 shows a typical example of the import by isolated chloroplasts of a protein targeted to the thylakoid lumen.

FIGURE 2 Import of a precursor thylakoid protein by isolated chloroplasts. A radiolabeled precursor protein consisting of the presequence of Rubisco small subunit fused to the cytochrome *f* precursor was incubated with isolated chloroplasts, and the sample was treated as described in Procedure B. Annotations: Molecular weight markers (M), translation product (T), washed chloroplast fraction (wC), protease-treated chloroplast fraction (pC), stromal fraction (S), washed thylakoid fraction (wTh), protease-treated stromal fraction (pS), protease-treated thylakoid fraction (pTh), and kilodaltons (K). The precursor, intermediate, and mature forms of the radiolabeled protein are denoted P, I, and M, respectively.

D. Import of Proteins by Isolated Thylakoids

Solutions

1. *10 mM HEPES, pH 8/ 60 mM EDTA:* To make 10 ml, mix 1 ml 100 mM HEPES, pH 8, 0.6 ml 1 M EDTA, pH 8, and 8.4 ml sterile distilled water. Store at 4°C.

2. 1 M EDTA, 100 mM methionine, 100 mM MgATP, thermolysin, 10 mM HEPES, pH 8, and SB (SDS sample buffer) are prepared as described in Volume 2, Mould and Gray, "Preparation of Chloroplasts for Protein Synthesis and Protein Import."

Steps

1. Set up translation reactions as in Procedure B.

2. Prepare isolated chloroplasts as in Volume 2, Mould and Gray, "Preparation of Chloroplasts for Protein Synthesis and Protein Import," and resuspend to 1 mg/ml chlorophyll with 1× SRM.

3. Pellet chloroplasts by microcentrifugation at half speed (2000–3000 *g*) for 3 min at 4°C and discard the supernatant.

4. Lyse chloroplasts by resuspending pellet to 1 mg/ml chlorophyll in 10 mM HEPES, pH 8. Pipette until the mixture becomes homogenous, taking care not to bubble air through the sample. Leave on ice for 5 min.

5. Microcentrifuge the sample at full speed for 5 min at 4°C.

6. Reserve the supernatant (stromal fraction) on ice.

7. Resuspend the pellet in 1 ml 10 mM HEPES, pH 8, and microcentrifuge at full speed for 5 min at 4°C.

8. Resuspend the thylakoid pellet in stromal fraction to 1 mg/ml chlorophyll (see Comment 5).

9. In a microcentrifuge tube, gently mix 25.0 μl thylakoid suspension, 2.5 μl methionine (100 mM stock), 5.0 μl translation product, 5.0 μl MgATP (100 mM stock), and 12.5 μl 10 mM HEPES, pH 8. Mixture totals 50.0 μl.

10. Incubate at 22–26°C for 40 min in an illuminated water bath (100 μmol photons m^{-2} sec^{-1}).

11. Add 1 ml ice-cold 10 mM HEPES, pH 8.

12. Microcentrifuge at full speed for 5 min at 4°C.

13. Resuspend pellet to 200 μl in 10 mM HEPES, pH 8.

14. Divide into two aliquots of 100 μl and place on ice.

15. Add 5 μl thermolysin (4 mg/ml stock) to one aliquot and incubate on ice for 40 min.

16. Add 1 ml 10 mM HEPES, pH 8/60 mM EDTA to each aliquot.

17. Microcentrifuge at full speed for 5 min at 4°C.

18. Discard supernatants and resuspend each pellet to 25 μl in 10 mM HEPES, pH 8.

19. Add 25 μl SB to each sample. Store samples at −20°C.

20. Heat samples to 80°C for 2 min before analyzing by SDS–polyacrylamide gel electrophoresis followed by fluorography. Figure 3 shows a typical example of the import of a thylakoid lumen protein by isolated thylakoids.

IV. COMMENTS

1. We generally linearize plasmids by restriction digests at a suitable site 3′ to the cDNA stop codon; although not essential for efficient transcription, it prevents run-on of transcription.

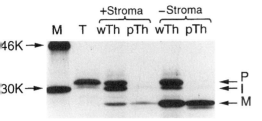

FIGURE 3 Import of a nuclear-encoded thylakoid protein by isolated thylakoids. The radiolabeled precursor form of the 23-kDa subunit of the photosystem II oxygen-evolving complex was incubated with thylakoids resuspended in stroma and treated as described in Procedure D (lanes labeled "+ Stroma"). A second sample was treated the same way except that the radiolabeled protein was incubated with thylakoid resuspended in 10 mM HEPES, pH 8 (lanes labeled "− Stroma"). Annotations: Molecular weight markers (M), translation product (T), washed thylakoids (wTh), protease-treated thylakoids (pTh), and kilodaltons (K). The precursor, intermediate, and mature forms of the radiolabeled protein are denoted P, I, and M, respectively. (See Comment 5.)

2. Transcription products can be visualized by electrophoresis of 5 μl reaction products and 0.5 μl DNA template on a 1% agarose gel stained with ethidium bromide.

3. Translation reactions sometimes produce several major polypeptides, usually due to premature termination of translation. Visualizing translation products by SDS–PAGE followed by fluorography enables translation conditions to be selected which produce the least truncated products.

4. The choice of radiolabeled amino acid used for the translation reaction is dependent on the protein sequence. We generally use [³⁵S]methionine but for some proteins a mixture of [³⁵S]methionine and [³⁵S]cysteine significantly increases the number of labeled residues in the translation product. The chosen amino acid must be present in the mature form of the protein generated after any processing by chloroplast enzymes. For proteins lacking methionine and cysteine, [³H]leucine is used.

5. Proteins translocated across the thylakoid membrane via the ΔpH-dependent pathway tend to be imported more efficiently by isolated thylakoids resuspended in 10 mM HEPES, pH 8, rather than in stroma (Mould *et al.*, 1991; see Fig. 3). However, because stromal extract is essential for the import of proteins translocated by the CPSecA-dependent pathway, we recommend the inclusion of stromal extract in preliminary import studies.

6. In our experience, some nuclear-encoded thylakoid proteins are imported efficiently *in vitro* by isolated chloroplasts and are targeted to the thylakoid network, but are imported much less efficiently by isolated thylakoids. Therefore we recommend Procedure C for preliminary studies.

V. PITFALLS

1. If import fails, positive controls should be used in parallel import reactions to test *in vitro* import assay conditions. The precursor of the small subunit of Rubisco from pea is a nuclear-encoded stromal protein that is imported efficiently by isolated chloroplasts (Anderson and Smith, 1986); the precursor of the 23-kDa subunit of the photosystem II oxygen-evolving complex from wheat is a thylakoid lumen protein that is imported efficiently by isolated chloroplasts and isolated thylakoids (James *et al.*, 1989).

References

Anderson, S., and Smith, S. M. (1986). Synthesis of the small subunit of ribulose-bisphosphate carboxylase from genes cloned into plasmids containing the SP6 promoter. *Biochem. J.* **240**, 709–714.

Cline, K., and Henry, R. (1996). Import and routing of nucleus-encoded chloroplast proteins. *Annu. Rev. Cell Dev. Biol.* **12**, 1–26.

James, H. E., Bartling, D., Musgrove, J. E., Kirwin, P. M., Herrmann, R. G., and Robinson, C. (1989). Transport of proteins into chloroplasts: Import and maturation of precursors to the 33-, 23-, and 16-kDa proteins of the photosynthetic oxygen-evolving complex. *J. Biol. Chem.* **264**, 19573–19576.

Keegstra, K., and Yousif, A. E. (1986). Isolation and characterization of chloroplast envelope membranes. *In* "Methods in Enzymology" (A. Weissbach and H. Weissbach, eds.), Vol. 118, pp. 316–325. Academic Press, San Diego.

Knott, T. G., and Robinson, C. (1994). The SecA inhibitor, azide, reversibly blocks the translocation of a subset of proteins across the chloroplast thylakoid membrane. *J. Biol. Chem.* **269**, 7843–7846.

Li, X., Henry, R., Yuan, J., Cline, K., and Hoffman, N. (1995). A chloroplast homolog of the signal recognition particle subunit SRP54 is involved in the posttranslational integration of a protein into thylakoid membranes. *Proc. Natl. Acad. Sci. USA* **92**, 3789–3793.

Michl, D., Robinson, C., Shackleton, J. B., Herrmann, R. G., and Klösgen, R. B. (1994). Targeting of proteins to the thylakoids by bipartite presequences: CFoII is imported by a novel, third pathway. *EMBO J.* **13**, 1310–1317.

Mould, R. M., and Robinson, C. (1991). A proton gradient is required for the transport of two lumenal oxygen-evolving proteins across the thylakoid membrane. *J. Biol. Chem.* **266**, 12189–12193.

Mould, R. M., Shackleton, J. B., and Robinson, C. (1991). Transport of proteins into chloroplasts: Requirements for the efficient import of two lumenal oxygen-evolving complex proteins into isolated thylakoids. *J. Biol. Chem.* **266**, 17286–17289.

Schnell, D. J., and Blobel, G. (1993). Identification of intermediates in the pathway of protein import into chloroplasts and their localisation to envelope contact sites. *J. Cell Biol.* **120**, 103–115.

Yuan, J., Henry, R., McCaffery, M., and Cline, K. (1994). SecA homolog in protein transport within chloroplasts: Evidence for endosymbiont-derived sorting. *Science* **266**, 796–798.

Peroxisomes

A Permeabilized Cell System to Study Peroxisomal Protein Import

Martin Wendland and William M. Nuttley

I. INTRODUCTION

The development of permeabilized cell systems has facilitated the reconstitution and biochemical characterization of a variety of intracellular protein trafficking steps. This article describes how coverslip-attached cells are permeabilized with streptolysin O (SLO) and incubated with exogenously added peroxisomal proteins. The protein import into peroxisomes is analyzed by a simple immunofluorescence technique. The procedure is described for CHO cells but it can be used for a variety of mammalian cells. Furthermore, the system can be applied to study the import of proteins into other subcellular organelles.

SLO is a bacterial cytolysin that binds to cholesterol in the plasma membrane and forms pores large enough to allow the exchange of cytosolic components of up to 150 kDa or even larger (Ahnert-Hilger *et al.*, 1989, see Volume 4, Ahnert-Hilger and Weller, "α-Toxin and Streptolysis O as Tools in Cell Biological Research"). It has been shown that treatment of cells with SLO results in the specific perforation of the plasma membrane while intracellular membranes remain intact (Miller and Moore, 1991; Wendland and Subramani, 1993a).

The import of proteins into peroxisomes is directed by peroxisomal targeting signals (PTSs) (Subramani, 1993). A carboxy-terminal tripeptide with the sequence of SKL (in the one-letter amino acid code) represents a prototype PTS. Cross-linking of a peptide ending in SKL to a carrier proteins creates an artificial substrate for the peroxisomal import machinery (Walton *et al.*, 1992).

II. MATERIALS AND INSTRUMENTATION

CHO wild-type cells are from the American Type Culture Collection. Luciferase from *Photinus pyralis* (Cat. No. L-5256), ATP (Cat. No. A-5394), creatine phosphate (Cat. No. P-6915), and creatine phosphokinase (Cat. No. C-3755) are from Sigma Chemical Company. Human serum albumin (Cat. No. 126658) and Mowiol (Cat. No. 475904) are from Calbiochem Novabiochem International. SLO (Cat. No. MR16) prepared by Wellcome Diagnostic is from Murex Diagnostics. *m*-Maleimidobenzoyl-*N*-hydroxy sulfosuccinimide ester (Sulfo-MBS, Cat. No. 22312) is from Pierce. Pronectin F (Cat. No. 205001) is from Stratagene. Rabbit reticulocyte lysate (Cat. No. L4151) is from Promega. Rabbit anti-HSA antibodies (Cat. No. 65-051) are from ICN Biopharmaceuticals Inc. Rabbit antiluciferase antibodies are from Dr. S. Subramani (University of California, San Diego). Guinea pig anti-HSA and rabbit antitubulin antibodies are from Dr. S.J. Singer (University of California, San Diego). Rabbit anticatalase antibodies are from Dr. P.B. Lazarow (Mount Sinai School of Medicine, New York). Rhodamine-conjugated goat anti-rabbit IgG (Cat. No. 111-025-003), rhodamine-conjugated goat anti-guinea pig IgG (Cat. No. 106-025-003), FITC-conjugated goat anti-rabbit IgG (Cat. No. 111-095-003), and FITC-conjugated goat anti-guinea pig IgG (Cat. No. 106-095-003) are from Jackson ImmunoResearch Laboratories. A synthetic peptide with the amino acid sequence NH_2-CRYHLKPLQSKL-COOH is from Multiple Peptide Systems. Centricon-30 microconcentrators (Cat. No. 4208) are from Amicon. All other reagents are purchased from standard sources. Immunofluorescence microscopy is performed on a Zeiss photoscope II microscope.

III. PROCEDURES

A. Coating of Coverslips with Pronectin F

Coat coverslips with pronectin F to strengthen cell attachment. This is vital for allowing maximum cell recovery after permeabilization.

Solution

1. *Pronectin F:* Dissolve pronectin F powder in the diluent supplied by the manufacturer to obtain a 1-mg/ml stock solution and store at room temperature. For immediate use, dilute the stock solution in sterile phosphate-buffered saline (PBS) to 20 μg/ml.

Steps

1. Autoclave coverslips (18 mm circular) in a glass petri dish.

2. Spot drops of 50–100 μl of diluted pronectin F solution (20 μg/ml) in tissue culture dishes and invert coverslips over the drops.

3. Leave for 2 hr at room temperature, invert the coverslips (coated side up), and rinse them three times with sterile PBS. The coverslips are now ready for the immediate seeding of cells or can be stored for later use at room temperature for up to 4 months.

B. Preparation of Cells and Seeding onto Coverslips

1. Cell Culture

Steps

Grow CHO cells in α-MEM containing 10% fetal calf serum (v/v), 200 U/ml penicillin G, and 200 μg/ml streptomycin. Maintain the cultures in a humidified incubator at 37°C with 5% CO_2. For cell passaging use standard trypsinization procedures. It is recommended that cultures never be overgrown.

2. Seeding of Cells onto Coverslips

Steps

The seeding of cells should be performed at least 36 hr before the import assay. Place up to 12 coated coverslips, coated side facing up, into a 10-cm tissue dish. Trypsinize cells from a confluent dish and seed at 1×10^6 cells/10-cm dish for permeabilization with 0.2 U/ml SLO and at 3×10^6 cells/10-cm dish for permeabilization with 2.0 U/ml SLO. Avoid cell clumping because it impairs the permeabilization with SLO. Two hours before the import assay, change the medium.

C. Permeabilization of Cells with SLO

Solutions

1. *Transport buffer:* To make 1 liter, dissolve 4.76 g of HEPES (20 mM), 10.79 g of potassium acetate (110 mM), 0.41 g of sodium acetate (5 mM), 0.43 g of magnesium acetate (2 mM), and 0.38 g of EGTA (1 mM) in 800 ml distilled water, adjust pH to 7.3 with 10 N potassium hydroxide, and bring the total volume to 1000 ml. Autoclave and store at room temperature. Before use add dithiothreitol (DTT) to 2 mM (0.308 g/liter).

2. *Streptolysin O stock solution:* Vials containing 40 U of SLO as lyophilized powder are supplied by the manufacturer. Dissolve the powder of one vial in 10 ml of distilled water to obtain a 4 U/ml stock solution. Aliquot and store at −70°C. Stock solutions should not be freeze–thawed more than twice. For immediate use, dilute SLO stock in transport buffer/DTT to 0.2 or 2.0 U/ml or any other desired dilution.

Two differing cell permeabilization procedures are described that allow the study of peroxisomal protein import when it is either independent (0.2 U/ml SLO) or dependent (2.0 U/ml SLO) on externally added cytosol.

1. Permeabilization with 0.2 U/ml SLO

Steps

1. The cells should be at subconfluent density. Dilute the SLO stock solution in transport buffer/DTT to 0.2 U/ml; prepare 40 μl per coverslip. Two minutes before permeabilization, prewarm the diluted SLO solution to 30°C.

2. Before treatment with SLO the culture medium has to be rinsed off the cells. A major problem encountered is the loss of cells during the multiple wash steps required throughout the assay. It has been found that careful dip washing of the coverslips allows the highest cell recoveries. Fill three beakers with 50–100 ml of transport buffer (room temperature) and, with the use of forceps, carefully dip the coverslips sequentially into beakers 1–3. Subsequently remove excess fluid by blotting the edge of each coverslip on Kimwipe tissue, but do not dry completely. Invert coverslips cell side down over SLO drops (see step 3).

3. Carry out the permeabilization and subsequent incubation with transport mix in a humidified floating box. A household plastic box with a lid works well for this purpose. Insert paper towels on box bottom, soak with water, and place a layer of Parafilm on top. Preincubate the chamber in a water bath at 30°C. Spot drops of 40 μl of diluted SLO solution on top of Parafilm and invert coverslips cell side down over SLO drops.

4. Incubate for 5–10 min at 30°C. Dip wash coverslips as before and blot off excess fluid. The cells are now permeabilized and ready for the import assay. By this time, have the transport mix spotted onto Parafilm (see Procedure D, step 1).

2. Permeabilization with 2.0 U/ml SLO

Steps

1. Cells should be at confluent density, ideally forming a contiguous monolayer. However, the formation of cell layers is detrimental.

2. The permeabilization procedure is the same as just described, except that the cells are permeabilized with 2.0 U/ml SLO and the incubation with SLO is extended to 10 min.

3. After incubation with SLO, wash the coverslip and incubate them in 1 ml of cold transport buffer for 15 min to ensure the leakage of endogenous cytosol. We use 18-mm circular coverslips which fit nicely into the wells of a 12-well tissue culture plate.

4. Dip wash the coverslips and blot off excess fluid. The coverslips are now ready for incubation with transport mix.

Note that the described conditions were developed for a specific preparation (lot) of SLO and that there may be variations in the use of different cell lines. We recommend starting with a serial dilution of SLO and assaying SLO-treated cells for perforation of the plasma membrane and leakage of endogenous cytosol. The permeabilization and access to the cell interior can be assayed by incubating SLO- or mock-treated cells with antibodies against the cytoskeleton as is shown in Fig. 1. Leakage of endogenous cytosol can be tested by comparing the lactate dehydrogenase activity (a cytosolic marker enzyme of 135 kDa) of SLO-permeabilized cells with nonpermeabilized cells as described by Wendland and Subramani (1993a).

D. Import Reaction

Solutions

1. *ATP-regenerating mix*

    a. ATP stock (100 mM): Dissolve 60 mg of ATP in 800 μl of H_2O and add about 150 μl of 1 N NaOH to neutralize the pH to 7.0–8.0. Check pH and adjust the volume to 1 ml. Aliquot and store at $-20°C$ or, for long-term storage, at $-70°C$.

    b. Creatine phosphate stock (250 mM): Dissolve 64 mg of creatine phosphate in 1 ml of H_2O to make a 250 mM stock. Aliquot and store as described for ATP.

    c. Creatine phosphokinase stock (1000 U/ml): Dissolve lyophilized powder of creatine phosphokinase in 50% glycerol/H_2O (v/v) to obtain a stock of 1000 U/ml. Aliquot and store at $-20°C$ in a freeze-thaw cycle-free freezer.

    d. ATP-regenerating mix: For use in the import assay, mix 1 vol of ATP stock (100 mM) with 2 vol of creatine phosphate stock (250 mM) and 2 vol of creatine phosphokinase stock (1000 U/ml). Mix well and add 3 μl of this ATP-regenerating mix to 60 μl final volume of the transport mix (see Procedure D3).

2. *Reporter molecules for the import into peroxisomes*

    a. HSA-SKL stock solution: Coupling of human serum albumin (HSA) with a 12-mer peptide (sequence: NH_2-CRYHLKPLQSKL-COOH) is performed by using the bifunctional cross-linker m-maleimidobenzoyl-N-hydroxysulfosuccinimide ester as described by Harlow and Lane (1988). Determine the protein concentration and adjust to 3 μg/μl. For the import assay, use 1–2 μl per 60 μl transport mix.

    b. Luciferase stock solution: Dissolve the lyophilized luciferase powder in transport buffer to obtain a 1-mg/ml solution. For maximum solubility, leave on ice for 30 min with periodic gentle agitation. Remove insoluble material by centrifugation (5 min at 10,000 g). Dialyze the supernatant overnight against transport buffer, concentrate in a Centricon-30 microconcentrator to 1/10th vol, and determine the protein concentration. The stock solution should contain 5–10 μg/μl protein. For the import assay, use 1–2 μl per 60 μl transport mix.

3. *Transport mix:* Prepare at least 60 μl of transport mix per coverslip (18 mm circular) to prevent the cells from drying out during the import reaction.

    a. Transport mix without addition of cytosol: Mix on ice 56 μl of transport buffer/DTT with 3 μl of ATP-regenerating mix and 1–2 μl of the HSA-SKL stock or 1–2 μl of the luciferase stock. Leave on ice until use.

    b. Transport mix with addition of cytosol: Mix 30 μl of rabbit reticulocyte lysate with 26 μl of transport buffer/DTT, 3 μl of the ATP-regenerating mix, and 1–2 μl of the HSA-SKL stock or 1–2 μl of the luciferase stock.

Steps

1. Prewarm a humidified chamber, as described, to 37°C in a water bath. By the time the SLO permeabilization is finished, 60 μl of transport mix should be spotted on the Parafilm.

2. Invert SLO-treated coverslips (cell side down) over transport mix and incubate for 45–60 min at 37°C.

3. After incubation, carefully lift coverslips, dip wash, and incubate for 30 min at room temperature in 1 ml of transport buffer containing 4 % formaldehyde to fix the cells.

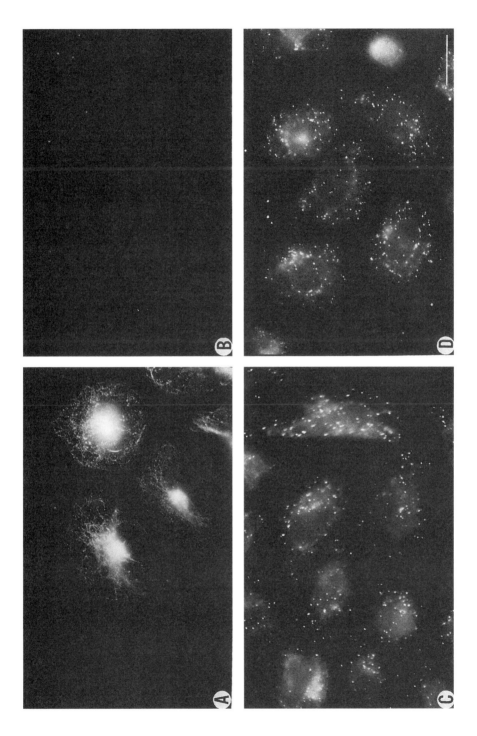

FIGURE 1 Permeabilization of CHO cells with SLO and reconstitution of peroxisomal protein import. Cells were treated with (A) or without (B) 0.2 U/ml SLO, fixed, and incubated with antitubulin antibodies followed by rhodamine-coupled secondary antibodies. Cells permeabilized with 0.2 U/ml SLO were incubated for 45 min at 37°C with HSA-SKL (C) or luciferase (D), and the localization of the exogenously added substrates was analyzed by immunofluorescence. Bar, 20 μm. [Reproduced, by copyright permission of the Rockefeller University Press, from Wendland and Subramani (1993a).]

E. Immunofluorescence Analysis

Solutions

1. *Antibodies for immunofluorescence analysis:* All antibody dilutions are made in PBS. A serial dilution of the antibodies is recommended to find the optimal dilution. Prepare 30 μl of antibody solution per coverslip. Rabbit anti-HSA antibodies are diluted to 1:1000 and rabbit antiluciferase antibodies to 1:50. FITC-conjugated secondary antibodies are diluted to 1:50 and rhodamine-conjugated secondary antibodies to 1:100. Because of a better signal-to-background ratio, FITC-conjugated secondary antibodies are preferred over rhodamine-conjugated antibodies.

2. *Mowiol (mounting medium):* Prepare as described by Harlow and Lane (1988) (see also article by Herzog *et al.*). Instead of DABCO we use 0.1% *p*-phenylenediamine as antifade compound.

Steps

1. Incubate the coverslips for 5 min in PBS containing 1% Triton X-100. Subsequently dip wash coverslips.

2. Blot off excess fluid and invert the coverslips cell side down over a drop of primary antibody solution spotted on Parafilm. Incubate for 20–30 min.

3. Dip wash coverslips, blot off excess fluid, invert over a drop of secondary antibody solution, and incubate as above.

4. Dip wash coverslips and blot off excess fluid. By this time have slides prepared with drops of Mowiol and mount coverslips cell side down onto slides. The slides are now ready for microscopic analysis. Mowiol will harden when left at room temperature overnight (keep in the dark). Subsequently the slides can be stored at −20°C and can be kept for a month or longer.

Note that the SLO treatment only permeabilizes the cell membrane; the membranes of peroxisomes remain intact as judged by the impermeability to antibodies (Wendland and Subramani, 1993a). To allow access of antibodies to peroxisomal matrix proteins, permeabilize the peroxisomal membrane by treatment with Triton X-100.

F. Step-by-Step Procedure

1. Seed cells on coated coverslips 40–48 hr before the import assay (see Procedure B2).

2. Rinse coverslips and invert over a drop of SLO solution. Incubate in a humidified chamber for 5–10 min at 30°C (see Procedure C).

3. Rinse coverslips and invert over a drop of transport mix. Incubate in a humidified chamber for 45–60 min at 37°C (see Procedure D).

4. Rinse coverslips and incubate for 30 min in transport buffer/4% formaldehyde. Treat coverslips for 5 min with PBS/1% Triton X-100 and subject to immunofluorescence analysis (see Procedure E).

IV. COMMENTS

When SLO-permeabilized cells are incubated with exogenously added proteins such as HSA-SKL or luciferase, these substrates are imported into vesicular structures (Figs. 1C and 1D). These vesicles are identified as peroxisomes by colocalization with the peroxisomal marker protein catalase (Figs. 2A and 2B). Depletion of endogenous cytosol from SLO-permeabilized cells abolishes the competence to import proteins into peroxisomes, but import is reconstituted by the addition of external cytosol (Figs. 2C and 2D). Use of the permeabilized cell system has allowed the characterization of several other features of the peroxisomal import event such as signal, energy, time, and temperature dependence (Wendland and Subramani,

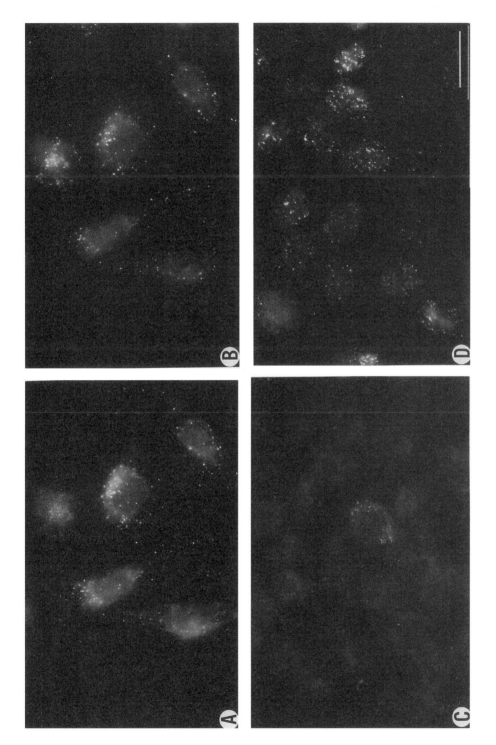

FIGURE 2 Colocalization of imported HSA-SKL with catalase and cytosol dependence of peroxisomal protein import. Cells permeabilized with 0.2 U/ml SLO were incubated with HSA-SKL, and processed for double labeling indirect immunofluorescence. Cells were incubated simultaneously with guinea pig anti-HSA (A) and rabbit anticatalase antibodies (B) followed by species-specific secondary antibodies coupled with FITC (A) and rhodamine (B). Cells were permeabilized with 2.0 U/ml SLO and incubated with HSA-SKL in the absence (C) or presence (D) of external cytosol in the form of rabbit reticulocyte lysate. Bar, 20 μm. [Reproduced, by copyright permission of the Rockefeller University Press, from Wendland and Subramani (1993a).]

1993a). Various factors and compounds can be added to or omitted from the transport mix to study their role in peroxisomal protein import. This system is also able to evaluate the role of various known proteins in peroxisomal protein import by the use of antibody inhibition studies. Wiemer *et al.* (1995) used antibody inhibition to demonstrate the essential function of the PTS1 receptor in PTS1-dependent import. This technique has also been used to study the involvement of members of the hsp70 family in peroxisomal protein import (Walton *et al.* 1994; W. M. Nuttley, unpublished observations). Another application of the import assay is its use in characterizing cell lines from human patients displaying peroxisomal disorders such as Zellweger syndrome. Wendland and Subramani (1993b) have used the assay to characterize several cell lines from patients for cytosolic versus organelle-associated defects.

V. PITFALLS

1. The source and quality of SLO is of critical importance. SLO preparations from certain suppliers did not work reliably.

2. The treatment of cells with SLO, especially with higher SLO concentrations, significantly reduces cell attachment and cells are easily washed off the coverslips. Avoid vigorous washing. Confluent cell densities strengthen cell attachment.

3. When blotting off fluid from coverslips, do not dry coverslips completely as this may damage the cells.

4. A titration of antibody solution is necessary to find the best signal-to-background ratio. Also, compare the use of FITC- and rhodamine-conjugated secondary antibodies for improvement of signal detection.

References

Ahnert-Hilger, G., Mach, W., Fohr, K. J., and Gratzl, M. (1989). Poration by alpha-toxin and streptolysin O: An approach to analyze intracellular processes. *Methods Cell Biol.* **31**, 63–90.

Harlow, E., and Lane, D. (1988). "Antibodies: A Laboratory Manual." Cold Spring Harbor Laboratory, Cold Spring Harbor, NY.

Miller, S. G., and Moore, H.-P. H. (1991). Reconstitution of constitutive secretion using semi-intact cells: Regulation by GTP but not calcium. *J. Cell Biol.* **112**, 39–54.

Subramani, S. (1993). Protein import into peroxisomes and biogenesis of the organelle. *Annu. Rev. Cell Biol.* **9**, 445–478.

Walton, P. A., Gould, S. J., Feramisco, J. R., and Subramani, S. (1992). Transport of microinjected proteins into peroxisomes of mammalian cells: Inability of Zellweger cell lines to import proteins with the SKL tripeptide peroxisomal targeting signal. *Mol. Cell. Biol.* **12**, 531–541.

Walton P. A., Wendland, M., Subramani, S., Rachubinski, R. A., and Welch, W. J. (1994). Involvement of 70-kD heat-shock proteins in peroxisomal import. *J. Cell Biol.* **125**, 1037–1046.

Wendland, M., and Subramani, S. (1993a). Cytosol-dependent peroxisomal protein import in a permeabilized cell system. *J. Cell Biol.* **120**, 675–685.

Wendland, M., and Subramani, S. (1993b). Presence of cytoplasmic factors functional in peroxisomal protein import implicates organelle-associated defects in several human peroxisomal disorders. *J. Clin. Invest.* **92**, 2462–2468.

Wiemer, E. A. C., Nuttley, W. M., Bertolaet, B. L., Li, X., Francke, U., Wheelock, M. J., Anne, U. K., Johnson, K. R., and Subramani, S. (1995). Human peroxisomal targeting signal-1 receptor restores peroxisomal protein import in cells from patients with fatal peroxisomal disorders. *J. Cell Biol.* **130**, 51–65.

Nuclear Transport

Assay of Nuclear Protein Import in Permeabilized Cells Using Flow Cytometry

Bryce M. Paschal

I. INTRODUCTION

Molecular trafficking between the cytoplasm and the nucleus is mediated by the nuclear pore complex (NPC), a supramolecular assembly of proteins ($\sim 125 \times 10^6$ Da) that spans the double membrane system of the nuclear envelope (reviewed by Pante and Aebi, 1995). Small molecules (<30 kDa) can freely diffuse through ~ 10-nm channels in the NPC. The sizes of most proteins and ribonucleoprotein particles, however, preclude efficient diffusion-based transport. These macromolecules are actively transported through a ~ 25-nm central gated channel by mechanisms that involve cytosolic transport factors (reviewed by Melchior and Gerace, 1995).

One of the milestones in the study of nuclear protein import came with the discovery of the targeting signals that mediate protein import, known as nuclear localization sequences (NLSs). The most thoroughly characterized NLS is that of SV40 large T antigen, which consists of the sequence PKKKRKV (Kalderon, et al., 1984; Lanford and Butel, 1984). Importantly, this highly charged sequence is both necessary and sufficient to mediate protein import. This observation led to the use of reporter proteins coupled with synthetic peptides containing the SV40 large T antigen NLS to monitor nuclear protein import (Goldfarb et al., 1986; Lanford et al., 1986). Analysis of the distributions of NLS-containing reporter proteins microinjected into *Xenopus* oocytes or mammalian tissue culture cells provided some of the first information about the behavior of the nuclear protein import apparatus *in vivo*. This included the finding that nuclear protein import is a saturable process that could be mediated by cellular receptors (Goldfarb et al., 1986).

A second advance in the study of nuclear protein import came with the development of *in vitro* assays that reconstitute nuclear import of NLS-containing reporter proteins. The major technical challenge in developing an assay that reconstitutes nuclear protein import derives from the difficulty in isolating intact nuclei from cultured cells and tissues. Thus, the conditions used for extraction and homogenization typically compromise the structural integrity of the nuclear envelope. Forbes and co-workers showed that *Xenopus* egg extracts could be used to reseal nuclei isolated from rat liver and that these nuclei support protein import (Newmeyer et al., 1986). These investigators also showed that import-competent nuclei could be assembled by combining naked chromatin with cytosolic and membrane fractions from *Xenopus* eggs (Newmeyer et al., 1986).

Gerace and co-workers developed an assay that reconstitutes nuclear protein import in tissue culture cells grown on glass coverslips (Adam et al., 1990). The principle involved selective perforation of the plasma membrane with digitonin, a detergent that binds to cholesterol. Intracellular membranes, such as the nuclear envelope, remain intact because they contain proportionally less cholesterol than the plasma membrane. Treatment with digitonin results in the release of many soluble proteins, and it was found that these permeabilized cells required exogenous cytosol to reconstitute nuclear protein import of an NLS-containing reporter protein. The permeabilized cell assay reproduces the known features of nuclear protein import observed *in vivo*. Specifically, protein import in permeabilized cells requires a functional NLS, is nucleotide and temperature dependent, and is inhibited by reagents that bind to O-linked NPC proteins (Adam *et al.,* 1990).

The microscopy-based assays involving *Xenopus* egg extracts (Newmeyer *et al.,* 1986) and cells grown on coverslips (Adam *et al.,* 1990) have been extremely useful for characterizing some of the proteins required for nuclear protein import. In practice, however, these assays have technical limitations. The *Xenopus* egg extract assay system requires the addition of cytosol for *de novo* nuclear assembly around chromatin or for resealing rat liver nuclei damaged during biochemical isolation. Thus, it is difficult to distinguish between factors whose function is related to nuclear structure and factors whose function is related to nuclear protein import. In the permeabilized cell assay system, nuclear protein import levels vary with cell density, cells can detach during the permeabilization and washing steps, and the perimeter of the coverslip is subject to artifacts due to evaporation during the transport reaction. Both assay systems are relatively labor-intensive as they require examination of each sample by fluorescence microscopy and quantitation of import using image analysis software.

We have developed a flow cytometry-based method for measuring nuclear protein import in permeabilized cells that circumvents the technical problems associated with microscopy-based assays (Paschal and Gerace, 1995). The use of suspension culture HeLa cells allows the entire assay to be performed in a test tube, and analysis by flow cytometry provides a rapid and highly accurate quantitation of protein import. This high throughput version of the permeabilized cell assay enables the user to reliable measure protein import in 50 samples in 2–3 hr (Fig. 1).

II. MATERIALS AND INSTRUMENTATION

HeLa S3 cells (CCL-2.2) are from the American Type Culture Collection. Joklik's modified SMEM (Cat. No. 22300), newborn calf serum (Cat. No. 16010-159), and penicillin–

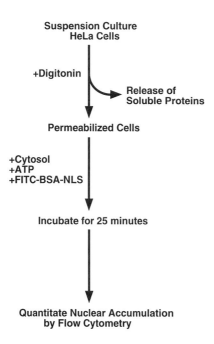

FIGURE 1 Flow diagram of the nuclear protein import assay.

streptomycin (Cat. No. 15140-122) are from GIBCO-BRL. ATP (Cat. No. A-2383), creatine phosphate (Cat. No. P-7936), creatine phosphokinase (Cat. No. C-7886), sodium bicarbonate (Cat. No. S-4019), HEPES (Cat. No. H-7523), potassium acetate (Cat. No. P-5708), magnesium acetate (Cat. No. M-2545), EGTA (Cat. No. E-4378), dithiothreitol (DTT, Cat. No. D-5545), NaCl (Cat. No. S-7653), potassium chloride (Cat. No. P-9333), sodium phosphate, dibasic (Cat. No. S-9763), potassium phosphate, monobasic (Cat. No. P-0662), phenylmethyl-sulfonyl fluoride (PMSF, Cat. No. P-7626), dimethyl sulfoxide (DMSO, Cat. No. D-8779), hydroxylamine (Cat. No. H-9876), sodium carbonate (Cat. No. S-7795), Sephadex G-25 (Cat. No. G-25-150), PD-10 columns (Cat. No. 5-4805), and trypan blue (Cat. No. T-8154) are from Sigma. High-purity bovine serum albumin (BSA, Cat. No. Cat. No. 238031), aprotinin (Cat. No. 236 624), leupeptin (Cat. No. 1017 101), and pepstatin (Cat. No. 258 286) are from Boehringer-Mannheim. The FITC Isomer I (Cat. No. F-1906) and the sulfo-SMCC (Cat. No. 22322) are from Molecular Probes and Pierce, respectively. Centricon filters (30 kDa cutoff, Cat. No. 4208) are from Amicon. High-purity digitonin (Cat. No. 300410) is from Calbiochem. The polystyrene tubes used for the transport assays (Cat. No. 2058) are from Fisher. The instrument used for flow cytometric analysis is the FACScan unit from Beckton-Dickinson.

The following equipment required for growing and harvesting HeLa cells are from Bellco Glass. The 250-ml spinner flask (Cat. No. 1965-00250) and stir plate (Cat. No. 7760-0600) are used for the continuous culture of HeLa cells. Additional spinner flasks are required to scale up the preparation to 15 liters. These include 1000-ml (Cat. No. 1965-01000), 3000-ml (Cat. No. 1965-03000), and 15-liter (Cat. No. 1964-15000) spinner flasks. The 15-liter spinner flask requires an overhead drive unit (Cat. No. 7764-00110), cap assembly (Cat. No. 7764-10100), and Teflon paddle assembly (Cat. No. 1964-30015). The cell harvest is carried out using four 780-ml conical glass centrifuge bottles (Cat. No. 3045-0080).

The equipment for centrifugation includes the JS5.2 swinging bucket rotor and J6B centrifuge, the JA-20 fixed angle rotor and J2 centrifuge, and the type 60Ti fixed angle rotor and L7 centrifuge, all from Beckman. Cytosol dialysis is carried out using the colloidion vacuum dialysis apparatus (Cat. No. 253310) and 10,000-Da cutoff membranes (Cat. No. 27110) available from Schleicher and Schuell. The homogenizer used for cell disruption is a 0.02-mm clearance stainless-steel unit (Cat. No. 885310-0015) available from Kontes.

III. PROCEDURES

A. Preparation of Cytosol

A significant feature of the permeabilized cell assay is the option to use cytosols prepared from a variety of cells and tissues. We routinely use suspension culture HeLa cells for preparing cytosol. Cytosol prepared from these cells has the highest specific activity of any tested in our laboratory. Only 2–5 mg/ml of HeLa cell cytosol is required to obtain a maximum level of protein import in the assay. This could reflect the relatively high activity of the nuclear protein import machinery in cells undergoing exponential growth.

Solutions

1. *10× transport buffer:* 200 mM HEPES, pH 7.4, 1.1 M potassium acetate, 20 mM magnesium acetate, and 5 mM EGTA. To make 1 liter, dissolve 47.6 g HEPES, 107.9 g potassium acetate, 4.8 g magnesium acetate, and 1.9 g EGTA in 800 ml distilled water. Adjust the pH to 7.4 with 10 N NaOH, and bring the final volume to 1 liter. Sterile filter and store at 4°C.

2. *1× transport buffer:* 20 mM HEPES, pH 7.4, 110 M potassium acetate, 2 mM magnesium acetate, and 0.5 mM EGTA. To make 1 liter, add 100 ml of 10× transport buffer to 900 ml distilled water.

3. *1 M HEPES stock, pH 7.4:* To make 500 ml, dissolve 119.1 g HEPES (free acid) in 400 ml distilled water. Adjust the pH to 7.4 with 10 N NaOH, and bring the final volume to 500 ml. Sterile filter and store at 4°C.

4. *1 M potassium acetate stock:* To make 500 ml, dissolve 49 g potassium acetate in 400 ml distilled water. Bring the final volume to 500 ml, sterile filter, and store at 4°C.

5. *1 M magnesium acetate stock:* To make 500 ml, dissolve 107.2 g magnesium acetate (tetrahydrate) in 400 ml distilled water. Bring the final volume to 500 ml, sterile filter, and store at 4°C.

6. *0.2 M EGTA stock:* To make 500 ml, dissolve 38 g EGTA (free acid) in 400 ml distilled water. Adjust the pH to ~7.0 with 10 N NaOH, and bring the final volume to 500 ml. Store at 4°C.

7. *Cell lysis buffer:* 5 mM HEPES, pH 7.4, 10 mM potassium acetate, 2 mM magnesium acetate, and 1 mM EGTA. To make 500 ml, combine 2.5 ml 1 M HEPES, pH 7.4, 5 ml 1 M potassium acetate, 1 ml 1 M magnesium acetate, 2.5 ml 0.2 M EGTA, and 489 ml distilled water. Store at 4°C.

8. *PBS:* To make 1 liter, dissolve 8 g sodium chloride, 0.2 g potassium chloride, 1.44 g sodium phosphate (dibasic), and 0.24 g potassium phosphate (monobasic) in 900 ml distilled water. Adjust the pH to 7.4, and bring the final volume to 1 liter. Store at 4°C.

Steps

1. Grow HeLa cells at a density of $2–7 \times 10^5$ cells per milliliter in a spinner flask (30–50 rpm) in a 37°C incubator (CO_2 is not required). The medium is Joklik's modified SMEM, containing 2.0 g sodium bicarbonate and 2.38 g HEPES per liter of media. Adjust the pH to 7.3, sterile filter, and store at 4°C. Before use, supplement the medium with 10% newborn calf serum and 1% penicillin–streptomycin. The cells should have a doubling time of approximately 18 hr, making it necessary to dilute the culture with fresh, prewarmed medium every 1–2 days.

2. HeLa cells from a 250-ml culture provide the starting point for scaling up the preparation to 15 liters. This is carried out by sequential dilution of the culture into larger spinner flasks. The culture should not be diluted to a density below 2×10^5 cells/ml. The spinner flasks used for the scaling up the preparation are 250 ml (1 each), 1 liter (1 each), 3 liters (2 each), and 15 liters (1 each). This process generally takes 5 days.

3. Perform the cell harvest and subsequent steps at 0–4°C. Collect the cells by centrifugation ($300\,g \times 15$ min) in 780-ml conical glass bottles in a Beckman J6B refrigerated centrifuge equipped with a JS5.2 swinging bucket rotor. The cell harvest takes about 1 hr.

4. Wash the cells by sequential resuspension and centrifugation. Two washes are carried out in ice-cold PBS (1 liter each), and one wash is carried out in 1× transport buffer containing 2 mM DTT.

5. Resuspend the cells in a total volume of 100 ml in 1× transport buffer containing 2 mM DTT, transfer to 50-ml polypropylene centrifuge tubes (with graduations), and collect by centrifugation. The yield from a 15-liter culture should be approximately 40 ml of packed cells.

6. Resuspend the cell pellet using 1.5 vol of lysis buffer, supplemented with 3 μg/ml each aprotinin, leupeptin, pepstatin, 0.5 mM PMSF, and 5 mM DTT. Allow the cells to swell on ice for 10 min.

7. Disrupt the cells by two to three passes in a stainless-steel homogenizer. Monitor the progress of homogenization by trypan blue staining and phase-contrast microscopy. The goal is to obtain ~95% cell disruption. Excessive homogenization should be avoided because it results in nuclear fragmentation and the release of nuclear contents into the soluble fraction of the preparation.

8. Dilute the homogenate with 0.1 vol of 10× transport buffer and centrifuge in a fixed-angle rotor such as the Beckman JA-20 ($40,000\,g \times 30$ min).

9. Filter the resulting low-speed supernatant fraction through four layers of cheesecloth. Subject the filtered low-speed supernatant fraction to ultracentrifugation in a fixed angle rotor such as the Beckman type 60 Ti ($150,000\,g \times 60$ min).

10. Dispense the resulting high-speed supernatant fraction (~50 ml, protein concentration ~5 mg/ml) into 1- and 4-ml aliquots, flash freeze in liquid N_2, and store at $-80°C$ indefinitely.

11. HeLa cell cytosol is generally subjected to a rapid dialysis step before use in transport reactions. Thaw a 4-ml aliquot of cytosol at $0–4°C$ and dialyze for 3 hr in 1× transport buffer containing 2 mM DTT and 1 μg/ml each of aprotinin, leupeptin, and pepstatin (two buffer changes). We use a vacuum apparatus and a colloidion membrane to achieve a twofold concentration of the sample. Dispense the dialyzed, concentrated cytosol into 100-μl aliquots, flash freeze in liquid N_2, and store at $-80°C$.

B. Preparation of NLS Transport Ligand

Synthetic peptides containing an NLS can be used to direct the nuclear import of a variety of fluorescent reporter proteins. We use FITC–BSA–NLS as a reporter protein in the flow cytometry assay. The average size of the protein conjugate (>70 kDa) is too large to diffuse through the NPC, and it displays neglible nonspecific binding to the permeabilized cell. Preparation of FITC–BSA–NLS is carried out in two steps: fluorescent labeling of BSA using isothiocyanate chemistry and attachment of NLS peptides using the heterobifunctional cross-linker sulfo-SMCC. FITC-labeled proteins are readily detected by both flow cytometry and fluorescence microscopy, making the transport ligand useful for both types of studies. Sulfo-SMCC provides a covalent linkage between primary amines on BSA and a cysteine present on the N terminus of the NLS peptide.

Solutions

1. *PBS:* To make 1 liter, dissolve 8 g sodium chloride, 0.2 g potassium chloride, 1.44 g sodium phosphate (dibasic), and 0.24 g potassium phosphate (monobasic) in 900 ml distilled water. Adjust the pH to 7.4, and bring the final volume to 1 liter. Store at 4°C.

2. *0.1 M sodium carbonate:* To prepare 250 ml, dissolve 3.1 g sodium carbonate in 200 ml distilled water, adjust the pH to 9.0, and bring the final volume to 250 ml. Store at 4°C.

3. *1.5 M hydroxylamine:* To prepare 100 ml, dissolve 10.4 g in a total volume of 100 ml distilled water. Store at room temperature.

4. *10 mg/ml FITC:* Add 1 ml DMSO to 10 mg FITC in an amber vial and vortex to dissolve.

5. *20 mM sulfo-SMCC:* Prepare a 20 mM stock of sulfo-SMCC in the following manner. Preweigh a microfuge tube on a fine balance, and use a small spatula to add approximately 1–2 mg of sulfo-SMCC to the microfuge tube. Reweigh the microfuge tube containing the sulfo-SMCC, and add DMSO for a final concentration of 8.7 mg/ml.

Steps

1. Dissolve 10 mg high-purity BSA in 1 ml sodium carbonate buffer, pH 9.0.

2. Stir the BSA solution in a glass test tube with a micro-stir bar, and add 0.1 ml of 10 mg/ml FITC dropwise using a P200 pipetman. Cover with foil and stir for 60 min at room temperature.

3. Stop the reaction by adding 0.1 ml 1.5 M hydroxylamine.

4. Separate FITC-labeled BSA from unincorporated FITC by desalting on a Sephadex G-25 column (dimensions: 1.5 × 20 cm) equilibrated in PBS. Run the column at ~0.5 ml/min and collect 0.5-ml fractions. The bright yellow FITC–BSA will elute in the void volume of this column.

5. Pool the four or five most concentrated fractions, dispense into 1-mg aliquots, and freeze in foil-wrapped microfuge tubes at $-20°C$.

6. Combine 1 mg of FITC–BSA with 50 μl of freshly prepared 20 mM sulfo-SMCC and mix end over end for 45 min at room temperature.

7. Separate the sulfo-SMCC-activated FITC–BSA from unincorporated sulfo-SMCC by desalting on a disposable PD-10 column (Sephadex G-25) equilibrated in PBS. After loading the sample, fill the buffer reservoir of the column with PBS and collect 0.5-ml fractions. The bright yellow FITC–BSA will elute in the void volume as before.

8. Pool the three most concentrated fractions of sulfo-SMCC-activated FITC–BSA and combine with 0.3 mg of NLS peptide (CGGGPKKKRKVED). Mix end over end in a foil-wrapped microfuge tube overnight at 4°C.

9. Remove unincorporated NLS peptide by subjecting the sample to four cycles of centrifugation and resuspension in 1× transport buffer using a 2-ml 30-kDa cutoff Centricon filter. Follow the manufacturerĩOs recommendations for centrifugation conditions.

10. Adjust the FITC–BSA–NLS conjugate to a final concentration of 2 mg/ml, dispense into 50-μl aliquots, flash freeze in liquid N_2, and store at −80°C.

C. Nuclear Protein Import Assay

Solutions

1. *10× transport buffer:* 200 mM HEPES, pH 7.4, 1.1 M potassium acetate, 20 mM magnesium acetate, and 5 mM EGTA. To make 1 liter, dissolve 47.6 g HEPES, 107.9 g potassium acetate, 4.8 g magnesium acetate, and 1.9 g EGTA in 800 ml distilled water. Adjust the pH to 7.4 with 10 N NaOH, and bring the final volume to 1 liter. Sterile filter and store at 4°C.

2. *1× transport buffer:* 20 mM HEPES, pH 7.4, 110 M potassium acetate, 2 mM magnesium acetate, and 0.5 mM EGTA. To make 1 liter, add 100 ml of 10× transport buffer to 900 ml distilled water.

3. *Complete transport buffer:* 1× transport buffer containing 1 μg/ml each aprotinin, leupeptin, pepstatin, and 2 mM DTT.

4. *10% digitonin:* To make 2 ml, add 0.2 g high-purity digitonin to 1.7 ml DMSO and dissolve by vigorous vortexing. Dispense into 20-μl aliquots and freeze at −20°C.

5. *100 mM MgATP:* To make 5 ml, add 0.5 ml 1 M magnesium acetate and 0.1 ml 1 M HEPES, pH 7.4, to 4 ml distilled water. Add 275.1 mg ATP, dissolve by vortexing, and bring the final volume to 5 ml. Dispense into 20-μl aliquots and freeze at −80°C.

6. *250 mM creatine phosphate:* To make 5 ml, add 0.32 g creatine phosphate to 4 ml distilled water, dissolve by vortexing, and bring the final volume to 5 ml. Dispense into 20-μl aliquots and freeze at −20°C.

7. *2000 U/ml creatine phosphokinase:* To make 5 ml, dissolve 10,000 U creatine phosphokinase in 20 mM HEPES, pH 7.4, containing 50% glycerol. Dispense into 1-ml aliquots and store at −20°C.

Steps

1. Harvest the suspension culture HeLa cells (50 ml at a density ~5 × 10^5/ml) by centrifugation (300 g × 5 min) in a clinical-type centrifuge equipped with a swinging bucket rotor.

2. Wash the cells by gentle resuspension of the pellet in 25 ml 1× transport buffer and collect by centrifugation.

3. Resuspend the washed cell pellet with complete transport buffer in a total volume of 1 ml. Measure the cell density using a hemacytometer, and dilute the sample to 5 × 10^6 cells/ml using complete transport buffer.

4. Add digitonin to a final concentration of 50 μg/ml. Gently pipette the cells two to three times and place on ice for 5 min.

5. Dilute the cells 10-fold in complete transport buffer and collect by centrifugation.

6. Carefully pour off the supernatant fraction and resuspend the cell pellet in a total volume of 0.25 ml in complete transport buffer. The final concentration should be ~2.5 × 10^7 cells/ml.

7. Assemble the transport reactions in polystyrene snap-top tubes on ice. Each reaction contains (final concentration) FITC–BSA–NLS (25 μg/ml), MgATP (1 mM), creatine phosphate (5 mM), creatine phosphokinase (20 U/ml), digitonin-permeabilized cells (5 × 10^6/ml), HeLa cell cytosol (2 mg/ml), and complete transport buffer to 40 μl final volume.

8. Transfer the tubes to a 30°C water bath and incubate for 25 min.

9. Transfer the tubes to a ice bath and add 4 ml transport buffer to each tube.

10. Collect the cells by centrifugation, aspirate the supernatant fraction, and gently resuspend the small (but visable) cell pellet in ~0.25 ml transport buffer using a P1000 pipetman.

11. Measure the fluorescence of 10^4 cells from each sample by flow cytometry, using an instrument such as the Becton-Dickinson FACScan instrument. Histograms depicting the cell-associated fluorescent signal resulting from the nuclear import of FITC–BSA–NLS are shown in Fig. 2.

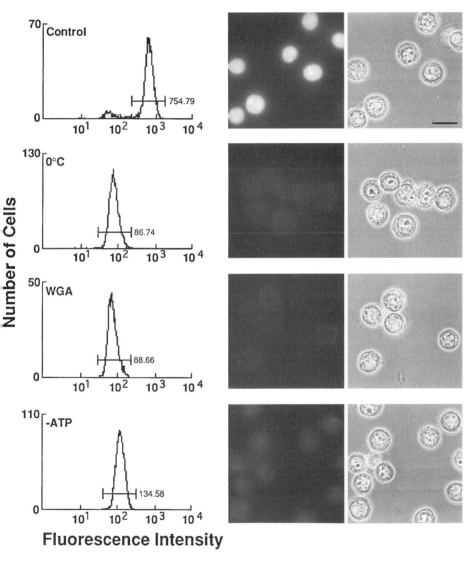

FIGURE 2 Analysis of nuclear protein import by flow cytometry and fluorescence microscopy. Nuclear transport reactions were carried out under standard reactions (control) or under conditions that inhibit nuclear protein import *in vivo* and *in vitro* (0°C, WGA, or -ATP). The fluorescence output generated from the flow cytometry is plotted as a histogram, with the mean fluorescence of the bracketed region indicated (left). An aliquot of each sample was also photographed by epifluorescence (middle) and phase-contrast microscopy (right). Bar, 20 μm. Reprinted by permission from the *J. Cell Biol.*, Rockefeller University Press.

IV. COMMENTS

We typically acquire fluorescence data from 10^4 cells per sample, which should take about 15 sec depending on the cell density and flow rate of the instrument. Examples of the fluorescence output, presented as histograms, are shown (Fig. 2, left). In this experiment, the nuclear protein import supported by HeLa cell cytosol was assayed under standard conditions (control) or under conditions that inhibit nuclear protein import *in vitro* and *in vivo* (0°C, WGA, −ATP). Because the flow cytometer meaures total cell-associated fluorescence, it is important to confirm that FITC−BSA−NLS accumulates within the nucleus of the permeabilized cells. This is done by applying a small aliquot of cell suspension to a polylysine-coated coverslip and viewing the cells by epifluorescence and phase-contrast microscopy (Fig.2, middle and right). Microscopic examination is not routinely performed, although it should be employed when testing new preparations of FITC−BSA−NLS to confirm that the cell-associated fluorescence is within the nucleus. On this note, we have observed that transport ligands coupled with an excessive number of NLSs peptides display nonspecific interactions with membranes and and cytoskeletal elements (see below).

Flow cytometry is a rapid and convenient method for measuring nuclear protein import in tissue culture cells. The advantages of this method over coverslip-based assays include the ability to analyze large numbers of samples, ease of quantitation, and reproducibility. For example, it is not essential to perform duplicate assays because the standard deviation between identical samples is usually ~2%. Because the flow cytometer can measure cell-associated fluorescence at multiple wavelengths simultaneously, it is possible to quantitate protein import and monitor parameters such as DNA content in the same cells. This could provide an approach for exploring the biochemical basis of the regulation of the nuclear protein import machinery during the cell cycle. Finally, it should be possible to use permeabilized cells and flow cytometry to measure protein import into the endoplasmic reticulum, mitochondria, or peroxisomes by using the appropriate targeting signals.

V. PITFALLS

There are three major technical concerns related to the nuclear protein import assay. The first is that unfractionated cytosols from a variety of sources have inhibitory activities that are manifest when too much cytosol is used in the assay. It is important, therefore, to perform a simple titration experiment with each batch of cytosol to determine the concentration that will support maximal protein import.

The second concern is that the extent of digitonin permeabilization depends on the purity of the detergent and the growth state of the cells. This concern can be alleviated by testing several concentrations of digitonin in the range of 25−100 µg/ml, thereby defining the concentration that yields ~95% permeabilization as assessed by trypan blue staining.

The third concern is that artificial NLS−ligands are conjugated with multiple lysine-rich peptides, resulting in a highly charged macromolecule. If the molar ratio of peptide to carrier is much greater than 20:1, the ligand can display nonspecific interactions with the permeabilized cells. It is critical, therefore, to use the approprate concentration of cross-linker for the specified length of time. Furthermore, transports reactions using a new batch of transport ligand should be examined by fluorescence microscopy to confirm the nuclear localization.

References

Adam, S. A., Sterne-Marr, R. E., and Gerace, L. (1990). Nuclear protein import in permeabilized mammalian cells requires soluble cytoplasmic factors. *J. Cell Biol.* 111, 807−816.

Goldfarb, D. S., Gariepy, J., Schoolnik, G., and Kornberg, R. D. (1986). Synthetic peptides as nuclear location signals. *Nature* 326, 641−644.

Kalderon, D., Roberts, B. L., Richardson, W. D., and Smith, A. E. (1984). A short amino acid sequence able to specify nuclear location. *Cell* 39, 499−509.

Lanford, R. E., and Butel, J. S. (1984). Construction and characterization of an SV40 mutant defective in nuclear transport of T antigen. *Cell* 37, 801−813.

Lanford, R. E., Kanda, P., and Kennedy, R. C. (1986). Induction of nuclear transport with a synthetic peptide homologous to the SV40 T antigen transport signal. *Mol. Cell Biol.* **8**, 2722–2729.

Melchior, F., and Gerace, L. (1995). Mechanisms of nuclear protein import. *Curr. Opin. Cell Biol.* **7**, 310–318.

Newmeyer, D. D., Finlay, D. R., and Forbes, D. J. (1986). *In vitro* transport of a fluorescent nuclear protein and exclusion of non-nuclear proteins. *J. Cell Biol.* **103**, 2091–2102.

Pante, N., and Aebi, U. (1995). Toward a molecular understanding of the structure and function of the nuclear pore complex. *Intl. Rev. Cytol.* 225–255.

Paschal, B. M., and Gerace, L. (1995). Identification of NTF2, a cytosolic factor for nuclear import that interacts with nuclear pore complex protein p62. *J. Cell Biol.* **129**, 925–937.

Motility Assays for Motor Proteins and Other Motility Models

Microtubule Motility Assays

R. A. Cross

I. INTRODUCTION

As the pivotal role of microtubule (MT) motors in the cell cycle is more widely recognized, so the demand for motility assays will increase. This article provides a set of robust and straightforward protocols, based on experience in the author's laboratory. Readers seeking to set up more specialized assays should refer to one of the excellent methodological compendia that are available (Inoué, 1986; Cross and Kendrick Jones, 1991; Scholey, 1993). The protocols and software described here are available from webpage http://194.82.114.11/motorhome.html.

II. MATERIALS AND INSTRUMENTATION

A. Hardware

1. Microscope

Unstained MTs are most conveniently visualized by video microscopy using computer-enhanced differential interference contrast (DIC). MTs assembled from fluorescently labeled tubulin can be visualized using epifluorescence. This article concentrates on these two contrast modes. It is also possible to visualize microtubules by dark field, which gives a high contrast image of unstained microtubules, but this mode is very susceptible to dirt in the solutions and so may be inconvenient for routine work. Interference reflection produces higher contrast than DIC and gives information in the Z direction, but again is susceptible to dirt.

For all modes of contrast the main requirement of the microscope is that it should be as simple as possible. It should also be heavy to give it some vibration resistance. Uprights are cheaper, optically straightforward, and more convenient if the bathing solution is to be exchanged during observation (which is often the case). Inverted scopes are more stable and provide much better access to the specimen if other inputs (micromanipulators, uncaging lasers, evanescence prisms) are envisioned.

Human eyes work in color, have a higher dynamic range, a better spatial resolution, and a larger view field than a video camera. It is very useful to be able to use them to find focus. To do this the microscope needs to incorporate some sort of device to switch the light between the eyepieces and the camera.

2. Illumination

Because 100-W Hg lamps generate large amounts of heat, it is necessary to remove this heat from the illuminating light. Glass heat filters may be used, but the best way to filter out heat is to use a hot mirror, a mirror which transmits heat but reflects light (Technical Video).

DIC optics are optimized for visible wavelengths, and the conventional method uses a green interference filter to give narrow band illumination. In practice the improvement this brings is too slight to be obvious by the naked eye, and removing the green filter is a convenient way to brighter illumination. The UV emitted by Hg lamps is substantially removed by optical (lead) glass, but for DIC an in-line UV filter is nonetheless a sensible precaution.

A useful improvement in both fluorescence and DIC image quality can be achieved by sending the illuminating light through a fiber-optic light scrambler (Technical Video). Light from the lamp is focused into one end of the fiber (Fig. 1) and the other end emits a gaussian disc of light that is used to illuminate the microscope.

3. Optical Train

For maximum image quality in all contrast modes, it is advisable to reduce the number of optical components between the objective and the camera to a minimum. The objective is the most critical component. The light intensity transmitted by a lens is proportional to the square of its numerical aperture (NA) and the resolution rises linearly with the NA. It is important, therefore, to use a 1.4 NA (therefore oil immersion) 60× or 100× planapo objective in order to maximize light-gathering power and resolution; this is particularly so in epifluorescence, where the objective doubles as the condensor.

4. Antivibration Hardware

Vibration will degrade the highly magnified image. Low frequencies (people walking across the floor) will cause the image to bounce around, whereas high frequencies will be averaged

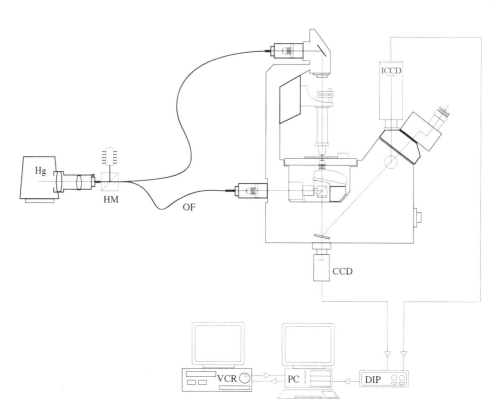

FIGURE 1 Video microscope. An inverted microscope set up for both fluorescence and DIC video microscopy. Hg, mercury lamp; HM, hot mirror; OF, optical fiber; CCD, charge-coupled device camera (for DIC); ICCD, intensified CCD (for fluorescence); DIP, digital image processor; PC, personal computer; VCR, video cassette recorder.

out by the video framing rate, causing the image to blur. The extent of this problem is, however, often overemphasized. Before purchasing expensive and awkward vibration-damping equipment, try placing a few layers of bubble wrap under the microscope base plate.

5. Temperature Control Hardware

Motility rates for several motors are extremely temperature sensitive in the range of room temperature, and temperature control is consequently important if measurements taken at different times are to be compared. The best way is to temperature clamp the entire microscope. If you have air conditioning, this will already be happening. We do not, so we use water jackets for the objective and stage. These are readily fashioned from flexible, narrow-bore copper tubing.

6. Camera

For fluorescence of moving objects, two types of low-light level/high-contrast/high-framing rate cameras are suitable: intensified charge-coupled device (ICCD) and intensified silicon-intensified tube (ISIT) cameras. ICCD cameras are better for current purposes because ISIT cameras, although more sensitive, introduce spatial and intensity distortions across the view field that are tedious to correct for. For DIC, a nonintensified scientific grade CCD is fine.

7. Camera Coupling/Magnification

The ideal magnification sets four or more camera pixels across the width of the MT. For a 760×512-pixel (2/3 in.) CCD, this corresponds to a square field with sides of $20–25$ μm. Zoom couplings are not a good idea because they absorb a lot of light.

8. Image Processor

The just-described cameras all come with a hardware box providing real-time analog enhancement of the video signal (any or all of gain, back-off, and shading correction). A digital video processor is able to digitize incoming video frames, perform frame averaging, contrast enhancement, background subtraction, and caption overlay, and then reencode the image as an analog video signal, all in real time. The Hamamatsu Argus 10/20 is so well thought out that it is virtually standard equipment for video microscopy laboratories. The best and most flexible arrangement is to perform analog enhancement and then feed the signal to the Argus and to the computer and VCR (see Fig. 1).

9. Monitor

A high-quality, 14-in. multiformat monitor is worth the investment. Larger monitors look impressive but are not helpful.

10. Video Recorder

At the time of writing, the only practical way to store large amounts of video is on tape. Recording to VCRs inevitably involves some degradation of the image (loss of spatial resolution, noise, contrast effects). For practical purposes the resolution loss is potentially the most serious problem. The effective resolution following recording can be visualized by recording and replaying a test card image having black and white lines at various spatial frequencies. SVHS VCRs record considerably more faithfully than standard VHS. The more expensive domestic models are as good as the "professional" versions, in our experience. Models with RS232 interfaces (e.g., Panasonic AG 5700) allow software control of the VCR. We have evolved a video-recording strategy that offers maximum flexibility: Time-lapse digital recording of video frames to a computer hard disc (with no resolution loss) and simultaneous SVHS recording to a VCR. The VCR runs uninterrupted in the background for 3 hr per tape, generating an archive. The operator is free to go back to this archive at a later date and recapture interesting sequences for analysis.

Video disks [optical memory disk recorders (OMDRs)] are convenient for data analysis because they offer random access to frames, but media are too expensive to use for routine work. Many laboratories record to VCRs and then transfer interesting sequences to OMDRs because the random access to frames is useful for analysis. In our experience, this is more conveniently and inexpensively done now using a large computer hard disk.

Currently, different video formats unfortunately operate in different countries. In the United States and Japan, the NTSC format applies (620 × 480 lines; 30 frames/sec). European countries use PAL (SECAM, France), which has a higher spatial resolution but a lower time resolution (768 × 512; 25 frames/sec). It is worthwhile buying a VCR that can play back multiple formats so that tapes provided by visiting scientists can be played. Editing VCRs that can convert formats are available, but the investment is probably not worthwhile. On the rare occasions when this is necessary, video services can do it.

11. Computing

The most convenient way to analyze motility is to capture a sequence of frames into computer memory and to track objects using a mouse-driven cursor. It helps to have a hard disk big enough to hold 2 day's work (more is dangerous because of the temptation not to backup) and enough hard memory to hold the stack of captured frames. A typical 20 frame stack uses about 8 Mb of application memory. If larger stacks are needed, then more memory is needed. Processed stacks can conveniently be archived to removable disks. We use 230-Mb magneto-optical disks. CD writers are becoming less expensive and may soon be worth considering if a permanent archive is required. It is best to avoid video compression protocols (JPEG, MPEG), as all involve some data loss.

12. Frame Grabber Card

Large numbers of video grabbing cards are available. Only a few are supported by RETRAC and *NIH Image* (See Section II, B). The situation is fluid, so please check the software documentation for a list of supported cards.

B. Software

On the Mac, the best route for analysis is *NIH Image*, which can be customized using macros to track objects and output the data in a spreadsheet-compatible format. A macro for basic tracking through *NIH Image* stacks is available for downloading from our website (see Section I). *NIH Image* is currently being ported to the PC by Scion Corp., who intend to make the program freely available.

RETRAC for the PC is to the author's knowledge the best thing available for either platform because it is purpose written. The program is written in assembler and consequently runs very quickly under DOS or Windows 95. The latest version supports software control of a VCR, time-lapse frame grabbing from either VCR or live video, autofocus, autocontrast, tracking (including drift correction) spatial filtration, and magnification. A freeware version of the program and support documentation is available from our website. A pro version of the program with an integrated file manager is in preparation. Figure 2 (color plate) shows a screenshot during tracking.

C. Glassware

The type of slide used does not matter. The type of coverslip does. The thickness of the coverslip should be matched to the objective. The objective will be marked appropriately [e.g., 60/planapo DIC 1.4 0.17/160 means a 60× objective selected as strain free for DIC aplanatic (flat field); apochromatic (low chromatic aberration for blue yellow and green) optimized for cover glasses 0.17 mm thick and with a 160-mm focal length]. We use Chance 22 × 22-mm No. 1.5 coverslips, without any special cleaning treatment. Experience suggests

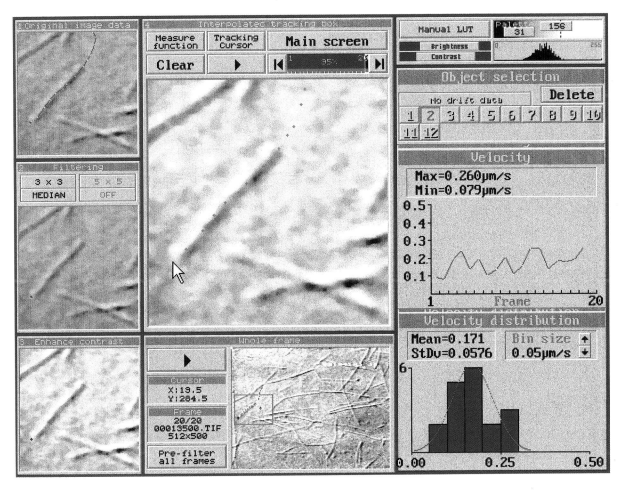

FIGURE 2   Example of a RETRAC screen.

hat "good" coverslips do not improve on cleaning and that "bad" coverslips cannot by cleaning be made "good."

II. PROCEDURES

A. Taxol-Stabilized Microtubules

Solutions

1. *1 M K-PIPES:* PIPES dissolves around its isoelectric point of about pH 6.5. Add 65 g solid KOH and then slowly add 302 g PIPES buffer (Sigma P-6757) to 500 ml water. Once everything is dissolved, monitor pH and roughly adjust by adding more KOH pellets as necessary. Allow the warm solution to cool and then fine adjust pH using 5 M KOH. Be careful not to overshoot, as there is no way back.

2. *100 mM NaGTP stock solution:* Nucleoside triphosphates such as GTP and ATP undergo rapid hydrolysis at acidic pH, so efforts should be made to control pH when dissolving and storing them. Dissolve 1 g NaGTP (Sigma G-8877) in 15 ml 10 mM K-PIPES, pH 6.9, monitoring pH. Rapidly reneutralize pH by titrating in 5 M KOH. Finely adjust pH, then make volume up to 19.1 ml. Store frozen at −20°C in aliquots of 5–2000 μl. Do not add MgCl$_2$ to the stock solution (it precipitates).

3. *100 mM MgATP stock solution:* Dissolve 5.87 g NaATP (Sigma A-7699 ATP ultra or Boehringer 519 987) in 60 ml 10 mM K-PIPES, pH 6.9, continuously monitoring pH and holding as close as possible to neutral using concentrated KOH. Once the ATP is dissolved, add 10 ml of 100 MgCl$_2$ and readjust pH to 6.9. Adjust volume to 100 ml and freeze in aliquots of 5–5000 μl.

4. *10 mM taxol stock solution:* Wear gloves and work in the fume hood. Inject 2.93 ml anhydrous dimethyl sulfoxide (DMSO) (Aldrich 27685-5) into a 25-mg bottle of taxol (Sigma T-7402). Dissolve by vortexing and store as 2- to 20-μl aliquots at −20°C. Taxol is stable in DMSO but unstable in water. It is insoluble in aqueous buffers above about 18 μM. DMSO is explosive if it gets wet.

5. *0.2 M NaEGTA:* Dissolve 15.2 g EGTA (Sigma E-4378) in 190 ml water. Adjust pH to neutral by adding concentrated NaOH, then make volume to 200 ml. Store at room temperature.

6. *1 M MgCl$_2$:* 20.33 g MgCl$_2 \cdot$6H$_2$O to 100 ml water. Sterile filter and store at room temperature.

7. *BRB 80 (Brinkley reassembly buffer):* 80 mM K-PIPES, 1 mM MgCl$_2$, and 1 mM EGTA, pH 6.9. Make up as a 10× stock, store at 4°C, and dilute freshly for use.

8. *Purified tubulin (100 μM):* Purified tubulin in BRB80 should be flash frozen in 10- to 25-μl aliquots in the presence of 30% glycerol by immersion in liquid nitrogen and stored either at −70°C or preferably in liquid nitrogen.

Steps

1. Thaw an aliquot of tubulin (typically 200 μM), and add stock 100 mM NaGTP to 1 mM and MgCl$_2$ to 2 mM. Warm to 37°C and incubate for 20 min.

2. After 20 min, add taxol from a 10 mM stock in DMSO to 20 μM final. Dilute microtubules 1000-fold for use using BRB80 buffer supplemented with 20 μM taxol.

B. Preparation of Sample Cells

Steps

1. Apply single-sided Scotch tape to the long edges of a microscope slide such that the strip of glass surface between the two pieces of tape is 8–10 mm wide. Trim away overhangs with a razor blade.

2. Extrude two parallel stripes of Apiezon M grease, along the inner edges of the tape strips, from a syringe with a squared-off wide-bore needle.

3. Tap a clean coverslip onto the grease. The volume of the flow cell can be adjusted by spacing the grease strips apart and/or by placing spacers between the coverslip and the slide. Single-sided Scotch magic tape is about 50 μm thick, giving a flow cell of about 10 mm $\times$ 5 mm $\times$ 50 μm, or 25 μl. Thinner metal or cellophane foils can be used to make a shallower flow cell and conserve sample. It is helpful to make the flow cell shallow because the microtubules below the top surface scatter light and reduce contrast. For inverted scopes, it is convenient to arrange flow crosswise (see Fig. 3).

C. Surface Adsorption of Motor

Solution

1. *Motility buffer:* BRB80 plus 1 mM MgATP. For fluorescence work only, degas and add 1% of:

> **a.** 100$\times$ antibleach mix (GOC): 100 mg/ml glucose oxidase (Sigma G-7016), 18 mg/ml catalase (Sigma C-100), and 300 mg/ml glucose (Sigma G-7528) in BRB80 plus 50% glycerol. When aliquoting, fill tubes to exclude oxygen, cap, and store at $-20°C$.
>
> **b.** 100$\times$ diluted MTs in motility buffer or motility buffer plus GOC.

Steps

1. Place the flow cell flat. Inject into the cell using a Gilson 1 chamber volume of motor solution. The solution is drawn into the cell by capillarity. Incubate the slide in a moisture chamber for 2–5 min at 20°C to allow the motor to adsorb to the glass.

2. Wash the cell with 2 chamber volumes of assay buffer, applying the solution to one end of the chamber using a micropipette and drawing the solution gently through the cell using the capillary action of the torn edge of a strip of Whatman 3MM, placed at the exit of the chamber

3. Flow in 1 vol of MTs in motility buffer + taxol, and mount the slide on the microscope stage, oiling the condensor to the bottom of the slide (it may be possible for quick-and-dirty assays to use a dry condensor).

4. For fluorescence work: Degas some BRB80. To 10 ml, add 100 μl of 100$\times$ GOC. Take another 3 ml and add MgATP to 2 mM. Fill and cap tubes to exclude oxygen and hold buffers on ice. Add taxol to 20 μM freshly before use.

A

B

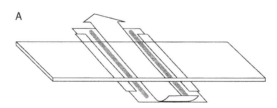

FIGURE 3 Flow cells. (A) Flow cell for inverted microscope. (B) Flow cell for upright microscope.

D. Microscope Setup

1. For DIC

Setup for video fluorescence and DIC microscopy is described elsewhere in this volume (See Volume 3, Section 8B, "Video and Digital Fluorescence Microscopy" for additional information), and I make only brief recommendations here. Before the day's work, align the microscope roughly using a test specimen (a slide made using a suspension of plastic beads provides a stable and realistic test specimen). Switch on the lamp and allow a few minutes for the arc to stabilize. Rack down the objective and oil it to the slide. Insert some neutral density filtration to protect your eyes from the intensely bright light, focus roughly on the top surface of the grease at the edge of the chamber, and then drive the stage to center the sample below the objective. Find some beads attached to the undersurface of the cover slip. Open the condensor aperture and close the field aperture. Obtain Koehler illumination by focusing and centering the condensor so that a sharp image of the field diaphragm appears in the viewfield. Open the field diaphragm again and adjust DIC sliders close to extinction.

Focusing on MTs in the experimental flow cell is also best done using the grease surface as a guide. Focus as above, then remove neutral density filters and switch in the video system. Adjust fine focus to image the surface. Adjust light intensity to almost saturate the camera (this is the point where signal to noise is maximal). With the contrast on the Argus set to max, microtubules should be visible without background subtraction. Defocus slightly, collect a background image, and subtract. Microtubules should now be clearly visible.

2. For Epifluorescence

A test sample of multispectral fluorescent beads is very useful (Molecular Probes multispeck M-7900). Switch on the arc lamp and allow a few minutes for the arc to stabilize. Once the lamp is stable, align the microscope for epifluorescence: Remove an objective and place a piece of paper on the stage. Inset some neutral density filtration. Close the field diaphragm slightly, focus, and center the image of the lamp filament that appears on the paper. Replace the objective.

Focusing on microtubules in the experimental cell is much easier with dark-adapted eyes. Using the full intensity of the mercury lamp, rack the objective down until MTs are visible, first as a dim red glow and then as a sharply defined bright red lines on a black background. Immediately reduce the illumination intensity to protect against photobleaching, switch in the intensified camera, and start recording.

E. Recording Data

The most flexible arrangement for data recording is to set up time-lapse digital recording of video frames to computer hard disk (with no resolution loss) and simultaneous SVHS recording to VCR. The VCR runs uninterrupted in the background for 3 hr and generates an archive. The operator is free to go back to this archive at a later date and recapture interesting sequences for analysis.

F. Analyzing Data

Calibration is by imaging a stage graticule, a slide with etched lines at 1- or 10-μm intervals (from microscope manufacturers). It is important to calibrate both in X and Y; simply rotate the camera 90°. Most systems will give you a different number of pixels per micron in X and Y. Tracking software compensates for this effect.

The best way to track is to follow the tip of a moving microtubule; tracking the centroids, as is common in cell tracking, for example, will give you the wrong answer as soon as the microtubule bends. For maximum accuracy, the time lapse between frames should be adjusted to minimize the effects of operator error when tracking using the mouse. In practice we try to collect 20 frames, adjusting the time lapse so that the microtubules move across the full field (22 μm) during this time.

IV. COMMENTS

A. Archiving Data

It is very important to have a formal system for identifying every video frame on every tape. In this way there is no possibility of confusing data sets. The simplest way to do this is to time and date stamp the frames as they are generated, using the overlay feature of the Argus. As ever, keeping careful written notes also helps immensely. For complex experiments it can be useful to speak notes onto the audio track of the tape. Currently, digital video clips are most conveniently archived to 230-Mb magneto-optical disks.

B. Imminent Technology

As computers get quicker, it is realistic to start recalculating images in real time. Autocontrast is one interesting possibility, whereby the pixels of each incoming frame are parsed and the look-up table is stretched to optimize contrast. It will be some time before we can dispense with the VCR. Real-time full resolution recording to disk is pushing the limits at present, but sufficiently fast sustained data transfer rates will soon be available. This is not the real problem, however. One frame of PAL video is 768×512 pixels, which with 8 bit (256 greys) data means each frame is 384 Kb. Real-time recording to hard disk fills the disk up at about 0.5 Gb per minute, and it soon becomes necessary to archive data to video tape.

C. Workstation Ergonomics

It is worth paying some attention to the ergonomics of your microscope workstation. Microscope focus, mouse, keyboard video, and contrast–adjustment electronics all need to be within reach of a seated operator. Screens should be visible with only a slight turn of the head. It is very helpful to have a foot switch to dim the room lights, and blinds on any windows.

D. Best Practice

Because of inherent uncertainties about the way a particular protein attaches to a particular glass, motility assays are at their strongest when used to measure the relative motility in different treatments of samples. It is commonly assumed that motility assays measure motor-driven microtubule sliding under zero load. It is probably more correct to assume that an unspecified, variable (but low) load applies.

V. PITFALLS

1. The most common fault in video microscopy is to overprocess an indifferent optical image. Too much processing can seriously degrade the amount of information in the image. A good primary image has high spatial resolution (sharpness), high contrast, and low background noise. Obtaining a good image is (1) a function of specimen preparation, and (2) of microscope setup.

2. In DIC, a troublesome problem is sudden variations in light intensity caused by the arc of the mercury lamp wandering. These are not noticeable in normal modes of microscopy, but with electronic amplification of contrast they become annoying. The only solution is to change the lamp. Cooling the lamp using a fan may help. Mercury lamps typically need changing after 100 hr because their intensity then drops fairly rapidly.

3. Some motor proteins bind better to the glass surface than others. Erratic motility may be due to the protein denaturing on the glass or binding in such a way that its force-generating conformational change is inhibited. Areas of uncoated glass can also bind microtubules and inhibit sliding. Increase the motor concentration if possible or try infusing the motor twice over and/or reducing or eliminating the wash step prior to infusing microtubules. Including casein at 0.1–1 mg ml in the assay buffer efficiently protein coats glass. Motor activity can also be sensitive to thiol oxidation, so try including 5 mM dithiothreitol in the motility buffer.

References

Cross, R. A., and Kendrick, Jones, J. (1991). Motor proteins. *J. Cell Sci.* Suppl. 14.

Inoué, S. (1986). "Video Microscopy." Plenum Press, New York.

Kron, S. J., Toyoshima, Y. Y., Uyeda, T. Q. R., and Spudich, J. A. (1991). Assays for actin sliding movement over myosin-coated surfaces. *In* "Methods in Enzymology" (R. B. Vallee, ed.), Vol. 196, pp. 399–416. Academic Press, San Diego.

Scholey, J. M. (ed.) (1993). Motility assays for motor proteins. "Methods in Cell Biology," Vol. 39.

In Vitro Assays for Mitotic Spindle Assembly and Function

Rebecca Heald, Régis Tournebize, Isabelle Vernos, Andrew Murray, Tony Hyman, and Eric Karsenti

I. INTRODUCTION

Since the early 1980s, cell biology has entered a phase in which technology is so powerful that fundamental questions concerning the morphogenesis and function of cellular organelles can be addressed. One essential and beautiful structure is the mitotic spindle. The work of Lohka and Maller (1985), followed by that of a few other laboratories (Murray and Kirschner, 1989; Sawin and Mitchison, 1991; Shamu and Murray, 1992), has opened up a novel approach to studying such a complex and dynamic structure. Instead of purifying individual spindle components and putting them back together, hoping that a spindle will assemble, the idea is to open up the cell and prepare a cytoplasmic extract as crude and concentrated as possible to keep the conditions close to the *in vivo* situation. At first sight, this approach seems uninformative: one merely mimics *in vitro* what happens *in vivo*. In fact, it has proven extremely powerful as demonstrated by recent experiments that combine this *in vitro* system with video microscopy, the use of simple DNA templates bound to beads, and specific reagents (Boleti *et al.*, 1996; Heald *et al.*, 1996; Sawin *et al.*, 1992; Vernos *et al.*, 1995). Thus, this system can be used both to analyze the mechanism of spindle assembly and to evaluate the role of individual molecules in the process by adding specific inhibitors such as antibodies or dominant negative constructs. In this way, both complexity and specificity are combined at the molecular level in one reconstituted system.

II. COMMENTS ABOUT PROTOCOLS

Good protocols detailing the preparation of Xenopus egg extracts and sperm nuclei have already been published (Murray, 1991). This article describes only the preparation of cytostatic factor (CSF) extracts as optimized in our laboratory for spindle assembly reactions. For the preparation of DNA beads, plasmid DNA is linearized and filled in with nucleotides so that one end contains biotinylated bases and the other end contains thionucleotides to inhibit exonuclease activity (Fig. 1). The DNA is then coupled to 2.8-μm magnetic beads. Spindle assembly reactions are based on published protocols (Sawin and Mitchison, 1991; Shamu and Murray, 1992). Spindle assembly around sperm nuclei is accomplished by cycling the extract through interphase in order to allow DNA replication and centrosome duplication before the extract is induced back into mitosis (Fig. 2a). Spindle assembly around DNA beads is achieved by first assembling chromatin on the beads in an extract that is cycled

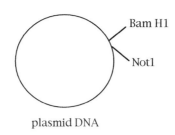

. restriction digest

plasmid DNA

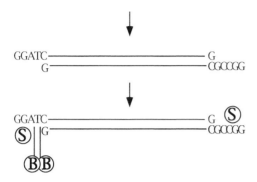

. Klenow fill in with
biotin-dATP, -dUTP,
thio-dCTP, -dGTP.

gel filtration

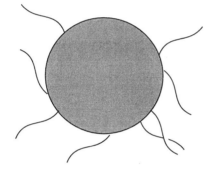

. Couple to Dynabeads
M-280 streptavidin

FIGURE 1 Steps in preparing DNA beads.

through interphase. The beads are then retrieved and resuspended in fresh mitotic extract
Fig. 2b). Microtubules are visualized by the addition of fluorochrome-labeled tubulin. For
data collection, small samples of the reaction can be squashed with fixative between a slide
and coverslip or observed live by video microscopy. Alternatively, samples can be diluted
and spun onto coverslips to allow immunofluorescent analysis. A protocol is also given for
the analysis of specific antibodies or dominant negative fusion proteins, which can simply be
added to the spindle assembly reaction.

II. MATERIALS AND INSTRUMENTATION

A. Preparation of *Xenopus laevis* Egg Extracts

Equipment includes an incubator at 16°C, clinical centrifuge, DuPont Sorvall RC-5 centrifuge, HB-4 rotor with rubber adaptors (Sorvall Cat. No. 00363), Beckman ultraclear SW50
tubes (Cat. No. 344057), Sarstedt 13-ml adaptor tubes (Cat. No. 55.518), 1-ml syringes, 18-
and 27-gauge needles, and glass Pasteur pipettes.

Mature female frogs are from African Reptile Park. Pregnant mare serum gonadotropin
PMSG, Intergonan) is from Intervet. Human chorionic gonadotropin (HCG; Cat. No. CG-
0), L-cysteine (Cat. No. C-7755), cytochalasin D (Cat. No. C-8273), EGTA (Cat. No. E-

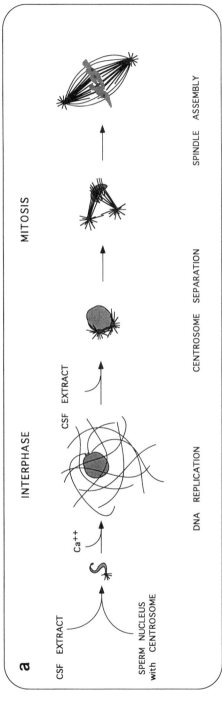

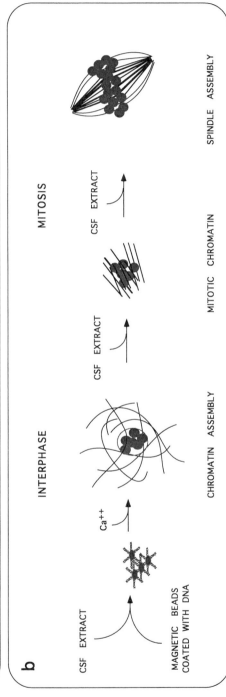

FIGURE 2 Spindle assembly in Xenopus egg extracts. (a) Successive steps in spindle assembly around sperm nuclei. (b) Steps in spindle assembly around DNA beads.

378), 4.9 M $MgCl_2$ (Cat. No. 104.20), and ATP (Cat. No. A-2383) are from Sigma. $CaCl_2$ (Cat. No. 2382.0500), KCl (Cat. No. 1.04936.1000), NaCl (Cat. No. 1.06404.1000), and sucrose (Cat. No. 1.07653.1000) are from Merck. Chymostatin (Cat. No. E16), leupeptin (Cat. No. E18), and pepstatin (Cat. No. E19) are from Chemicon. HEPES is from Biomol. Creatine phosphate (Cat. No. 127 574) is from Boehringer-Mannheim.

B. Preparation of DNA Beads

Biotin-21-dUTP (Cat. No. 5201-1) is from Clonetech. Biotin-14-dATP (Cat. No. 520-95245A) is from GIBCO-BRL. Thio-dCTP (Cat. No. 27-7360), thio-dGTP (27-7370), and G-50 gel filtration (NICK) columns (17-0855-01) are from Pharmacia. Restriction enzymes are from Boehringer-Mannheim. Klenow (Cat. No. M2201) is from Promega. A magnetic particle concentrator (MPC) and a Kilobase BINDER kit containing Dynabeads M-280 streptavidin, washing, and binding solutions can be obtained from Dynal. Trizma base (Cat. No. T-1503) is from Sigma. EDTA (Cat. No. 1.08418.1000) is from Merck.

C. Spindle Assays

Freshly prepared CSF extract, sperm nuclei (3000/μl) (Murray, 1991), rhodamine-labeled tubulin (20–30 mg/ml) (see Volume 2, Ashford *et al.*, "Preparation of Tublin from Bovine Brain" for additional information; Hyman, *et al.*, 1991; Hyman, 1991), 76 × 26-mm microscope slides, 22 × 22-mm coverslips, and a fluorescence microscope are needed. PIPES (Cat. No. P-6757), Triton X-100 (Cat. No. T-8787), and Hoechst dye (bisbenzimide, Cat. No. H-33342) are from Sigma. Glycerol (Cat. No. 1.04093.2500) and formaldehyde (Cat. No. 1.04003) are from Merck. For sedimentation and immunofluorescence experiments, corex tubes (15 ml) are equipped with plastic adaptors (homemade; see Evans *et al.*, 1985) to support 12-mm round coverslips. The tubes are centrifuged in an HB-4 rotor containing rubber adaptors. pGEX expression vectors are from Pharmacia.

IV. PROCEDURES

A. Preparation of *Xenopus laevis* CSF Egg Extracts

Solutions

1. *PMSG (200 units/ml):* Add 5 ml sterile distilled water to vial containing 1000 units. Store at 4°C.

2. *HCG (500 units/ml):* Add 10 ml sterile distilled water to vial containing 10,000 units. Store at 4°C.

3. *MMR:* 100 mM NaCl, 2 mM KCl, 1 mM $MgCl_2$, 2 mM $CaCl_2$, 0.1 mM EDTA, and 5 mM HEPES, pH 7.8. Prepare 500 ml 1× MMR from 10× stock by combining 50 ml 10× and 450 ml distilled water. To prepare 1 liter of 10× stock, combine 58.4 g NaCl, 1.49 g KCl, 2.04 ml of 4.9 M $MgCl_2$ stock solution (Sigma), 2.94 g $CaCl_2$, 372 mg EDTA, 11.9 g HEPES, and 900 ml distilled water. Adjust pH to 7.8 with 6 M NaOH and bring volume to 1 liter. Autoclave and store at room temperature.

4. *Dejellying solution:* 2% L-cysteine, pH 7.8. To prepare 250 ml, dissolve 5 g cysteine in 250 ml water. Adjust pH to 7.8 with 6 M NaOH. Prepare fresh just before use.

5. *Protease inhibitors (LPC) (10 mg/ml):* To prepare 1 ml, dissolve 10 mg each of leupeptin, pepstatin, and chymostatin in 1 ml dimethy sulfoxide (DMSO). Store in 50-μl aliquots at −20°C.

6. *Cytochalasin D (10 mg/ml):* To prepare 1 ml, dissolve 10 mg cytochalasin D in 1 ml DMSO. Store in 50-μl aliquots at −20°C.

7. *XB:* 100 mM KCl, 0.1 mM $CaCl_2$, 1 mM $MgCl_2$, 50 mM sucrose, and 10 mM K-HEPES, pH 7.7. Prepare fresh from stock solutions of 20× XB salts, 1 M HEPES, and 1.5 M sucrose. To prepare 20× XB salts (2 M KCl, 20 mM $MgCl_2$, 2 mM $CaCl_2$), dissolve 74.6

g KCl, 147 mg $CaCl_2$, and 2.04 ml of 4.9 M $MgCl_2$ in 500 ml of distilled water. Filter sterilize and store at 4°C. To make 1 M HEPES, dissolve 59.6 g HEPES in 200 ml distilled water. Adjust pH to 7.8 with concentrated KOH and volume to 250 ml. Filter sterilize and store in 5-ml aliquots at −20°C. To prepare 1.5 M sucrose, dissolve 256.7 g sucrose in distilled water. Store in 15-ml aliquots at −20°C. For an extract preparation to make spindles, prepare 250 ml of XB, add 12.5 ml 20× XB salts, 2.5 ml 1 M HEPES, and 8.3 ml of 1.5 M sucrose to 200 ml of distilled water. Adjust pH to 7.7 with KOH if necessary and bring volume to 250 ml.

8. *CSF-XB:* 100 mM KCl, 0.1 mM $CaCl_2$, 2 mM $MgCl_2$, 5 mM EGTA, 50 mM sucrose, 10 μg/ml LPC, and 10 mM HEPES, pH 7.7. Prepare by adding $MgCl_2$, EGTA, and LPC stock solutions to XB solution (above). To prepare 0.5 M EGTA, dissolve 19.02 g EGTA in 100 ml distilled water and adjust pH to 7.7 with 10 M NaOH. Store in 1-ml aliquots at −20°C. To make 50 ml of CSF-XB, add 10.2 μl of 4.9 M $MgCl_2$, 0.5 ml of 0.5 M EGTA, and 50 μl of 10 mg/ml LPC to 49.5 ml of XB. Check the pH and adjust if necessary to pH 7.7 with 1 M KOH.

9. *Energy mix:* The 20× stock contains 150 mM creatine phosphate, 20 mM ATP, 2 mM EGTA, and 20 mM $MgCl_2$. Combine 55.1 mg ATP, 32.7 mg creatine phosphate, 10 μl 1 M $MgCl_2$, 20 μl 0.5 M EGTA, and 5 ml distilled water. Store in 100-μl aliquots at −20°C.

Steps

1. Inject two to four frogs subcutaneously with 0.5 ml (100 units) PMSG each using a 1 ml syringe and a 27-gauge needle at least 4 days before planning to make an extract. They should be used within 2 weeks after the priming injection. The number of frogs required depends on the quantity and quality of eggs. To obtain 1 ml of extract, approximately 5 ml of eggs must be used (1 SW50 tube)

2. Twelve to 18 hr before use, inject frogs subcutaneously with 0.5 ml (500 units) HCG. Place the frogs in individual boxes containing 500 ml MMR at 16°C.

3. Prepare all solutions before starting. Rinse all glassware with distilled water (eggs stick to plastic dishes). Cut off the end of a glass Pasteur pipette and fire-polish it to make a wide mouth pipette.

4. Collect laid eggs in MMR. Frogs can also be squeezed, which often gives the highest quality eggs. Keep eggs from different frogs in separate batches in 400-ml beakers. Discard batches of eggs containing more than 5% of lysed, ugly, or stringy eggs.

5. Pour off MMR and add 50–100 ml dejellying solution. When laid, eggs are enveloped in a transparent jelly coat and do not pack closely together. Swirl the beaker frequently, and change the cysteine solution two to three times. After removal of the jelly coat, eggs pack together. This takes about 5 min. Eggs left for too long in cysteine will lyse.

6. Pour off the cysteine solution and add 50–100 ml MMR. Repeat the rinse several times. After removal of the jelly coat, the eggs become fragile. They lyse easily and can activate if in contact with air. They must always remain immersed in buffer. Remove all bad looking eggs: white and puffy, flattened, or activated ones (darker pole retracted) and those with mottled pigmentation. Wash again in MMR.

7. Wash three times with 50–100 ml XB.

8. Remove as much buffer as possible, keeping all eggs immersed. Wash twice with CSF-XB.

9. Transfer eggs to SW50 tubes containing 1 ml CSF-XB plus 10 μl cytochalasin D (100 μg/ml). Always immerse the pipette tip in solution before expelling eggs to prevent contact with air. Transfer the SW50 tubes to 13-ml Sarstedt adaptor tubes, which contain 0.5 ml of water to prevent the tubes from collapsing.

10. Centrifuge in a clinical centrifuge at 150 g for 30 sec, then at 700 g for 30 sec at 16°C. Then remove all excess buffer from the top of the packed eggs. Removal of buffer is critical to obtain a concentrated cytoplasm.

11. Place the tubes in an HB-4 rotor containing rubber adaptors. Centrifuge at 10,000 rpm for 15 min at 16°C to crush the eggs.

12. Place the tubes on ice. A yellow lipid layer is at the top of the tube. Underneath is the cytoplasmic layer, then heavy membranes, and yolk particles at the bottom of the tube. Wipe the sides of the tubes with a tissue before piercing with an 18-gauge needle at the bottom of the cytoplasmic layer. Slowly and carefully remove it, using a 1-ml syringe with the needle opening facing upwards. Remove the needle from the syringe before expelling the extract into a 1.5-ml eppendorf tube.

13. Add LPC and cytochalasin D to 10 μg/ml (1:1000 dilution of stocks). Add energy mix (1:20 dilution of stock). Mix gently. The extract can be kept on ice up to 6 hr before use.

Comments on Sperm Nucleus Preparation

The preparation is based on Gurdon (1976) and is modified by Murray (1991). Our only modification is to mash fragments of testes between two frosted (rough) slides before filtering through a cheesecloth mesh.

B. Preparation of DNA Beads

Solutions

1. *TE:* 10 m*M* Tris and 1 m*M* EDTA, pH 8. To prepare 500 ml, dissolve 0.61 g Tris base and 0.15 g EDTA in 450 ml distilled water. Adjust pH to 8.0 with 2 *M* HCl and bring volume to 500 ml. Autoclave and store at room temperature.

2. *Washing and binding solutions:* Included with Kilobase BINDER kit.

3. *Bead buffer:* 2 *M* NaCl, 10 m*M* Tris, and 1 m*M* EDTA, pH 7.6. To prepare 250 ml, dissolve 29.2 g NaCl, 0.30 g Tris base, and 75 mg EDTA in 200 ml distilled water. Adjust pH to 7.6 with 2 *M* HCl and bring volume to 250 ml. Store at room temperature.

Steps

1. Prepare plasmid DNA by Qiagen column purification. Although the sequence of the DNA is not important, the plasmid should be more than 5 kb to effectively induce chromatin assembly. Cut 50 μg of the DNA with two restriction enzymes that have unique sites in the polylinker to produce one short and one long DNA fragment. One end of the long fragment should terminate in an overhang containing Gs and Cs, whereas the other should contain only As and Ts (e.g., *Not*I, *Bam*HI). See Fig. 1.

2. Ethanol precipitate the DNA and resuspend in 25 μl TE. Quantify recovery by OD$_{260}$ measurement.

3. Prepare fill-in reaction in 70 μl containing 1× Klenow buffer, 30 μg DNA, 50 μ*M* nucleotides (biotin-dATP, biotin-dUTP, thio-dCTP, and thio-dGTP), and 20 units Klenow. Incubate for 2 hr at 37°C.

4. Remove unincorporated nucleotides, following instructions supplied with Pharmacia Nick columns. Elute the DNA in a large volume (400 μl); the recovery is better than with spin columns. Quantify recovery by OD$_{260}$ measurement.

5. Prepare coupling mix by combining 400 μl biotinylated DNA and 400 μl binding solution. Set aside 25 μl of the coupling mix for later evaluation of coupling efficiency.

6. Prepare 4 μl of streptavidin Dynabeads for each microgram of DNA recovered (120 μl for 30 μg). Retrieve beads using the MPC (magnet) and wash once with 5 vol of binding solution (600 μl for 120 μl beads). Retrieve the beads and resuspend them in coupling mix containing DNA.

7. Incubate bead/coupling mixture for several hours (or overnight) on a rotator at 20°C.

8. Retrieve the beads and save the supernatant. Compare the OD$_{260}$ of the supernatant to the sample taken before coupling to determine the amount of DNA immobilized. Typically two-thirds of the DNA is coupled.

9. Wash beads twice with washing solution and then twice with bead buffer. After the last wash, resuspend the beads in bead buffer so that the final concentration of immobilized DNA is 1 μg/5 μl of beads. Store at 4°C.

C. Spindle Assembly *in Vitro*

Solutions

1. *400 mM CaCl$_2$ stock solution:* To make 100 ml, dissolve 5.9 g CaCl$_2$·2H$_2$O in 100 ml distilled water. Store in aliquots at −20°C.

2. *4.9 M MgCl$_2$ stock solution (Sigma)*

3. *Calcium solution:* 4 mM CaCl$_2$, 100 mM KCl, and 1 mM MgCl$_2$. To prepare 100 ml combine 1 ml 400 mM CaCl$_2$ stock solution, 20.4 μl of 4.9 M MgCl$_2$ stock solution, and 0.75 g KCl and bring volume to 100 ml with distilled water. Store in aliquots at −20°C.

4. *10× MMR:* See Procedure A.

5. *Hoechst dye solution:* 10 mg/ml bisbenzimide. To prepare 1 ml, add 10 mg to 1 ml distilled water. Store in the dark at 4°C.

6. *Spindle fix:* For 1 ml, combine 600 μl of 80% glycerol, 300 μl of 37% formaldehyde 100 μl of 10× MMR, and 0.5 μl of 10 mg/ml Hoechst dye. Always prepare fresh on day of use.

7. *5× BRB80:* 0.4 M PIPES, 5 mM MgCl$_2$, and 5 mM EGTA, pH 6.8. To prepare 250 ml, add 30.2 g PIPES, 0.26 ml of 4.9 M MgCl$_2$, and 0.48 g EGTA to 200 ml distilled water. While stirring, add KOH pellets until the PIPES dissolves. Adjust final pH to 6.8 with 10 M KOH and bring volume to 250 ml. Sterilize by filtration and store at 4°C.

8. *Dilution buffer:* 30% glycerol, 1% Triton X-100, in BRB80. For 100 ml, combine 30 ml glycerol, 1 ml Triton X-100, 20 ml 5X BRB80, and 49 ml distilled water. Store at room temperature.

9. *Cushion:* 40% glycerol in BRB80. For 500 ml, combine 200 ml glycerol, 100 ml 5× BRB80, and 200 ml distilled water. Store at 4°C.

10. *0.2 M Tris, pH 8:* To prepare 100 ml, dissolve 2.42 g Tris base in 80 ml distilled water. Adjust pH to 8.0 with concentrated HCl and bring volume to 100 ml. Autoclave and store at room temperature.

11. *Mounting medium:* 90% glycerol and 10% 0.2 M Tris–HCl, pH 8. For 10 ml, combine 9 ml glycerol and 1 ml 0.2 M Tris–HCl, pH 8.

Steps for Sperm DNA Spindle Assembly (Fig. 2)

1. On ice, add 0.5 μl rhodamine tubulin and 1.25 μl sperm nuclei to 50 μl CSF extract in a 1.5-ml tube (about 75 sperm/μl extract).

2. Incubate for 10 min at 20°C, then release extract into interphase by adding 5 μl calcium solution. Mix by pipetting with a cut-off tip.

3. Incubate for 80 min at 20°C. Check that the extract is in interphase by transferring 1.2 μl to a microscope slide, using a cut-off tip. Carefully place 6 μl of fixation solution on top of the drop of extract, and squash gently by placing a 22 × 22-mm coverslip on top. If the sample is to be saved, seal the coverslip to the slide with nail polish. By fluorescence microscopy, nuclei should appear large, round, and uniform. Microtubules should be long and abundant.

4. At 90 min postcalcium addition, add 0.5 vol (25 μl) of fresh CSF extract to the reaction Continue incubation at 20°C.

5. Take samples at 15, 30, 45, 60, and 90 min after adding fresh extract to assess the spindle assembly reaction.

6. For immunofluorescent analysis of samples, transfer 10–20 μl of the spindle assembly reaction to a 1.5-ml Eppendorf tube and add 1 ml of dilution buffer. Layer the mixture over 5 ml of cushion in a 15-ml modified corex tube containing a 12-mm coverslip. Centrifuge tubes for 15 min at 10,000 rpm in an HB-4 rotor. Aspirate supernatant and cushion before removing the coverslip. Postfix coverslips in −20°C methanol for 5 min, transfer to phosphate-buffered saline (PBS), and stain with primary and secondary antibodies. Stain the DNA with 5 μg/ml Hoechst for 2 min. After washing, place the coverslips upside down on a 4-μl drop of mounting medium and seal with nail polish.

Steps for Chromatin Bead Spindle Assembly (Fig. 2)

1. Transfer 3 μl of DNA beads (about 0.5 μg DNA) to a 0.5-ml Eppendorf tube and place on magnet. Remove supernatant, and wash beads by resuspending them in 20 μl of extract.

2. Retrieve beads and resuspend in 100 μl CSF extract. Transfer to a 1.5-ml Eppendorf tube and incubate at 20°C.

3. After 10 min, release the CSF extract into interphase by adding 10 μl of calcium solution. Incubate for 2 hr at 20°C.

4. Return the extract containing the beads to mitosis by adding 50 μl of fresh CSF extract. Incubate for an additional 30 min at 20°C.

5. Incubate the bead mixture on ice for several minutes. Retrieve the beads on ice over 10–15 min. Due to the viscosity of the extract, bead retrieval is slow. Pipette the mixture every several minutes, keeping the tube on the magnet.

6. Remove the supernatant and resuspend the beads in 150 μl of fresh CSF extract containing 1.5 μl rhodamine-labeled tubulin.

7. Incubate at 20°C. Monitor the spindle assembly by taking 1.2-μl samples and squashing with fixative as described earlier. Spindle assembly requires between 30 and 90 min, depending on the extract.

D. Assay of Reagents Interfering with Spindle Assembly

Solutions

1. *CSF-XB:* See Procedure A.

2. *PBS:* Dissolve 8 g NaCl, 0.2 g KCl, 1.44 g Na_2HPO_4, and 0.2 g KH_2PO_4 in 800 ml of distilled water. Adjust pH to 7.2 and volume to 1 liter. Sterilize by autoclaving and store at room temperature.

Steps for Using Antibodies to Study the Function of a Given Protein in Spindle Assembly

1. Prepare a concentrated solution (at least 2 mg/ml) of an affinity-purified polyclonal or monoclonal antibody in PBS.

2. Follow the protocol for spindle assembly *in vitro* as described earlier with the following modifications.

3. In the first step, in addition to rhodamine tubulin and sperm nuclei or DNA beads, add the affinity-purified antibody in a volume corresponding to 1/10th of the total volume of CSF extract. Mix well and incubate on ice for 10 to 20 min. Two control samples should be run in parallel, one containing a similar amount of a control antibody and another containing the same volume of PBS. It is also a good idea to try different concentrations of antibody (ranging from 200 to 400 μg/ml).

4. Proceed with the spindle assembly protocol. Use one-half of the treated CSF extract for sperm DNA reactions. Keep the rest of the treated extract on ice to add it to the reaction at step 4.

5. Analyze samples taken at the different time points and compare the phenotypes obtained in the antibody-treated samples with the control samples.

---

6. Samples diluted and spun onto coverslips can be analyzed by immunofluorescence using an FITC-conjugated secondary antibody to localize the added antibody.

Steps for Using Recombinant Proteins to Study the Function of a Given Protein in Spindle Assembly

1. Prepare constructs expressing different fragments of the protein under study fused to GST. Purify the fusion proteins by affinity on glutathione beads as described (Smith and Johnson, 1988) and dialyze them extensively against CSF-XB.

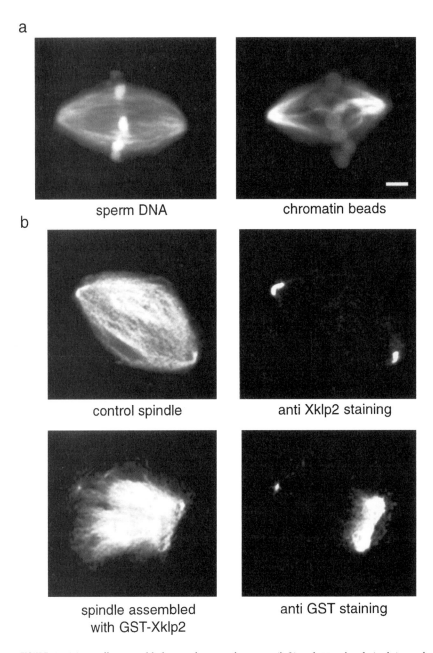

a

sperm DNA   chromatin beads

b

control spindle   anti Xklp2 staining

spindle assembled   anti GST staining
with GST-Xklp2

FIGURE 3 (a) Spindles assembled around sperm chromatin (left) and DNA beads (right) visualized by the addition of rhodamine-labeled tubulin. Scale bar is 5 μm. (b) Assay of GST fusion protein containing the C terminus of the kinesin-like protein Xklp2. (Top) Control sperm DNA spindle (left) and localization of the endogenous Xklp2 protein by immunofluorescence (right). (Bottom) Spindle reaction in the presence of GST–Xklp2. Distorted spindles predominate in which centrosome separation is blocked, leading to the formation of a half spindle containing one pole (left). Localization of the GST–Xklp2 to the pole by immunofluorescence with anti-GST antibodies (right).

2. Proceed with the spindle assembly protocol as described for the use of antibodies. Add fusion protein to extract reactions in a range of concentrations close to that of the endogenous protein. Controls in this case are samples containing only the GST protein and samples containing the same volume of CSF-XB.

3. Analyze samples as for the antibody-treated samples.

4. In addition, some samples can be used for immunofluorescent analysis as described earlier. Coverslips are stained with an anti-GST antibody to examine the localization of the fusion protein on spindles (see Fig. 3).

5. The exact proportion of fusion versus endogenous protein can be analyzed by Western blot. After addition of the fusion protein to the CSF extract, add a sample to Laemmli sample buffer, run on an SDS–PAGE gel of appropriate acrylamide concentration, and blot. Probe the blot with an antibody that recognizes both the endogenous and the recombinant proteins. A correlation can then be made between the proportion of added protein and its effect on spindle assembly.

V. PITFALLS

The biggest pitfall in studying spindle assembly with this system is the problem of reproducibility in Xenopus egg extracts. To obtain good eggs, the frog colony must be healthy and well cared for. This requires a substantial commitment on the part of the laboratory. Even with healthy frogs, there is seasonal variation in the quality of eggs, with summer being the off season. Furthermore, even experienced extract makers do not manage to prepare functional extracts every time, which can be frustrating. Therefore, experiments must be repeated several times to ensure a valid interpretation.

References

Boleti, H., Karsenti, E., and Vernos, I. (1996). Xklp2, a novel Xenopus centrosomal kinesin-like protein required for centrosome separation during mitosis. *Cell* **84**, 49–59.

Evans, L., Mitchison, T. J., and Kirschner, M. W. (1985). Influence of the centrosome on the structure of nucleated microtubules. *J. Cell Biol.* **100**, 1185–1191.

Gurdon, J. B. (1976). Injected nuclei in frog oocytes: Fate, enlargement, and chromatin dispersal. *J. Embryol. Exp. Morphol.* **36**, 523–540.

Heald, R., Tournebize, R., Blank, T., Sandaltzopoulos, R., Becker, P., Hyman, A., and Karsenti, E. (1996). Self organization of microtubules into bipolar spindles around artificial chromosomes in Xenopus egg extracts. *Nature* **382**, 420–425.

Hyman, A., Drechsel, D., Kellogg, D., Salser, S., Sawin, K., Steffen, P., Wordeman, L., and Mitchison, T. (1991). Preparation of modified tubulins. *In* "Methods in Enzymology" (R. B. Vallee, ed.), Vol. 196, pp. 478–485. Academic Press, San Diego.

Hyman, A. A. (1991). Preparation of marked microtubules for the assay of the polarity of microtubule-based motors by fluorescence. *J. Cell Sci.* (Suppl. 14), 125–127.

Lohka, M., and Maller, J. (1985). Induction of nuclear envelope breakdown, chromosome condensation, and spindle formation in cell-free extracts. *J. Cell Biol.* **101**, 518–523.

Murray, A. (1991). Cell cycle extracts. *In* "Methods in Cell Biology" (B. K. Kay and H. B. Peng, eds.), Vol. 36, pp. 581–605. Academic Press, San Diego.

Murray, A. W., and Kirschner, M. W. (1989). Cyclin synthesis drives the early embryonic cell cycle. *Nature* **339**, 275–280.

Sawin, K. E., LeGuellec, K., Philippe, M., and Mitchison, T. J. (1992). Mitotic spindle organization by plus-end-directed microtubule motor. *Nature* **359**, 540–543.

Sawin, K. E., and Mitchison, T. J. (1991). Mitotic spindle assembly by two different pathways *in vitro*. *J. Cell Biol.* **112**, 925–940.

Shamu, C. E., and Murray, A. W. (1992). Sister chromatid separation in frog egg extracts requires DNA topoisomerase II activity during anaphase. *J Cell Biol.* **117**, 921–934.

Smith, D., and Johnson, K. (1988). Single step purification of polypeptides expressed in *Escherichia coli* as fusions with glutathione S-transferase. *Gene* **67**, 31–40.

Vernos, I., Raats, J., Hirano, T., Heasman, J., Karsenti, E., and Wylie, C. (1995). Xklp1, a chromosomal Xenopus kinesin-like protein essential for spindle organization and chromosome positioning. *Cell* **81**, 117–127.

In Vitro Motility Assays with Actin

James R. Sellers and He Jiang

I. INTRODUCTION

The interaction between actin and myosin has been studied for years using a variety of techniques, including ultracentrifugation, light scattering, chemical cross-linking, fluorescence, and measurement of the effect of actin on the MgATPase activity of myosin. The sliding actin *in vitro* motility assay constitutes a relatively recent technique for studying actin–myosin interaction. This assay, developed by Kron and Spudich (1986), takes advantage of the ability to image rhodamine–phalloidin-labeled actin filaments by fluorescence microscopy as they interact with and are translocated by myosin bound to a coverslip surface. The sliding actin *in vitro* motility assay is among the most elegant biochemical assays, reproducing the most fundamental property of a muscle, the ability of myosin to translocate actin using only the two highly purified proteins. It is a close *in vitro* correlate of the maximum unloaded shortening velocity of muscle fibers (Homsher *et al.*, 1992). As will be seen below, it is simple to set up, reproducible, quantitative, and utilizes as little as 1μg of myosin per assay. The assay is now used routinely in a large number of laboratories studying myosin and actin biochemistry.

This article discusses the design of the assay, describes the equipment required for its setup, and deals with methods for quantification and presentation of the results. Because different myosins exhibit a range of actin translocation speeds from 0.04 to 60 μm/sec (Sellers and Goodson, 1995), it will be necessary to discuss modification of the experimental setup for fast and slow myosins. We will describe the instrumentation that we use in our system and elaborate on other options where applicable.

II. MATERIALS AND INSTRUMENTATION

The following reagents are from Sigma Chemical Company: MOPS (Cat. No. M-5162), EGTA (Cat. No. E-4378), ATP (Cat. No. A-5394), glucose (Cat. No. G-7528), glucose oxidase (Cat. No. G-6891), catalase (Cat. No. C-3155), and methylcellulose (Cat. No. M-0512) Rhodamine–phalloidin is from Molecular Probes, Inc. (Cat. No. R415). Nitrocellulose (Superclean grade, Cat. No.11180) is from Ernest F. Fullam, Inc. Bovine serum albumin (BSA Cat. No. 160069) and dithiothreitol (DTT, Cat. No. 856126) are from ICN. The following items are from Thomas Scientific: microscope slides (Cat. No. 6684-H30), microscope 18 mm² No. 1 thickness coverslips (Cat. No. 6667-F24), 24 × 60-mm No. 0 thickness coverslips (Cat. No. 6672-A08), Apiezon grease-M (Cat. No. 7893-T40), and a diamond scribe (Cat

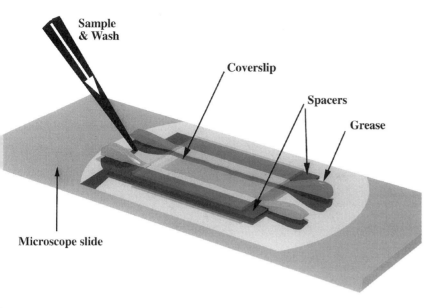

Sample & Wash

Coverslip

Spacers

Grease

Microscope slide

FIGURE 1 Schematic of the flow cell.

No. 5680-D60). Eppendorf combitips (Cat. No. 22 26130-4) are from Brinkman Instruments. Sony sVHS videotapes (ST120) are from local suppliers.

III. PROCEDURES

A. Construction of Flow Cells

Steps

Figure 1 shows a schematic of a flow cell produced by the following method.

1. Prepare nitrocellulose-coated coverslips by first placing one drop of a 1% solution of nitrocellulose in amylacetate into a 10 cm round dish filled with deionized water. Gently place a No. 1 thickness 18-mm^2 coverslip on the film formed after evaporation of the solvent. Using forceps, tear away excess film and remove coverslips by submersion and underwater inversion while lifting the coated coverslip out of the water. Air dry with the film side up. Alternatively, 2–3 μl of a 0.1% solution of nitrocellulose in amylacetate is spread directly on a coverslip and air dried to create the film. Some investigators use silicon coating of the coverslips (Fraser and Marston, 1995).

2. Prepare spacers that are 24 × 3–4 mm by cutting a No. 0 thickness 24 × 60-mm coverslip with a diamond scribe.

3. Using a 5-ml Eppendorf combitip syringe filled with Apiezon M grease, place two parallel tracks of Epizoon M grease about 10 mm apart and 25 mm long on a 25 × 75-mm glass microscope slide. On the outside of the grease tracks place two coverslip spacers. Place a nitrocellulose-coated coverslip with the coated side down onto the tracks. Tap gently with forceps to create a tight seal. The volume of such a flow cell is typically 10–30 μl. Alternatively, fingernail polish or some other type of glue can be used to fix the nitrocellulose-coated coverslip to the spacers in the absence of grease.

B. Preparation of Rhodamine–Phalloidin-Labeled Actin

1. Place 60 μl of 3.3 μM (in methanol) rhodamine–phalloidin into an Eppendorf tube and dry using a Speed-Vac concentrator.

2. Redissolve the rhodamine–phalloidin powder in 3–5 μl of methanol.

3. Add 85 μl of 20 mM KCl, 20 mM MOPS (pH 7.4), 5 mM MgCl$_2$, 0.1 mM EGTA, and 10 mM DTT (buffer A). To make 100 ml of buffer A, add 1 ml of 2 M KCl, 1 ml of 2 M MOPS (pH 7.4), 0.5 ml of 1 M MgCl$_2$, 0.02 ml of 0.5 M EGTA, and 15.4 mg DTT; bring to 100 ml with H$_2$O and adjust pH to 7.4.

4. Add 10 μl of a freshly diluted 20 μM F-actin solution (in buffer A) and incubate overnight on ice.

5. Centrifuge for 15 min at 435,000 g in a TL-100 ultracentrifuge (Beckman Instruments), remove the supernatant, and gently resuspend the pink rhodamine–phalloidin-labeled actin pellet in buffer A using a pipettor tip that has been cut to widen the bore size.

6. The rhodamine–phalloidin-labeled actin solution is stable for several weeks on ice.

C. Preparation of the Sample for Motility Assay

Solutions

1. *Wash solution:* 80 mM KCl, 20 mM MOPS (pH 7.4), 5 mM MgCl$_2$, 0.1 mM EGTA, and 5 mM DTT; to make 100 ml, add 4 ml of 2 M KCl, 1 ml of 2 M MOPS, (pH 7.4), 0.5 ml of 1 M MgCl$_2$, 0.02 ml of 0.5 M EGTA, and 77 mg of DTT. Bring volume to 100 ml with H$_2$O and pH to 7.4.

2. *Blocking solution:* 1 mg/ml BSA in 0.5 M NaCl, 20 mM MOPS (pH 7.0), 0.1 mM EGTA, and 1 mM DTT solution. To make 100 ml of blocking solution, add 10 ml of 5 M NaCl, 1 ml of 2 M MOPS (pH 7.0), 0.02 ml of 0.5 M EGTA, 15.4 mg of DTT, and 100 mg of BSA. Bring to 100 ml with H$_2$O and pH to 7.0.

3. *Rhodamine–phalloidin-labeled actin solution:* 20 nM rhodamine–phalloidin-labeled actin. To make 1 ml, take 10 μl of 2 μM rhodamine–phalloidin-labeled actin, 10 μl of 500 mM DTT, and 980 μl of wash solution.

4. *ATP–actin wash:* 1 mM ATP and 5 μM F-actin (unlabeled) in wash solution. To make 1 ml, add 10 μl of 0.1 M ATP and 50 μl of 100 μM F-actin to 940 μl of wash solution.

5. *4× stock solution:* 80 mM MOPS (pH 7.4), 20 mM MgCl$_2$, and 0.4 mM EGTA. To make 100 ml, add 16 ml of 2 M KCl, 4 ml of 2 M MOPS (pH 7.4), 2 ml of 1 M MgCl$_2$, and 0.08 ml of 0.5 M EGTA. Bring volume to 100 ml with H$_2$O and pH to 7.4.

6. *1.4% methycellulose solution:* Dissolve 1.4 g of methylcellulose in a final volume of 100 ml of H$_2$O by stirring overnight. Occasionally it is necessary to homogenize the solution with a glass–Teflon homogenizer to aid in solubilization. Dialyze the dissolved methylcellulose against H$_2$O overnight. Divide into 10-ml aliquots and store frozen at −20°C.

7. *Motility buffer:* 80 mM KCl, 20 mM MOPS (pH 7.4), 5 mM MgCl$_2$, 0.1 mM EGTA, 1 mM ATP, 50 mM DTT, 0.7% methylcellulose, 2.5 mg/ml glucose, 0.1 mg/ml glucose oxidase, and 0.02 mg/ml catalase. To make 1 ml, add 250 μl of 4× stock solution, 10 μl of 0.1 M ATP, 40 μl of 2 M KCl, 100 μl of 0.5 M DTT (prepare fresh each day by adding 77 mg DTT to 1 ml of H$_2$O), 20 μl of 125 mg/ml glucose, 20 μl of 5 mg/ml glucose oxidase, 2 μl of 1 mg/ml catalase, 58 μl of H$_2$O, and 0.5 ml of 1.4 % methylcellulose. Mix well by pipetting up and down repeatedly.

Steps

1. Apply 0.2 mg/ml of myosin in 0.5 M NaCl, 20 mM MOPS (pH 7.), 0.1 mM EGTA, and 1 mM DTT to fill the flow chamber. Wait 1 min.

2. Wash with 75 μl of blocking solution. Wait 1 min.

3. Wash with 75 μl of wash solution, followed by 75 μl of ATP–actin wash solution. Wait 1 min. This step is optional.

4. Wash with 75 μl of wash solution, followed by 75 μl of rhodamine–phalloidin-labeled actin solution. Wait 1 min.

5. Initiate reaction by the addition of 75 μl motility buffer. This is a very viscous solution and it helps to use a piece of filter paper placed at the outflow of the flow chamber as a wick.

Comments

In the protocol just described, myosin is bound to the surface as monomers. If myosin is to be bound as filaments, it is necessary to block with BSA that is in a low ionic strength solution, such as the wash solution. Alternatively, heavy meromyosin or a soluble myosin such as myosin I can be applied to the flow chamber at either low or high ionic strength. In some cases, myosin or HMM can be attached to the surface via specific antibodies against their carboxyl-terminal sequence (Winkelmann *et al.*, 1995; Kelley *et al.*, 1992; Cuda *et al.*, 1993), which may also serve the purpose of further purifying the desired isoform of myosin.

Although each myosin has its own characteristic velocity, the velocity of a given myosin can vary with ionic and assay conditions. In general the velocity tends to increase as the ionic strength is raised from 20 to 100 m*M* and increases with temperature. At higher ionic strengths the actin filaments typically begin to become weakly associated with the myosin-coated surface and move erratically. The velocity of actin filament translocation by some myosins such as vertebrate smooth muscle myosin and *Limulus*-striated muscle myosin is markedly increased (two to four times) by the inclusion of 200 n*M* tropomyosin in the motility buffer (Wang *et al.*, 1993; Umemoto and Sellers, 1990). However, tropomyosin inhibits the movement of brush border myosin I (Collins *et al.*, 1990).

Step 3 is optional. It is often included to improve the "quality" of movement by binding unlabeled actin to noncycling myosin heads that would otherwise bind and tether the labeled actin added subsequently. This step can be omitted if the myosin is capable of moving actin filaments smoothly without this step.

D. Recording and Quantifying Data

Steps and Equipment

A schematic diagram of the equipment setup is shown in Fig. 2. The following describes the equipment used in our laboratory. There is a wide selection of video microscopy equipment represented by many manufacturers.

1. The microscope slide is place under a 100×, 1.4 NA Plan-Neofluor objective in an Axioscope microscope (both from Carl Zeiss, Germany) equipped for epifluorescence. Other microscopes, including ones with an inverted format, are also suitable and a variety of objectives can be used, but note that high numerical apertures are required for maximal brightness.

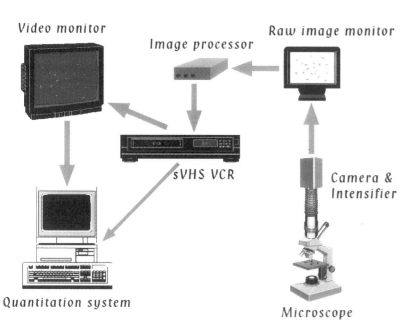

FIGURE 2 Schematic of the video setup.

Illumination is via a 100-W mercury lamp. An IR filter should be placed between the light source and the sample to attenuate heat; neutral density filters are useful to attenuate the light intensity if needed. A filter set designed for rhodamine fluorescence measurements should be utilized in the filter cube. For quantitative work it is necessary to control the temperature of the assay. This can be accomplished in several ways. The most inexpensive way is to create an air curtain using a hair dryer. A more practical approach is to manufacture a water jacket for the objective and to use a circulating water bath to regulate temperature. In our experience, commercial stage heaters are not sufficient if oil immersion objectives are used as the objective acts as a large heat sink.

2. Actin filaments are imaged using a VS 2525 image intensifier (Videoscope Internation) coupled with a C2400 Newvicon camera (Hamamatsu Photonics). Other low-light imaging systems are possible, such as an intensified charged couple device (CCD) camera or a silicon-intensified target (SIT) or intensified SIT camera. Several manufacturers sell this equipment. The raw camera image can be fed to a standard black-and-white video monitor (TR 124MA, Panasonic).

3. It is useful to process the raw image using an image processor such as the Argus 10 (Hamamatsu Photonics) to perform frame averaging and/or background substraction. The processed image is displayed on another video monitor and is recorded on an sVHS tape via a AG7350 sVHS recorder (Panasonic).

4. The movement of individual actin filaments is determined using an automated tracking system equipped with a VP110 digitizer from Motion Analysis.

Comments

Given the range of motility rates of different myosins, there is no standard number of frames to average in order to get a good image. If the myosin is moving at 5 μm/sec, a 2 or 4 frame average is used along with high illumination levels that can be tolerated because of the short exposure time need to define a filament path. With slow myosins that may move at rates of less than 0.1 μm/sec, it is possible to average 64 frames and to reduce the light intensity so that longer recording periods are possible.

Other formats are available for recording data such as VHS, U-matic, or optical memory disk recorders, but each of these has a disadvantage in terms of poorer resolution (VHS) or cost (U-matic and optical memory disk recorders). (See Volume 2, Cross, "Microtubule Motility Assays" for additional information.)

The quantification of the rate of actin filament sliding is perhaps the most difficult part of the motility assay. The method just described requires a fairly expensive apparatus, but is accurate, very fast, and can give unbiased results (for an extensive discussion of quantification of data, see Homsher et al., 1992). The user tells the computer what frame rate and for how long of a time to collect data from either the live image or a prerecorded image. The computer grabs the specified frames, performs a gray-level threshold to define the actin filament image, determines the centroid position of each actin filament in each frame, connects the centroids to form paths, calculates the incremental velocity between each successive data point in a path, and, finally, calculates a mean $\pm$SD for each filament path. This process takes less than a minute for a field of 25–50 actin filaments. Several investigators use commercial frame grabbers and write their own software for semiautomated tracking of actin filaments (Work and Warshaw, 1992). There is at least one commercial source of semiautomatic tracking software on the market (RETRAC, by Nick Carter, Institute for Applied Biology, University of York, York YO15DD, UK).

E. Presentation of Data

Figure 3 shows three frames taken at 1-sec intervals of actin filaments moving over a myosin-coated surface as it would appear on the video monitor. Data from such an experiment are most commonly presented as the mean $\pm$SD of the velocity of the population of actin filaments. In general, the SD is typically 15–20% of the mean. There are two cases where merely reporting this number does not always accurately describe what is happening in the

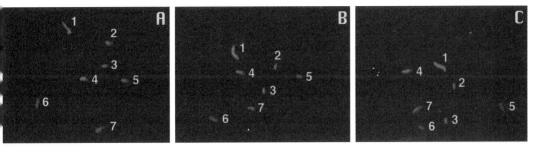

FIGURE 3 Three time sequences from a typical field. Rhodamine–phalloidin-labeled actin filaments moving over a surface coated with rabbit skeletal muscle myosin. The video sequences were taken at about 1-sec intervals.

assay. One such case is when something (perhaps a regulatory protein) is affecting the number of filaments that are moving. If, in the absence of the regulatory protein, >95% of the actin filaments are moving at 1 μm/sec whereas only 5% of the filaments move at any velocity in the presence of the regulatory protein, reporting only the mean value for the velocity in each case does not reflect the difference that is observed in the assay between the two conditions. A better method for data display for this example is to display all data in the form of a histogram so that one can see that most of the actin filaments are not moving in the presence of the regulatory protein. This display also allows the reader to see whether the regulatory protein affects the speed of movement of the few actin filaments that remain moving. The other case where more complex data display is necessary is if the filaments are moving erratically. Here the mean velocity will underestimate the "instantaneous" velocity and will have a considerably larger standard deviation than for smoothly moving filaments. One way to graphically display these data is to show a path plot in which the centroid position of the moving actin filaments is plotted in two-dimensional space as a function of time (Fig. 4).

Often the best way to present data (although not practical for the print medium) is by playing the videotape sequence directly to the audience. The presenter should be aware, however, of the many different and incompatible video formats in use around the world! For example, a sVHS tape will not play properly in a VHS machine, and within the VHS family, there are several formats such as PAL, NTSC, and SECAM that are not mutually compatible.

IV. PITFALLS

1. Actin filaments are moving erratically or only a fraction of the actin filaments are moving. The cause of this phenomena is usually noncycling heads in the preparation. Using the actin–ATP wash solution described in step 3C usually helps or eliminates this problem, but if the erratic motility persists, do the following. Bring the myosin or HMM solution to 0.5 M in NaCl and add actin to a final concentration of 10 μM, ATP to 2 mM, and MgCl$_2$ to 5 mM. Immediately sediment at 435,000 g for 15 min in a Beckman TL-100 ultracentrifuge. Remove the supernatant and use for motility assay.

2. Actin filaments shear quckly into small dots. Several things can contribute to this phenomena. Poor quality myosin containing a significant number of noncycling heads might be a problem. See Pitfall 1 for advice on how to remove these. Decreasing the density of myosin heads on the surface and/or increasing the ionic strength of the assay solution also sometimes helps decrease the light intensity. In general, myosin bound to the surface as monomers tends to shear less than myosin bound as filaments.

3. Actin filaments appear wobbly when they move or are moving in a back and forward type manner. If the assay does not contain methylcellulose, any portion of the actin filament that is not bound along its length by myosin will experience Brownian motion and appear very wobbly. Even though these filaments may be moving, their movement will be erratic and difficult to quantitate. Increasing the density of myosin on the surface, decreasing the ionic strength of the assay, or using methycellulose in the assay buffer usually helps. The back and forward motion of the actin filaments seen in the presence of methycellulose is

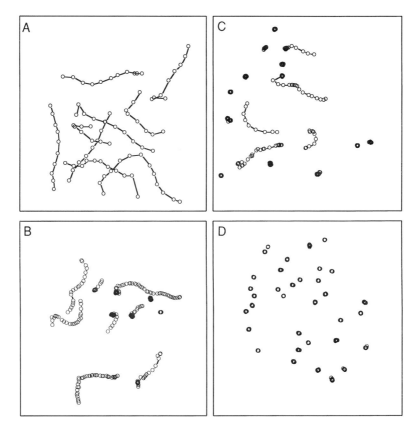

FIGURE 4 Path plots as a means to display data. Path plots are constructed by plotting the centroid of the actin filaments at constant time intervals. In this hypothetical example, four plots are shown. (A) Path plots of a smooth and fast-moving field of actin filaments. (B) Path plots in a case where the movement of the actin filaments by the myosin is slower. Note that a few filaments did not move as evidenced by the large dot representing the overlaying of all the centroids in that path. (C) A case where the movement of many of the actin filaments is erratic and where many are not moving at all. (D) A case where all the actin filaments are nonmotile.

merely Brownian motion in the presence of the viscous solution where the actin filament is restricted to move mostly along its long axis. If the actin filament is not bound by myosin it will move back and forward. Increasing the density of myosin or decreasing the ionic strength of the solution should help.

4. Actin filaments photobleach rapidly. Decrease the light intensity if possible and use image processing to do frame averaging to improve the signal-to-noise ratio. Make sure that the glucose, glucose oxidase, and catalase components of the motility buffer are good. The addition of 50 mM DTT also aids in preventing photobleaching.

5. Actin filaments leave comet tail-like images as they move. If you are frame averaging, merely decrease the number of frames averaged. If not, the problem is likely to be encountered when the actin filaments are moving fast and a non-CCD type camera is used. The streaking or persistence in this case is related to the fact that the tube cameras effectively average about four frames in producing their image. The persistence can be attenuated by increasing the light level or by switching to a lower magnification objective.

Acknowledgments

The authors thank Debra Silver for comments on the manuscript and Earl Homsher for helpful discussions.

References

Collins, K., Sellers, J. R., and Matsudaira, P. (1990). Calmodulin dissociation regulates brush border myosin I (110-kD-calmodulin) mechanochemical activity in vitro. *J. Cell Biol.* 110, 1137–1147.

Cuda, G., Fananapazir, L., Zhu, W.-S., Sellers, J. R., and Epstein, N. D. (1993). Skeletal muscle expression and abnormal function of β-myosin in hypertrophic cardiomyopathy. *J. Clin. Invest.* **91,** 2861–2865.

Fraser, I. D. C., and Marston, S. B. (1995). *In vitro* motility analysis of smooth muscle caldesmon control of actin-tropomyosin filament movement. *J. Biol. Chem.* **270,** 19688–19693.

Homsher, E., Wang, F., and Sellers, J. R. (1992). Factors affecting movement of F-actin filaments propelled by skeletal muscle heavy meromyosin. *Am. J. Physiol. Cell Physiol.* **262,** C714–C723.

Kelley, C. A., Sellers, J. R., Goldsmith, P. K., and Adelstein, R. S. (1992). Smooth muscle myosin is composed of homodimeric heavy chains. *J. Biol. Chem.* **267,** 2127–2130.

Kron, S. J., and Spudich, J. A. (1986). Fluorescent actin filaments move on myosin fixed to a glass surface. *Proc. Natl. Acad. Sci. USA* **83,** 6272–6276.

Sellers, J. R., and Goodson, H. V. (1995). Motor proteins 2: Myosin. *Protein Profile* **2,** 1323–1423.

Umemoto, S., and Sellers, J. R. (1990). Characterization of in vitro motility assays using smooth muscle and cytoplasmic myosins. *J. Biol. Chem.* **265,** 14864–14869.

Wang, F., Martin, B. M., and Sellers, J. R. (1993). Regulation of actomyosin interactions in *Limulus* muscle proteins. *J. Biol. Chem.* **268,** 3776–3780.

Winkelmann, D. A., Bourdieu, L., Kinose, F., and Libchaber, A. (1995). Motility assays using myosin attached to surfaces through specific binding to monoclonal antibodies. *Biophys. J.* **68,** 2444–2453.

Work, S. S., and Warshaw, D. M. (1992). Computer-assisted tracking of actin filament motility. *Anal. Biochem.* **202,** 275–285.

Use of Acrosomal Processes in Actin Motility Studies

Sergei A. Kuznetsov and Dieter G. Weiss

I. INTRODUCTION

The movement of organelles in cells is dependent on two types of motility systems, one actin filament based and the other microtubule based. The movement along microtubules is driven by kinesins and dyneins. The movement along actin filaments is driven by myosins. Evidence shows that the newly discovered unconventional myosins serve as motors for the movement of membranous organelles in eukaryotic cells (Langford, 1995).

Organelle movement thus far observed on actin filaments is unidirectional. The directionality of myosins can be determined by an assay involving the differential assembly of skeletal muscle actin filaments on the barbed (+) and pointed (−) ends of acrosomal processes.

Acrosomal processes of the horseshoe crab sperm consist of very stable bundles of parallel actin filaments that have uniform polarity. The plus, barbed end of all filaments is oriented toward the distal end of the bundle (Fig. 1a).

Isolated acrosomal fragments can be used as a seeds for polymerization of actin from skeletal muscle. At the proper actin concentration polymerization occurs only on the preferred, barbed ends (Figs. 1a, 1c, and 1d). All newly assembled filaments have the same polarity.

The use of polymerized actin filaments on the ends of acrosomal processes has proven to be very useful method in determining the directionality of newly identified myosins. This assay shows that actin-dependent motors move axoplasmic organelles toward the barbed end or fast-growing end of the actin filaments (Langford et al., 1994). Wolenski et al. (1995) used this assay to determine that myosin V is a barbed-end directed motor.

II. MATERIALS AND INSTRUMENTATION

A. Reagents

Tris (Cat. No. T-6791), EGTA (Cat. No. E-4378), phalloidin–TRITC-labeled (Cat. No. P-1951), K-ATP (Cat. No. A-8937), dithiothreitol (DTT, Cat. No. D5545), dimethyl sulfoxide (DMSO, Cat. No. D-587), and Triton X-100 (Cat. No. T-6878) are from Sigma. Magnesium chloride (Cat. No. 5833) is from Merck.

B. G-actin

Monomeric actin from rabbit skeletal muscle is prepared according to the method of Spudich and Watt (1971) and further purified by gel filtration chromatography on Sephacryl S-300

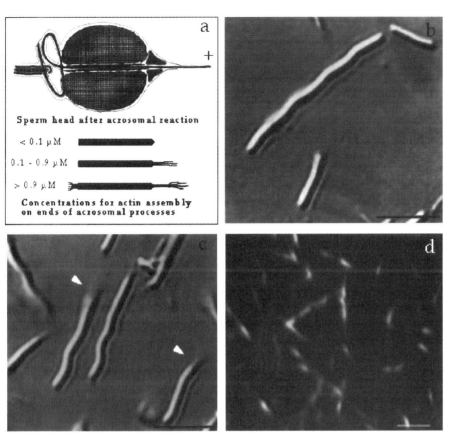

FIGURE 1 (a) Schematic of sperm head after acrosomal reaction and appearance of newly polymerized actin filaments. VEC-DIC microscopy of an acrosome fragment without (b) and with polymerized skeletal muscle actin filaments at one end. (c) At an actin concentration of 0.5 μM, polymerization occurs only on the preferred, barbed ends. After actin polymerization, each bundle had visible structures on one end only (arrowheads). (d) Video-intensified fluorescence microscopy of the same preparation as in c. Note that there is weak fluorescence in acrosomes and strong fluorescence in newly polymerized actin bundles. Bar: 5 μm.

By electrophoretic analysis the absence of myosin and other contaminating proteins should be controlled. A stock solution of monomeric actin (30 μM) can be stored for up to 2 weeks at 0°C in 2 mM Tris–HCl buffer (pH 8.0) with 1 mM Na–ATP, 0.05 mM MgCl$_2$, and 0.2 mM DTT.

C. Materials for Sample Preparation

Materials needed include glass slides and coverslips (No. 0, 80–120 μm thick) (O. Kindler); VALAP sealant, mixture of equal parts by weight of vaseline, lanolin, and paraffin (MP 51–53°C); cotton-tip applicators; and Scotch 3M tape.

D. DIC and Fluorescence Microscopy

Acrosomal processes with assembled skeletal muscle actin filaments can only be observed by video-enhanced contrast, differential interference contrast (VEC-DIC) microscopy (Allen and Allen, 1983; Weiss *et al.*, 1990, see Volume 3, Weiss, "Video-Enhanced Contrast Microscopy" for additional information) and video-intensified fluorescence microscopy. We use a Zeiss Axiophot microscope (C. Zeiss) equipped with an oil immersion condenser (NA 1.4) and a 100× DIC Plan Neofluar oil objective (NA 1.32). The mercury arc lamp (HBO 100) for DIC is required to ensure that the Hamamatsu C2400-07 Newvicon camera works at optimal saturation to acquire DIC images and a Hamamatsu C2400-97. An intensified CCD camera system or a silicon intensifier target (SIT) camera (Hamamatsu Photonics Inc.) is used for fluorescence images.

The analog and digital processing of fluorescence and DIC signals can be performed in parallel using two ARGUS 10 real-time image processors (Hamamatsu Photonics Inc.). A Mitsubishi BV-1000 SVHS or other SVHS video recorder is used to record processed images.

III. PROCEDURES

Solutions

1. *100 mM ATP stock solution:* For 1 ml stock, weigh out 61.4 mg ATP into 700 μl H_2O. Adjust pH to 6.8 with 1 N KOH and make up to 1 ml with H_2O. Store in aliquots at $-20°C$.

2. *1 M DTT stock solution:* For 1 ml stock, weigh out 154.2 mg DTT into 1 ml H_2O. Store in aliquots at $-20°C$.

3. *200 μM TRITC-labeled phalloidin stock solution:* Dissolve 0.1 mg TRITC-labeled phalloidin into 0.3 ml DMSO, add DMSO to 0.385 ml. Store in aliquots at $-20°C$.

4. *0.2 M EGTA stock solution:* For 100 ml stock, weigh out 7.61 g EGTA into 80 ml H_2O, adjust pH to 7.0–7.2 with 1 N KOH, add H_2O to 100 ml, and store at 4°C.

5. *1 M $MgCl_2$ stock solution:* For 100 ml stock, weigh out 20.3 g $MgCl_2 \cdot 6H_2O$, dissolve, add H_2O to 100 ml, and store at 4°C.

6. *10% Triton X-100 stock solution:* For 10 ml stock, weigh out 1 g Triton X-100 and add 9 ml H_2O. Store at 4°C.

7. *0.25 M Tris stock solution, pH 7.5:* For 100 ml stock, weigh out 3 g Tris into 80 ml H_2O, adjust pH to 7.5 with 1 N HCl, and make up to 100 ml with H_2O. Store at 4°C.

8. *0.25 M Tris stock solution, pH 8.0:* For 100 ml stock, weigh out 3 g Tris into 80 ml H_2O, adjust pH to 8.0 with 1 N HCl, and make up to 100 ml with H_2O. Store at 4°C.

9. *Acrosome buffer (buffer A):* 25 mM Tris buffer with 3 mM $MgCl_2$, and 1% Triton X-100, pH 8.0, at room temperature. For 10 ml, add the following components to 5 ml H_2O: 1 ml of 0.25 M Tris (pH 8.0), 33.3 μl of 1 M $MgCl_2$, and 1 ml of 10% Triton; make up to 10 ml with H_2O.

10. *Washing buffer (buffer B):* 25 mM Tris, and 2 mM $MgCl_2$; pH 7.5 at room temperature. For 100 ml, add the following components to 50 ml H_2O: 10 ml of 0.25 M Tris (pH 7.5) and 200 μl of 1 M $MgCl_2$; make up to 100 ml with H_2O.

11. *Working buffer (buffer C):* 25 mM Tris, 1 mM EGTA, 2 mM $MgCl_2$, 1 mM ATP, and 2 mM DTT, pH 7.5 at room temperature. For 10 ml, add the following components to 5 ml H_2O: 1 ml of 0.25 M Tris (pH 7.5), 50 μl of 0.2 M EGTA, 20 μl of 1 M $MgCl_2$, 100 μl of 100 mM ATP, and 20 μl of 1 M DTT; make up to 10 ml with H_2O.

Steps

Preparation of Acrosomal Processes

1. Horseshoe crabs (*Limulus polyphemus*) can be obtained from the Department of Marine Resources of the Marine Biological Laboratory. Carry out all steps on ice or at 2°C.

2. Collect sperm from horseshoe crabs as described previously (Tilney, 1975): induce animals to discharge their sperm by rubbing gently near the gonadopores and collect the sperm with a Pasteur pipette. A number of animals must be milked repeatedly in order to obtain sufficient quantities of sperm. Sperm can be collected throughout the year.

3. Wash the sperm twice in sea water and pellet at 700 *g* for 5 min.

4. Resuspend the pellet in 5 vol of buffer A to induce the false discharge of acrosomal processes (Fig. 1a) and to remove membranes (Tilney, 1975).

5. Gently shear acrosomal processes into short fragments (2–10 μm long) by several passages through the tip of a Pasteur pipette.

6. Spin the sheared acrosomal processes for 5 min at 3000 g to remove nuclei and attached axonemes. After centrifugation, the supernatant contains primarily short fragments of acrosomal processes and an occasional flagellar axoneme as determined by video microscopy.

7. Pellet the acrosomal processes by spinning for 10 min at 10,000 g and then resuspend in buffer B to remove the detergent.

8. Repeat the centrifugation procedure five times and resuspend the acrosomal processes in buffer C.

Barbed-End Assembly of Actin Filaments on Acrosomal Processes

1. To a preparation of acrosomal processes resuspended in buffer C, add monomeric actin from skeletal muscle to a final concentration of 0.5 μM.

2. Incubate this mixture for 1 hr at room temperature followed by overnight incubation on ice.

3. After assembly of the actin filaments, dilute an aliquot 10-fold into WB containing 0.5 μM TRITC–phalloidin to stabilize and stain the actin filaments (Figs. 1b–d).

Sample Preparation for Microscopy

1. After 1 hr of staining (a longer staining time can produce more highly fluorescent acrosomal processes), place a 4-μl droplet of the solution containing acrosomal processes with polymerized actin filaments (or without polymerized actin filaments for control experiments) on a plastic surface (20 × 13-mm rectangle of 3M Scotch tape) and gently cover with a No. 0 coverslip (22 × 26 mm) (Fig. 2).

2. Incubate the coverslip in the solution for 5 min in a humidified chamber in the dark to allow adsorption of the acrosomal processes to the glass surface. Before use, rinse with the same buffer as used for later in the assay.

3. Apply 25 μl of buffer C (control) or a probe containing organelles to be tested (Fig. 3) to the slide between two spacer strips of Scotch tape.

4. After the addition of buffer C or a probe to be tested, apply the glass coverslip with adsorbed acrosomal processes to the slide and seal the preparation with valap (60°C, apply with cotton-tip applicator)

VEC-DIC and Fluorescence Microscopy

By VEC-DIC microscopy, the isolated acrosomal fragments appear as highly refractive, wavy bundles (Fig. 1b). After actin polymerization, each bundle has visible structures only on one

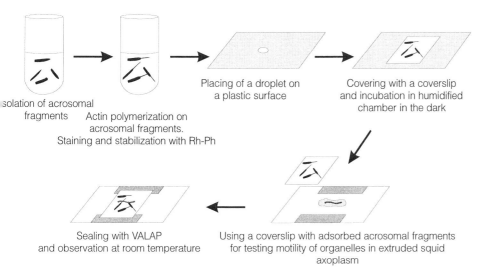

FIGURE 2 Procedure of sample preparation for testing squid axoplasmic organelle movement on acrosomal processes.

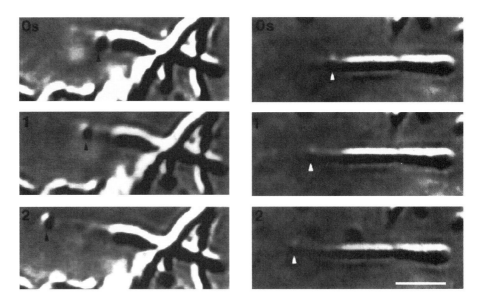

FIGURE 3 Squid axoplasmic organelles (arrowheads) move in proximo-distal direction on actin filaments assembled at the barbed end of acrosomal fragments. This demonstrates the presence of a barbed-end directed motor. Time in seconds. Bar: 5 μm.

end (Fig. 1c). The newly polymerized actin filaments stabilized and labeled with rhodamine–phalloidin can be detected as bright spots only on this end by straightforward video-intensified fluorescence microscopy (Weiss and Maile, 1993) (Fig. 1d).

For VEC-DIC the video signals should first be subjected to analog contrast enhancement (Allen and Allen, 1983; Weiss *et al.*, 1990), followed by real-time digital image processing in the following steps: subtraction of an out-of-focus background (mottle pattern), accumulation or averaging of images to increase signal-to-noise ratio, and finally selection of the desired range of gray levels (Allen and Allen, 1983; Weiss and Maile, 1993). The detailed steps are described elsewhere in this volume (Weiss).

IV. COMMENTS

This motility assay can be used to test native motors on their organelles. To this end the probe to be tested consists of a postnuclear supernatant fraction of a cell extract. Similarly extruded cytoplasm from squid giant axon (Fig. 3) (Weiss *et al.*, 1990) or plant cells (*Chara* or *Nitella*) can be used. In these cases, native actin filaments and microtubules must be removed. We use 100 μM nocodazole (methyl-(5-[2-thienyl-carbonyl]-1*H*-benzimidazol-2-yl)carbamate, Sigma Cat. No. M-1404) and 50 μM cytochalasin B (Sigma Cat. No. C-6762). At this concentration, only the phalloidin-stabilized acrosomal actin filaments are not depolymerized (Langford *et al.*, 1994).

If purified motors are to be tested, these have to be adsorbed to latex beads. In order to retain enzyme activity, proper orientation of the motor molecules may be essential, as was achieved by Wolenski *et al.* (1995) using antibody coupling.

IV. PITFALLS

1. *G-actin titration.* The critical concentration for actin polymerization at just one end of acrosomes may vary depending on the batch of monomeric actin. It is, therefore, required to verify by microscopy that polymerization has occurred only at one (the barbed) end. Otherwise the G-actin concentration should be adjusted accordingly.

2. *Too high particle abundance.* The quality of VEC-DIC images is severely reduced by too large amounts of spherical or highly phase-retarding objects that depolarize the illumination light or prevent full contrast enhancement, respectively. In this case, particle concentration has to be greatly diminished. If particles are so scarce that they rarely get into contact with actin filaments, laser tweezers may be required in order to place them manually (Wolenski *et al.*, 1995).

3. *Photodamage.* In order to reduce potential photo damage during observation by fluorescence microscopy, we usually search for and observe a given field in the preparation by VEC-DIC microscopy preferentially at a wavelength that is not absorbed or damaging before employing fluorescence microscopy. In addition, bleaching of the fluorescent-labeled actin filaments can be minimized by depleting molecular oxygen in the preparations by adding 2.5 mg/ml glucose (Sigma Cat. No. G-7528), 0.1 mg/ml glucose oxidase (Sigma Cat. No. G-7016), and 0.02 mg/ml catalase (Sigma Cat. No. C-100).

References

Allen, R. D., and Allen, N. S. (1983). Video-enhanced microscopy with a computer frame memory. *J. Microsc.* **129**, 3–17.

Langford, G. M. (1995). Actin- and microtubule-dependent organelle motors: Interrelationships between the two motility systems. *Curr. Opin. Cell Biol.* **7**, 82–88.

Langford, G. M., Kuznetsov, S. A., Johnson, D., Cohen, D. L., and Weiss, D. G. (1994). Movement of axoplasmic organelles on actin filaments assembled on acrosomal processes: Evidence for a barbed-end directed organelle motor. *J. Cell Sci.* **107**, 2291–2298.

Owen, C., and DeRosier, D. (1993). A 13-Å map of the actin-scruin filament from the *Limulus* acrosomal process. *J. Cell Biol.* **123**, 337–344.

Spudich, J. A., and Watt, S. J. (1971). The regulation of rabbit skeletal muscle contraction. I. Biochemical studies of the interaction of the tropomyosin–troponin complex with actin and the proteolytic fragments of myosin. *J. Biol. Chem.* **246**, 4866–4871.

Tilney, L. G. (1975). Actin filaments in the acrosomal reaction of *Limulus* sperm. *J. Cell Biol.* **64**, 289–310.

Weiss, D. G., and Maile, W. (1993). Principles, practice, and applications of video-enhanced contrast microscopy. *In* "Electronic Light Microscopy" (D. M. Shotton, ed.), pp. 105–140. Wiley-Liss, New York.

Weiss, D. G., Meyer, M. A. and Langford, G. M. (1990). Studying axoplasmic transport by video microscopy and using the squid giant axon as a model system. *In* "Squid as an Experimental Animal" (D. L. Gilbert, W. J. Adelman, Jr., and J. M. Arnold, eds.), pp. 303–321. Plenum Press, New York.

Wolenski, J. S., Cheney, R. S., Mooseker M. S., and Forscher P. (1995). *In vitro* motility of immunoadsorbed brain myosin V using a *Limulus* acrosomal process and optical tweezer-based assay. *J. Cell Sci.* **108**,1489–1496.

Use of *Xenopus* Egg Extracts for Studies of Actin-Based Motility

Julie A. Theriot and David C. Fung

I. INTRODUCTION

The amoeboid or crawling motility of animal cells and free-living soil amoebae has fascinated cell biologists since it was first observed well over 200 years ago. In recent decades we have learned that amoeboid motility is dependent on the dynamic behavior of the cell's actin cytoskeleton, and the individual roles of different cytoskeletal components are beginning to be elucidated, although biochemical research into this process has been relatively slow compared to other cell biological behaviors. A major barrier to the detailed biochemical analysis of actin-dependent whole-cell motility has been the intimate role of the cell plasma membrane in the physical organization in addition to the biochemical organization of protrusive cellular structures such as lamellipodia and filopodia; lamellipodial motility cannot be reconstituted in a cytoplasmic extract. Within the past decade, a large number of cell biology laboratories have begun using an unexpected model system for whole-cell actin-based motility: the actin-dependent "rocketing" movement of intracellular bacterial pathogens, including *Listeria monocytogenes* and *Shigella flexneri*. These organisms grow directly in the cytoplasm of infected host cells and induce the formation of an actin-rich "comet tail" consisting of numerous short filaments of host cell actin cross-linked together. Elongation of actin filaments at the proximal end of the stationary comet tail results in rapid movement of the bacterium through the host cell cytoplasm. Actin filament dynamics in comet tails associated with moving bacteria are strikingly similar to dynamics in the lamellipodium (Theriot, 1994). Indeed, the bacteria appear to be undergoing a form of protrusive motility where the body of the bacterium itself is imitating a small fragment of the lamellipodial leading edge. Since the bacteria are in a sense substituting for the plasma membrane, this system raised the possibility that protrusive actin-dependent motility could be reconstituted in a cell-free cytoplasmic extract.

Cytoplasmic extracts made from the eggs of the clawed frog *Xenopus laevis* have been extremely useful for the reconstitution of a number of complex cell biological processes, including studies of DNA replication, chromosome architecture, nuclear envelope breakdown and assembly, mitotic spindle assembly and dynamics, cell cycle regulation, and microtubule dynamics [see Murray (1991) and other articles in the same volume of "Methods in Cell Biology"; also Volume 4, Matthews, "Preparation and Use of Translocating Cell-Free Translation Extracts from *Xenopus* Eggs" for additional information]. *Xenopus* eggs are rich sources of all the factors that the organism needs to develop from the point of fertilization to the midblastula transition, where zygotic transcription begins. Because of their large size, it is simple to prepare essentially undiluted cytoplasm by crushing the eggs in a low-speed centrifuge spin. With minor modifications of earlier protocols, it is possible to reconstitute the

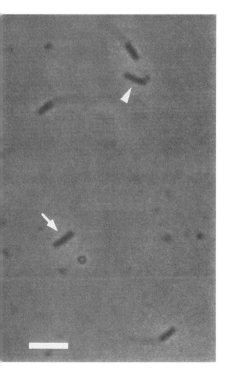 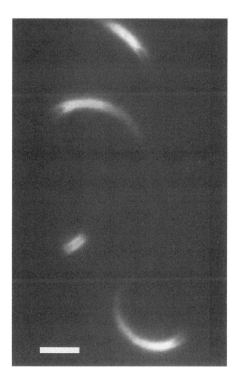

FIGURE 1 Actin-based motility of *L. monocytogenes* strain SLCC-5764 in a *Xenopus* egg cytoplasmic extract. (Left) Phase-contrast image. (Right) Epifluorescence image showing distribution of rhodamine–actin. Three moving bacteria associated with actin-rich "comet tails" are shown, along with one stationary bacterium surrounded by an actin filament cloud (white arrow) and one stationary bacterium that has not associated with actin filaments (white arrowhead). Bars, 4µm.

actin-based rocketing motility of intracellular bacterial pathogens in *Xenopus* egg cytoplasmic extracts (see Fig. 1). Movement of bacteria in the extracts appears to accurately mimic their actin-based motility inside infected cells. This reconstitution in a biochemically tractable system may help not only to elucidate the molecular mechanism of bacterial actin-based motility, but also the mechanisms involved in whole-cell protrusive activity and amoeboid movement.

II. MATERIALS AND INSTRUMENTATION

Disodium EDTA (Cat. No. E-5134), HEPES (Cat. No. H-3375), EGTA (Cat. No. E-3889), sucrose (Cat. No. S-1888), cysteine (free base, Cat. No. C-7755), leupeptin (Cat. No. L-2884), pepstatin (Cat. No. P-4265), chymostatin (Cat. No. C-7268), pregnant mare serum gonadotropin (PMSG, Cat. No. G-4527), human chorionic gonadotropin (HCG, Cat. No. C-0434), creatine phosphate (Cat. No. P-7936), ATP (disodium salt, Cat. No. A-3377), Nonidet P-40 (Cat. No. N-6507), catalase (Cat. No. C-40), glucose oxidase (Cat. No. G-7016), glucose (Cat. No. G-7021), ampicillin (Cat. No. A-9518), Tris base (Cat. No. 1503), dithiothreitol (DTT, Cat. No. D-5545), and iodoacetic acid (Cat. No. I-7768) are all from Sigma. Dimethyl sulfoxide, anhydrous (DMSO, Cat. No. 27, 685-5), is from Aldrich. Brain heart infusion (BHI, Cat. No. 0037-17-8), Bacto-agar (Cat. No. 0140-01), tryptic soy broth (TSB, Cat. No. 0370-17-3), tryptone (Cat. No. 0123-17-3), and yeast extract (Cat. No. 0127-17-9) are from Difco. Tetramethylrhodamine iodoacetamide (TMR-IA, Cat. No. T-6006) is from Molecular Probes. Formaldehyde, 16% solution (Cat. No. 18505), is from Ted Pella. Versilube oil (Cat. No. F-50) is from Andpak-EMA. Gravid female *X. laevis* are from Nasco.

Any upright or inverted microscope equipped with phase-contrast and epifluorescence optics may be used. We use a Nikon Diaphot-300 inverted microscope with epifluorescence filters for rhodamine from Chroma Technology. An intensified charge-coupled device video camera (ICCD, Dage-MTI GenIISys/CCD-c72) is hooked up to a Pentium-based data acquisition computer running MetaMorph (Universal Imaging Corporation), which also controls shutters that switch between transmitted and epifluorescence light sources. For details of motion analysis, see Fung and Theriot (1997).

Bacterial strains described in this article may be obtained from the authors by written request. Please note that postal regulations may not permit shipping of human pathogens to all countries. Researchers wishing to work with *L. monocytogenes* should consult their biosafety office.

III. PROCEDURES

A. Preparation of *Xenopus* Egg Extract

Cytostatic factor-arrested extracts are prepared essentially according to the method of Murray (1991) with the important modification that cytochalasin is omitted. These extracts are arrested in the cell cycle at metaphase of meiosis II.

Solutions

1. *Marc's modified Ringer's (MMR):* 100 mM NaCl, 2 mM KCl, 1 mM MgCl$_2$, 2 mM CaCl$_2$, 0.1 mM EDTA, and 5 mM NaHEPES, pH 7.8 (Murray, 1991). To make 1 liter of a 10× stock, add 58.5 g NaCl, 1.5 g KCl, 2.0 g MgCl$_2$·6H$_2$O, 2.9 g CaCl$_2$·2H$_2$O, 0.4 g EDTA disodium salt, and 11.9 g HEPES to distilled water, adjust to pH 7.8 with 10 N NaOH, and complete volume to 1 liter. Autoclave and store at room temperature. Dilute to 1× with distilled water immediately before use.

2. *Xenopus extract buffer (XB):* 100 mM KCl, 0.1 mM CaCl$_2$, 2 mM MgCl$_2$, 5 mM EGTA, 10 mM KHEPES, pH 7.7, and 50 mM sucrose (Murray, 1991; there called CSF-XB). Prepare a 20× salt stock by adding 151.1 g KCl, 0.3 g CaCl$_2$·2H$_2$O, 8.1 g MgCl$_2$·6H$_2$O, 38.0 g EGTA, and 47.7 g HEPES to distilled water, adjust to pH 7.7 with 10 N KOH, and complete volume to 1 liter. Filter sterilize and store at room temperature. Prepare XB immediately before use by mixing 50 ml of 20X salt stock with 50 ml of 1 M sucrose stock and adding distilled water to 1 liter.

3. *2% cysteine, pH 7.8:* To make 500 ml, add 10 g cysteine (free base) to distilled water and adjust pH to 7.8 by adding 2.25 ml of 10 N NaOH. Make fresh immediately before use.

4. *Protease inhibitor cocktail stock solution (1000×):* 10 mg/ml each leupeptin, pepstatin, and chymostatin in dry DMSO. Store in small aliquots (10 μl) at −20°C.

5. *1 M sucrose stock solution:* To make 1 liter, dissolve 342.3 g sucrose in distilled water and complete volume to 1 liter. Filter sterilize and store at room temperature.

Steps

1. Prime frogs 3 to 5 days in advance by injecting 50 units PMSG. Use sterile 1-cc tuberculin syringes for all injections. Inject the hormones underneath the loose back skin into the dorsal lymph sac. As with handling any animal, it is best to learn this procedure from an investigator who is familiar with the proper location and technique for injection. Never use the same syringe for more than one frog or put a used syringe back in the hormone stock solution; this can spread diseases among the animals. Twenty to 24 hr prior to extract preparation, inject 150 units HCG into primed frogs. Put each frog into a separate bucket with 1 liter MMR and keep at 16°C overnight to lay out eggs.

2. Remove frogs from buckets and return to tank (frogs may be reused after 6 to 8 months). Discard all the eggs from any frog where a large number of the eggs are dead, white, or exploded. Pour off excess MMR and remove chunks of debris, skin, etc. Collect the good eggs in a beaker and rinse several times in MMR by swirling, allowing the eggs to settle, and pouring off excess liquid. Remove as many bad eggs as possible using a Pasteur pipette whose end has been broken off and fire polished.

3. Pour off the MMR and dejelly the eggs with 2% cysteine. Gently swirl the eggs at intervals. Dejellying takes about 5–10 min. The eggs will pack together very tightly (apparent volume will decrease at least threefold).

4. Pour off the cysteine and wash the eggs four times in XB. Wash twice more in XB with protease inhibitors (final concentration 10 $\mu g/ml$ each). Using the fire-polished Pasteur pipette, transfer the eggs in a minimum volume of buffer to centrifuge tubes (e.g., SW50 tubes) containing 1 ml of XB with protease inhibitors. Suck eggs into pipette, allow to settle to the bottom, and drip out slowly in order to minimize the amount of buffer transferred. Remove excess XB. *Optional:* Overlay eggs with 1 ml of Versilube oil.

5. Spin the eggs in a clinical centrifuge at room temperature for 60 sec at 1000 rpm and then for 30 sec at 2000 rpm to pack them into the tube. Remove all the XB and oil from the top of the tube with a Pasteur pipette or pipetman. The eggs should still be intact at this stage.

6. Crush the eggs in a swinging bucket rotor by spinning at 10,000 rpm for 10 min at 16°C. To use a Sorvall HB4 rotor, put the SW50 tubes inside 15-ml corex tubes or polypropylene tubes, and put these inside rubber adapters.

7. The extract will be in three layers: lipid, cytoplasm, and yolk, from top to bottom. The cytoplasm is the straw-colored or orange layer near the middle. Collect it by puncturing the side of the tube and drawing it out with a needle and syringe.

8. Add protease inhibitors to a final concentration of 10 $\mu g/ml$ each, do a second clarifying spin if necessary, and add 1/10 vol sterile 1 M sucrose (final sucrose concentration 150 mM). Use immediately or freeze in small aliquots (50–200 μl) in liquid nitrogen. Extract can be stored at −80°C for at least 2 months. A typical yield is 1–2 ml extract per frog.

9. *Optional:* ATP-regenerating system. 20× "energy mix" (Murray, 1991) is 150 mM creatine phosphate, 20 mM ATP, pH 7.4, 2 mM EGTA, and 20 mM MgCl$_2$. To use, add 1/20 vol energy mix to cytoplasmic extract either before freezing or immediately before use. This step is not necessary for the observation of bacterial motility, but it does help maintain consistent activity of the extract over several hours (Marchand *et al.*, 1995).

10. *Optional:* Extract clarification. Crude *Xenopus* egg cytoplasmic extracts contain numerous phase-dense membraneous particles (see Fig. 1). The addition of 0.5% Nonidet P-40 and 1 mM EGTA (pH 7.7) together results in the disappearance of almost all of the vesicles from the extract without affecting actin-based bacterial motility. The greater optical clarity of the extracts is particularly useful when thick chambers are being used. Nonidet P-40 and EGTA should be added to extract after thawing and immediately prior to use.

B. Preparation of Bacteria

Media

1. *Brain heart infusion (BHI):* Prepare liquid medium according to manufacturer's instructions. For plates, add 15 g Bacto-agar per liter of broth, autoclave, and pour about 30 ml per 10-cm petri dish.

2. *Tryptic soy broth (TSB):* Prepare liquid medium according to manufacturer's instructions. For plates, add 15 g Bacto-agar per liter of broth, autoclave, and pour about 30 ml per 10-cm petri dish. For ampicillin plates, add drug after TSB has cooled below 60°C.

3. *Luria–Bertani medium (LB):* Add 10 g tryptone, 5 g yeast extract, and 10 g NaCl per liter of broth and sterilize by autoclaving. For plates, add 15 g Bacto-agar per liter of broth, autoclave, and pour about 30 ml per 10-cm petri dish.

4. *Ampicillin stock solution (500×):* Dissolve 500 mg ampicillin in 10 ml distilled water. Store in 1-ml aliquots at −20°C.

Listeria Monocytogenes. *L. monocytogenes* causes meningitis in newborns and immunocompromised people, and miscarriages in pregnant women. It is a level 2 biohazard. The infectious dose is high for healthy adults, and it can be handled safely on the benchtop with appropriate precautions. Among these is that all waste which has come into contact with infectious bacteria should be treated with bleach, where possible, and autoclaved before disposal. General advice on obtaining and handling microbial pathogens can be found in Russell (1994) or any medical microbiology laboratory manual.

1. Streak *L. monocytogenes* strain SLCC-5764 onto a BHI-agar plate and incubate at 37°C until colonies are 1–2 mm in diameter (about 1 day). The plate can be wrapped in Parafilm and stored at 4°C for up to 1 month.

2. Put 2 ml BHI broth in a 15-ml sterile disposable test tube. Scrape a colony off the plate with a sterile toothpick or flamed metal loop and innoculate the broth. Grow overnight with rapid shaking at 37°C. It is important that the bacteria be grown at 37°C; *L. monocytogenes* grown below 30°C develop flagella which interfere with the initiation of actin-based movement.

3. When the culture has reached stationary phase, transfer 1 ml into an Eppendorf centrifuge tube and pellet the bacteria by spinning 1–2 min at full speed in a microcentrifuge. Remove supernatant.

4. Resuspend bacterial pellet in 1 ml XB (see Procedure A) by gently pipetting up and down (do not vortex). Pellet bacteria again and resuspend in 100 μl XB. The bacterial suspension can be stored on ice for at least 8 hr with no loss of activity.

5. If it is not practical or desirable to grow fresh bacteria for each day of experiments, a stock can be prepared as follows. After rinsing bacteria in 1 ml XB, resuspend in 1 ml XB containing 10 mM iodoacetic acid. Mix gently (rotate tube) for 20 min at room temperature. Pellet bacteria and rinse twice in XB. These killed bacteria are fully functional for actin-based motility and can be stored in 50% glycerol at −20°C for several months.

***Escherichia coli* Expressing IcsA from *Shigella flexneri*.**   Researchers wishing to avoid handling human pathogens may use nonpathogenic bacteria engineered to express the virulence factors required for actin-based motilty. We routinely use *E. coli* expressing the IcsA (VirG) protein from the gram-negative pathogen *S. flexneri*. *S. flexneri* causes bloody diarrhea in humans with a low infectious dose, whereas the *E. coli* expressing IcsA is avirulent.

1. Streak *E. coli* strain MBG263 (pHS-3199) (Goldberg and Theriot, 1995) onto a TSB- or LB-agar plate containing 100 μg/ml ampicillin. Incubate at 37°C until colonies are 1–2 mm in diameter (overnight).

2. Innoculate a single colony from the plate into 2 ml of LB or TSB with 100 μg/ml ampicillin in a sterile disposable test tube. Grow overnight at 37°C with rapid shaking. In the morning, back-dilute 1:100 (i.e., add 20 ml of overnight culture to 2 ml fresh medium with ampicillin) and grow to an optical density at 600 nm of about 0.2 to 0.3; this will take approximately 2 hr in TSB and approximately 3 hr in LB.

3. Remove 1 ml of the culture and spin the bacteria down in an Eppendorf tube at low speed (Eppendorf microfuge at about 400 rpm or bench-top picofuge) for about 1 min. Remove the supernatant and resuspend the pellet by pipetting up and down in 1 ml of XB with 1% formaldehyde. Do not vortex. Incubate the tube with gentle rocking at room temperature for 10 min.

4. Spin down as before and resuspend the bacterial pellet in 1 ml of XB (see Procedure A). Spin again and resuspend pellet in 100 μl XB. Do not vortex. These fixed bacteria can be stored in the refrigerator for at least 2 weeks without losing activity. For long-term storage they may be kept in 50% glycerol at −20°C.

C. Preparation of Rhodamine–Actin

Addition of rhodamine–actin to the *Xenopus* egg cytoplasmic extracts is not necessary for the reconstitution of bacterial actin-based motility. However, we have found it to be extremely useful for rapid visualization of the actin-rich comet tails associated with moving bacteria and it greatly facilitates quantitative analysis of actin dynamics during movement. Actin can be purified from a variety of sources. We use rabbit skeletal muscle actin purified essentially according to the method of Pardee and Spudich (1982). Actin may be covalently labeled on lysine residues using tetramethylrhodamine *N*-hydroxysuccinimidyl ester (TMR-NHS) or on cysteine residues using tetramethylrhodamine iodoacetamide (TMR-IA). In our hands, actin labeled with TMR-IA retains more activity after freeze–thawing than does actin labeled with TMR-NHS. Rhodamine-labeled rabbit skeletal muscle actin (Cat. No. AR05) and rhodamine-labeled human platelet actin (Cat. No. APHR) can be purchased from Cytoskeleton. We have not tested these products for use in motility assays. Here we describe our standard procedure for labeling rabbit skeletal muscle actin with TMR-IA; however, any actin isoform labeled using any standard protocol should work as a tracer.

Solutions

1. *G buffer:* 5 mM Tris, pH 8.0, 0.2 mM $CaCl_2$, 0.2 mM ATP, and DTT as needed (10 μM or 0.2 mM; see details in protocol below). Prepare stock solutions of 1 M Tris, pH 8.0, 0.1 M $CaCl_2$, 0.1 M ATP, pH 7.0, and 1 M DTT. For 1 M Tris stock solution, dissolve 121 g Tris base in distilled water, adjust pH to 8.0 with concentrated HCl, adjust volume to 1 liter, and store at room temperature. For 0.1 M $CaCl_2$, add 14.7 g $CaCl_2 \cdot 2H_2O$ to 1 liter distilled water and store at room temperature. For 0.1 M ATP, dissolve 2.8 g in distilled water, adjust pH to 7.0 with 10 N NaOH, complete volume to 50 ml, and store in small aliquots at $-20°C$. For 1 M DTT, dissolve 18 g in 100 ml distilled water and store in small aliquots at $-20°C$. Prepare G buffer by combining 5 ml of 1 M Tris stock, 2 ml of 0.1 M $CaCl_2$ stock, 2 ml of 0.1 M ATP stock, and 200 μl DTT stock (for 0.2 mM) or 10 μl DTT stock (for 10 μM) and completing volume to 1 liter with distilled water. Salts may be mixed in advance but add ATP immediately before use.

2. *G to F conversion buffer (5×):* 80 mM Tris, pH 8.0, 250 mM KCl, 1 mM ATP, and 2 mM $MgCl_2$. Use stocks prepared above. For 100 ml, combine 8 ml of 1 M Tris, pH 8.0, 1 ml of 0.1 M ATP, 1.86 g KCl, 0.04 g $MgCl_2 \cdot 6H_2O$, and complete volume to 100 ml. Salts may be mixed in advance but add ATP immediately before use.

3. *TMR-IA stock solution:* Dissolve 10 mg TMR-IA in 180 μl dry DMSO for a 100 mM stock solution. This should be stored in the dark at $-20°C$. Immediately prior to actin labeling, add 8 μl of the 100 mM stock solution to 92 μl dry DMSO to make 100 μl of an 8 mM "working stock."

Steps

1. Purify actin from rabbit skeletal muscle according to standard protocols (e.g., Pardee and Spudich, 1982). Purified actin should be stored in monomeric form in G buffer containing 0.2 mM DTT at $-80°C$.

2. Dialyze purified actin into G buffer with 10 μM DTT at 4°C for 4 hr. Higher DTT concentrations interfere with iodoacetamide labeling. Measure actin concentration and dilute to 5 mg/ml in G buffer with 10 μM DTT. One to 2 ml of 5 mg/ml actin is a good amount for small-scale labeling.

3. Warm actin solution to room temperature and transfer it into a polypropylene tube that can hold at least twice the reaction volume. Add 50 μl of 8 mM TMR-IA working stock per milliliter of actin; hold the open tube on a vortexer with one hand and pipette the dye into the actin solution with the other hand in order to assure rapid mixing of dye and protein. After thorough mixing, seal the polypropylene tube and wrap it in aluminum foil. Stir gently (rotate tube) at 4°C overnight.

4. Pour gel filtration column (P-10 or G-25 resin) and equilibrate in G buffer with 0.2 mM DTT (from now on, always use 0.2 mM DTT). Prepacked columns such as PD-10 from Pharmacia are convenient for separating 1-ml reactions. A homemade column should be about 1 cm in diameter, with at least 10 ml resin per milliliter of reaction solution. Centrifuge the reaction solution for 10 min at top speed in an Eppendorf microcentrifuge to remove precipitates before loading on column. Carefully load the column and run with G buffer with 0.2 mM DTT. Labeled protein will elute in void volume, whereas unreacted label will remain in resin.

5. Pool pink protein-containing fractions and measure the total volume. Polymerize actin by adding 1/4 vol 5× G to F conversion buffer and mixing thoroughly. Let sit at 4°C overnight.

6. Spin out filaments at 100,000 g for 1 hr at 4°C. Pellet should be glassy and dark pink. Remove supernatant, overlay pellet with a small amount of G buffer with 0.2 mM DTT, and let sit on ice for 1 hr to soften. Resuspend pellet with a sawed-off yellow tip in a minimal volume of G buffer. Sonicate to break up filaments (optional). Dialyze against G buffer overnight to depolymerize. Clarify in centrifuge. Freeze in small aliquots (2–5 μl) in liquid nitrogen and store at −80°C. For ease of use in motility assays, the actin may be diluted to an appropriate working concentration (e.g., 0.5 mg/ml) before freezing. Labeling stoichiometry may be determined by comparing protein concentration to fluorophore concentration (measured by absorption at 540 nm; $\epsilon \approx 60,000$ for rhodamine covalently attached to protein). A good yield is 30% with a labeling stoichiometry of 0.8 mol fluorophore per mole protein. Individual filaments should be visible by epifluorescence microscopy with a 60 or 100× objective.

D. Motility Assay

1. In a small (0.5 ml) Eppendorf tube, combine 10 μl *Xenopus* egg extract (prepared under Procedure A), 1 μl washed bacterial suspension in XB (prepared under Procedure B), and 0.5 μl of a 0.5-mg/ml solution of rhodamine–actin (prepared under Procedure C). Mix by pipetting gently up and down. Do not vortex. Incubate tube on ice for 30 min to 2 hr.

2. Remove 1 μl of assay mix and spot it onto the middle of a clean glass microscope slide. Gently overlay the droplet with a clean glass 22 × 22-mm coverslip. Press down on the coverslip until the droplet has spread to the edges. Seal the edges of the coverslip with melted beeswax or melted VALAP. VALAP is a 1:1:1 mixture of vaseline, lanolin, and paraffin that has a lower melting point than beeswax and can be easily applied to the edge of the coverslip with a cotton-tipped applicator stick (Q-tip). Incubate the sealed slide in the dark at room temperature for about 30 min.

3. Examine the slide on an upright or inverted microscope equipped with phase-contrast and epifluorescence optics (see Fig. 1). Actin-rich comet tails associated with moving bacteria will be easily observed in the rhodamine channel using a 40× or higher power objective. Bacteria will be visible as phase-dense rods. Comet tails are often not easily seen under phase contrast but can sometimes be detected as phase-dense or phase-lucent streaks behind the moving bacteria. Bacteria and comet tails can be visualized using differential interference contrast microscopy, but we have found phase contrast to be more useful for this assay. Movement can best be appreciated by recording time-lapse video sequences (Fung and Theriot, 1997). Typical movement rates range from about 3 to 30 μm/min.

IV. COMMENTS

Like all *in vitro* motility assays, the reconstitution of bacterial actin-based motility described here has a variety of uses. Since 1994, when this assay was first published, the reconstitution has proven useful for the examination of the roles of individual host proteins (Marchand *et al.*, 1995; Theriot *et al.*, 1994) and bacterial proteins (Goldberg and Theriot, 1995; Kocks *et al.*, 1995; Smith *et al.*, 1995) involved in actin-based motility. We have found *Xenopus* egg cytoplasmic extracts to be an excellent starting material for the purification of host factors involved in motility (M. C. Sanders, W. B. Simms, P. Farazi, D. C. Fung and J. A. Theriot, manuscript in preparation). We have also found that the *in vitro* assay for bacterial motility presents advantages over movement in infected tissue culture cells for the semiautomated tracking of bacterial trajectories (Fung and Theriot, 1997) and quantitative examination of actin filament dynamics during motility (D. C. Fung and J. A. Theriot, manuscript in preparation).

This form of actin polymerization-dependent movement is not observed only with pathogenic bacteria and is likely to represent a general type of actin-based motility. The rocketing movement of endosomes and polycationic beads has been observed in uninfected cells.

Similarly, in *Xenopus* egg extracts, vesicles formed from cellular membranes occasionally induce comet tail formation and can be seen to move rapidly in an actin-dependent fashion (see, e.g., Marchand *et al.*, 1995). This may be of particular interest to researchers wishing to study the actin dynamics involved in rocketing motility but who are not interested in bacterial pathogens per se. A rich source of movement-competent vesicles is the yellowish lipid layer that forms at the top of the centrifuge tube after the *Xenopus* egg-crushing spin (step 7 in Procedure A). This material can be resuspended in XB and added to the cytoplasmic extract like the bacterial suspensions recommended earlier to generate numerous comet tails and impressive vesicular movement (D. C. Fung and J. A. Theriot, unpublished observations). Intriguingly, the active component in the vesicles may be lipid rather than protein, as pure lipid vesicles containing PIP_2 or PIP_3 also exhibit dramatic actin-based motility in *Xenopus* egg cytoplasmic extracts (L. Ma and M. W. Kirschner, manuscript submitted).

The general approach described in this article may be used to reconstitute bacterial actin-based motility in several different extract types. We have used *Xenopus* egg cytoplasmic extracts arrested in the cell cycle in metaphase of meiosis II for all our published work, as these cytostatic factor-arrested extracts are simple to prepare and require minimal manipulation of the eggs (Murray, 1991). However, bacterial actin-based motility can also be reconstituted in interphase extracts prepared from activated eggs (Marchand *et al.*, 1995). Movement is generally slightly faster in interphase extracts and the actin-rich comet tails appear longer (J. A. Theriot, unpublished observations); these effects are due to cell cycle-dependent changes in actin filament dynamics. Shortly after the original description of reconstitution of movement in *Xenopus* egg extracts was published in 1994 (Theriot *et al.*, 1994), M. D. Welch and T. J. Mitchison (personal communication) succeeded in reconstituting bacterial actin-based motility in concentrated extracts prepared from human platelets, with movement rates similar to those observed in *Xenopus* egg extracts. Researchers who do not have access to a colony of *X. laevis* may find platelets easier to obtain as outdated batches from blood banks. Platelet extracts are also an excellent starting material for purification of host cell factors involved in movement (see Volume 2, Laurent and Carlier, "Use of Platelets Extracts for Actin-Based Motility of *Listeria monocytonenes*" for additional information; Welch *et al.*, 1997).

V. PITFALLS

1. The choice of bacterial strain is extremely important. For *L. monocytogenes,* the protein ActA must be expressed at high levels on the bacterial surface when the organisms are used in the motility assay. However, most wild-type strains of *L. monocytogenes* express only low levels of ActA on their surfaces when grown in broth and do not upregulate ActA expression until after infection. We therefore use a "hyperhemolytic" strain, SLCC-5764, that expresses high levels of all virulence factors (including ActA) in an apparently nonregulated fashion. Alternatively, ActA can be overexpressed in a nonhyperhemolytic strain by introducing a high copy-number plasmid carrying the *actA* gene under an active promoter (see Marchand *et al.*, 1995). For *E. coli* expressing IcsA (VirG), at least three strains have been described in the published literature that present robust actin-based motility (Goldberg and Theriot, 1995; Kocks *et al.*, 1995). Results with other strains cannot be predicted.

2. Pathogenic bacteria may lose virulence when maintained in the laboratory. We recommend that fresh plates be prepared from frozen stock at least once a month and that all liquid cultures be innoculated from a plate rather than from a previous liquid culture.

3. Extract quality is extremely important. In our hands, crude cytoplasmic extracts cannot be diluted more than two- or threefold without losing activity in the motility assay. Extracts prepared from unhealthy eggs may be inactive for bacterial actin-based motility.

4. Concentrated cytoplasmic extracts from any source containing actin filaments and myosin II bipolar filaments have a tendency to contract over time. In order to prevent this from happening, protocols for preparing *Xenopus* egg extracts for most purposes recommend the addition of cytochalasin (e.g., Murray, 1991; see Matthews, this volume). Because cytochalasin completely inhibits bacterial actin-based motility both in infected tissue culture cells

and in cytoplasmic extracts, it must be omitted for these assays. Contraction may be minimized by keeping the extract on ice, but contraction cannot be completely avoided with a crude extract and will render the extract useless for motility assays a few hours after warming.

5. Neither the concentration of rhodamine–actin nor the number of bacteria used in the motility assay has any significant effect on activity. The amounts we have suggested here are convenient for most of our experiments, but optimum amounts should be determined empirically for any particular application.

6. Like all fluorophores, tetramethylrhodamine can be bleached by excessive illumination. Photobleaching in a motility assay may be minimized by the addition of free radical scavengers to the extract. We have used a mixture of 0.05 mg/ml catalase, 0.1 mg/ml glucose oxidase, 2.5 mg/ml glucose, and 0.5 mM DTT (after Kron *et al.*, 1991); some photobleaching will still occur. The addition of these anti-bleaching abents may affect bacterial speeds and the rate of actin depolymerization. For long-term recordings, epifluorescence illumination must be shuttered.

Acknowledgements

We are grateful to Matt Welch, Tim Mitchison, Le Ma, and Marc Kirschner for the communication of results prior to publication. Our work is supported by grants from the National Institutes of Health and the W. M. Keck Foundation.

References

Fung, D. C., and Theriot, J. A. (1997). Movement of bacterial pathogens driven by actin polymerization. *In* "Motion Analysis of Living Cells" (D. R. Soll, ed.). Wiley, New York, in press.

Goldberg, M. B., and Theriot, J. A. (1995). *Shigella flexneri* surface protein IcsA is sufficient to direct actin-based motility. *Proc. Natl. Acad. Sci. USA* **92**, 6572–6576.

Kocks, C., Marchand, J. B., Gouin, E., d'Hauteville, H., Sansonetti, P. J., Carlier, M. F., and Cossart, P. (1995). The unrelated surface proteins ActA of *Listeria monocytogenes* and IcsA of *Shigella flexneri* are sufficient to confer actin-based motility on *Listeria innocua* and *Escherichia coli* respectively. *Mol. Microbiol.* **18**, 413–423.

Kron, S. J., Toyoshima, Y. Y., Uyeda, T. Q., and Spudich, J. A. (1991). Assays for actin sliding movement over myosin-coated surfaces. *In* "Methods in Enzymology" (R. B. Vallee, ed.), Vol. 196, pp. 399–416. Academic Press, San Diego.

Marchand, J. B., Moreau, P., Paoletti, A., Cossart, P., Carlier, M. F., and Pantaloni, D. (1995). Actin-based movement of *Listeria monocytogenes*: Actin assembly results from the local maintenance of uncapped filament barbed ends at the bacterium surface. *J. Cell Biol.* **130**, 331–343.

Murray, A. W. (1991). Cell cycle extracts. *Methods Cell Biol.* **36**, 581–605.

Pardee, J. D., and Spudich, J. A. (1982). Purification of muscle actin. *In* "Methods in Enzymology" (D. W. Frederiksen and L. W. Cunningham, eds.), Vol. 85, pp. 164–181. Academic Press, San Diego.

Russell, D. G. (1994). Obtaining and maintaining microbial pathogens. *Methods Cell Biol.* **45**, 1–4.

Smith, G. A., Portnoy, D. A., and Theriot, J. A. (1995). Asymmetric distribution of the *Listeria monocytogenes* ActA protein is required and sufficient to direct actin-based motility. *Mol. Microbiol.* **17**, 945–951.

Theriot, J. A. (1994). Actin filament dynamics in cell motility. *Adv. Exp. Med. Biol.* **358**, 133–145.

Theriot, J. A., Rosenblatt, J., Portnoy, D. A., Goldschmidt-Clermont, P. J., and Mitchison, T. J. (1994). Involvement of profilin in the actin-based motility of *L. monocytogenes* in cells and in cell-free extracts. *Cell* **76**, 505–517.

Welch, M. D., Iwamatsu, A., and Mitchison, T. J. (1997). Actin polymerization is induced by Arp 2/3 protein complex at the surface of *Listeria monocytogenes*. *Nature* **385**, 265–269.

Use of Platelet Extracts for Actin-Based Motility of *Listeria monocytogenes*

Valérie Laurent and Marie-France Carlier

I. INTRODUCTION

The process of local actin assembly in cells is thought to be the motor of changes in cell shape and locomotion. The molecular basis for site-directed actin assembly in response to extracellular ligands, however, and the components of the machinery involved are far from being understood.

A small group of intracellular pathogens, including bacteria *Listeria monocytogenes* (Tilney and Portnoy, 1989), *Shigella flexneri* (Bernadini *et al.*, 1989), and *Rickettsia* and also viruses like vaccinia virus (Cudmore *et al.*, 1995), are able to interact with the cellular machinery leading, under normal life conditions, to local actin assembly in response to stimuli and to induce polymerization of actin present in the host cytoplasm. The actin meshwork continuously built at the surface of the pathogen rapidly aquires the rigidity and the viscous drag sufficient to cause propulsion of the bacteria or virus at a rate equal to the rate of actin polymerization (Theriot *et al.*, 1992). This process eventually allows the bacteria to spread from cell to cell. The propulsive movement of *Listeria* therefore is mechanistically similar to other actin-based motile processes such as the extension of the lamella at the leading edge of locomoting cells. Movement of *Listeria,* initially observed in cultured macrophages (Tilney *et al.*, 1992), could be reconstituted *in vitro* in acellular extracts of *Xenopus* eggs (Marchand *et al.*, 1995; Theriot *et al.*, 1994). The development of the *in vitro* motility assay is expected to allow the biochemical identification of the components of the cascade involved in site-directed actin assembly (Fig. 1).

Cytoplasmic extracts of *Xenopus* egg contain a wealth of cellular functions and have been useful tools in the study of microtubule dynamics and regulation of the cell cycle, but they cannot be considered as a specialized system for the actin-based motile response. Motile blood cells such as platelets, neutrophils, and macrophages, in contrast, are actin-rich cells known to display actin assembly in response to stimulation and are easily available in amounts suitable for biochemical studies. This article describes a protocol for the observation of actin-based movement of *L. monocytogenes* in cytoplasmic extracts of human platelets (Tables I and II). Complementary information concerning the use of a cellular extracts in *Listeria* motility assays can be found elsewhere (Rosenblatt *et al.*, 1997; Carlier *et al.*, 1997).

II. MATERIALS AND INSTRUMENTATION

Brain–heart infusion (BHI) culture medium is from Difco Laboratories. NaCl (Cat. No. 1.06404), KCl (Cat. No. 1.04936), NaH$_2$PO$_4$ (Cat. No. 1.06346), NaHCO$_3$ (Cat. No. 1.06329),

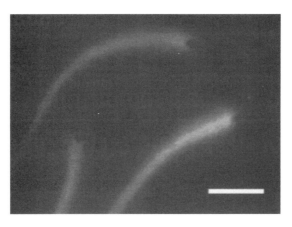

FIGURE 1 Movement of *Listeria monocytogenes* in a platelet extract. This field was observed simultaneously in fluorescence and in phase contrast. Bar: 10 μm.

MgCl$_2$ (Cat. No.1.05833), glucose (Cat. No.1.04074), Tris (Cat. No. 1.08382), sucrose (Cat. No. 1.07687) and CaCl$_2$ (Cat. No. 1.02382) are from Merck. EGTA (Cat. No.E-4378), bovine serum albumin (BSA, Cat. No. A-6003), phenylmethylsulfonyl fluoride (PMSF, Cat. No.P-7626), methylcellulose (Cat. No. M-0512), HEPES (Cat. No. H-3375), erythromycin (Cat. No. E-6375), chloramphenicol (Cat. No. C-0378), leupeptin (Cat. No. L-2884), pepstatin (Cat. No. P-4265), and chymostatin (Cat. No. C-7268) are from Sigma. 5-(and 6)-Carboxyte-tramethylrhodamine succinimidyl ester (NHSR) (Cat. No. C-1171) is from Molecular Probes. Dithiothreitol (DTT, Cat. No. EU0006B) is from Euromedex. ATP (Cat. No. 519 987), glucose oxidase (Cat. No. 646 423), and catalase (Cat. No. 106 810) are from Boehringer-Mannheim. Paraffin (Cat. No. 26 152 294) and lanolin (Cat. No. 24 485 291) are from Prolabo.

L. monocytogenes strain Lut12 (pactA3) overexpressing ActA was used for motility assays.

Actin was purified from rabbit back muscles as described (Spudich and Watt, 1971; McLean-Flechter and Pollard, 1980). Actin was rhodamine labeled with NHSR as previously

TABLE I Effect of the Concentration of Methylcellulose on the Rate of Propulsion of *Listeria* Movement[a]

Methyl cellulose (%)	Average rate $\pm$ SD (μm/min)
0	8.2 $\pm$ 1.3
0.05	8.3 $\pm$ 1.1
0.1	7.7 $\pm$ 0.5
0.25	6.9 $\pm$ 1.5
0.5	7.1 $\pm$ 0.9

[a] The presence of methylcellulose in the assay avoids the lateral drift of the bacteria with their actin tail bound, which results from the low viscosity of the diluted platelet extracts. The rates of at least 10 bacteria were measured at different concentrations of methylcellulose (in %, w/v). The percentage of motile bacteria was 30% in the presence of 0.1% methylcellulose in a 40-fold diluted platelet cytoplasm. Four hundred bacteria were counted for this estimation. Only bacteria with comet tails were counted as motile bacteria. Data show that the rate of movement is independent of the concentration of methylcellulose.

| Platelet cytoplasm dilution | Average rate ± SD (μm/min) | |
	Measured in the first half hour after preparation of the sample	Measured 1 hr after preparation of the sample
1/8	—	—
1/12	—	≤1.5
1/18	12.1 ± 2.6	8 ± 0.5
1/40	9.1 ± 1.3	7.3 ± 1.4
1/80	4.6 ± 0.7	—

[a] Average rates were measured as described under Table I, in the presence of 0.1% methycellulose. The dilution factor of the platelet cytoplasm in the final assay medium (left column) was calculated assuming that the platelet cytoplasm contains 280 μM monomeric actin (Weber *et al.*, 1992) and measuring the concentration of actin in the platelet extract by the DNase I inhibition method (for details see Carlier *et al.*, 1997). The rates of *Listeria* movement were measured immediately after preparing the sample or 1 hr later. Data show that it took a longer time for bacteria to start moving when the platelet cytoplasm was more concentrated. At a 8-fold dilution of the cytoplasm, bacteria were surrounded by an actin cloud only after 90 min of incubation. The clouds did not rearrange into actin tails and no movement was detectable even after 4 hr of incubation. At a 12-fold dilution of the cytoplasm, few bacteria were initially surrounded by a "cloud" of rhodamine–F-actin. After 90 min, short comet tails were observed and a larger number of bacteria surrounded by an actin "cloud" were observed. After 3.5 hr, rates not exceeding 1.5 μm/min were measured. Data show that the rate of movement measured at higher dilutions tended to decrease with the incubation time.

described (Isambert *et al.*, 1995). Rhodamine–G-actin (60 μM) was stored at −80°C as 25-μl aliquots. Once thawed, rhodamine–G-actin can be kept on ice and used for up to 3 weeks.
All observations were performed using a phase-contrast and fluorescence microscope.

III. PROCEDURES

A. Preparation of Platelet Extracts

Solutions

1. *2.25 M NaCl stock solution:* Dissolve 6.59 g of NaCl in 50 ml of distilled water. Store at 4°C.

2. *0.135 M KCl stock solution:* Dissolve 0.50 g of KCl in 50 ml of distilled water. Store at 4°C.

3. *0.595 M NaHCO$_3$ stock solution:* Dissolve 2.50 g of NaHCO$_3$ in 50 ml of distilled water. Store at 4°C.

4. *0.018 M NaH$_2$PO$_4$ stock solution:* Dissolve 0.12 g of NaH$_2$PO$_4$ in 50 ml of distilled water. Store at 4°C.

5. *1 M MgCl$_2$ stock solution:* Dissolve 10.16 g of MgCl$_2$ in 50 ml of distilled water. Store at 4°C.

6. *0.5 M EGTA stock solution:* Dissolve 9.51 g of EGTA in 50 ml of distilled water. Store at 4°C.

7. *0.275 M glucose stock solution:* Dissolve 2.72 g of glucose in 50 ml of distilled water. Store at 4°C.

8. *30% BSA stock solution:* Dissolve 1.5 g of albumin in 50 ml of distilled water. Store at 4°C.

9. *0.5 M Tris, pH 7.5:* To make 50 ml, add 3.03 g of Tris in distilled water and adjust pH to 7.5 with 12 N HCl.

10. *Washing buffer:* 135 mM NaCl, 2.7 mM KCl, 11.9 mM NaHCO$_3$, 0.36 mM NaH$_2$PO$_4$, 2 mM MgCl$_2$, 0.2 mM EGTA, 5.5 mM glucose, and 0.3% albumin, pH 6.5. To make 250 ml, add 15 ml of 2.25 M NaCl stock solution, 5 ml of 0.135 M KCl stock solution, 5 ml of 0.595 M NaHCO$_3$ stock solution, 5 ml of 0.018 M NaH$_2$PO$_4$ stock solution, 0.5 ml of 1 M MgCl$_2$ stock solution, 0.1 ml of 0.5 M EGTA stock solution, 5 ml of 0.275 M glucose stock solution, and 2.5 ml of 30% albumin stock solution. Adjust pH to 6.5 with 1 N HCl and adjust to 250 ml with distilled water. Store at 4°C.

11. *Sonication buffer:* 10 mM Tris–HCl, pH 7.5, 10 mM EGTA, and 2 mM MgCl$_2$. To make 50 ml, add 1 ml of 0.5 M Tris, pH 7.5, 2 ml of 0.5 M EGTA stock solution, and 0.1 ml of 1 M MgCl$_2$ stock solution. Adjust to 50 ml. Store at 4°C.

12. *Leupeptin (2 mg/ml):* Dissolve 2 mg of leupeptin in 1 ml of 50% methanol in water. Aliquot by 25 μl and store at −20°C.

13. *Pepstatin (1 mg/ml):* Dissolve 1 mg of pepstatin in dimethyl sulfoxide (DMSO). Aliquot by 25 μl and store at −20°C.

14. *Chymostatin (1 mg/ml):* Dissolve 1 mg of chymostatin in DMSO. Aliquot by 25 μl and store at −20°C.

15. *PMSF (50 mM):* Dissolve 87 mg of PMSF in 10 ml of pure ethanol.

16. *0.2 M DTT stock solution:* To prepare 25 ml of solution, dissolve 0.77 g of DTT in distilled water. Store at −20°C.

17. *0.2 M ATP stock solution:* To prepare 25 ml of solution, dissolve 3.02 g of ATP in distilled water. Adjust pH to 7 with 6 N NaOH. Store at −20°C.

18. *2 M sucrose stock solution:* To prepare 10 ml of solution, dissolve 6.84 g in distilled water. Store at −20°C.

Steps

A typical preparation starting with 5 units of outdated unstimulated human platelets is described below.

1. Share platelet-rich plasma in 30-ml plastic tubes. It is necessary to use plastic tubes and plastic material to avoid platelet stimulation. Sediment cells by centrifugation at 1600 g for 15 min at 18°C in a swinging bucket rotor (e.g., Sorvall HS4 rotor). Platelets sediment as a whitish superior layer on top of the deep red erythrocyte pellet.

2. Carefully remove the supernatant and gently resuspend platelets in 1 ml washing buffer. Avoid resuspending erythrocytes.

3. Collect resuspended platelets in six conical, plastic tubes (15-ml, Falcon). Disperse aggregates by gentle pipetting with a 1-ml pipette. Adjust the volume to 14 ml with washing buffer. Wash the cells by gentle overturning of the tubes.

4. Sediment cells by centrifugation at 1600 g for 15 min at 18°C in a swinging bucket rotor. Repeat steps 2 to 4 until erythrocyte contamination gets negligible, while decreasing the volume of the platelet suspension to 30 ml (two Falcon tubes of 15 ml).

5. After the last centrifugation, carefully remove the supernatant. The volume of pelleted platelet must be about 3 ml total, in two tubes. Each pellet is resuspended in 1 ml of sonication buffer.

6. Add protease inhibitors to a final concentration of 10 μg/ml: 12.5 μl of 2 mg/ml leupeptin stock solution, 25 μl of 1 mg/ml pepstatin stock solution, 25 μl of 1 mg/ml of chymostatin stock solution, and 5 μl of PMSF (final concentration of 1 mM) to each tube.

7. Sonicate resuspended platelets 6 × 10 sec, cooling tubes on ice between each treatment.

8. Sediment membrane fragments by centrifugation at 100,000 g for 45 min at 4°C (Beckman TL100).

9. Supplement supernatants (2×2.5 ml) with 1 mM ATP, 1 mM DTT, and 150 mM sucrose.

10. Aliquot the cytoplasm platelet extract by 25 or 50 μl and freeze the aliquots in liquid nitrogen. Store at $-80°C$.

B. *Listeria monocytogenes* Culture

Grow bacteria at $37°C$ in BHI in the presence of chloramphenicol (7 μg/ml) and erythromycin (5 μg/ml). Dilute the overnight culture (final $OD_{560\,nm} = 0.3$) in 5 ml of fresh medium and grow to stationary phase ($OD_{560\,nm} = 1.5$ to 2). Freeze 100-μl aliquots at this point in liquid nitrogen and keep as suspensions of cells at $-80°C$ in 30% glycerol.

C. Motility Test

Solutions

1. *10 mM Tris–HCl, pH 7.5:* To prepare 500 ml of solution, add 10 ml of 0.5 M Tris–HCl, pH 7.5 stock solution to distilled water. Complete the volume to 500 ml with distilled water. Store at $4°C$.

2. *2% methylcellulose solution stock:* Heat 250 ml of 10 mM Tris–HCl, pH 7, in a beaker to $60°C$ and add 5 g of methylcellulose. Stir vigorously. Cool the beaker on ice and stir with a glass rod until the solution gets thick. Store at room temperature.

3. *0.2% methylcellulose:* Add 1 ml of the 2% methylcellulose stock solution to 9 ml of 10 mM Tris–HCl, pH 7.5.

4. *1 mg/ml glucose oxidase stock solution:* Dissolve 1 mg of glucose oxidase in 1 ml of distilled water. Store at $4°C$.

5. *3 mg/ml catalase stock solution:* Add 1.5 ml of catalase to 10 ml distilled water. Store at $4°C$.

6. *1 M glucose stock solution:* Dissolve 0.19 g of glucose in 1 ml distilled water. Store at $4°C$.

7. *Free radical scavenger solution:* For 10 motility tests, mix 5 μl of glucose oxidase stock solution, 5 μl catalase stock solution, and 5 μl glucose stock solution in an Eppendorf tube. Keep on ice.

8. *80 mM ATP, 0.1 M DTT, 0.1 M MgCl$_2$, pH 7 stock solution:* Dissolve 0.48 g ATPNa$_2$ and 0.15 g DTT in distilled water (8 ml) and add 1 ml of 1 M MgCl$_2$ stock solution. Adjust pH to 7 with 6 N NaOH. Adjust to 10 ml. Aliquot by 0.5 ml and freeze at $-20°C$.

9. *1% BSA stock solution:* Dissolve 0.1 g of BSA in 10 ml of distilled water. Store at $4°C$.

10. *1 M KCl stock solution:* Dissolve 37 g of KCl in 500 ml of dstilled water. Store at $4°C$.

11. *1 M MgCl$_2$ stock solution:* Dissolve 10.16 g of MgCl$_2$ in 50 ml of distilled water. Store at $4°C$.

12. *1 M CaCl$_2$ stock solution:* Dissolve 7.35 g of CaCl$_2$ in 50 ml of distilled water. Store at $4°C$.

13. *2 M sucrose stock solution:* Dissolve 6.84 g in distilled water. Store at $-20°C$.

14. *X$_b$ buffer, pH 7.7:* 100 mM HEPES, pH 7.7, 100 mM KCl, 1 mM MgCl$_2$, 0.1 mM CaCl$_2$, and 50 mM sucrose. To make 50 ml, add 1.2 g of HEPES powder, 5 ml of 1 M KCl stock solution, 50 μl of 1 M MgCl$_2$ stock solution, 5 μl of 1 M CaCl$_2$ stock solution, and 2.5 ml of 2 M sucrose stock solution in distilled water. Adjust pH to 7.7 with 10 M KOH. Complete the volume to 50 ml. Store at $4°C$.

15. *Valap:* Mix vaselin, lanolin, and solid paraffin in equal amounts and homogenize at $50°C$.

Video-Microscopy Motility Assay

1. For 10 motility tests, coat 10 slides and 10 coverslips with the 1% BSA stock solution. Let BSA adsorb on the glass surface for a few minutes. Blow the glass surfaces dry using air or nitrogen under pressure. BSA-coated slides and coverslips can be stored for several days.

2. Thaw a 100-μl bacteria sample, sediment bacteria by a short centrifugation at 4°C for 5 min at 13,000 rpm in a microfuge. Remove the supernatant and resuspend the pellet in 25 μl X_b buffer, pH 7.7, to obtain about 6×10^9 bacteria/ml. Keep bacteria on ice.

3. At room temperature, in an Eppendorf microtube, mix 2 μl of cytoplasm platelet extract, 8 μl of 0.2% methylcellulose, 1.5 μl of free radical scavenger solution, 1 μl of 80 mM ATP, 0.1 M DTT, 0.1 M $MgCl_2$, pH 7, stock solution, 0.6 μl of rhodamine–G-actin and 0.5 μl of bacteria. Vortex or homogenize thoroughly with a 20-μl pipette.

4. Place a 2.5-μl aliquot of the above mixture on a BSA-coated slide and cover with a 22 $\times$ 22-mm BSA-coated coverslip.

5. Seal preparations with valap before observations.

6. The observations are carried out at room temperature as soon as the slide is ready. Movements of bacteria were observed on a Zeiss III RS microscope equipped with a silicon-intensified camera (LHESA LHL4036), which allowed observation of bacteria by phase contrast and of fluorescent actin tails simultaneously, using a 100-W tungsten–iodine lamp for phase contrast and a 200-W mercury arc for fluorescence. Images were recorded in real time or in time lapse on a video tape recorder (Panasonic). All image analysis was performed using a Hamamatsu, Argus-10 image processor.

IV. COMMENTS

L. monocytogenes strain Lut12 (pacta3) is a pathogenic strain that must be handled with precautions. Wear gloves and cautiously decontaminate all materials that have been in contact with *Listeria* using bleach.

Coating slides and coverslips with BSA is essential because total protein concentration is only 2 mg/ml in the motility assay. BSA coating prevents the adsorption of proteins from the extract on glass.

The average rates of *Listeria* varied in the range of 3 to 10 μm/min from platelet extract to platelet extract, reflecting the variability of the cytoplasm of platelets.

Acknowledgments

We thank Dr. Matt Welch for private communication on the use of platelet extracts for *Listeria* motility assays.

References

Bernadini, M. L., Mounier, J., d'Hauteville, H., Coquis-Rondon, M., and Sansonetti, P. J. (1989). Identification of IcsA, a plasmid locus of *S. flexneri* that governs bacterial intra- and intercellular spread through interaction with F-actin. *Proc. Natl. Acad. Sci. USA.* **86**, 3867–3871.

Carlier, M.-F., Laurent, V., Santolini, J., Melki, R., Didry, D., Xia, G.-X., Hong, Y., Chua, N.-M., and Pantaloni, D. (1997). Actin depolymerizing factor (ADF/Cofilin) enhances the rate of filament turnover: Implication in actin-based motility. *J. Cell. Biol.* **136**, 1307–1322.

Cudmore, S., Cossart, P., Griffiths, G., and Way, M. (1995). Actin-based motility of *Vaccinia* virus. *Nature* **378**, 636–638.

Isambert, H., Venier, P., Maggs, A. C., Fattoum, R., Kassab, R., Pantaloni, D., and Carlier, M.-F. (1995). Flexibility of actin filaments derived from thermal fluctuations. *J. Biol. Chem.* **270**, 11437–11444.

Marchand, J.-B., Moreau, P., Poaletti, A., Cossart, P., Carlier, M.-F., and Pantaloni, D. (1995). Actin-based movement of *Listeria monocytogenes:* Actin assembly results from the local maintenance of uncapped filament barbed ends at the bacterium surface. *J. Cell. Biol.* **130**, 331–343.

McLean-Flechter, S., and Pollard, T.D. (1980). Identification of a factor in conventional muscle actin preparation which inhibits actin filament self association. *Biochem. Biophys. Res. Commun.* **96**, 18–27.

Rosenblatt, J., Agnew, B. J., Abe, H., Bamburg, J. R., and Mitchison, T. J. (1997). Xenopus actin depolimerizing factor/cofilin (XAC) is responsible for the turnover of actin filaments in *Listeria monocytogenes* tails. *J. Cell. Biol.* **136**, 1323–1332.

Theriot, J. A., Mitchison, T. J., Tilney, L. G., and Portnoy, D. A. (1992). The rate of actin-based motility of intracellular *Listeria monocytogenes* equals the rate of actin polymerization. *Nature* 357, 257–260.

Theriot, J. A., Rosenblatt, J., Portnoy, D. A., Goldschmidt-Clermont, P. J., and Mitchison, T. J. (1994). Involvement of profilin in the actin-based motility of *Listeria monocytogenes* in cells and in cell-free extracts. *Cell* 76, 505–517.

Tilney, L. G., and Portnoy, D. A. (1989). Actin filaments and the growth, movement and spread of the intracellular bacterial parasite, *Listeria monocytogenes. J. Biol. Chem.* 109, 1597–1608.

Tilney, L. G., DeRosier, D. J., Weber, A., and Tilney, M. S. (1992). How *Listeria* exploits host cell actin to form its own cytoskeleton? II. Nucleation, actin filament polarity, filament assembly and evidence for a pointed end capper. *J. Cell. Biol.* 118, 83–93.

Spudich, J. A., and Watt, S. (1971). The regulation of rabbit skeletal muscle contraction: Biochemical studies of the interaction of the tropomyosin-troponin complex with actin and the proteolytic fragments of myosin. *J. Biol. Chem.* 246, 4866–4871.

Weber, A., Nachmias, V. T., Pennise, C. R., Pring, M., and Safer, D. (1992). Interaction of Tβ_4 with muscle and platelet actin: Implication for actin sequestration in resting platelet. *Biochemistry* 31, 6179–6185.

Isolation of Actin Comet Tails of *Listeria monocytogenes*

Antonio Sechi and J. Victor Small

I. INTRODUCTION

Many microorganisms have developed the ability to invade eukaryotic cells. Among them, *Listeria monocytogenes* has gained particular attention since the mid-1980s (Tilney and Portnoy, 1989). This gram-positive human pathogenic bacterium can enter cells and recruit their motile machinery to propel itself through the cytoplasm and, eventually, into neighboring cells (Tilney and Tilney, 1993) at the tip of a comet tail of actin filaments.

It has been suggested that the mechanism(s) underlying the movement of *Listeria* is similar to that used for lamellipodia protrusion; thus this invasive pathogen has been adopted as a model to clarify the general mechanism(s) of actin-based motility (Theriot and Mitchison, 1991; Theriot *et al.*, 1992; see Volume 2, Theriot and Fung, "Use of *Xenopus* Egg Extracts for Studies of Actin-Based Motility" for additional information).

As one approach toward defining the requisite components of the comet tails, as well as their structural organization, this article describes procedures used to isolate them from infected cells.

II. MATERIALS AND INSTRUMENTATION

Minimum essential medium (MEM) with Earle's salts (Cat. No. 21090-022), sodium pyruvate (Cat. No. 11360-039), MEM nonessential amino acids (Cat. No. 11140-035), L-glutamine (Cat. No. 25030-024), gentamicin (Cat. No. 15750-037), and penicillin–streptomycin (Cat. No. 15070-022) are from GIBCO-BRL. Fetal calf serum (Cat. No. 40378) is from SEBAK (Vienna). Sodium chloride (Cat. No. 1.06404), disodium hydrogen orthophosphate (Cat. No. 1.06380), potassium dihydrogen orthophosphate (Cat. No. 4873), magnesium chloride hexahydrated (Cat. No. 5833), glucose (8337) are from Merck. Triton X-100 (Cat. No. T9284), MES (Cat. No. M8250), and bacitracin (Cat. No. B-0125) are from Sigma. Phalloidin was a generous gift from Professor H. Faulstich (Heidelberg), but can be obtained from Sigma (Cat. No. P-2141). Brain heart infusion (BHI) medium (Cat. No. 0037-17-8) is from Difco Laboratories. EGTA (Cat. No. 03780) is from Fluka. Falcon tissue culture flasks (Cat. No. 3108) and Falcon petri dishes are from Becton-Dickinson. Electron microscopy grids (G300HEXC3; Cu) are from Science Services. Glutaraldehyde (EM grade Cat. No.R1020) and formvar are from Agar Scientific Ltd. Sodium silico tungstate (SST, Cat. No. S019) and EM forceps (Nos. 4 and 5) are from TAAB. Filter paper (Whatman No. 1) and disposable syringe filters (0.20/0.45 μm cutoff) are from Sartorius. The table-top centrifuge Hettich

Universal was equipped with a swing-out rotor (type No. 1323), capable of taking petri dishes of 3.5 cm in diameter.

III. PROCEDURES

A. Preparation of Dishes for Infection and Tail Isolation

Steps

1. To prepare the plastic adaptors (Fig. 1b, 1), cut a plastic tube of 12 mm inner diameter into slices 5 mm thick.

2. Glue four adaptors to the bottom of a 3.5-cm petri dish (Fig. 1b, 2).

3. Fill the space between the inner wall of the dish and the outer perimeter of the adaptors with plasticine to give more stability to the whole (Fig. 1b, 3 and 4).

B. Growth of *Listeria monocytogenes*

1. *Bacterial medium:* Dissolve 37 g of BHI medium in 1 liter of distilled water and autoclave at 121–124°C for 15 min. The medium can be stored at room temperature.

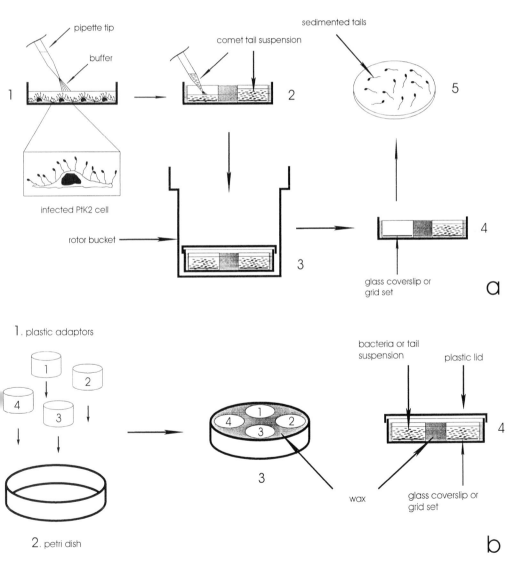

FIGURE 1 (a) Basic steps of tail isolation. (b) Schematical representation of petri dish assembly used for the sedimentation of tails (steps 2–4 in Procedure A). For more details, see text.

2. *Bacterial culture:* The hemolytic strain *L. monocytogenes* wild type is routinely grown in BHI at 37°C with shaking.

C. Tissue Culture Cells

PtK2 cells (American Type Culture Collection CCL 56) are routinely cultured in minimum essential medium supplemented with 10% heat-inactivated fetal calf serum (FCS), 1 mM sodium pyruvate, 1% nonessential amino acids, 2 mM L-glutamine, 100 U/ml penicillin, and 100 μg/ml streptomycin in a CO_2 incubator (5% CO_2, 95% humidified air) at 37°C.

D. Infection

See Fig. 2.

Solutions

1. *Infection medium:* Standard PtK2 medium lacking antibiotics and serum (infection medium) and standard PtK2 medium lacking FCS (postinfection medium).

2. *Phosphate-buffered saline (PBS):* Dissolve 7.65 g of NaCl, 0.724 g of Na_2HPO_4, and 0.210 g of KH_2PO_4 in 800 ml of distilled water. Adjust the pH to 7.2–7.4 at room temperature with 1 M NaOH and bring the volume to 1 liter. Filter through a 0.45-μm filter and store at 4°C.

Steps

1. Plate a total of 2.4×10^5 PtK$_2$ cells in 3.5-cm petri dishes the day before the experiment

2. Grow the bacteria to an optical density of 1.4–1.5 at 600 nm, corresponding to around 1×10^9–1×10^{10} colony-forming units (CFU) per milliliter.

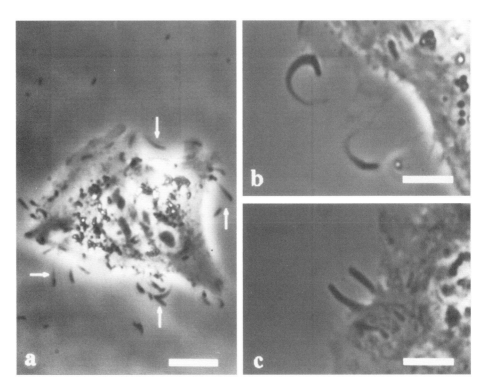

FIGURE 2 Phase-contrast pictures showing *Listeria*-infected cells. (a) Arrows point at *Listeria*-induced protrusions (b and c) Protrusions in different stages of formation. Bars: a, 15 μm; b and c, 6 μm.

3. Wash bacteria twice with PBS. After the final wash, resuspend the pellet in infection medium to give a final bacterial concentration of $1 \times 10^6 – 1 \times 10^7$ (CFU)/ml.

4. Wash the PtK_2 cells three times with prewarmed (37°C) infection medium.

5. Add 1 ml of the bacterial suspension to each dish and centrifuge the bacteria onto PtK_2 cells at 700 g for 5 min at room temperature. This step is critical to obtain 100% infection and to synchronize the infection process.

6. Return the PtK_2 cells to the incubator for 1 hr to permit entry of bacteria.

7. Remove free bacteria by washing three times with prewarmed postinfection medium. Add gentamicin to a final concentration of 15 μg/ml to prevent extracellular bacterial growth.

8. Return the PtK_2 cells to the incubator for at least 5 hr (the optimal time depending on the type of cells and on the bacterial strain used).

E. Isolation of Comet Tails

See Fig. 1.

Solution

1. *Cytoskeleton buffer (CB)*: 10 mM MES, 150 mM NaCl, 5 mM EGTA, 5 mM $MgCl_2 \cdot 6H_2O$, and 5 mM glucose, pH 6.1, at room temperature. To make 1 liter, dissolve the following components in 800 ml of water: 1.95 g MES, 8.76 g NaCl, 1.90 g EGTA, 1.02 g $MgCl_2 \cdot 6H_2O$, and 0.90 g glucose. Adjust the pH to 6.1 with 1 M NaOH and fill up to 1 liter. Filter through a 0.20-μm filter. Store at 4°C.

Steps

1. Wash *Listeria*-infected cells gently three times with CB. To minimize the mechanical manipulation, which might induce the loss of tails, aspirate solution gently from the side of the dish with one hand while adding new solution gently to the opposite side with the other hand.

2. After the last wash, add 1 ml of fresh CB to the dish. Then, using a 1-ml automatic pipette, repeatedly and gently pipette the buffer onto the cell surface to detach the protruding comet tails. Do this at least 10 times (Fig. 1a, 1).

3. Collect the buffer containing the tails and transfer it to the petri dish wells containing 10-mm-round glass coverslips or a set of formvar-coated grids (Fig. 1a, 2) (for the preparation of grid sites, see Volume 3, Small and Sechi, "Whole-Mount Electron Microscopy of the Cytoskeleton: Negative Staining Methods").

4. Centrifuge at 2000 g for 5 min at room temperature to sediment the protrusions onto the coverslips (Fig. 1a, 3).

5. Discard the supernatant and wash three times in CB. The protrusions can now be processed for immunofluorescence microscopy or electron microscopy (Fig. 1a, 4 and 5). Alternatively, the tails can be harvested for biochemical analysis.

F. Electron Microscopy

See Fig. 3 and Volume 3, J. V. Small and A. Sechi, "Whole-Mount Electron Microscopy of the Cytoskeleton: Negative Staining Methods" for additional information.

Solutions

1. *Negative stain solution:* 3% SST. Weigh out 3 g of SST and dissolve in 80 ml of water. Adjust pH to 7.2–7.3 with 1 M NaOH. Bring the volume to 100 ml . Filter through a 0.20-μm syringe filter and store at 4°C.

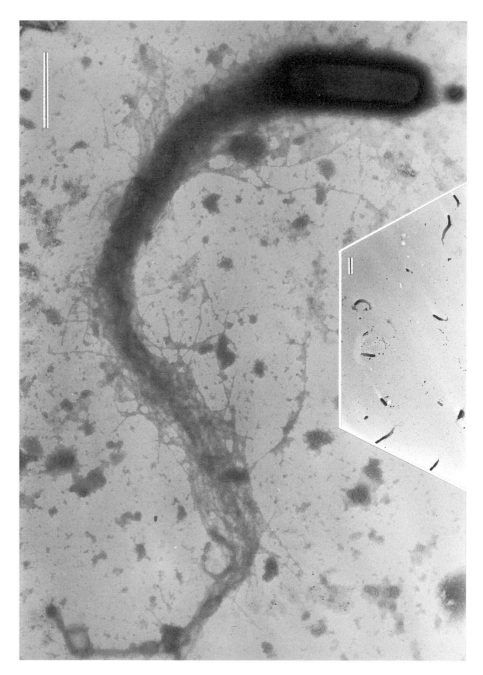

FIGURE 3 Electron micrograph of an isolated actin comet tail. Bar: 1 μm. (Inset) Low magnification picture showing field of actin tails sedimented onto filmed grid. Bar: 4 μm.

2. *Spreading solution:* 100 μg/ml bacitracin in water with 0.1% amyl alcohol. Prepare just before use.

Steps

1. Extract the isolated protrusions with 0.5% Triton X-100 in CB for 60 sec at room temperature. Alternatively, the comets may be treated with a mixture of GA–Triton X-100 (J. V. Small and A. Sechi, volume 3) for 60 sec at room temperature to decrease the extent of the extraction.

2. Wash gently three times in CB.

3. Add CB containing 10 μg/ml of phalloidin. Incubate for at least 30 min at room temperature to stabilize the actin filaments.

4. Repeat step 2.

5. Fixation. Add 2.5% GA and incubate for 10 min at room temperature.

6. Repeat step 2.

7. Rinse grid with water and negatively stain with 3% SST as described by Small and Sechi (this volume).

IV. COMMENTS AND PITFALLS

Our estimation of the average yield of isolated tails is 3% of the total protrusions of infected cells. Increasing centrifugational force with the tail suspension does not produce an increase in the number of isolated tails, only an increase of background (cells debris, free bacteria, etc.) that makes the analysis of the filament organization more difficult. The amount of background also depends on the extent and intensity of pipetting. In our hands, too vigorous pipetting and too many passages produce an increase of the background debris without increasing the number of isolated tails.

References

Sechi, A. S., Wehland, J., and Small, J. V. (1997). The isolated comet tail pseudopodium of *Listeria monocytogenes*: A tail of two actin filament populations, long and axial and short and random. *J. Cell Biol.* **137**, 155–167.

Theriot, J. A., and Mitchison, T. J. (1991). Actin microfilaments dynamics in locomoting cells. *Nature* **352**, 126–131.

Theriot, J. A., Mitchison, T. J., Tilney, L. G., and Portnoy, D. A. (1992). The rate of actin-based motility of intracellular *Listeria monocytogenes* equals the rate of actin polymerization. *Nature* **357**, 257–260.

Tilney, L. G., and Portnoy, D. A. (1989). Actin filaments and the growth, movement, and spread of the intracellular bacterial parasite, *Listeria monocytogenes*. *J. Cell Biol.* **109**, 1597–1608.

Tilney, L. G., and Tilney, M. S. (1993). The wily ways of a parasite: Induction of actin assembly by *Listeria*. *Trends Microbiol.* **1**, 25–31.

Preparation and Fixation of Fish Keratocytes

K. I. Anderson and J. V. Small

I. INTRODUCTION

The fish keratocyte is a fast crawling epithelial cell, first described by Goodrich (1924) and more recently adopted as a useful model system for studying cell motility (Euteneuer and Schliwa, 1984; Sheetz *et al.*, 1989; Theriot and Mitchison, 1991; Lee *et al.*, 1994; Small *e. al.*, 1995). This article describes a procedure for the preparation of primary cultures of these cells for microscopic observation, as well as techniques for their fixation and staining.

II. MATERIALS

Plastic petri dishes, 30 and 90 mm diameter; culture chambers (50-mm petri dishes with a 20-mm hole drilled through the center, see Fig. 1); round coverslips, 30 or 15 mm diameter washed extensively in distilled water and dried on lint free paper (velin tissue, Taab Laboratories, No. V056); glass slides (76 × 26); 5 ml syringe filled with Dow Corning high vacuum silicon grease; Whatman filter paper No. 1, 90 mm; Parafilm; forceps; freshly killed trou (*Salmo alpinus*); Dulbecco's modified Eagles medium (DMEM), GIBCO Cat. No. 042-02501; Sigma type V collagenase, (Cat. No. C-9263; chicken serum (Sigma Cat. No. C-5405) dissecting microscope; paraformaldehyde; Triton X-100; rhodamine-labeled phalloidin (Sigma Cat. No. P 1951); and inverted microscope equipped for phase-contrast and fluorescence microsopy.

III. PROCEDURES

Solutions

1. *Steinberg medium (100 ml of 10× stock):* 3.04 g NaCl, 0.07 g Ca(NO$_3$)$_2$ · 4H$_2$O, 0.04$^-$ g KCl, and 0.197 g MgSO$_4$ · 6H$_2$O.

2. *Fish Ringers (500 ml of 5× stock):* 6.54 g NaCl, 0.15 g KCl, 0.20 g NaHCO$_3$, 0.15 g CaCl$_2$ · 2H$_2$O, and 0.12 g Tris.

3. *Start Medium:* 20 ml of DMEM, 14 ml of 5× fish Ringers, 2 ml of chicken serum, 1 ml of 10× Steinberg medium, and 1 ml of 1 *M* PIPES, pH 7.0; make up to 100 ml with distilled water and store frozen as 5-ml aliquots.

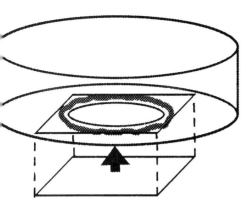

4. *Running medium:* 20 ml of 5× fish Ringers, 1 ml of 10× Steinberg medium, and 1 ll of 1 *M* PIPES, pH 7.0; make up to 100 ml with distilled water.

5. *Collagenase:* Dissolve collagenase V at 0.5 mg/ml in fish Ringers; adjust pH to around .5 and freeze in 1-ml aliquots. Dilute one part collagenase with two parts start medium mmediately before use.

Preparing the Primary Culture

ption 1: Scale Explant

teps

acrifice fish, place in aluminum foil or plastic wrap, and rinse generously with DMEM. .eep on ice.

1. Under the dissecting microscope, harvest scales from the general area along the middle dge behind the gills; these scales are typically the largest. Grasp scales at the side with orceps to avoid damaging the tissue at the scale tip. Notice that there is a metallic sheen n the bottom side of the scale (facing toward the fish) and that the tissue is attached to the op, outer scale surface. Place scales on ice in a 30-mm petri dish containing a few milliliters f DMEM.

2. In order to help loosen debris and slime from the scales, shake the dish gently on ice n an orbital or other suitable shaker for 10–20 min.

3. Arrange 30-mm coverslips in inverted 90-mm petri dishes, five to seven coverslips/ ish.

4. Cut the end off of a 1-ml pipette tip and rinse scales thoroughly by repeatedly drawing 1em into the tip and expelling. Drain DMEM at least once and replace with fresh DMEM r start medium. After the final wash step, transfer scales to start medium on ice.

5. Place a small drop of start medium in the middle of each coverslip within the petri ish. Remove scales from start medium under the dissecting microscope and determine 'hich side the tissue is on. Carefully place two scales in each drop with the lump of tissue acing up, metallic sheen down. This is important for proper adhesion: the scale will adhere oward the side with the metallic sheen.

6. Place a round 15-mm coverslip on top of each drop to form a sandwich. There should e sufficient medium between the coverslips so that the tissue will not dry out overnight, ut not so much that the scales float freely, out of contact with the lower coverslip. To check nis, inspect the coverslip sandwiches on an inverted phase-contrast microscope under low 1agnification (16×). Tap the stage lightly: scales that are floating freely will vibrate indepen- ently of the stage and bottom coverslip.

7. Tear a piece of Whatman No. 1 filter paper into four equal pieces. Place one piece

on the inside of the top of the dish, dampen with DMEM, and seal the dish with Parafilm
Incubate at 4°C overnight.

8. Inspect dishes the following morning on an inverted microscope and locate coverslip
on which the cells have spread to form a monolayer halo around the scale (Fig. 2).

9. Prepare 500 μl of the start–collagenase mixture per coverslip.

10. Grease a ring around the hole in the bottom of each culture chamber (Fig. 1). La
selected coverslips out on a paper towel. Center culture chambers over the top coverslip am
press the greased ring onto the bottom coverslip to form a water-tight seal.

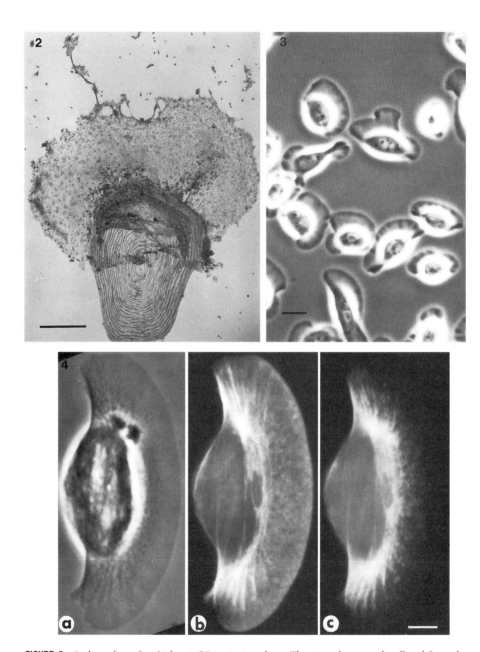

FIGURE 2 Scale explant after 24 hr at 4°C in start medium. The tissue has spread well and formed a monolaye
halo about the scale. Some individual cells (arrowheads) and groups of cells (arrows) have already broken free fro
the explant. Bar: 0.5 mm.

FIGURE 3 Individual keratocytes, released from the tissue monolayer through collagenase treatment and engage
in rapid locomotion. Bar: 10 μm.

FIGURE 4 Phase-contrast (a) and fluorescence microscopy images (b and c) of a fish epidermal keratocyte doub
labeled for actin (b, rhodamine–phalloidin) and muscle type myosin II (c, antibody to smooth muscle myosin S-1
Bar: 5 μm.

11. Carefully add 500 μl collagenase mix to each chamber so that the top coverslips float up on top of the medium. It may be helpful to first place a drop of collagenase mix at the side of the top coverslip and wait for the liquid to be drawn underneath. Remove top coverslip. Incubate for 15 min at room temperature. This step facilitates the dissociation of the tissue into single cells.

12. Add 5 ml running buffer per chamber. Incubate in fridge or at room temperature for at least 2 hr to allow adequate cell spreading before observation.

Option 2: Explant from Isolated Tissue

Steps

1. Follow steps 1, 2, and 4 listed in Option 1.

2. Drain DMEM from the petri dish containing the scales and replace with 1 ml 0.5 mg/ml collagenase in fish Ringers. Incubate for 7 min at room temperature. Rinse once with DMEM and return scales to start medium. Leave scales to "mature" for around 1 hr at 20°C, after which time the tissue can be more readily removed from the scale. Return the petri dish to an ice box.

3. Under a dissecting microscope, transfer a scale to a drop of start medium in a 90-mm petri dish. Using the flat edge of a fine hypodermic needle and a pin, scrape the translucent epithelial tissue from the scale surface. Wash the tissue by transferring it through one or two fresh drops of start medium and then into a further drop on a water-washed 30-mm coverslip. Place a 15-mm coverslip over the tissue and proceed according to steps 6 and 7 listed in Option 1, but not at 4°C.

4. Incubate petri dish at room temperature (below 22°C). Cells will normally move out spontaneously from these explants within a few hours and can be used on the same day. Accelerated outgrowth may be achieved by treating extended monolayers of cells with 0.2% trypsin (in Ca- and Mg-free Hanks solution) for 1 min or as long as necessary to disrupt cell–cell contacts (Fig. 3).

B. Fixation and Staining of Keratocytes

Solutions

1. *Phosphate-buffered saline (PBS) (1 liter):* 0.36 g NaH_2PO_4, 1.1 g Na_2HPO_4, and 9.0 g NaCl, pH 7.2–7.4.

2. *3% paraformaldehyde (PFA/PBS):* Prepare by heating 100 ml PBS to 80°C. While stirring, add 3.0 g PFA. After it dissolves, cool and store as frozen aliquots.

3. *0.5% Triton X-100 in PBS.*

4. *Phalloidin:* Dilute rhodamine–phalloidin to 0.1 μg/ml in PBS immediately prior to use.

5. *0.1% glutaraldehyde in PBS:* This is optional, see Section IV.

Steps

1. Carefully remove coverslips from their culture chambers. Wipe away excess vacuum grease with a piece of tissue paper and arrange the coverslips with the cells up in a 90-mm petri dish.

2. Flood the dish with 3% PFA/PBS and incubate for 20 min at room temperature.

3. Fill a 50-ml beaker and another 90-mm petri dish with PBS. Remove coverslips one by one from the 3% PFA/PBS, rinse by immersing in the beaker filled with PBS, and transfer to the petri dish with PBS.

4. Fill a 25-ml beaker with 0.5% Triton/PBS. Extract each coverslip individually by holding it with forceps in 0.5% Triton/PBS for 45 sec. Rinse in PBS and transfer to a fresh petri dish filled with PBS.

5. Drain the PBS and add rhodmaine–phalloidin in PBS. Cover the dish, seal with Parafilm, and incubate overnight at 4°C. Alternatively, process cells for immunofluorescence microscopy as described by Mies *et al.* (volume 2, "Multiple Immunofluorescence Microscopy of the Cytoskeleton") (Fig. 4).

6. Mount coverslips cell side down in a drop of Gelvatol on a standard (76 × 26 mm) glass slide and allow to dry.

IV. COMMENTS

Once the tissue has spread around the scale (Fig. 2), a generous amount of start medium may be added and the coverslip sandwiches may be kept in the fridge for several days prior to treatment with collagenase. Because cell velocity is proportional to temperature, spreading can be encouraged by incubation at room temperature. However, trout cells are sensitive to warmer temperatures and prolonged exposure to temperatures above 20°C should be avoided. Cells can also be fixed and stained while the coverslips are still mounted in their chamber by adding solutions directly into the chamber. The method described here is useful for preparing primary cultures from a wide variety of fish types (goldfish, guppy, various trout, etc.). Cells from different sources differ subtly in morphology, cytoskeletal organization, and speed and nature of motility. Cells may also be grown on formvar-coated electron microscope grids (nickel, 150 mesh) by a simple modification of this method (Small *et al.*, 1995). Paraformaldehyde fixation may cause membrane blebbing and result in irregular fixation of the cytoskeleton. These effects can be minimalized by fixing cells in 0.1% glutaraldehyde (25% EM grade, Agar Sceintific Ltd.) in PBS for 30 sec prior to PFA fixation.

V. PITFALLS

The amount of medium in the coverslip sandwiches is of crucial importance. If the scale is not pressed against the lower coverslip, the cells will not attach and spread during the overnight incubation. If this occurs the cells are most likely still viable. Drain some of the start medium from the sandwich using a pipette or piece of filter paper and incubate at room temperature.

References

Anderson, K. I., Wang, Y.-L., and Small, J. V. (1996). Coordination of protrusion and translocation of the keratocyte involves rolling of the cell body, *J. Cell Biol.* **134**, 1209–1218.

Euteneuer, U., and Schliwa, M. (1984). Persistent, directional motility of cells and cytoplasmic fragments in the absence of microtubules. *Nature* **310**, 58–61.

Goodrich, H. B. (1924). Cell behavior in tissue cultures. *Biol. Bull. (Woods Hole)* **46**, 252–262.

Lee, J., Leonard, M., Oliver, T., Ishihara, A., and Jakobson, K. (1994). Traction forces generated by locomoting keratocytes. *J. Cell Biol.* **127**, 1957–1964.

Sheetz, M. P., Turney, S., Qian, H., and Elson, E. L. (1989). Nanometer-level analysis demonstrates that lipid flow does not drive membrane glycoprotein movements. *Nature* **340**, 284–288.

Small, J. V., Herzog, M., and Anderson, K. (1995). Actin filament organization in the fish keratocyte lamellipodium. *J. Cell Biol.* **129**, 1275–1286.

Theriot, J. A., and Mitchison, T. J. (1991). Actin microfilament dynamics in locomoting cells. *Nature* **352**, 126–131.

PART 5

Antibodies

Production of Antibodies

Production of Polyclonal Antibodies in Rabbits

Christian Huet

I. INTRODUCTION

The major problem encountered in the production of antibodies is the amount of antigen available. This amount determines the method of injection. Subcutaneous injection along the spine can be used when 1 mg of antigen is available. Usually, however, the amount of purified antigen is as small as 100 μg. In this case the best location for injection is the popliteal lymph nodes in the hind legs. Even a smaller amount of antigens can be efficient in raising antibodies but probably with lower serum titers.

On the basis of 200 μg of antigen, the method of immunization to be described here implies injections in both lymph nodes and subcutaneously along the spine. Usually the lymph nodes are buried in fat and are rather difficult to isolate; that is why young animals should be used. To aid in their location, we inject Evans blue into the two hind paws of the rabbit 2 hr before injecting the antigen. The dye is carried via the lymphatic vessels to the popliteal lymph nodes, making them more clearly visible.

About a third of the rabbits have endogenous antikeratin antibodies, which can give a false reaction in immunocytochemistry. It has been suggested that the false reaction is due to chicken feathers (composed largely of keratin) used as a filler in commercial rabbit chow. This emphasizes the importance of taking preimmune serum from a rabbit that will be used to prepare antibodies against a particular antigen.

II. MATERIALS AND INSTRUMENTATION

A. Rabbits

Animals weighing approximately 2–3 kg should be chosen. White rabbits are preferable because the node, after injection of the blue dye, is easier to detect through the leg skin.

B. Syringes

1-ml (Cat. No. BS-01T), 2-ml (Cat. No. BS-H2S), 10-ml (Cat. No. BS-10ES), and 50-ml (Cat. No. BS50ES) plastic syringes from Terumo Europe NV

2 × 3-ml Multifit syringes with Luer-lok from Becton Dickinson (Cat. No. SMBDL 1020F)

0.45 × 12-mm (Cat. No. NN-2613R), 0.6 × 25-mm (Cat. No. NN-2325R), and 1.2 × 40-mm (Cat. No. NN-1838R) nonreusable needles are from Terumo Europe NV

Two-headed 18-gauge needles (adjusted as shown in Fig. 1 and used to bridge two glass syringes when preparing the antigen–adjuvant emulsion) are usually obtained from the local machine shop or from Aubry

C. Tools

1 pair of scissors with straight heavy and blunt blades

1 pair of microscissors with sharp delicate blades

1 pair of dissection forceps, straight

1 pair of dissection forceps, curved

Clamp "mosquito"

1 MikRon Autoclip Applier (9 mm) and MikRon wound clips (Cat. No. 7631, distributed by Clay Adams–Becton Dickinson)

D. Chemicals

Complete Freund's adjuvant (Cat. No. 0638-60-7) and incomplete Freund's adjuvant (Cat. No. 0639-60-6) from Difco Laboratories

Instamed phosphate-buffered saline (PBS, Cat. No. L 182-10) prepared by Seromed

Evans blue (Cat. No. E-2129), sodium azide (Cat. No. S-2002), sodium chloride (Cat. No. S-9625), and lysine (Cat. No. L-6001) from Sigma

Glutaraldehyde EM grade (Cat. No. G003) from TAAB

Hemocyanin (Cat. No. 37 48 05) from Calbiochem

Fentanyl (which is a regulated substance) from Janssen

Sephadex G-10 and G-50 from Pharmacia

E. Miscellaneous Items

50-ml Falcon plastic tubes (Cat. No. 2098) from Becton Dickinson

Dialysis tubing

70% alcohol in a squeeze bottle and in a beaker large enough to submerge instruments

0.22-μm (Cat. No. SLGV 025 BS) and 0.45-μm (Cat. No. SL HV 025 LS) filters from Millipore

Mortar (Cat. No. 01 447 341) from Prolabo

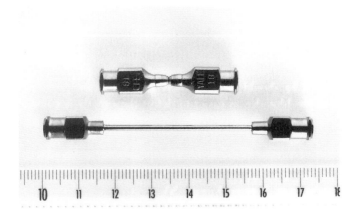

FIGURE 1   Two two-headed 18-gauge needles.

Rabbit restraining box, razor blades, lab coat and kitchen towel, animal grooming shears, gauze, liquid nitrogen, and ice bucket

II. PROCEDURES

Rabbits should be kept in the animal house for at least a week before surgery to ensure that healthy rabbits have been obtained and that they are adjusting properly to their new conditions of living.

Tools can be either autoclaved (dry) or soaked in 70% alcohol to sterilize.

A. Injection of Dye to Display Popliteal Lymph Node

Solutions

1. *Sterile 0.15 M NaCl solution:* To make 100 ml, dissolve 0.9 g of NaCl in distilled water and adjust to a total volume of 100 ml in a cylinder. Filter on a 0.22-μm Millipore filter adjusted on a 50-ml syringe and store in a sterile 50-ml Falcon tube.

2. *2.5% Evans blue solution:* Add 1.25 g Evans blue powder to 50 ml of saline solution and filter on 0.45-μm Millipore filter. Store at 4°C.

Materials

1. Needles, 0.6×2 mm
2. Syringe, 2 ml
3. Lab coat

Steps

1. Spread out a lab coat (with a knotted sleeve) on a bench or table. Bring the rabbit from its cage, and place it on the lab coat. Put its head in the knotted sleeve of the lab coat to immobilized it as much as possible. One person should hold the two hind legs while another injects the dye.

2. Spread the two middle toes apart, and squirt the area with 70% alcohol from a squeeze bottle. This has the advantage of cleaning the area and also clumps the hairs together, rendering the injection easier.

3. Insert the needle into the skin between the two toes, pass it subcutaneously for about a centimeter, and inject 0.2–0.5 ml of the dye solution. Repeat for the other hind paw. Hold the rabbit very firmly when doing these two painful injections because the animals usually react very nervously.

Dye should be injected at least 2 hr before injection of antigen into the lymph node (the antigen solution can be prepared in the meantime). The staining remains visible for a couple of days. One should not be surprised to find, later, a blue bunny exhibiting pale blue ears in the cage (however, we never found him in a hurry shaking its pocketwatch).

B. Preparation of the Antigen Solution

The antigen should be in solution in a sterile and nonimmunogenic solvent such as saline or PBS (100–500 μg/ml). About 1 ml of antigen solution will be made into an emulsion with an equal amount of complete Freund's adjuvant. Then, 2 ml of emulsion will be injected into the rabbit, about 1 ml into the two popliteal lymph nodes and the rest subcutaneously along the spine.

1. Solution of a Proteic Antigen

Solutions

1. *0.15 M NaCl or PBS:* 0.15 M NaCl in 10 mM NaPO$_4$ in sterile water. Prepare saline as directed in Section A.

2. *PBS 10× stock solution:* Dissolve one flask of Seromed PBS powder in 1 liter distilled water.

Step

1. Prepare about 1 ml of purified antigen in a saline solution or in PBS (100–500 μg/ml final).

2. Antigen in Bands on Nitrocellulose Blots

Very often the pure antigen is not isolated but is visualized on nitrocellulose blots in which the protein bands are stained with Ponceau S. The frozen isolated band is pulverized with a mortar and pestle cooled down with liquid nitrogen.

Solution

1. *0.15 M NaCl or PBS:* 0.15 M NaCl in 10 mM NaPO$_4$ in sterile water. See Section B1.

Materials

1. Mortar and pestle

2. Liquid nitrogen

3. Spatula or scalpel

4. Glass syringe, 3 ml

5. Needle: 1.2 × 40 mm

Steps

1. Pour liquid nitrogen into the mortar (with the pestle in it) and let it cool.

2. Cut the nitrocellulose band into small pieces with scissors, and let the pieces fall into the liquid nitrogen in the mortar.

3. When the nitrogen is essentially gone, though the surfaces are still near liquid nitrogen temperature, grind vigorously with the pestle to pulverize the nitrocellulose.

4. If there are several bands of sample antigen, then scrape the powder from the surfaces of the mortar (using a scalpel or a spatula) and transfer into a 5-ml glass vial (plastic snap cap). After adding 1 ml of saline solution (see above), vigorously mix with a Vortex mixer. Take up 1 ml antigen solution from the vial into a 3-ml glass syringe with a 1.2 × 40-mm needle.

5. If the antigen band is sparse, then warm the mortar above 0°C and add 1 ml of saline solution into the mortar. Take up 1 ml of antigen suspension directly into a 3-ml glass syringe fitted with a 1.2 × 40-mm needle.

3. Synthetic Peptides Coupled to KLH as Antigens

Synthetic peptides may be a good source of antigens. A good size is a sequence of 10–15 amino acids. These amino antigens must be coupled to a carrier. Keyhold limpet hemocyanin (KLH) is most commonly used. The peptide is covalently crosslinked to KLH by adding glutaraldehyde.

Solutions

1. *PBS:* See Section B1.
2. *5% glutaraldehyde:* Dilute the 25% glutaraldehyde stock to 5% in water.
3. *1 M lysine:* Dissolve 1 g of lysine in 10 ml water.

Materials

1. Sephadex G-10 column
2. Sephadex G-50 column or dialysis tubing

Steps

1. Desalt peptide over a Sephadex G-10 column (column volume should be about 9 × the sample volume) in PBS buffer.

2. Desalt KLH either by dialyzing or by running through a Sephadex G-50 column in PBS buffer. Adjust concentration to 5–10 mg/ml (OD_{280} = 1).

3. Add 20–40 molar excess of peptide to the KLH solution. (The molecular weight of KLH is considered to be 100,000.)

4. At 30-min intervals, add five aliquots of 5% glutaraldehyde solution (MW = 100). Final concentration is 10 mM. Leave the reaction to complete overnight with agitation in the cold.

5. Block the potential free unreacted aldehyde groups by adding 1 M lysine to a final concentration of 25 mM.

6. To immunize, use the complete mixture, including occasional aggregates.

C. Producing the Emulsion

Solutions

Use one of the antigen solutions as prepared in Section B.

Tools

1. Two 3-ml glass syringes
2. Two-headed needle
3. Needles, 1.2 × 40, 0.6 × 25, and 0.45 × 12 mm

Steps

1. Take up 1 ml of complete Freund's adjuvant in a 3-ml glass syringe with a 1.2 × 40-mm needle.

2. Remove the needle and replace it with a two-headed needle. Push the syringe plunger until adjuvant appears at the open end of the needle.

3. Attach the syringe containing the antigen solution (100 μg in 1 ml) on that end. Make sure that no air is trapped and that both ends are tightly fitted. Hold one syringe in each hand and run the solutions back and forth between the two syringes with your thumbs. Continue until the emulsion becomes rather firm, then push two or three more times and stop (if you continue the emulsion may become too stiff).

4. When a stable emulsion is achieved, push the emulsion into one of the two glass syringes and remove the other one. Place a 0.45 × 12-mm needle on the syringe. Use a 0.6 × 25-mm needle to avoid clogging if the antigen is on powdered nitrocellulose. Remove any bubbles by tapping the syringe on the edge of a bench until the bubbles rise, then expel them.

5. Push the plunger until the emulsion begins to emerge at the tip of the needle, then put the plastic cover on the needle and put the syringe in an ice bucket, oriented vertically with the covered needle and emulsion in the ice. Keeping the syringe on ice should prevent phase separation of the emulsion.

D. Anesthetizing and Shaving the Rabbit

Solution

1. *Fentanyl citrate (50 μg/ml) stock solution.*

Tools

1. Plastic syringe, 10 ml
2. Needle, 0.6 × 25 mm
3. Animal grooming shear
4. Lab coat and towel

Steps

1. Weigh the rabbit.

2. Place the rabbit on a towel on a table and inject the anesthetic intramuscularly into the leg muscle. (Do not let rabbits stand on a slippery surface; they hate it. Put a piece of cloth under their paws.) It is not necessary to cover the rabbit's head. With a 10-ml syringe and a 0.6 × 25-mm needle, inject about 5 ml/3 kg body weight of the anesthetic: First inject into each leg three-fourths of the total amount and then the remainder about 10–15 min later.

3. Wait about 20 min to see the effect. The rabbit is somewhat calmed by the anesthesia (but not necessarily unconscious).

4. Shave (against the grain of the fur) a strip along the back (roughly 5 cm wide). Gather the shavings and throw them in a wastepaper basket (do not wait until the end of the procedure).

5. Shave the backs of both legs, extending from midthigh to the heel, and passing roughly 3 cm forward on the outside of the leg.

E. Popliteal Lymph Node Isolation

Solution

1. *70% alcohol*

Tools

1. Surgical tools (autoclaved or imersed in alcohol)
2. Clean kitchen towel
3. Gauze

Steps

1. The rabbit should be unconscious from the anesthesia, although there may still be some reflex movements.

2. Lay it on the clean kitchen towel (the lab coat used before is presumably dirty and full of fur hair).

3. To find the node on the left leg, grasp the knee with the left hand and feel with the thumb the area where the node would be expected. [The node is at the back (posterior) of

the bend of the knee, somewhat to the outward (lateral) side.] You will feel a mass about 1 cm in diameter; it may appear bluish through the skin and moves as you manipulate it with your thumb.

4. Apply 70% alcohol (from a squeeze bottle) to the general area.

5. Use small surgical scissors to make an incision in the skin approximately 2 cm long over the node and arranged roughly parallel to the axis of the leg. Carefully make the incision deeper until the node (appearing blue because of the previous dye injection) bulges out, under pressure from the side by your thumb (Fig. 2).

6. Isolate the lymph node using the bent forceps to dilacerate the connective tissue. (This is safer than using sharp tools because of the proximity of the artery.) In some cases the node is just below the artery, which makes access to the node difficult. The node often appears smaller than might be expected.

F. Injecting the Antigen

Solution

1. *Antigen emulsion:* Prepare as described in Section B and keep in the glass syringe.

Materials

Autoclip Applier and wound clips (immersed in 70% alcohol)

The immunization schedule is given in Table I.

Steps

1. Remove the syringe containing the emulsion from the ice bucket, and take the plastic cover off the needle.

2. Insert the needle into the node, reaching approximately the middle. Carefully expel the emulsion until the node seems full. You will see white spots here and there as the node fills, and some may leak out (especially if the needle goes too far or is otherwise ill-placed). Each node can take approximately 0.3–0.5 ml.

3. Close the wound with the surgical clip applier and spray 70% alcohol in the surrounding area. Repeat for the popliteal node in the other leg.

4. Apply 70% alcohol to the shaved portion of the back, using a squeeze bottle.

5. Inject the remaining emulsion (approximately 1 ml) subcutaneously into numerous small depots arranged along the spine. To do so, insert the needle through the skin and pass it horizontally under the skin for a short distance. Then expel the emulsion to make a small bump roughly 0.5 cm in diameter.

6. The rabbit can now be returned to its cage.

G. Boosters

Solutions

1. *Incomplete Freund adjuvant.*

2. *Stock antigen solution in saline or PBS.*

Materials

1. Needles, 0.45 × 12 and 0.6 × 25 mm

2. Plastic syringes, 1 ml

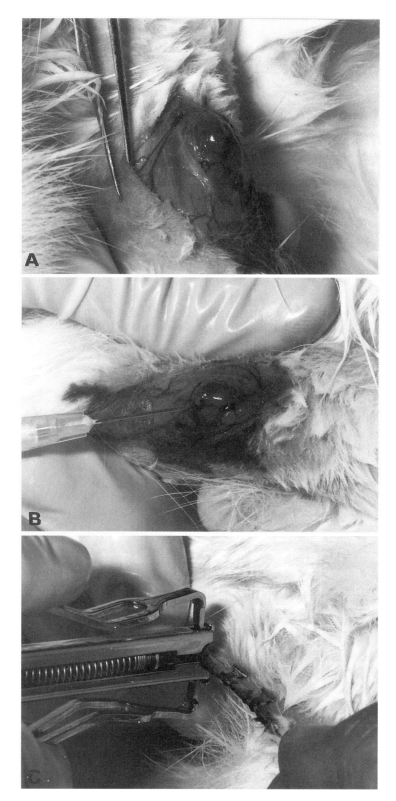

FIGURE 2   (**A**) Exposing the lymph node. (**B**) Injecting into the lymph node. (**C**) Stapling the wound.

Steps

1. *Three weeks later,* detect the subscapular cavity by palpation between the spine and the scapulae. Inject approximately 100 μg of protein mixed with an equal volume of incomplete Freund's adjuvant (1–2 ml can be injected). Use a 0.45 × 12-mm needle on the syringe.

TABLE I Immunization and Bleeding Schedule

Immunization		
Day	Injection mode	Amount
−10	*Rabbit in the animal house*	
0	Lymph node	20–100 μg + complete Freund's adjuvant
21	Subscapular	40–100 μg + incomplete Freund's adjuvant
28	Intramuscular	50–100 μg in PBS
29	Intravenous	20–50 μg in PBS

	Bleedings	
	Day	Bleeding
	36	Fasting
	37	First bleeding
	41	Fasting
	42	Second bleeding
	48	Fasting
	49	Third bleeding
	55	Fasting
	56	Fourth bleeding

2 weeks rest

Boosters		
Day	Injection mode	Amount
0	Subscapular	20–50 μg in PBS
8–10	Intramuscular	20 μg in PBS
9–11	Intravenous	20 μg in PBS

Bleedings

Same schedule as above

2. *Seven to 10 days later,* inject protein (approximately 50 μg) in 1 ml saline solution (*no adjuvant*) into the gluteus maximus (leg muscle). Use a 0.45 × 12-mm needle.

3. *One day later,* inject intravenously (into the lateral ear vein) approximately 50 μg (0.5 to 1 ml) of protein (*no adjuvant*). Use a 0.6 × 25-mm needle

H. Bleedings

Ten days after the last booster, obtain 40–50 ml from the central artery of the dorsal face of the ear. To clear the blood of lipids, it is advisable to fast the rabbit overnight before the procedure. The bleeding schedule is summarized in Table I.

Materials

1. Restraining box

2. Razor blades

3. Needles, 1.2 × 40 mm

4. Falcon tubes, 50 ml

5. Clamp "mosquito"

6. Gauze

Steps

1. Place the rabbit in the restraining box on the bench.

2. With a razor blade shave about 1 cm of ear skin above the central artery of the ear. Rub the skin on that area to obtain clearly visible vasodilation (do not use solvent for this purpose as it severely damages the skin and renders further bleedings difficult).

3. Insert a large (1.2 × 40-mm) needle about 0.5 cm into the skin and parallel to the artery. To insert the needle into the blood vessel, push it to the right or to the left with your thumb, depending on the ear and the place of insertion you have chosen. This method, although a little more difficult than direct insertion into the artery, has the advantage of maintaining the needle by the skin during the bleeding and of preventing its rejection due to struggling by the rabbit.

4. Keep rubbing the ear base with one finger during the bleeding to keep the artery fully dilated while holding the needle and the plastic tube with the other hand. Collect the blood (one can obtain 40–45 ml) into a 50-ml plastic tube.

5. Remove the needle and stop blood by firmly holding gauze on the wound (use a mosquito clamp). Make absolutely sure that bleeding completely stops before returning the rabbit to its cage.

6. Repeat the bleeding three times at 1-week intervals.

I. Serum Storage

Solution

1. *1% NaN$_3$ stock:* Dissolve 0.5 g sodium azide in 50 ml distilled water. Store at 4°C.

Materials

1. Wooden sticks or glass rods

2. Falcon tubes, 50 ml

3. Plastic blood containers, 25 ml

Steps

1. Collect the blood at room temperature and leave on the bench for about 60 min.

2. Carefully detach the clot from the sides of the plastic tube with a wooden stick or a glass rod. The clot is fully detached from the walls when it turns freely in the tube.

3. Let it stand overnight at 4°C to allow complete contraction of the clot.

4. Pour the serum out of the tube into another Falcon tube and spin it for 10 min at 3000 rpm.

5. Store the serum in 25-ml plastic containers in the fridge at 4°C (never in a freezer). Add sodium azide (0.1% final) to prevent contaminant growth.

6. When used, keep the serum container continuously on an ice bucket and never warm up to prevent growth of contaminants. From time to time sodium azide may be added.

IV. PITFALLS

1. The nitrocellulose powder must be very fine. If not fine enough it will clot the injection needle. The size of the particles is also important for proper uptake by the macrophages and for proper processing and presentation of the antigen to the lymphocytes. If the powder is not fine enough, the solution can be poured back into the mortar, frozen with liquid nitrogen, and ground again with the pestle. After thawing, it can be taken back into the syringe.

2. Unsuccessful coupling is generally due to the presence of amino groups in the solutions. Normally it comes from ammonium sulfate used to precipitate the proteins or peptide or from the use of Tris buffer. Desalt or dialyze very carefully. Also make sure that no amino groups are present in the water used.

3. If the emulsion is unstable, it will separate within a few minutes into aqueous and lipid phases (e.g., if the antigen solution contains a little detergent). In that case you can leave the freshly prepared emulsion for about 5 min in a −20°C freezer. Mix again by pushing the pestle back and forth before injecting.

4. Tetanization of the rabbit may occur; do not panic. The animal can easily recover with strong and prolonged cardiac massage. Inject the anesthetic in several aliquots (normally two sets of injections at 10 to 15-min intervals are sufficient). If tetanization occurs during the surgery or while you are injecting your most precious and rare antigen, simply place the tools on the bench and start to activate the cardiac reflexes by pressing on the thorax. During the massage you will be able to feel the heart beats and monitor the rabbit's recovery.

5. Very often, insertion of a needle into the ear artery produces severe vasoconstriction and the artery turns white. Proceed by rubbing the ear base of the rabbit until blood flow resumes.

Suggested Reading

Avrameas, S., and Terninck, T. (1969). The cross linking of proteins with glutaraldehyde and its use for the preparation of immunoadsorbents. *Immunocytochemistry* **6**, 53–66.
Goudie, R. B., Horne, C. H. W., and Wilkison, P. A. (1966). A simple method for producing antibodies specific to a single selected diffusible antigen. *Lancet* **2**, 1224.
Kreis, T. E. (1986). Microinjected antibodies against the cytoplasmic domain of vesicular stomatitis virus glycoprotein block its transport to the cell surface. *EMBO J.* **5**, 931–941.
Louvard, D. (1987). Production of polyclonal antibodies. *In* "EMBO Practical Course: Antibodies in Cell Biology." EMBO, Heidelberg.

Production of Mouse Monoclonal Antibodies

Ariana Celis, Kurt Dejgaard, and Julio E. Celis

I. INTRODUCTION

The aim of this article is to illustrate the various steps involved in the production of mouse monoclonal antibodies (Köhler and Milstein, 1975; Galfre *et al.*, 1977; Goding, 1987; Milstein, 1986; Harlow and Lane, 1988; Liddel and Cryer, 1991; Asai, 1993). First, the mouse is immunized with an appropriate antigen and then cell lines are created *in vitro* by fusing spleen lymphocytes with a myeloma cell line (e.g., P3-X63-Ag8, HGPRT−, immortal) that confer immortality to the somatic cell hybrid (hybridoma). Spleen and myeloma cells are fused and plated in a special medium containing hypoxanthine, aminopterin, and thymidine (HAT) that allows the growth of some of the hybrids but not of the parent cells. Selected hybridomas that are capable of endless reproduction produce only one kind of antibody that is specific for one kind of antigen.

II. MATERIALS AND INSTRUMENTATION

Freund's complete and incomplete adjuvants are from the Statens Serum Institute. Polyethylene glycol (M_r 6000) is from Koch–Light, and the 50× stock HAT solution (Cat. No. 03-080-1B) is from Biological Industries. The fluid plaster is from Aerosols.

The 96-well plates (Cat. No. 655180) are from Greiner and the 1-ml sterile plastic syringes (Cat. No. 308400) are from Becton Dickinson, Dublin. The gauze (Medical gauze EUR.PH, 100% cotton, 28 threads/cm^2) is from Smith & Nephew, the Gilson tips (C20, Cat. No. G 23800) are from Gilson, and the 20G1 sterile needles (Cat. No. 304827450) are from Becton-Dickinson Fabersanitas.

All other reagents, materials, and equipment are as described in the article "General Procedures for Tissue Culture" by Celis and Celis in Volume 1.

III. PROCEDURES

A. Immunization

Solutions

1. *Hanks' buffered saline solution (HBSS) without Ca^{2+} and Mg^{2+}, 10× stock solution:* To make 1 liter, weigh 4 g of KCl, 0.6 g of KH_2PO_4, 80 g of NaCl, and 0.621 g of Na_2H-$PO_4 \cdot 2H_2O$. Complete to 1 liter with distilled water.

2. *1× HBSS without Ca²⁺ and Mg²⁺:* To make 1 liter of 1× solution, take 100 ml of the 10× stock and complete to 1 liter with distilled water. Autoclave and keep at 4°C.

3. *Freund's complete adjuvant:* Sterile. Keep at 4°C.

4. *Freund's incomplete adjuvant:* Sterile. Keep at 4°C.

5. *Fluid plaster.*

6. *70% ethanol.*

Steps

1. Suspend the antigen in 0.25 ml of HBSS and mix with an equal volume of complete Freund's adjuvant. Make an emulsion by aspirating the solution up and down with the aid of a 1-ml syringe connected to a 20G1 needle. If you have enough antigen, emulsify the solution as described by Huet (1993) (see Volume 2, Huet, "Production of Polyclonal Antibodies in Rabbits" for additional information). The Freund's complete adjuvant should be handled with care as it contains killed *Mycobacterium tuberculosis* bacteria. For soluble proteins we routinely use 20–50 μg of protein for injection. See also the article by Huet (in this volume) concerning antigens contained in acrylamide gels or nitrocellulose blots.

2. Select an adult Balb/C mouse, squirt the area with 70% ethanol, and inject 0.5 ml of the emulsion intraperitoneally and subcutaneously.

3. Four weeks after the first injection, inject 0.5 ml of the antigen suspended in HBSS and mix with an equal volume of Freund's incomplete adjuvant. Emulsify before injection. On week 5, cut a little piece of the mouse tail with a sterile blade and collect 30–50 μl of blood in a Gilson tip closed at the pointed end by flaming. Massage the tail to increase the blood flow. Stop the bleeding by pressing the tail firmly for a few seconds with tissue paper. Spray with a liquid plaster.

4. Leave the blood to coagulate and test the plasma (diluted 1:40 in HBSS).

5. Repeat injections every 2 weeks until the result of the immunization is satisfactory as judged by immunofluorescence and or ELISA tests.

6. Usually the spleen is removed for fusion 4 to 5 days after the boost injection. To boost, inject the soluble protein (in the absence of the adjuvant) into the tail vein of the mouse.

B. Preparation of Mouse Myeloma Cells (P3-X63-Ag8)

Solution

1. *Complete Dulbecco's modified Eagle's medium (DMEM):* DMEM medium supplemented with antibiotics (penicillin, 100 U/ml; streptomycin, 100 μg/ml), glutamine (2 m*M*) and 10% fetal calf serum (FCS). To make 500 ml, mix 440 ml of DMEM medium, 5 ml of a 100× stock solution of glutamine, 5 ml of a 100× stock solution of penicillin–streptomycin, and 50 ml of FCS.

Steps

1. Grow P3-X63-Ag8 cells in several 75-cm² flasks until they reach a density of 5×10^5 cells/ml (see Volume 1, Celis and Celis, "General Procedures for Tissue Culture" for additional information).

2. The night before the fusion, remove the culture supernatant by aspiration and add fresh DMEM medium.

3. The day of the fusion, centrifuge the cells at 300–400 g in a bench centrifuge for 3 min and wash two times with DMEM without FCS. Resuspend at a final concentration of 2×10^6 cells/ml in DMEM without FCS and keep in the CO_2 incubator until use. You need a total of 5 ml (1×10^7 cells).

C. Preparation of Mouse Macrophages

Solution

1. *DMEM without FCS:* As in Section B but add 50 ml of sterile water instead of FCS.

Steps

1. Sacrifice an adult Balb/C mouse by cervical dislocation.

2. Disinfect the abdominal region with 70% ethanol and inject intraperitoneally 5 ml of DMEM without FCS using a 5-ml syringe connected to a 20G1 needle. Carry out the operation in the sterile hood.

3. Rock the mouse (massage the abdomen) and slowly aspirate the injected medium (Fig. 1A). Observe under a microscope and check for the presence of blood and feces. The medium should look clear.

4. Centrifuge the suspension in a bench centrifuge at 300–400 g and resuspend in 10 ml of DMEM without FCS. Count the cells in a hemocytometer (see Volume 1, Celis and Celis, "General Procedures for Tissue Culture" for additional information). A total of about 1×10^7 macrophages are needed. Keep cells in a 37°C humidified 8% CO_2 incubator.

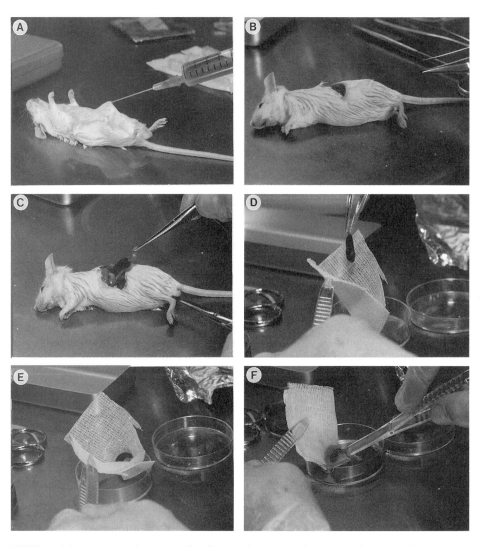

FIGURE 1 (A) Intraperitoneal injection of medium to obtain macrophages. (B and C) Surgical steps to remove the spleen. (D–F) Disaggregation of the spleen.

D. Preparation of Spleen Cells from the Immunized Mouse

Solutions

1. *DMEM without FCS:* As in Section B but add 50 ml of sterile water instead of FCS.

2. *DMEM with 20% FCS:* Prepare as described in Section B but containing 20% FCS.

3. *50× stock HAT:* Aliquot and keep at −20°C.

4. *DMEM–HAT:* To 500 ml of DMEM containing 20% FCS, add 10 ml of a 50× stock HAT solution.

Steps

1. Clean the left side of the immunized mice with 70% ethanol and make a lateral incision (left side, Fig. 1B).

2. Cut first the skin and then the muscle layer. The spleen is easy to localize due to its elongated shape and dark red color (Fig. 1C).

3. Remove the spleen aseptically and place it in a 60-mm tissue culture dish containing 7 ml of DMEM without serum. Trim off the adherens and wash two times with DMEM without serum.

4. Place the spleen inside a pocket of sterile gauze (6 × 6 cm) placed in a 60-mm tissue culture dish (Fig. 1D) containing 4 ml of DMEM without serum. The gauze, folded in four, is placed in aluminum foil and sterilized for 30 min in an oven at 160°C.

5. Disaggregate the tissue with the aid of the flat handle of a disposable sterile plastic forceps (Fig. 1F). Scrape gently against the tissue. Wash the gauze with 6 ml of DMEM without serum and collect the cells in a 10-ml conical centrifuge tube.

6. Wash the splenocytes once in DMEM without serum. Centrifuge the suspension for 3 min at 300–400 g and resuspend the cells at a final concentration of 2×10^7/ml in DMEM without serum. You need 5 ml (1×10^8) for the fusion. Keep the cells in the 37°C humidified 8% CO_2 incubator until use.

E. Cell Fusion

Solutions

1. *50% polyethylene glycol (PEG):* To make 5 ml, weigh 5 g of PEG 6000 and add 5 ml of 2× HBSS. Warm carefully in a microwave oven until it melts. The temperature should not exceed 56°C. Rock it carefully until all the particles dissolve. Allow the solution to cool and filter-sterilize. The solution should be made fresh. Keep at 37°C.

2. *2× HBSS without Ca^{2+} and Mg^{2+}:* To make 100 ml, take 20 ml of 10× HBSS and complete to 100 ml. Filter-sterilize and keep at 4°C.

Steps

1. Remove the flask containing the P3-X63-Ag8 cells (5 ml, total of 1×10^7 cells) from the CO_2 incubator. Have a stopwatch ready.

2. Mix 5 ml of the spleen cell suspension (10^8 cells) with 5 ml of the P3-X-63-Ag8 cells (10^7 cells).

3. Centrifuge the suspension at 300–400 g for 2 min in a bench centrifuge.

4. Discard the supernatant by aspiration.

5. Disperse the pellet by gently tapping the tube (important), and rotate the tube to achieve a fine dispersion of the cells.

6. Add 0.6 ml of 50% PEG kept at 37°C. Rotate the tube and rock for 1.5 min at room temperature.

7. Slowly add 8 ml of DMEM–HAT (kept 37°C) and resuspend by gently pipetting up

and down. Centrifuge at 300–400 g for 2 min. Aspirate the supernatant and tap the pellet gently (important).

8. Repeat step 7 twice.

9. Resuspend the pellet in 30 ml of DMEM–HAT and place in a 25-cm^2 tissue culture flask.

10. Centrifuge the macrophages and resuspend in 20 ml of DMEM–HAT.

11. Add the macrophages to the fused cells. Mix.

12. Transfer to a 175-cm^2 tissue culture flask containing 500–600 ml of DMEM-HAT.

13. Plate 0.25 ml of the cell suspension in microtiter plates (96 wells). Wrap the plates with Saran wrap to avoid evaporation (Fig. 2A) and place in a 37°C humidified 8% CO$_2$ incubator.

14. It is optimal to have single colonies growing after 8–10 days (Fig. 2B). Screen the plates for the production of specific antibody using ELISA (see Volume 2, Perlmann and Perlmann, "Enzyme-Linked Immunoabsorbent Assay" for additional information) or immunofluorescence (see other relevant articles in this volume).

F. Cloning by Limiting Dilution

Steps

1. Count and dilute cells in DMEM–HAT to a concentration of about 300 cells per 100 ml. Shake the cell suspension to ensure homogeneous dispersion.

2. Plate 0.3 ml in 96-well microtiter plates.

3. Wrap the plates with Saran wrap and place in the 37°C humidified 8% CO$_2$ incubator.

4. Select wells having a single colony (Fig. 2B). Repeat cloning. Screen once again for the production of specific antibodies.

5. Expand and freeze (see Volume 1, Celis and Celis, "General Procedures for Tissue Culture" for additional information).

G. Determination of the Immunoglobulin Subtype

Several kit-based procedures (Serotec, Amersham International, Behring Diagnostics, and others) are available commercially. Follow the vendor's instructions.

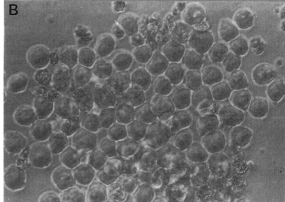

FIGURE 2   (A) Plating the fused cells in microtiter wells and mitrotiter plates wrapped in Saran wrap. (B) Hybridoma colony.

H. Production of Antibody in Ascitic Fluid

Steps

1. Centrifuge exponentially growing hybridomas and resuspend in 0.5 ml of DMEM lacking serum at a concentration of 10^7 cells/ml.

2. Inject intraperitoneally 0.5 ml of the cell suspension per mouse. Observe the animals regularly.

3. Following abdominal swelling, kill the animal by cervical dislocation. Disinfect the abdomen with 70% ethanol and aspirate the fluid with a syringe connected to a 20G1 needle. Work in the sterile hood.

4. Centrifuge the ascites at 300–400 g for 5 min. Store the supernatant at −80°C. Freeze the cells as described in the article by Celis and Celis, "General Procedures for Tissue Culture" in Volume 1.

5. If the cells do not produce ascites, inject the mouse intraperitoneally with pristane (2, 6, 10, 14-tetramethylpentadecane) (0.5 ml) 2 weeks prior to injection of the cells.

IV. PITFALLS

1. Following the addition of PEG and onward, the cells are very sensitive and should be handled with care.

2. Clone hybridomas as soon as possible as they may be overgrown by nonproducing cells.

3. If the hybrids do not grow well, try another batch of fetal calf serum.

4. When cloning by limiting dilution, it is advisable to add macrophages or spleen cells to help growth.

References

Asai, D. J. (ed.) (1993). "Antibodies in Cell Biology." Academic Press, San Diego.

Galfre, G., Howe, S. C., Milstein, C., Butcher, G. W., and Howard, J. C. (1977). Antibodies to major histocompatibility antigens produced by hybrid cell lines. *Nature* **266**, 550–552.

Goding, J. W. (1987). "Monoclonal Antibodies: Principles and Practice," 2nd Ed. Academic Press, London.

Harlow, E., and Lane, D. (1988). "Antibodies: A Laboratory Manual." Cold Spring Harbor Laboratory Press, Cold Spring Harbor, NY.

Huet, C. (1993). Production of polyclonal autibodies in rabbits. *In* "Cell Biology: A Laboratory Handbook" (J. E. Celis, ed.), Vol. 2, pp. 245–252. Academic Press, San Diego.

Köhler, G., and Milstein, C. (1975). Continuous cultures of fused cells secreting antibody of predefined specificity. *Nature* **256**, 495–497.

Liddel, E., and Cryer, A. (eds.) (1991). "A Practical Guide to Monoclonal Antibodies." John Wiley, Chichester.

Milstein, C. (1986). From antibody structure to immunological diversification of the immune response. *Science* **231**, 1261–1269.

Rapid Procedures for Preparing Monoclonal Antibodies and Identifying Their Epitopes

Kirsten Niebuhr, Andreas Lingnau, Ronald Frank, and Jürgen Wehland

I. INTRODUCTION

The generation of monoclonal antibodies is a process normally requiring several months, as most standard immunization protocols alone take 7 weeks on the average (see Volume 2, Celis *et al.*, "Production of Mouse Monoclonal Antibodies" for additional information). Moreover, the probability of obtaining specific antibodies is dependent on the immunogenicity of the antigen. Often it is difficult to achieve a sufficient immune response against proteins that are highly conserved between different species and thus often not recognized as antigens (e.g., cytoskeletal proteins). This article describes a very rapid and effective immunization method that takes only 17 days and, by eliciting a local immune response and by using lymph nodes for fusion, gives very good results also for poor antigens. In our experience it is even practicable to immunize with a mixture of different antigens as the resulting antibodies can be screened in Western blots and classified by the molecular weights of the respective antigen. In addition, a method for the identification of antibody epitopes using immobilized synthetic peptides is described.

II. MATERIALS AND INSTRUMENTS

1. 1-ml syringes with 0.45×13-mm or equivalent needles (e.g., Microlance 3, Becton-Dickinson), adjuvant [e.g., Alu-Gel-S (No. 12261, Serva), or Freund's complete/incomplete adjuvant (F-5881/F-5506, Sigma)].

2. Surgical instruments (fine scissors and tweezers).

3. 24-well plates (No. 146485), 6-well plates (No. 150229), and 10-cm petri dishes (No. 172958) are from Nunc.

4. Microscopic glass slides with frosted ends.

5. Optional: conical tubes in 96-well systems (Linbro Tubes-96, No. 61-226-C2, ICN), 24-well storage system for supernatant samples (Linbro Tubes-24, No. 61-238-05, ICN), miniblotting apparatus (Miniblotter MN16, No. 014-300, Biometra), and multitest slides (No. 60-408-05, ICN).

6. For the characterization of the epitopes: Immobilized synthetic peptides covering the primary sequence of the protein antigen synthesized according to the SPOT method (Frank *et al.*, 1992); commercially available at "Jerini Bio Tools" or "Genosys Biotechnologies."

7. All other reagents, materials, and equipment are as described in Volume 2, Celis *et al.*, "Production of Mouse Monoclonal Antibodies."

III. PROCEDURES

A. Immunization

This procedure is modified from Brodsky (1985) and Kubagawa *et al.* (1982).

Steps

1. We routinely immunize three Balb/C mice in parallel. The first injection, consisting of antigen suspended in phosphate-buffered saline (PBS) mixed with an equal volume of complete Freund's adjuvant, is given 17 days before the fusion is scheduled. A total volume of 200 μl containing 10–30 μg antigen is sufficient for injecting three mice. Adjuvants such as Alu-Gel S (Serva) or Hunter's Titer Max (Cyt Rx) can be substituted for Freund's adjuvant in the procedure.

2. Anesthesize the mice with ether or Metofane (Janssen) and inject the antigen subcutaneously into both hind legs using a 1-ml plastic syringe with a 0.45×13-mm or equivalent needle. In areas where animal protection committees discourage footpad immunizations, local subcutaneous injections in the vicinity of the popliteal lymph node can be used alternatively.

3. Three days later (day 14 before the fusion) give the first booster injection, in incomplete Freund's adjuvant. For the subsequent booster injections on days 10, 7, 3, and 1, dissolve the antigen in PBS.

4. On "day 0" sacrifice all three mice and remove their popliteal lymph nodes for the fusion. Because only a local immune response is elicited, the antibody titer in the blood may be low. Therefore, no blood sampling or serum testing is necessary before the fusion.

B. Preparation of Lymphocytes from Popliteal Lymph Nodes

Solutions

1. *70% ethanol.*

2. *Sterile, prewarmed PBS.*

3. *Prewarmed DMEM containing 25 m*M *HEPES (No. 22320-022; Life Technologies) without supplements.*

Steps

1. Kill the mice by cervical dislocation. Dip them in a beaker with 70% ethanol and place on a clean dissecting board under the sterile hood.

2. The popliteal lymph nodes are located in the hollow of the knee. Due to the adjuvant, their size should be greatly increased. Remove the skin by making a cut around the thigh and then pulling it down like a stocking.

3. The lymph nodes are visible as pale, yellowish spheres with a diameter of about 2–4 mm. They can be localized by pushing up the hollow of the knee from the back as shown in Figs 1A and 1B. Clean the hind legs with 70% ethanol, remove the tissue containing the lymph node with a generous cut, and wash it with PBS. Take care that the lymph nodes are not damaged at this stage.

4. Transfer the lymph nodes into a petri dish with DMEM–HEPES and remove the adherent muscle tissue using sterile scissors and tweezers (Fig. 1C). Place the lymph nodes into another petri dish with medium, cut them in half with scissors, and mince them between the frosted ends of two sterile glass microscope slides (Fig. 1D).

5. Collect the cells in a 50-ml centifuge tube and allow the larger debris to sediment. Transfer the supernatant containing the cells into a new 50-ml tube and wash the cells with HEPES–medium.

C. Cell Fusion

Solutions

1. *DMEM containing 25 mM HEPES.*

2. *OPTIMEM 1 medium (No. 31985-047; Life Technologies):* This medium is supplemented with 5% FCS, glutamine, antibiotics, and hypoxanthine/azaserine (HA) (Sigma A-9666). Test the appropriate concentration for your myeloma line; we use 1.5 vials per 500 ml OPTIMEM.

3. *50% PEG:* See Volume 2, Celis *et al.*, "Production of Mouse Monoclonal Antibodies" for additional information.

Steps

1. Three days before the fusion is scheduled, prepare feeder cells and seed them into fifteen 24-well plates containing OPTIMEM–HA selection medium. Thus at the time of the

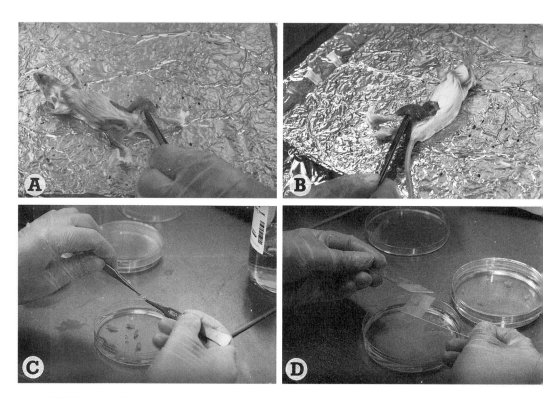

FIGURE 1 (A and B) Identification of the popliteal lymph nodes by pushing up the hollow of the knee from the back. (C) Removal of the adherent muscle tissue . (D) Mincing of the lymph nodes between the frosted ends of glass microscope slides.

fusion the medium is already conditioned by the macrophages and in addition it serves as a control that the macrophages have been prepared aseptically. The use of 24-well plates reduces the number of samples to 360 and yields sufficient amounts of supernatant (2 ml) for screening in ELISA, Western blot, and immunofluorescence. We routinely prepare peritoneal macrophages from three mice (one mouse per five 24-well plates) according to the protocol described by Ariana Celis, Kurt Dejgaard, and Julio E. Celis, this volume.

2. Prepare the mouse myeloma cells as described by Ariana Celis, Kurt Dejgaard, and Julio E. Celis, this volume. For six lymph nodes we use a total of about 4×10^8 P3-X63-Ag8 cells (ATCC CRL 1580), which corresponds to 10 to 12 densely grown petri dishes. Harvest the cells, resuspend in DMEM–HEPES, and transfer to a 50-ml tube.

3. Centrifuge the lymphocytes and myeloma cells for 5 min at 200 g. Aspirate off the supernatants, resuspend the pellets in HEPES medium, and then combine and mix the cells.

4. Centrifuge the cells and aspirate off the supernatant.

5. Gently tap the tube to disperse the cell pellet. Add 1 ml of prewarmed (37°C) PEG over a period of 1 min (drop by drop) while gently shaking the centifuge tube. Continue shaking for another minute.

6. Carefully dilute the PEG by adding prewarmed DMEM–HEPES. Start adding the medium dropwise (1 ml within 30 sec), increase the speed of addition (3 ml within 30 sec), and finally fill to a total volume of 20 ml. Place the tube in a 37°C water bath for 5 min.

7. Centrifuge the cells, resuspend them in OPTIMEM containing HA, aliquot them into the plates containing the prepared macrophages, and place them in the incubator. After 3–4 days the abundant myeloma cells will die; 8–10 days after the fusion, hybridoma colonies should be ready for screening. We usually obtain around 10 clones per well.

D. Rapid Screening and Propagation of Positive Clones

Steps

1. Take a 1-ml sample of supernatant from each well. For this purpose a system that consists of removable 1-ml tubes in a holder containing 8×12 tubes corresponding to a microtiter plate (Linbro Tubes-96, ICN) is very useful. When the supernatants are collected in this way, they can be transferred onto ELISA plates with a multichannel pipette.

2. Supernatants that give promising results in ELISA (see article by Hedvig Perlmann and Peter Perlmann, this volume) can be further tested in Western blots (see article by Julio E. Celis, Jette B. Lauridsen, and Bodil Basse, this volume). An efficient method to test many supernatant samples simultaneously is to use a device that allows the parallel application of different antibodies on a single membrane (Miniblotter MN16, Biometra). Separate the antigen (mixture) to be probed on SDS–polyacrylamide gels without slots, which results in protein bands over the total width of the gel. After transfer onto a membrane, apply the supernatants in "lanes" and compare their signals directly.

3. Rapid screening in immunofluorescence is facilitated by the use of multiwell slides (ICN), which allows 8–12 samples to be tested simultaneously.

4. Select the supernatants which give the most promising results. Resuspend the cells in the corresponding wells and transfer them into six-well dishes containing OPTIMEM–HA and macrophages as feeder cells. When they have reached a sufficient density, freeze the cells as described in Volume 1, Celis and Celis, "General Procedures for Tissue Culture." We usually freeze four vials per well. At the same time, collect the supernatant (6–7 ml) in a convenient storage system (Tubes-24, Linbro, ICN) for further testings. Add sodium azide at a final concentration of 0.02% (w/v) for preservation.

5. Perform subcloning by limiting dilution as soon as possible (see Volume 2, Celis *et al.*, "Production of Mouse Monoclonal Antibodies" for additional information). After the clones have reached a sufficient size, select wells with a single colony and test the supernatant. Expand positive clones, freeze a stock of the cells, and repeat the subcloning once.

Pitfalls

The most difficult part of the procedure is the identification of the lymph nodes. If they are not visible as shown in Fig. 1, excise the area under the knee generously, as most of the lymphocytes will be lost when the lymph nodes are damaged at this stage. After a short time in the prewarmed medium, the lymph nodes can be distinguished from the muscle tissue as yellowish, compact stuctures. Use forceps to tear apart the remaining muscle tissue and remove it as efficiently as possible.

E. Identification of Antibody Epitopes

For this assay the primary sequence of the polypeptide under investigation is subdivided into overlapping peptides that are synthesized on a membrane support according to the SPOT method (Frank, 1992). It is useful to choose a peptide length of 15 amino acid residues and an offset of 3 to 5 amino acids (i.e., a sequence overlap of 12 to 10 amino acids). The antibody-binding assay itself is relatively fast and easy: the membrane can be probed like a dot or immunoblot. Moreover, the peptide sheet can be reused several times by applying stringent washing and stripping procedures.

A prerequisite for this method is that the site of interaction with the antibody is linear as it is unlikely that binding domains dependent on the three-dimensional structure of the protein being analyzed (i.e., conformational epitopes) will be displayed by the relatively short synthetic peptides.

Solutions

1. *100% ethanol.*

2. *Blocking solution:* Whichever solution you prefer for Western blots.

3. *Tris-buffered saline:* 20 mM Tris base, 140 mM NaCl, pH adjusted to 7.6, with 0.1% (v/v) Tween 20 (=TBS-T); TBS-T with 0.5 M NaCl and TBS-T with 0.5% (v/v) Triton X-100.

3. Appropriate secondary antibodies (preferably developed with a chemoluminescence kit that is usually very sensitive and avoids colored spots on the sheet, e.g., ECL (enhanced chemoluminescence kit, Amersham) for horseradish peroxidase-coupled detection reagents).

4. *"Stripping buffers"*: (A) 8 M urea, 1% (w/v) SDS, and 0.5% (v/v) ß-mercaptoethanol; (B) 10% (v/v) acetic acid and 50% (v/v) ethanol.

NOTE

Check the secondary antibodies first to make sure that they do not give an unspecific signal.

Steps

1. If the peptide sheet is dry, moisten it with ethanol. Wash three times (5 min each) with TBS-T and saturate the filter sheet with blocking buffer for 2 hr at room temperature or, preferably, overnight at 4°C.

2. Incubate the sheet with antibody in blocking reagent for 1 hr at room temperature and wash at least 5 min each with TBS-T, TBS-T + NaCl, TBS-T + Triton, and TBS-T.

3. Incubate the peptide sheet for 1 hr at room temperature with an appropriate dilution of secondary antibody. After washing, perform the enzyme reaction (preferably chemoluminescence) to detect the bound antibody. The respective binding site can be determined from the sequence that the detected synthetic peptides have in common. Generally the smaller the offset of the overlapping peptides, the more precise the binding site can be delineated.

4. With a stringent washing procedure the membrane can be regenerated and reprobed several times. Perform each washing step three times for at least 10 min at room temperature with water, "stripping buffer" A (see earlier), "stripping buffer" B (see earlier), and finally with ethanol. Afterwards dry the peptide sheet and store at −20°C or saturate and probe with another antibody after checking the remaining background using secondary antibody.

Pitfalls

The stringent stripping procedure usually removes the antibodies quantitatively. In some cases, however, the affinity of the antibody to be tested is rather high and traces are still left on respective spots. These might be recognized by the secondary antibodies in a subsequent assay. Freezing and thawing of the peptide sheet usually reduce this background.

If the enzyme reaction is developed with substrates that give a colored precipitate on the membrane, start the stripping procedure as follows: wash with water twice, wash with *N,N*-dimethylformamide three times, and sonicate. Then continue as described earlier.

References

Brodsky, F. M. (1985). Clathrin structure characterized with monoclonal antibodies. I. Analysis of multiple antigenic sites. *J. Cell Biol.* **101**, 2047–2054.

Frank, R. (1992). Spot-synthesis: An easy technique for the positionally addressable, parallel chemical synthesis on a membrane support. *Tetrahedron* **48**, 9217–9232.

Kubagawa, H., Mayumi, M., Kearney, J. F., and Cooper, M. D. (1982). Immunoglobulin V_H determinants defined by monoclonal antibodies. *J. Exp. Med.* **156**, 1010–1024.

In Vitro Production of Monoclonal Antibodies

Matthew Holley

I. INTRODUCTION

This article describes a simple, efficient method of *in vitro* immunization for the production of monoclonal antibodies (Kohler and Milstein, 1975; Galfre *et al.*, 1977; Reading, 1982). Some essential features of the method have been described previously (see Volume 2, Celis *et al.*, "Production of Mouse Monoclonal Antibodies" for additional information). The specific method (Cel-Prime) has been developed by Immune Systems Ltd (UK) to provide a reliable system for making monoclonal antibodies with the minimum level of expertise and equipment. The method is also designed to enhance the yield of IgG class antibodies by culturing splenocytes with antigen-presenting cells that have previously been exposed to the antigen. The splenocytes are then fused with a myeloma cell line and screened, cloned, and characterized by standard procedures.

One of the attractions of the method is that it requires only small quantities of antigen, which can be presented in many different forms. The recommended dose of antigen is 60 μg, although both larger and smaller quantities can be used. Successful results have been obtained from tissue pellets dissolved in SDS buffer (Holley and Richardson, 1994; Holley and Nishida, 1996), bands of protein removed from polyacrylamide gels, unconjugated peptides composed of as few as nine residues, and suspensions of cells cultured with splenocytes during immunization.

II. MATERIALS AND EQUIPMENT

All media and plasticware for immunization were purchased economically in kit form from Immune Systems Ltd. (Cat. No. H.P2). Each kit is sufficient for two immunizations and includes support cells (antigen-presenting cells), basic media for culturing support cells and splenocytes, an immunization medium, culture media to test antigen sterility, and a positive control antigen/antibody system. Plasticware includes 2 × 25-ml vented culture flasks, one pack of 35-mm petri dishes, 6 × 15-ml test tubes, and two homogenizers for splenocyte dissociation. Ten- to 12-week-old BALB/c mice of either sex were obtained from B&K Universal. Plastic pipettes, 1 (Cat. No. 40101), 5 (Cat. No. 40105), and 10 ml (Cat. No. 40110), are from Bibby Sterilin Ltd. Equipment included an RS Biotech incubator, an Heraeus Class II laminar flow hood, an Olympus CK2 inverted microscope with phase contrast, and a Burkard Fugette benchtop centrifuge.

Media and plasticware for two fusions can be purchased as a convenient kit (Cat. No.

(SP.1) from Immune Systems Ltd., including SP$_2$/O-Ag-14 myeloma cells, myeloma cell culture medium (T20-5007), hybridoma growth medium (T20-5008), Doma-Drive hybridoma growth-enhancing supplement (T31-0403), and a polyethylene glycol solution (fusion medium Cat. No. T30-1222). The α modification of Eagle's medium with glutamine but without ribosides or deoxyribosides (Cat. No. 22561-021), the α modification of Eagle's medium with glutamine, ribosides, and deoxyribosides (Cat. No. 22571-020), fetal calf serum (FCS, Cat. No. 105020-37), and HAT (hypoxanthine, aminopterin, and thymidine at 50×; Cat. No. 21060-017) are from Life Technologies Ltd. Glutamine (Cat. No. G-7513), penicillin (Cat. No. P-3032), streptomycin (Cat. No. S-9137), and dimethyl sulfoxide (DMSO, Cat. No. D2650) are from Sigma. Cell strainers (70-μm pore size; Cat.No. 2350), 25-cm^2 vented flasks (Cat. No. 3108), and 15 (Cat. No. 2097)- and 50-ml (Cat. No. 2070) centrifuge tubes are from Falcon (Becton Dickinson UK Ltd.). Forty-eight-well culture plates (Cat. No. 3548) and U-bottom 96-well plates (Cat. No. 3799) are from Costar UK Ltd. Twelve-well glass slides (Cat. No. 6041205) are from ICN Biomedicals Ltd. PAP pens (Cat. No. S2002) are from Dako UK Ltd.

II. PROCEDURES

A. Immunization

Solutions

1. *Support cell medium.* Supplied ready to use.
2. *Immunization medium.* Supplied ready to use.
3. *Eagle's Medium without ribosides and deoxyribosides.*

Steps

See Fig. 1.

1. Check your timing to ensure that you have sufficient SP2 myeloma cells for fusion following immunization. Expansion of myeloma cells may take up to 2 weeks if the cells are initially frozen, but the immunization will be complete within 7–8 days.

2. Thaw the frozen support cells and culture with 5 ml support cell medium in a 25-cm^2 gas exchange flask at 37°C and 6% CO$_2$ for 24 hr.

3. Discard the support cell medium and replace with 5 ml of fresh medium mixed with the chosen antigen. At this stage the density of support cells is very low and they should be cultured undisturbed for 48 hr.

4. Prepare splenocytes from the 10- to 12-week-old BALB/c mouse following the method described by Celis *et al.* (Volume 2, "Production of Mouse Monoclonal Antibodies"). Dissociate the cells in Eagle's medium, first with fine forceps and then by reflux of the tissue through a sterile plastic pipette tip that is about 0.5 mm in diameter. Pipette the cells through the strainer into a 50-ml plastic centrifuge tube. Wash with further Eagle's medium if necessary. Spin the cells at 800 g for 5 min in a bench centrifuge. Resuspend the cells in 5 ml of immunization medium previously warmed to 37°C.

5. Remove the medium from the support cell culture and gently wash the attached support cells with 5 ml of fresh support cell medium. Add the splenocytes, the remaining immunization medium (total volume 30 ml), and fresh antigen and incubate at 37°C and 6% CO$_2$ for 3–4 days (Fig. 2). During this period the support cells should form a confluent monolayer and the medium will become the color of straw.

B. Preparation of Myeloma Cells

Solution

1. *Culture medium for SP2 cells:* Myeloma growth medium or Eagle's medium *with* ribonucleosides and deoxyribonucleosides, 10% FCS, 1% glutamine (unless already present

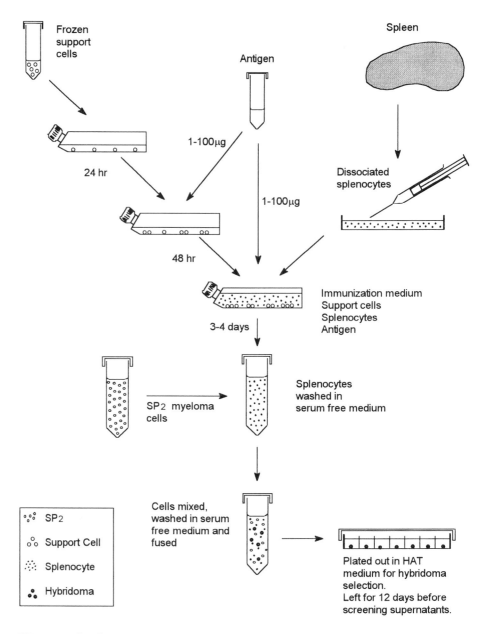

FIGURE 1 A flow diagram representing the major steps in the *in vitro* immunization procedure.

in the saline), 200 U/ml penicillin, and 200 mg/ml streptomycin. Store at 4°C for up to 2 weeks.

Steps

1. Thaw a frozen vial of myeloma cells and transfer them to 10 ml of SP2 medium previously equilibrated to 37°C with 6% CO_2 in a 75-cm^2 culture flask and incubate for 1–2 days. The ideal seeding concentration is 10^5 cells per milliliter.

2. Allow cells to approach confluence and then subculture them at a density of 10^5 cells per milliliter. Repeat this procedure until you have a total of 10^7 cells for each fusion of immunized splenocytes. Excess SP2 cells can be frozen in FCS with 10% DMSO as described previously (see Volume 1, Celis and Celis, "General Procedures for Tissue Culture" for additional information).

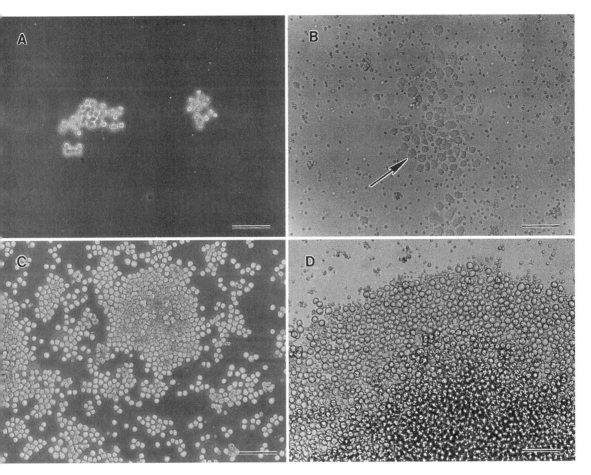

FIGURE 2 (A) Two groups of support cells after 3 days in culture. The initial density of cells is very low and the clones are sparsely distributed. (B) Supports cells (arrow) and residual blood cells and splenocytes after cells have been removed from the culture flask for fusion. (C) Myeloma cells. (D) The edge of a hybridoma colony 10 days after fusion. Scale bars: 100 μm.

C. Fusion

Solutions

1. *Eagle's medium.*

2. *Fusion medium:* 50% weight by volume of polyethylene glycol in Eagle's medium.

3. *HAT growth medium:* Eagle's medium with 10% FCS, 10% Doma-Drive, 1% glutamine, and HAT medium (1×).

Steps

See Fig. 1.

1. Pipette 0.5 ml of HAT growth medium into each well of 8 × 48-well plates and place in the incubator.

2. Collect the splenocytes from the immunization flask, dislodging attached cells from the flask by running a jet of Eagle's medium across the growth surface, and transfer them to a 50-ml centrifuge tube. Spin the cells at 550 g for 5 min, remove the medium, resuspend the cells in 5 ml fresh Eagle's medium without calf serum, and transfer them to a 10-ml centrifuge tube.

3. Suspend the SP2 cells and calculate the volume of medium that contains a total of 10^7 cells. Follow the cell counting procedure described previously (see Volume 1, Celis and

Celis, "General Procedures for Tissue Culture" for additional information). Spin these cells at 300 g for 5 min, remove the medium, resuspend them in 5 ml of Eagle's medium, and mix them with the splenocytes. Spin the suspension at 550 g for 5 min and then remove the medium. Repeat this procedure twice.

4. After removing the final supernatant, gently resuspend and stir the pellet in 0.8 ml of fusion medium. Agitate the suspension by shaking the tube. Using a stopwatch, add and mix HAT growth medium in the following volumes at the given times: 0.5 ml at 75 sec; 1 ml at 2, 3, and 4 min; and 5 ml at 5 min.

5. Spin the cells at 100 g for 5 min, remove the medium, and resuspend the cells in 5 ml of fresh HAT growth medium. Transfer the cells to a flask containing 25 ml of HAT growth medium and leave to stand at room temperature for 15 min.

6. Mix the medium gently to ensure that the cells are suspended at an even density and place a drop into each well of the 8 × 48-well plates prepared earlier. If necessary, add further drops until the medium is fully distributed among the wells. The plates should then be incubated undisturbed for 12 days at 37°C and 6% CO_2.

7. Remove supernatant from wells that contain clones visible to the naked eye and screen them by the appropriate method, e.g., against antigen by ELISA, immunoblotting, or immunofluorescence labeling of isolated cells or histological sections. Replace the supernatants with fresh HAT growth medium and return the plates to the incubator for 1–2 days during screening. Select wells that contain desired supernatants and clone the hybridoma cells by limiting dilution in fresh HAT growth medium in U-bottomed 96-well plates (see Volume 2, Celis et al., "Production of Mouse Monoclonal Antibodies" for additional information). Most hybridoma cell lines are robust and can be frozen during log-phase growth in FCS with 10% DMSO (see Volume 1, Celis and Celis, "General Procedures for Tissue Culture" for additional information).

IV. COMMENTS

The kits provided by Immune Systems have been assembled to provide all the media and plastics, excluding pipettes, to complete the procedures. Additional media for myeloma and hybridoma cell growth are available in complete form using the catalog numbers listed in this article. Doma-Drive is also available separately and is an invaluable supplement for enhancing hybridoma cell growth.

Each fusion produces at least 350 hybridoma cell lines, equivalent to the number of culture wells used. Because many of these wells support multiple clones the total number of hybridoma lines probably exceeds 500. The majority of antibodies are class IgM, and the number of IgG class antibodies varies widely between different types of antigen. Although IgMs are not favored by some researchers, they have produced very good results in both light and electron microscopy and they can be used profitably with IgGs for double labeling.

When making antibodies for use as cell markers, we have developed an efficient screening system. Hybridoma supernatants are screened either on frozen sections prepared on a cryostat or on dissociated cells dried onto multiwell glass slides coated with gelatin. The tissue is fixed with cold acetone, and each well is encircled by a hydrophobic barrier drawn with a PAP pen. Wells are then briefly washed in saline, incubated overnight in supernatant, and labeled with fluorescent anti-mouse immunoglobulins for 2 hr. The IgM and IgG classes are initially distinguished by different fluorochromes bound to anti-mouse IgM or IgG class antibodies. Within 24 hr it is thus possible to define the cellular staining pattern for a given antibody, its likely cross-reactivity with other cells, and its class. More rigorous testing is necessary to confirm antibody isotypes, but in practice it is possible to identify IgG class antibodies quickly and to clone the appropriate hybridoma cells without delay.

When the desired clones have been selected, we withdraw media from all wells in the original 48-well plates, replace them with freezing medium (FCS with 10% DMSO), and place them in a −80°C freezer. This method allows clones to be reviewed within 3–6 months if necessary.

V. PITFALLS

1. The initial density of support cells is alarmingly low but this is necessary to avoid excess growth during immunization.

2. Media in the corner wells of 48-well plates tend to evaporate during the 12 days from fusion. They can be filled with media but not seeded with hybridomas. The problem can also be reduced by creating higher humidity in the incubator and by using culture plates fitted with low evaporation lids.

3. The hybridoma cells grow vigorously once the medium is replaced during screening and sometimes it is useful to transfer them to 24-well plates.

Acknowledgments

MCH is a Royal Society University Research Fellow supported by The Wellcome Trust. I thank Steve Bourne and Chris Popham of Immune Systems Ltd. for their support in the preparation of this article.

References

Galfre, G., Howe, S. C., Milstein, C., Butcher, G. W., and Howard, J. C. (1977). Antibodies to major histocompatibility antigens produced by hybrid cell lines. *Nature (London)* **266**, 550–552.

Holley, M. C., and Nishida, Y. (1995). Monoclonal antibody markers for early development of the stereociliary bundles of mammalian hair cells. *J Neurocytol.* **24**, 853–864.

Holley, M. C., and Richardson, G. P. (1994). Monoclonal antibodies specific for endoplasmic membranes of mammalian cochlear outer hair cells. *J. Neurocytol.* **23**, 87–96.

Kohler, G., and Milstein, C. (1975). Continuous cultures of fused cells secreting antibody of predefined specificity. *Nature (London)* **256**, 495–497.

Reading, C. L. (1982). Theory and models for immunization in culture and monoclonal antibody production. *J. Immunol. Meth.* **53**, 261–291.

Rapid Production of Antibodies in Chicken and Isolation from Eggs

Harri Kokko, Ilpo Kuronen, and Sirpa Kärenlampi

I. INTRODUCTION

Rabbits have previously been preferred as producers of polyclonal antibodies. It was found, however, already in 1962 that immunoglobulin concentration in the yolk is equal to or greater than that found in hen's serum (Patterson et al., 1962). To date, hens are recognized as convenient and inexpensive sources of antibodies. According to the increasing number of publications, the antibodies produced in hens are useful in many applications, including immunotherapy and immunodiagnostics (Schade et al., 1991; Akita and Nakai, 1992). The increasing preference of hens as antibody producers is partly based on the development of efficient methods for the purification of antibodies from egg yolk (Jensenius et al., 1981; Akita and Nakai, 1992, 1993). In some cases, hens, as more distant relatives to mammals, offer a good alternative to rabbits in producing antibodies against mammalian antigens (Stuart et al., 1988).

We have produced antibodies in hens against several different types of antigens, e.g., proteins, synthetic peptides (Kokko and Kärenlampi, 1992; Kuronen et al., 1993), plant viruses (Kokko et al., 1996), and fungal and bacterial cell walls. According to our experiences, hens are very effective producers of specific antibodies. Using the protocol described in this article, homogeneous antibodies can be recovered monthly from egg yolks in amounts corresponding to 300 ml of high-titer rabbit antiserum. The antibody titer peaks shortly after 2 weeks from the first antigen exposure, staying high for at least 100 days. We have found that high pH combined with NaCl are superior in the affinity purification of chicken egg antibodies. These conditions can be successfully used also for the purification of antibodies from mammalian origin to which low pH is generally recommended.

II. MATERIALS AND INSTRUMENTATION

Freund's complete (Cat. No. F-4258) and incomplete (Cat. No. F-5506) adjuvants, Trizma base (Cat. No. T-1503), and p-nitrophenyl phosphate, sodium salt, Sigma104 (Cat. No. 104-0) are from Sigma. Alkaline phosphatase-labeled anti-chicken IgG (AP–anti-chicken IgG, Cat. No. 61-3122) is from Zymed Laboratories. Dextran sulfate, sodium salt (MW $\approx$ 500,000, Cat. No. 17-0340-01) is from Pharmacia. Anhydrous Na_2SO_4 (Cat. No. 6649), $CaCl_2 \cdot 2H_2O$ (Cat. No. 2382), NaCl (Cat. No. 6604), $MgCl_2 \cdot 6H_2O$ (Cat. No. 5833), $Na_2HPO_4 \cdot 2H_2O$ (Cat. No. 6580), KH_2PO_4 (Cat. No. 4873), 25% glutaraldehyde solution (Cat. No. 4239), iethylamine (Cat. No. 803010), sodium azide (Cat. No. 6688), and glycerol (87%, Cat. No. 094) are from Merck. Fat-free milk powder (food grade) is from the local grocery store.

Water is purified by a Milli-Q apparatus (Millipore). Dialysis tubing with a 12,000 molecular weight cutoff (Spectra/Por 4, Cat. No. 3787-D22) is provided by Spectrum. The 96-well inyl assay plates (ELISA plates, Cat. No. 2596) are from Costar. ELISA reader is from SLT abinstruments.

II. PROCEDURES

A. Immunization of Hens

Steps

1. House the hens in individual cages to avoid mixing of the eggs.

2. You may want to use an assistant in the handling of the hen (Fig. 1). Grab the hen with your left hand from the back by locking the wings. With the right hand, grab the feet. Put the hen on the table and move your left hand over its head. In this position, the hen is

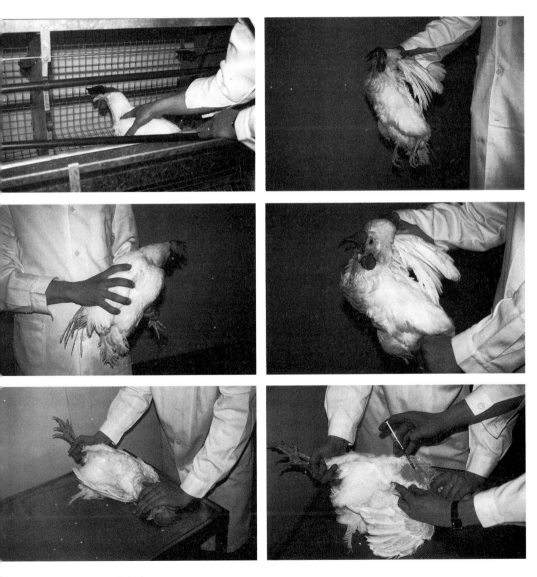

FIGURE 1 Immunization of chicken.

calm and easy to immunize. Let your assistant do the immunization. You can also immuniz[e]
the hen by yourself. Lift the hen from the back with your right hand. Grab the feet with th[e]
left hand. Push the hen's head under your left arm. Immunize with your right hand.

3. For the primary immunization, dilute the antigen (maximal concentration of the adju[-]
vant is 50%, v/v) in Freund's complete adjuvant and emulsify in an Eppendorf tube using [a]
syringe with a needle (18 to 21 gauge).

4. Give one to two injections of the antigen emulsion into the pectoralis muscles of th[e]
hen (Fig. 1). The maximal volume to one muscle is about 500 μl. The injected amount (10[–]
1000 μg) depends on the type of the antigen.

5. Repeat the immunization every 2 to 4 weeks by the same procedure but using Freund['s]
incomplete adjuvant as an emulgator.

6. Collect the eggs daily. Label the eggs by an identification number and date using [a]
water-proof pen and store the eggs refrigerated at 4°C until the isolation of the immunoglobu[-]
lins. Yolk antibodies are stable *in situ* in the refrigerator for at least 1 year.

B. Isolation of Egg Immunoglobulins

Several different methods have been published for the isolation of immunoglobulins from
eggs. Two alternative methods are provided here. The dextran sulfate method (Procedur[e]
B1), modified from Jensenius *et al.* (1981), is classic and works well. The water dilutio[n]
method (Procedure B2), adapted from Akita and Nakai (1993), is simpler and less expensive[.]
It has also worked well in larger scales.

1. Dextran Sulfate Method

This method is modified from Jensenius *et al.* (1981).

Solutions

1. *10× Tris-buffered saline (TBS) stock solution:* To make 1 liter of 1.4 M NaCl–10[0]
mM Tris–HCl solution, add 81.82 g of NaCl and 12.11 g of Trizma base to approximatel[y]
900 ml of purified water, adjust pH to 7.4 with 6 M HCl, and bring to a total volume of [1]
liter with purified water. Store at room temperature. Dilute 1 to 10 with purified wate[r]
before use.

2. *10% (w/v) dextran sulfate solution:* To make 100 ml, solubilize 10 g of dextran sulfat[e]
in purified water and adjust to a total volume of 100 ml. You may add 0.05% sodium azid[e]
for a prolonged storage at room temperature.

3. *1 M calcium chloride solution:* To make 100 ml, solubilize 14.7 g of CaCl$_2$·2H$_2$O i[n]
purified water and adjust to a total volume of 100 ml. Store at room temperature.

4. *36% (w/v) sodium sulfate solution:* To make 100 ml, solubilize 36 g of anhydrou[s]
Na$_2$SO$_4$ in purified water and adjust to a total volume of 100 ml. Store at room temperature[.]
If precipitates appear, make a fresh solution.

5. *10× phosphate-buffered saline (PBS) stock solution:* To make 2 liters, dissolve 136 [g]
of NaCl, 37.6 g of Na$_2$HPO$_4$·2H$_2$O, and 4 g of KH$_2$PO$_4$ to purified water and bring to [a]
total volume of 2 liters. Store at room temperature. Dilute 1 to 10 with purified water be[-]
fore use.

Steps

Carry out the entire procedure at room temperature:

1. Separate the egg yolk from the white (Fig. 2).

2. Add 1× TBS on the yolk to bring the volume to 50 ml and mix. (From this mixtur[e]

IGURE 2 Separation of egg yolk from the white.

ou can take a small sample, dilute it 1:100 with 2% MP–TBS, and titer the antibody in ELISA or Western blot.)

3. Centrifuge the diluted yolk for 10 min at 2000 g. Discard the pellet and save the upernatant.

4. Mix the supernatant with 3 ml of 10% dextran sulfate solution and 7.5 ml of 1 M alcium chloride solution and incubate for 30 min. Centrifuge as before. If the resulting upernatant is cloudy, add an additional 1–3 ml of dextran sulfate solution and recentrifuge. Save the clear supernatant.

5. Slowly add solid sodium sulfate (a total of 20 g/100 ml) to the supernatant and incubate or 20 min. Centrifuge as before. Save the pellet.

6. Dissolve the pellet in 10 ml of 1× TBS. Separate the dissolved immunoglobulins from nondissolved material by centrifugation for 10 min at 2000 g. Save the clear supernatant.

7. Add 6.2 ml of 36% sodium sulfate solution to the supernatant. Centrifuge as before and save the pellet.

8. Dissolve the immunoglobulin pellet in 5 ml of 1× TBS or 1× PBS.

9. Dialyze the immunoglobulin solution at 4°C overnight with constant stirring against 1 iter of 1× TBS or 1× PBS in Spectra/Por 4 dialysis tubing.

10. Repeat step 9 again.

11. Divide the dialyzed immunoglobulin solution into small aliquots and store at −20°C. mmunoglobulins can also be stored at 4°C for short periods. Lyophilized preparations can e stored at 4°C for several years.

2. Water Dilution Method

This method is adapted from Akita and Nakai (1993).

Solutions

1. *10× TBS:* See Procedure B1.

2. *10× PBS:* See Procedure B1.

Steps

1. Separate the egg yolk from the white carefully so as not to break the yolk and rinse it once with purified water (Fig. 2).

2. Dilute the yolk by adding 9 vol of purified water (i.e., one yolk to about 100 ml). Adjust the pH to 5.0–5.2 with 1 M HCl (about 1 ml) and incubate for 6 hr at 4°C.

3. Centrifuge for 25 min at 10,000 g at 4°C. Save the supernatant.

4. Add solid sodium sulfate (a total of 20 g/100 ml) to the supernatant that has been prewarmed to room temperature. Incubate for 20 min at room temperature. Centrifuge for 20 min at 2000 g. Save the pellet.

5. Dissolve the pellet in 5–10 ml of 1× TBS or 1× PBS, depending on the intended use of your antibody.

6. You may further purify the immunoglobulins by ultrafiltration, alcohol precipitation, ion-exchange chromatography, gel chromatography, affinity chromatography, etc.

C. Immunoaffinity Purification of the Antibody

Solutions

1. *10× TBS:* See Procedure B1.

2. *5 M NaCl stock solution:* To make 100 ml, dissolve 29.22 g of NaCl in purified water and adjust to a total volume of 100 ml. Store at room temperature.

3. *Tris–NaCl buffer, pH 7.4:* To make 100 ml of 10 mM Tris–290 mM NaCl, dilute 1 ml of 10× TBS and 3 ml of 5 M NaCl to purified water. Store at room temperature.

4. *0.1% diethylamine–100 mM NaCl buffer, pH 11.0:* To make 1 liter, pipette 1 g of diethylamine into a small amount of purified water to prevent its evaporation. Add 20 ml of 5 M NaCl stock solution and purified water to about 900 ml. Adjust the pH to 11.0 with M HCl and bring to a total volume of 1 liter with water. Store at 4°C. After a prolonged storage, check the pH. See also Comment 6.

5. *1 M Tris–HCl, pH 7.0:* To make 100 ml, dissolve 12.11 g of Trizma base to 90 ml of purified water. Adjust the pH to 7.0 with 6 M HCl and bring to a total volume of 100 ml. Store at room temperature.

6. *1% sodium azide stock solution:* Dissolve 0.5 g NaN$_3$ in 50 ml purified water. Store at 4°C.

7. *50% (w/v) glycerol–TBS:* To make 100 ml, add 50 g glycerol and 10 ml 10× TBS stock solution into purified water.

Steps

1. Prepare the affinity column by one of the standard procedures using the antigen of interest (for protein antigens, see Huet, 1994).

2. Wash the column with 1× TBS.

3. Adsorb the clear (filtrate if necessary) antibody preparation or antiserum (1–2.5 mg immunoglobulin/ml TBS) to the column.

4. Wash the column with Tris–NaCl buffer, pH 7.4, until A_{280} is below 0.01.

5. Elute the antibody as 1.0-ml fractions with diethylamine–NaCl buffer into tubes containing 100 μl 1 M Tris–HCl, pH 7.0, to neutralize the eluate.

6. Measure absorbance at 280 nm from each fraction and pool the ones containing the immunoglobulins.

7. For storage and further use of the column, rinse it with TBS, add 0.02% sodium azide and store at 4°C.

8. Dialyze the affinity-purified antibody against 1× TBS. Add 0.05% sodium azide.

9. Store the antibody in 50% glycerol–TBS at −20°C.

D. Screening and Titering of the Antibody

Solutions

1. *10× Tris–NaCl buffer stock solution, pH 9.5:* To make 100 ml of 1 M Tris–1 M NaCl, dissolve 12.11 g of Trizma base and 5.84 g of NaCl to 90 ml of purified water. Adjust the pH to 9.5 with 6 M HCl and bring to a total volume of 100 ml. Store at room temperature.

2. *200× MgCl₂ stock solution:* To make 100 ml of 1 M MgCl₂, add 20.33 g of MgCl₂·6H₂O to 100 ml of purified water. Store at room temperature.

3. *Alkaline phosphatase substrate solution:* 1 mg of *p*-nitrophenyl phosphate/ml of 100 mM Tris, 100 mM NaCl, and 1 mM MgCl₂, pH 9.5. Weigh 10 mg of *p*-nitrophenyl phosphate, add 1 ml of 10× Tris–NaCl buffer stock solution and 50 μl of 200× MgCl₂ stock solution, and bring to a total volume of 10 ml with purified water. Prepare just before use.

4. *0.5% (v/v) glutaraldehyde solution:* To make 10 ml, dilute 50 μl of 25% glutaraldehyde solution to 10 ml with purified water. Prepare just before use.

5. *10× TBS:* See Procedure B1.

6. *2% fat-free milk powder (MP–TBS):* To make 50 ml, dissolve 1 g of fat-free milk powder to 50 ml of 1× TBS. This is stable for 1 day.

Steps

Carry out the entire procedure at room temperature. You may use a plate shaker for the incubations.

1. Treat the ELISA plate by adding 50 μl of 0.5% (v/v) glutaraldehyde in each well for 30 min to improve the binding of a peptide antigen. For other antigens, you may need other methods (follow the instructions of ELISA plate manufacturers; see also Harlow and Lane, 1988). Wash the wells with 200 μl of 1× TBS. Shake the solution off the plate. Dry the plate by pressing against a paper towel.

2. Binding of the antigen: Add the antigen in a volume of 50 μl (in case of peptide antigen: 1 mg of peptide/50 ml of 1× TBS) to each well and incubate for at least 1 hr (you may incubate the plate overnight). Remove the solution.

3. Blocking: Add 250 μl of 2% MP–TBS and incubate for 30 min. Wash with 250 μl of 1× TBS.

4a. For screening of the yolk preparations, add the immunoglobulin solution (from Procedure B1, step 2 diluted 1 to 100 with 2% MP–TBS) in a volume of 50 ml (with some immunoplates, it is better to use 100 μl), incubate for 2 hr and wash thoroughly (five times) with 1× TBS.

4b. For titering, make stepwise dilutions of 1:3 from the antibody preparation in 2% MP–TBS and incubate as in step 4a.

5. Add 50 μl of the second antibody (AP–anti-chicken IgG, 1:1000 dilution in 2% MP–TBS) and incubate for 1 to 2 hr. Wash the wells five times with 1× TBS.

6. Developing: Add 150 ml of alkaline phosphatase substrate solution and incubate for 10 to 20 min. Stop the reaction by adding 50 ml of 0.1 M NaOH and measure the absorbance at 405 nm with an ELISA reader.

IV. COMMENTS

1. Every antigen elicits a unique response. Peptides and haptens require binding to a carrier protein (BSA, ovalbumin) to become sufficiently antigenic. The response may also depend on the animal species or individual that is immunized. Therefore, only general instructions can be given as to the amount of antigen that should be used in the immunization.

2. When producing antibodies against synthetic peptides representing a known protein, it is useful to screen first the protein for hydrophilicity and antigenicity indexes by a suitable computer program (commercially available sequence-analysis software can be obtained from GCG, PROSIS, and Intelligenetics).

3. Hens produce useful antibodies already in 2 weeks after the first immunization. The response does not necessarily rise dramatically after second or third immunizations. This suggests that in chicken the booster immunizations mainly maintain the IgY production elicited by the primary immunization.

4. The antigen response can be tested by several methods. Different types of antigens (proteins, peptides, haptens, bacterial and viral particles, etc.) may require different methods for binding to the immobilizing surface (ELISA plate, Western blot). For details, see Perlmann and Perlmann (1994) Celis *et al.* (1994); and related articles in this volume. The binding of peptides to a polyvinyl ELISA plate can be improved by glutaraldehyde treatment.

5. The immunoglobulin concentration can be approximated using an A_{280} value of 1.4 as equal to 1 mg of immunoglobulin/ml. Following the technique of Procedure B1, approximately 100 to 190 mg of immunoglobulin per egg can be recovered.

6. In the immunoaffinity purification, the antibody–antigen affinity varies. We have varied the composition of the elution buffer as follows: diethylamine, 0.1–0.4%, NaCl, 0–250 mM and pH 11.0–11.5. NaCl may prevent the precipitation of the immunoglobulins at high pH

V. PITFALLS

1. Glycoproteins are good immunogens but often give rise to antibodies reacting with a wide variety of glycoproteins.

2. Every egg should be regarded as an individual. There is a great day-to-day variation and successive eggs may contain very different amounts of the specific antibody. Therefore, it is important to screen all the eggs before pooling the antibody preparations.

3. All lipids (cloudiness) have to be carefully removed by dextran sulfate (Procedure B1, step 4) before the precipitation of the IgY with sodium sulfate. Before the immunoaffinity chromatography, filtrate the antibody preparation through a 0.45-μm filter if cloudy. The immunoglobulin yield will otherwise decrease and the amount of contaminants increase.

4. Precipitation of immunoglobulin with sodium sulfate in cold forms an insoluble precipitate.

5. No specific binding proteins are available for chicken immunoglobulins (protein A and protein G do not bind IgY).

6. If you want to label your antibody via amino groups, remove all other reactive substances such as Tris (TBS) or sodium azide (preservative). You may dissolve the final precipitate (Procedure B1, step 8) in PBS instead of TBS.

References

Akita, E. M., and Nakai, S. (1992). Immunoglobulins from egg yolk: Isolation and purification. *J. Food Sci.* **57,** 629–634.

Akita, E. M., and Nakai, S. (1993). Comparison of four purification methods for the production of immunoglobulins from eggs laid by hens immunized with an enterotoxigenic *E. coli* strain. *J. Immunol. Methods* **160,** 207–214.

Celis, J. E., Lauridsen, J. B., and Basse, B. (1994). Determination of antibody specificity by Western blotting and immunoprecipitation. *In* "Cell Biology: A Laboratory Handbook" (J. E. Celis, ed.), Vol. 2, pp. 305–313. Academic Press, San Diego.

Harlow, E., and Lane, D. (1988). "Antibodies, a Laboratory Manual. Cold Spring Harbor Laboratory, Cold Spring Harbor, NY.

Huet, C. (1994). Purification of immunoglobulins. *In* "Cell Biology: A Laboratory Handbook" (J. E. Celis, ed.), Vol. 2, pp. 291–296. Academic Press, San Diego.

Jensenius, J. C., Andersen, I., Hau, J., Crone, M., and Koch, K. (1981). Eggs: Conveniently packaged antibodies. Methods for purification of yolk IgG. *J. Immunol. Methods* **46,** 63–68.

Kokko, H., and Kärenlampi, S. O. (1992). Antibody from hen's eggs against a conserved sequence of the gametophytic self-incompatibility proteins of plants. *Anal. Biochem.* **201,** 311–318.

Kokko, H. I., Kivineva, M., and Kärenlampi, S. O. (1996). Single-step immunocapture RT-PRC in the detection of raspberry bushy dwarf virus. *Biotechniques* **20,** 842–846.

Kuronen, I., Kokko, H., and Parviainen, M. (1993). Production of monoclonal and polyclonal antibodies against human osteocalcin sequences and development of a two-site ELISA for intact human osteocalcin. *J. Immunol. Methods* **163,** 233–240.

atterson, R., Youngner, J. S., Weigle, W. O., and Dixon, F. J. (1962). Antibody production and transfer to egg yolk in chickens. *J. Immunol.* **89**, 272–278.

erlmann, H., and Perlmann, P. (1994). Enzyme-linked immunosorbent assay. In "Cell Biology: A Laboratory Handbook" (J. E. Celis, ed.), Vol. 2, pp. 322–328. Academic Press, San Diego.

chade, R., Pfister, C., Halatsch, R., and Henklein, P. (1991). Polyclonal IgY antibodies from chicken egg yolk: An alternative to the production of mammalian IgG type antibodies in rabbits. *ATLA Altern. Lab. Anim.* **19**, 403–419.

tuart, C. A., Pietrzyk, R. A., Furlanetto, R. W., and Green, A. (1988). High affinity antibody from hen's eggs directed against the human insulin receptor and the human IGF-I receptor. *Anal. Biochem.* **173**, 142–150.

Purification of Immunoglobulins

Purification of Immunoglobulins

Christian Huet

INTRODUCTION

Purification of immunoglobulins is recommended when one wants to achieve good quality in immunocytochemistry and is necessary when one wants to prepare antibodies labeled with fluorochromes, enzymes, or electron-dense reagents. The purification of antibodies may be performed at two levels. Either one needs to obtain the bulk fraction of immunoglobulins or highly purified monospecific antibodies have to be obtained by affinity purification.

The immunoglobulin fraction is obtained by salt precipitation or on an ion-exchange resin. The affinity purification presented here may be used to purify either the antigen (using purified immunoglobulins) or the antibody (if the purified antigen is available).

MATERIALS AND INSTRUMENTATION

Glycerol (Cat. No. 4095) from Merck

Plastic syringes, 1 ml (Cat. No. BS-01T) and 2 ml (Cat. No. BS-H2S) from Terumo Europe NV

DE-52 preswollen anion exchanger from Whatmann

Ultrogel AcA 22 and ACT-Ultrogel AcA 22 (Cat. Nos. 230 121 and 249 203) from Sepracor

Sephadex G-50 (Cat. No. 17 0045 01) and CNBr–Sepharose 4B (Cat. No. 17 04 30 01) from Pharmacia

Ammonium chloride (Cat. No. A-4915), sodium azide (Cat. No. S-2002), sodium bicarbonate (Cat. No. S-6014), sodium chloride (Cat. No. S-9625), lysine (Cat. No. L-6001), $K_2HPO_4 \cdot 3H_2O$ (Cat. No. P-5504), and KH_2PO_4 (Cat. No. P-5379) from Sigma

Glutaraldehyde EM grade (Cat. No. G 003) from TAAB

Instamed phosphate-buffered saline (PBS, Cat. No. L 182-10) prepared by Seromed

III. PROCEDURES

A. Precipitation of Immunoglobulins with Ammonium Sulfate

Solutions

1. *Antiserum.*
2. *$(NH_4)_2$ SO_4-saturated solution:* 53.1 g $(NH_4)_2$ SO_4/100 ml saturated solution at 20°C.
3. *1 M $K_2HPO_4 \cdot 3H_2O$:* 228.2 g/liter H_2O.
4. *1 M KH_2PO_4:* 136.1 g/liter H_2O.
5. *1 M $KH_2 PO_4$/K_2HPO_4, pH 7.8:* Mix 8.5 vol of 1 M KH_2PO_4 with 91.5 vol of 1 M K_2HPO_4. Check the pH before use.
6. *1 M KH_2PO_4/K_2HPO_4, pH 7.4:* Mix 19 vol of 1 M KH_2PO_4 with 81 vol of 1 M K_2HPO_4. Check the pH before use.

Steps

1. Spin the antiserum at 10,000 rpm for 10 min to clear the serum.
2. Cool the serum on an ice bucket.
3. Keep the serum at 4°C under constant agitation with a magnetic stirrer.
4. Slowly add dropwise the saturated solution of ammonium sulfate to a final concentration of 40% saturation. This should be done in about 30 min (volume of ammonium sulfate added = 0.66 × volume of serum).
5. Let solution sit for 1 hr on ice.
6. Spin at 10,000 rpm for 10 min at 4°C and save the pellet.
7. Dissolve the precipitate in about half of the initial volume of the serum.
8. Dialyze against 2 liters PBS for 2 days (four changes) or against 10 mM phosphate buffer (pH 7.8) if the purification is to be continued on a DE-52 column.
9. Measure the absorbance at 280 nm. The absorbance of a 1% (w/v) IgG solution is 14.0. Usually a 1:20 dilution in PBS is convenient.

$$\text{total mg protein} = \frac{A_{280} \times \text{dilution} \times \text{volume}}{1.4}.$$

10. Continue purification or store at −20°C in 50% glycerol.

B. Purification on DEAE Column

Solutions

1. *1 M KH_2PO_4/K_2HPO_4, pH 7.8.*
2. *1 mM $KH_2 PO_4$/K_2HPO_4, pH 7.8.*
3. *PBS 10× stock solution:* Dissolve one flask of Seromed PBS powder in 1 liter distilled water.
4. *1% NaN_3:* Dissolve 0.5 g NaN_3 in 50 ml distilled water. Store at 4°C.
5. *DE-52 preswollen anion exchanger.*
6. *1 M NaCl:* Dissolve 5.8 g in 100 ml distilled water.

Steps

1. Stir the resin with 1 M KP_i, pH 7.8, for about 15 min. About 15–30 ml of buffer i used for every dry gram of cellulose or about 6 ml per gram of wet anion-exchange exchanger
2. Adjust the pH to 7.8 while stirring with 1 M K_2HPO_4.

3. Allow the slurry to settle and decant the supernatant containing the fines. The stock resin can be stored at 4°C in the presence of 0.01% azide.

4. Pack the resin in a plastic syringe (nothing fancy): 1 ml packed volume of DEAE can adsorb 10–20 mg of immunoglobulins, or 2.5 ml DEAE can retain immunoglobulins from ml of serum. Column volume = mg protein/15.

5. Wash the column with 10 vol of 10 mM KP$_i$, pH 7.8.

6. Allow the serum or ammonium sulfate-precipitated IgG to run through the resin column.

7. Elute the column with 10 mM KP$_i$. Discard the dead volume (V_0 = one-third of the column volume) and collect 1.5 the volume of the column.

8. Read the OD of the fractions and pool all fractions having an OD higher than 0.2.

9. Elute with salt. For mouse or guinea pig serum, use 70 mM NaCl in 10 mM KP$_i$, and or rabbit serum use 40 mM NaCl in 10 mM KP$_i$.

10. Read the OD of the fractions and pool all fractions having an OD higher than 0.2.

11. Dialyze against PBS. If necessary, concentrate the pooled fraction by dialyzing against PBS–50% glycerol. Store at 4°C in 0.01% sodium azide or at −20°C in 50% glycerol.

C. Immunoadsorbent Preparation and Affinity Purification

Solutions

Gel Activation with Glutaraldehyde

1. *1 M KP$_i$, pH 7.4.*
2. *25% Glutaraldehyde stock solution.*
3. *1% NaN$_3$:* Dissolve 0.5 g NaN$_3$ in 50 ml distilled water. Store at 4°C.

Coupling the Protein to the Gel (Glutaraldehyde as Cross-linker)

1. *0.1 M KP$_i$, pH 7.4.*
2. *Protein solution:* 2–5 mg/ml in phosphate buffer.

Coupling the Protein to Sepharose 4B (Cyanogen as Cross-linker)

1. *1 M NaCO$_3$:* Dissolve 8.4 g in 100 distilled water.
2. *1 M lysine:* Dissolve 1 g of lysine in 10 ml water.
3. *1 mM HCl.*
4. *1 M NaCl:* Dissolve 5.8 g in 100 ml distilled water.
5. *1 M CH$_3$COONa/HCl, pH 4:* 1 M CH$_3$COONa is made by dissolving 27.22 g in 200 ml water. To 50 ml of 1 M CH$_3$COONa, add 40 ml of 1 N HCl solution, made up to 250 ml.

Affinity Purification

1. *PBS.*
2. *1 M K$_2$PO$_4$.*
3. *1 N HCl/2 M lysine buffer:* To make 2 M lysine, dissolve 2 g of lysine in 20 ml of water. Add the lysine slowly to the HCl, measuring the pH until it reaches 2.2.

Steps

Gel Activation with Glutaraldehyde

This step is presented here although activated gels are commercially available. If you obtaine[d] such an activated gel, go to step 1 under Coupling the Protein to Act Ultrogel AcA 22.

1. Wash AcA 22 Ultrogel on a Buchner funnel, porosity C, with double-distilled wat[er] (10–20 vol).

2. Wash the gel with 0.1 M KP$_i$ buffer, pH 7.4.

3. Activate the gel by mixing 20 parts gel, 10 parts 0.1 M KP$_i$, pH 7.4, 20 parts 25[%] glutaraldehyde, and 50 parts H$_2$O in a flask. Keep rotating the flask at 37°C for 18 hr or [at] 4°C for 48 hr.

4. Remove the flask from the agitator, let it sit for a while, then remove the fines generate[d] during incubation. Add 2 vol double-distilled H$_2$O to the gel, mix, and remove the fine[s] (repeat three times).

5. Wash the gel with double-distilled H$_2$O thoroughly (i.e., until you do not smell th[e] glutaraldehyde at all).

6. Add NaN$_3$ (to 0.01%) to the washed activated gel and store in a 50-ml tube in th[e] cold. If you want to store for a long period (over a month), add 0.1% glutaraldehyde [to] the beads.

Coupling the Protein to Act Ultrogel AcA 22 (Glutaraldehyde as Cross-linker)

1. The concentration of dialyzed protein in 0.1 M KP$_i$ must be 2–4 mg/ml.

2. Wash the gel with 0.1 M KP$_i$ buffer thoroughly (i.e., you do not smell glutaraldehyde [at] all).

3. Mix equal volumes of protein and gel, incubate at 37°C for 18 hr (or at 4°C for 48 hr) and keep rotating.

4. Spin the beads at 100 g for 2 min. Remove and save supernatant for step 5.

5. Wash the uncoupled protein from the gel with 10 vol PBS at room temperature. Combine the supernantants and calculate coupling efficiency.

6. Add equal volume of 0.1 M lysine in PBS, pH 7.4, to the protein-coupled gel and incubate at 37°C (keep rotating) for at least 2 hr.

7. Wash the gel with PBS at room temperature.

Coupling the Protein to CNBr-Activated Sepharose 4B (Cyanogen Bromide as Cross-linker)

You can use Sepharose 4B that has been activated with cyanogen bromide, a ready-to-us[e] stabilized derivative.

1. Weigh out the required amount of CNBr-activated Sepharose 4B (1 g gives about 3.5-ml packed column).

2. Wash and reswell on a Büchner funnel (porosity 6) with 200 ml of 1 mM HCl pe[r] gram of Sepharose powder.

3. Dissolve protein or peptide to be coupled in 0.2 M NaHCO$_3$ and 0.5 M NaC[l] pH 8.5.

4. Wash with the coupling buffer.

5. Mix protein solution (use 1–10 mg of antigen/ml of gel) with gel suspension an[d] incubate for 2 hr at room temperature and then overnight under agitation (rotate, d[o] not stir).

6. Spin out beads (100 g for 2 min) and remove supernatant.

7. Wash uncoupled protein from beads with 10 vol of coupling buffer.

8. Combine supernatants and read the OD_{280}. Calculate coupling efficency.

9. Block free unreacted cross-linker groups by incubating with 0.1 M lysine in coupling ffer for 2 hr.

10. Wash excess lysine and nonconvalently bound protein by cycles of coupling buffer llowed by acetate buffer 0.5 M NaCl, 0.1 M CH_3 COONa/HCl, pH 4.

11. Resuspend beads, add 0.01% sodium azide, and store at 4°C.

finity Purification

1. Incubate the gel (keep rotating with serum or antiserum for 2 hr at 37°C or for 18 hr 4°C or pass the serum through a packed column). If the volume of serum is larger than e void volume of the column, recycle the serum.

2. The volume of antiserum to be used depends on the titer of antiserum. Usually, the aximum capacity of the column is equal to 70% of the equivalent point; 10 mg IgG retains tween 30 and 50 mg anti-IgG.

3. Wash the gel with PBS until the OD_{280} is less than 0.02.

4. Pack the column.

5. Prepare fraction tubes, adding 0.3 ml of 0.1 M K_2PO_4 to each of them.

6. Elute the column in the cold with 0.2 M HCl/2 M lysine, pH 2.2.

7. Collect 30 drops per tube.

8. Read the OD_{280} of each tube. (Mix well before reading the OD.)

9. Pool the fractions in which the OD is greater than 0.1. Adjust pH to 7–7.4.

10. When the OD of the fraction is lower than 0.08, neutralize the column by eluting th 1 M K_2PO_4 immediately.

11. Rinse the column immediately with PBS, add NaN_3 to 0.01%, and store in the cold. he column may be reused several times if carefully stored.)

12. Dialyze eluted protein in the cold against 100 vol of 0.01% NaN_3, PBS for 48 hr hree or four changes).

13. Concentrate protein up to 3–5 mg/ml and store in 50% glycerol–PBS at −20°C.

Purification of Protein A–Sepharose Columns

rotein A is a 42,000 MW protein extracted from *Staphylococcus aureus* and has the ability bind two immunoglobulin molecules. It has a strong affinity for the Fc region of the munoglobulins.

Protein A–Sepharose CL-4N resins are commercially obtained from Pharmacia Biotech-ology. One milliliter of resin has bound some 2 mg of protein A and this can bind about) mg of IgG.

Binding and eluting conditions are available from the suppliers for the different subclasses antibodies and for different animal species.

. PITFALLS

1. Polyclonal affinity-purified rabbit antibodies are very difficult to concentrate higher an 2–3 mg/ml. Furthermore, they precipitate slowly when stored at 4°C. Storage in 50% ycerol PBS at −20°C is recommended. Purified monoclonal antibodies stay well in solution concentrations as high as 10 mg/ml.

2. Never use a magnetic stirrer to agitate gel; it will break the beads. Any other method f smoothly agitating will do (rotating, rocking).

3. Unsuccessful coupling is generally due to the presence of amino groups in the solutions.

Normally it comes from the ammonium sulfate used to precipitate the proteins or peptide or from the use of Tris buffer. Desalt or dialyze very carefully, with volumes of buffer large enough. Also ensure that no amino groups are present in the water used (storage in plastic containers may be a source of contamination).

4. Other eluants may be used to desorb the affinity-bound immunoglobulins: raising lowering pH of the initial solvent, addition of polarity-reducing agents (i.e., ethylene glycol dioxane), denaturing agents (i.e., urea, guanidine), or chaotropic agents (i.e., thiocyanate trifluoroacetate).

5. As denatured immunoglobulins may be obtained from the affinity column, it is critical to elute at low pH and to elute quickly and at a carefully maintained low temperature.

Suggested Reading

Ey, P., Prowse, S., and Jenkin, C. (1978). Isolation of pure IgG1, IgG2a and IgG2b immunoglobulins from mouse serum using protein A–Sepharose. *Biochemistry* **15**, 429–436.

Harlow, E., and Lane, D. (1988). "Antibodies: A Laboratory Manual." Cold Spring Harbor Laboratory Press, Cold Spring Harbor, NY.

Kreis, T. (1987). Purification of antibodies. *In* "EMBO Practical Course: Antibodies in Cell Biology." EMBO, Heidelberg.

Mäkelä, O., and Seppälä, I. (1986). Haptens and carriers. *In* "Handbook of Experimental Immunology" (D. Weir, ed.), Vol. 1. Blackwell Scientific, London.

Ternynck, T., and Avrameas, S. (1976). Polymerization and immunobilization of proteins using ethylchloroformate and glutaraldehyde. *Scand. J. Immunol.* **3**, 29.

Ternynck, T., and Avrameas, S. (1987). "Techniques Immunoenzymatiques." Editions INSERM, Paris.

Antibody Specificity

Determination of Antibody Specificity by Western Blotting and Immunoprecipitation

Julio E. Celis, Jette B. Lauridsen, and Bodil Basse

I. INTRODUCTION

Immunoblotting (Towbin *et al.,* 1979; Symington, 1984; Harlow and Lane, 1988; Otto and Lee, 1993) and immunoprecipitation (Harlow and Lane, 1988; Otto, 1993; see Volume 4, Lukas and Bartek, "Immunoprecipitation of Proteins under nondenaturing conditions" for additional information) are among the most common techniques used to determine the specificity of an antibody. As some antibodies work well in immunoblotting but not in immunoprecipitation and vice versa, it is important to set up both techniques in the laboratory.

II. MATERIALS AND INSTRUMENTATION

Trizma base (Cat. No. T-1503), glycine (Cat. No. G-7126), bovine hemoglobin (Cat. No. H2625), and Bovine Serum Albumin (BSA, Cat. No. A-4503) are from Sigma. EDTA (Cat. No. 8418), 30% hydrogen peroxide (Cat. No. 7209), and sodium deoxycholate (Cat. No. 1065040100) are from Merck. Protein A–Sepharose CL-4B (Cat. No. 17-0780-01) is from Pharmacia Biotech and Nonidet P-40 (NP-40, Cat. No. 56009 2L) is from BDH. Peroxidase-conjugated rabbit anti-mouse immunoglobulins (Cat. No. P0260), rabbit immunoglobulins to mouse immunoglobulins (Cat. No. Z0259), and the mouse monoclonal antibody against proliferating cell nuclear antigen (PCNA) (Cat. No. M0879) are from DAKO. The monoclonal antibody against MEK 2 is from Transduction Laboratories (Cat. No. M24520). Dehydrated bacto skim milk (Cat. No. 0032173) is from Difco Laboratories. HRP color development reagent (Cat. No. 170-6534) is from Bio-Rad. Tissue culture media and supplements are as described in the article by Celis and Celis, Volume 1 "General Procedures for Tissue Culture."

Nitrocellulose Hybond-C (Cat. No. RPN 203C) and the ECL kit are from Amersham and Filter-Count (Cat. No. 6013149) is from Packard. Plates (96 wells, Cat. No. 655180) are from Greiner. Other sterile plasticware are as described in the article by Ariana Celis and Julio E. Celis in Volume 1. Rectangular (24 × 19 cm) pie dishes are from Corning (Cat. No. PX 38567), and X-ray films (X-Omat DS; 18 × 24 cm, Cat. No. 508 7838) are from Kodak. The Trans-Blot electrophoretic transfer cell is from Bio-Rad (Cat. No. 170-3910) and the orbital shaker (Red Rotor PR75) is from Pharmacia. The power supplies (EPS 500/400) are from Pharmacia.

III. PROCEDURES

A. Two-Dimensional Gel Immunoblotting

The procedure is illustrated using Simian virus 40 (SV40)-transformed human keratinocyt (K14) proteins (whole extracts) separated by isoelectric focusing (IEF) two-dimensional (2 D) gel electrophoresis and blotted to a nitrocellulose membrane (Towbin *et al.,* 1979). Blot are analyzed using the HRP color development reagent and the ECL protocol.

1. Blotting
Solution

1. *TGM (Tris, glycine, methanol):* 25 mM Tris, 192 m*M* glycine, and 20% (v/v) methano To make 5 liters, weigh 15.14 g of Tris base and 72.07 g of glycine. Add 1 liter of methano and complete to 5 liters with distilled water.

Steps

1. Run 2-D gels as described in the article by Celis, *et al.,* Volume 4, "High-Resolutio Two-Dimensional Gel Electrophoresis of Proteins: Isoelectric Focusing (IEF) and Nonequi librium pH Gradient Electrophoresis (NEPHGE). Applications to the Analysis of Culture Cells and Mouse Knockouts." In this particular case we have run an IEF 2-D gel of huma K14 keratinocyte proteins. Mix the unlabeled extract in O'Farrel lysis solution with a sma amount of [^{35}S]methionine-labeled K14 proteins (about 500,000 cpm; for labeling of cel see article by Celis, *et al.,* Volume 4, "High-Resolution Two-Dimensional Gel Electrophores of Proteins: Isoelectric Focusing (IEF) and Nonequilibrium pH Gradient Electrophores (NEPHGE). Applications to the Analysis of Cultured Cells and Mouse Knockouts" in orde

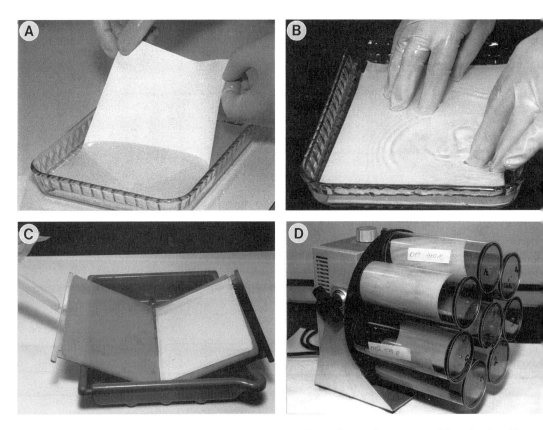

FIGURE 1 (A) Wetting the nitrocellulose membrane. (B) Placing the membrane on top of the gel. (C) Holder wit fiber gel pads. (D) Rotating roller system (Navigator).

facilitate the immunodetection of the antigen(s). For 1-D blots, resuspend the samples in Laemmli's buffer (Laemmli, 1970).

2. Place a piece of 3MM Whatmann paper (a bit larger than the size of the gel) in a rectangular glass pie dish (24 × 19 cm) containing about 100 ml of TGM. Place the gel on top of the paper.

3. Equilibrate for 5 min at room temperature.

4. Wet the fiber gel pads in TGM.

5. Open the gel holder (Fig. 1C) and place one fiber gel pad in each side.

6. Wet the nitrocellulose membrane (14 × 16 cm) in TGM by capillary action as shown in Fig. 1A. Use gloves to handle the membranes.

7. Place the wet nitrocellulose membrane on top of the gel (Fig. 1B). The operation should be done under the buffer. Rub the membrane from one end to the other to eliminate bubbles.

8. Place another wet 3MM Whatmann paper on top of the nitrocellulose membrane. Rub the paper carefully to avoid bubbles. Lift the "sandwich" (same side up) and place it on top of the fiber gel pad located on the black holder (Fig. 1C).

9. Close the gel holder and place it in the Trans-Blot tank containing about 750 ml of TGM. The black side of the holder should face the cathode side (indicated with black in the tank). Up to three holders can be inserted in the tank. If only one cassette is used, insert in the middle track.

10. Fill the tank with TGM, connect the electrodes, and run at room temperature for 4 hr at 130 mA.

11. Following protein transfer, dry the membranes and expose to an X-ray film to assess the quality of the transfer (Fig. 2A). Dry sheets can be kept for extended periods of time without significant changes in the reactivity of the proteins. Blots that do not contain radioactivity can be stained with amido black (see Volume 4 Hoffmann et al., "Calcium Overlay Assay" for additional information).

Immunodetection

HRP color development reagent

Solutions

1. *Hank's buffered saline solution (HBSS) without* Ca^{2+} *and* Mg^{2+}: 10× stock solution. To make 1 liter, weigh 4 g of KCl, 0.6 g of KH_2PO_4, 80 g of NaCl, and 0.621 g of Na_2H-$PO_4 \cdot 2H_2O$. Complete to 1 liter with distilled water.

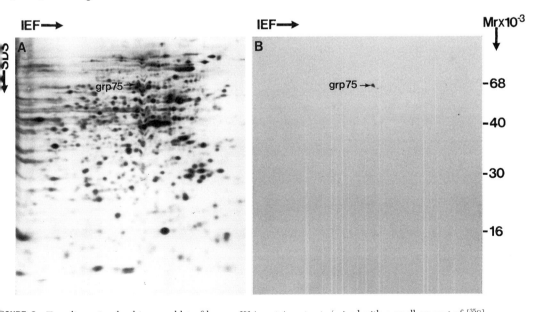

FIGURE 2 Two-dimensional gel immunoblot of human K14 protein extracts (mixed with a small amount of [^{35}S]-methionine-labeled sample) blotted to a nitrocellulose membrane (A) and reacted with mAB 30A5 which recognizes the glucose regulated protein 75, grp 75 (B). SDS, sodium dodecyl sulfate; IEF, isoelectric focusing.

2. *1× HBSS without Ca²⁺ and Mg²⁺:* To make 1 liter of HBSS, use 100 ml of the 10×
stock solution and complete to 1 liter with distilled water.

3. *2× TBS stock:* 40 mM Tris and 1 M NaCl, pH 7.5. To make 1 liter, add 4.84 g of Tris
base and 58.44 g of NaCl. Adjust to pH 7.5 with HCl and complete to 1 liter with distilled
water. Store at 4°C.

4. *Dehydrated Bacto skim milk:* 50 mg/ml in HBSS.

5. *mAB 30A5(culture supernatant) or any other antibody.*

6. *Peroxidase-conjugated rabbit anti-mouse immunoglobulins:* To make 10 ml of a 1:200
dilution, add 50 μl of peroxidase-conjugated anti-mouse immunoglobulins to 10 ml of HBSS.

7. *HRP color development solution:* To make 60 ml, dissolve 30 mg of HRP color development reagent in 10 ml of methanol (protect from light). Mix 30 ml of 2× TBS and 20 ml of
distilled water. Just before use add 30 μl of ice-cold 30% H_2O_2 to the TBS and mix the two
solutions.

Steps

1. Cut the appropriate area of the blot with a scalpel using the X-ray film as reference
and wet by capillarity in the skim milk solution. Incubate with shaking for 1 hr at room
temperature (or overnight in the cold room) in the same solution. Incubation can be done
in a rectangular plastic dish (15 × 18.5 cm; need a minimum volume of 40 ml per dish) or
a rotating roller system (Navigator Model 128, BIOCOMP, Fig. 1D). In the latter case, the
volume needed is considerably less (8 ml). For 1D gel strips we use the chamber shown in
Fig. 3 (Pierce). The following immunodetection procedure is illustrated using the rotating
roller system.

2. Wash the blots three times for 10 min each in HBSS.

3. Add 8 ml of mAB 30A5 (culture supernatant) or a similar antibody diluted 1:10 in
HBSS. Roll for 2 hr at room temperature.

4. Wash three times for 10 min each with HBSS.

5. Add 8–10 ml of a 1:200 dilution of peroxidase-conjugated rabbit anti-mouse immunoglobulins in HBSS. Roll for 2 hr at room temperature.

6. Transfer the blot to a rectangular plastic dish and wash three times for 10 min each
in HBSS.

7. Prepare the HRP color development solution just before use.

8. Aspirate the HBSS and add 30 ml of the HRP solution. Shake at room temperature
until staining appears (Fig. 2B). Discard the HRP color developer according to the safety
regulations enforced in your laboratory.

FIGURE 3 Chamber for incubating one-dimensional gel strips with antibodies.

9. Rinse well with demineralized or tap water and dry. Superimpose the dry blot and the -ray film with the aid of the radioactive marks.

Figure 2 shows immunoblots of human K14 proteins extracts (mixed with a small amount [^{35}S]methionine-labeled sample) blotted to nitrocellulose (autoradiography, Fig. 2A) and acted with mAB 30A5 (specific for grp 75) (Fig. 2B). Ideally, 2-D gel protein blots [IEF d nonequilibrium pH gradient electrophoresis (NEPHGE)] should be used to determine e specificity of the antibody. The procedure described here has been used in no less than)00 blots using hundreds of polyclonal and monoclonal antibodies (Celis *et al.*, 1991).

ECL detection

he procedure has been slightly modified from the ECL Western blotting protocol described Amersham.

olutions

1. *10× TBS, pH 7.6 (Tris-buffered saline) stock:* To make 2 liters, add 48.4 g Tris base .20 *M*) and 160 g NaCl (1.37 *M*). Adjust to pH 7.6 with HCl (37% fuming) and complete 2 liters with distilled water. Store at 4°C.

2. *TBS–Tween 0.05%:* To make 1 liter, add 100 ml of 10× TBS, 0.5 ml of Tween 20, d complete to 1 liter with distilled water.

3. *Blocking solution:* 5% skim milk; 5 g per 100 ml of TBS–Tween.

4. *mAB anti-MEK 2.*

teps

1. Wet the nitrocellulose blot with TBS–Tween in a rectangular plastic dish and transfer it to a bottle (fitting the Navigator system, Fig. 1D) containing 10 ml of blocking solution.

2. Block for 1 hr at room temperature or overnight in the cold room.

3. Wash 3× for 10 min in TBS–Tween or until the washing buffer is clear.

4. Add 10 ml of the primary antibody (mAB anti-MEK 2) diluted 1:2500 in TBS–Tween and incubate for 1.5 hr at room temperature.

5. Wash 3× for 10 min in TBS–Tween.

6. Add 10 ml of peroxidase-conjugated secondary antibody diluted 1:1000 in TBS–Tween for 1 hr at room temperature.

7. Transfer the membrane blot to a plastic dish and wash 3× for 10 min in TBS–Tween.

ne next steps are carried out in the dark room:

8. Mix 20 ml of detection solution 1 with 20 ml of detection solution 2. This is sufficient for at least eight blots of 13 × 15 cm.

9. Lift the blot carefully and dry it gently by touching a piece of paper towel. Place the blot in a clean plastic dish and add the detection solution. Leave for 1 min (no shaking).

0. Let the solution run of the blot as described earlier and put it on top of a piece of plastic in a film cassette (protein side up). Cover it quickly with Saran wrap and carefully smooth out air pockets with a piece of paper.

1. Turn of the light (red safety light is allowed). Place an X-ray film on top of the membrane and close the cassette. Expose for 5, 15, 30 sec up to 15 min if necessary.

Figure 4 shows an IEF 2-D gel blot of [³⁵S]methionine-labeled protein from K14 cell reacted with anti-MEK 2 antibodies.

C. Immunoprecipitation under Denaturing Conditions

This procedure is illustrated using a [³⁵S]methionine-labeled extract from transformed human amnion (AMA) cells and a mouse monoclonal antibody directed against the PCNA. See also Harlow and Lane (1988).

Solutions

1. *1 M Tris–HCl, pH 7.4 (stock solution):* To make 100 ml, dissolve 12.11 g of Tris base in distilled water. Adjust the pH to 7.4 with HCl and complete to 100 ml with distilled water.

2. *1 M NaCl (stock solution):* To make 500 ml, weigh 29.22 g of NaCl and complete to 500 ml with distilled water.

3. *0.5 M EDTA (stock solution):* To make 50 ml, take 9.31 g of EDTA and add about 40 ml of distilled water. Stir and slowly add dropwise 10 N NaOH until the EDTA goes into solution. Complete with distilled water.

4. *NP-40 (stock solution), 10% (w/v):* To make 100 ml, weigh 10 g of NP-40 and complete to 100 ml with distilled water.

5. *Sodium deoxycholate (stock solution), 2.5%:* To make 500 ml, weigh 12.5 g sodium deoxycholate and complete to 500 ml with distilled water.

6. *10% SDS (stock solution):* To make 50 ml, weigh 5 g of SDS and complete to 50 ml with distilled water.

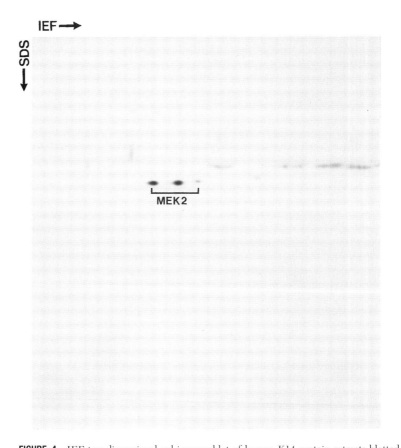

FIGURE 4 IEF two-dimensional gel immunoblot of human K14 protein extracts blotted to a nitrocellulose membrane, reacted with an antibody against MEK 2, and developed using the ECL procedure. The position of MEK 2 and its variant is indicated. SDS, sodium dodecyl sulfate; IEF, isoelectric focusing.

7. *Buffer C:* 50 mM Tris–HCl, pH 7.4, 150 mM NaCl, 5 mM EDTA, 0.5% NP-40, 0.5% sodium deoxycholate, and 0.1% SDS. To make 1 liter, add 50 ml of 1 M Tris, pH 7.4, 150 ml of 1 M NaCl, 10 ml of 0.5 M EDTA, 200 ml of 2.5% sodium deoxycholate, 10 ml of 10% SDS, and 50 ml of 10% NP-40. Adjust pH to 7.28 and complete to 1 liter with distilled water. Keep at 4°C.

8. *Buffer C with bovine serum albumin (BSA):* To make 100 ml, weigh 200 mg of BSA and add 100 ml of buffer C.

9. *Protein A–Sepharose:* To make 1 ml, weigh 100 mg of protein A–Sepharose and add 1 ml of buffer C containing BSA.

10. *Mouse monoclonal antibody against PCNA:* Tissue culture supernatant, protein concentration = 15.1 g/liter.

11. *Rabbit anti-mouse immunoglobulins.*

Steps

1. Label AMA cells with [^{35}S]methionine as described in the article by Celis *et al.*, Volume 4, "High-Resolution Two-Dimensional Gel Electrophoresis of Proteins: Isoelectric Focusing (IEF) and Nonequilibrium pH Gradient Electrophoresis (NEPHGE). Applications to the Analysis of Cultured Cells and Mouse Knockouts." Use a 35-mm tissue culture dish instead of a microtiter well. One needs about 0.6 ml of labeling medium to cover the monolayer. Use [^{35}S]methionine at a concentration of 500 μCi/ml. Cover the tissue culture dish with Saran wrap and place in a 37°C humidified 5% CO_2 incubator.

2. Aspirate the radioactive media and wash the cells twice with HBSS.

3. Add 1.5 ml of buffer C containing BSA and rock the culture dish at room temperature. If the antigen is known to be proteolyzed easily, add protease inhibitors (phenylmethylsulfonyl fluoride, aprotinin, leupeptin, and pepstatin A) to buffer C.

4. Bend the end of a Pasteur pipette under the flame and use it to remove the cells.

5. Place the cell suspension in a plastic conical tube and centrifuge at 6000 g in the SS-34 rotor of the Sorvall centrifuge (4°C).

6. Take 0.1 ml of the supernatant and add it to a 1.5-ml Eppendorf microtest tube. Add 10 μl of the anti-PCNA monoclonal antibody.

7. Incubate with rotation for 1 hr at 4°C (cold room).

8. Add 5 μl of rabbit anti-mouse immunoglobulins. Incubate with rotation for 1 hr at 4°C (cold room). This step is not necessary for polyclonal antibodies.

9. Add 100 μl of the protein A solution and incubate with rotation at 4°C for 1 hr (cold room).

10. Centrifuge in an Eppendorf centrifuge for 30 sec at 10,000 g and wash the pellet four times with buffer C.

11. After the last centrifugation, remove all the liquid and resuspend in 50 μl of O'Farrel lysis solution (O'Farrel, 1975; see also Volume 4, Celis *et al.*, "High-Resolution Two-Dimensional Gel Electrophoresis of Proteins: Isoelectric Focusing (IEF) and Nonequilibrium pH Gradient Electrophoresis (NEPHGE). Applications to the Analysis of Cultured Cells and Mouse Knockouts." Leave for 60 min at room temperature.

12. Centrifuge in an Eppendorf centrifuge and run a fraction of the supernatant in IEF and NEPHGE gels as described in the article by Celis *et al.*, Volume 4, "High-Resolution Two-Dimensional Gel Electrophoresis of Proteins: Isoelectric Focusing (IEF) and Nonequilibrium pH Gradient Electrophoresis (NEPHGE). Applications to the Analysis of Cultured Cells and Mouse Knockouts."

Figure 5 shows IEF and NEPHGE gels of PCNA immunoprecipitates. In addition to PCNA, one can also detect keratins and vimentin (arrowheads). The latter are also detected in control immunoprecipitates (results not shown). Ideally, the immunoprecipitates should be analyzed by 2-D gel electrophoresis (IEF and NEPHGE).

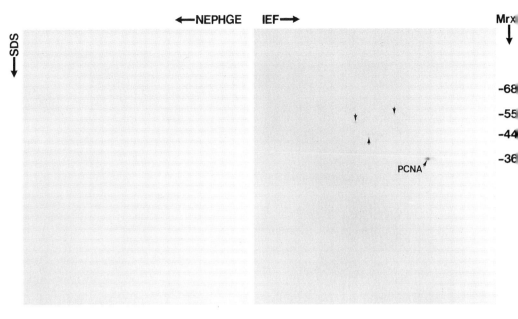

FIGURE 5 Autoradiograms of IEF and NEPHGE two-dimensional gels of [^{35}S]methionine-labeled proteins immunoprecipitated with PCNA antibodies.

IV. PITFALLS

A. Western Blotting

1. It is a good idea to remove excess vaseline floating in the buffer with a tissue.

2. Avoid air bubbles when making the "sandwich" and lifting it to the fiber gel pad.

3. For 2-D blotting it is important to run a range of concentrations of the protein mixture. Choose a protein concentration that does not give streaking. Never run more protein than is necessary.

4. Membrane proteins may streak and it may be necessary to use detergents such as Triton X-114.

5. H_2O_2 is unstable. Check the expiration date on the bottle.

B. Immunoprecipitation

1. Variable levels of keratins and vimentin are usually observed in the immunoprecipitates (short arrows in Fig. 5). Make sure you run a proper control.

2. The amount of antibody needed for immunoprecipitation varies from antibody to antibody. It is advisable to try a range of concentrations. Recommended initial volumes are 10–20 μl for culture supernatant and 5 μl for sera.

3. Label the cells with as much radioactive isotope as possible to enrich for low-abundancy proteins.

4. In case of no signal, use other buffers to solubilize the sample (see Harlow and Lane, 1988; Otto and Lee, 1993). Try a different amino acid for labeling. If still no signal, try immunoblotting.

References

Celis, J. E., *et al.* (1991). The master two-dimensional gel database of human AMA cells proteins: Towards linking protein and genome sequence and mapping information. *Electrophoresis* **12**, 765–801.

Glenney, J. R., Jr., Glenney, P., and Weber, K. (1982). Erythroid spectrin, brain fodrin, and intestinal brush border proteins (TW-260/240) are related molecules containing a common calmodulin-binding subunit bound to a variant cell type-specific subunit. *Proc. Natl. Acad. Sci. USA* **79**, 4002–4005.

Harlow, E., and Lane, D. (1988). "Antibodies: A Laboratory Manual." Cold Spring Harbor Laboratory, Cold Spring Harbor, NY.

Laemmli, U. K. (1970). Cleavage of structural proteins during the assembly of the head of bacteriophage T4. *Nature (London)* **227**, 680–685.

O'Farrell, P. H. (1975). High resolution two-dimensional electrophoresis of proteins. *J. Biol. Chem.* **250**, 4007–4021.

O'Farrell, P. Z., Goodman, H. M., and O'Farrell, P. H. (1977). High resolution two-dimensional electrophoresis of basic as well as acidic proteins. *Cell* **12**, 1133–1142.

Otto, J. J. (1993). Immunoblotting. *In* "Antibodies in Cell Biology" (D. J. Asai, ed.), pp 105–117. Academic Press, San Diego.

Otto, J. J., and Lee, S. (1993). Immunoprecipitation methods. *In* "Antibodies in Cell Biology" (D. J. Asai, ed.), pp 119–127. Academic Press, San Diego.

Symington, J. (1984). *In* "Two-Dimensional Gel Electrophoresis of Proteins: Methods and Applications" (J. E. Celis and R. Bravo, eds.), pp 126–168, Academic Press, New York.

Taylor-Papadimitriou, J., Purkis, P., Lane, E. B., McKay, I. A., and Chang, S. E. (1982). Effect of SV40 transformation on the cytoskeleton and behavioural properties of human keratinocytes. *Cell. Diff.* **11**, 169–180.

Towbin, H., Staehelin, T., and Gordon, J. (1979). Electrophoretic transfer of proteins from polyacrylamide gels to nitrocellulose sheets: Procedure and some application. *Proc. Natl. Acad. Sci. USA* **76**, 4350–4354.

Enzyme-Linked Immunosorbent Assay

Hedvig Perlmann and Peter Perlmann

I. INTRODUCTION

The enzyme-linked immunosorbent assay (ELISA) (Engvall and Perlmann, 1971) is a highly versatile and sensitive technique that can be used for qualitative or quantitative determinations of practically any antigen or antibody. Reagents are stable, nonradioactive, and, in many cases, commercially available. The use of 96-well microtiter plates as the solid phase and the availability of microtiter plate readers have added to the usefulness of the method. In one of its simplest forms, the assay involves immobilization of one reagent (e.g., antigen) on a plastic surface, followed by the addition of test antibodies specific for the antigen and, after washing, enzyme-conjugated second antibodies against the test antibodies. The addition of substrate giving colored, fluorescent, or luminescent reaction products makes it possible to determine the concentrations of the reactants at very low levels. Commercially available micromodifications of the assay (Amersham Life Science) permit the determination of antigens (e.g., certain cytokines) at concentrations <1 pg/ml. Of several enzymes suitable for ELISA, alkaline phosphatase (ALP) and horseradish peroxidase (HRP) are most commonly used. Instead of preparing enzyme-conjugated antibodies (or antigens), a useful and more versatile system that also amplifies the reaction is to biotinylate the reagent and let it bind enzyme-conjugated streptavidine (O'Sullivan and Marks, 1981; Ternynck and Avrameas, 1990).

This article provides three examples of ELISA protocols: indirect ELISA to determine antibodies, competitive ELISA for defining antigenic specificities, and sandwich ELISA to detect antigens. For the many modifications and applications the literature should be consulted (e.g., see Hornbeck, 1991; Kemeny, 1992; Maloy et al., 1991; Mark-Carter, 1994; Ravindranath et al., 1996; Zielen et al., 1996).

One important application of ELISA is ELISPOT (Czerkinsky et al., 1983; Sedgwick and Holt, 1983), which makes it possible to enumerate immunoglubulin or antibody-secreting cells on a single cell basis (Kawabata, 1995). It is also being used to detect cells that secrete cytokines or various antigens (Klinman and Nutman, 1994). As it is possible to detect secreting cells, even when they are very few, it is a very sensitive method. This article gives the protocol for a cytokine ELISPOT.

II. MATERIALS AND INSTRUMENTATION

Flat-bottomed microtiter plates: Maxisorp from Nunc or High Binding from Costar
Round-bottomed microtiter plates for preparation of dilutions.

Micropipette, multichannel pipette, and disposable pipette tips (Finnpipette, Lab Systems OY)

V_{max} kinetic microplate reader with computer program SOFTmax from Molecular Devices Corporation

Microplate washer (Titertek, ICN Biomedicals)

96-well nitrocellulose plates: Multiscreen-HA, Millipore Corp.

Round-bottomed tissue culture tubes (5 ml) (Falcon 2058, Becton Dickinson Labware)

Millipore vacuum control machine (Millipore Corp.)

Cell incubator with 5% CO_2 atmosphere

Dissection microscope ($\times 40$)

II. PROCEDURES

A. Alkaline Phophatase Conjugation

See O'Sullivan and Marks (1981).

Solutions

1. *ALP:* Type VII-S, Sigma P-5521 (10,000 U = 3–5 mg protein).
2. *PBS 10$\times$ concentrated stock solution:* 40 g $Na_2HPO_4 \cdot 12H_2O$, 5 g KH_2PO_4, 81 g NaCl, and H_2O to 1000 ml.
3. *Glutardialdehyde:* From Merck, for electron microscopy.
4. *Tris buffer, pH 8.0:* 4.44 g Tris–HCl, 2.65 g Tris base, 203 mg $MgCl$, 200 mg NaN_3, and H_2O to 1000 ml.

Steps

1. Centrifuge (2000 rpm) 3–4 mg ALP. Discard supernatant.
2. Add 1 mg of affinity-purified polyclonal antibodies or monoclonal antibodies purified by ammonium sulfate precipitation to the pellet.
3. Dialyse against PBS for 4 hr at room temperature.
4. Weigh dialysate and add 8 μl 25% glutardialdehyde per gram of solution.
5. Mix on roller drum at 4°C overnight.
6. Dialyze against PBS at 4°C overnight and then against Tris buffer, pH 8.0.
7. Add 0.5% BSA and store at 4°C.

B. Biotinylation of Immunoglobulin

See Ternynck and Avrameas (1990).

Solutions

1. *0.1 M NaHCO$_3$:* 84.01 g/1000 ml H_2O.
2. *DMF:* N,N-dimethylformamide (BDH).
3. *Biotin:* D-biotinyl-ϵ-aminocaproic acid N-hydroxysuccinimide ester (Boehringer Mannheim, Cat. No. 1008960).

Steps

1. Dialyze affinity-purified polyclonal antibodies or monoclonal antibodies purified by ammonium sulfate precipitation against 0.1 M NaHCO$_3$, at 4°C overnight and adjust t 1 mg/ml.

2. Dissolve 1 mg biotin in 100 μl ice-cold DMF in a glass tube (keep dark). Immediately before the next step add 900 μl ice-cold, filtered 0.1 M NaHCO$_3$.

3. Mix 1 ml antibody (1 mg/ml) with 200 μl biotin (1 mg/ml). Keep the tube dark in an ice bath for 2 hr and then at 4 hr at room temperature with occasional shaking.

4. Dialyze overnight at 4°C against PBS containing 0.02% NaN$_3$.

5. Add 0.5% BSA and store at 4°C.

C. Optimal Reagent Concentrations

1. *Concentration of coating reagent:* The plastic solid phase is practically saturated whe coated with the reagent at a concentration of 10 μg/ml. Try 10 μg/ml down to 1 μg/ml coating buffer, pH 9.6. (Some proteins adhere better to the plate at a lower pH.) For unknow antigens or monoclonals, try PBS or a buffer of pH ~5.

2. *Concentration of test reagent:* If possible, use one known positive, one known negativ and, if available, a standard. Find the concentration where the known positive sample give a good positive reading and responds to dilution.

3. *Concentration of developing reagent:* Commercial enzyme conjugates usually have recommended concentration. Try a few dilutions around that, e.g., 1/500, 1/1000, 1/200 and 1/4000. Choose the lowest concentration assuring excess; the only limiting factor in th setup should be the amount of test reagent.

D. Elisa Protocols

Solutions

1. *Coating buffer, pH 9.6:* 1.59 g Na$_2$CO$_3$, 2.93 g NaHCO$_3$, 200 mg NaN$_3$, and H$_2$O to 1000 ml.

2. *Incubation (diluent) buffer:* 100 ml PBS stock (10×), 5 g BSA, 0.5 ml Tween 20, 1 ml 20% NaN$_3$, and H$_2$O to 1000 ml.

3. *Washing buffer:* 45 g NaCl, 2.5 ml Tween 20 (0.05%), and H$_2$O to 5000 ml.

4. *Enzyme substrate buffer:* 97 ml diethanolamine, 800 ml H$_2$O, 1 ml 20% NaN$_3$, and 1C mg MgCl$_2$·6H$_2$O (should be added last). Adjust finally to pH 9.8 with 1 M HCl (~10 ml).

5. *Substrate for ALP:* NPP (p-nitrophenyl phosphate, 5-mg tablets (Sigma Chemical Co. 1 tablet/5 ml of enzyme substrate buffer.

6. *Streptavidin/ALP:* From Sigma Chemical Co.

1. Indirect ELISA for Screening of Specific Antibodies in Serum or Hybridoma Supernatants

Reactants

1. *Antigen.*

2. *Antibody specific for test antigen.*

3. *Anti-immunoglobulin enzyme conjugate.*

4. *Substrate.*

Steps

1. Coat plate with 50 μl/well of antigen, diluted in coating buffer, pH 9.6, overnight at 4°C.

2. Block with 100 μl/well of incubation buffer for 1 hr at 37°C. Wash four times with washing buffer.

3. Add 50 μl/well of immune serum, e.g., diluted 1/1000, or hybridoma supernatant, e.g., 1/5, in incubation buffer and incubate for 1 hr at 37°C. Wash four times with washing buffer.

4. Add 50 μl/well of ALP-conjugated anti-immunoglobulin (e.g., goat anti-human γ-chains, rabbit anti-mouse Ig, etc.) diluted in incubation buffer. Incubate for 1 hr at 37°C. Wash four times with washing buffer.

5. Develop with fresh NPP, 50 μl/well, and read absorbance at 405 nm.

2. Competitive ELISA for Defining Antigenic Specificities and Determining Possible Antigenic Cross-Reactivity between Antigens

Reactants

1. *Antigen.*

2. *Serial dilutions of test antigen + antibody specific for test antigen.*

3. *Anti-immunoglobulin enzyme conjugate.*

4. *Substrate.*

Steps

1. Coat plate with 50 μl/well of antigen, diluted in coating buffer, pH 9.6, overnight at 4°C.

2. Block with 100 μl/well of incubation buffer for 1 hr at 37°C. Wash four times with washing buffer.

3. Mix 40 μl of serial dilutions of test antigen in incubation buffer and 40 μl/well of an immune serum with known specificity, diluted in incubation buffer in round-bottomed microtiter plates. Incubate for 1 hr at 37°C and transfer 50 μl of the mixtures to the coated plate; incubate for 4 hr at room temperature. Wash four times with washing buffer.

4. Add 50 μl/well of relevant ALP-conjugated anti-immunoglobulin diluted in incubation buffer. Incubate for 1 hr at 37°C. Wash four times with washing buffer.

5. Develop with fresh NPP, 100 μl/well, and read absorbance at 405 nm.

3. Sandwich ELISA for Detecting Antigens

Reactants

1. *Capture antibody specific for test antigen.*

2. *Antigen.*

3. *Biotinylated antibody specific for test antigen:* See Section IV.

4. *Streptavidin/ALP.*

5. *Substrate.*

Steps

1. Coat plate overnight at 4°C with 50 μl/well of antigen-specific monoclonal antibody diluted in PBS or antigen-specific polyclonal antibody diluted in coating buffer.

2. Block with 100 μl/well of incubation buffer for 1 hr at 37°C. Wash four times with washing buffer.

3. Add 50 μl/well of test reagent (cell culture supernatants or serum dilutions). Incubate for 1 hr at 37°C. Wash four times with washing buffer.

4. Add 50 μl/well of biotinylated monoclonal antibody specific for a different determinant of the antigen or biotinylated polyclonal antibody of the same antigen specificity as used for coating (see Section IV) diluted in incubation buffer. Incubate for 1 hr at 37°C. Wash four times with washing buffer.

5. Add 50 μl/well of ALP-conjugated streptavidin and 1/2000 in incubation buffer for 1 hr at 37°C. Wash four times with washing buffer.

6. Develop with fresh NPP, 50 μl/well, and read absorbance at 405 nm.

E. Isolation of Mononuclear Cells (PBMC) from Peripheral Blood

See Volume 1, Celis and Celis, "General Procedures for Tissue Culture" and Volume 1, Hokland *et al.*, "Isolation of Mononuclear Cells from Human Blood and Bone Marrow and Identification of Leukocyte Subsets by Multiparameter Flow Cytometry" for additional information.

F. Elispot Protocol

Solutions

1. *Coating buffer, pH 9.6:* Millipore filtered (0.45 μm), 1.59 g Na_2CO_3, 2.93 g $NaHCO_3$, 200 mg NaN_3, and H_2O to 1000 ml.
2. *PBS Millipore filtered (0.45 μm) (10× concentrated stock solution:* 40 g $Na_2HPO_4 \cdot 12$ H_2O, 5 g KH_2PO_4, 81 g NaCl, 1.0 ml 20% NaN_3, and H_2O to 1000 ml.
3. *Streptavidin/ALP.*
4. *Substrate for ALP:* Kit is from Bio-Rad.

Interleukin ELISPOT

Reactants

1. *Antibody specific for the test interleukin.*
2. *Biotinylated anti-interleukin antibody:* Of noncompeting specificity.
3. *Substrate.*

Steps

1. Dilute PBMC in tissue culture medium to 2×10^6 cells/ml and set up 0.5–1 ml cultures in round-bottomed 5-ml Falcon tubes.
2. Add antigen and/or polyclonal activator [e.g., 1–10 μg/ml of phytohemagglutinin (PHA)] for cell stimulation.
3. Incubate the tubes (leaning, 45° angle) for 4 hr at 37°C in a cell incubator (humid atmosphere containing 5% CO_2).
4. Coat nitrocellulose plates with 100 μl/well of monoclonal antibodies against test interleukin diluted to 15 μg/ml in filtered coating buffer overnight at 4°C.
5. Wash the plates six times with 200 μl/well of filtered PBS. The PBS is flicked out except for the last wash which is sucked through using the vacuum control machine.
6. Add 100 μl/well of PBMC (20,000–200,000 cells) in quadruplicates and incubate plates for 40–42 hr in a cell incubator.
7. Wash the plates as described earlier. Avoid cells remaining on the filter surface.
8. Add 100 μl/ml of biotinylated anti-interleukin antibody, 1 μg/ml, in filtered PBS. Incubate for 2–4 hr at room temperature.

9. Wash as described earlier and add 100 μl streptavidin/ALP, diluted 1/1000 in filtered PBS. Incubate for 1.5 hr at room temperature.

0. Wash as described previously, add substrate, and incubate at room temperature until dark spots emerge (1–2 hr).

1. Stop color development by washing with 3 $\times$ 200 μl/well of tap water.

2. Leave plates to dry; count spots in a dissection microscope.

V. COMMENTS

A. Controls

Four to eight wells in every plate are used for blanks (substrate, but no reagents) and the mean value is subtracted from test values. All samples are set up in duplicates. Negative controls with incubation buffer replacing the reagents should always be included and should give OD readings well below OD 0.100. In the two-site sandwich applications, it is especially important that the capture antibody does not bind to the second antibody and vice versa. For screening of unknown samples, and comparison between different runs, include a known positive sample as a reference. For estimation of background, include expected negatives, e.g., nonimmune sera, cell culture medium, or irrelevant antigen.

B. Blocking

After coating, vacant protein-binding sites on the plastic surface should be blocked. BSA, casein, milk powder, or gelatine are commonly used. To avoid cross-reactive antibody binding to the blocking protein, the same protein used for dilution of the reagents should be included in the incubation (diluent) buffer.

C. Incubation Times

Coating the plates overnight is often practical, and the coated plates can be stored for several weeks at 4°C, wrapped in plastic film. Coating for 3–4 hr at room temperature or 1 hr at 37°C may often be enough. The same holds true for the specific binding of the reagents. The signal may be significantly increased by longer incubations, but shorter incubations at 37°C may suffice as well. The development of color with the substrate varies in time, but requires usually between 10 and 60 min.

D. Amplification

Amplification of the signal may be obtained, e.g., by use of an extra layer such as enzyme-conjugated anti-immunoglobulin as the developing reagent or biotinylated antibody followed by streptavidin–enzyme conjugate as the developing reagent.

E. Sandwich ELISA

Antibody sandwich ELISAs are sensitive and very useful for the detection of antigen, e.g., cytokines in cell culture supernatants. As an example, for determining human IL-4, we use monoclonal antibodies of two different specificities against IL-4. For isotype determinations in human serum or lymphocyte culture supernatants, we use goat or rabbit antibodies made highly Fc specific by affinity purification. The same polyvalent antibody preparation can be used both as a capture antibody and as an enzyme-conjugated or biotinylated second antibody. For IgG subclass determinations, we use monoclonal antibodies as capture antibodies and Fc-specific, affinity-purified (depleted of anti-mouse Ig reactivity) goat anti-human IgG as the second antibody. For quantitation of isotypes or IgG subclasses, standard immunoglobulin preparations or myeloma protein solutions of known concentrations are commercially available.

F. Quantitation

For quantitation of specific antibodies, a standard curve with serial dilutions (e.g., 300, 100, 30, 10, 3, and 1 ng/ml) of a relevant standard immunoglobulin is prepared in wells coated with affinity-purified anti-immunoglobulin instead of the antigen. The linear range from a log–log curve is used for interpolation of the experimental values. Similarly, the amount of antigen can be determined in a sandwich ELISA with the help of a standard curve with known amounts of the antigen run in parallell. One standard curve can be used for several plates if care is taken that all plates are developed for the same time.

G. Synthetic Peptides

When the antigen is a short peptide (<20 amino acids), it may have to be coupled to a carrier protein for coating. This protein should also be used for blocking and in the incubation (diluent) buffer.

H. ELISPOT

To be able to discriminate between real ELISPOTs and artifacts, e.g., caused by cells or debris remaining on the nitrocellulose filter, requires some training. The artifactual spots are usually smaller and lack the diffuse rim that is characteristic for real spots. It is important that the plate is not moved or shaken during incubation in the cell incubator (see also Fig. 1).

V. PITFALLS

A. High Backgrounds

Purity and specificity of the reagents are the basic requirements for reliable ELISA determinations. Furthermore the sensitivity of the assay depends on low background. Therefore low OD readings of the negative controls are absolutely essential. With appropriate controls it is

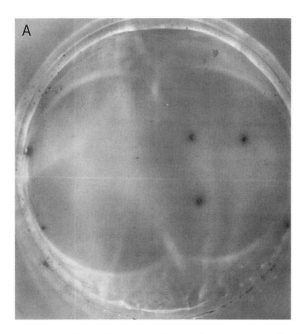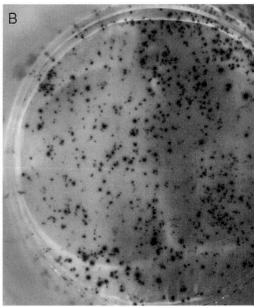

Figure 1 Spleen cells (2×10^6 mononuclear cells/ml) from a mouse immune to the malaria parasite *Plasmodium vinckei were incubated with lysates of infected erythrocytes in tubes for 2 days and thereafter in anti-IL4-coated (A) or anti-IFNγ coated wells (B) (2×10^5 cells/well) in nitro-cellulose (for details see H. Perlmann et al., 1995).*

possible to identify reagents giving rise to unwanted binding of the enzyme conjugate and to remove the possible cross-reacting antibodies by affinity purification and/or neutralization.

B. Competitive Antibodies

When determining antigen-specific antibodies of a certain isotype or IgG subclass, problems may arise if antibodies of other isotypes with higher affinity compete for the same antigenic sites. If the relative concentration of the test antibody is high enough, that problem may be solved by coating with a capture antibody specific for its isotype or subclass, followed by addition of the sample and then the antigen conjugated with enzyme or biotin.

C. Adsorption-Induced Protein Denaturation

Loss of functional activity of antibodies (as much as 90% for polyclonal antibodies and all for some monoclonal antibodies) due to adsorption to polystyrene requires serious consideration (Butler et al., 1992). Changes in the conformation of antigens at coating may lead to masking of native epitopes as well as to exposure of "new" epitopes. Of interest in this context is the finding of Jitsukawa et al. (1989) that by mixing antigen or antibody with a stabilizing protein such as BSA (10 μg/ml), a considerably increased coating efficiency may be achieved.

References

Butler, J. E., Ni, L., Nessler, R., Joshi, K. S., Suter, M., Rosenberg, B., Chang, J., Brown, W. R., and Cantarero, L. A. (1992). The physical and functional behavior of capture antibodies adsorbed on polystyrene. J. Immunol. Methods 150, 77–90.

Czerkinsky, C. C., Nilsson, L. A., Nygren, H., Ouchterlony, O., and Tarkowski, A. (1983). A solidphase enzyme linked imusospot (ELISPOT) assay for enumeration of specific antibody secreting cells. J. Immunol. Methods 65, 109–121.

Engvall, E., and Perlmann, P. (1971). Enzyme-linked immunosorbent assay (ELISA): Quantitative assay of immunoglobulin G. Immunochemistry 18, 871–874.

Hornbeck, P. (1991). Enzyme-linked immunosorbent assays. In "Current Protocols in Immunology" (J. E. Coligan, A. M. Kruisbeek, D. H. Margulies, E. M. Shevach, and W. Strober, eds.), pp. 2.1.1.–2.1.22. Greene Publishing Associates and Wiley-Interscience, New York.

Jitsukawa, T., Nakajima, S., Sugawara, I., and Watanabe, H. (1989). Increased coating efficiency of antigens and preservation of original antigenic structure after coating in ELISA. J. Immunol. Methods 116, 251–257.

Kawabata, T. T. (1995). Enumeration of antigen specific antibody forming cells by the enzyme-linked immunospot (ELISPOT) assay. Meth. Immunol. 1, 125–135.

Kemeny, D. M. (1992). Titration of antibodies. J. Immunol. Methods 150, 57–76.

Klinman, D. M., and Nutman, T. B. (1994). ELISPOT assay to detect cytokine secreting murine and human cells. In "Currrent Protocols in Immunology" (J. E. Coligan, A. M. Kruisbeek, D. H. Margulies, E. M. Shevach, and W. Strober, eds.), pp. 6.19.1–6.19.8. Greene Publishing Associates and Wiley-Interscience, New York.

Maloy, W. L., Coligan, J. E., and Paterson, Y. (1991). Indirect ELISA to determine antipeptide antibody titer. In "Current Protocols in Immunology" (J. E. Coligan, A. M. Kruisbeek, D. H. Margulies, E. M. Shevach, and W. Strober, eds.), pp. 9.4.8.–9.4.11. Greene Publishing Associates and Wiley-Interscience, New York.

Mark-Carter, J. (1994). Epitope mapping of a protein using the Geysen (Pepscan) procedure. In "Methods in Molecular Biology" (B. M. Dunn and M. W. Pennington, eds.), Vol. 36, pp. 207–223.

O'Sullivan, M. J. and Marks, V. (1981). Methods for preparation of enzyme–antibody conjugates for use in enzyme immuno assay. In "Methods in Enzymology" (J. G. Langone and H. Van Vunakis, eds.), Vol. 73, pp. 147–166. Academic Press, San Diego.

Perlmann, H., Kumar, S., Vinetz, J. M., Kullberg, M., Miller, L. H., and Perlmann, P. (1995). Cellular mechanisms in the immune response to malaria in Plasmodium vinckei-infected mice. Infection and Immunity 63, 3987–3993.

Ravindranath, M. H., Ravindranath, R. M. H., Morton, D. L., and Graves, M. C. (1994). Factors affecting the fine specificity and sensitivity of serum antiganglioside antibodies in ELISA. J. Immunol. Methods 169, 257–272.

Sedgwick, J. D., and Holt, P. G. (1983). A solid phase immuno enzymatic technique for the enumeration of specific antibody-secreting cells. J. Immunol. Methods 57, 301.

Ternynck, T., and Avrameas, S. (1990). Avidin–biotin system in enzyme immunoassays. In "Methods in Enzymology" (M. Wilchek and E. A. Bayer, eds.), Vol. 184, pp. 469–581. Academic Press, San Diego.

Zielen, S., Bröker, M., Strnad, N., Schwenen, L., Schön, P., Gottvald, G., and Hofmann, D. (1996). Simple determination of polysaccharide specific antibodies by means of chemically modified ELISA plates. J. Immunol. Methods 193, 1–7.

Epitope Characterization by Combinatorial Phage Display Analysis

Baruch Stern, David Enshell-Seijffers, and Jonathan M. Gershoni

I. INTRODUCTION

The epitope is that aspect of the antigen that is recognized by its corresponding antibody. Obviously, therefore, when working with monoclonal antibodies (mAbs) as research tools or therapeutics, one would like to map and characterize the specific relevant epitopes.

Mapping is simply delineating the region within the antigen that the antibody specifically contacts. Often a contiguous linear peptide can be defined within a protein and such epitopes have been referred to as linear, continuous, or nonconformational. Mapping discontinuous epitopes presents a more challenging problem.

Characterization of the epitope goes one step beyond mapping. One can identify within the epitope those residues that are critical, e.g., those that may be actually contacting the CDR loops of the antibody. There are also the residues that may dictate the three-dimensional configuration of the epitope. Finally, residues may lie within the boundries of the epitope but actually have no contact or role in recognition; extensive moderation could be tolerated at their positions.

Thus characterization of an epitope requires its mapping and the identification of essential structural requirements for antibody binding. Alternatively, in characterizing an epitope, one strives to learn the degree of flexibility and tolerance the antibody may have for variation within the limits of the epitope. The latter thus indicates the degree of potential cross reactivity the mAb may be able to demonstrate.

Combinatorial phage display epitope libraries have been developed to map epitopes, as has been reviewed by many authors (e.g., Dower, 1992; Lane and Stephen, 1993; Cortese et al., 1994; Burritt et al., 1996). This article describes how these libraries can be used to characterize epitopes as well. In principle, one screens a library with an antibody of interest and selects a *collection* of phages that bind with various affinities. The random inserts of the phages are compared among themselves, thus revealing a structural motif that represents the characteristic common denominator of requirements for the mAb being studied.

II. MATERIALS AND INSTRUMENTATION

In all procedures, water is double distilled and the reagents are of analytical grade.

Trizma-base (Cat. No. T-1503), glycine (Cat. No. G-7126), potassium phosphate monobasic (KH_2PO_4) (Cat. No. P-5379), potassium phosphate dibasic (K_2HPO_4) (Cat. No. P-8281)

kanamycin (Cat. No. K-4000), polyethylene glycol 8000 (PEG) (Cat. No. P2139), and bovine serum albumin (BSA) (Cat. No. A4503) are from Sigma.

Bacto-gelatin (Cat. No. 0143-01), Bacto-agar (Cat. No. 0140-01), Bacto-yeast extract (Cat. No. 0127-17-9), and Bacto-tryptone (Cat. No. 0123-17-3) are from Difco.

Six-well cluster plates (Cat. No. 3506) and flat-bottom, 96-well plates (Cat. No. 3596) are from Costar. U-bottom, sterile 96-well plates (Cat. No. 25850-96) are from Corning. Petri dishes (Cat. No. 101VR20) and 100 × 16-mm round-bottom, screw-capped tubes (Cat. No. 142AS) are from Bibby Sterilin, Ltd.

Nitrocellulose membrane filters (0.45 μm, NC45) are from Schleicher and Schuell (Cat. No. 401-169). The TMB membrane peroxidase substrate system (Cat. No.50-77-03) is from Kirkegaard & Perry Laboratories, Inc. ECL immunodetection (Cat. No.RPM2106) is from Amersham International plc.

AffiniPure rabbit anti-mouse IgG Fc fragment specific (RbaMIgGFc) (Cat. No.315-005-008) and horseradish peroxidase–goat anti-mouse whole IgG (GtaMIgG-HRP) (Cat. No.115-035-003) are from Jackson ImmunoResearch Laboratories, Inc.

To probagate the phages, we use bacterial strain K91KAN, a kind gift from Dr. George Smith. Alternatively, any *Escherichia coli* strain that is F+/Hfr is suitable.

We constructed and used an epitope phage display library based on the fuse5 vector [kindly provided by Dr. George Smith (Scott and Smith, 1990)]. Our library contains a random 20-mer insert in the NH$_2$ terminus of the pIII protein (Stern and Gershoni, 1997). The protocol described in this article can be performed using any phage display library published in the literature (see also Devlin *et al.*, 1990; Cwirla *et al.*, 1990; Kay *et al.*, 1993) and obtained either as a gift or purchased [e.g., Ph.D. phage display peptide library kit (Cat. No. 8100) from New England Biolabs].

For the dot blot assay we use a MilliBlot-S system (MilliPore Corporation, Cat. No. MBBDS0480) and quench the filters with evaporated spray dried skim milk 1.5% fat (Marvel).

III. PROCEDURES

A. Plate Coating

Here an anti-Fc antibody is immobilized on the bottom of the wells of a six-well cluster plate and used to capture the mAb of interest via its Fc domain. Often the immobilization of the antibody has been achieved by using biotinylated antibodies and streptavidin (e.g., Scott and Smith, 1990). The use, however, of anti-Fc is intended to allow uniform orientation of the antibody on the plate and has been found to be very efficient.

Solutions

1. *2 M Tris–HCl:* Add 242.2 g Trizma-base to 500 ml water and titer to pH 7.5 with HCl; make up to 1 liter with water.

2. *10× TBS:* Add 87.6 g NaCl to 250 ml 2 M Tris–HCl and make up final volume to 1 liter with water. For coating plates, a 1× TBS autoclaved solution is used and can be stored at room temperature.

3. *Blocking solution A:* TBS/0.25% gelatin. Add 0.25 g gelatin to 100 ml TBS and autoclave. Immediately after autoclaving, swirl the mixture to homogeneity and store at room temperature.

Steps

1. Pipette 700 μl (see Comment 1) of TBS containing 35 μg of RbaMIgFc (see Comment 2) onto the bottom of a 35-mm tissue culture, six-well cluster plate. Place the plate in a humidified box (any plastic container with a wet paper towel) at 4°C overnight on a rocker.

2. The next day discard the excess solution and immediately add blocking solution A, completely filling the wells. Incubate the plate for 2 hr at room temperature.

3. Wash the dish rapidly five times using TBS. Fill each well half way, swirl the liquid in the plate, and pour the contents into a sink. Slap the plate face down on a clean piece of paper towel to remove residual fluid.

4. Add 700 μl of blocking solution A diluted 1:10 in TBS containing 10 μg of a specific mAb to each well and put the plate in a humidified box, rocking gently at room temperature for 4 hr.

B. Affinity Selection

The phages that are recognized by the immobilized mAb are captured.

Solutions

1. *Elution buffer:* Prepare 0.1 N HCl with autoclaved water (about 50 ml) adjust the pH to 2.2 with solid glycine, and add 1 mg BSA/ml. Filter sterilize and store at 4°C.

2. *Neutralizing buffer:* Add 12.11 g of Trizma-base to 80 ml water, adjust the pH to 9.1 with HCl, and make up to a final volume of 100 ml. Autoclave and store at room temperature.

Steps

1. Wash the coated plates rapidly six times with TBS, slapping the plate face down each time on a clean piece of paper towel. Add 700 μl of TBS containing 10^{11} phages. Put the plates in a humidified box and incubate the box at 4°C overnight, shaking gently on a rocker.

2. The next day, remove the solution containing the phages and wash the plate rapidly 10 times in TBS (as described in step 1).

3. To elute specifically bound phages, add 400 μl elution buffer and shake gently on a rocker for 10 min at room temperature to dissociate the immunocomplexed mAb/phage and/ or mAb/anti-Fc fragment antibodies (see Comment 3).

4. Transfer the eluate containing the phages into an Eppendorf tube containing 75 μl of neutralizing solution (see Comment 3) and mix by pipetting up and down.

5. To improve the yield of phages, steps 3 and 4 can be repeated.

6. Pool both phage eluates.

C. Determining the Titer of Phages

Solutions

1. *LB medium:* Add 10 g Bacto-tryptone, 10 g NaCl, and 5 g yeast extract to a final volume of 1 liter in water and adjust the pH to 7.0 with a few drops of a saturated solution of NaOH. Autoclave and store at room temperature.

2. *Kanamycin (Kan) stock solution (50 mg/ml):* Dissolve 0.5 g kanamycin in 10 ml water and filter sterilize.

3. *LB/Kan medium:* Dilute the kanamycin stock solution 1:1000 in LB medium.

4. *LB/Kan agar plates:* Prepare 500 ml of LB medium in a 1-liter flask and add 10 g of Bacto-agar and a spin bar. Autoclave the mixture (the agar wil go into solution during the autoclavization). Meanwhile, set up empty petri dishes (about 20–30). Let the autoclaved solution cool while mixing gently (so to avoid bubbles) on a magnetic stirrer to about 50°C (just hot enough to touch) and add 500 μl of the stock kanamycin (to give a final concentration of 50 μg/ml). Pour the hot medium into the petri dishes, adding at least enough to cover the bottom of each.

Steps

1. Inoculate 2.5 ml of LB/Kan medium with a single colony of *E. coli* K91Kan. Grow in a 37°C shaker incubator at 225 rpm overnight.

2. Prepare 0.5% agarose dissolved in water, bring to a boil, and cool to 50°C.

3. In a 100 × 16-mm, round-bottom screw cap tube, add 200 μl of the overnight culture plus 3.5 ml of the agarose. Roll the test tube briefly between both hands and quickly pour its contents onto a prewarmed (37°C) LB/Kan plate.

4. Leave the plates on the bench to dry for 15 min at room temperature.

5. Prepare three 10-fold dilutions of the eluate containing the phages and spot three 3-μl drops of each dilution onto the plate.

6. After the drops dry, incubate the plates at 37°C overnight, by which time tiny turbid plaques will become visible. Calculate the titer of the phages.

D. Selecting for Positive Phages (see comment 4)

Solutions

1. *Potassium–phosphate buffer:* Dissolve 2.31 g KH_2PO_4 (anhydrous) and 12.54 g K_2HPO_4 (anhydrous) in 90 ml water, adjust volume to 100 ml, and autoclave.

2. *"Terrific broth":* Add 12 g Bacto-tryptone, 24 g yeast extract, and 4 ml (ca. 5.0 g) glycerol to water and adjust the final volume to 900 ml. Divide into 225-ml volumes in separate 500-ml glass bottles and autoclave. Before use, add 25 ml of separately prepared autoclaved potassium–phosphate buffer.

3. *Blocking soultion B:* Dissolve 5 g evaporated milk in 100 ml TBS.

4. *PEG–NaCl:* Dissolve 192.85 g NaCl in 600 ml hot water. Then dissolve 330 g PEG in the salt solution. Adjust the volume to 1 liter and let the turbid solution cool while mixing until it clears. Divide the solution into three 500-ml glass bottles and autoclave. The hot autoclaved solutions form two phases. Mixing the solutions while cooling them in a cool water bath regenerates the uniform and clear cooled PEG–NaCl solutions. Store at 4°C.

Steps

1. Prepare an overnight culture as described in Section C, step 1.

2. Inoculate 10 ml of prewarmed Terrific broth in a 125-ml flask with 200 μl of the overnight culture of cells and shake at 225 rpm, 37°C. Stop shaking when the OD_{600} of the 1:10 dilution reaches 0.2 and allow the sheared F-pili to regenerate for 15 min at 37°C (see Comment 5).

3. To plate the phages, remove and place the cap of a 100 × 16-mm round-bottom screw cap tube on the bench facing upwards and lean the tube on the cap, almost horizontally. Lay 200 μl of the cells 1 cm from the tube opening and add to the drop of cells an aliquot of the eluate containing 300–500 phages. Raise the tube to a vertical position and the drop will slide to the bottom of the tube. Replace the cap and incubate for 15 min at room temperature (see Comment 6).

4. Prepare 0.5% agarose as described in Section C, step 2.

5. Add 3.5 ml of the agarose to the tube, roll it briefly between both hands (so to mix the content), and quickly pour the agarose/infected bacteria mixture onto a prewarmed LB/Kan plate. Leave the plate on the bench to dry for 15 min at room temperature.

6. Incubate the plate at 37°C overnight.

7. Meanwhile, inoculate 2.5 ml of LB/Kan with a single colony of *E. coli* K91Kan. Grow in a 37°C shaker incubator at 225 rpm overnight.

8. The next day, fill the wells of a U-bottom, sterile 96-well plate with 200 μl of Terrific broth containing a 100-fold dilution of the *E. coli* K91Kan overnight culture. Stab single plaques using sterile toothpicks and inoculate the wells of the plate by dipping the tip of the toothpick into the media. Secure the plates in a humidified box and shake overnight at 175 rpm, 37°C (see Comment 7).

9. Centrifuge the plates at 3000 rpm (1500 *g*) for 20 min at room temperature. Using a multichannel pipettor, transfer, in a sterile manner (avoiding the bacterial pellet), 125 μl well of the supernatant to a flat-bottom, 96-well plate already containing 50 μl/well of PEG NaCl solution and mix by raising and lowering the solution in the tip several times. Incubate the plates at 4°C for 2 hr and save the original plates containing the bacterial pellet. The later will be used as master plates and should be sealed with Parafilm and stored at 4°C.

10. Centrifuge the plates at 3000 rpm (1500 g) for 20 min at room temperature. To remove the bulk of the fluid, invert the plate into a biohazard bag, collecting the waste for disposal. Remove the residual fluid by slapping the plate gently face down on several layers of paper towels. Then resuspend the pellet in a total of 100 μl TBS.

11. Prepare nitrocellulose membrane blots by applying 80-μl aliquots of the phages from each well to a MilliBlot-S system using a vacuum transfer system.

12. Block membranes with blocking solution B by rocking them for 1 hr at room temperature (see Comment 8).

13. Wash briefly in TBS and incubate the membrane overnight in blocking solution B diluted 1:10 in TBS containing 1 μg/ml mAb of interest at 4°C, with gentle rocking (see Comment 9).

14. Wash the membrane for 5 min in TBS and repeat five times. Add blocking solution B diluted 1:10 in TBS, add GtaMIgHRP diluted 1:5000, and incubate for 1 hr at room temperature, with gentle rocking (see Comments 8 and 9).

15. Wash the membrane as described in step 14.

16. The positive signals can be detected either by the TMB membrane peroxidase substrate system or by ECL immunodetection (see Fig. 1).

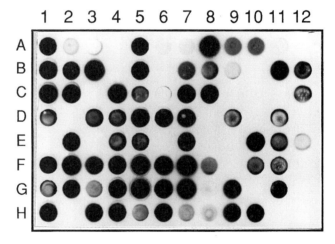

FIGURE 1 Analysis of the epitope corresponding to mAb GV4H3. GV4H3 is a murine mAb anti-gp 120 of HIV-1$_{IIIB}$. It was used to screen the 20-mer pIII phage display library, and after one round of panning, the phages were used to prepare the dot blot (top). The signals were obtained using the ECL reaction. As negative control phages of the fd wild type were spotted at positions 12G and 12F. (Bottom) Eight of the positive phages were sequenced and their inserts aligned with gp120. For each phage, the flanks of the vector are indicated in lowercase (as and sl). The core of the epitope is emphasized in bold. Note that some variation in the core is tolerated.

E. Amplification for Next Round of Biopanning

The method that has been described thus far involves a single step of biopanning. In many cases, this will detect numerous phages that are capable of reacting with a specific antibody with various affinities. However, in order to find the phages with the highest affinities, a second and third round of panning are recommended (Scott and Smith, 1990). For this, amplification is necessary (see also Comment 10).

Solutions

1. *Azide stock solution (5%):* Azide is extremely toxic and should be handled with care! Dissolve 2.5 g sodium azide in 50 ml water and store at 4°C.

2. *TBS/azide:* Add 200 μl of the azide stock solution to 50 ml TBS.

Steps

1. Amplifying the phages after a round of biopanning is done by plating the phages at a density of $2-3 \times 10^4$ on a 90-mm LB/Kan plate (plating of phages was done as described in Section D, steps 1–6).

2. Remove the large plaques with the tip of a Pasteur pipette (see Comment 11).

3. Collect the soft agarose layer by scraping it off the bottom nutrient layer using a microscope slide. Pulverize the collected agarose by injecting it through a 18-gauge, 1.5 in. needle connected to a 10-ml syringe directly into a 50-ml hinge cap test tube containing 20 ml TBS. Wash the syringe with an additional 10 ml TBS.

4. Seal the closed tube with a layer of Parafilm so to ensure that it does not accidentally open and place it horizontally on a rocker at 4°C overnight.

5. Centrifuge at 6000 g for 20 min at 4°C. Carefully collect the supernatant and precipitate the phages with 0.4 vol of a PEG/NaCl solution. Mix well and incubate on ice for 4 hr.

6. Centrifuge in a SS34 rotor at 27,000 g for 20 min at 4°C. Discard the supernatant carefully, maintaining the integrity of the pellet. Resuspend the pellet in 5 ml TBS/azide. If several plates are used, phages can be pooled at this stage.

7. Centrifuge at 5000 g for 5 min at 4°C and transfer the cleared supernatant to a clean tube. Add 0.4 vol of PEG/NaCl, mix by inverting several times, and place at 4°C for 4 hr or overnight.

8. Centrifuge at 27,000 g for 20 min at 4°C. Discard the supernatant and resuspend the pellet in 1 ml of TBS/azide.

9. For the next round of panning (as described previously), use 25% of the phage solution and store the rest (4°C) for future use.

F. Verification of Positive Phages

This section simply confirms that the candidate positive phages are indeed correct.

Steps

1. Prepare an overnight culture of K91Kan cells as described in Section C, step 1.

2. Dilute the overnight culture 1:100 in "Terrific broth" and aliquot it out into 100 × 16-mm round-bottom, screw-capped tubes (2 ml/tube).

3. Stab each positive well of the "master plate" (see Section D, step 9) with a sterile toothpick and use it to inoculate one of the tubes prepared in step 2.

4. Grow overnight at 37°C while shaking at 225 rpm.

5. The next day, transfer the medium containing the bacteria and phages to 2-ml microfuge tubes and spin down at maximal speed for 10 min at room temperature.

6. Transfer 1.5 ml supernatant to a new 2-ml microfuge tube, carefully avoiding the pellet, and add 0.6 ml PEG/NaCl. Mix by inverting the tube several times and leave on ice for 20 min.

7. Centrifuge as in step 5.

8. Pour off the supernatant and tap the tube upside down on a paper towel so as to remove excess medium and recentrifuge briefly, maintaining the orientation of the pellet so as to concentrate any residual fluid that can then be aspirated.

9. Resuspend the pellet with 200 μl TBS.

10. Prepare dot blots by applying 50 μl of each resuspended pellet to a nitrocellulose membrane filter using a Miliblot-S vaccum manifold system. Triplicate samples should be used.

11. Continue the procedure as is described in Section D, steps 12–16.

12. Sequence the insert of the positive phages using the appropriate primers and any standard single-strand sequencing protocol.

G. Analysis

Once a collection of sequences representing the inserts of the selected phages is produced, one must analyze them so to learn the characteristics of the epitope corresponding to the mAb of interest. This is done by simply aligning the sequences for their greatest degree of homology. Often it is sufficient to peruse the inserts and identify even a single common dipeptide represented in most phages. One then aligns the inserts using this dipeptide as a point of reference and then searches for further homologies upstream and downstream to the dipeptide (not necessarily contiguous with it). In doing so, conservative exchanges that emphasize a motif rather than the need for strict sequence homology can be recognized. Obviously, once a motif is identified, then one should attempt to align it to a corresponding region within the original antigen.

A few points should be made. By performing such an analysis, one can define the core of an epitope. In Fig. 1 it is clear that the epitope of the mAb GV4H3 lies in the region of the sequence APAGFAIL within the sequence of the HIV-1 gp120 envelope protein. It also becomes clear that the core is the tripeptide GFA. Moreover, it is evident that the F to W exchange is permissible.

One can learn more by ranking the phages according to their apparent affinity for the mAb, thus discovering the specific importance and relative contribution of each residue of the epitope toward establishing high affinity.

One should also particularly pay attention to *discontinuous* motifs. In using a library that has a relatively long insert (10–20 residues), it is possible to identify motifs with a periodicity. Thus, for example, the anti HIV-1 gp120 mAb GV1A8 identifies the motif HxxIxxLW, where x represents any amino acid. This motif is characteristic for an α-helical conformation in the epitope. Thus by characterizing epitopes rather than simply mapping them, one can reveal protein secondary configurations in the antigen.

IV. COMMENTS

1. This is just enough to cover the bottom of the well.

2. When using antibodies from other organisms instead of murine mAbs, then one must use the appropriate corresponding anti-Fc antibody.

3. Check the pH of both the elution buffer (pH 2.2) and the neutralizing solution (pH 9.1) before use. The mixture of the two comes out to be pH 7–8.5 and should be confirmed. One can use pH indicator paper for this test.

4. Some of the phages that are eluted are the result of nonspecifc adsorption to the well plastic, mAb, or other. One must therefore identify those phages that are genuinely positive, i.e., are recognized specifically by the mAb.

5. When the 10-ml culture becomes quite turbid (after approximately 2.5 hr), start reading the OD_{600} of 1/10 dilutions.

6. The purpose of these "acrobatics" is to keep the culture sterile while infecting the bacteria with the phages.

7. Plaques will be tiny and extremely difficult to see. When picking the isolated plaques we found it easiest to hold the plate up to the light in order to see them clearly.

8. One can add NP-40 (final concentration 0.1%) to improve the signal-to-background ratio in these assays. This can also be done in the TBS washing steps.

9. Nondiluted blocking solution B sometimes is better for reducing the background.

10. When carrying out more than a single round of biopanning one must also be aware that a library can contain phages that for no obvious reason, are able to multiply much faster than the rest, thereby enabling them to dominate in a liquid cell culture. We were able to overcome this problem by growing the phages as plaques on a plate.

11. Large plaques are most probably those derived from phages that replicate fastest and it is advisable to separate them out and analyze them later if need be (or simply throw them away).

References

Burritt, J. B., Bond, C. W., Doss, K. W., and Jesaitis, A. J. (1996). Filamenous phage diplay of oligopeptide libraries. *Anal. Biochem.* **238**, 1–13.

Cortese, R., Felici, F., Galfre, G., Luzzago, A., Monaci, P., and Nicosia, A. (1994). Epitope discovery using peptide libraries displayed on phage. *Tibtech* **12**, 262–267.

Cwirla, S. E., Peters, E. A., Barret, R. W., and Dower, W. J. (1990). Peptides on phages: A vast library of peptides for identifying ligands. *Proc. Natl. Acad. Sci. USA* **87**, 6378–6382.

Devlin, J. J., Panganiban, L. C., and Devlin, P. E. (1990). Random peptide libraries: A source of specific protein binding molecules. *Science* **249**, 404–406.

Dower, W. J. (1992). Phage power. *Curr. Biol.* **2**, 251–253.

Kay, B. K., Adey, N. B., He, Y-S., Manfredi, J. P., Mataragnon, A. H., and Fowlkes, D. M. (1993). An M13 phage library displaying random 38-amino-acid peptides as a source of novel sequences with affinity to selected targets. *Gene* **128**, 59–65.

Lane, D. P., and Stephen, C. W. (1993). Epitope mapping using bacteriophage peptide libraries. *Cur. Opin. Immunol.* **5**, 268–271.

Scott, J. K., and Smith, G. P. (1990). Searching for peptide ligands with an epitope library. *Science* **249**, 386–390.

Stern, B., and Gershoni, J. M. (1997). Construction and use of a 20-mer phage display epitope library. *Methods in Molecular Biology*, Vol. 87. Combinatorial Peptide Library Protocols (S. Cabilly, ed.) Humana Press Inc. Totowa, NJ.

PART 6

Immunocytochemistry

Conjugation of Fluorescent Dyes to Antibodies

Benjamin Geiger and Tova Volberg

I. INTRODUCTION AND GENERAL CONSIDERATIONS

For more than four decades, fluorescence microscopy has been the leading method for the sensitive and specific localization of a large variety of molecules in cells and tissues. Such molecular mapping has over the years provided much information, not only on the distribution of the molecules of interest but also on their functional properties. The first and most commonly used flurophore-conjugated probes are antibodies, which enable direct or indirect localization of a very wide variety of antigenic molecules. The list of specific probes that have been used for fluorescence microscopic analysis has expanded and presently includes such probes as lectins, toxins, hormones, and growth factors. It is beyond the scope of this article to discuss extensively the various aspects of fluorescence microscopy and the chemistry of fluorophore conjugation. It should nevertheless be pointed out that the quality and fidelity of immunofluorescence localization depend on multiple factors, all of which should be optimized for each experimental system. These include:

1. Proper processing of the specimen (i.e., fixation and permeabilization when necessary).

2. High affinity and specificity of the probes (i.e., antibodies, lectins).

3. Optimal choice of fluorophore and conjugation procedures.

4. Appropriate labeling procedure.

5. Availability of a microscopic system suitable for examination of the labeled specimens.

Although each of these variables may significantly affect the quality of labeling, this article focuses on only one of these factors: the conjugation of various fluorophores to antibodies. The high demand for immunofluorescent regents has resulted in the introduction of many different fluorescent dyes suitable for conjugation to proteins. Some of the new fluorophores offer a broader excitation–emission range than previously attainable, allow for multiple labeling of specimens, and are often less susceptible to photobleaching. Moreover, in addition to the "classical" amine-reactive fluorophores (such as the isothiocyanate derivatives), there are now batteries of fluorescent reagents, suitable for coupling to other functional groups such as thiols, hydroxyls, and carboxylates. The procedures described in this article are, however, restricted to fluorophores bearing amine-reactive groups, including sulfonyl chloride, isothiocyanate, and dichlorotriazinyl. For most immunocytochemical purposes these fluorophores and conjugation procedures appear to be satisfactory. There are, however, some general considerations affecting the choice of suitable fluorophores that are broadly applicable and should be pointed out:

1. The choice of fluorophores should take into account factors such as photobleaching, autofluorescence of the specimen, and the filter sets available in the microscope. Usually, rhodamine-based dyes are preferable to fluorescein-based dyes due to their lower susceptibility to photobleaching during microscopic examination.

2. For most conjugation procedures it is recommended not to use highly diluted IgG solutions. Optimal conjugation is obtained using 1–5 mg/ml IgG solutions.

3. It is important to monitor carefully the extent of conjugation. Usually, conjugates containing two to five fluorophores per antibody molecule are optimal. A lower level of modification results in weak signal whereas overmodification may lead to high nonspecific background labeling, therefore removal of under- and overmodified probes is advisable.

4. The solution of fluorescent antibodies should be protected from intense light and from frequent changes in temperature. Storage of labeled antibodies at 4°C or frozen in liquid nitrogen in small aliquotes is recommended.

Another practical consideration of many researchers is whether they should carry out the conjugation themselves or, alternatively, purchase conjugated antibodies from commercial sources. There are no general rules that apply to all commercially available fluorophore-conjugated antibodies, yet they are usually quite expensive, are often of variable quality, and are commonly available only as secondary ("anti-IgG") reagents for indirect ("sandwich") labeling.

II. MATERIALS AND INSTRUMENTATION

Lissamine–rhodamine B (sulforhodamine B, Polysciences Inc., Cat. No. 0643)

Dichlorotriazinyl amino fluorescein (DTAF 2096D-1, Research Organics, Inc., Sigma Cat. No. D-0531, or Molecular probes)

Fluorescein isothiocyanate (FITC, Calbiochem Cat. No. 34323 or Sigma Cat. No. F-1628)

Phosphorus pentachloride (PCL5, BDH)

Sephadex G-50 (Cat. No. A-0043-01, Pharmacia)

DEAE-cellulose (DE-52, Cat. No. 4057050, Whatman)

Carbonate buffer, phosphate buffer, sodium chloride, acetone, ethanol, and glycine

Pestle and mortar, small funnels, Whatman No. 1 filter paper, ice bucket, 15-ml glass tubes, microfuge tubes (Eppendorf 3810), small stirring bar, magnetic stirrer, vortex, centrifuge, and spectrophotometer

III. PROCEDURES

A. Conjugation of Rhodamine B 200 Sulfonyl Chloride (RB200SC) to Antibodies

1. Preparation of RB200SC

This procedure follows that of Brandtzaeg (1973).

NOTE

Caution: The entire procedure should be performed in a fume hood! Avoid direct contact with PCl$_5$ or activated rhodamine.

Steps

1. Weigh in fume hood 1 g of lissamine–rhodamine B and 2 g PCl$_5$.

2. Mix the two powders thoroughly with pestle and mortar for 5 min at room temperature.

3. Add 10 ml acetone (or dioxane) to the activated rhodamine and mix occasionally for 5 min to maximally dissolve the activated fluorophore.

4. Filter the solution through Whatman No. 1 filter paper, into a tube kept on ice.

5. Distribute 100- to 200-μl aliquots into small stoppered bottles or microfuge tubes, and store at $-70°C$.

6. Determine the concentration of RB200SC by diluting it 1:10 in acetone and further 1:100 in phosphate-buffered saline (PBS). Measure the absorbance of the solution at 565 nm and calculate the concentration of RB200SC using an extinction coefficient $E_{565\,nm}^{1\%}$ of 1265.

NOTE

Do not refreeze a fluorophore solution once it has been thawed and opened.

2. Conjugation of RB200SC to Antibodies and Fractionation of the Conjugate

Solutions

1. *1.0 M carbonate buffer, pH 9.0:* Mix 1.0 *M* sodium carbonate and 1.0 *M* sodium bicarbonate until the pH is 9.0.

2. *1 M glycine (7.5 g in 100 ml distilled water).*

3. *0.1 M phosphate buffer, pH 8.0:* Mix 0.1 *M* sodium phosphate with sodium hydrogen phosphate until the pH is 8.0.

4. *0.25 M NaCl (1:20 dilution of 5 M NaCl stock) in 10 mM phosphate buffer, pH 8.0 (1:10 dilution of 0.1 M stock solution).*

Steps

1. Working at room temperature, mix 1 vol of 1.0 *M* carbonate buffer, pH 9.0, with 4 vol of the antibody solution in PBS in a 15-ml glass tube.

2. While stirring the antibody solution, add 20–30 mg RB200SC/mg of antibody in 3 to 4 aliquots over a period of 30 min. Stir the solution for an additional 20 min and stop the reaction by adding 0.1 vol of 1 *M* glycine.

3. Load the reaction mixture on a Sephadex G-50 column (5–6 bed volumes per sample volume), preequilibrated with 10 m*M* phosphate buffer, pH 8.0. Collect the first colored (protein conjugated) peak.

4. Load the fluorescent protein solution on a DEAE-cellulose column (1-ml bed volume per mg of antibody), preequilibrated with 10 m*M* phosphate buffer pH 8.0.

5. Wash the column with the same buffer and elute the labeled protein with 0.25 *M* NaCl in 10 m*M* phosphate buffer, pH 8.0.

6. Read the absorbance at 280 and 575 nm and calculate the protein concentration and the fluorophore/protein ratio (*F/P*), according to the following equation and Fig. 1:

$$\text{Rhodamine-labeled antibodies (mg/ml)} = \frac{OD_{280} - (0.32 \times OD_{575})}{1.4}. \tag{1}$$

7. Store the labeled antibodies at 4°C or freeze small aliquots in liquid nitrogen.

B. Labeling of IgG with Dichlorotriazinyl Amino Fluorescein (DTAF)

Solutions

1. *DTAF (5 mg/ml in ethanol).*

2. *1 M carbonate buffer.*

3. *10 mM phosphate buffer.*

4. *0.25 mM NaCl in 10 mM phosphate buffer.*

5. *1 M glycine in water.*

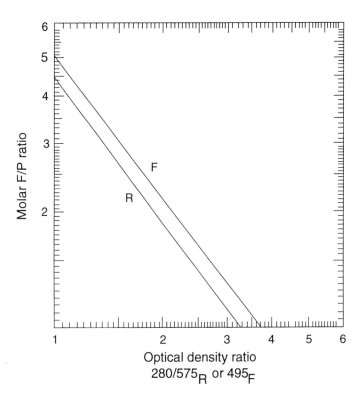

FIGURE 1 Calculation of *F/P* ratio of rhodamine- and fluorescein-conjugated antibodies.

Steps

1. Prepare a fresh stock solution of DTAF (5 mg/ml in ethanol) in a 15-ml glass tube (heat the solution slightly to completely dissolve the powder).

2. Mix the antibody solution in PBS, with 1/4 vol of 1 *M* carbonate buffer, pH 9.0.

3. Add 40 μg DTAF (8 μl) per milligram of IgG and mix well.

4. Incubate the mixture on ice for 30 min with gentle stirring, then add 1:10 vol of the 1 *M* glycine solution.

5. Separate the IgG-bound DTAF from the unbound dye on Sephadex G-50 and save the first colored peak.

6. Load the fluorescent protein solution on a DEAE-cellulose column (1 ml bed volume/mg antibody), preequilibrated with the 10 m*M* phosphate buffer, pH 8.0.

7. Wash the column with the same buffer and elute the labeled protein with 0.25 *M* NaCl in 10 m*M* phosphate buffer, pH 8.0.

8. Read the absorbance at 280 and 495 nm and calculate the protein concentration and the fluorophore/protein ratio (*F/P*), according to the following equation and Fig. 1:

$$\text{Fluorescein-labeled antibodies (mg/ml)} = \frac{\text{OD}_{280} - (0.35 \times \text{OD}_{495})}{1.4}. \tag{2}$$

9. Store the labeled antibodies at 4°C or freeze small aliquots in liquid nitrogen.

C. Coupling of Fluorescein Isothiocyanate (FITC) to Antibodies

Solutions

1. *1 M carbonate buffer.*

2. *10 mM phosphate buffer.*

3. *0.25 mM NaCl in 10 mM phosphate buffer.*

Steps

1. Mix the IgG solution with 1/4 vol of 1 M carbonate buffer, pH 9, at room temperature in a 15-ml glass centrifuge tube.

2. Weigh out Celite–FITC (0.5 mg powder/mg of IgG).

3. Add the Celite–FITC to the IgG solution and immediately mix by Vortex or centrifuge briefly.

4. Add the stir bar and mix continuously to suspend the Celite–FITC. Stir for 20 min at room temperature.

5. Separate the labeled IgG from the Celite by a brief centrifugation.

6. Separate the IgG-bound FITC from the unbound dye on Sephadex G-50 and save the first colored peak.

7. Load the fluorescent protein solution on a DEAE-cellulose column (1 ml bed volume/mg antibody), preequilibrated with 10 mM phosphate buffer pH 8.0.

8. Wash the column with the same buffer and elute the labeled protein with 0.25 M NaCl in 10 mM phosphate buffer, pH 8.0.

9. Read the absorbance at 280 and 495 nm and calculate the protein concentration and the fluorophore/protein ratio (*F/P*), according to Eq. 2 and Fig. 1.

10. Store the labeled antibodies at 4°C or freeze small aliquots in liquid nitrogen.

Reference

Brandtzaeg, P. (1973). *Scand. J. Immunol.* Suppl. 2, 273.

Immunofluorescence Microscopy of Cultured Cells

Mary Osborn

I. INTRODUCTION

Immunocytochemistry is the method of choice for locating an antigen to a particular structure or subcellular compartment provided that an antibody specific for the protein under study is available. Immunofluorescence is a sensitive method requiring only one available antigenic site on the protein. Usually the indirect technique is used. In this technique the first antibody is unlabeled and can be made in any species. After it has bound to the antigen a second antibody, made against IgGs of the species in which the first antibody is made and coupled to a fluorochrome such as fluoroscein isothiocyanate, is added. The distribution of the antigen can then be viewed in a microscope equipped with the appropriate filters.

Immunofluorescence as a method to study cytoarchitecture and subcellular localization gained prominence with the demonstration that antibodies can be produced to actin even though it is a ubiquitous components of cells and tissues (Lazarides and Weber, 1974). Cytoskeletal structures visualized in cells in immunofluorescence microscopy include the three filamentous systems: microfilaments, microtubules, and intermediate filaments [Figs. 1–4 (see color plate) and micrographs in the article by Mies *et al.*, Volume 2, "Multiple Immunofluorescence Microscopy of the Cytoskeleton"]. In addition, proteins can be located to other cellular subcompartments and organelles, e.g., the plasma or nuclear membranes (Fig. 5), the Golgi apparatus, or the endoplasmic reticulum (Fig. 6), or to other cellular structures such as mitochondria (Fig. 7) and vesicles. Other proteins can also be localized to subcompartments of the nucleus or even of the nucleolus. In addition to its use in identifying cytoskeletal structures and organelles, immunocytochemistry has proved useful in building up a biochemical or protein chemical anatomy of a structure. Examples include the location of the microfilament-associated proteins to the stress fiber and the description of the biochemical anatomy of such structures as microvilli and stereo cilia. A third and very important use of the technique has been to demonstrate heterogeneity in mixed cultures, e.g., of neuronal cultures, or of other primary cell cultures [cf. Fig. 4 (see color plate) and Fig. 8].

The micrographs that accompany this article show not only the beauty of some of the structures, but also some of the advantages of the technique. First, only the arrangement of the particular protein against which the antibody is made is visualized. Second, for those proteins that form part of the supramolecular structures, the arrangement of those structures throughout the cell is revealed. Third, numerous cells can be visualized at the same time, and therefore it is relatively easy to determine how the structures under study vary under particular conditions. Immunofluorescence microscopy is also a useful method to establish appropriate conditions to study a structure at higher resolution in the electron microscope.

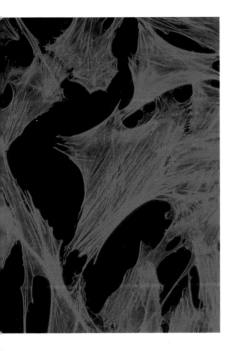

FIGURE 1 Actin stress fibers in rat mammary cell line revealed by staining with rhodamine-labeled phalloidin (×400).

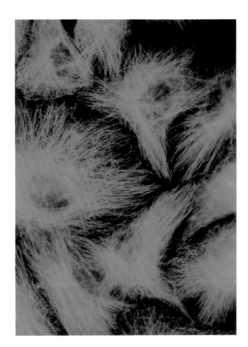

FIGURE 2 Microtubules in the monkey CV-1 cell line revealed by staining with antibodies to tubulin (×400).

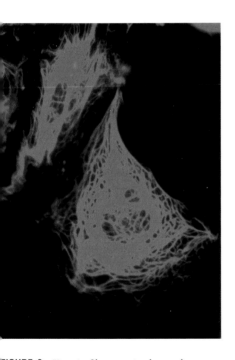

FIGURE 3 Keratin filaments in the rat kangaroo PtK2 cell line revealed by staining with antibodies to keratin (×600).

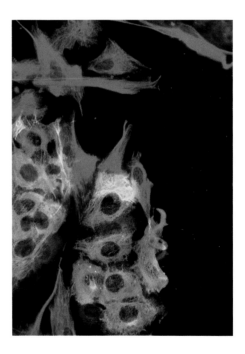

FIGURE 4 Artificial mixture of cells from the human breast carcinoma cell line MCF-7 stained with an antibody to keratin and an FITC-labeled second antibody (in green) and the human fibroblast cell line HS27 stained with the V9 antibody to vimentin (in red). Note that each cell type contains only a single intermediate filament. The yellow color results from MCF-7 and HS27 cells that lie over each other (×150).

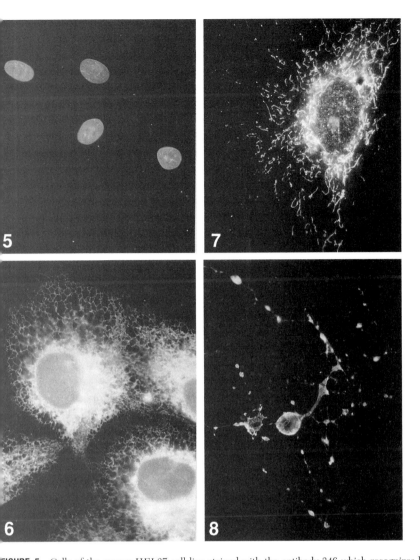

FIGURE 5 Cells of the mouse HEL37 cell line stained with the antibody 346 which recognizes lamins A/C. Only the nuclear membrane is stained. ×400.

FIGURE 6 6V-1 cells stained with the ID3 antibody against protein disulfide isomerase to reveal the endoplasmic reticulum (cf. Vaux *et al.*, 1990). ×600.

FIGURE 7 Mouse 3T3 cell stained with an antibody to cytochrome oxidase to show the arrangement of mitochondria. ×600.

FIGURE 8 An oligodendrocyte in a rat optic nerve primary culture stained with antibodies to galactoceroboside. Other cells present in the same field are not stained (cf. Raff *et al.*, 1978). ×400.

A 1:1 correspondence has been shown for a parallel-processed or even the same specimen when studied under fluorescence and under the electron microscope. Electron microscopic methods not only allow location of the antigen to a particular structure at higher resolution, but may also allow the determination of interactions between a structure that is immunolabeled and other unlabeled structures in the cell.

Other reviews of immunofluorescence of cultured cells that may be helpful include those of Osborn (1981) and Osborn and Weber (1981), and for live cells, Wang and Taylor (1989). For an overview of the different cytoskeletal and motor proteins, see Kreis and Vale (1993).

II. MATERIALS AND INSTRUMENTATION

A. Antibodies

Antibodies to many cellular proteins can now be purchased commercially. Firms offering a variety of antibodies to cytoskeletal and other proteins include Amersham, Biomakor, Dako,

Sigma, and Novocastra; other firms have put together specialized collections emphasizing one or another narrower area.

Primary antibodies may be monoclonal antibodies made in mice or polyclonal antibodies made in species such as guinea pigs and rabbits. The appropriate dilution is established by a dilution series. Monoclonal antibodies supplied as hybridoma supernatants can often be diluted 1:1 to 1:20 for immunofluorescence or even more if other more sensitive immunocytochemical procedures are used (see also Volume 2, Osborn and Isenberg, "Immunocytochemistry of Frozen and of Parrafin Tissue Sections" for additional information). Monoclonal antibodies supplied as ascites fluid can be diluted in the range of 1:100 to 1:1000. Polyclonal antibodies supplied as sera should be diluted in the range of 1:20 to 1:100. Note that many rabbits have relatively high levels of autoantibodies against keratins and/or other cellular proteins, so check presera. Affinity purification in which the antigen is coupled to a support and the polyclonal antibody is then put through the column usually results in a dramatic improvement in the quality of the staining patterns. Affinity-purified antibodies should work in the range of 5–20 μg/ml.

Secondary antibodies directed against IgGs of the species in which the first antibody is made are usually purchased already coupled to FITC or rhodamine (for list of suppliers, see above). The working dilution for the secondary antibody is established by running a dilution series. Usually 1:50 to 1:150 dilutions of the commercial products are appropriate. An essential control is to check that the second antibody is negative when used alone. If nonspecific staining is present, it can sometimes be removed by absorbing the antibody on fixed monolayers of cells or on an acetone cell powder.

Antibodies other than IgMs should be stored in the freezer ($-70°C$ for valuable primary antibodies and affinity-purified antibodies, $-20°C$ for the rest). Antibodies should be stored in small aliquots and repeated freezing/thawing should be avoided. IgMs may be inactivated by freezing/thawing and are better kept in 50% glycerol in a freezer set at -20 to $-25°C$. If dilutions are made in a suitable buffer [e.g., phosphate-buffered saline (PBS), 0.5 mg/ml bovine serum albumin, 10^{-3} M sodium azide], diluted antibodies are stable for several months at 4°C.

B. Reagents and Other Useful Items

Methanol is of reagent grade. Formaldehyde can be diluted 1:10 from a concentrated 37% solution (e.g., Analar-grade BDH Chemicals). As such solutions usually contain 11% methanol, it may be better to make the formaldehyde solution from paraformaldehyde. In this case, heat 18.5 g paraformaldehyde in 500 ml PBS to 60°C and filter through a 0.45-μm filter. Store at room temperature.

PBS contains per liter 8 g NaCl, 0.2 g KCl, 0.2 g KH_2PO_4, and 1.15 g Na_2HPO_4, adjusted to pH 7.3 with NaOH. With respect to mounting medium, polyvinyl alcohol-based mounting media have the advantage in that although they are liquid when the sample is mounted, they solidify within several hours of application. In addition, the fluorescence is stable if the sample is held in the dark and at 4°C. Samples can be reexamined and photographed after months or even years. Commonly used mounting media include Elvanol 51-05 (Serva) and Mowiol (4-88) (Hoechst). Place 6 g analytical-grade glycerol in a 50-ml plastic conical centrifuge tube, add 2.4 g of Mowiol 4.88 or Elvanol 51-05, and stir for 1 hr to mix. Add 6 ml distilled water and stir for a further 2 hr. Add 12 ml of 0.2 M Tris buffer (2.42 g Tris/100 ml water, pH adjusted to 8.5 with HCl as FITC has maximal fluorescence emission at this pH), and incubate in a water bath at 50°C for 10 min, stirring occasionally to dissolve the Mowiol. Clarify by centrifugation at 1200 g for 15 min and aliquot. Store at $-20°C$; unfreeze as required. Once unfrozen, the solution will be stable for several months at room temperature.

Other useful items include round (12-mm) or square (12 $\times$ 12-mm) glass coverslips (thickness 1$\frac{1}{2}$). Ten round coverslips fit in a petri dish of 5.5 cm diameter. For screening purposes or when a large number of samples are needed (e.g., for hybridoma screening), microtest slides that contain 10 numbered circles 7 mm in diameter (Flow Labs, Cat. No. 6041505) are useful. Tweezers (e.g., Dumont No. 7) are used to handle the coverslips. Ceramic racks into which coverslips fit (Cat. No. 8542E40, A. Thomas) and matching glass containers are also needed. Glass beakers (30 ml) are used to wash the specimens. Cells growing in suspension can be firmly attached to microscope slides using a cytocentrifuge such as the Cytospin (Shandon Instruments).

C. Equipment

The essential requirement is access to a microscope equipped with appropriate filters to visualize the fluorochromes in routine use. The Zeiss inverted microscope, Photomicroscope III, or Axiophot is suitable, as are models from other manufacturers. An automatic camera, with spot and whole field options for photography, is a useful accessory. Epifluorescence, an appropriate high-pressure mercury lamp (HBO 50 or HBO 100), and appropriate filters (so that specimens doubly labeled with fluorescein and rhodamine can be visualized) are basic requirements. Lenses should also be selected carefully. The depth of field of the lens will decrease as the magnification increases. Round cultured cells will be in focus only with a 25× or 40× lens, whereas flatter cells can be studied with a 63× or 100× lens. To enable phase and fluorescence to be studied on the same specimen, some lenses should have phase optics. Only certain lenses transmit the Hoechst DNA stain (e.g., Neofluar lenses) and this stain also requires a separate filter set.

Confocal microscopy is being increasingly used, particularly for rounded cells and for structures in or near the nucleus.

III. INDIRECT IMMUNOFLUORESCENCE PROCEDURE

Steps

1. Trypsinize cells 1–2 days prior to the experiment onto glass coverslips or on multitest slides that have been washed in 100% ethanol and oven sterilized. For most applications, choose coverslips or multitest slides on which cells are two-thirds or less confluent. Drain coverslip or slide on filter paper to remove excess medium, but do not allow it to dry.

2. Place coverslips in ceramic rack and multitest slides in metal racks, and immerse in methanol precooled to −10 to −20°C. Leave for 6 min.

3. Make a wet chamber by lining a 13-cm-diameter (for coverslips) or a 24 × 24-cm square petri dish (for slides) with two or three sheets of filter paper, and add sufficient water to moisten the filter paper.

4. Wash the fixed specimens briefly in PBS, remove excess PBS by touching to dry filter paper, and place cell side up over the appropriate number in the wet chamber.

5. Add 5–10 µl of an appropriate dilution of the primary antibody with an Eppendorf pipette. Use the tip to spread the antibody over the coverslip without touching the cells. Replace the top of the wet chamber, transfer to a 37°C incubator with humidity, and incubate for 45 min.

6. Wash by dipping each coverslip individually three times into each of three 30-ml beakers containing PBS. Wash slides by replacing slides in metal rack and transferring through three PBS washes (180 ml each, leave for 2 min in each). Remove excess PBS with filter paper.

7. Replace specimens in wet chamber. Add 5–10 µl of an appropriately diluted second antibody carrying a fluorescent tag. Return to 37°C incubator for a further 30–45 min.

8. Repeat step 6.

9. Identify microscope slides with small adhesive labels on which date, specimen number, antibody, or other information is written. Place slides in cardboard microslide folders (e.g., 6708-M10 Thomas Scientific). Mount two coverslips per slide by inverting each coverslip and placing cell side down on a drop of mounting medium placed on the slide, with a disposable ring micropipette. Cover with filter paper and press gently to remove excess mounting medium. For samples on multitest slides, use 6 × 2.5-cm glass coverslips on which a drop of mounting medium has been placed. Secure the coverslips with nail polish.

10. Photograph. Use a fast film (e.g., Kodak 35-mm Tri-X) and push the development e.g., with Diafine (Acufine).

IV. COMMENTS

Specimens should not be allowed to dry out at any stage in the procedure. If coverslips are accidentally dropped, the side on which the cells are can be identified by focusing on the cells under an upright microscope and scratching gently with tweezers.

A. Fixation

The procedure just given results in good results with many cytoskeletal and other antigens however the optimal fixation protocol depends on the specimen, the antigen, and the location of the antigen within the cell. Three requirements are needed. First, the fixation procedure must retain the antigen within the cell. Second, the ultrastructure must be preserved as far as possible without destroying the antigenic determinents recognized by the antibody. Third the antibody must be able to reach the antigen; i.e., the fixation and permeabilization steps must extract sufficient cytoplasmic components so that the antibodies can penetrate the fixed cells. In the procedure just given, fixation and permeabilization are achieved in a single step i.e., with methanol. Alternative fixation methods include the following:

1. Formaldehyde–methanol: 3.7% formaldehyde in PBS for 10 min (to fix the cells), then methanol at $-10°C$ for 6 min (to permeabilize the cells).

2. Formaldehyde–Triton: 3.7% formaldehyde in PBS for 10 min, then PBS with 0.2% Triton X-100 for 1 min at room temperature (see Volume 2, Mies *et al.*, "Multiple Immunofluorescence Microscopy of the Cytoskeleton" for additional information).

3. Glutaraldehyde: Fix in 1% glutaraldehyde (electron microscopic grade) in PBS for 15 min, then methanol at $-10°C$ for 15 min. Immerse in fresh sodium borohydride solution (0.5 mg/ml in PBS) for 3 × 4 min. Wash with PBS 2 × 3 min each. Note that the sodium borohydride step is necessary to reduce the unreacted aldehyde groups; without this step the background will be very high.

Note that formaldehyde treatment destroys the antigenicity of many antigens. Alternatively in a very few cases positive staining may be observed only after formaldehyde fixation. Very few antigens react after glutaraldehyde fixation.

B. Special Situations

1. Fluorescently labeled phalloidin, a phallotoxin that binds to filamentous actin, is usually used to reveal the distribution of filamentous actin in cells (Fig. 1; see color plate). To obtain good staining patterns, fix cells for 10 min in 3.7% formaldehyde in PBS. Wash with PBS. Incubate for 1 min in 0.2% Triton X-100 in PBS and wash with PBS. Incubate with an appropriate dilution of rhodamine-labeled phalloidin (e.g., Sigma Cat. No. P-1951) for 30 min at 37°C, wash with PBS, and mount in Mowiol.

NOTE

Caution: Phalloidin is extremely poisonous.

2. To stain endoplasmic reticulum, use either an antibody, e.g., ID3 against a sequence in the tail region of protein disulfide isomerase (Vaux *et al.*, 1990) (Fig. 6), or the lipophilic cationic fluorescent dye $DiOC_6$ (3,3-dihexyloxacarbocyanine iodide, Kodak 14414) (Terasaki *et al.*, 1984). To stain with dye, fix for 5 min in 0.25% glutaraldehyde in 0.1 *M* cacodylate and 0.1 *M* sucrose buffer, pH 7.4. Wash. Stain for 80 sec with dye, mount in buffer, and observe using a 63× or 100× lens and the fluorescein filter. Reticular structures should be apparent. Note that mitochondria will also be stained. To stain only mitochondria, use either an antibody, e.g., to cytochrome oxidase (Fig. 7), or the dye rhodamine 123.

3. Special fixation procedures may also be needed for other membrane structures in cells In addition, lectins can be used to stain carbohydrate-containing organelles, e.g., staining of Golgi apparatus with fluorescently labeled wheat germ or other agglutinins.

4. To stain DNA, pipette 100 μl Hoechst dye (33258, Cat. No. B2883, Sigma) on to coverslip for 1–2 min, drain excess dye, and mount directly in Mowiol or stain with Hoechst between steps 8 and 9 in the immunofluorescence procedure (shock solution 1 mM, working solution 4 μl/ml PBS).

5. Some cellular structures, such as microtubules, are sensitive to calcium. In this case add 2–5 mM EGTA to the 3.7% formaldehyde solution in Section IVA and to the methanol in Section III, step 2.

6. Sometimes for cell surface components it may be advantageous to stain live cells. Expose such cells to antibody for 25 min and proceed with steps 6–8 in Section III. Then fix cells in 5% acetic acid/95% ethanol for 10 min at −10°C (Fig. 8, cf. Raff *et al.*, 1979).

7. It is often advantageous to visualize two or three antigens in the same cell. See Volume 2, Mies *et al.*, "Multiple Immunofluorescence Microscopy of the Cytoskeleton" for methods to do this.

C. Stereomicroscopy

Fluorescence microscopy gives an overview of the whole cell. With practice, specimens can be seen in three dimensions when looking through the microscope. Stereo micrographs can be made using a simple modification of commercially available parts (Osborn *et al.*, 1978a). Alternatively, confocal microscopy is increasingly being used and is particularly useful for round cells, which are not in focus with the higher-power 63 or 100 lenses, or to document arrangements and obtain greater resolution at different levels in the cell (cf. Fox *et al.*, 1991).

D. Limit of Resolution

Theoretically this is ~200 nm when 515-nm wavelength light and a numerical aperture of 1.4 are used. Objects with dimensions above 200 nm will be seen at their real size. Objects with dimensions below 200 nm can be visualized provided they bind sufficient antibody but will be seen with diameters equal to the resolution of the light microscope (cf. visualization of single microtubules in Osborn *et al.*, 1978b). Thus, objects closer together than 200–250 nm cannot be resolved by fluorescence microscopy, e.g., microtubules in the mitotic spindle or ribosomes.

V. PITFALLS

Occasionally no specific structures are visualized, even though the cell is known to contain the antigen. This may be because:

1. Antibodies can be species specific. This can be a particular problem with monoclonal antibodies, which, for instance, may work with human but not with other species. If in doubt check the species specificity with the supplier before purchase.

2. The fixation procedure may inactivate the antigen. For instance, many intermediate filament antibodies no longer react after fixation protocols such as those in Section IVA.

3. The antigen can be poorly fixed or extracted by the fixation procedure.

4. The antibody may not be able to gain access to the antigen; e.g., antibodies to tubulin often do not stain the intracellular bridge.

5. The specimens may be generally fluorescent and it can be hard to decide whether this is due to specific or nonspecific staining.

References

Fox, M. H., Arndt-Jovin, D. J., Jovin, T. M., Baumann, P. H., and Robert-Nicoud, M. (1991). Spatial and temporal distribution of DNA replication sites localized by immunofluorescence and confocal microscopy in mouse fibroblasts. *J. Cell Sci.* **99**, 247–253.

Kreis, T., and Vale, R. (1993). "Guidebook to the Cytoskeletal and Motor Proteins." Oxford University Press, London/New York.

Lazarides, E., and Weber, K. (1974). Actin antibody: The specific visualization of actin filaments in non-muscle cells. *Proc. Natl. Acad. Sci. USA* **71**, 2268–2272.

Osborn, M. (1981). Localization of proteins by immunofluorescence techniques. *Techniq. Cell. Physiol.* **P107**, 1–28.

Osborn, M., Born, T., Koitzsch, H.-J., and Weber, K. (1978a). Stereo immunofluorescence microscopy. I. Three-dimensional arrangement of microfilaments, microtubules and tonofilaments. *Cell* **13**, 477–488.

Osborn, M., and Weber, K. (1981). Immunofluorescence and immunochemical procedures with affinity purified antibodies. *In* "Methods in Cell Biology," Vol. 23. Academic Press, New York.

Osborn, M., Webster, R. E., and Weber, K. (1978b). Individual microtubules viewed by immunofluorescence and electron microscopy in the same PtK2 cell. *J. Cell Biol.* **77**, R27–R34.

Raff, M. C., Mirsky, R., Fields, K. L., Lisak, R. P., Dorfman, S. H., Pilbenberg, D. H., Gregeon, N. A., Leibowitz, S., and Kennedy, M. C. (1978). Galactocereboside is a specific cell surface antigenic marker for oligodendrocytes in culture. *Nature* **274**, 813–816.

Terasaki, M., Song, J., Wong, J. R., Weiss, M. J., and Chen, L. B. (1984). Localization of endoplasmic reticulum in living and glutaraldehyde fixed cells with fluorescent dyes. *Cell* **38**, 101–108.

Vaux, D., Tooze, J., and Fuller, S. (1990). Identification by anti-idiotype antibodies of an intracellular membrane protein that recognizes a mammalian endoplasmic reticulum retention signal. *Nature* **345**, 495–502.

Wang, Y.-L., and Taylor, D. L. (1989). "Methods in Cell Biology," Vols. 29 and 30. Academic Press, New York.

Multiple Immunofluorescence Microscopy of the Cytoskeleton

Brigitte Mies, Klemens Rottner, and J. Victor Small

I. INTRODUCTION

Cells possess an extensive scaffolding of fibrillar elements, collectively referred to as the cytoskeleton. The components of the cytoskeleton are involved in diverse cellular functions ranging from mitosis to cell motility to signal transduction. In cultured cells, the three primary components of the cytoskeleton, actin filaments, microtubules, and intermediate filaments, can be readily visualized using immunofluorescence microscopy. This article describes a protocol for double and triple labeling of the cytoskeleton that can also be applied to the localization of proteins putatively associated with one or more of its components.

The use of antibodies to localize proteins within cells requires that the cells be chemically fixed and rendered permeable to the antibody molecules. The pitfalls of this approach are as numerable as the different techniques employed. The compromises that must, by necessity, be made should thus be recalled when drawing conclusions about the results obtained. In general, one aims to achieve optimal structural preservation combined with intense antibody labeling. But the properties of many antibodies do not allow us this luxury. As a rule, stronger fixation, giving better structural preservation, leads to weaker antibody labeling. Inorganic solvents (normally acetone or methanol) serve as weak fixatives and aldehydes (formaldehyde and glutaraldehyde) as stronger fixatives, and various recipes employing these alone or in combination have been described in studies of the cytoskeleton (e.g., Fujiwara and Pollard, 1980; Lazarides, 1982; Osborn and Weber, 1982; see also Volume 2, Osborn "Immunofluorescence Microscopy of Cultured Cells" for additional information). Permeabilization of the cell membrane is achieved either with inorganic solvents or with anionic detergents. We have had good experience with techniques involving the use of aldehyde–detergent mixtures that effect simultaneous penetration of the cell and fixation of the cytoskeleton and the structures bound to it. The same protocols yield good preservation of the cytoskeleton also in the electron microscope (e.g., Small, 1988), but at the expense of a major loss of the membrane-bound organelles.

In this update of a previous article (Herzog *et al.*, 1994), we emphasize the labeling characteristics of a set of commercially available antibodies and the types of compromises that are necessarily imposed when multiple labeling is required.

II. MATERIALS AND REAGENTS

1. *Coverslips:* Round glass coverslips 10 or 12 mm in diameter, cleaned in 60% ethanol/40% HCl (10 min), rinsed with H_2O (2×5 min), drained, cleaned with lint-free paper, and sterilized for tissue culture by exposure to ultraviolet light in the culture dish.

2. *Humid chamber:* Large petri dish 14 cm in diameter, or similar container with lid containing a glass plate (around 9 cm²) coated with a layer of Parafilm and supported on a moistened piece of filter paper on the bottom of the dish. A few drops of water on the glass plate facilitate spreading and flattening of the Parafilm.

3. *Washing reservoir:* Two multiwell dishes, 24 wells each (e.g., Falcon or Nunc).

4. *Rotary shaker:* Rotating–tilting table for washing (optional).

5. *Filter paper:* Whatman No. 1, 9 cm in diameter.

6. *Forceps:* Dumont No. 4 or No. 5 or equivalent watchmaker forceps.

7. *Pipettes:* Set of automatic pipettes (0–20 µl, 20–200 µl, 50–1000 µl) or capillary pipettes for diluting and aliquoting antibodies. Pasteur pipettes.

8. *Phalloidin:* Rhodamine (TRITC)- or fluorescein-conjugated phalloidin from Sigma. Store as 0.1-mg/ml stocks in methanol at −20°C.

9. *Secondary antibodies:* Commercial secondary antibodies carrying fluorescein, Texas red, Cy-3, or coumarin conjugates are in our experience of generally good quality. Biotinylated secondary antibodies in subsequent combination with avidin/streptavidin conjugates are an interesting alternative (see also Section VII). We generally store our antibodies in a refrigerator at 4°C. The antibodies and other probes used here are listed in Table I.

10. *Gelvatol, Vinol:* The basic ingredient of the mounting medium is polyvinyl alcohol (MW 10,000, around 87% hydrolyzed) that comes under various trade names: Elvanol, Mowiol, Gelvatol, etc. We use Vinol 203 from Air Products and Chemicals Inc.

III. PROCEDURES

Solutions

1. *0.5 M EGTA stock solution:* For 500 ml stock, weigh out 95.1 g EGTA (Sigma E 4378) into 400 ml H₂O, Adjust pH to 7.0 with 1 N NaOH and make up to 400 ml with H₂O. Store at room temperature in a plastic bottle.

TABLE I Triple Label Combinations

	Fixation d	Fixation c
Combination	α-Actinin Vimentin Myosin	α-Actinin Vinculin Phalloidin
First antibody mix	a-α-actinin IgM (1:100)[a] a-vimentin IgG (1:5)[b] a-nonmuscle myosin (1:25)[c]	a-α-actinin IgM (1:100) a-vinculin IgG (1:200)[g]
Second antibody mix	FITC a-mouse IgM (1:100)[d] AMCA a-mouse IgG (1:100)[e] TRITC a-rabbit Igs (1:60)[f]	CY3 a-mouse IgM (1:100)[h] Biotinylated a-mouse IgG (1:100)[i]
Third step		AMCA avidin D (1:100)[j] FITC–phalloidin (1:100)[k]

[a] Mouse anti-α-actinin IgM (BM 75.2, Sigma).
[b] Mouse antivimentin IgG (V9 Cat. No. V-6630, Sigma).
[c] Polyclonal anti-nonmuscle myosin (BT-561, Biomedical Technologies, Inc.).
[d] FITC (fluorescein) anti-mouse IgM (µ-chain specific) (Cat. No. F5259, Sigma).
[e] Coumarin-conjugated, affinity-purified goat anti-mouse IgG (H + L) (Jackson).
[f] TRITC (rhodamine) swine anti-rabbit Igs (Cat. No. R156, Dako).
[g] Mouse antivinculin IgG (clone hVin1 Cat. No. V-9131, Sigma).
[h] CY3 anti-mouse IgM (Zymed).
[i] Biotinylated anti-mouse IgG (Southern Biotech. Associates, Cat. No. 1030-08).
[j] Coumarin, AMCA avidin D (Cat. No. A-2008, Vector Laboratories).
[k] FITC phalloidin (Cat. No. P-8543, Sigma).

2. *1 M MgCl₂ stock solution:* For 500 ml stock, weigh out 101.6 g $MgCl_2$, add H_2O to 500 ml, dissolve, and store at 4°C.

3. *Cytoskeleton buffer (CB):* 10 mM MES (Sigma M-8250), 150 mM NaCl, 5 mM EGTA, 5 mM $MgCl_2$, and 5 mM glucose. For 1 liter, add the following amounts to 800 ml H_2O: MES, 1.95 g; NaCl, 8.76 g; 0.5 M EGTA, 10 ml; 1 M $MgCl_2$, 5 ml; and glucose, 0.9 g. Adjust pH to 6.1 with 1 N NaOH and fill up to 1 liter. Store at 4°C. For extended storage, add 100 mg streptomycin sulfate (Sigma S-6501).

4. *Tris-buffered saline (TBS), 10× concentrated stock solution:* 200 mM Tris (Merck Cat. No. 8382), 1.54 M NaCl, 20 mM EGTA, 20 mM $MgCl_2$, pH 7.5 at room temperature. For 1 liter, add the following components to 800 ml H_2O: Tris, 24.2 g; NaCl, 89.9 g; 0.5 M EGTA, 40 ml; and 1 M $MgCl_2$, 20 ml. Adjust pH to 7.5 with 1 N HCl and make up to 1 liter. Store at 4°C. Dilute required amount 1:9 with H_2O before use.

5. *Phosphate-buffered saline (PBS) working solution:* 137 mM NaCl, 2.7 mM KCl, 4.3 mM $Na_2HPO_4 \cdot 7H_2O$, and 1.4 mM KH_2PO_4, pH 7.4.

6. *Modified CB solution (CB, pH 7.0):* 10 mM PIPES, 150 mM NaCl, 5 mM EGTA, 5 mM glucose, 5 mM $MgCl_2$, and 100 µg/ml streptomycin. For 1 liter, add 10 ml of 1 M PIPES, pH 7.0, 8 ml of 4 M NaCl or 8.76 g, 10 ml of 0.5 M EGTA, 5 ml of 1 M glucose or 0.9 g, 5 ml of 1 M $MgCl_2$ or 1.0165 g, and 100 µg streptomycin.

7. *Triton X-100 (T):* Make up 10% aqueous stock and store at 4°C.

8. *Glutaraldehyde (GA) stock:* Make up 2.5% solution of glutaraldehyde by diluting 25% glutaraldehyde EM grade (Agar scientific Ltd, Cat. No. R 1020 or equivalent) in CB. Readjust pH to 6.1 and store at 4°C.

9. *Paraformaldehyde (PFA) stock:* Make up a stock 3 or 4% solution of paraformaldehyde in PBS or CB (see fixative mixtures) (analytical grade Merck Cat. No. 4005). To make 100 ml: heat 80 ml of CB (or PBS) to 60°C, add 3g paraformaldehyde, and mix for 30 min. Add a few drops of 10 M NaOH until the solution is clear, cool, adjust pH (see appropriate mixture), and make up to 100 ml. Store in aliquots at −20°C.

10. *Fixative mixtures:* Aldehyde fixative mixtures are made up using the stock solutions above to give the combinations listed under step 3.

11. *Methanol:* Store at −20°C.

12. *Blocking solution:* 1% bovine serum albumin and 5% horse serum in TBS.

13. *Antibody mixtures:* These are made up in TBS or in the blocking solution without serum. To remove any unwanted particles, centrifuge (10,000 g for 10 min) the diluted mixtures before use. The antibody combinations used for this article are listed in Table I.

14. *Mounting medium:* Mix 2.4 g of polyvinyl alcohol with 6 g glycerol (87%) and then with 6 ml H_2O. After at least 2 h at room temperature, add 0.2 ml 0.2 M Tris–HCl, pH 8.5, to the mixture and further incubate the solution for 10 min at 60°C. Remove any precipitate by centrifugation at 17,000 g for 30 min. Store in aliquots at −20°C. [As antibleach agents added to mounting medium, additives are available that considerably reduce bleaching and thus enable multiple pictures to be taken of the same cells. We use *n*-propyl gallate (Giloh and Sedat, 1982) at 5 mg/ml or phenylenediamine (Johnson *et al.*, 1982) at 1–2 mg/ml in the mounting medium. After dissolving the additive, degas mounting medium before storage.]

Steps

1. Seed the cells onto coverslips in the petri dish and allow them to attach and spread for 4–48 hr in an incubator at 37°C.

2. Aspirate growth medium and rinse dish gently with CB (warmed to room temperature), avoiding shifting of the coverslips over each other. Aspirate CB and replace with one of the following fixative solutions:

    a. Methanol (−20°C) for 5 min

    b. 4% PFA in PBS 10 min; 0.5% T in PBS 30 sec

    c. 0.25% GA, 0.5% T in CB (pH 6.1), 1 min; wash CB (pH 7.0); 4% PFA in CB (pH 7.0), 10 min

    d. 0.25% T, 0.5% GA in CB (pH 6.1), 1 min; wash CB; 1% GA in CB (pH 6.1), 10 min

    e. 3% PFA, 0.3% T, 0.1% GA, in CB (pH 6.1), 15 min

    f. 3% PFA, 0.3% T, 0.2% GA in CB (pH 6.1), 15 min

    g. 3% PFA, 0.3% T, 0.05% GA in CB (pH 6.1), 15 min

3. Rinse twice with CB, 2×10 min.

4. Block: Invert each coverslip onto a 20-μl drop of blocking solution on Parafilm in the humid chamber. Before transfer to drop, dry the back side of the coverslip by holding it briefly on filter paper with a pair of forceps, taking care not to allow the cell side to dry. Drain any excesss solution from the cell side by touching the edge of the coverslip to the filter paper. Incubate on blocking solution for 15 min or until first antibody mixtures are prepared. (Back side of coverslip should not be wet or else coverslip will sink during the washing step.)

5. Apply drops (10–15 μl) of first antibody mixture to unused part of Parafilm and transfer coverslips to appropriate drops after draining excess blocking solution on filter paper. Replace lid on petri dish and leave at room temperature for 30 min.

6. Wash: To ease removal of coverslips for washing, pipette 10 μl TBS under their edge to lift them up from the Parafilm. Using forceps, transfer coverslips to multiwell dish in which the wells are filled to the brim with TBS so that the liquid surface is flat. The coverslips will float well, cell side down, as long as the back side remains dry. For efficient washing, transfer dish gently to a tilting rotating table for 10 min. Repeat washing steps after transfer of coverslips to a second dish containing fresh TBS.

7. Change Parafilm in humid chamber and apply drops of second antibody mixture. Transfer coverslips to drops after briefly draining excess TBS with filter paper and incubate for 30 min.

8. Wash as described in step 6.

9. Apply third antibody as described in step 7.

10. Wash as described in step 6.

11. Mount: Add a small drop of mounting medium to a cleaned glass slide using, for example, a plastic disposable pipette tip. Drain excess TBS from coverslip and gently invert onto drop. *Note:* The mounting medium dries quite fast so the drops should be applied singly and not in batches. If necessary, remove excess medium after mounting by applying small pieces of torn filter paper to the coverslip edge.

12. Observe directly in a fluorescence microscope with a dry lens. An oil immersion lens can be used the next day when the mounting medium has solidified.

V. CHOICE OF FIXATION FOR MULTIPLE LABELING

Different antibodies commonly require different fixation protocols to give optimal labeling. It is therefore important to test different fixation conditions for each antibody to determine the best compromise fixation for multiple labeling. Table II lists the characteristics of the commercial antibodies used in this article in terms of the intensity of label obtained with each of the seven fixation protocols described. Note that these data refer to only two cell types: Swiss 3T3 cells and a primary human fibroblast cell line (provided by Professor J.E. Celis). In general, you should establish the strongest fixation protocol that your antibodies can tolerate and draw your conclusions accordingly.

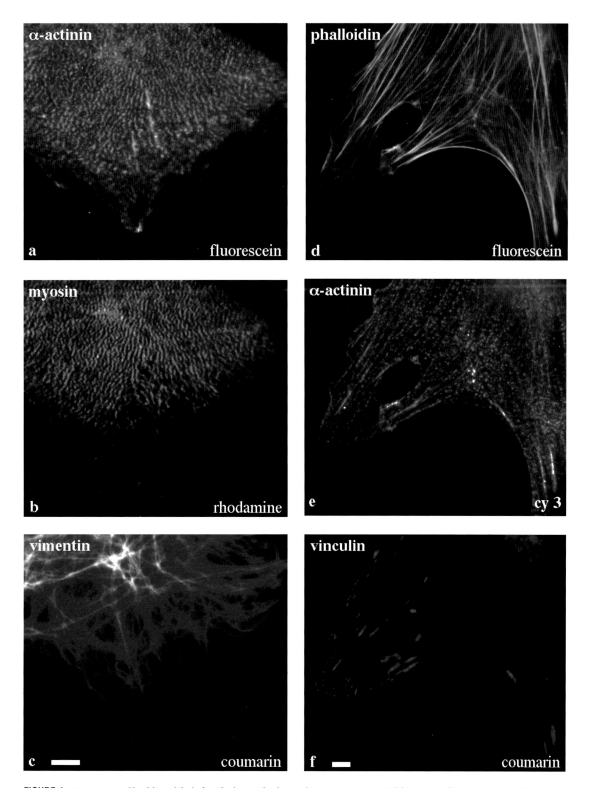

FIGURE 1 Human ear fibroblasts labeled with the antibody combinations given in Table I, using fixation mixture d (a,b,c) or fixation mixture c (d,e,f) given in the text. Bars: 5μm.

TABLE II Labeling Characteristics of the First Antibodies Used in this Article[a]

Antibody fixative	a	b	c	d	e	f	g
M-a-α-actinin IgM BM 75.2 (Sigma)	++++[b]	+	+++	++	+++	+++	++
M-a-α-tubulin clone DM1A (Sigma)	+	+	+++	++++	+++	+++	+++
M-a-human vinculin h-Vin1 (Sigma)	−	+++	++	−	−	−	−
M-a-β-actin AC 15 (Sigma)	++++	+	+	+++	+++	++++	+++
M-A-vimentin V9 (Sigma)	+	+	++	+++	+++	+++	+++
R-a-NM myosin (BTI)	+++	++	++	++	+++	++	+++
Phalloidin	−	+++	+++	+++	+++	+++	+++

[a] a–f correspond to fixation options given in the text.
[b] +, weakly positive; ++, moderate label; +++, strong label; ++++, very good label; −, negative.

V. COMMENTS

A. General Remarks

We have generally aimed for fixation protocols that preserve the actin cytoskeleton best. Although stress fibers are easily preserved with most fixative protocols, the delicate peripheral lamellipodia are normally distorted or lost after methanol or formaldehyde fixation. This is why glutaraldehyde is included in several of the present mixtures. A stronger glutaraldehyde fixation than what we have used here is best for lamellipodia (see, e.g., Small, 1988) but cannot be used with several of the antibodies described. So again we have had to compromise. Osborn and Weber (1982) recommend using sodium borohydride to reduce free aldehyde groups after glutaraldehyde fixation and thereby the autofluorescence introduced by this fixative. In the current study we have obviated the need for the borohydride step by reducing the glutaraldehyde treatment to below the level at which autofluorescence sets in. If autofluorescence is observed, for example, in the region of the nucleus, this can be eliminated by a brief treatment (5 min) of the coverslips in ice-cold cytoskeleton buffer containing freshly dissolved sodium borohydride (0.5 mg/ml), prior to immunolabeling.

B. Specific Characteristics of the Antibodies Worth Noting

1. α-Actinin (BM 75.2) shows the strongest label after methanol fixation (Fig. 2d) and for the aldehyde mixtures when more glutaraldehyde is present (Fig. 1; see color plate).

2. For tubulin (DM1A), microubules are best labeled in cells fixed with glutaraldehyde. Both methanol and formaldehyde fixations give discontinuously stained microtubules (Fig. 3).

3. For human vinculin (h-Vin 1), the best results are seen with formaldehyde fixation, but it does tolerate low concentrations of glutaraldehyde.

4. New lots of β-actin (AC15) show different characteristics as compared to original ones (Gimona *et al.*, 1994). The total actin cytoskeleton is labeled after methanol (Fig. 2c), but only the lamellipodia are labeled after aldehyde fixation (Fig. 2a). We presume that there is a more easy access of epitopes in lamellipodia due to the absence of certain actin-binding proteins.

5. Vimentin (V9) does not label 3T3 cells, but does show same characteristics as the tubulin antibody with human cells, i.e., only continuous fibers after stronger aldehyde fixation (Fig. 1c; see color plate).

6. Nonmuscle myosin (NM myosin) is a good label with all fixative combinations.

7. Phalloidin does not stain after methanol fixation. Extended incubation in phalloidin (e.g., overnight) gives an improved label.

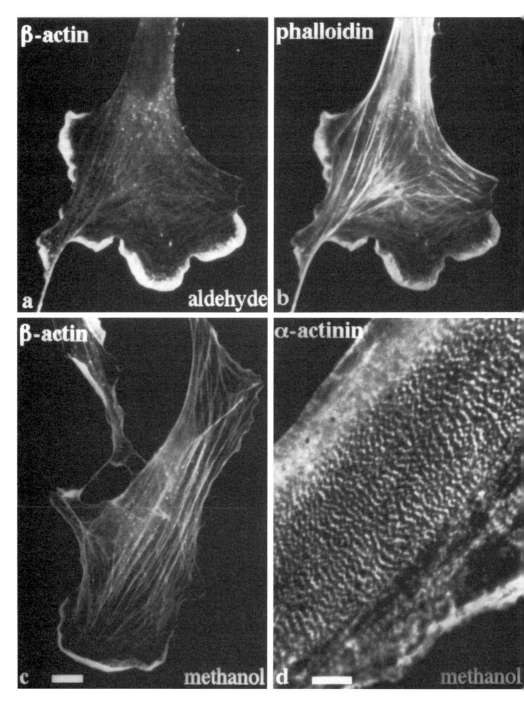

FIGURE 2 (a–c) Comparison of the labeling by the monoclonal b–actin antibody in 3T3 fibroblasts fixed in aldehyde fixation mixture f (a) or in methanol (c). The phalloidin label in b shows that aldehyde fixation drastically reduces (in 'a') the labeling of the stress fibers, leaving only lamellipodia strongly stained. Bar: 10 μm. (d) Labeling with the monoclonal a–actinin antibody in a human ear fibroblast fixed with methanol. Bar: 5 μm.

As far as secondary antibodies are concerned, we have found that those conjugated with Cy3 give the strongest label. When using Cy3 (or Cy2) on streptavidin, we have generally observed a punctate background label that can be problematic.

VI. FILTERS

For double immunofluorescence microscopy with FITC and rhodamine (or Texas red, Cy3), standard filter sets are available. If bleed-through of one color into the wrong channel is

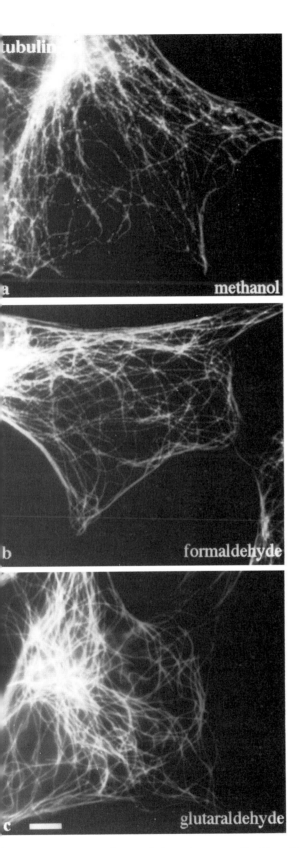

FIGURE 3 Appearance of microtubules in human ear fibroblasts (labeled with DMIA antibody) after fixation in methanol (a), formaldehyde fixation mixture b (b), or glutaraldehyde fixation mixture d (c). Bar: 5 μm.

observed, the filter combinations should be rechecked by the manufacturer. When using coumarin in a triple combination, an ultraviolet excitation filter is used together with a barrier filter that excludes green fluorescence arising from UV excitation of fluorescein. A filter combination supplied on request by Zeiss that is suitable for coumarin comprises G365 for excitation and BS 450–490 as a barrier filter (see, e.g., Small *et al.*, 1988).

VII. PITFALLS

If problems arise from sinking of coverslips during washing, use another washing protocol, e.g., immersion of coverslips cell side up in separate petri dishes containing TBS. Damaged cells normally arise from inadvertently allowing the coverslip to dry at any stage of the procedure or by touching the cell side with filter paper. Labeling with phalloidin can be improved by including this probe also in the first antibody. Successful double or triple immunofluorescence labeling requires that the individual antibody combinations each produce intense staining with a clean background, when used alone.

References

Brandtzaeg, P. (1973). Conjugates of immunoglobulin G with different fluorochromes. I. Characterization by anionic exchange chromatography. *Scand. J. Immun.* 2, 273–290.

Fujiwara, K., and Pollard, T. D. (1980). Techniques for colocalizing contractile proteins with fluorescent antibodies. *In* "Current Topics in Developmental Biology," Vol. 14, pp. 271–296. Academic Press, New York.

Giloh, H., and Sedat, J. W. (1982). Fluorescence microscopy: Reduced photobleaching of rhodamine and fluorescein protein conjugates by *n*-propyl gallate. *Science* 217, 1252–1255.

Gimona, M., Vandekerckhove, J., Goethals, M., Herzog, M., Lando, Z., and Small, J. V. (1994). A β-actin specific monoclonal antibody. *Cell Motil. Cytoskel.* 27, 108–116.

Herzog, M., Draeger, A., Ehler, E., and Small, J. V. (1994). Immunofluorescence microscopy of the cytoskeleton. *In* "Cell Biology Laboratory Handbook" (J. E. Celis, ed.), pp. 355–360. Academic Press, San Diego.

Johnson, G. D., Davidson, R. S., McNamee, K. C., Russell, G., Goodwin, D., and Holborow, F. J. (1982). Fading of immunofluorescence during microscopy: A study of the phenomena and its remedy. *J. Immunol. Meth.* 55, 231–242.

Lazarides, E. (1982). Antibody production and immunofluorescent characterization of actin and contractile proteins. *In* "Methods in Cell Biology" (L. Wilson, ed.), Vol. 24, pp. 313–331. Academic Press, New York.

Osborn, M., and Weber, K. (1982). Immunofluorescence and immunocytochemical procedures with affinity purified antibodies: Tubulin-containing structures. *In* "Methods in Cell Biology" (L. Wilson, ed.), Vol. 24, pp. 97–132. Academic Press, New York.

Small, J. V. (1988). The actin cytoskeleton. *Electron. Microsc. Rev.* 1, 155–174.

Small, J. V., Zobeley, S., Rinnerthaler, G., and Faulstich, H. (1988). Coumarin–phalloidin: A new actin probe permitting triple immunofluorescence microscopy of the cytoskeleton. *J. Cell Sci.* 89, 21–24.

Immunofluorescence Microscopy of Yeast Cells

Kathryn R. Ayscough and David G. Drubin

I. INTRODUCTION

Yeast has been used extensively as an experimental organism for genetic and biochemical studies. However, for a range of technical reasons, cytological studies were originally mostly limited to electron microscopy, which was too laborious for routine use. The biochemical and genetic characterization of an increasing number of proteins in yeast provided a tremendous impetus to develop immunofluorescence microscopy methods to permit easier and more rapid protein localization.

The major difficulties facing the researcher wishing to do immunofluorescence microscopy in yeast are the small cell size (3–10 μM), the near round shape of the cell, and the presence of a thick cell wall that must be removed to allow access of antibodies to the cell milieu. These problems have been overcome (Kilmartin and Adams, 1984), and a slew of proteins in yeast have been successfully immunolocalized. These proteins include components of the actins and microtubule cytoskeletons and proteins residing in various organelles such as the nucleus, mitochondria, peroxisomes, Golgi apparatus, endoplasmic reticulum, and vacuoles (see Pringle *et al.*, 1991 and references therein). One technical point emerging from these localization studies has been that there is not a single procedure that is always useful. Rather, a general methodology can be used with appropriate modifications.

This article presents a general protocol that has been successfully used to immunolocalize many proteins in yeast, followed by a number of modifications that have been found to be of use in particular cases. The article then discusses various aspects of performing immunofluorescence in yeast and indicates stages of the procedure that could be modified to optimize staining obtained. Examples of immunofluorescence staining in yeast are shown in Fig. 1. It should be noted that due to space constraints the methods given here are those developed for use in the yeast *Saccharomyces cerevisiae*. More specific methods for immunofluorescence in *Schizosaccharomyces pombe* are available elsewhere (Alfa *et al.*, 1993).

II. MATERIALS AND INSTRUMENTATION

A. Chemical Reagents

Formaldehyde (F79-500), methanol (A452-4), acetone (A18-4), Tween 20 (BP337-100), yeast extract (DF0127-17-9), Bacto-peptone (DF0118-17-0), and malt extract (DF0186-17-7) are from Fisher. Bovine serum albumin (BSA, A-7906), β-mercaptoethanol (M-6250), 4′6′-diamidino-2-phenylindole dihydrochloride (DAPI, D-9542), glucose (G-8270), glycerol (G33-

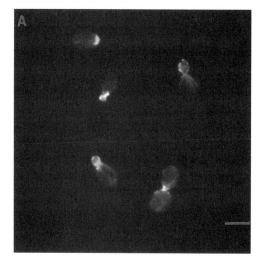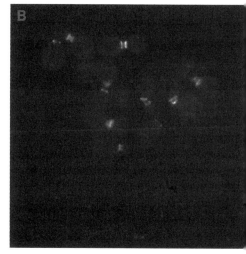

FIGURE 1 Immunofluorescence staining in yeast cells. (A) Staining of the actin cytoskeleton. Note the highly polarized staining of the cortical actin patches in the small and medium budded cells, and the staining at the septal region in large budded cells. (B) Staining of the septin ring at the mother-bud neck using antibodies to the septin protein, Cdc11p (a gift from John Pringle, University of North Carolina, Chapel Hill). Bar: 5 μm.

1), n-propylgallate (P-3130), p-phenylenediamine (P-6001), poly-L-lysine (P-1524), KH_2PO_4 (P-5379), K_2HPO_4 (P-8281), SDS (L-5750), NaH_2PO_4 (S-8282), Na_2HPO_4 (S-0876), NaCl (S-7653), and sorbitol (S-3889) are from Sigma. Zymolyase-20T (32-092-1) and Zymolyase-100T (32-093-1) are from ICN Pharmaceuticals. Vectashield antifade mounting solution (H-1000) is available from Vector Laboratories.

B. Antibodies

Tubulin, YOL1/34, monoclonal rat antibody is from Serotec (MCA 78S). Actin, C4, antibody is from Boehringer-Mannheim (1 378 996) and from ICN (69-100-1). Anti-Myc-tag antibodies are from Santa Cruz Biotechnology Inc. (9E10 monoclonal—Cat. No. SC-40; A14 polyclonal—Cat. No. SC-789). Monoclonal anti-HA tag antibodies are from BAbCO (MMS101P). Secondary antibodies, fluorescein–isothiocyanate (FITC)-conjugated goat anti-rabbit (55646), FITC goat anti-mouse (55493), FITC X anti-rat (55745), tetramethylrhodamine–isothiocyanate (TRITC) goat anti-rabbit (55666), and TRITC goat anti-mouse (55527), are from Cappel/Organon Technika Inc. CY3-conjugated sheep anti-rabbit (C-2306) is from Sigma. The CY3 fluorophore is observed using the normal rhodamine filter on a microscope. It is very bright and has been of particular use when trying to observe signals that are weak with other secondary antibodies. However, it should be noted that the background signal is also increased.

C. Instruments

Centrifugation steps for washing cells are performed in a CL2, clinical centrifuge from IEC. Multiwell slides (60-408-05) are from ICN Pharmaceuticals. Coverslips (12-518-105E) are from Fisher. Microscopy is performed using a Zeiss Axioskop fluorescence microscope with a HB100 W/Z high-pressure mercury lamp and a Zeiss 100X Plan-Neofluar oil immersion objective.

III. PROCEDURES

Solutions

1. *YPD growth medium:* For the general immunofluorescence procedure, grow yeast cells in YPD medium to log phase ($0.5–1 \times 10^7$ cells/ml). To make 1 liter, add 10 g yeast extract, 20 g Bacto-peptone, and 950 ml distilled H_2O; separately make a solution of 40% glucose (20 g glucose in 50 ml distilled H_2O). Autoclave both solutions, then combine the two solutions to obtain YPD medium.

2. *Spheroplast recovery medium:* For 500 ml, add 100.2 g sorbitol, 10 g glucose, 2.5 g yeast extract, 2.5 g malt extract, and 5 g Bacto-peptone. Make volume up to 500 ml and autoclave.

3. *Zymolyase cell wall digestion mix:* To make a stock solution of zymolyase-20T, add 10 mg zymolyase-20T, 1 ml of 1 M potassium phosphate buffer, pH 7.5 [18 ml 1 M KH_2PO_4 (68.0 g/500 ml), 82 ml 1 M K_2HPO_4 (87.1 g/500 ml)], and 9 ml distilled H_2O. Store at $-20°C$ as 1-ml aliquots. For the working digest mix, take 1 ml buffer A (below), add 40 μl zymolyase stock solution and 2 μl β-mercaptoethanol (0.5 ml of this mix is required for each sample being processed).

4. *Phosphate-buffered saline (PBS) buffer (10× stock solution):* 40 g NaCl, 0.2 g KCl, 1.44 g Na_2HPO_4, and 0.24 g KH_2PO_4, pH with NaOH to pH 7.2, adjust volume with distilled H_2O to 500 ml, autoclave, and store at room temperature.

5. *Buffer A:* 0.1 M potassium phosphate buffer, pH 7.5, and 1.2 M sorbitol. Take 18 ml 1 M KH_2PO_4 (68.0 g/500 ml) and 82 ml 1 M K_2HPO_4 (87.1 g/500 ml) to make a 1 M KPi, pH 7.5, stock. Make a 2 M sorbitol stock (182 g/500 ml). To make 100 ml of buffer A, take 10 ml of 1 M KPi, pH 7.5, 60 ml sorbitol stock, and 30 ml distilled H_2O. Filter sterilize.

6. *Buffer B (PBS + BSA):* For 100 ml, take 10 ml 10× PBS stock, 90 ml distilled H_2O, and 100 mg BSA. Filter sterilize. Sodium azide can be added to this solution to 1 mM to prevent deterioration.

7. *Buffer C:* 20% (w/v) dried skim milk in PBS. For 100 ml take 20 g Carnation milk powder and add to 10 ml 10× PBS; adjust volume to 100 ml with distilled H_2O.

8. *Buffer D:* 2% milk and 0.1% Tween 20 in PBS. Take 2 g milk powder, 0.5 ml of a 20% Tween 20 stock, and 10 ml 10× PBS; adjust volume to 100 ml with distilled H_2O.

9. *0.5% SDS in PBS:* To make a 10% stock solution of SDS, take 10 g SDS and make up to 100 ml with distilled H_2O. For 1 ml of solution, take 50 μl 10% SDS, 100 μl 10× PBS, and 850 μl distilled H_2O.

10. *Mounting solution:* We usually make our mounting solution in the lab using an antifade reagent, *p*-phenylenediamine. As this chemical is toxic (LD_{50} 80 mg/kg) and reported to be carcinogenic, we prefer to weigh it out infrequently and make large batches that can be stored at $-80°C$ for long periods of time. Add 100 mg *p*-phenylenediamine to 10 ml PBS. If the pH is below 9.0, bring it to pH 9.0 by adding NaOH while stirring. Add 90 ml glycerol and stir until homogeneous. If desired, add 2.25 μl of DAPI (1 mg/ml in water), which will allow nuclei to be visualized. Store mounting medium at $-80°C$ in the dark. The solution should be discarded when it loses its clear color and turns brown. An alternative antifade solution that is less toxic than *p*-phenylenediamine is a solution of *n*-propylgallate (LD_{50} 3.8 g/kg). We have not had experience with this antifade solution for immunofluorescnce in yeast but it can be made as follows: 0.2 g *n*-propylgallate in 10 ml of 70% glycerol made up in PBS, pH 9.0 (Giloh and Sedat, 1982). DAPI can be added to visualize nuclei at the same concentration as above. In addition, various antifade mounting solutions are commercially available, e.g., Vectashield from Vector Laboratories.

A. General Immunofluorescence Protocol

This is the procedure that has been most widely used and is very similar to that described by Pringle *et al.* (1991). It has been used in our laboratory for visualizing the actin cytoskeleton and several actin-binding proteins (e.g., Abp1p and cofilin), for the neck filament proteins, Cdc10p and Cdc11p, for nuclear proteins (e.g., Anc1p), and for microtubules. See Fig. 1 for examples of immunofluorescence staining in yeast.

Steps

1. Fix 5 ml of an actively growing population of log-phase cells (1×10^7 cells/ml) by adding 0.67 ml 37% formaldehyde and allow to stand at room temp for at least 1 hr.

2. Spin down cells for 2 min at 700 g (2500 rpm) and wash twice with 2.5 ml o buffer A.

3. Resuspend cells in 0.5 ml cell wall digestion mix. Incubate for 35 min at 37°C.

4. During this incubation time the slides can be prepared for cell mounting by placing 15 μl of 1 mg/ml poly-L-lysine on each well of a multiwell slide (precleaned by immersing in distilled H_2O, then in 95% EtOH and air drying). Allow to sit for 5 min and then wash each well three times with distilled water and air dry.

5. Put 15 μl of cell suspension on each well. Allow to settle for 10 min and then aspirate off gently. If obtaining enough cells for observation is a problem, the suspension can be spun down and resuspended in a smaller volume just before placing on the slide.

6. Place the slide in −20°C methanol for 6 min and then in −20°C acetone for 30 sec This step flattens and permeabilizes the cells.

7. After air drying the slides completely, wash each well 10 times with blocking buffer B by gently aspirating off the wash buffer and then reapplying from a pipette. From this point on it is important not to let the cells dry out. Keep the slides in a humid environment e.g., in a covered petri dish with a damp tissue.

8. Remove the final buffer B wash from the well and place 15 μl of the primary antibody onto the cells. The antibody should be diluted as necessary in buffer B. In general, antibodies are used at a greater concentration (about 10-fold greater) for immunofluorescence than for Western blotting. Incubate for 1 hr.

9. Wash each well 10 times with buffer B and add 15 μl of diluted secondary antibody We use 1/1000 dilution of Cappel secondary antibodies or 1/200 dilution of Sigma CY3-conjugated antibodies. Incubate for 1 hr in a dark place to prevent bleaching of the fluorophore.

10. Wash again 10 times with buffer B, aspirate the buffer, and then put 5 μl of the mounting solution on each well. Cover gently with coverslip, lowering from one side to exclude bubbles. Then seal around the edges with nail polish. Slides can be viewed immediately or stored at −20°C in the dark for several months.

B. Modification to General Immunofluorescence Procedure

This method follows many of the same steps as the general protocol except that the methanol acetone step is replaced with an incubation using SDS. This adaptation of the original procedure has permitted the localization of several proteins, particularly those that seem to be embedded in large complexes of proteins, such as those at the presumptive bud site, or that need some denaturation to expose an epitope. Proteins localized using this method include Cdc42p (Ziman *et al.*, 1993), Sec4p (Novick and Brennwald, 1993), and Smy1p (Lillie and Brown, 1994).

Steps

1–5. Same as for general protocol.

6. Aspirate off the excess suspension from each well and add 10 μl 0.5% SDS in PBS for 5 min. This step permeabilizes the cells but they will not be flattened as much as in the previous procedure. Due to the presence of SDS the solution loses surface tension. To avoid spreading of solution from the wells, the volume of solution used in this permeabilization stage and in all subsequent wash steps and incubation procedures should be reduced to 10–12 μl rather than 15 μl.

7. After the incubation aspirate the SDS gently and wash the cells 10 times with buffer B. Do not allow the slide to dry in between the SDS step and washes.

8–10. Same as for general protocol.

C. Immunofluorescence Protocol 2

This procedure allows the preservation of the spindle pole body and has been used successfully for localizing proteins to these structures. The protocol given here is that of Rout and Kilmartin (1990) with adaptations by Stirling et al. (1994). The main difference from the more general protocol is that the cell wall is digested prior to cell fixation.

Steps

1. Prepare spheroplasts from log-phase cells (1×10^7 cells/ml) by digestion with 40 μg/ml zymolyase-100T in 1.1 M sorbitol for 20 min at 28°C.

2. Allow the spheroplasts to recover for 30–60 min at room temperature by incubating in spheroplast recovery medium.

3. After gently washing in 1.1 M sorbitol, apply spheroplasts to poly-L-lysine coated slides and allow cells to settle for 5–10 min. Aspirate off excess cells and immerse slide in methanol for 5 min at −20°C, followed by acetone for 30 sec at room temperature.

4. After air drying the slides, add blocking buffer C to each well for 1 min.

5. Remove the blocking buffer from the well and place 15 μl of the primary antibody onto the cells. The antibody should be diluted as necessary in buffer D. Incubate for 1 hr at room temperature.

6. Wash each well 10 times with buffer D and add 15 μl of diluted secondary antibody. Incubate for 1 hr at room temperature. The incubation should be in the dark to prevent bleaching of the fluorophore.

7. Wash again 10 times with buffer D and put 5 μl of mounting solution on each well. Cover with coverslip, exclude bubbles, and seal around the edges with nail polish.

IV. COMMENTS

As already mentioned, optimization of a few key variables is likely to be important for successful localization of a protein of interest. In general, start with the approach that seems most appropriate. For example, if a protein binds actin *in vitro,* try a procedure that has given success in actin localization. Alternatively, if it is thought that the protein might be a component of the spindle pole body, it is likely that a rather different procedure optimized for other spindle pole body components will be the best starting place. If you have little or no idea where the protein resides within the cell, we recommend that you start with the general protocol given in Section IIIA and then try some of the modifications suggested in Table I.

A. Yeast Strains and Growth Conditions

The first factors to be considered in any immunofluorescence experiment are the yeast strain itself and the conditions under which the cells are grown. The choice of yeast strain might be constrained by the questions being asked. For example, studies on proteins involved in pseudohyphal growth in yeast must be in a strain that is capable of undergoing this morphogenetic change. For localizing the majority of proteins, however, most lab strains are satisfactory, although it is often helpful for cytological studies if diploids can be used because they are larger than haploid cells, facilitating observation of cell structures.

As with the choice of strain, the conditions of growth will vary according to the biological questions being asked. In general, a population of exponentially growing cells in rich medium is preferable. One reason for this choice of growth conditions is that nonexponential phase cells often have thicker cell walls, making cell wall digestion and permeabilization more difficult. However, if the protein of interest has a role in sporulation, for example, then clearly modifications to the methodology will have to be made.

Sometimes it may be necessary to express a protein of interest from a plasmid. In such

TABLE I Stages of Immunofluorescence in *S. cerevisiae* and Guide to Variables Used to Optimize Staining for a Specific Antiserum[a]

Stage	Standard protocol	Suggested modification
1.	Fix cells for 60 min with 4% formaldehyde	Reduce time of fixation Use paraformaldehyde or glutaraldehyde–formaldehyde fixations Alternatively, omit this step altogether
2.	Digest cell wall for 35 min	Vary time of digestion if cells are not grown in rich media or if they cannot be fixed in exponential growth phase
3.	MeOH/acetone treatment	Replace this stage with a 5-min incubation with 0.1–0.5% SDS Omit this step altogether
4.	Block, by 10× washes, with PBS/BSA solution	If staining looks nonspecific, block for longer or change blocking solution
5.	Primary antibody (1 hr)	If the staining is nonspecific then the antibody could be further affinity purified and/or preabsorbed against cells deleted for the gene of interest Vary dilution of the antibody
6.	Secondary antibody	If the signal is very weak, try a different fluorophore, e.g., CY3 If the signal is nonspecific, try further dilution of the antibody
7.	Observation using fluorescence microscope	If the signal is weak, find the best microscope available or use an electronic system to enhance the signal

[a] Details of the standard method and of many of the modifications suggested here are given in the text.

cases, cells will be grown on synthetic media with appropriate selection. Growth on minimal media often leads to thicker cell walls, which may require a slightly longer cell wall digestion period than in the more general protocol given in this article. In addition, the exponential phase for cell growth in minimal media occurs at a lower cell density than with cells grown in rich media. Despite these considerations, immunolocalization of Sac6p (Drubin *et al.,* 1988) was better in cells grown on minimal media then in cell grown in rich media. It should be noted that the expression of proteins from plasmids is not optimal as the level of protein within the cell is usually higher than normal. If a protein of interest is expressed from a plasmid, it is preferable to use a centromere-based plasmid (three to five copies per cell) and to use the gene's own promoter. The use of high expression promoters, such as that for the *GAL1* and *GAL10* genes, may help in observation of a protein, but it may be difficult to demonstrate whether the localization is real or whether it occurs just when there is too much of the protein. It is likely that many low-level proteins can be observed without overexpression by using high-titer antibodies and optimizing key variables in the immunofluorescence procedure.

B. Fixation

In general, concentrated formaldehyde can be added from a 37% stock to a final concentration of 3.7–5.0%. Our lab routinely adds the formaldehyde to cells while still in the growth medium and incubates in fixative for about 1 hr at room temperature. The addition of formaldehyde to the medium is preferable as the fixation is rapid. In cases where cells have been centrifuged, rinsed in buffer, and then resuspended in a buffer containing formaldehyde, changes to organelle morphology have been noted (Pringle *et al.,* 1991). Following fixation the cells can be washed and cell wall digestion can be performed. We have found that after washing, cells can be stored fixed at 4°C for several hours, although leaving them overnight is usually not optimal. Longer storage may be hard to avoid when performing a time course, although this problem may be circumvented by initiating cultures at different times so samples will be fixed simultaneously.

Two variables with regard to fixation are the time of fixation and the fixative itself. Sometimes a short fixation in formaldehyde is best (5–20 min). This is presumably because longer mes increase the level of cross-linking to a level where the appropriate epitopes are no longer accessible. Reduction of fixation time has proved useful in cases where proteins are thought to be part of large complexes, such as the spindle pole body.

Although almost all immunofluorescence in yeast reported to date has used formaldehyde-fixed cells, there may be cases in which this is not optimal. A number of alternative protocols have been reported in which the conditions of fixation vary. We have not had experience with these methods, but if proteins cannot be localized using the protocols given, these might provide useful variations. First, the use of paraformaldehyde is widely reported (Roberts et al., 1991; TerBush and Novick, 1995). Cells are fixed for 20 min using 3.7% formaldehyde, they are then spun, washed, and resuspended in buffer containing freshly prepared 4% paraformaldehyde for a further 60 min. An alternative protocol involves fixation with a low concentration (0.025%) of glutaraldehyde added to the growth medium, followed by a further fixation of 30 min with 3.5% formaldehyde (Pringle et al., 1991). Finally, in some cases, it has been found preferable to avoid the use of aldehyde fixatives altogether. For example, fixation using just methanol and acetone at −20°C has also been used (C. Beh, personal communication).

C. Antibodies

There are commercially available antibodies for a few proteins, e.g., for tubulin, YOL1/34; and for actin, C4. In other cases an antibody to a protein of interest may be available from a fellow researcher. However, it is most common that antibodies to a protein have not been raised.

For most immunofluorescence procedures, it appears that purified polyclonal antibodies are as good as, or superior to, monoclonal antibodies. A preference for raising polyclonal antibodies derives from the probability that the serum will contain multiple antibody species that recognize different epitopes on the protein of interest. Thus, if a particular epitope is sensitive to fixation or is deeply embedded in the native protein, it is also probable that others will not be. In addition, if the antiserum is raised to a large portion of the protein, it is possible that several antibodies could bind simultaneously, thus increasing the strength of an immunofluorescence signal. However, when generating a polyclonal antiserum, it is important to recognize that the serum is likely to contain a plethora of antibodies that might react with other antigens in yeast cells. To ensure specificity, antibodies should be affinity purified. Microaffinity purification of antibodies using antigen bound to nitrocellulose and larger scale, column-based techniques have been reported for the affinity purification of antibodies (Pringle et al., 1991). Both approaches have been successfully used in our laboratory. Another stage of purification that can be useful is to preabsorb the antiserum against fixed yeast cells in which the gene encoding the protein of interest has been deleted.

A more recent development in immunofluorescence microscopy has been the advent of numerous peptide tags that can be attached via recombinant DNA manipulations to the protein of interest. Proteins can then be visualized using commercially available antibodies. One advantage of using a tagged protein is that, providing the antibody preparation is pure, only the protein carrying the tag will be detected by immunofluorescence. This is of particular importance when studying one member of a family of highly related proteins. A number of tags are available, and proteins have been successfully localized in yeast using both the myc tag (Evan et al., 1985; Terbush and Novick, 1995) and the HA tag (Bogerd et al., 1994). It should be noted that in most cases multiple copies of the tag (usually three to six copies) need to be appended to the protein before the signal is sufficiently strong for immunofluorescence.

The use of larger tags, such as glutathione S-transferase, has also proved successful in the immunolocalization of proteins in yeast (J. Pringle, personal communication). However, the large size of the tag (27.5 kDa in this case) is a concern. If such a tag is to be used, it is important to demonstrate that the tagged protein is functional, e.g., showing that it can rescue a phenotype associated with a mutation in the protein.

The use of positive control antibodies is very valuable when attempting immunofluorescence for a protein of unknown cell localization. This will determine whether the technique has at least been successful for proteins that should be localized in a specific way. Once a staining pattern has been observed, it is imperative that the localization is then demonstrated to be specific. In this regard, showing a lack of staining in a strain in which the gene encoding the protein of interest is deleted is a suitable control. Another good control when using an affinity-purified antiserum is to preabsorb the serum against the antigen and show that specific staining pattern is no longer obtained. Ultimately, there is no perfect control to show that the staining pattern reflects normal localization and it is important to confirm results by complementary approaches. These approaches might include coimmunoprecipitation to show binding to another component of the structure to which the protein of interest is localized and cofractionation *in vitro*.

D. Double Labeling

To demonstrate the spatial relationships of two different proteins in the same cell, double labeling is often valuable. This can be relatively straightforward if primary antibodies from different types of animals are available. Appropriate secondary antibodies can then discriminate the antibody types to allow the staining patterns of each protein to be observed. It should be noted that the primary and secondary antibodies can be incubated with cells either sequentially or simultaneously. If the proteins show colocalization, it is important to demonstrate that this is not due to cross-species reactivity of the secondary antibodies used. Thus, control samples should be set up in which one or the other primary antibody is left out of the incubation. Both secondary antibodies should then be added. A specific signal in the absence of the relevant primary antibody would indicate cross-reactivity. To remedy this, cross-reacting antibodies can be eliminated by preabsorbing the antiserum with immobilized antibodies from the second. A second concern with regard to double labeling is that of crossover fluorescence. For example, sometimes illumination for FITC can result in some rhodamine fluorescence from a double-labeled sample. This can usually be circumvented by using different fluorophores (e.g., Texas red or CY3) or by using different filter sets on the microscope.

E. Microscopes and Recording Images

The quality of the microscope itself can make a tremendous difference when doing immuno-fluorescence. We would always advise finding the best microscope available before investing time in trying to improve the brightness of your samples. When we switched to a better fluorescence microscope, we could see that some actin mutants previously thought to lack actin cables actually contained faint cables that could only be seen with the better optics. One technique that can be tried to increase the signal is to use tertiary and quaternary antibody "sandwiching" of fluorescent antibodies. However, our experience is that the background noise increases rapidly so seeing the specific signal may not be much easier. If using a sandwiching technique, it is important to carefully optimize the concentration of each antibody. The wider use of electronic systems such as CCD cameras can help with weak signals. It is preferable to obtain a weak but accurate signal rather than obtaining a brighter signal that is only achievable by overexpression of the protein.

Acknowledgments

We thank Keith Kozminski, Lisa Belmont, Nathan Machin, and James Close for helpful comments and for critical reading of the manuscript. This work was supported by grants to David Drubin from the National Institute of General Medicine Sciences (GM-42759) and the American Cancer Society (CB-106, FRA-442). Kathryn Ayscough is an International Prize Traveling Fellow of the Wellcome Trust (038110/Z/93/Z).

References

Alfa, C., Fantes, P., Hyams, J., McLeod, M., and Warbrick, E. (1993). "Experiments with Fission Yeast: A Laboratory Course Manual." Cold Spring Harbor Press, Cold Spring Harbor, NY.
Bogerd, A. M., Hoffman, J. A., Amberg, D. C., Fink, G. R., and Davis, L. I. (1994). *nup1* mutants exhibit pleiotropic defects in nuclear pore complex function. *J. Cell Biol.* **127**, 319–332.
Drubin, D. G., Miller, K. G., and Botstein, D. (1988). Yeast actin binding proteins: Evidence for a role in morphogenesis. *J. Cell Biol.* **107**, 2551–2561.

van, G. I., Lewis, G. K., Ramsey, G., and Bishop, J. M. (1985). Isolation of monoclonal antibodies specific for human c-*myc* proto-oncogene product. *Mol. Cell Biol.* **5**, 3610–3616.

iloh, H., and Sedat, J. W. (1982). Fluorescence microscopy: Reduced photobleaching of rhodamine and fluorescein protein conjugates by n-propyl gallate. *Science* **217**, 1252.

ilmartin, J. V., and Adams, A. E. M. (1984). Structural rearrangements of tubulin and actin during the cell cycle of the yeast, *Saccharomyces. J. Cell Biol.* **98**, 922–939.

illie, S. H., and Brown, S. S. (1994). Immunofluorescence localization of the unconventional myosin, Myo2p, and the putative kinesin-related protein, Smy1p to the same regions of polarized growth in *Saccharomyces cerevisiae. J. Cell Biol.* **125**, 825–842.

ovick, P., and Brennwald, P. (1993). Friends and family: The role of the Rab GTPases in vesicular traffic. *Cell* **75**, 597–601.

ringle, J. R., Adams, A. E. M., Drubin, D. G., and Haarer, B. K. (1991). Immunofluorescence methods for yeast. *In* "Methods in Enzymology" (C. Guthrie and G. R. Fink, eds.), Vol. 194, pp. 565–602. Academic Press, San Diego.

oberts, C. J., Raymond, C. K., Yamashiro, Y. K., and Stevens, T. H. (1991). Methods for studying the yeast vacuole. *In* "Methods in Enzymology" (C. Guthrie and G. R. Fink, eds.), Vol. 194, pp. 644–661. Academic Press, San Diego.

out, M. P., and Kilmartin, J. V. (1990). Components of the yeast spindle and spindle pole body. *J. Cell Biol.* **111**, 1913–1927.

tirling, D. A., Welch, K. A., and Stark, M. J. R. (1994). Interaction with calmodulin is required for the function of Spc110p, an essential component of the yeast spindle pole body. *EMBO. J.* **13**, 4329–4342.

erBush, D. R., and Novick, P. (1995). Sec6, Sec8, and Sec15 are components of a multisubunit complex which localizes to small bud tips in *Saccharomyces cerevisiae. J. Cell Biol.* **130**, 299–312.

iman, M., Preuss, D., Mulholland, J., O'Brien, J. M., Botstein, D., and Johnson, D. I. (1993). Subcellular localization of Cdc42p, a *Saccharomyces cerevisiae* GTP-binding protein involved in the control of cell polarity. *Mol. Biol. Cell* **4**, 1307–1316.

Immunocytochemistry of Frozen and of Paraffin Tissue Sections

Mary Osborn and Susanne Isenberg

I. INTRODUCTION

Immunocytochemistry of tissue sections can yield valuable information as to the location of antigens. Thus it can determine whether the antigen is ubiquitous or is present only in certain tissues. It can further determine whether all cells in a given tissue are positive for a given antigen, or whether the antigen is restricted to one or a few specialized cell types within the tissue. Its uses are not limited to normal tissues, and testing of tumor tissues in immunocytochemistry can yield information important for determination of tumor type. As in the article on immunocytochemistry of cultured cells (by Osborn, this volume), choice of antibodies and of fixation method can be of critical importance. New antibodies should be tested on both frozen sections and on paraffin sections of a variety of tissues in which the antigen is thought to be present. In general, more antibodies will react on frozen sections than on paraffin sections; however, morphology is better preserved in the paraffin sections.

For other reviews of methods, see Denk (1987) and Sternberger (1979), and for overviews of the use of these methods in histopathology, see Osborn and Weber (1983), Tubbs *et al.* (1986), Jennette (1989), and Osborn and Domagala (1990).

II. MATERIALS AND INSTRUMENTATION

A. Antibodies

See the article by Mary Osborn in this volume. Dilute monoclonal antibodies 1:5 to 1:30, ascites fluid 1:100 to 1:1000, and polyclonal antibodies 1:20 to 1:40. Dilute both primary and secondary antibodies as far as possible to save money and to avoid unspecific reactions.

B. Cryostat

Use, e.g., a Reichert-Jung Cryostat Frigocut Model CM:3050 (Leica). A useful accessory is a freezing head with variable temperature control. Use a C-knife and resharpen when necessary.

Tissue-tek

CT compound from Miles Laboratories.

Paraffin Embedding

Automatic machines are useful only for laboratories that process a large number of samples. Check equipment for paraffin embedding in the local pathology department.

Sliding Microtomes

These are relatively inexpensive, e.g., Reichert-Jung Model HN40, and can be used with disposable knives.

Water Baths for Histology

These are thermostatically controlled and relatively shallow, e.g., 20 cm in diameter and 4 cm deep.

Microscopes

For a microscope to view immunofluorescence specimens see the article by Mary Osborn this volume. To view peroxidase- or streptavidin–biotin-stained specimens only a simple light microscope is required.

II. PROCEDURES

A. Cryostat Sections

Steps

1. Freezing of Tissue Blocks

1. Fill the inner beaker of the freezing apparatus with isopentane and the outer beaker with liquid nitrogen about 30 min before freezing tissues (Fig. 1). Use reagent-grade isopentane. Wait until isopentane reaches -120 to $-130°C$. At $-155°C$ the isopentane will freeze.

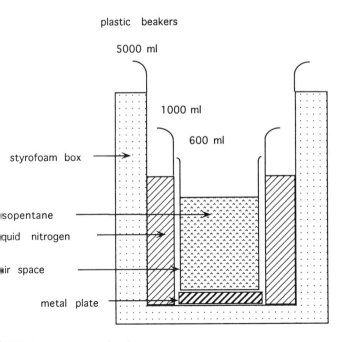

plastic beakers

5000 ml

1000 ml

600 ml

styrofoam box

isopentane

liquid nitrogen

air space

metal plate

FIGURE 1 Apparatus used to freeze tissues.

2. Dissect tissues. Cut into small blocks (~4–7 mm) using a scalpel. Place block on prenumbered square of paper, with the surface that will be sectioned furthest away fro the paper. For small specimens, e.g., vessels, place a drop of Tissue-tek on the paper, the add the tissue.

3. Drop tissue blocks into iospentane. Leave for at least 30 sec. Remove blocks wi plastic tweezers and transfer directly to plastic vials (scintillation vials ~6 × 2.5 cm wo well) or metal cans (~3 × 3 cm) with screw tops that have been precooled on dry ice. Clo vials and store in a −70°C freezer. Tissue blocks are stable for several years.

2. Cutting Cryostat Sections

1. Wash microscope slides by dipping in acetone, air-dry, and store at room temperatur Store plastic tweezers and brush in the cryostat. Precool cryostat and freezing head.

2. Place Tissue-tek on precooled freezing head, and mount block so that the larger sic of the cut section is at 90° to the knife blade.

3. The optimal cutting temperature is different for different tissues. Most tissues cut we at −15 to −20°C. For liver, use −10°C. If the sections wrinkle as they are cut and loc mushy, decrease the temperature. If the sections have cracks and look brittle, increase tl temperature.

4. To cut sections use the C-knife. First trim the block to get a good cutting surfac Then adjust the section thickness to 5 μm, and use the automatic advance. Now cut thre or four sections so that the preset section thickness is achieved. Then bring the antiroll pla on the knife to stop the section from rolling up and to keep it flat on the knife. The antiro plate must be parallel to the knife edge and should only protrude very slightly over the edg

5. For optimal sections the knife and the antiroll plate must be kept clean (use a so cloth dipped in acetone). Always clean the knife in the cutting direction and never tl reverse, so as not to damage the cutting edge of the knife.

6. Remove the antiroll plate and hold the microscope slide over but not touching the cu section. The section should now spring on to the slide because of the difference in tempera ture between knife and slide. The quality of the section can be checked using toluidine blu or hematoxylin–eosin staining (see below).

7. Dry the sections at room temperature for 30 min. Then either use directly or plac in a slide box and put in −70°C freezer. Cut sections are stable for months at −70°C.

8. Sections from a few tissues may not stick firmly enough to slides and may come o during subsequent processing. If this happens try coating slides with 0.1% polylysine (e.g Sigma Cat. No. 8920) in water.

B. Paraffin Sections

Steps

1. Embedding Tissues in Paraffin

Human material is often received from the clinic already embedded in paraffin. Protocol vary depending on the clinic, with time of fixation in formaldehyde being very variable (e.g 4 hr to over the weekend). To embed animal tissue in the lab, cut into 4 to 7-mm block and place it for 4–8 hr in 3.7% formaldehyde in water or phosphate-buffered saline (PBS 1 hr in 50% ethanol, 2 × 1 hr in 70% ethanol, 2 × 1 hr in 96% ethanol, 2 × 1 hr in 100% ethanol, 1 × 1 hr in xylene, 1 × 2 hr in xylene, and 2 × 2 hr in Paraplast Plus (Shandon)

2. Cutting Paraffin Sections

1. To obtain very thin sections (1–2 μm), put the paraffin blocks in a freezer at −20°C for around 30 min. Mount the block in the holder of a sliding microtome. If the block cut well do not use the automatic advance, but rely instead on the natural expansion of the bloc as it warms up to advance the block. If the block is not easy to cut use the automatic advanc set at a thickness of 1–2 µm. Correct adjustment of the knife and of the cutting angle is very important. Use an inclination angle (β) of 15° (Fig. 2). If the inclination angle is less than 10° the knife will not cut the block, and if it is greater than 15° the block will break.

2. Trim the block until the cutting surface is optimal. Then cut a section, using a paint brush to draw the section onto the knife so that it does not roll up.

3. Dip a second paint brush in water so that the section will adhere to it, and move the section to a water bath held at 40–45°C. If the section is placed with the shiny smooth surface touching the water the warmth will smooth out the section and wrinkles will vanish!

4. Place a microscope slide under the section, and using a brush position the section on the slide.

5. Dry the sections. For immunohistochemical methods, dry overnight at 37°C. For normal histological methods, set drying oven to 60°C so that the paraffin melts in part during the drying step, and dry for 1–2 hr.

3. Deparaffinization

Immerse sections 2 × 10 min in xylene, 1 × 3 min in 100% ethanol, 1 × 3 min in 95% ethanol, and air-dry.

4. Trypsinization

Some laboratories routinely use trypsinization or other proteolytic treatment of formalin-fixed, paraffin-embedded tissues prior to immunohistochemistry, e.g., 5 min in 0.1% trypsin (Sigma Cat. No. T-8128) in PBS at room temperature. As stressed by Ordonez *et al.* (1988) this can enhance the staining by certain antibodies but may also result in false-negative staining with other antibodies. Thus, control trypsinization conditions carefully.

5. Treating Sections in a Microwave Oven

Cut sections onto Superfrost or similar slides. After deparaffinization immerse slides in citrate buffer (2.1 g citric acid monohydrate/liter adjusted to pH 6.0 with NaOH). Microwave for two cycles of 5 min each at a setting of 650 or 700 W. Add more buffer between cycles so slides stay covered during the microwave step. Cool to room temperature (cf. Cattoretti *et al.*, 1992).

C. Histologic Staining of Sections

Stain frozen sections directly. Deparaffinize paraffin sections (see earlier), substituting a wash with distilled water for the air-drying step.

Solutions

1. *Toluidine blue:* Immerse sections in 1% toluidine blue for approximately 1 min. Wash with distilled water. Mount in water-soluble embedding medium, e.g., Glycergel (Dako Cat. No. C0563).

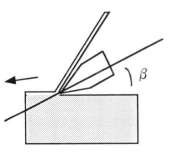

FIGURE 2 Correct adjustment of knife and of cutting angle to cut paraffin sections.

2. *Hemotoxylin–eosin:* Immerse sections for 10 min in Mayer's Hemalum solution (Merck). Wash for 10 min under running tap water, for 5 min in eosin (Merck) and twice with distilled water, and then run through an alcohol series (e.g. 75%, 95% for 2 min each then 100% for 5 min, then 2 × 5 min in xylol) and mount in Eukitt (Riedel de Haën Cat No. 33949) or Entellan (Merck Cat. No. 1.07961.0100).

D. Immunocytochemistry

Three methods are described in detail: immunofluorescence, the immunoperoxidase method and the more sensitive streptavidin–biotin method. The two latter methods have the advantage that nuclei can be counterstained with hemotoxylin and that only a simple light microscope is required to visualize the stain; however, fluorescence generally gives greater resolution.

As noted for cells, many antibodies that react well on cryostat sections may not react on the same tissue after it has been fixed in formaldehyde and embedded in paraffin. In such a case it may be advantagous to try fixing tissue, e.g., in B5, Bouin's or Zenker's fixative or alcohol, prior to paraffin embedding. An interesting alternative is to use sections of formaldehyde-fixed, paraffin-embedded material that have been treated in a microwave oven

For all methods mark the position of the section after fixation; either use a diamond penci or circle the section with a water-repellent marker (Dako Cat. No. S2002). Remove excess buffer after rinsing steps with Q-Tips. Use 10 μl of antibody per section. Apply with an Eppendorf pipette and use the pipette tip to spread the antibody over the section without touching the section. Several manufacturers (e.g., Dako) produce excellent protocol sheets for each immunocytochemical method.

Steps

1. Immunofluorescence

1. Fix cryostat sections or paraffin sections deparaffinized as described earlier for 10 min in acetone at $-10°C$. Air-dry.

2. Use steps 2–9 of the protocol given in the article by Mary Osborn in this volume for multitest slides. Nuclei can be counterstained with Hoechst dye (see Osborn's article). Positively stained cells will be green (see Fig. 3, color plate), if an FITC-labeled second antibody is used, or red, if a rhodamine-labeled second antibody is selected.

2. Peroxidase Staining

1. Fix cryostat sections for 10 min in acetone at $-10°C$, air-dry, and wash in PBS.

2. Deparaffinize paraffin sections, and incubate for 30 min at room temperature in 100 ml methanol containing 100 μl H_2O_2 to block endogenous peroxidase activity. Wash in PBS.

3. Incubate for 10 min at 37°C with normal rabbit serum diluted 1:10 in bovine serum albumin (PBS), 0.5 mg/ml BSA. Drain but do not wash after this step.

4. Incubate with primary antibody (e.g., mouse monoclonal) for 30 min at 37°C.

5. Wash three times in PBS.

6. Incubate with second antibody coupled to peroxidase, e.g., rabbit anti-mouse for a monoclonal first antibody (Cat. No. P0260 from Dako diluted 1:10 to 1:20).

7. Wash three times in PBS and once in Tris buffer (6 g NaCl, 6 g Tris/liter, pH 7.4).

8. Develop for 10 min at room temperature using freshly made solutions (e.g., 0.06 g diaminobenzidine, Fluka Cat. No. 32750 in 100 ml Tris buffer, 0.03 ml H_2O_2). *Note*: Diaminobenzidine is a carcinogen; handle with care.

9. Wash in tap water.

10. Apply a light counterstain by immersing the slide in Hemalum for 1 to 10 sec. Remove when staining reaches the required intensity.

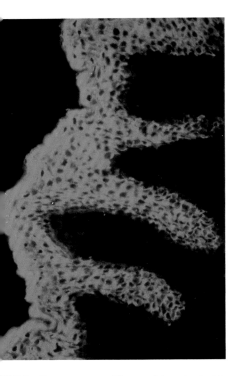 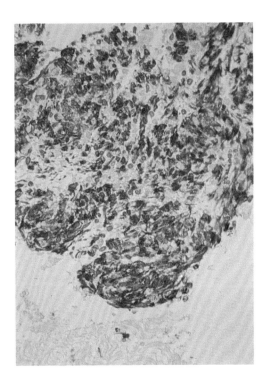

URE 3 Frozen section of human skin stained with 1 keratin antibody with FITC-labeled second antidy. Only the epidermis is stained (×150).

FIGURE 4 Frozen section of human rhabdomyosarcoma stained after microwave treatment with desmin antibody DEB 5 and with peroxidase-labeled second antibody. Brown tumor cells are positive for desmin. Nuclei are counterstained blue (×160).

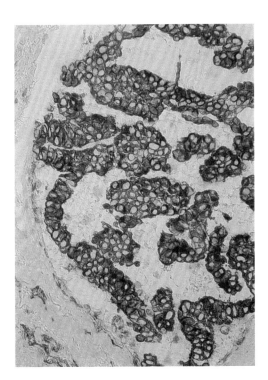

FIGURE 5 Frozen section of human breast carcinoma stained with the keratin KL1 antibody and with peroxidase-labeled second antibody. Brown tumor cells are positive for keratin. Nuclei are counterstained blue (×150).

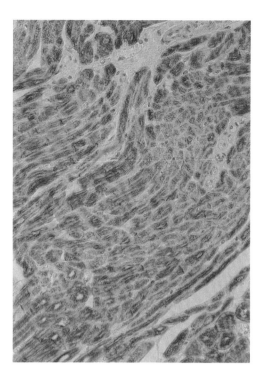

FIGURE 6 Paraffin section of human heart stained with antibody after microwave fixation with the desmin DER 11 antibody in the streptavidin–biotin technique. Note striations. Nuclei are counterstained blue (×150).

FIGURE 7 Paraffin section of human breast carcinoma stained after microwave treatment with keratin K antibody in the streptavidin–biotin technique. Tumor cells are keratin positive. Nuclei are counter stained blue (×150).

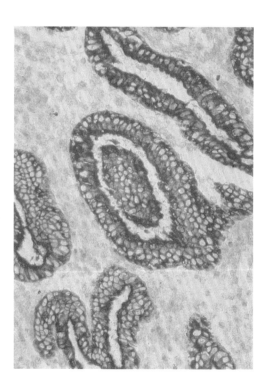

FIGURE 8 Frozen section of human uterus is stained with KL1 keratin antibody in the alkaline phosphatase–antialkaline phosphatase technique. Only the epithelial cells are positive (×150).

11. Wash in tap water.

12. Mount in Glycergel (Dako Cat. No. C0563).

Notes

1. Structures that are positively stained will be dark brown, whereas nuclei will be light blue (Figs. 4 and 5; see color plates).

2. The method can be made more sensitive by using an additional step with a peroxidase–antiperoxidase complex (see Sternberger, 1979).

3. Streptavidin–Biotin Stain

Buy the reagents separately or use the Histostain-SP kit from Zymed Laboratories (Cat. No. 95-6543 for mouse primary antibody, Cat. No. 95-6143 for rabbit primary antibody). These kits are based on the strong binding between streptavidin, a 60,000-kDa protein isolated from *Streptomyces avidinii*, and biotin, a water-soluble vitamin (MW 244, $K_d = 10^{-15}\ M$). Instructions are given for the mouse kit.

1. Fix cryostat sections for 10 min in acetone at $-10°C$ and air-dry. Sections can be treated with 0.23% periodate for 45 sec. Go to step 3.

2. Deparaffinize paraffin sections and air-dry. Incubate 10 min in PBS at room temperature, then 10 min in H_2O_2 solution (9 parts methanol to 1 part 30% H_2O_2 in water). Wash 3×2 min in PBS.

3. Reduce nonspecific background staining by blocking for 10 min in 10% goat serum at room temperature. Then drain but do not wash.

4. Incubate with primary antibody, e.g., mouse monoclonal antibody, in wet chamber for 30 min at 37°C.

5. Wash 3×2 min in PBS.

6. Add biotinylated second antibody, e.g., goat anti-mouse, for 10 min at room temperature and repeat step 5.

7. Add enzyme conjugate streptavidin–peroxidase diluted 1:20 for 5 min at room temperature. This binds to the biotin residues on the second antibody.

8. Develop by adding substrate–chromogen mixture for 5 min at 37°C or 15 min at room temperature. The enzyme peroxidase catalyzes the substrate hydrogen peroxide and converts the chromogen aminoethylcarbazole to a red, colored deposit.

9. Wash 3×2 min in distilled water.

10. Counterstain with Hemalum between 1 and 10 sec.

11. Wash 7 min in tap water.

12. Mount in Glycergel.

Notes

1. Structures that are positively stained will be red, whereas nuclei will be light blue (Figs. 6 and 7; see color plates).

2. Optimal dilution of primary antibody has to be determined for each antibody.

3. The streptavidin–peroxidase conjugate is the same for all species; the blocking serum and second antibody vary according to the species in which the first antibody is made.

Other Methods:

1. The alkaline phosphatase–antialkaline phosphatase technique is also a very sensitive method (Fig. 8; see color plate).

2. Both fluorescent and chromogenic signals can be enhanced up to 1000× with Tyramide

Signal Amplification Technology (New England Nuclear). With this technique primarily antibodies can be diluted in some instances more than 10,000 fold.

IV. Comments

If possible, include (1) a positive control, i.e., a section of a tissue known to contain the antigen; (2) a negative control, i.e., a section known not to have the antigen; and (3) a reagent control, i.e., a section stained with nonimmune serum instead of the primary antibody. Ideally, (1) should be positive and (2) and (3) negative. If high backgrounds are obtained, it may help to adjust the antibody concentrations by increasing the length of time for washing steps or by including 0.5 M NaCl or 5% BSA in the antibody solutions.

References

Cattoretti, G., Becker, M. H. G., Key, G., Duchrow, M., Schlüter, C., Galle, J., and Gerdes, J. (1992). Monoclonal antibodies against recombinant parts of the Ki-67 antigen (MIB 1 and MIB 3) detect proliferating cells in microwave processed formalin-fixed paraffin sections. *J. Pathol.* **168**, 357–363.

Denk, H. (1987). Immunohistochemical methods for the demonstration of tumor markers. *In* "Morphological Tumor Markers" (G. Seifert, ed.), pp. 47–70. Springer-Verlag, Berlin.

Jenette, J. C. (1989). "Immunohistology in Diagnostic Pathology." CRC Press, Boca Raton, FL.

Ordonez, N. G., Manning, J. T., and Brooks, T. E. (1988). Effect of trypsinization on the immunostaining of formalin-fixed, paraffin-embedded tissues. *Am. J. Surg. Pathol.* **12**, 121–129.

Osborn, M., and Domagala, W. (1997). Immunocytochemistry. *In* "Comprehensive Cytopathology," Second Edition (M. Bibbo, ed.), pp. 1033–1074. Saunders, Philadelphia.

Osborn, M., and Weber, K. (1983). Tumor diagnosis by intermediate filament typing: A novel tool for surgical pathology. *Lab. Invest.* **48**, 372–394.

Sternberger, L. A. (1979). "Immunohistochemistry," 2nd Ed. Wiley, New York.

Tubbs, R. R., Gephardt, G. N., and Petras, R. E. (1986). "Atlas of Immunohistology." American Society of Clinical Pathologists Press, Chicago.

PART 7

Vital **S**taining of **C**ells

Labeling of Endocytic Vesicles Using Fluorescent Probes for Fluid-Phase Endocytosis

Nobukazu Araki and Joel A. Swanson

I. INTRODUCTION

Endocytosis occurs by the invagination of plasma membrane to form an intracellular vesicle. Endocytic vesicles, and the intracellular organelles they communicate with, can be labeled by the inclusion of fluorescent, membrane-impermeant molecules in the extracellular medium. Such probes are enclosed in the vesicles as they form and remain contained in endocytic compartments. Ultimately, these fluid-phase endocytic probes either accumulate in lysosomes, where they may be degraded, or are returned to the extracellular medium by recycling vesicles. Fluid-phase endocytosis is also called pinocytosis, which is sometimes subdivided into macropinocytosis and micropinocytosis, the formation of large and small pinosomes, respectively (Swanson, 1989a; Swanson and Watts, 1995).

A variety of fluorescent probes can be used to label endocytic organelles in living cells. This article describes methods for labeling and observing these compartments in living cells and for immunofluorescence microscopy of similarly labeled cells. In addition, a method for the quantitative measurement of fluid-phase pinocytosis using fluorescent probes is described. These methods are optimized for the study of macrophages, which are actively endocytic cells. It should be kept in mind that most other kinds of cells exhibit lower rates of endocytosis and different kinetics of delivery from endosomes to lysosomes. One may therefore have to extend incubation times or increase probe concentrations to obtain strong fluorescent signals in the microscope or fluorometer. For any new cell type, we recommend that compartment identities be determined empirically, by comparing fluorescent probe distributions in cells fixed after various pulse–chase intervals with the immunofluorescent localization of known markers for endocytic compartments (e.g., Racoosin and Swanson, 1993). Monoclonal antibodies recognizing some of these marker proteins are available from the Developmental Studies Hybridoma Bank.

II. MATERIALS AND INSTRUMENTATION

The fluorescent fluid-phase probes, fluorescein dextran, MW 3000 (FDx3, Cat. No. D-3305), MW 10,000 (FDx10, Cat. No. D-1821), and lysine-fixable fluorescein dextran, MW 10,000 (Cat. No. D-1820), are from Molecular Probes. Lucifer yellow CH (Cat. No. 86150-2) is from Aldrich. Fluorescein dextran, MW 150, 000 (FDx150 Cat. No. FD-150S), paraformaldehyde (Cat. No. P-6148), bovine serum albumin (BSA, Fraction V. Cat. No. A-9647), Triton X-100 (Cat. No. T-9284), HEPES (Cat. No. H-3375), Trizma base (Cat. No. T-1503), and *p*-phenylenediamine (Cat. No. P-6001) are from Sigma. A primary antibody, rabbit anti-cathepsin D serum, was a gift from Dr. S. Yokota, Yamanashi Medical School. Texas red-labeled anti-rabbit IgG (goat) (Cat. No. TI-1000) is from Vector Lab. Dulbecco's modified essential medium (DMEM, Cat. No. 31600-034), fetal bovine serum (FBS, Cat.16000), and goat serum (Cat. No. 16210) are from GIBCO-BRL. Circular glass coverslips of 12 and 25 mm in diameter (No. 1 thickness, Cat. No. 12-545-102) and silicon oil (Cat. No. 5159-500) are from Fisher Scientific. Twenty-four-well (Cat. No. 430262) and 6-well (Cat. No.430343) plates are from Corning Costar.

An epifluorescence microscope (Axiophoto FL Carl Zeiss) is used for observation of both living and fixed cells (see Volume 3, Tanke, "Fluorescence Microscopy" for additional information). A lucifer yellow filter set was obtained from Omega Optical (Set No. XF-14), in addition to the usual fluorescein filter set. For high-resolution observation, a 100× Planapo lens, numerical aperture (NA) 1.4, or a 100× Plan-Neofluar lens, NA 1.3, was used. An inverted-type fluorescence microscope (Carl Zeiss IM-35) equipped with nuvicon video and a multichannel plate intensifier (Video Scope Int'l) and thermo-controlled stage (Medical Systems Corp.) were used for longer observations of living cells. A personal computer installed with image analysis software (MetaMorph 2.0, Universal Imaging Co.) was used for collecting images from the intensified video signal and for making time-lapse movies.

III. PROCEDURES

A. Fluorescence Microscopy of Endocytic Compartments Labeled with Fluorescent Probes

Solutions

1. *Ringer's buffer + BSA (RB):* 155 mM NaCl, 5 mM KCl, 2 mM CaCl₂, 1 mM MgCl₂, 2 mM NaH₂PO₄, 10 mM HEPES, 10 mM D-glucose, pH 7.2, plus 0.05% BSA. To make 1 liter, add 9.1 g of NaCl, 0.37 g of KCl, 0.275 g of NaH₂PO₄ · H₂O, 2.38 g of HEPES, 1.8 g of D-glucose, 0.22 g of CaCl₂, 0.2 g of MgCl₂ · 6H₂O, and 0.5 g BSA to 950 ml distilled water, adjust to pH 7.2 with 1 N NaOH, and bring the volume to 1 liter. Sterilize with 0.22-μm filter and store at 4°C.

2. *Labeling medium:* To make 1 ml, dissolve 0.5 mg of lucifer yellow, FDx3, FDx10, or FDx150 in 1 ml RB. The concentrations of probes may be changed depending on cell types. One can differentially label macropinosomes and micropinosomes using different-sized probes (Fig.1). To label both macropinosomes and micropinosomes, use low molecular weight probes such as lucifer yellow and FDx3. To label primarily macropinosomes, use larger probes, such as FDx150 (Araki *et al.*, 1996). Warm to 37°C before adding to cells.

3. *8% paraformaldehyde stock solution:* To make 50 ml, add 4 g paraformaldehyde to 30 ml distilled water and heat to 70°C while stirring. Add a few drops of 1 N NaOH so that the mixture becomes clear. Bring the final volume to 50 ml with distilled water, and filtrate with paper filter. This solution may be kept in aliquots at −20°C.

4. *80 mM HEPES stock solution, pH 7.2:* To make 100 ml, add 1.91 g of HEPES to 70 ml of distilled water. Adjust pH to 7.2 with 1 N NaOH while stirring. Bring the final volume to 100 ml with distilled water. Store at 4°C.

5. *Fixative:* 4% paraformaldehyde in 40 mM HEPES buffer, pH 7.2, containing 6.8% sucrose. To make 10 ml, add 5 ml of 8% paraformaldehyde solution and 0.68 g of sucrose to 5 ml of 80 mM HEPES, pH 7.2. Warm to 37°C before applying to cells.

6. *Phosphate-buffered saline (PBS):* To make 5 liters, dissolve 40 g of NaCl, 1 g of KCl, 5.75 g of Na₂HPO₄, and 1 g of KH₂PO₄ in 4 liters of distilled water. Bring the final volume to 5 liters with distilled water.

7. *Mounting medium:* To make 10 ml, put 10 mg *p*-phenylenediamine, 1 ml of PBS, and 9 ml of glycerol in a 15-ml tube. Wrap the tube with aluminum foil to protect from light and mix overnight on a rotating wheel. Divide into aliquot tubes and store at −20°C.

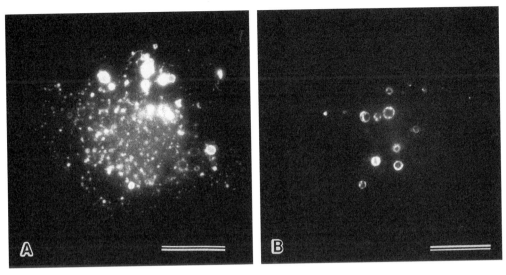

FIGURE 1 Differential labeling of macropinosomes and micropinosomes with lucifer yellow (A) and FDx150 (B). Macrophages were incubated for 5 min in labeling medium containing lucifer yellow or FDx150, then briefly washed and fixed immediately. A low molecular weight probe, lucifer yellow (MW 457), labels both macropinosomes and micropinosomes (A). A larger probe, FDx150 (MW 150,000), labels predominantly macropinosomes (B). Bars: 10 μm.

Steps

1. Culture cells on 12-mm, circular, No. 1 thickness coverslips in 24-well culture dishes in DMEM with 10% heat-inactivated FBS.

2. Replace the culture medium with prewarmed labeling medium. Swirl the dishes and incubate cells in labeling medium for various times at 37°C. Then quickly rinse by changing warm RB to remove fluorescent probe and chase in RB as necessary. For mouse macrophages, a 2- to 5-min labeling incubation without chase will label early endosomes, including micro- and macropinosomes (Fig. 1). A subsequent chase in the absence of the probe for 5 to 15 min should label late endosomes. A 30- to 60-min labeling incubation labels all endocytic compartments, including early and late endosomes and lysosomes, and a 30-min labeling followed by a chase longer than 30 min should label only lysosomes (Fig. 2).

3. Fix the cells in 4% paraformaldehyde in 40 m*M* HEPES, pH 7.2, containing 6.8% sucrose for 30 to 60 min at 37°C.

4. Rinse 3 × 5 min with PBS.

5. Mount the coverslip, cell side down, on a slide using mounting medium. Seal the coverslip and the slide with nail polish.

6. Observe the slide with an epifluorescence microscope. We can observe living cells without fixation for short periods using a simple microscope culture chamber. This method has been described previously in detail (Swanson, 1989b; Raccosin and Swanson, 1994). Briefly, assemble a chamber on a slide using small coverslip fragments to support the coverslip. RB should be added to fill the space between the slide and the coverslip. Seal the coverslip to the slide using a heat-melted paraffin-based compound (Swanson, 1989b). The method for longer observations of living cells is described next.

B. Observation of Endocytic Compartments in Living Cells

Step

1. Plate cells on 25-mm No. 1 coverslips in a 6-well culture dish containing tissue culture medium.

2. Replace the culture medium with labeling medium. Incubate in labeling medium for various times at 37°C, as described earlier, to label the endocytic compartments.

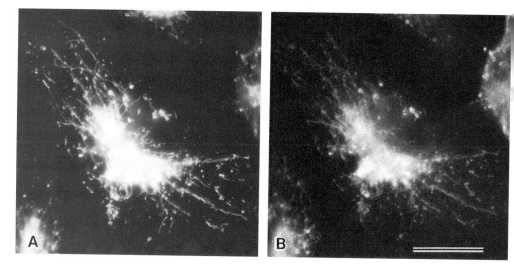

FIGURE 2 Immunofluorescence of cathepsin D in endocytic compartments labeled with endocytic probes. Macrophages were incubated for 30 min in labeling medium containing 0.5 mg/ml lysine-fixable FDx10 and chased for 30 min. Cells were then fixed and processed for immunofluorescence using a primary antibody against cathepsin D and Texas red-conjugated secondary antibody. (A) Fluorescein image shows that FDx10 labels tubular lysosomes. (B) Texas red image shows that tubular lysosomes, which are labeled with FDx10, are positive for cathepsin D. Bar: 10 μm.

3. Wash away fluorescent probe from the coverslip with RB.

4. Assemble the coverslip in a Leiden chamber (Medical System Corp.), fill with 1 ml of RB, and slowly add a small amount of silicon oil to cover the surface of RB in the chamber.

5. Put the chamber on the thermo-controlled stage at 37°C.

6. Observe the coverslip with an inverted fluorescence microscope using the lowest light exposure possible, as intense excitation light may cause not only photobleaching but also photochemical damage to living cells. Under optimal conditions of labeling, we can observe cells for up to an hour under low-intensity illumination conditions. This can be extended by inserting a shutter into the light path to control exposures.

7. Collect time-lapse images using a silicon-intensified target (SIT) or other sensitive video camera and MetaMorph 2.0 software (Universal Imaging Co.) on a personal computer. These images may be recorded on laser optical disks (Panasonic) or electronic storage media for later processing into movies of living cells.

C. Immunofluorescence Identification of Endocytic Compartments Labeled with a Fluorescent Probe

Solutions

1. *0.25% NH₄Cl in PBS (NH₄Cl/PBS):* To make 100 ml, dissolve 0.25 g of NH$_4$Cl in 100 ml PBS.

2. *0.25% Triton X-100 in PBS (Tx/PBS):* To make 300 ml, add 0.75 g of Triton X-100 to 300 ml PBS while stirring.

3. *2% heat-inactivated goat serum in Tx/PBS (HIGS-Tx/PBS):* To make 10 ml, add 0.2 ml of heat-inactivated goat serum to 9.8 ml of Tx/PBS.

4. *Primary antibody:* Dilute serum or antibody with HIGS-Tx/PBS. In the example shown, we diluted rabbit anti-cathepsin D serum at 1:500.

5. *Secondary antibody:* Dilute Texas red-labeled anti-rabbit IgG with HIGS-Tx/PBS at 1:250–500.

Steps

1. Incubate the cells with labeling medium as described earlier. Lysine-fixable FDx or lucifer yellow should be used when the other markers are to be localized by immunofluorescence. Nonfixable FDx would be lost during permeabilizing of cells.

2. Rinse in PBS to remove excess fluorescent probe, unless the cells were chased in RB without fluorescent probe.

3. Fix in 4% paraformaldehyde in 40 mM HEPES, pH 7.2, containing 6.8% sucrose for 1 hr at 37°C.

4. Rinse with PBS 3 × 5 min and further immerse in NH$_4$Cl/PBS for 10 min to quench free aldehyde.

5. Treat cells with HIGS-Tx/PBS for 2 × 5 min for cell permeabilization and blocking.

6. Put Parafilm in a container with a moist paper. Place one 40-μl drop of primary antibody on the Parafilm for each coverslip.

7. Wipe the cell-free side of the coverslip with a Kimwipe. Place the coverslip cell side down on a drop of primary antibody. Incubate with the primary antibody for 1 hr at room temperature in a moist chamber.

8. After incubation, put the coverslip back into the well and wash three times for 5 min each with Tx/PBS.

9. Using the same method as for the primary antibody, incubate with secondary antibody, e.g., Texas red-conjugated anti-rabbit IgG diluted in HIGS-Tx/PBS at a concentration 1:500, for 1 hr at room temperature.

10. Wash the coverslip with Tx/PBS twice and PBS once for 5 min each.

11. Mount the coverslip on a slide using the mounting medium. Seal the coverslip with nail polish. Observe the specimens with an epifluorescence microscope using fluorescein and rhodamine filter sets.

D. Quantitative Fluorometric Analysis of Endocytic Compartments Labeled with Fluorescent Probes

Solution

1. *Lysis buffer:* 0.1% Triton X-100 in 50 mM Tris, pH 8.5. To make 100 ml, dissolve 0.6 g of Trizma base and 0.1 g of Triton X-100 in 80 ml of distilled water. Adjust pH to 8.5 with 1 N HCl. Bring the volume to 100 ml.

2. *0.1% BSA/PBS:* To make 2 liters, dissolve 2 g of BSA in 2 liters of PBS.

3. *Standard solutions of fluorescent probes:* Dilute the labeling medium to concentrations 0, 1, 5, 10, and 20 ng probe/ml in lysis buffer. Each solution should be more than 2 ml.

Steps

1. Plate the cells at high density (e.g., 2 × 10^5 cells/well) in a 24-well culture dish. Triplicate experiments are desirable.

2. Replace the culture medium with labeling medium containing fluorescent probes. Dual labeling with FDx and lucifer yellow is possible (Berthiaume *et al.*, 1995). Incubate at 37°C for various times. A 0-min incubation should be done as a control to determine the background level.

3. Discard the labeling medium and rinse the culture dish twice by dipping into a 1-liter beaker filled with ice-cold 0.1% BSA/PBS for 5 min each. Repeat with another beaker filled with cold PBS for 5 min.

4. Drain PBS and aspirate remaining PBS completely.

5. Put 0.5 ml of lysis buffer into each well and leave it at least 30 min to complete cell lysis.

6. Fill disposable 1-cm plastic cuvettes with 0.75 ml of lysis buffer. Add 0.4 ml of cell lysate into the plastic cuvette and dilute it with another 0.75 ml of lysis buffer so that the final volume is 1.9 ml.

7. Measure the fluorescence of lysate in a spectrofluorometer (SLM/Aminco 500C). Fluorescein can be measured at excitation (exc.) 495 nm, emission (em.) 514 nm. Lucifer yellow is exc. 430 nm, em. 580 nm. These wavelength allowed selective measurement of FDx and lucifer yellow when the cells are labeled with both probes. Lucifer yellow alone is best measured at exc. 430 nm, em. 540 nm.

8. Measure the protein concentration of lysates remaining in wells using a BCA protein assay kit (Pierce Chemical Co.).

9. Prepare the standard solutions, 0, 1, 5, 10, and 20 ng probe/ml, in lysis buffer and measure in the spectrofluorometer to obtain a standard curve.

10. Calculate the amount of probes from the standard curve and express the value as ng probe/mg protein.

IV. PITFALLS

1. A prolonged exposure to intense excitation light may cause release of fluorescent probes from endocytic vesicles into cytoplasm, especially in living cells. To avoid this, reduce the intensity of the excitation light or the exposure time as much as possible.

2. Lucifer yellow can be seen using some fluorescein filter sets with a wide band pass (e.g., Olympus BP490), but not with some others (e.g., Zeiss No. 09). Choose an appropriate filter set for lucifer yellow (e.g., Omega Optical Set No. XF-14, Zeiss No. 05).

3. Many kinds of fluorescent-conjugated probes are commercially available; however, some are not good fluid-phase probes. Texas red albumin is very efficiently taken up by adsorptive endocytosis, although it is sometimes used as a fluid-phase probe. Lysine-fixable Texas red dextran may bind nonspecifically to cells and coverslips.

4. For quantitative analysis, plate the cells at a higher density to increase the sensitivity for fluorescence.

References

Araki, N., Johnson, M. T., and Swanson, J. A. (1996). A role for phosphoinositide 3-kinase in the completion of macropinocytosis and phagocytosis by macrophages. *J. Cell Biol.* **135**, 1249–1260.

Berthiaume, E. P., Mediana, C., and Swanson, J. A. (1995). Molecular size-fractionation during endocytosis in macrophages. *J. Cell Biol.* **129**, 989–998.

Racoosin, E. L., and Swanson, J. A. (1994). Labeling of endocytic vesicles using fluorescent probes for fluid-phase endocytosis. *In* "Cell Biology: A Laboratory Handbook" (J. E. Celis, ed.), pp. 375–380. Academic Press, San Diego.

Racoosin, E. L., and Swanson, J. A. (1993). Macropinosome maturation and fusion with tubular lysosomes in macrophages. *J. Cell Biol.* **121**, 1011–1020.

Swanson, J. A. (1989a). Phorbol esters stimulate macropinocytosis and solute flow through macrophages. *J. Cell Sci.* **94**, 135–142.

Swanson, J. A. (1989b). Fluorescent labeling of endocytic compartments. *In* "Methods in Cell Biology" (Y.-L. Wang and D. L. Taylor, eds.), Vol. 29, pp. 137–151. Academic Press, New York.

Swanson, J. A., and Watts, C. (1995). Macropinocytosis. *Trends Cell Biol.* **5**, 424–428.

Labeling of the Endoplasmic Reticulum with DiOC$_6$(3)

Mark Terasaki

I. INTRODUCTION

The endoplasmic reticulum (ER) is the site of synthesis of membrane proteins, secreted proteins, and membrane lipids, and it is the principal intracellular calcium store. The ER can be localized in living or fixed cells or in cell-free membrane preparations with simple procedures using the fluorescent dye DiOC$_6$(3) (Terasaki *et al.*, 1984). The dye is particularly easy to use because it permeates through the plasma membrane and because its fluorescence is bright and stable.

DiOC$_6$(3) is a general membrane dye, i.e., DiOC$_6$(3) is not a specific dye for the ER. When the technique is successful, the ER is recognized by its characteristic morphology and continuity. When the technique is not successful, it is often because the ER cannot be distinguished from the other organelles or because the ER has a dense three-dimensional structure that cannot be resolved by the light microscope. It is very important to remember the nonspecificity of staining by DiOC$_6$(3). If the ER is not imaged clearly, then one cannot claim that the fluorescence image "shows" the distribution of ER; the fluorescence could just as well be due to mitochondria, lysosomes, or other organelles. These issues are discussed in more detail elsewhere (Terasaki and Reese, 1992).

The best results have been obtained with cells that have regions where the ER has a two-dimensional organization. Examples are certain fibroblasts or epithelial cells in culture, which have a large thin peripheral region, or certain plant cells, which have a distinct cortical ER network adjacent to the plasma membrane. The technique is not as useful in cells or regions of cells in which the ER has a three-dimensional distribution, though in a few cases, confocal microscopy has been useful in these regions.

General directions on how to stain with DiOC$_6$(3) are given in this article. More details about staining mechanisms were reported in an early review (Terasaki, 1989), and a later review contains more recent references (Terasaki, 1993). Koning *et al.* (1993) describe how to stain living yeast cells with DiOC$_6$(3). A different method for labeling ER, which requires microinjection of DiI (a long-chain dicarbocyanine dye), is described elsewhere (Terasaki and Jaffe, 1991; Jaffe and Terasaki, 1993; Terasaki and Jaffe, 1993). An even more specific way of labeling the ER uses green fluorescent protein targeted to the ER (Hampton *et al.*, 1996; Terasaki *et al.*, 1996), although this requires special reagents and methods for getting the coding DNA or mRNA into cells.

II. MATERIALS AND INSTRUMENTATION

$DiOC_6(3)$ can be obtained from Molecular Probes (Cat. No. D-273) or Eastman Kodak (Cat. No. 136 8141). A stock solution of 0.5 mg/ml in ethanol, protected from light, can be stored at room temperature for at least 1 year.

Rhodamine 6G and hexyl ester of rhodamine B (Molecular Probes, Cat. Nos. R-634 and R-648) have very similar staining properties as $DiOC_6(3)$ except that they are viewed with rhodamine optics (Terasaki and Reese, 1992). Their staining is not more specific for the ER, as was stated in the 1993 Molecular Probes catalog. The 1996 catalog suggests that they are less toxic.

To image the fluorescence staining, a high numerical aperture lens (1.2–1.4) is required. A standard fluorescein filter works with this dye. To document the staining in fixed cells, 35-mm photographic film is sufficient. To document changes of the ER in living cells, it is best to have an electronic imaging system such as an SIT camera with imaging processing hardware or confocal microscope.

A convenient test specimen is onion epithelium (Knebel *et al.*, 1990). Briefly, cut a 1-cm square out of an onion layer, peel off the inner epithelium, and mount it in 0.5 μg/ml $DiOC_6(3)$ in spring water (distilled water does not work). See Terasaki and Dailey (1996) for a drawing of this preparation.

III. PROCEDURES

A. Living Cells

Two alternate methods are used: Observe cells in a low concentration of $DiOC_6(3)$ or stain cells briefly with a high dye concentration then observe them in dye-free medium.

1. First Method

Solution

1. 0.5 μg/ml $DiOC_6(3)$ in growth medium, diluted from stock $DiOC_6(3)$ in ethanol (see Section II).

Steps

1. Mount the cells in growth medium containing $DiOC_6(3)$.

2. After about 5 min, observe staining, without washing out the dye. If there is only mitochondrial staining, wait longer or make another sample with a higher dye concentration. If the cells start to round up and detach, make another sample with a lower dye concentration.

2. Second Method

Solution

1. 2.5 μg/ml $DiOC_6(3)$ in growth medium, diluted from stock $DiOC_6(3)$ in ethanol (see Section II).

Steps

1. Incubate the cells in 2.5 μg/ml $DiOC_6(3)$ for 5 min.

2. Wash the coverslip and mount it in dye-free medium.

3. Observe the staining, and change the dose or length of incubation if required.

A potential disadvantage of the first method is that the dye in the medium will contribute too much background fluorescence, and a potential disadvantage of the second method is that the dye will eventually leak out of the cells, reducing staining intensity.

For whole tissues, immerse the tissue in 0.5 μg/ml DiOC$_6$(3) and observe after 5 min. Change the concentration if staining is not optimal. Alternatively, stain the tissue in 2.5 μg/ml for 5–10 min, wash, mount, and observe.

B. Fixed Cells

For fixation, glutaraldehyde is required because formaldehyde often vesiculates the ER. It might be best to use EM-grade glutaraldehyde from a freshly opened vial, but less high-quality glutaraldehyde seems to work well also; the main criterion is whether the ER is vesiculated. For mammalian cultured cells, 0.25% glutaraldehyde has worked well. This concentration is sufficient to fix well but is relatively low so that autofluorescence takes longer to develop. It is important to pay attention to the osmolarity of the buffer, even after fixation. Twice hypertonic buffers cause the ER to shrink. Twice hypotonic buffers seem to leave the ER intact, but more dilute buffers cause the ER to vesiculate. For cultured cells, a buffer consisting of 100 mM sucrose and 100 mM sodium cacodylate, pH 7.4 has worked well; HEPES or phosphate buffers will probably work well also. Finally, it is important not to let the coverslip dry out at any time.

Once the cells have been stained, the coverslip must be mounted in some way to allow microscopic observations. For living cells, the main consideration is to keep the conditions so that the cells are as healthy as possible. For fixed cells, glycerol (or any detergents) and nonpolar solvents, such as those in nail polish used to seal coverslips, should be avoided because these extract the staining. Another consideration for mounting coverslips is that there must be good access for a high numerical aperture (i.e., oil immersion) objective lens. One way to mount fixed or living cells on coverslips is to use a chamber made of silicon rubber gasket material (Ronsil). Cut a small hole in it with a cork borer. Fill the chamber with buffer and quickly place the coverslip over the hole, pressing it down so as to make a seal. Then wipe off the excess with a Kimwipe. This is not airtight, and small bubbles will start to appear at the edge of the chamber after 30 min to 1 hr. As the staining quality also decreases with time, this is usually not a problem.

Solutions

1. *Sucrose cacodylate buffer:* 100 mM sucrose and 100 mM sodium cacodylate, pH 7.4. Adjust sucrose concentration for correct osmolarity of the cells being stained. This can be stored at room temperature.

2. *0.25% glutaraldehyde in sucrose cacodylate buffer.*

3. *2.5 μg/ml DiOC$_6$(3) in sucrose cacodylate buffer:* Use the stock DiOC$_6$(3) in ethanol (see Section II).

Steps

1. Fix the cells in the glutaraldehyde solution for 3–5 min.

2. Stain the cells in DiOC$_6$(3) in sucrose cacodylate buffer for 10 sec.

3. Wash the cells and mount them in dye-free sucrose cacodylate buffer.

4. Look at cells immediately because staining becomes less optimal after about 10–20 min and then slowly deteriorates. The deterioration consists of development of autofluorescence and dye redistribution to some large vesicles at the expense of the ER.

C. Cell Fractions

Solution

1. 0.5 μg/ml DiOC$_6$(3) in the buffer used to isolate the cell fractions.

Step

1. Mount the fractions in the DiOC$_6$(3)-containing buffer. Try to use fractions dilute enough so that the membranes can be easily distinguished from each other.

D. Photography

What to aim for is a print that looks as good or almost as good as what your eyes see in the eyepiece. If the image in the microscope eyepiece looks blurry or dim to your eyes, then photography will probably not improve it. Use a 100× lens if possible, with a 63× lens as second choice. Figure 1 was taken with a 100× lens.

DiOC$_6$(3) fluorescence is bright, and it should be possible to obtain a fairly good image on photograph film. To get an optimum image, it may be necessary to experiment with the photographic conditions. Films with higher ASA are more sensitive to light but are more grainy. Try a black and white ASA 400 film first. Push the processing about one stop to increase the sensitivity (e.g., for TMAX film, develop for 10 min instead of 7.5 min). In this author's opinion, because there is a greater range of adequate exposure times for fluorescence than for transmitted light, the light exposure meter is less necessary for obtaining good exposures of fluorescence images. Try exposures of 1, 2, 4, or 8 sec. Try not to expose longer than about 10 sec because vibrations during long exposures can blur the image. Once a good exposure time has been determined, take as many photographs as possible to increase the chance of getting good images. Try to work fast so that there is less bleaching. With fixed cells, it is also good to work fast because the staining quality seems to deteriorate gradually after about 10–20 min. If possible, try to develop the film soon to get feedback on what adjustments in the photographic procedures can be done. When printing negatives, recall that film has greater dynamic range than paper, so that details that can be seen in the brightest and darkest regions of the negative can sometimes be reproduced on print paper only in the brightest or the darkest regions but not both.

E. Video Microscopy

With video imaging, it is possible to obtain many images of the same cell, although video image quality is usually poorer than photographic image quality. To collect useful data, it is important to optimize the amount of illumination with respect to photodynamic damage versus image quality.

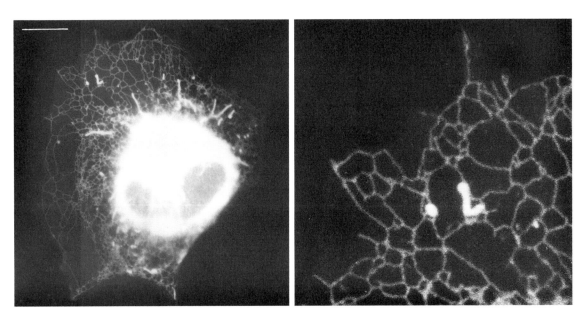

FIGURE 1 A CV-1 cell (monkey kidney epithelial cell line) fixed in glutaraldehyde, then stained with DiOC$_6$(3). (Left) The ER is the network in the thin spread periphery of the cell. Several mitochondria (which are much thicker and usually unbranched) are seen. No details in the central region can be seen because of the overlap of many organelles. (Right) Higher magnification of the same cell (printed from the same negative) shows more details of the ER network. Two brightly stained objects are near the center of the image; the longer one is a mitochondrion, and the round one to the left is either a small mitochondrion or a lysosome. Bar = 10 μm.

F. Double Labeling Using Immunofluorescence

Localization of intracellular antigens requires permeabilization, but $DiOC_6(3)$ staining is extracted by the usual methods of permeabilizing cells (detergents, methanol). Double labeling has been accomplished in a procedure where cells are stained with $DiOC_6(3)$, photographed, permeabilized, processed for immunofluorescence, and then photographed again (Terasaki et al., 1986; Terasaki and Reese, 1992). Probably the most difficult part of this rather arduous effort is to find the same cells again. One way to do this is as follows. Scrape a cross on the coverslip using a rubber policeman or any other thin implement, such as a pipetman tip. Break off a small corner of the coverslip to keep track of orientation. Using a phase-contrast inverted microscope at low magnification (i.e., 10–16×), draw the positions of cells near the cross. Under high magnification, it is possible to find the intersection of the cross by scanning the coverslip, finding a scraped pathway, and following it to the center of the coverslip where it intersects with the other pathway. Record on the map which cells were photographed, and use this map to find the same cells after immunofluorescence.

Photodynamic damage is thought to be caused by reaction of an excited fluorophore with molecular oxygen to form a harmful free radical. To reduce photodynamic damage, one can add a 1:100 dilution of EC oxyrase to the medium (Waterman-Storer et al., 1993). This product is an enzyme system from the respiratory chain of Escherichia coli and acts to eliminate the molecular oxygen (Oxyrase).

IV. COMMENTS

To troubleshoot problems or to devise procedures for staining in new situations, it is useful to know about staining mechanisms of $DiOC_6(3)$. $DiOC_6(3)$ is a lipophilic cationic molecule that is permeable to the plasma membrane. In living cells exposed to low concentrations of dye, $DiOC_6(3)$ accumulates only in the mitochondria, due to the large negative membrane potential there. At higher concentrations, the dye stains the ER and other organelles, presumably because the mitochondria are saturated with the dye. In fixed cells, the dye stains mitochondria, ER, and other organelles at all concentrations. $DiOC_6(3)$ very probably is in the membrane bilayer of stained cells. Any treatment that extracts bilayers will also extract the dye.

It can also be useful to compare fixation for $DiOC_6(3)$ staining with fixation for conventional electron microscopy and for immunofluorescence of intracellular proteins. The aldehydes glutaraldehyde and formaldehyde are chemical fixatives that permeate through the plasma membrane. For electron microscopy, glutaraldehyde is preferred because formaldehyde tends to vesiculate membranes. Both do not "fix" membranes because lipids are still able to diffuse in the bilayer. Osmium does fix membranes, but in the process, prevents membranes from being stained with $DiOC_6(3)$. Osmium also destroys fluorescence in prestained cells. For immunofluorescence, formaldehyde is used more often because glutaraldehyde-fixed cells develop autofluorescence. Antibodies do not cross membranes of aldehyde-fixed cells. Therefore, either aldehyde cells must be permeabilized with detergents such as Triton X-100 or living cells can be fixed and permeabilized by using other agents such as cold methanol. Both these treatments extract membranes and make it impossible to use $DiOC_6(3)$ staining.

References

Hampton, R. Y., Koning, A., Wright, R., and Rine, J. (1996). In vivo examination of membrane protein localization and degradation with green fluorescent protein. Proc. Natl. Acad. Sci. USA 93, 828–833.

Jaffe, L. A., and Terasaki, M. (1993). Structural changes of the endoplasmic reticulum of sea urchin eggs during fertilization. Dev. Biol. 156, 556–573.

Knebel, W., Quader, H., and Schnepf, E. (1990). Mobile and immobile endoplasmic reticulum in onion bulb epidermis cells: Short- and long-term observations with a confocal laser scanning microscope. Eur. J. Cell Biol. 52, 328–340.

Koning, A. J., Lum, P. Y., Williams, J. M., and Wright, R. (1993). DiOC6 staining reveals organelle structure and dynamics in living yeast cells. Cell Motil. Cytoskel. 25, 111–128.

Terasaki, M. (1989). Fluorescent labeling of endoplasmic reticulum. *In* "Methods in Cell Biology" (Y.-L. Wang and D. L. Taylor, eds.), Vol. 29, pp. 125–135. Academic Press, San Diego.

Terasaki, M. (1993). Probes for endoplasmic reticulum. *In* "Fluorescent Probes of Living Cells: A Practical Manual" (W. T. Mason, ed.), pp. 120–123. Academic Press, London.

Terasaki, M., Chen, L. B., and Fujiwara, K. (1986). Microtubules and the endoplasmic reticulum are highly interdependent structures. *J. Cell Biol.* **103**, 1557–1568.

Terasaki, M., and Dailey, M. E. (1995). Confocal microscopy of living cells. *In* "Handbook of Biological Confocal Microscopy" J. Pawley, ed.), 2nd Ed., pp. 327–246.

Terasaki, M., and Jaffe, L. A. (1991). Organization of the sea urchin egg endoplasmic reticulum and its reorganization at fertilization. *J. Cell Biol.* **114**, 929–940.

Terasaki, M., and Jaffe, L. A. (1993). Imaging of the endoplasmic reticulum in living marine eggs. *In* "Methods in Cell Biology" (B. Matsumoto, ed.), Vol. 37, Academic Press, Orlando, FL.

Terasaki, M., Jaffe, L. A., Hunnicutt, G. R., and Hammer, J. A., III. (1996). Structural change of the endoplasmic reticulum during fertilization: Evidence for loss of membrane continuity using the green fluorescent protein. *Dev. Biol.* **179**, 320–328.

Terasaki, M., and Reese, T. S. (1992). Characterization of endoplasmic reticulum by co-localization of BiP and dicarbocyanine dyes. *J. Cell Sci.* **101**, 315–322.

Terasaki, M., Song, J. D., Wong, J. R., Weiss, M. J., and Chen, L. B. (1984). Localization of endoplasmic reticulum in living and glutaraldehyde fixed cells with fluorescent dyes. *Cell* **38**, 101–108.

Waterman-Storer, C. M., Sanger, J. W., and Sanger, J. M. (1993). Dynamics of organelles in the mitotic spindles of living cells: Membrane and microtubule interactions. *Cell Motil. Cytoskel.* **26**, 19–39.

Use of Fluorescent Analogs of Ceramide to Study the Golgi Apparatus of Animal Cells

Richard E. Pagano and Ona C. Martin

I. INTRODUCTION

Studies of lipid traffic at the Golgi complex have been made possible using fluorescent analogs of ceramide (Cer), N-[7-(4-nitrobenzo-2-oxa-1,3-diazole)]-6-aminocaproyl D-*erythro*-sphingosine (C_6-NBD-Cer) and N-[5-(5,7-dimethyl BODIPY)[1]-1-pentanoyl]-D-*erythro*-sphingosine (C_5-DMB-Cer) (reviewed in Pagano, 1989, 1990; Rosenwald and Pagano, 1993). These molecules are vital stains for the Golgi apparatus (Lipsky and Pagano, 1985a; Pagano *et al.*, 1991). In living cells they are metabolized to the corresponding fluorescent analogs of sphingomyelin (SM) and a glycolipid, glucosylceramide (GlcCer). These fluorescent metabolites are subsequently transported to the plasma membrane from the Golgi complex by a vesicle-mediated process analogous to the transport of newly synthesized membrane and secretory proteins (Lipsky and Pagano, 1983, 1985b; Kobayashi and Pagano, 1989). Using fluorescence video imaging, C_6-NBD-Cer has been used to study the dynamics of the Golgi apparatus in cultured astrocytes (Cooper *et al.*, 1990). C_6-NBD-Cer also stains the Golgi apparatus of fixed cells, most likely through an interaction(s) with endogenous lipids and cholesterol, and serves as a *trans*-Golgi marker for both light and electron microscopy (Pagano *et al.*, 1989). In polarized cells, C_6-NBD-Cer is metabolized to fluorescent analogs of SM and GlcCer, and the latter is preferentially delivered to the apical cell surface (van Meer *et al.*, 1987; van't Hof and van Meer, 1990).

This article describes procedures for the (1) preparation and storage of fluorescent Cer/BSA complexes for incubation with cells; (2) incubation of living cells with fluorescent Cer analogs; (3) incubation of fixed cells with fluorescent Cer analogs; and (4) examination of the distribution of the fluorescent Cer and its metabolites at the electron microscopic (EM) level.

II. MATERIALS AND INSTRUMENTATION

C_6-NBD-Cer (Cat. No. N-1154) and C_5-DMB-Cer (Cat. No. D-3521) are from Molecular Probes, Inc. The lipids are stored at $-70°C$ in chloroform/methanol (19:1, v/v) and should appear as single fluorescent spots following thin-layer chromatography in chloroform/methanol/15 mM $CaCl_2$ (60/35/8, v/v/v). Defatted BSA (Cat. No. A6003) is from Sigma Chemical

---

[1]The BODIPY fluorophore has an approximately two- to threefold higher fluorescence yield and greater photostability than NBD (Johnson *et al.*, 1991). The BODIPY Cer analog is also useful because its fluorescence emission spectrum is dramatically red-shifted as the probe concentrates in membranes. Thus, with the appropriate microscope filters, the Golgi apparatus can be observed within living cells without interfering fluorescence from other intracellular membranes.

Company. 3,3'-Diaminobenzidine tetrahydrochloride (DAB) is from Polysciences, preweighed in 10-mg aliquots in sealed serum vials. Unless indicated, all other materials are from Sigma Chemical Company.

Microscopy is performed with a conventional fluorescence microscope. We routinely use a Zeiss IM-35 (Carl Zeiss, Inc.) equipped with a Planapo 100× (1.3 NA) objective and an electronic shutter at the mercury lamp (100 W) housing. To minimize exposure to the exciting light, the samples are briefly focused with a neutral density filter (1 OD) in the light path. The filter is then removed, and the sample is exposed for 2–4 sec using full illumination for photomicroscopy. For NBD-labeled specimens, samples are excited at 450–490 nm and the fluorescence is observed at 520–560 nm. For BODIPY-labeled specimens, samples are excited at 450–490 nm and the fluorescence is observed at either ≥520 nm (green + red wavelengths) or at ≥590 nm (red wavelengths) (see Pagano et al., 1991). Black and white photomicrographs are obtained using Tri-X film (Eastman Kodak Co.) and processed at ASA 1600 with Diafine developer (Acufine, Inc.); color photomicrographs are obtained using either Ektachrome 400 or Kodachrome 200 film which must be processed by a Kodak film laboratory.

III. PROCEDURES

A. Preparation of NBD- or BODIPY-Cer/BSA Complexes

These complexes are used for subsequent labeling of living or fixed preparations of cells (see below). We routinely prepare them as either dilute (5 nmol/ml) or concentrated (0.5 nmol/μl) stock solutions.

1. Dilute (5 nmol/ml) Stock Solutions

Solutions

1. Approximately 1 mM C_6-NBD-Cer or C_5-DMB-Cer stock solution in chloroform/methanol (19/1, v/v).
2. 10 ml serum-free balanced salt solution[2] containing 0.34 mg defatted BSA/ml in a 50-ml plastic centrifuge tube.
3. 500 ml serum-free balanced salt solution for dialysis.

Steps

1. Dispense 50 nmol C_6-NBD-Cer or C_5-DMB-Cer in chloroform/methanol into a small glass test tube and dry, first under a stream of nitrogen, and then in vacuo for at least 1 hr.
2. Dissolve dried Cer in 200 μl absolute ethanol.
3. Inject Cer into the 10 ml BSA solution (while vortex mixing).
4. Rinse Cer/ethanol tube with a little of the Cer/BSA solution and combine with the Cer/BSA complex.
5. Dialyze overnight at 4°C against 500 ml serum-free balanced salt solution.
6. Recover in 10 ml balanced salt solution and store in a plastic tube at −20°C.

2. Concentrated (0.5 nmol/μl) Stock Solutions

Solutions

1. Approximately 1 mM C_6-NBD-Cer or C_5-DMB-Cer stock solution in chloroform/methanol (19/1, v/v).

[2]We use HMEM, 10 mM 4-(2-hydroxyethyl)-1-piperazineethane sulfonic acid-buffered minimal essential medium, pH 7.4, without indicator.

2. 450 μl serum-free balanced salt solution[3] containing 250 nmol DF-BSA in a small glass test tube

Steps

1. Dispense 250 nmol C_6-NBD-Cer or C_5-DMB-Cer from the chloroform/methanol stock solution into a small glass test tube and dry down, first under a stream of nitrogen and then *in vacuo* for at least 1 hr.

2. Add 50 μl ethanol to the dried Cer. Vortex mix to completely dissolve the sample

3. Using a micropipette, add the ethanol solution of the fluorescent Cer to the DF-BSA solution while vortex mixing.

4. Rinse the tube that contained the ethanol solution with an aliquot of the fluorescent Cer/DF-BSA complex.

5. Transfer the Cer/DF-BSA complex to a plastic conical centrifuge tube and store at $-20°C$.

B. Staining the Golgi Apparatus with Fluorescent Ceramides

1. Living Cells

Solutions

1. 5 nmol/ml or 0.5 nmol/μl C_6-NBD- or C_5-DMB-Cer/BSA complex.

2. HMEM for cell incubations.

Steps

1. Rinse cells grown on glass coverslips or on plastic tissue culture dishes in HMEM and transfer to an ice water bath at 2°C.

2. Incubate the cells for 30 min at 2°C with 5 nmol/ml C_6-NBD-Cer/BSA or C_5-DMB-Cer/BSA in HMEM.

3. Rinse the samples several times with ice-cold HMEM, transfer to 37°C, and further incubate for 30 min.

4. Wash the samples in HMEM and observe under the fluorescence microscope. Prominent labeling of the Golgi apparatus and weaker labeling of other intracellular membranes by the fluorescent Cers should be seen (see Fig. 1).

2. Fixed Cells

Solutions

1. HMEM for rinsing cells.

2. 0.5% glutaraldehyde/10% sucrose/100 mM Pipes, pH 7.0.

3. HCMF.

4. *Optional:* Freshly prepared $NaBH_4$ in HCMF (0.5 mg/ml).

5. 5 nmol/ml or 0.5 nmol/μl C_6-NBD-Cer/BSA complex (do not use C_5-DMB-Cer/BSA complex; see Section IV).

6. 3.4 mg/ml DF-BSA in HCMF.

[3]We use HCMF, 10 mM 4-(2-hydroxyethyl)-1-piperazineethane sulfonic acid-buffered Puck's saline without calcium and magnesium.

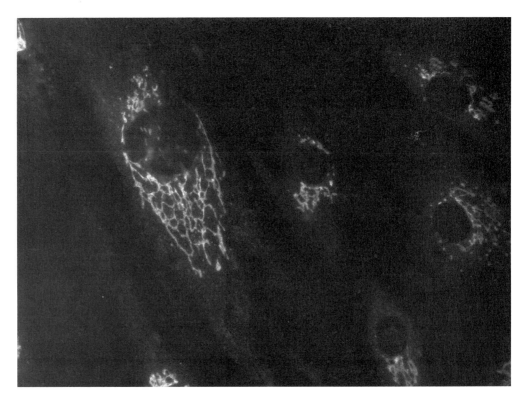

FIGURE 1 Living human skin fibroblasts were incubated with C_6-NBD-Cer as described in Section IIIB. Note prominent fluorescence at the Golgi apparatus and weaker labeling at other intracellular membranes.

Steps

1. Rinse cells grown on glass coverslips or on plastic tissue culture dishes in HMEM and fix for 5–10 min at room temperature in 0.5% glutaraldehyde/10% sucrose/100 mM Pipes, pH 7.0.

2. Wash the cells in HCMF. [*Optional:* The samples can be transferred to an ice water bath and incubated (3 × 5 min) with $NaBH_4$ in ice-cold HCMF to reduce glutaraldehyde-induced autofluorescence. For most cell types and fixation conditions, the staining of the Golgi apparatus is so prominent and autofluorescence is so low that this step is not necessary.]

3. Rinse the samples several times with ice-cold HCMF, transfer to an ice water bath, and incubate for 30 min at 2°C with 5 nmol/ml C_6-NBD-Cer/BSA complex.

4. Rinse the samples several times with HCMF and incubate at room temperature (4 × 30 min) with 3.4 mg/ml defatted BSA in HCMF. [This incubation serves to remove ("back-exchange") excess C_6-NBD-Cer from the fixed cells (Pagano *et al.*, 1989)].

5. Wash the samples in HCMF and observe under the fluorescence microscope. Prominent labeling of the Golgi apparatus by C_6-NBD-Cer is seen.

C. Method for Examining Distribution of the Fluorescent Cer and Its Metabolites at the EM Level

See Pagano *et al.* (1989, 1991).

Solutions

1. 0.1 M Tris, pH 7.6.

2. 1.5 mg DAB/ml 0.1 M Tris, pH 7.6, freshly prepared and kept on ice.

3. 0.1 *M* Na cacodylate, pH 7.4.

4. 1% OsO4 in 0.1 *M* Na cacodylate, pH 7.4.

Steps

1. Cells should be grown in 35-mm-diameter plastic tissue culture dishes, not on glass coverslips.

2. Label living or fixed cells according to desired protocol (e.g., as in Section IIIB). Living cells should be fixed after labeling (see Section IIIB). Always include a sample that has not been treated with the fluorescent lipid as a control.

3. Wash cells in 0.1 *M* Tris (pH 7.6) and add 0.9 ml DAB solution to the culture dish. Cover dish and place in the dark at room temperature for ≥10 min.

4. Irradiate sample for 30 min at room temperature using the 476.5-nm line of an argon laser operating at 50-mW power. To obtain a large area of irradiated cells, expand the laser beam to a line ~1 mm wide × 1 cm long using a cylindrical lens. [Alternatively, the specimen may be irradiated using a 6.3× objective and filters appropriate for NBD fluorescence (Zeiss Cat. No. 487717), although only a very small area of the culture dish (≤1 mm diameter) is irradiated.]

5. After irradiation, wash the sample five or more times in 0.1 *M* Tris (pH 7.6) and observe using phase optics for evidence of a DAB reaction product. Using a dissecting scope and a needle, circumscribe the region of cells that are DAB positive on the inside of the culture dish.

6. Rinse the sample in 0.1 *M* cacodylate buffer (pH 7.4) and treat with 1% OsO₄ in 0.1 *M* cacodylate buffer for 60 min at room temperature.

7. Wash in cacodylate buffer, dehydrate, and embed.

8. After polymerization is complete, the area of DAB-positive cells should be readily identified by the scratch made in step 6. Cut out this region of the dish and mount for thin section electron microscopy. A typical result using this procedure on human skin fibroblasts after treatment with C₆-NBD-Cer is shown in Fig. 2.

IV. COMMENTS

Using the protocols described in this article, the Golgi apparatus of cells can be readily stained. The prominent labeling of this organelle by fluorescent Cer analogs has been observed

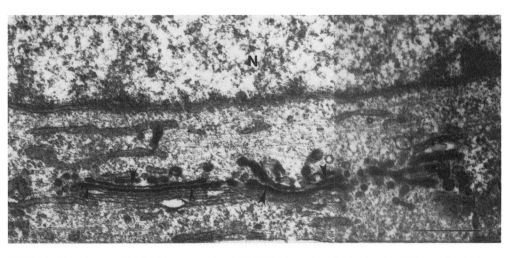

FIGURE 2 Living human skin fibroblasts treated with C₆-NBD-Cer as described in Section IIIB were fixed, photo-bleached in the presence of diaminobenzidine (Section IIIC), processed for thin section electron microscopy, and photographed. Note the black deposition product in the Golgi stacks and in Golgi-associated vesicles (at arrows). N, nucleus. Bar: 0.5 μm. [Reproduced from Pagano *et al.* (1989). *The Journal of Cell Biology*, **109**, 2067–2079 by copyright permission of The Rockefeller University Press.]

in all cell types tested to date. We believe this labeling results from the spontaneous transfer of the lipid into cells followed by molecular trapping at the Golgi apparatus (Pagano, 1989; Pagano *et al.*, 1989). Although C_6-NBD-Cer or C_5-DMB-Cer can be delivered to cells either from lipid vesicles (Lipsky and Pagano, 1983, 1985a,b) or from a BSA complex (Pagano and Martin, 1988; Pagano *et al.*, 1989), the use of fluorescent Cer/BSA complexes as described in this article is especially convenient because they can be prepared in advance of an experiment and stored frozen. In addition, repeated freezing and thawing does not affect cell labeling or the purity of the fluorescent lipids.

Although both C_6-NBD-Cer and C_5-DMB-Cer are vital stains for the Golgi apparatus, only C_6-NBD-Cer is suitable for the staining of fixed cells. In using C_6-NBD-Cer for staining fixed cells, we found that fixation time, temperature, and buffer composition are not critical. However, brief treatment of fixed cells with detergents or fixation with methanol/acetone at $-20°C$ eliminates labeling of the Golgi apparatus.

Visualization of C_6-NBD-Cer and C_5-DMB-Cer at the electron microscope level is performed using methods adapted from Maranto (1982) and Sandell and Masland (1988) in which cells treated with various fluorescent compounds were photobleached in the presence of DAB. The photooxidation products catalyze the polymerization of DAB to yield a high molecular weight osmiophilic compound that is visualized at the EM level. Double-label experiments indicate that both C_6-NBD-Cer and C_5-DMB-Cer label a subset of Golgi membranes that corresponds to the *trans*-Golgi elements (Pagano *et al.*, 1989, 1991).

References

Cooper, M. S., Cornell-Bell, A. H., Chernjavsky, A., Dani, J. W., and Smith, S. J. (1990). Tubulovesicular processes emerge from *trans*-Golgi cisternae, extend along microtubules, and interlink adjacent *trans*-Golgi elements into a reticulum. *Cell* 61, 135–145.

Johnson, I. D., Kang, H. C., and Haugland, R. P. (1991). Fluorescent membrane probes incorporating dipyrromethendboron difluoride fluorophores. *Anal. Biochem.* 198, 228–237.

Kobayashi, T., and Pagano, R. E. (1989). Lipid transport during mitosis: Alternative pathways for delivery of newly synthesized lipids to the cell surface. *J. Biol. Chem.* 264, 5966–5973.

Lipsky, N. G., and Pagano, R. E. (1983). Sphingolipid metabolism in cultured fibroblasts: Microscopic and biochemical studies employing a fluorescent analogue of ceramide. *Proc. Natl. Acad. Sci. USA* 80, 2608–2612.

Lipsky, N. G., and Pagano, R. E. (1985a). A vital stain for the Golgi apparatus. *Science* 228, 745–747.

Lipsky, N. G., and Pagano, R. E. (1985b). Intracellular translocation of fluorescent sphingolipids in cultured fibroblasts: Endogenously synthesized sphingomyelin and glucocerebroside analogs pass through the Golgi apparatus *en route* to the plasma membrane. *J. Cell Biol.* 100, 27–34.

Maranto, A. R. (1982). Neuronal mapping: A photooxidation reaction makes lucifer yellow useful for electron microscopy. *Science* 217, 953–955.

Pagano, R. E. (1989). A fluorescent derivative of ceramide: Physical properties and use in studying the Golgi apparatus of animal cells. *Methods Cell Biol.* 29, 75–85.

Pagano, R. E. (1990). The Golgi apparatus: Insights from lipid biochemistry. *Biochem. Soc. Trans.* 18, 361–366.

Pagano, R. E., and Martin, O. C. (1988). A series of fluorescent *N*-(Acyl)-sphingosines: Synthesis, physical properties, and studies in cultured cells. *Biochemistry* 27, 4439–4445.

Pagano, R. E., Martin, O. C., Kang, H. C., and Haugland, R. P. (1991). A novel fluorescent ceramide analog for studying membrane traffic in animal cells: Accumulation at the Golgi apparatus results in altered spectral properties of the sphingolipid precursor. *J. Cell Biol.* 113, 1267–1279.

Pagano, R. E., Sepanski, M. A., and Martin, O. C. (1989). Molecular trapping of a fluorescent ceramide analog at the Golgi apparatus of fixed cells: Interaction with endogenous lipids provides a *trans*-Golgi marker for both light and electron microscopy. *J. Cell Biol.* 109, 2067–2079.

Rosenwald, A. G., and Pagano, R. E. (1993). Intracellular transport of ceramide and its metabolites at the Golgi complex: Insights from short-chain ceramides. *Adv. Lipid. Res.* 26, 101–118.

Sandell, J. H., and Masland, R. H. (1988). Photoconversion of some fluorescent markers to a diaminobenzidine product. *J. Histochem. Cytochem.* 36, 555–559.

van Meer, G., Stelzer, E. H. K., Wijnaendts-van-Resandt, W., and Simons, K. (1987). Sorting of sphingolipids in epithelial (Madin–Darby canine kidney) cells. *J. Cell Biol.* 105, 1623–1635.

van't Hof, W., and van Meer, G. (1990). Generation of lipid polarity in intestinal epithelial (Caco-2) cells: Sphingolipid synthesis in the Golgi complex and sorting before vesicular traffic to the plasma membrane. *J. Cell Biol.* 111, 977–986.

Staining of Mitochondria

Martin Poot

I. INTRODUCTION

After mitochondria have been visualized successfully by vital staining with rhodamine 123 (Johnson *et al.*, 1980), numerous other fluorescent stains that target mitochondria have become available (Table I). The first group of mitochondrial stains consists of fluorescent cations such as rhodamine 123 and the MitoTracker CMXRos dye (Fig. 1). They target mitochondria by virtue of their interior negative membrane potential. The second group of mitochondrial stains accumulates in mitochondria by specifically binding to a structural component of the organelle, e.g., nonyl acridine orange binds to cardiolipin (Petit *et al.*, 1992). The amount of fluorescence obtained with this dye, therefore, may reflect the amount of mitochondrial material in a cell. The third group consists of reduced dyes that become fluorescent after oxidation inside the mitochondrion (e.g., dihydrorhodamine 123, CMXRos-H_2). Table I lists the stains used most frequently and may serve as a dye selection guide. This article describes a general protocol for the staining of animal cells with mitochondrial dyes. In a second protocol, a cell fixation procedure for use after mitochondria were stained with vital stains is described.

II. MATERIALS

Rhodamine 123 (Cat. No. R-302), dihydrorhodamine 123 (Cat. No. D-632), MitoTracker Green FM (Cat. No. M-7514), MitoFluor Green (Cat. No. M-7502), nonyl acridine orange (Cat. No. A-1372), MitoTracker Red CMXRos (Cat. No. M-7512), MitoTracker Red CMXRos-H_2 (Cat. No. M-7513), propidium iodide (Cat. No. P-3566), and SYTOX Green nucleic acid stain (Cat. No. S-7020) are from Molecular Probes, Inc. or from Molecular Probes Europe B.V. Phosphate-buffered saline solution (PBS; Cat. No. D-5527) and HBSS powder (Cat. No. H-1387) can be obtained from Sigma Chemical Co. A 37% stock solution of formaldehyde (Cat. No. 47608) can be obtained from Fluka Chemical Corporation. Standard equipment for tissue culture, a fluorescence microscope, and a flow cytometer equipped with either an arc lamp or a laser as an excitation light source (optional) are also needed.

TABLE I   Features of Mitochondrial Stains

Dye	Excitation max. (nm)	Emission max. (nm)	Sensitivity toward functional state	Fixability with aldehydes	Photostability
Rhodamine 123	506	530	+[a]	−	−
MitoTracker Red CMXRos	594	608	+	+	+
MitoTracker Green FM	480	516	−	+	+
MitoFluor Green	480	516	−	−	+
Nonyl acridine orange	497	519	−	−	+

[a] +, photographable with ordinary skill; −, difficult to photograph.

III. PROCEDURES

A. Vital Staining of Mitochondria

The procedure is adapted from that of Poot and co-workers (1996).

Solutions

1. *Cell culture medium:* Use the same medium as used to grow the cells in. If powder medium is used, reconstitute this according to the manufacturer's recommendations and supplement with 10% fetal bovine serum (FBS).

2. *Dye stock solutions:* Depending on the purpose, 0.5 or 0.25 mM stock solutions of the dye(s) should be prepared in dry dimethyl sulfoxide or absolute ethanol. In cases where a reduced dye (e.g., dihydrorhodamine 123) is used, the stock solution should be flushed with an inert gas (e.g., nitrogen, argon, helium). As fluorescent dyes generally decompose if illuminated, it is necessary to keep stock solutions in well-sealed dark reagent bottles. All dye stock solutions can be kept at −20°C.

3. *HBSS-FBS:* Reconstitute from powder according to the manufacturers recommenda-

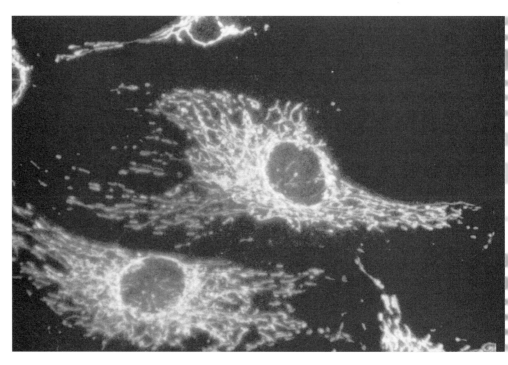

FIGURE 1   MRC-5 cells (human fetal lung fibroblasts) grown on a coverslip and stained live with the MitoTracker Red dye CMXRos according to the procedure described in the text.

tions for Hank's balanced salt solution (HBSS) and supplement nine parts of HBSS with one part of FBS.

Steps

For Microscopy

1. Culture adherent cells overnight on sterilized coverslips. Take coverslips with cells out of the cell culture dish, rinse coverslips once with prewarmed cell culture medium, place them in a small cell culture dish (35-mm-diameter dishes fit 18×18-mm^2 coverslips) with 1 ml of prewarmed cell culture medium, and maintain the dishes at 37°C. If larger dishes are being used, the volume of the staining and other solutions has to be adjusted.

2. Thaw out dye solutions at room temperature and protect from light (e.g., in a drawer).

3. Add 1 μl of 0.5 mM dye stock solution (or 2 μl of 0.25 mM dye stock solution) to the dish with the coverslip (containing 1 ml of warm cell culture medium) and swirl immediately to distribute the dye solution evenly. Incubate for 15 to 30 min at 37°C in the dark or at subdued light. After staining, briefly rinse the coverslip three times with prewarmed HBSS-FBS.

4. Invert coverslip and mount onto a slide in HBSS-FBS, leaving some clearance. Do NOT apply pressure. Seal the coverslip by a method regularly used in the laboratory (e.g., melted wax, nail polish).

5. To observe rhodamine 123, MitoTracker Green, MitoFluor Green, or nonyl acridine orange-stained cells by fluorescence microscopy, use "fluorescein-type" excitation (Ex.) and emission (Em.) filters (Ex.: 465–495 nm; Em.: 515–555 nm); CMXRos-stained cells are best observed by using "Texas Red" excitation and emission filters (Ex.: 540–580 nm; Em.: 600–660 nm).

For Flow Cytometry

1. Harvest cultured cells by standard procedures in 15-ml screw-capped centrifuge tubes and centrifuge for 5 min at 200 g at room temperature. Resuspend the cell pellet at 0.5 to 1.0×10^6 cells/ml in prewarmed cell culture medium. Leave cell suspensions in a water bath at 37°C for at least 5 min. This incubation preserves the functional state of the mitochondria.

2. Thaw out dye solutions at room temperature and protect from light (e.g., in a drawer).

3. Aliquot cell suspensions into 12×75-mm polypropylene tubes. Add 1 μl of 0.5 mM dye stock solution (or 2 μl of 0.25 mM dye stock solution) to 1 ml of prewarmed cell suspension. Mix immediately by briefly vortexing at maximal speed. Incubate for 15 to 30 min at 37°C in the dark or at subdued light. After staining, put tubes with cell suspensions in a melting ice bath. Optionally, dead cells can be excluded by costaining with 5 μM propidium iodide (if rhodamine 123, MitoTracker Green, MitoFluor Green, or nonyl acridine orange is used) or with 100 nM SYTOX Green nucleic acid stain (if CMXRos or CMXRos-H$_2$ is used).

4. Set up and optimize the flow cytometer. If the cells were stained with rhodamine 123, MitoTracker Green, MitoFluor Green, or nonyl acridine orange dyes, the argon laser should be tuned to the 488-nm line. In case of CMXRos staining, the argon laser can be tuned to the 488- or 514-nm line; alternatively the 543- or the 594-nm line of a HeNe laser can be used. To collect fluorescence from rhodamine 123, MitoTracker Green, MitoFluor Green, or nonyl acridine orange staining, use a band pass filter centered around 530 nm; for CMXRos use a band pass filter centered around 610 nm. Due to the wide variation in the cellular contents of mitochondria, it is advisable to use logarithmic signal amplification for the signal channels collecting mitochondria-related fluorescence. Carefully resuspend the cell sample by gently pipetting up and down a few times immediately before analysis.

B. Fixation of Stained Cells

Solutions

1. *3.7% formaldehyde in PBS:* Dilute formaldehyde from a stock solution into PBS immediately before use.

For Microscopy

1. After cells on coverslips have been stained with a fixable mitochondrial dye (see Table I), slowly add 1 ml of 3.7% formaldehyde in PBS at 37°C dropwise to the staining solution in the small dish in which the coverslip is stained. Incubate for at least 15 min at room temperature. Alternatively, the staining solution can be poured off and replaced by PBS at 37°C. Pour off the PBS and add 1 ml of prewarmed formaldehyde. It is critical to add prewarmed fixative to the cells in order to fully conserve their morphology.

2. After cell fixation, pour off the formaldehyde and either view cells directly in a drop of HBSS-FBS (for viewing procedure, see Procedure A, steps 4 and 5) or wash coverslips twice with PBS at room temperature. At this stage, cells can be counterstained with a dye of a different fluorescence emission wavelength. For instance, cells stained with CMXRos can be counterstained with 100 nM SYTOX Green nucleic acid stain for at least 15 min at room temperature; cells stained with MitoTracker Green can be counterstained with 5 μM propidium iodide for at least 15 min at room temperature.

3. For antibody labeling, pour the PBS off the formaldehyde-fixed cells and add 1 ml of acetone at −20°C to the coverslips in the staining dish. Incubate the coverslips for at least 15 min at room temperature. To give antibodies full access to intracellular epitopes, it is necessary to permeabilize all cell membranes. Acetone preserves cellular morphology the best.

4. Pour the acetone off and wash the coverslips three times with PBS at room temperature. At this stage, coverslips can be treated with "blocking solution" and then incubated with the antibody of choice.

For Flow Cytometry

1. Centrifuge cells stained with a fixable mitochondrial dye (see Table I) for 5 min at 200 g at room temperature. Discard most of the staining solution and resuspend the cell pellets by gently tapping in a small amount of residual staining solution. While spinning on a vortex at medium speed, add 5 to 10 ml of 3.7% formaldehyde in PBS at 37°C dropwise. Incubate for at least 10 min at room temperature. This procedure minimizes the formation of cell clumps.

2. Centrifuge fixed cell samples for 5 min at 200 g at room temperature. Discard most of the fixative and resuspend the cell pellets by gently tapping in a small amount of residual fixative. Add 1 ml of PBS. At this stage, cells can either be analyzed directly by flow cytometry or be counterstained with a dye of different fluorescence emission wavelength. For instance, cells stained with CMXRos can be counterstained with 100 nM SYTOX Green nucleic acid stain for at least 15 min at room temperature; cells stained with MitoTracker Green can be counterstained with 5 μM propidium iodide for at least 15 min at room temperature.

3. If fixed cells are to be labeled with an antibody, the resuspended cells should be permeabilized with 5 to 10 ml of acetone at −20°C for at least 15 min. After acetone permeabilization, centrifuge cell samples for 5 min at 200 g at room temperature. Discard most of the acetone and resuspend the cell pellets by gently tapping in a small amount of residual acetone.

4. Add 5–10 ml of PBS to the acetone-treated cell samples and incubate at room temperature for at least 15 min. After PBS treatment (rehydration), centrifuge cell samples for 5 min at 200 g at room temperature. Discard most of the PBS and resuspend the cell pellets by gently tapping in a small amount of residual PBS. At this stage, cell suspensions can be treated with "blocking solution" and then incubated with the antibody of choice.

IV. COMMENTS

The procedures described in this article have been used with a variety of cultured animal cells. No data exist on their possible use with intact animal or plant tissues or with yeast or cultured plant cells.

V. PITFALLS

1. As analysis of mitochondria by flow or image cytometry is intended to generate information on the physiological state of the cells under study, it is paramount to prepare fresh staining solutions and to prewarm these to 37°C at all times. A "cold shock" may cause mitochondria to "wrinkle" rapidly, affecting cell staining accordingly.

2. Dye solutions can be subject to photodegradation. It is, therefore, advisable to store all dye solutions in a freezer (not a "No-Frost" freezer) in the dark. Shortly before use, dye solutions can be thawed out at room temperature while being kept in the dark (e.g., in a drawer). Solutions of reduced dyes have to be flushed with inert gas (e.g., nitrogen, argon) to prevent oxidation.

3. Dye concentrations in the range of 0.5 to 1 μM are recommended, as nonmitochondrial staining may occur at higher dye concentrations.

4. For flow cytometric analysis, it is critical to obtain a suspension of single cells. The procedures described in this protocol were found to best achieve this goal. During the staining period, cells tend to clump; to obtain meaningful data on a per cell basis, it is essential to resuspend cells immediately before analysis.

5. If little or no fluorescence is found with stained cells, check whether the excitation wavelength and output power of the laser or arc lamp are compatible with the dye used; check filter combinations in front of the photomultiplier tubes. If those parameters meet the features of the dye and still little or no signal is obtained, increase signal amplification; the sensitivity of the detection system of the flow or image cytometry system may vary. If the signal is still weak, check cells by fluorescence microscopy to see whether cell staining is dim or bright.

References

Johnson, L. V., Walsh, M. L., and Chen, L. B. (1980). Localization of mitochondria in living cells with rhodamine 123. *Proc. Natl. Acad. Sci. USA* **77**, 990–994.

Petit, J.-M., Maftah, A., Ratinaud, M.-H., and Julien, R. (1992). 10 N-Nonyl-acidine orange interacts with cardiolipin and allows the quantification of this phospholipid in isolated mitochondria. *Eur. J. Biochem.* **209**, 267–273.

Poot, M., Zhang, Y.-Z., Krämer, J., Wells, K. S., Jones, L. J., Hanzel, D. K., Lugade, A. G., Singer, V. L., and Haugland, R. P. (1996). Analysis of mitochondrial morphology and function with novel fixable fluorescent stains. *J. Histochem. Cytochem.* **44**, 1363–1372.

Appendix: Internet Resources

Julian A. T. Dow

I. INTRODUCTION

Progress in any discipline depends critically on awareness and appraisal of the work that has gone before. In cell biology, as in any contemporary life science, this requirement is becoming increasingly demanding, as both the pace and the quality of (at least some!) research has accelerated hugely over the last two decades. To have a hope of keeping abreast of this information explosion, a researcher must have certain generic and specific skills. The purpose of this article is to outline the use of the Internet in selecting relevant information, and so contains pointers to more than 130 recommended sites; and because the technology of the Internet is itself advancing so quickly, to provide generic strategies that will allow information to be identified in the future.

II. WHAT IS THE INTERNET?

The Internet is a global network of computers that has evolved from smaller networks such as the American ARPAnet defense network. Every computer on the network, from desktop PC to mainframe, has a unique "IP address," of the form 123.456.789.012, and usually an accompanying name, of the form personalname.sitename.class. These identities are interconvertible because of global "phone books" held on domain name servers (DNS). In this way, it is possible to find an informative name from a numerical IP address, or vice versa. Because of the hierarchical structure of these names, it is possible to direct packets of information from any one computer to any other computer in the world, usually via several intermediate machines. A passing acquaintance with these conventions is important because all the resources described in this article are accessed through this naming convention, and so it is helpful in understanding the error messages that all too frequently occur.

III. CONNECTING TO THE INTERNET

For academic users in most countries, connection to the Internet is freely available and is the norm. For users in more remote locations or with uncooperative computing service departments, it will be necessary to purchase a modem and enter into an agreement with a commercial Internet provider (IP). The cost is usually a flat rate of $10–20 per month, together with the cost of a call to a local number run by the company. In the former case, the connection to the Internet is patent at all times; in the latter, it lasts for as long as the phone call.

IV. INTERNET SERVICES

The Internet gives access to a range of services, each governed by their own protocols. Before concentrating on the World Wide Web (WWW), it is important to review the other protocols.

A. Electronic Mail (email)

The simplest and most widely used is electronic mail, or email, and use of this service is nearly compulsory for scientists! Messages, attached files, and even binaries (computer programs) can be sent around the world in a few minutes. It is important to distinguish email (personal) addresses from Internet (machine) addresses: the former generally have an "@" sign separating the individual's account number from the IP name of the machine on which the individual's mail is held. For example, my email address is GBAA02@udcf.gla.ac.uk, from which it can be deduced that (i) my mail is held on a machine with the name udcf.gla.ac.uk, (ii) on that machine, my user identifier is GBAA02, and (iii) I am an academic user from the United Kingdom (ac.uk).

It is easy to send email to someone whose address is known, but how can it be found out? Methods that rely on the WWW are discussed later, but a guessing strategy can surprisingly often yield results. Although most mail accounts are held under arcane user identifiers, almost all mail systems allow the adoption of multiple aliases, and the worldwide conventions are to set up either firstname.lastname or initials.lastname. Similarly, machine names are usually abbreviated. So j.a.t.dow@gla.ac.uk and julian.dow@gla.ac.uk would both be creditable guesses if you knew that I worked in Glasgow. If these approaches fail, you will usually receive an automatically generated message from the computer you were trying to reach, which may list email addresses of people with similar names at that site. If not, you may send an email to postmaster@machine.name, where machine.name is your best guess of a host machine's address. By Internet rules, such mail must be seen and handled by a postmaster, and so a request for an email address will usually give a prompt and courteous reply.

B. File transfer protocol (ftp)

This is a protocol whereby files can be copied from remote computers onto your computer, and vice versa. Although there are explicit programs to help this transfer (such as "Fetch" on the Macintosh), ftp is now usually performed implicitly, by attaching a file to an email message, or because modern WWW browsers can handle ftp connections automatically.

C. Gopher

Again, this is a cutting-edge protocol of only a few years ago, which has now largely been superseded by the WWW. Gopher browsers provided a familiar nested-folder-and-file metaphor (rather like the Mac or Windows desktops) for providing information. Again, although dedicated gopher client software (such as "TurboGopher") exists, the same job is now achieved transparently with WWW browsers.

D. News

The usenet newsgroups are a valuable resource. Of several thousand newsgroups, those likely to be of most relevance are those starting with "bionet," which have biological content, and those starting with "comp.sys," which have computer-based content. Unfortunately, the obvious newsgroup (bionet.cellbiol) is pretty moribund; however, the related newsgroup, bionet.molbio.methds-reagnts, is easily the most active part of bionet and is an excellent place to seek advice with problems in molecular biology. Although WWW browsers (see below) can now handle news reading, at present specialist programs (like "Newswatcher" on the Mac) are easier to use. This will probably change over the next few years.

E. World Wide Web

This terminology is used rather loosely to cover the hypertext transfer protocol (http) that has come to dominate Internet information transfers. In essence, the metaphor is of text and graphics files that contain embedded hypertext links to other Internet resources, i.e., a document can contain text, graphics, animations, and sound files, and by clicking a mouse on a particular piece of text, it is possible to jump to a related item. Text that links to another resource is usually identifiable as blue text, or underlined.

Strictly, http is only one of the protocols that make up the WWW. So the annoying "http://" that precedes most addresses is important because most Internet browsers can also handle other addresses.

Prefix	Protocol
http:	Hypertext transfer protocol (i.e., "normal" pages)
ftp:	File transfer protocol
gopher:	Gopher protocol
mailto:	email protocol
news:	USENET newsgroups

V. CHOICE OF BROWSER

The preceding section showed that modern "web browsers" are multiply talented creatures, which explains the prodigious demands that are made on both hard disc and RAM. Despite their complexity, browsers are generally licensed free of charge to academics and at nominal cost to others. This is because the browser manufacturer can provide buttons (e.g., "what's cool") linking to sites that pay for the privilege, thus providing significant returns to subsidize the free software.

Early text-only (e.g., "Lynx") and graphical (e.g., "NCSA Mosaic" and "MacWeb") browsers are struggling against the brand leader, "Netscape." In turn, Netscape is battling against Microsoft's Internet explorer, although at the time of writing, Netscape users constituted 90% of all accesses to our local university site. Accordingly, this work will use Netscape as an example client, although it should be noted that fashions in browsers will change over time, the basic functionality and "look and feel" are likely to remain consistent.

WWW Addresses

Throughout this article, the facilities reviewed can be accessed from your computer by typing in the uniform resource locator (URL) or address that accompanies it. To do this, launch your browser, choose "Open location" under the "file" menu, and type in the address exactly as it appears. Sometimes, case of the letters is significant; these addresses never contain spaces.

The most common format of URL is made up of three parts: the type of the URL, usually "http:" followed by two slashes; the name or IP address of the machine that holds data you need, e.g., "www.ibls.gla.ac.uk"; and the directory path of the particular file you seek, e.g., "/news/index.html." Although constructions like this can make for quite demanding typing, the ability of all browsers to save bookmarks to frequently used pages means that you need only type in such an address once.

To help in the use of this article, the URLs described here are indexed on http://www.mblab.gla.ac.uk/~julian/trends.html. So this is the only URL that you need to type.

VI. GENERIC SEARCH ENGINES

Equipped with an Internet-capable computer and a contemporary WWW browser, how can sources of information be accessed? The simplest way, and one with utility that extends beyond molecular and cellular biology, is simply to click on the "Net search" button in the browser window and enter a term of interest.

The search engine will return with usually many matching entries, with hypertext links to each. It may seem surprising that, even for relatively arcane biomedical terms, multiple "hits" are readily obtained:

```
Lycos search: aquaporin
59,945,140 unique URLs
Found 29 documents with the words aquaporin (21), aquaporins (8)

1) http://www.med.jhu.edu/otl/9494.html
        Project Title:Water Channel Protein (JHU Ref. DM-9494) Inven
        aquaporin water channel (CHIP) is an integral membrane prote
        http://www.med.jhu.edu/otl/9494.html (2k)
        [100% relevant]

2) http://www.med.jhu.edu/otl/9918.html
        Project Title:Aquaporin 5 Water Channel (JHU Ref. DM-9918) I
        transmembrane water channel protein is isolated in highly purifie
        http://www.med.jhu.edu/otl/9918.html (2k)
        [100% relevant]

3) APStracts 2:0205F, 1995.
        APStracts 2:0205F, 1995. Bilateral ureteral obstruction downreg
        aquaporin-2 water channel in rat kidney. Frokiaer, Jorgen, David
        http://www.uth.tmc.edu/apstracts/1995/renal/December/205F.ht
        [99% relevant]

4) Homo sapiens - Biomembranes
        Homo sapiens - Biomembranes Homo sapiens - Biomembranes
        [R][W][F][D][T] cGMP-gated ion channel protein - human 207S
```

How does this work? There are tens of millions of WWW pages distributed on servers around the globe, but effectively all of these are scanned by automatic web browsers, called "robots," that read a page, index every word on it, and follow any links recursively to index all subpages. Most of these robots are run by commercial search companies, who subsidize their costs by selling advertising on their home pages, so the searches are free to the end user.

There is a continual war for user share among these companies, and so their relative merits may shift with time. However, the major players at the time of writing included:

Search site	URL
AltaVista	http://www.altavista.com
Hotbot	http://www.hotbot.com
Infoseek	http://guide.infoseek.com/
Lycos	http://www.lycos.com/
Magellan	http://www.mckinley.com/
Yahoo	http://www.yahoo.com/

The nature of the hits obtained with this strategy are likely to be highly specific: they will include institutional lists of Ph.D. theses, patent applications, and home pages of individual research groups. Frequently, this can be extremely useful; however, for a more general, "top-down" view of a whole subject area, it is necessary to employ another approach.

Project Title:
Aquaporin 5 Water Channel (JHU Ref. DM-9918)

Inventors:
Peter Agre, M.D.

Brief Description:
A transmembrane water channel protein is isolated in [
erythrocytes. An identical protein is also found in kidn
been isolated and its amino acid sequence determined.
channel protein has also been obtained from salivary g
lacrimal gland, cornea, and lung tissue. The amino aci
from the cDNA, and the protein has been designated A
protein sequence provided herein, the protein may be
Expression of the protein may be determined by either

Patent Status:
pending

Reference:
"Molecular Cloning and Characterization of an Aquap
Respiratory Tissues", Raina et al, J. Biol. Chem, 270

Potential Commercial Uses:

VII. META INDEXES

As the finding of links to useful pages can sometimes be time-consuming and haphazard, both commercial bodies and individuals have assembled indices of links that provide an entry point to whole subject areas. These in turn are assembled into pages with links to indices of indices, or "meta indices."

A good example is Yahoo (http://www.yahoo.com). The top level includes entry points to science:

- **Business and Economy [Xtra!]**
  Directory, Investments, Classifieds, ...

- **Computers and Internet [Xtra!]**
  Internet, WWW, Software, Multimedia, ...

- **Education**
  Universities, K-12, Courses, ...

- **Entertainment [Xtra!]**
  TV, Movies, Music, Magazines, ...

- **Government**

- **Recreation and Sports [Xtra!**
  Olympics, Sports, Games, Travel, Aut

- **Reference**
  Libraries, Dictionaries, Phone Numbers

- **Regional**
  Countries, Regions, U.S. States, ...

- **Science**
  CS, Biology, Astronomy, Engineering,

- **Social Science**

which leads to a page on biology:

- **Artificial Life** *(56)* NEW!
- **Astronomy** *(876)* NEW!
- **Aviation and Aeronautics** *(118)* NEW!
- **Biology** *(1525)* NEW!
- **Chaos** *(3)*

- **Meteorolog**
- **Museums a**
- **Nanotechno**
- **News** *(5*
- **Oceanograp**
- **Organizatio**
- **Paleontolog**

This has several dozen subheadings:

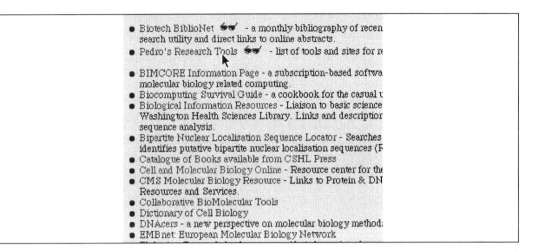

In turn, this yields both resources and indices of great relevance:

VIII. SUBJECT INDICES

To access data in the cell/molecular biology area, the most economical strategy is usually to set bookmarks to one or more pages of indices. The task of staying abreast of the field then devolves to the keeper of each page and so does not take up your valuable time. Good starting pages are:

- Yale University (http://www.med.yale.edu/library/sir/basic.html/) has a good selection of resources, including journals, organizations, model organisms, image libraries, and sequence analysis tools.

- The World Wide Web virtual library (http://golgi.harvard.edu/biopages.html) is curated from Harvard and has relevant sections on journals, software, biochemistry, molecular biology, biophysics, biotechnology, developmental biology, genetics, immunology, medicine, microbiology, and neurobiology.

- Yahoo (http://www.yahoo.com/Science/Biology/) has a small cell biology section, but has separate subindices for biochemistry, biotechnology, computational biology, developmental biology, immunology, journals, microbiology, molecular biology, neurosciences, and reproductive biology.

- Cell and Molecular Biology Online (http://www.tiac.net/users/pmgannon/) is a curated list of resources useful to cell biologists.

- The Biologist's Control Panel (http://gc.bcm.tmc.edu:8088/bio/bio_home.html) from Baylor College of Medicine is a curated list with an emphasis on sequence searching.

- BiowURLd (http://www.ebi.ac.uk/htbin/bwurld.pl) is an index searchable by topic.

- BioMedNet (http://www.biomednet.com/) is a commercial subscription-based service, although individual subscriptions were still free at the time of writing. It is targeted at biomedical scientists, and there are some good indices, described further in some sections below.

- Pedro's Biomolecular Research Tools (http://www.public.iastate.edu/~pedro/research_tools. html) is an indispensable collection of pointers to sequence analysis tools, journal contents, help and tutorial documents, and miscellany that you will find helpful at some time or another. This is an essential bookmark!

IX. SPECIFIC RESOURCES

This section considers the individual resources that are likely to be most useful to a cell biologist.

A. Sequence Databases

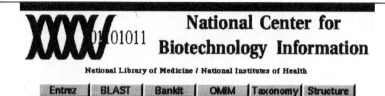

The most obvious use for computer technology in modern biology is in the analysis of DNA and derived peptide sequences. Classically, this has been the task of university mainframe computers, combined with institutional subscriptions to software packages and to databases. However, the development of powerful computers, efficient search algorithms, and the WWW has led to much of this utility being placed in the public domain. Leading the field is the American National Center for Biotechnology Information (**NCBI**) at http://www.ncbi.nlm.nih. gov/, which permits searching or browsing of several vital databases.

Resource	Description	Uses
Entrez	Multimedia combination of DNA and protein sequence databases with the subset of MEDLINE that refers to them	Literature search on gene families, related sequences
GenBank	Database of known DNA sequences and derived peptide sequences	Can be searched explicitly to retrieve a particular sequence, but its use is implicit in the other resources listed here
BLAST	Rapid sequence comparison tool, based on precompiled indices of blocks of sequence	Compare your novel DNA with all known sequences to see if it, or related genes, is already known
Bankit	Sequence deposition	Submit your DNA/protein sequence directly to GenBank
OMIM	Encyclopedic human genetic disease database with links to sequence and MEDLINE databases	Obtain a thorough description of any human disorder with a possible genetic component, including entry points into the literature

How are these resources used? The simplest use would be a direct access of GenBank. On reading a paper in which a novel sequence has been described, the sequence can be fetched from GenBank using the accession number published in the paper. As well as the DNA sequence, each record usually includes annotations such as the position of the open reading frame, the polyadenylation site, and citations of paper(s) that describe the sequence. Similarly, if you have characterized a new DNA sequence, you can provide it to the GenBank database directly by completing a series of forms on your browser, using the Bankit resource.

This simple use, however, is a small subset of the power of the resource. Other facets bring workers quickly up to speed on areas outside their main interest. For example, if one were interested in aquaporins, then one could use the Entrez database to find both MED-LINE articles and related proteins or genes.

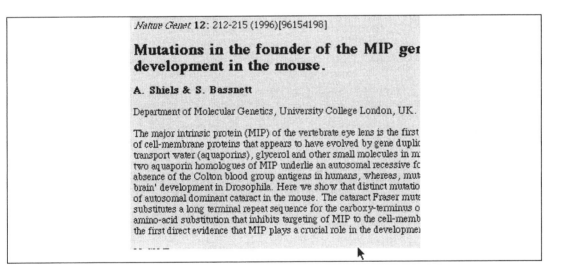

Your attention might be drawn to a paper linking mutations in MIP (a member of the aquaporin family) with cataracts:

It is possible to view an interesting abstract immediately:

By selecting related articles, it is possible to obtain a direct entry into the MIP/cataract literature.

By looking at nucleotide links from a particular paper, it is possible to switch to the GenBank database and pull out sequences of related genes as a first step to assembling a phylogenetic tree or to designing degenerate oligonucleotides for a polymerase chain reaction-based cloning strategy.

The Online Mendelian Inheritance in Man (OMIM) database provides an alternative approach, which is to focus on disease states. The entries are heavily curated, which means that the relevance of any links can be higher than the straightforward computer-based cross-indexing described earlier.

This can lead to informative minireviews on each disease related to the search topic. It usefully includes diseases that are no longer considered to be caused by mutations in a particular gene; this helps workers exclude candidate diseases that were mistakenly assigned to a particular locus.

Select Entries from OMIM --
Online Mendelian Inheritance in Man

11 entries found, searching for "aquaporin"

*107776 AQUAPORIN-1; AQP1
*107777 AQUAPORIN-2; AQP2
*600170 AQUAPORIN-3; AQP3
*600442 AQUAPORIN-5; AQP5
*600308 AQUAPORIN-4; AQP4
#110450 BLOOD GROUP--COLTON; CO
*125800 DIABETES INSIPIDUS, RENAL TYPE
*111000 BLOOD GROUP--KIDD SYSTEM; JK
#222000 DIABETES INSIPIDUS, RENAL TYPE, AUTOSOMAL RECESSIVE
*304800 DIABETES INSIPIDUS, NEPHROGENIC
*154050 MAJOR INTRINSIC PROTEIN OF LENS FIBER; MIP

DIABETES INSIPIDUS, NEPHROGENIC, AUTOSOMAL RECESSIVI

TABLE OF CONTENTS

- **TEXT**
- **REFERENCES**
- **CREATION DATE**
- **EDIT HISTORY**
- **CLINICAL SYNOPSIS**

Database Links

[MEDLINE]

Note: pressing the 📖 symbol will find the citations in the NCBI MEDLINE subset v
matches the text of the preceding OMIM paragraph, using the Entrez MEDLINE neig

TEXT

Ray et al. (1990) and Langley et al. (1991) reported on a family in which 2 sisters of
parents were affected with vasopressin-resistant (nephrogenic) diabetes insipidus. D
locus as a probe demonstrated that each sister had inherited different Xq28 regions fr

The **Medline** database (http://www.ncbi.nlm.nih.gov/PubMed) is also searchable directly, although the "added value" of the Entrez and OMIM presentations will probably make such use rare.

Another major use of this resource is to establish whether a novel sequence that you have just cloned is in fact already known. Sequences can be entered in the BLAST search form (http://www.ncbi.nlm.nih.gov/BLAST/) and compared with a daily updated database of all known sequence, including genome project expressed sequence tags (ESTs), in a matter of 2 or 3 min. Such a search should use the option "blastn" to compare nucleotides with nucleotides to answer the question, "Has any one published my DNA sequence before?" The option "blastx" may be used to ask the question, "Is my DNA sequence related to any known peptide?"

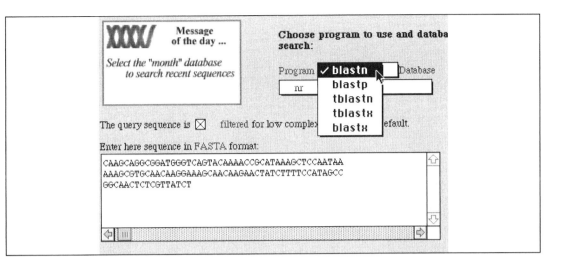

Other major databases, excellently indexed by **Pedro's Biomolecular Search Tools** (http://www.biophys.uni-duesseldorf.de/bionet/research_tools.html), now have several "mirror" sites worldwide to ease the pressure on the original. This includes pointers to servers that will take protein sequences, calculate multiple sequence alignments, and optionally display them, and programs that will scan sequences for defined motifs, e.g., **Prosite** (http://expasy.hcuge.ch/sprot/prosite.html). There are also databases of protein structural motifs, such as Brookhaven's Protein Data Bank (http://www.pdb.bnl.gov/) or the Cambridge SCOP (http://www.bio.cam.a-c.uk/scop/). There are also programs to model restriction digests on DNA and programs to generate phylogenetic trees from related sequences. Another good index is provided by the BCM **Biologists' Control Panel** (http://gc.bcm.tmc.edu:8088/bio/bio_home.html). A survey of available molecular biology databases is also available (http://www.ai.sri.com/people/pkarp/mimbd/rsmith.html).

Major database centers include **ExPASy** in Switzerland (http://www.expasy.ch/expasy-top.html), which is home to a SwissProt protein database, Prosite, and a 2D protein gel database, and the European Bioinformatics Institute (**EBI**) in the United Kingdom (http://www.ebi.ac.uk/), which carries the EMBL and Swissprot databases, together with several others, including a mirror of Flybase.

B. Genome Projects and Model Organisms

A related area is the curation of data into genome projects. In relatively few years, the complete genome of each of several important organisms (human, *Drosophila*, and yeast) will be known. The next step will be an elaborate detective story, as clues to the function of huge numbers of sequenced genes are pieced together. Genome project databases exist to facilitate this work. There is a good index to genome projects on http://golgi.harvard.edu/.

- The Flybase database (http://cbbridges.harvard.edu:7081) is an outstanding exemplar. It is possible to search by gene, function, author, or chromosomal localization. Graphical maps can be assembled, and Java scripting is used to put some intelligence in the user's workstation instead of interacting dumbly with the server. There is a related, but distinct, server run by the **Berkeley *Drosophila* Genome Project** (http://fruitfly.berkeley.edu/); both sites are indispensable to *Drosophila* workers.

- The *Arabidopsis thaliana* database (http://genome-www.stanford.edu/Arabidopsis/) has the full range of services, including EST searching, links to other databases, a graphical interface, and relevant meetings.

- Worm lovers can access the *Caenorhabditis elegans* WWW server on http://eatworms. swmed.edu/, or a database engine on http://probe.nalusda.gov:8300/other/index.html. This is a subset of the Agricultural Genome Information Server (**AGIS**) on http:// probe.nalusda.gov:8000/index.html.

- The **human database** has several manifestations, one run by the genome database (GDB) (http://gdbwww.gdb.org/). This has a very useful subpage (http://gdbwww.gdb.org/ gdb/hgpResources.html) of related resources, such as other genome centers, chromosome-specific databases, and mutation databases.

- The **mouse genome** database (http://www.informatics.jax.org/mgd.html) is run from the Jackson laboratory and includes a mammalian homolog query form.

- The *Saccharomyces cerevisiae* genome database (http://genome-www.stanford.edu/ Saccharomyces/) has the happy position of being the only model for which the genome has been completely sequenced. This means that workers in other species who identify novel genes can now seek yeast homologous sequences with some confidence.

C. Dictionary of Cell Biology

For quick reference, it is worthwhile to set a bookmark to the online version of the **Dictionary of Cell Biology** (curated by this author) on http://www.mblab.gla.ac.uk/dictionary. This corresponds to the text of the second edition of this Academic Press work, with subsequent electronic updates destined for the third edition. Although it can confidently be guaranteed that no dictionary is complete, the nearly 6000 hyperlinked entries cover a wide range of terms of use to cell biologists, and the dictionary receives about 2000 accesses per month.

D. Keratinocytes

The Internet is a happy hunting ground for those with interests in dermatology, and an excellent page (http://info.med.yale.edu/library/sir/medicine.html) curated by Yale lists relevant journals, reference resources, and organizations (including the **Internet Dermatology Society,** which has its own excellent home page on (http://www.telemedicine.org/ids.htm). Numerous resources are available for UV measurement and ozone matters. The **Dermatology WWW server** on http://www.uni-erlangen.de/docs/derma/ includes the **Dermatology Online atlas** and the **Project Dermatologic database.** There is also a site describing a proteome database of human heratenocytes on http://biobase.dk/cgi-bin/eehs.

E. Kidney Development

An innovative kidney development database, or "kidbase" (http://mbisg2.sbc.man.ac.uk/ kidbase/kidhome.html), essentially describes the distribution of markers in different stages of kidney development, with an emphasis on promoters and tissue-specific expression and a link to the Swiss-run **Homeobox** page (http://copan.bioz.unibas.ch/homeo.html), which has a vital consensus alignment to help you decide if you are working with one!

F. Neurosciences

There are several good index pages: the **WWW Virtual library** (http://neuro.med.cornell.edu/ VL/) or **Neurosciences on the Internet** (http://www.neuroguide.com/). The latter has pointers to newsgroups, meetings, and resources devoted to human neurodegenerative diseases, in addition to a range of useful teaching and research resources.

G. Receptor Studies

The G-protein-coupled receptor database (http://receptor.mgh.harvard.edu/GCRDBHOME. html) is a valuable assembly of key paper references, multiple protein alignments, and a list of diseases caused by mutations in receptors.

H. Teaching Resources

Hyperlinked text pages, combined with a quiz-setting engine, provide a near-perfect teaching environment when eyeball-to-eyeball staff interaction is limited. Accordingly, many cell biology courses are being written in html, and a fraction of these are published as a resource on the Internet. Indices of such pages are found in **Cell and Molecular Biology Online** (http://www.tiac.net/users/pmgannon/teaching.html) and in **virtual courses on the web** (http://lenti.med.umn.edu/~mwd/courses.html). **Medical courses** are indexed on http://www.scomm.net/~greg/med-ed/courses.html.

Some specific resources that deserve mention and that could be included easily in teaching courses include:

- **The Society for Developmental Biology** (http://sdb.bio.purdue.edu/SDBEduca/EducaToC. html), which includes confocal and animated images, as well as the excellent "the interactive fly" (http://sdb.bio.purdue.edu/fly/aimain/1aahome.htm). There is also a disturbing image of a gefilte fish embryo.

- The U.S. Department of Energy's **Primer on Molecular Genetics** (http://www.gdb.org/ Dan/DOE/intro.html) is an excellent teaching aid, with particular emphasis on those parts of molecular biology relevant to a genome project.

- **Cells alive !** (http://www.cellsalive.com/) includes some nice animations.

- **Virtual cell** (http://ampere.scale.uiuc.edu/~m-lexa/cell/cell.html) is the beginning of a virtual reality walk through of a (plant) cell, together with cross-sections and electron micrographs.

- A solid **course in cell biology** (http://lenti.med.umn.edu/~mwd/cell.html) could serve as a useful adjunct to an introductory level-taught course.

- A major resource is the **MIT Biology Hypertextbook** (http://esg-www.mit.edu:8001/ esgbio/). The section on cloning, for example, would be most useful to provide background reading during a laboratory.

- The **Visible Embryo project** (http://visembryo.ucsf.edu/) is another developmental teaching resource.

It is also possible to find material that covers the use of the Internet itself. The **Biocomputing Hypertext Coursebook** (http://www.techfak.uni-bielefeld.de/bcd/Curric/welcome.html) covers sequence searching and alignment, and the use of Internet tools such as BLAST. The **Big Dummy's Guide to the Internet** (http://www.td.anl.gov/InternetGuide.html) covers an introduction to the essentials of how to connect to the Internet and to the kind of services that are available once logged on.

I. Journal Contents

As academics increasingly spend their time chained in front of their computers, the ability to search journal articles or even view and print full text and graphics without leaving one's desk is a welcome enhancement to productivity. However, journals are not charities, and so money is involved. Frequently, a journal will have a web site in which abstracts can be viewed, but access to full text is password protected and granted to those who have a subscription to the paper version. Alternative strategies include university site licenses and the handling of charging through intermediates, such as BioMedNet (see below). Journal sites also allow publishers to display their instructions to authors and subscription details permanently.

There are several good indices of journal sites. One is maintained by the WWW virtual library (http://golgi.harvard.edu/journals.html). Another (http://www.princeton.edu/ ~beasley/journals.html) marks sites listed with useful information, such as whether registration or payment is required. The invaluable **Pedro's Biomolecular Tools** has a section on journals (http://www.public.iastate.edu/~pedro/rt_journals.html).

Addresses for particular journals:

Journal	URL
Annual Review of Cell and Developmental Biology	http://www.annurev.org/series/cell/cell.htm
Biochemical Journal	http://www.portlandpress.co.uk/bj/
Biochemistry	http://www.acsinfo.acs.org/journals/bichaw/index.html
Biochimica et Biophyshica Acta	http://www.elsevier.nl/locate/bba
Bioessays	http://www.gold.net/users/ag64/bioindex.htm
Cell	http://www.cell.com/cell/
Current Opinion in Cell Biology	http://www.cursci.co.uk/BioMedNet/cel/celinf.html
Development	http://www.gold.net/users/ag64/devindex.htm
Developmental Biology	http://www.apnet.com/www/journal/db.htm
EMBO Journal	http://www.informatik.uni-rostock.de/HUM-MOLGEN/journals/EMBO-J/
FEBS Letters	http://www.elsevier.nl/locate/febslet
Genes and Development	http://www.cshl.org/journals/gnd/
Human Molecular Genetics	http://www.oup.co.uk/jnls/list/hmg/
Journal of Biological Chemistry	http://www.jbc.org
Journal of Cell Biology	http://www.jcb.org
Journal of Cell Science	http://www.gold.net/users/ag64/jcsindex.htm
Journal of Molecular Biology	http://www.hbuk.co.uk/jmb
Molecular and Cellular Biology	http://www.asmusa.org/jnlsrc/mcb1.htm
Nature	http://www.nature.com/
Neuron	http://www.cell.com/neuron
Proceedings of the National Academy of Sciences USA	http://www.pnas.org
Science	http://www.sciencemag.org
Trends in Biochemical Sciences	http://www.elsevier.nl/locate/tibs
Trends in Cell Biology	http://www.elsevier.nl/locate/tcb

There are also related pages that are of use. For example, there is a very complete listing of **journal names** and their **standard abbreviations** (http://mgd.cordley.orst.edu/useful_tools/abbrev.html) and an index of **publishers' web addresses** and catalogs (http://www.lights.com/publisher/).

For those who are prepared to pay for the privilege of viewing published papers in most of the leading journals from their desktop and who are prepared to pass their credit card details over the Internet, it is worth investigating **BioMedNet** (http://www.Biomednet.com/). Although most publishers are experimenting with charging models for electronic access to their journals, this service has the advantage of being a "one-stop shop."

J. Societies

Sites operated by learned societies typically include mission statements, contact details, society newsletters, and conference registration details and programs. Sometimes it may even be possible to register for meetings online.

Useful indices of about 150 biological learned societies can be found at the University of Waterloo (http://www.lib.uwaterloo.ca/society/biol_soc.html). The World Wide Web virtual library also has a listing (http://golgi.harvard.edu/afagen/depts/orgs.html), as does Yahoo (http://www.yahoo.com/Science/Biology/Organizations/). There is a listing of **Electron Microscopy Societies** on http://cimewww.epfl.ch/EMYP/soc.html.

Individual Societies

Society	URL
American Cancer Society	http://www.cancer.org/
American Society for Biochemistry and Molecular Biology	http://www.faseb.org/asbmb/asbmb.html
American Society for Cell Biology	http://www.faseb.org/ascb/
Biochemical Society	http://www.biochemsoc.org.uk/
BioMedNet	http://BioMedNet.com/
British Society for Developmental Biology (BSDB)	http://www.ana.ed.ac.uk/BSDB/
British Society for Immunology (BSI)	http://immunology.org
European Molecular Biology Organisation (EMBO)	http://WWW.EMBL-Heidelberg.de/ExternalInfo/embo/index.html
Federation of American Societies for Experimental Biology (FASEB)	http://www.faseb.org/
Federation of European Biochemical Societies (FEBS)	http://ubeclu.unibe.ch/mci/febs/
International Union of Biochemistry and Molecular Biology (IUBMB)	http://ubeclu.unibe.ch/mci/iubmb/
Society for Developmental Biology	http://sdb.bio.purdue.edu/
Society for Experimental Biology	http://www.demon.co.uk/SEB/
Society for in Vitro Biology (SIVB)	http://www.sivb.org
UNESCO Global Network for Molecular and Cell Biology (MCBN)	http://morgoth.unibe.ch/mci/unesco/index.html

K. Conferences

An alternative strategy for identifying upcoming conferences is to look at indices. Yahoo holds one at http://www.yahoo.com/Science/Biology/Conferences/. The University of Rostock compiles a listing of **conferences,** courses, and workshops in biology (http://www.informatik.uni-rostock.de/HUM-MOLGEN/anno/meetings.html), searchable by date, continent, or subject area.

L. Software

There is much useful software available over the Web in the public domain. Generally, it covers molecular biology (DNA sequence manipulation, alignment, etc.), teaching, or the technology of the Internet (HTML authoring tools, Web browsers, compression utilities, etc). A very brief index of biological software can be fund on Yahoo (http://www.yahoo com/Computers_and_Internet/Software/Scientific/Biology/). A much more comprehensive page is kept by the Genome Data Bank (http://www.gdb.org/Dan/softsearch/softsearch.html), which holds a listing of over 80 sites worldwide that keeps biological software for downloading. It also has a search engine that covers the EMBL software archive.

M. Grant Agencies

To increase accountability, several grant agencies have home pages, from which it is usually possible to obtain contact addresses, research themes and initiatives, closing dates, and lists of currently funded projects. It is also possible to download templates for grant proposals. The National Science Foundation (**NSF**) (http://www.nsf.gov/) and the National

Institute of Health (**NIH**) (http://www.nih.gov/) are examples in the United States, while links to UK agencies can be found on http://www.nerc.ac.uk/joint_res_councils.html.

N. Physical Resources

It is possible to search catalogs of materials relevant to cell and molecular biology, and even to order such items over the Internet. For example, the **American Type Culture Collection** (http://www.atcc.org/) contains eukaryotic and prokaryotic cell lines, clones from genome projects, and hybridomas. **The European Collection of animal cultures** (http://www.gdb.org/annex/ecacc/HTML/ecacc.html) offers a similar service. A more focused resource is the National Institutes for Child Health and Human Development's **Developmental Studies Hybridoma Bank** (http://www.viowa.edu/~dshbwww/), which provides antibodies against specific developmentally significant proteins.

Several genome projects, e.g., **Flybase** (http://morgan.harvard.edu/), allow you to order stocks directly over the Internet.

O. Commercial Suppliers

A useful alternative to suppliers' catalogs are online guides. These are well indexed in the **Anderson's Timesaving Comparative Guides** on the Web (http://www.atcg.com/aguide/atcghome.htm). There are also indices of especially smaller commercial suppliers on Yahoo (http://www.yahoo.com/Business_and_Economy/Companies/Scientific/Biology/Biotechnology/). A list of biotechnology companies that sell **restriction enzymes** (http://www.gdb.org/Dan/rebase/comp.html) is provided by the Genome Databank. Subscribers to **BioMedNet** (http://www.biomednet.com/) can view advertisements from a wide range of suppliers.

Individual companies frequently have helpful "added value" pages with news and protocols to encourage readers to set bookmarks. Some well-known examples are:

Supplier	URL
Amersham International	http://www.amersham.co.uk/
Boehringer	http://biochem.boehringer-mannheim.com/
GIBCO-BRL	http://www.lifetech.com/
Invitrogen	http://www.invitrogen.com/
Molecular Probes	http://www.probes.com/
New England Biolabs	http://www.neb.com/
Pharmacia	http://www.biotech.pharmacia.se/
Promega	http://www.promega.com/
Sigma-Aldrich	http://www.sigma.sial.com/

As shown a good strategy for guessing a URL for a company is to use http://www.companyname.com/, and if this does not work, to search for the company using Lycos (http://lycos.cs.cmu.edu/) or a similar search engine.

X. KEEPING UP TO DATE

It is in the nature of the Internet that many of these sites will move or close down and that new ones will start elsewhere over the next few years. What strategies can allow you to adapt? First, the Internet has temporary problems almost every day, so you should try an address several times before giving up on it. Second, this article has tried to provide several indices to each section, so if an individual resource goes missing, you should look in the relevant index page. If a whole index shuts down, you can try other indices. If neither strategy works, try an Internet search using key words from the missing page; even if you do not find the original, you may turn up new, even better resources. You could also try posting an inquiry to the relevant Usenet group. It is really quite unlikely that you will fail to pick up a missing resource with this combination of strategies.

XI. FUTURE PROSPECTS

The prospects for life sciences are most exciting. It is already possible to draw on videos, animations, and interactive diagrams and to access catalogs and genome databases from around the world without log in or subscription. In contrast, those services that do charge money, such as offprint viewing or delivery services, must provide a high level of utility and convenience to justify the cost. The result is that an informed and judicious use of Internet resources can accelerate research progress to an extent that would have seemed incredible only a few years ago. With luck, this article will have provided some pointers to encourage both sceptics and the converted to use the Internet more productively.